HOW TO CONSTRUCT

Introduction to Civil Engineering, Structures and Construction

ISBN: 978-1-943605-06-4

AF854068

Publisher: Ahmet Tuter

Author, Editor, Illustrator, Photographs, Cover Design, various roles: Ahmet Tuter

Contact: ahmet1eng@outlook.com, atuter@constructmagazine.com

Disclaimer

The publisher / author disclaims any and all liability or damages of any nature, resulting from the use of, or reliance on this book directly or indirectly, in any way and for any purpose. Any person or entity who uses or relies on this book directly or indirectly in any way and for any purpose, assumes full responsibility of such use or reliance. Use information in this book at solely your own risk. The publisher / author does not express or imply any warranty regarding the information included in this book. The information presented in this book may contain errors, inaccuracies or may be incomplete.

No information in this book can be considered as advice, rendering service, guide, solution or instruction about any subject, including in engineering, construction or project management.

(continues on next page)

(continued from previous page)

This is an introductory and summary level book, to describe the concepts in general for informational, educational purposes only, based on long established civil engineering and construction knowledge, and the scope and coverage depth of subjects vary. For anything to be designed and built, applicable local codes and licensed, local professional engineers and architects must be consulted. Design and construction of structures involve knowledge at greater depth than that is introduced in this book. This book was also not written with the intent to cover any course or exam scope.

The information presented in this book are based on well established knowledge and practices in civil engineering, construction, structures and project management, which were selected, categorized and explained by the author in his own words. The figures / graphs in this book were drawn by the author for illustrative purposes, with the intent to give examples and or illustrate certain terms or concepts in a simplified / idealized / typical way, and the actual figures that apply may be different and will likely include more details. Numbers given in the text or figures / graphs are also for illustrative purposes, which may be approximate or incorrect, although to the best of author's knowledge they are as close as possible or equal to realistic, actual values.

Usage of masculine nouns does not necessarily refer to masculine gender only.

Any organization names, company names, brand names, product or model names, trademarks that may appear in this book are the properties of their respective owners and they are shown here only for informational purposes without intent to infringe. The publisher/author has no affiliation, relationship or collaboration with such entities or endorsement from them.

Health and safety aspects of construction or finished structures are not covered in this book. These aspects must be verified from proper references, and with qualified safety and health professionals.

If you do not agree with these terms, please do not use this book.

Foreword & User Guide

You can find many useful, very well written books in specific areas of civil engineering, construction and specific type of structures, which cover one or few specific aspects in these subjects and they are almost always for engineers or students only.

I wanted to write a book that would cover all these subjects together more broadly that still covers many essentials and can be understood by anyone regardless of background, civil engineer or not.

I tried to keep the language as plain as possible and explained the technical terms I used, so that anyone can follow. Same is true even for the concepts that may go well beyond entry level, to hopefully make them understandable by anyone interested.

If you are not an engineer and anywhere in the book you find something too technical or too detailed, just keep reading, provided that you do your best to pay some attention. This should not affect your ability to follow the rest of the book except in very rare, specific cases. Hopefully this book will be an excellent source for you to learn about the essentials of civil engineering, construction and structures. I recommend that you start from the beginning of Chapter 1 and follow the topics in the order I presented, especially for about the first half of Chapter 1 (until the end of soil mechanics section). You can also choose to start reading the book from anywhere, but in that case you may often need to refer back to Chapter 1 anyway and find yourself reading in the order that is given in the book sometimes.

For civil engineers or engineering students, hopefully this will be a useful summary / a refresher / an interesting read at the least, in well organized format covering the most essentials. Besides, probably not many engineers are specialized in multi disciplines of civil engineering and construction, and this book discusses essentials of many subjects from different sub branches of civil engineering, which can be another way that the book can be useful.

The information presented in this book are based on long-time established, commonly recognized and well known theory, knowledge, and industry practices in civil engineering, construction and project management, which I selected, and then organized and explained in my own words as a civil engineer.

By far the longest chapter is Chapter 1, which introduces concepts and provides foundation for the rest of the book. Chapters 2 to 12 are about different types of structures each, where their design and construction aspects are discussed further. The last one, Chapter 13 is a standalone but still an important chapter, which is about project management and scheduling.

Only civil engineering and its subdisciplines are discussed in this book. In other words, architectural, mechanical, electrical, plumbing... are mentioned very rarely. For materials, reinforced concrete, steel and masonry are covered but wood or cold formed steel are not.

I spent my best efforts and considerable time to write this book, which was a gigantic undertaking that lasted several years. I created this book entirely by myself. Deciding what to write, search, categorizing and describing concepts, editing and improving text, formatting, drawing the figures, organizing photos, preparing cover and so on... were each colossal tasks on their own but for me the important thing was to complete an enormous project to my satisfaction, not how long it took.

This book is in English but it is not my native language; my native language is Turkish. I studied, lived and worked many, many years in the US, so I think I am close to native speaking level in English language, and even before coming to US, I had studied it and spoken it adequately *(I got my bachelors degree in Civil Engineering in Middle East Technical University, Ankara, Türkiye, one of the top few universities in the country and one of several with English technical instruction).* I didn't use anyone's help for editing, proofreading or input so what you see everywhere is exactly how I wrote it, which might have sometimes caused deviations from how a native speaker would express something, but honestly I believe that the book, for its type and purpose, will get an overall passing grade even from native speakers as far as its level of English.

Often as a way of conveniently explaining things, I used "we", "us" or "our" in the text. This can mean two things. It can mean only the author, myself, if it is something about where or how something is written in the book. For example: "as we discussed", or "let's list these factors" and so on... In other words, it doesn't mean that multiple people wrote the book, it was just me. Other times when I used "we", "us" or "our" refers to us engineers or people in construction industry in general, if it is used about a design or construction related practice. For example: "we want the reinforcing steel to fail first" or "after we finish our structural model" and so on...

This book is intended for readers from any country and not just the US, but my guess is that a considerable part of the readers will be from USA, so I used metric and imperial units both, but mostly metric. When I said "codes", or "standards", it means codes or standards that may apply to any country, unless I specifically mention a source for that piece of info. Regardless of that, I was not aware of any conflicts with US based practices to the best of my knowledge anywhere in the book. Generally I looked at US based sources the most, but also the ones from other countries as well.

There are 5 levels of headings in this book. For example Level 1, is the chapter number and title which says "Chapter 1 - General Concepts on Page 1. Level 2 is as you see on that same page which says "Civil Engineering". Level 3 is as you see a couple pages after that, where it says "Role of Architects". Levels 4 and 5 have each successively smaller and different fonts. For example the first Level 4 subsection you see in the book is in Chapter 1>Strength of Materials>Bending>Double Integration Method. There are a lot of Level 3 and 4 subsections. Level 5, does not occur that much. The first level 5 subsection is at Chapter 1>Soil Mechanics>Strength of Soil>Total Stress...>Liquefaction. For example, if I used section and subsection numbering, Liquefaction would have been: Subsection 1.5.3.1.1, but to avoid crowding titles, I did not use section numbering, although for a book of this scope and size they could definitely have been used as well.

In some paragraphs there are underlined words. These are just for emphasizing the importance of those words in that location and context, not because they contained link to somewhere.

Often figures are mentioned only with their numbers. For example if we said Figure 4, and we are under Concrete section, obviously it means Figure - Concrete 4.

I did not see any book before this one which covered essentials of civil engineering, construction and structures together in general which was written for anyone, engineer or not. So in my opinion this book was needed. I read autobiographies of several successful people before, and the one common thing they all say is that no matter what you do, you must love it to succeed, which gives me hope that you may like this book, because I enjoyed writing and preparing it, which took several years of my time. Civil engineering and construction change slowly so I hope that many people will use this book in many years to come.

I dedicate this book to my dad, who is also an engineer (mechanical), and to my mom, and to anyone I met or didn't meet from whom I learned something.

Sincerely,

Ahmet Tüter

August 26th, 2024

About the Author:

Author has over 20 years of experience in construction and civil engineering with various construction, consulting and owner firms and as self work. He holds BS in Civil Engineering, MBA degrees, PE license, PMP, LEED and Construction Management certificates. His other interests include robotics, plants, languages.

Table of Contents

CHAPTER 1 - GENERAL CONCEPTS in CIVIL ENGINEERING & CONSTRUCTION

This chapter discusses many essential concepts in civil engineering, construction and design of structures, and forms the foundation for the rest of the chapters. The topics are ordered to introduce the most fundamentals first, at least up to the end of soil mechanics section. If a technical term appears first time, it is often very briefly explained right there, considering readers who are not engineers.

CIVIL ENGINEERING

Chapter 1 is mainly about introducing civil engineering and its concepts, so first let's define the term "engineering" and then talk about "civil engineering". Engineering is an applied science, which aims to make things that work (or make things work for us) in real life, by using scientific and mathematical principles as its foundation. So may be we can think of engineers as scientists and mathematicians who are specialized in applying science and mathematics plus the specific knowledge in their area for the purposes we have just mentioned. To do this, engineers follow established and or mandatory procedures and tasks set by codes, standards or guidelines, but they must also properly perform and rely on their own calculations and in addition, often exercise something called "engineering judgment". To make this judgment properly, the engineer must not only possess experience, but also understand the theory behind the subject and behavior and interaction of the components involved.

Engineering is separated into branches. Civil Engineering is the oldest of all, looking from historical perspective. Civil engineers make structures that serve us, which means building our civilization.

By constructing structures, we mean not only to design and build them against failure, but also function as intended, without causing any discomfort to us or undue interruptions in service. In this book, starting with Chapter 2, common types of structures are introduced.

Above we said that civil engineers design against failure. The word "failure" has very different meanings and levels for different structures and situations. Other than the first thing that comes to our minds, which is the collapse of a structure, failure in civil engineering can also mean many other things. It can also mean too much displacement that would cause a structure or system to function inadequately or not at all, even if it stands. For example if your living room slab sagged visibly, you would not feel safe at all, even if that slab would still not collapse for a hundred years. For a utility system, failure may mean flooding of pipes. For a suspension bridge, it may mean too much swinging during windy weather. For a highway, failure may mean frequent stop of traffic due to poor geometrical design of the road, or accidents happening around a hill due to poor sight distance, or water ponding at certain locations during rain. For a tunnel, it may mean the leaking of water from tunnel walls. For a dam, it may mean accumulation of sediments in the reservoir too rapidly, before the dam's economic design life is over. For a stabilized slope, it can mean the sliding of slope, even if it may be very slowly. For a foundation system, it can mean footings settling not in equal amounts or just too much, and so on.. Of course, for each of the systems mentioned above as examples, many more failure modes exist.

Civil Engineering Disciplines

We will group civil engineering into five main branches here, which is the most common and straightforward way as: Structural, Geotechnical, Transportation, Water Resources and Construction. Note that in different places you can also see different groupings and additional categories (i.e. Materials Science, Environmental, Earthquake) on top of the basic categories you see below. Civil engineers specializing in the categories below always coordinate with each other as necessary, as every structure requires the input of more than one branches listed below.

- Structural Engineering: It is the branch of civil engineering that specializes in the analysis and design of structures, and probably the first branch that comes to an ordinary person's mind, about civil engineers. Structural engineers are also involved in specialty components that are attached to or part of the structures such as facade, masonry, component support frames and many more. A civil engineer who is specialized in structural discipline is also called as "Structural Engineer". Earthquake engineering - from structures point of view - is also considered as a part of structural engineering and an inseparable part of it, in many areas of the world. Structural engineers make sure that the structures and the load carrying system members within are strong enough to carry all loads that will be demanded from them *(capacity design)*, and not only that but also will deform, deflect or rotate only within tolerances, throughout their design life and for their expected performance level *(service level design / design for deformations)*. A structure is similar to a "skeleton", that carries the weight of the structure itself plus the loads applied on it during its service life. The structure also receives nonstructural items on it, for that project to fully function and serve us. For example, a building structure receives architectural shell items that will cover the exterior of the building, interior partitions, finishes, mechanical, electrical and plumbing work and so on. A bridge structure will receive the paving, railings and all other miscellaneous items on it, to serve us properly. The structural engineer of a project may also be called as Structural Engineer of Record (Commonly abbreviated as SER or EOR), who designs the main structure. The EOR may report to the owner directly too but it is usually through the owner's representative or the architect who are hired by the owner. EOR produces the structural design drawings and structural specifications in coordination with the architect and other consultants as applicable. After that, the design of certain specialty items which are not part of the main structure but nevertheless require structural engineering input such as facade, masonry, steel frame connections, shoring, post-tensioned slabs and so on... are normally designed by specialty structural engineers who have greater expertise in those specific areas, who work for a general contractor (or subcontractors of general contractor). Specialty structural engineers produce submittals including shop drawings per the criteria given in the design drawings and specifications, that were produced by the architect and/or the EOR.
- Geotechnical Engineering: It deals with soil mechanics, all types of foundations, retaining structures, deep excavations, slope stability, rock mechanics, ground improvement (as far as

strength, impermeability and stiffness) and groundwater from foundations point of view. Civil engineers who specialize in this field are called "Geotechnical Engineer". Geotechnical engineers must be in close communication with all other disciplines listed here. Civil engineers in all disciplines exercise engineering judgment when necessary but for geotechnical engineers this may be more often than others.

- Transportation Engineering: It deals with the design and construction of roads, highways and their supplementary components. As for bridges and tunnels, the route, exact layout and traffic capacity would fall under transportation, while their structural and foundation design are performed by structural and geotechnical engineers. Civil engineers who specialize in this field are called as "Transportation Engineer".
- Water Resources Engineering: It deals with structures that we build in or around water or at least meant to handle water. Examples are dams, flood control structures, water distribution and collection utility systems, aqueducts, harbors, wharves, coastal protection structures and so on... Civil engineers who work in this branch are called as "Water Resources Engineer". A significant portion of civil engineering relates to transportation and water resources engineering, as far as the total monetary value (even if not the number) of projects.
- Construction Management / Engineering: The branches listed above deal with design and mainly work from office although they should also visit the jobsite as often as possible during construction phase. This branch however, as the name implies, is constructing those designs made by the first four (except for design of temporary construction items, which construction engineers might also perform). Civil engineers who work in construction management must still at least be aware of the main principles of design, but their primary focus is constructing the structures in the field. Civil engineers who work in this field can also be called as construction/site engineer or manager. All civil engineers working in branches mentioned above must understand project management too, but for construction engineers this is even more critical. In Chapter 13, we present a summary of project management. Construction engineers may also need to work closely with the disciplines listed above especially during the design phase, to make sure that the constructability aspects - such as logistics, access, means and methods *(and its corresponding cost and schedule impacts)* - are well taken into account.

Role of Architects In Relation To Civil Engineers

Many people may not entirely know the distinction between architects and civil engineers, so let's clarify. Centuries ago civil engineer and architect meant the same person, but as the engineering and architectural knowledge improved and accumulated, it was inevitable to view them as separate fields which started somewhere in the 19th century. For example the frame structures (will be introduced) that we see everywhere and take for granted today, which requires engineering input rather than architectural, only started to appear in the second half of the 19th century.

In this book we cover only the civil engineering aspects of construction and structures, but not architectural. Architecture, is important especially in <u>commercial and residential</u> construction which are mainly the buildings and similar structures. Architects design the size, shape, aesthetics, interiors, functions of these structures, and civil engineers design the structure to carry itself and the components, while resisting all other types of loads that will act on that structure during its life against failure and discomfort. The design of civil infrastructure however (bridges, roads, dams, harbors, utilities, tunnels, sea walls and many more...) are primarily the responsibility of civil engineers although in some cases and upto a point architectural input may still be used.

Determining a building's shape, function and aesthetics etc... doesn't mean that architects can be unaware of how the structure of a building or its foundation will work, although those parts fall on structural and geotechnical engineers. For example as you will see later under earthquake section, to ensure good earthquake resistance for a building, its overall shape, size and regularity has the biggest impact, which is the architect's responsibility. If, from structural point of view the building was not conceived properly by the architect in the first place, for example if there are a lot of things that cause structural irregularities (will be discussed) the engineers will try to work around it but the cost of structural design and construction will go up. Or for example the architect may prefer as slender and thin members as possible but this may make structural design, rebar or structural steel detailing and or placement difficult to impossible. In other words, in almost all cases engineers will be able to accommodate an architect's requests for a safe and serviceable structure, but this can often mean a cost and schedule impact.

For buildings and similar types of structures, architects normally serve as the lead consultant to the owner. On these type of projects, they prepare the preliminary drawings, and then coordinate other disciplines such as structural, geotechnical, mechanical, electrical, plumbing, information systems, interior design, landscaping etc... to come up with a complete, final design package. For example, a structural engineer will have the preliminary architectural drawings which progressed at least to a point to show the overall size and geometry and intended locations and sizes of members, when starting the design of the structure itself. Then, after he/she designs the structure, he/she forwards the structural drawings to architect who coordinates this with the architectural set, and also with other disciplines' drawings. A construction manager consultant should preferably also be employed during design stage, to look at the job from project management and constructability point of view, and to make sure that the design can achieve the required value and quality at a given budget and schedule. For large projects this route is followed, and the established owner firms also have construction management capabilities in house. After the design stage is over, the contractor is selected, who obtains a permit from local jurisdiction to start construction. In commercial or residential projects, civil engineers working in construction project management must also know about and coordinate architectural work everyday.

Roles of Owner, Consultants and Contractor w.r.t Each Other

Let's summarize the role of all parties with respect to each other, in order to construct the big picture in mind in the very beginning. First, a project stems from a need of the owner, who is willing to spare funds and time, to bring that project into reality. From this point, the rest of the path as far as the role and relationship of parties and type of contracts between each may vary but we will describe a common situation here.

Unless the owner person or organization has expertise and means to design and build that project, he/she must hire consultants to design the project and a general contractor to build it. If the owner does not have any knowledge about construction at all, instead of directly hiring the consultants and the contractor, the owner may chose to hire a representative, which is commonly called as the owner's representative or construction manager (CM), who is knowledgeable in construction and will represent the owner's interests. Then the consultants such as the architect and engineers (structural, geotechnical, mechanical, electrical, IT...) are hired, who design the project per the owner's needs. Finally, when the project drawings and specifications are at a sufficient level of completion, which is typically called a bid set, the general contractors (GC) place bids for the project and the one who is seen as the most appropriate is awarded the contract. The type and specifics of contracts between each of the parties above vary. For example between the owner and the GC, there are contact types such as such as Lump Sum or Guaranteed Maximum Price and more. The important thing is that the responsibilities and scopes of everyone are well defined.

After being awarded the project, the GC typically hires subcontractors, who are specialized firms in their own areas, such as structural steel, reinforced concrete, masonry, earthwork, facade, flooring, painting, electrical, plumbing, roofing and many more as applicable... For example the specialty structural engineers typically work for the specialty subcontractors, who are hired by the GC (but from the owner's point of view, they work for the GC). It is essential for example that the structural drawings, *(which are prepared by the structural engineer of the project)*, give well defined, clear criteria to allow the design of specialty items by the specialty engineers who work for the subcontractors, but at the same time should leave all the "specifics", which must be well defined, to the specialty engineers in order to have clear boundaries as far as the design scope and liability. The specialty engineers will prepare submittals which include shop drawings, material samples, product literature as applicable, and submit them to the GC, who reviews these for coordination with other trades and in general compliance to the contract documents and then submits to architect / consulting engineers, who can make more in-depth review for compliance to project drawings and specifications. After these are acceptable to the consultants, the GC distributes the approved submittals to not only to the subcontractor who prepared them but also to all other affected trades (subcontractors). Based on these approved submittals, the subcontractors can start the production / installation of their specialty items, under the coordination and supervision of the GC. The timing of everything described in this paragraph, must ultimately comply to the initial, high level *(i.e. without too specific details)* project schedule done by the owner / owner's representative. The construction schedule that the GC makes later therefore must comply with it

CONSTRUCTION

For any type and size of construction project you can imagine, there are various factors that affect the initial decision about the feasibility, scope and size of a project and its initial planning. These include:

- The required demand from the project
- The project budget and schedule
- Long term maintenance and operational costs, vs. the expected revenue
- Constructability issues

- Aesthetical concerns
- Presence of existing features and structures if any
- Technical capabilities
- Availability of materials
- Social, economical, administrative, political issues for selection and impacts after completion

Therefore, when discussing different types of structures in the book, don't forget that these will always be considered as applicable to any type of structure and project, even if we may not mention them each time we discuss a type of structure, but mention only specific factors applicable to that particular type of structure.

After the projects' initial planning and then design, they are built in the field, to function as intended. Design falls under planning, and construction falls under execution part of a project.

Construction phase is much more expensive than the design phase. Any cost savings that may be achieved by even the most efficient designs can easily disappear and then additional cost increases may accumulate on top, if the constructability issues were not adequately addressed during the design phase or there is poor construction management in the field.

Construction is a locomotive industry for any country's economy. This is because of the high amount of monies involved, which is because so many different industries ultimately make things or provide services for construction industry. The involvement of so many things also means construction is a process which can be affected from many things. That is why, when the overall economy is doing well, construction business also does well, and during the times of economic recession, construction is one of the first industries to be negatively affected, with some job sites stopping, workers and engineers being laid off, some contractors and subcontractors not being able to keep their cash flow running... Therefore, from monetary point of view, construction is a highly rewarding but also a risky business. The high risk of construction is also because the construction process itself involves many different procedures. This means, there are a lot of variables to which construction tasks are sensitive, which must be considered during planning (design, scheduling, budgeting) phase and, managed and coordinated well during project execution (construction) phase. These variables include:

- Effectiveness of initial project planning and technical design of scope (affects everything else that comes after, throughout the project).
- The quality, quantity, size, delivery dates, installation durations of materials provided by a wide range of suppliers, and the coordination of all these during design and construction.
- The project team's effectiveness in coordinating workforce from many firms, and those firms' ability to provide required workforce and management on the job to achieve target schedule, cost and quality. The field crews' efficiency and skill, who come from a variety of firms.
- The effectiveness and suitability of means and methods of construction/installation which may also need to be preplanned and coordinated during as early as the design phase.
- Weather, environmental or other external factor related delays.
- Construction safety issues.
- Temporary structures or measures that will be needed.

- Coordination of construction phases / progress with each variable.
- Site access and staging.
- Issues that were unforeseen, unknown or not detailed during design, which need clarification during construction. For example on large projects, there can be as much as several thousand requests for information (RFI)s from the general contractor (many of which are originated by subcontractors) to the owner and / or consultants.
- Construction equipment related issues.
- Administrative, financial and/or political issues.
- Coordination with neighboring facilities.

and more... Again, these apply to any project, which should be kept in mind while reading about any type of structure covered in this book.

The proper management and coordination of all these variables and more, the big budgets, makes construction stressful and a high risk business, but, for people and firms who love and know what they are doing, by following established procedures and proper project management practices it will often be very satisfactory and rewarding as well.

It is quite common to group construction into subcategories as follows:

- Civil Construction: It is the construction of dams, bridges, utilities, roads, aqueducts, water channels, tunnels etc... and all infrastructure to build our civilization, which is why it is called "civil" construction, and civil engineers are called as civil engineers in the first place. It can also be referred to as "heavy construction".
- Industrial Construction: It is the construction of industrial facilities such as power plants, factories, refineries etc..., thus, the construction of facilities that serve us by processing or manufacturing things. Apart from designing and building of the structures that will make up the facility, which is the job of civil engineers, industrial construction also involves a lot of high technology and highly specialized, sensitive items attached to the structures. Although those specialized systems and components fall into the realm of other engineering fields, in the end, these systems will reside on or in the structures designed and built by civil engineers, they will be fed by the utilities that will be designed and built by civil engineers.
- Commercial Construction: It involves the construction of facilities that will host people for other than residing purposes, such as malls, hospitals, offices, theaters, stadiums, parking structures... In this category of construction, architects take the lead consultant role, who coordinate other consultants such as civil engineers (structural and geotechnical), electrical, mechanical, IT engineers, landscaping architects, interior designers and so on....
- Residential Construction: It is the construction of buildings and homes, for residing purposes. Hotels can fall under this or commercial category. Architects' role in residential construction is the same as in commercial construction, being the lead consultant and coordinating everyone else, responsible either directly to the owner or the owner's representative.

Having described the terms civil engineering and construction, we can now start to discuss the concepts in civil engineering in more detail. Construction aspects will also be mentioned in this chapter as applicable within the context. From Chapter 2 and forward, you will see relatively more construction related content when we focus on a certain type of structure. So the rest of this chapter is mostly about introducing and discussing various terms and concepts in civil engineering.

STRUCTURAL BASICS: STATICS (FORCES)

Before we can talk about anything else in civil engineering, first we need to introduce forces, which are covered in this section, Statics, and then stresses and deformations, which are covered in the next section, Strength of Materials. So, here we start...

The most fundamental thing to learn in civil engineering is to understand the forces that act on objects. To understand force, we must first understand what a vector is.

A vector, if you are familiar from high school physics, has both magnitude and direction. It can be carried to another location, as long as we do not interfere with its direction. This is called the line of action. To show vectors, we use arrows, which are scaled in proportion to their magnitudes. Both of these are illustrated below. The figure on the left shows line of action concept. We can slide these vectors anywhere we want, as long as they stay in their lines of actions, and nothing about them will change, nor their turning effect on a point which is not on that line. The figure on the right illustrates showing of vectors with scale, in proportion to their magnitudes. For example see that the 30 Newton (N) force is 3 units, 40 N force is 4 units and 50 N force is 5 units long (as in a 3-4-5 triangle here).

For example, energy, temperature, distance or speed are not vectors but scalars, because they have magnitude but not direction. But force, such as your weight, is a vector. It has both magnitude and direction, which we can simplify and show as a single force from the center of mass of your body directed towards the center of mass of the earth. Although distance is scalar, displacement is a vector. It has both magnitude and direction. We can show a car's displacement from point A to point B with a vector, by indicating its magnitude and direction.

We can resolve vectors into their components in perpendicular directions. For example consider the force shown as 50 N in the figure below. We can resolve this force to its x and y components as 30 N and 40 N as in the figure, in which case we must then delete the 50 N force itself, as the 30 N and 40 N will have the equivalent overall effect. This fact is very useful for us to know as you will see.

The subject of statics is about forces that act on various masses, which we also call as bodies, such as structural members. One very important assumption in statics, is that these bodies never deform. We assume that they are infinitely rigid. In reality there is no rigid material. Everything,even the most stiff materials in the universe deform under the action of forces. But in statics, we assume that the bodies are rigid *(we will account for stiffness and deformation in next section, strength of materials).*

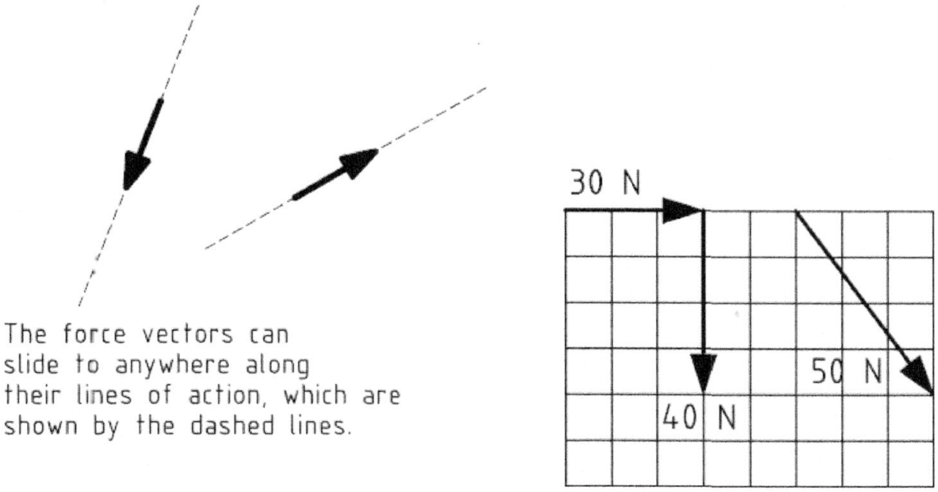

Figure - Statics 1

Types of Forces

Let's first list the types of forces. Forces can be as compressive, tensile or shear. Forces can also have turning effects, called moment.

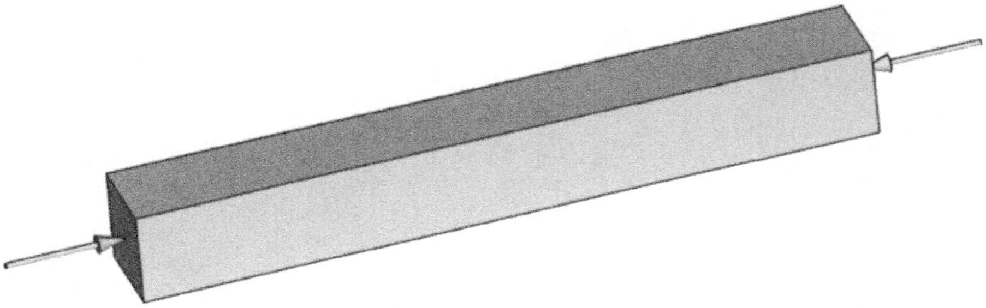

Figure - Statics 2

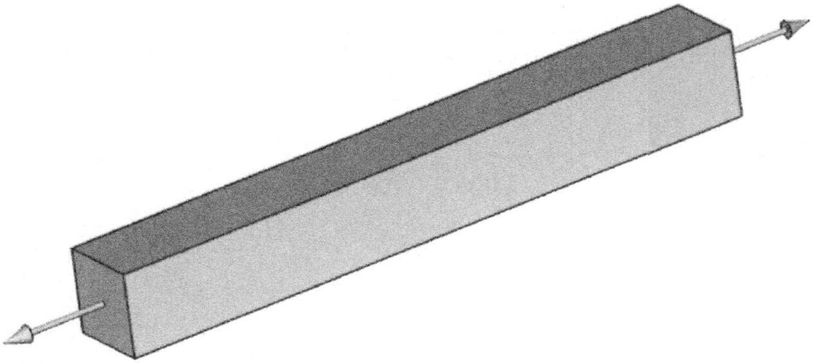

Figure - Statics 3

Forces can be in the form of:

- **Compression**
- **Tension**
- **Shear**

Forces can also cause turning effects such as:

- **Bending Moment (also called flexural moment or simply "Moment")**
- **Torsional Moment (also called as just "Torsion")**

Figure 2 above shows a member in compression under compressive forces, Figure 3 shows a member in tension under tensile forces, and Figure 4a and b below show shear force from different angles which causes two surfaces to slide against each other. For simplicity we show beams here but a member of any shape is possible, such as a column, deck, foundation or an element of any arbitrary shape.

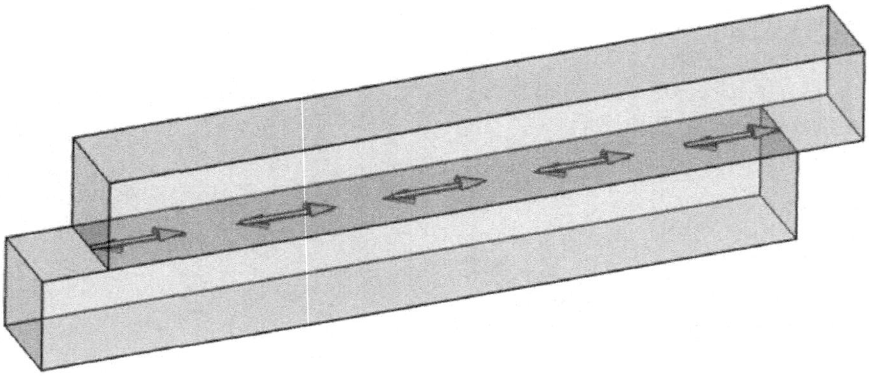

Figure - Statics 4a

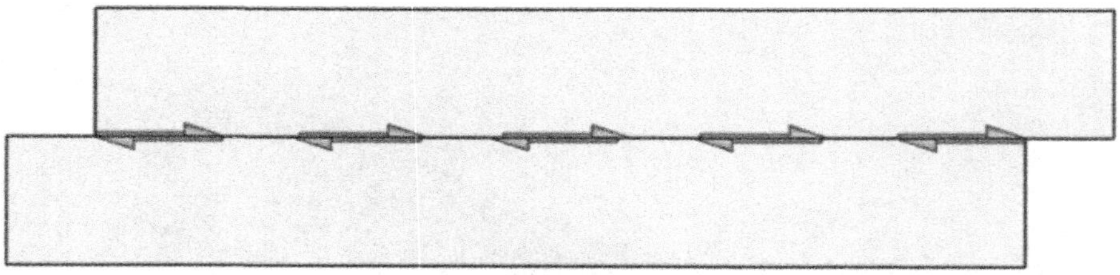

Figure - Statics 4b

For shear, we do not have such configurations as above in reality, in other words, we do not make two beams slide like this. This is just to show the concept of shear. In reality shear occurs within members, so for example you can think of the two beams above as adjacent fibers of a single beam.

In Figure 5 below, the concept of distributed load is shown, which is also called as "uniform load" if uniform everywhere. Here it is shown as linear, having one dimension, but it could also act on an area, in 2D, such as the snow load on a roof deck or the lateral earth pressure on a retaining wall.

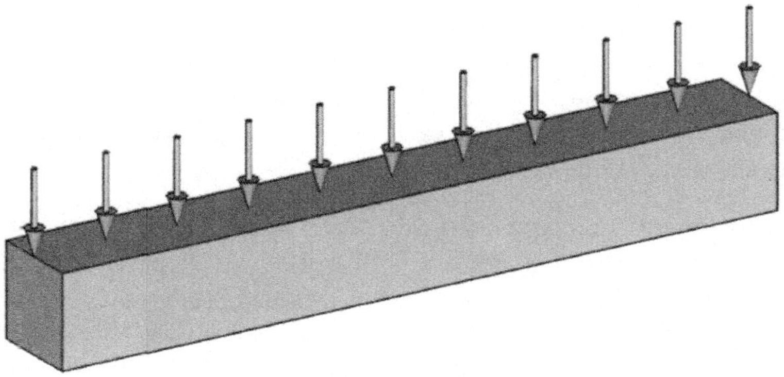

Figure - Statics 5

Figure 6 below shows moment, which is the turning effect of force, and found as:

$$\text{Moment} = \text{Force x distance} = M = F.d$$

By distance, we mear the shortest (perpendicular) distance from line of action of force to a point.

The two forces in Figure - 6 are equal in magnitude but opposite in direction and have a distance in between. This is called a "force couple", which creates a moment. And since we said the two forces are equal but opposite in direction, they cancel. Thus, all remaining is the moment as a result of the distance between them. So, the upper and lower parts of Figure 6 are equivalent for us. The moment of a force couple is found as in the formula above, multiplying the magnitude of one of the forces by the distance between the forces. If this beam was in space, it would rotate and its rotation speed would increase forever (rotational acceleration), in the absence of friction or any other outside force.

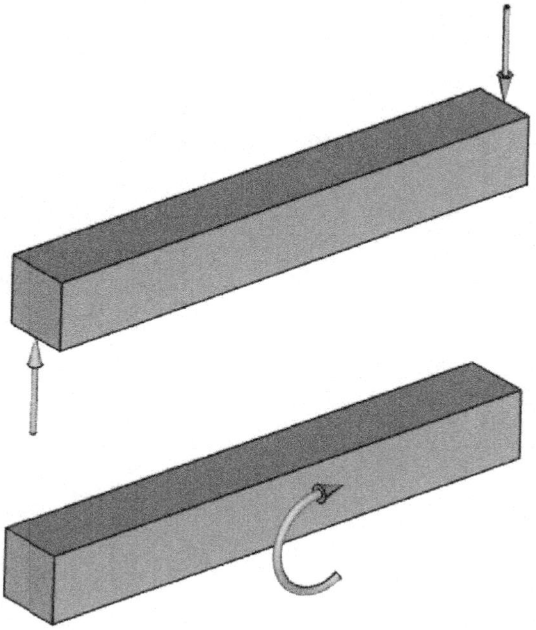

Figure - Statics 6

And what if the forces above were still opposite in direction but one of the forces was greater in magnitude? Then we would still have the moment, plus, the difference in force magnitudes as a resultant force, acting in the direction of the larger force. The moment in that case would be found by multiplying the smaller force with the distance. We said smaller since the remaining force from the larger one would remain in its original line as a force. If that beam was in space, it would still rotate with increasing rotation speed, but now it would also accelerate linearly in the direction of the net resultant force, with increasing linear speed (linear acceleration).

In the figure above the moment is shown on the side of the beam, but it is of no importance where we draw it, as long as we do not change its magnitude and around which axis it rotates. Wherever we draw this moment - at the end of the beam, middle of the beam, edge of our galaxy, it doesn't matter and it will have the same turning effect, in other words moment, on this beam. This is similar to what we said about forces earlier, which was, we could draw a force anywhere, as long as we do not alter its line of action. For example, if we had a force here and we changed its line of action, then its distance with respect to a point on the beam would change, and it would cause different moment on the beam, from M=F.d. This is why we can draw and show a moment anywhere as long as we do not change its rotation axis, and we can draw a force anywhere as long as we do not change its line of action.

And finally, Figure 7 shows torsional moment, simply called "torsion". It causes twisting, as seen below.

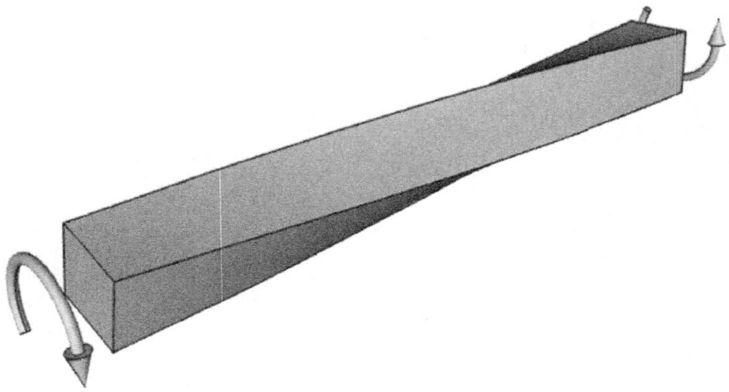

Figure - Statics 7

Free Body Diagram

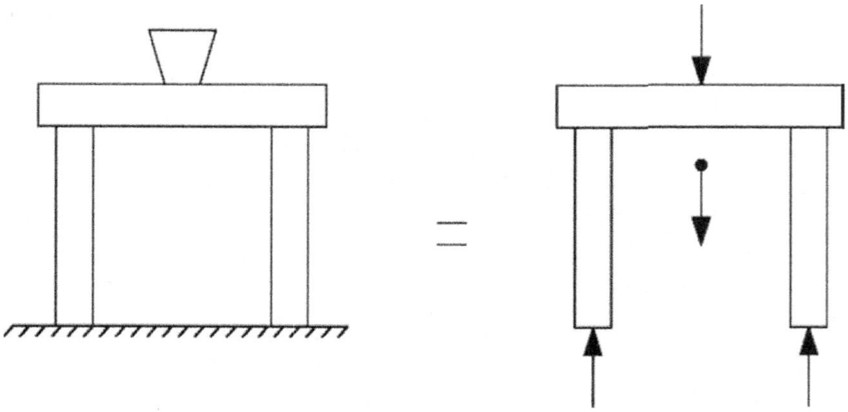

Figure - Statics 8

In structural analysis, in order to solve the forces acting on structures, we must first represent everything in a more understandable, solvable form, in terms of forces. Free body diagram (FBD) is used for this purpose. It is the representation of external forces on a body. Let's give an example.

In the figure above, on the left, a bowl is shown on a table. The figure on the right, shows the FBD of the table and for our purposes it is the same as the one on the left. The weight of the bowl acts from top, table's own weight acts at the center of mass of table, and the floor applies reaction forces.

Here in reality, the base of the bowl exerts uniform force on the table, and the table has uniform, distributed weight force everywhere, and the reaction forces from the base act on the cross section of the legs of the table as uniform 2D forces, but we simplify things and show them this way as single forces, in order to find external reactions, although we could definitely draw the FBD with uniform forces too, but it would be far more complicated to solve without any further benefits.

We can not always simplify forces this way. For example, to find the bending amount on a bridge deck, we could not just simplify all the vehicle loads and the self weight of the bridge deck as single forces.

Static Equilibrium of Bodies

The equilibrium of bodies concept is the starting point of all statics, thus structural or geotechnical analysis. It stems from Sir Isaac Newton's laws of motion which says that, if an object is not accelerating linearly or rotationally, then the net force and net moment acting on it must be zero. Thus, in F=ma and M=Iα, both F and M are zero, so, "a" (linear acceleration), and "α" (rotational acceleration), must also be both zero. Here "m" is the mass. "I" is the moment of inertia which we will explain under strength of materials. We will see F=ma later again in the book when studying earthquakes. And Newton's other law states that whenever there is a force or moment acting on a body, which is the action, there must be an equal and opposite force or moment countering it, which is the reaction. In that case, there is no net force or moment, thus no acceleration on the body, which means, the body is in static equilibrium.

We said "net" above. It means, there might be many forces and moments acting on the object, but, the bottom line is, they all add up to zero, so that our object will be in equilibrium, and remain "static", which is where the name of the topic we are studying right now comes from.

For example, table above represents an equilibrium condition. Nothing is accelerating because the sum of the weight of the bowl and the weight of the table, which are the action forces, were equally countered by the reaction forces that the floor was able to provide to the table's legs. So, the net force in vertical direction was zero. We also did not have any horizontal forces, so the net force in horizontal direction was also zero. We did not have any net moment acting anywhere either, so there was no rotational acceleration either. If we want to summarize, for equilibrium to exist:

Sum of forces in horizontal direction (x-axis) is zero: $\sum Fx=0$

Sum of forces in vertical direction (y-axis) is zero: $\sum Fy=0$

Sum of moment around z-axis is zero: $\sum Mz=0$

The Greek letter sigma means sum. The three equations above are called the equilibrium equations. These equations are what a structure must ultimately satisfy at all times, even if thousands of forces act on it every moment. We call by default the horizontal direction as x-axis and vertical as y-axis. And for any other inclined direction, we can always resolve forces, which are vectors, to their perpendicular components, as was shown in Figure 1. And we said z axis for the moment, because, if the paper plane you are looking at right now is x-y plane, the moment, in order to rotate on that plane, must be rotating around the z axis, which is perpendicular to x and y axes.

So, the three equilibrium equations above are for two dimensions, which is on the plane of paper you are looking at, in other words, the x-y plane. When studying structural analysis and solving manually, solving in 2D is sufficient for us in many cases. Even if we wanted to solve a 3D structural arrangement which is symmetric and regular, we could still solve it in 2D in many cases.

In three dimensions, we would have a total of six equations instead of three as above. The additional three equations will come from the equation for sum of forces along z axis, and, sum of moments around x and y axes, to bring the number of equations to 6.

In the figure below, a beam is fixed into a wall. On the free end of the beam, we apply forces. As we have just learned, for this beam to be in equilibrium, these forces must be countered by equal reaction forces. See the bottom part of the figure. The wall provided reaction forces of equal value, to counter these external forces. In addition, you can see that the wall also provided a reaction moment. This is because, if we did not have this reaction moment, the beam would rotate.

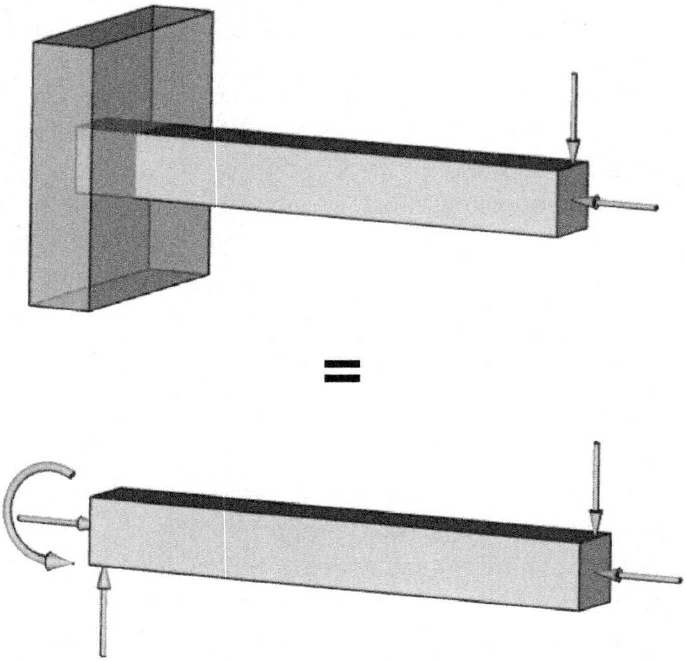

Figure - Statics 9

You can of course intuitively see this too in the figure above, but also see this by looking at the action and reaction forces more carefully. Look at the vertical action and reaction forces. Their magnitudes cancel as they are in opposite direction and of equal magnitude, but, their lines of action are not the same. There is a distance between their lines of action. So they create a moment. This is why, the wall must provide a reaction moment, in order to counter this moment on the beam. *(The horizontal action and reaction forces do not create a moment because they are on the same line of action, thus have zero distance between them.)*

In the figure below, you see almost the same example, but this time the beam is not fixed into the wall. Now we have a hinge (also called pin) type of support. Here, again the vertical and horizontal forces can be countered by equal reaction forces, but, the moment that the vertical force couple created can not be countered, because the hinge can not resist moment. It will allow rotation.

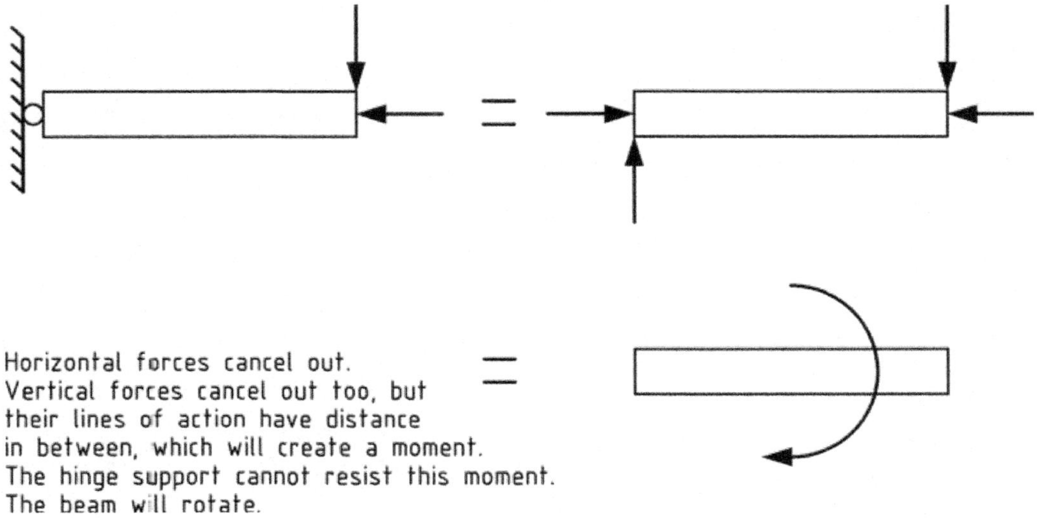

Figure - Statics 10

Therefore, the beam above would rotate. You can also intuitively see this from the figure. So, this beam is not in equilibrium and this is unacceptable in civil engineering.

Types of Structural Supports

First lets describe what "degree of freedom"means. It is a term imported from mathematics which means, the number of values in the final result of a calculation that are free to vary. In structural engineering, similarly, it means, the number of axis that a structural assembly is free to translate (move straight) or rotate. For example, a swinging door has only one degree of freedom. It can only rotate about the hinges, which is one axis, but it cannot move in x, y or z direction, and it cannot rotate around the two horizontal axis. A two dimensional system can have maximum 3 degrees of freedom, such as translation along x and y, and rotation around z axes. A 3D system can have 6 degrees of freedom, such as translation and rotation for all x y z axes. So, the supports we display below, each provide different degrees of freedom to structural members, or stated another way, each restrain translation or rotation in different directions, thus creating force or moment reactions along those.

Although for theoretical exercises there can be more, we have four common basic types of supports in structural analysis, as shown below. In the figure, all reactions can be in exact opposite directions too, which are not shown for clarity of the figure.

Fixed support, shown left, allows no rotation or movement (translation) in any direction, which means it can resist moment and forces in both directions, once external forces are applied.

Hinge (also called pin) type of support, can resist movements in both directions, but it will allow rotation. So it can create force reactions but not moment reaction. This is also useful. Sometimes we provide hinge supports in our structures, to make moments zero at certain locations, which would otherwise cause too much stress and the need for much stronger and expensive structure and or foundations. Or during earthquakes, we can deliberately allow hinges to form by themselves at certain locations in a structure for moments to be released.

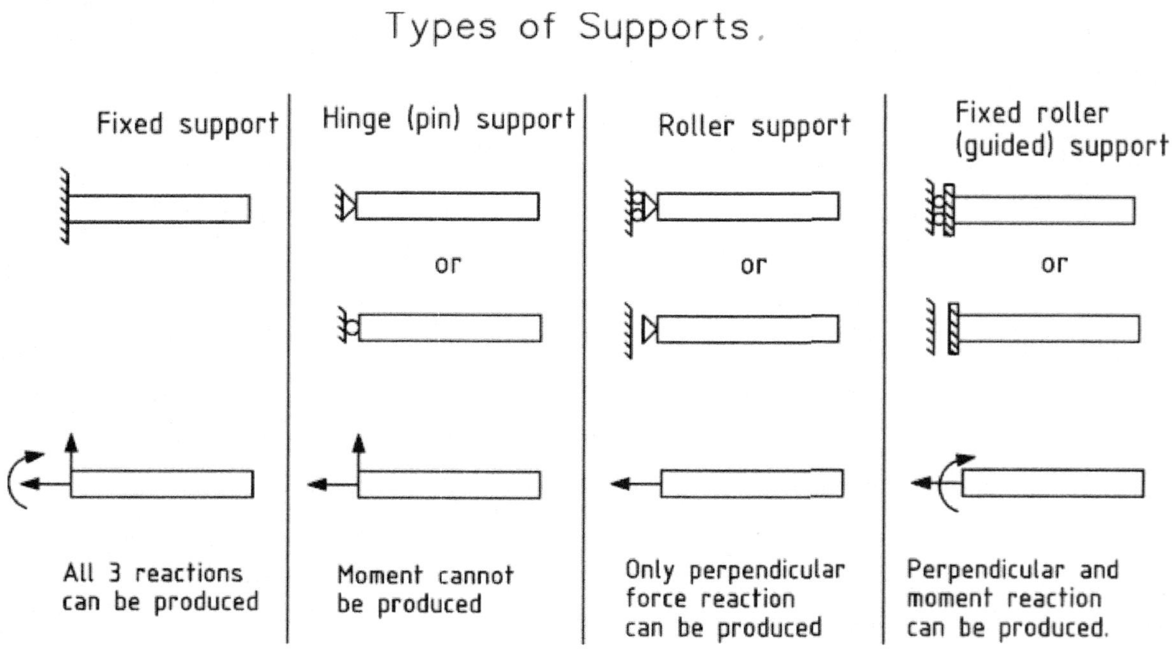

Figure - Statics 11

See the photo below for a hinge support, at the bottom of a center post of a widespan roof. The roof is already supported all around from its edges and here in the center the engineers wanted <u>no moment</u> is resisted at the bottom of the large center post, therefore hinge type of support was used.

The hinge support ensures that moment is released (zero) at this location.

The third one is a roller support. It can only produce a reaction force in one direction, which is perpendicular to its rolling axis. Again, this can be useful. For example, we may let one end of a bridge deck beam to rest on roller support, so during hot weather when the beam expands, it will not create any compression force or bending, as both are released there. Or, during earthquake, the fact that that end is free to move horizontally and rotate would protect the beam (and the pier it is connected on its other end) from those extra reaction forces that would otherwise be created.

And the last one, the fourth one from left is called a fixed roller support. Similar to roller support, it can only move (translate) in one direction but unlike roller support, it can also resist moment. This is encountered less often.

We don't necessarily have to construct these supports in the exact shapes shown above. The key point is that a support works as one of the supports above, no matter its or the connecting member's shape, orientation, in 2D or 3D. For example one edge of a truss system can simply sit on bearings *(as long as the truss' horizontal movement is restrained somewhere else otherwise it will be unstable horizontally)* which can effectively serve as pin or roller support, which allows rotations of truss, thermal movements and allows it to work separately from the supports below, except transferring gravity load.

Example: Solving Support Reactions

Based on what we covered so far, let's give a very simple example to show how we find reaction forces. In this book we have very few examples, and this is one of them, since this is a fundamental concept.

In the example below, we have a beam with its supports, with an inclined force acting externally. What you see below is actually a structural model (although the simplest), which is the representation of a structure in a mathematically solvable form. *(Note: We will discuss structural models later. All types of structures need to be represented as structural models first, so that we can analyse them, in other words, solve the reaction forces, and then find all stresses and displacements of members.)*

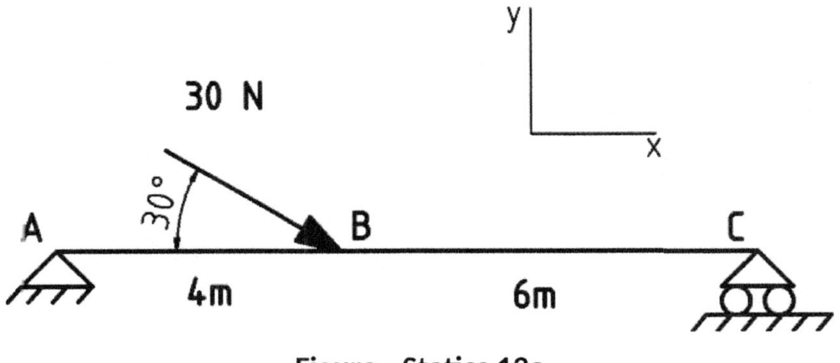

Figure - Statics 12a

Here we have a pin support at point A, and a roller support at C. This is called a simply supported beam, which is a term you see often. It is called simply supported, because these supports here are barely enough to provide sufficient reaction forces to keep the beam in equilibrium in both x and y directions and rotation around z axis. In this problem, we want to find the reaction forces at the supports A and C, as a result of applying the 30 N force as shown. First let's draw the FBD:

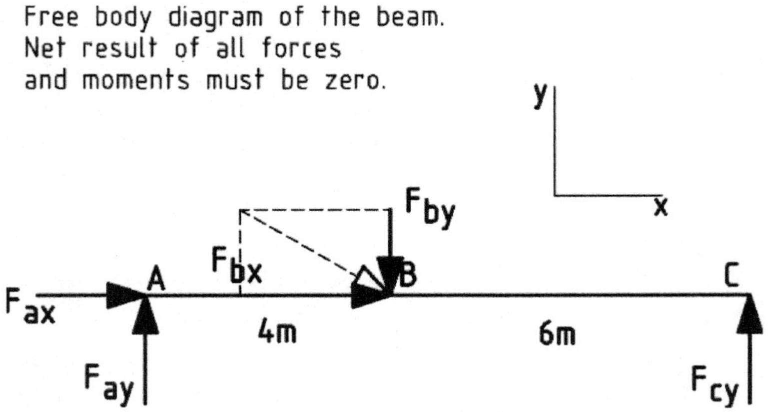

Figure - Statics 12b

As explained, to draw the FBD, we need to show all acting forces, whether action or reaction forces. So, we remove the supports and place the reaction forces they will provide instead, as in the figure above.

The hinge support at point A can create both vertical and horizontal reactions, while the roller support at C can only create vertical reaction. We show the vertical axis with y and horizontal axis with x. So for example, F_{ax} means, the reaction force F at point A, in x direction.

In the figure above, you can also see that we resolved the external force at B into its x and y components, to make our job easier, because it was inclined but we want to deal in only x and y directions, for convenience. Let's start with finding the components of force at B:

F_{bx} = F . cos30 = 30 x 0.866 = 25.98 N. Let's say 26 N.

F_{by} = F . sin30 = 30 x 0.5 = 15 N

Here we took the right and downwards directions as positive. We can choose any direction we want as positive. If we find the reactions as positive it means they act in the direction we chose. If we find negative, it means they act in the opposite direction of what we chose. The important thing is, once we choose a direction as positive, we need to keep it that way till the end.

Now we must remember the equilibrium conditions of: $\sum Fx=0$, $\sum Fy=0$ and $\sum Mz=0$.

In x direction, we have F_{ax} and F_{bx} only. Per the equilibrium condition above, their sum must be zero. So we can write them as: $F_{bx} + F_{ax} = 0$

We already found F_{bx} = 26N. So,

$F_{ax} + 26 = 0$, which gives us:

$F_{ax} = -26$ N

Now you see that the sign is minus, which means, it acts in the opposite direction to the positive direction that we chose before, which means, F_{ax} acts towards left. This makes sense, because F_{bx} was acting towards right, and since there is no other force in horizontal, they need to cancel each other to satisfy horizontal force equilibrium.

Now let's take a look at the vertical forces. To satisfy equilibrium in vertical, we can write them as:

$F_{ay} + F_{by} + F_{cy} = 0$

Here we know F_{by}, which we had found as 15 N. So,

$F_{ay} + 15 + F_{cy} = 0$

Now we have two unknowns and one equation here. But we still have one more equilibrium equation we can use, which is the one for moment. The sum of all moments must also be zero.

If the moment is zero, the beam is not rotating. If it is not rotating, then any point on the beam is also not rotating. Therefore, we can choose a convenient point, and take moments of forces with respect to it. We have F_{ay} and F_{cy} which we still do not know. So, if we take moment with respect to either point A or C, we can eliminate one of them from the equation, because a force cannot create any turning effect on the point it is acting. So let's choose point A and take moments of forces with respect to it. The forces at point A are eliminated since they already pass through point A. The horizontal force F_{bx} is also eliminated, because its line of action also passes through point A, so it cannot create any rotation about A. So all we have that can create moment on A are the forces F_{by} and F_{cy}. So let's write their moment with respect to point A, add them, and make it equal to zero, for moment equilibrium. Using M=F.d, we can write it as:

$F_{by} . 4 + F_{cy} . 10 = 0$

The distances of 4 and 10 meters can be seen from the FBD. This was the second equation we needed. Now we can insert F_{by}, into this equation, as we had found it as 15 N before. So it becomes:

$15 . 4 + F_{cy} . 10 + = 0$

From here, we find $F_{cy} = -6$ N

Again, the minus sign tells us that it is in opposite direction of positive, which we chose as downwards here, which means F_{cy} will act upwards.

Now we can put it in the vertical force equilibrium equation that we wrote on previous page, which was, $F_{ay} + 15 + F_{cy} = 0$, and insert $F_{cy} = -6$, into it, to find F_{ay}. So we will have,

$F_{ay} + 15 + (-6) = 0$

Which makes $F_{ay} = -9$ N

Again, its minus sign indicates that it acts upwards. Now we solved all reactions. See below:

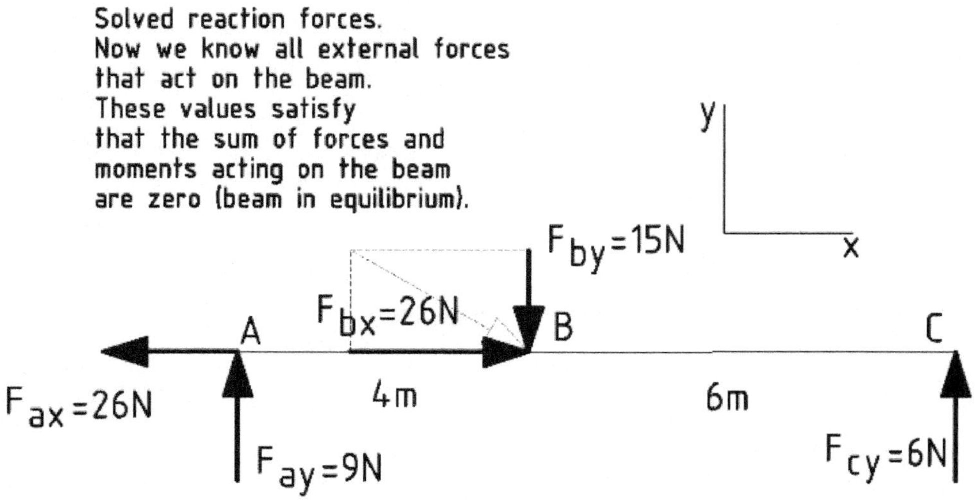

Figure - Statics 12c

Force and Moment Diagrams

Above, we saw how to find reaction forces provided by supports. But this is not enough. They are still external to the structure. We must also know the forces and moments internally, anywhere in the beam, which are then used to *see* if the beam will yield or how much it will axially deform or bend (covered under strength of materials topic next).

We can find internal forces by using the logic that, if the entire beam is in equilibrium, then any part of it must also be in equilibrium. This way, we can find forces and moments anywhere in the beam and then display the results on a graph. These are called axial force, shear force and bending moment diagrams (graphs). They show forces and moments anywhere in structural members at a glance and they are the first thing an engineer looks at, when he wants to see forces and moments in a member.

To keep things simple, let's only consider some significant locations on the beam, and then plot and connect their results on a graph. To see the big picture first, take a quick look at Figure Statics 13-c, a few pages forward. That is where we will reach. The steps below describe how to obtain that graph.

Again let's take x direction to the right as positive and point A as x = 0 for our beam, so point C will be at x = 10 meters. So, now let's consider significant points on the beam, and find internal forces and moments for those:

For x=0m (Point A):

Here we have the reaction forces from hinge support. We had found those as:

F_{ax} = -26 N

F_{ay} = -9 N. This is all we have at point A. Moment is zero here (which makes sense since there is a hinge here).

For x=1m:

See below. On the left we have the reaction forces at A. Since this portion of the beam must also be in equilibrium, on the right we must have forces to cancel them as 26N and 9N as shown. But this is not all. The vertical forces at point A and point x=1 create a force couple, which means moment. So, the moment at x=1 to counter this force couple will be:

M = 9N x 1 meter = 9 N.m

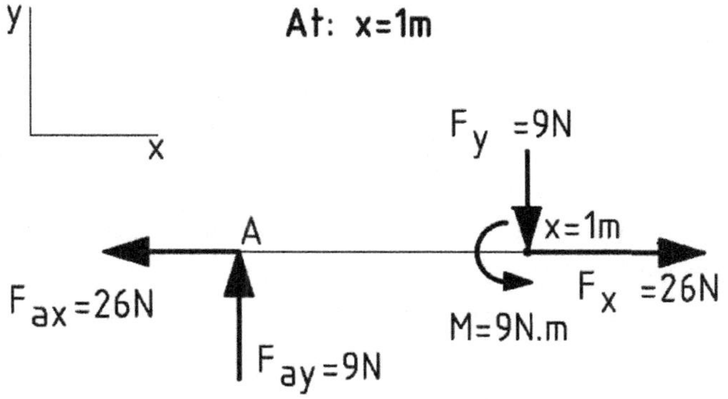

Figure - Statics 13a

Note that we could have found the same moment also by taking moment of forces with respect to anywhere on the beam, such as point A.

Horizontal forces are on the same line of action, therefore they do not cause any moment, which will also be true for the rest of the beam.

The forces in y direction are transverse forces here, as a shear force, because they are perpendicularly aligned to the beam's main axis. The forces in x direction are called as axial forces (tension or compression), as they act along beam's axis.

For x=4m (just to the left of B):

Now we are at x=4, but we stop slightly to the left of point B, before we reach to the external force at B. So, our reactions will still be the same, 26N and 9N.

This time we will have more moment because the distance between vertical forces increased:

M = 9N x 4 meters =36 N.m

You can also see this value in Figure Statics 13-c from moment graph.

For x=4m (just to the right of B):

Now we moved a little bit to the right, and we are still at x=4m but this time we already passed the point of application of force at B. So now we must also consider it. See Figure 13-b below.

We had found horizontal and vertical components of B as,

F_{bx} = 26 N and F_{by} = 15 N

Adding the forces in x and y directions to reach equilibrium, we get

$-26 + 26 + F_{(4m)x} = 0$

$-9 + 15 + F_{(4m)y} = 0$

So, at x=4 meters, just to the right of B, we find the forces in x and y direction as:

$F_{(4m)x} = 0$ N

$F_{(4m)y} = -6$ N

As you can see, the external force at point B suddenly affected the internal forces in x and y direction. On the graph in Figure Statics 13-c you can also see this sudden jump.

Pay attention that the horizontal force is now reduced to zero. And it will remain that way until point C, where we also have no horizontal reaction. So the only portion of our beam that has axial force is from point A to B (which acts as tension here). The value of shear (vertical) force also experienced a sudden change, in the amount of the vertical component of force at B.

The moment however will not be affected. This is because, at point B, although we have the force at B now, that also changed the value of the internal vertical force at beam's right face. So their difference is still the same which is 15 - 6 = 9 N pointing upwards, so the moment is still 9x4 = 36 Nm.

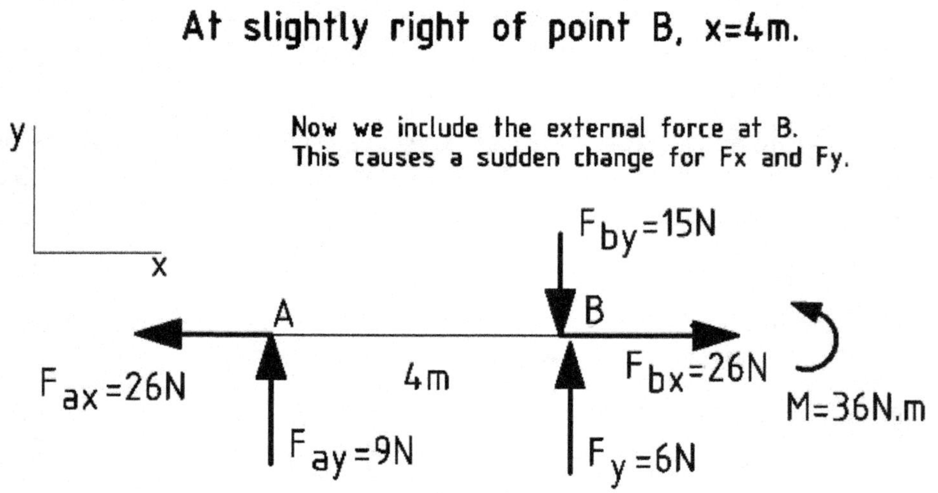

Figure - Statics 13b

<u>For x=5m:</u>

From x=4 until point C there is no external force, but let's show an intermediate point, such as at x=5. The x and y forces will remain the same. But moment will be affected because now the distance of internal vertical force of Fy changed, with respect to point A. Taking moments with respect to A:

$F_{by}.4 - F_{(5m)y}.5 + M_{(5m)z} = 0$

Inserting F_{by} and $F_{(5m)y}$ into above: 15x4 - 6x5 + $M_{(5m)z} = 0$ which gives us $M_{(5m)z} = 30$ N.m

So as you see, the moment started to decrease. This makes sense because the y component of external force B tries to turn the beam in clockwise direction but now the y component of force at x=5 tries to turn it in counter clockwise direction.

For x=10m (Point C):

This is where we have the roller support, which only provides vertical reaction force.

The vertical force in y direction, will be changed in the amount of vertical reaction force at C. So if we now add the vertical forces at A, B and C, we will get: -9 + 15 + (-6)=0. So, at point C, on the right face of the beam, no internal vertical force remained.

Moment also becomes zero here as we have roller support. You can also verify this by taking the moments of the three vertical forces with respect to anywhere on the beam.

Below, the Axial Force (N), Shear Force (V) and Bending Moment (M) diagrams are shown.

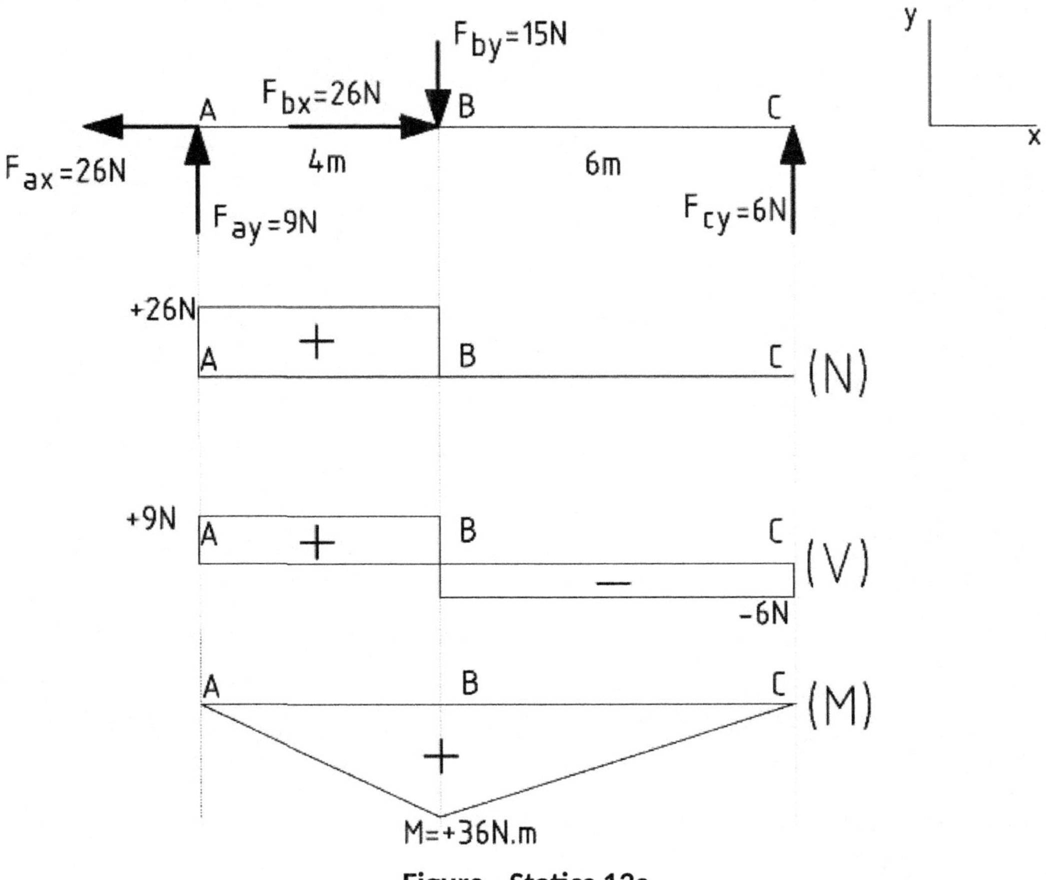

Figure - Statics 13c

Note that the moment diagram is shown downwards as positive, which is the conventional way, just to remind us that the bending of the beam this manner indicates positive moment. These internal forces can later be used during designing of the beam itself, such as its material and cross section, so that it will be able to withstand these internal forces and moment.

Now that we finished this exercise, a few useful side notes...

In the example above, if the load acted vertically, we would not have any axial force anywhere in the beam. But this doesn't mean that we would not have axial stress because there would be compressive and tensile stresses in the beam just because it bends. You will see this when we study bending stresses. This is another use of simply supported beam. By allowing the member to release moments at its ends, and also allowing one end to move freely, a simply supported beam allows only shear force and moment to develop in a beam, although we will still have axial stress in the beam due to bending but then we will know that that axial stress is purely from bending.

In the paragraph above we said that if the load acted vertically, we would not have any horizontal force anywhere. It means, the horizontal reaction that the pin support at point A is able to provide would also be zero. Then in such a case, you may ask, "What is the use of having pin support at A, when we only had vertical force acting on beam - we might as well have a roller support at A?". The answer is that, in such a case, the beam would be unstable. It is true that under the action of a perfectly vertical force, we would have zero horizontal forces, but, having that pin at point A ensures that our beam is stable against horizontal forces as well. For a structure to be stable, the supports of that structure must be able to satisfy all three equilibrium equations (speaking in 2 dimension), by providing reactions for them. Our beam provided both horizontal and vertical reactions, thus satisfied the two force equilibrium conditions, plus, although it did not have any fixed support to directly resist the moment, the supports at A and C worked together to keep the beam from rotating, so the moment equilibrium was also satisfied. For example a door is not an acceptable structure from a structural engineering point of view. It is not stable. We are easily able to rotate it, even if we cannot move it in either direction. That is called a "mechanism". A mechanism occurs, when something fails in a structure, it becomes unstable and moves like a mechanism, which is unacceptable.

Finally, do not confuse swinging rotation and bending flexure. The beam will bend in reality (although in statics we assume rigid members and do not yet deal with bending or axial deformations). Bending is different than swinging. Bending is the deforming of the member's body by a moment, which causes the member to cease being straight and start having a certain radius. Swinging happens around a hinge like a door, but while it happens, the member doesn't necessarily have to bend and may remain straight.

The force and moment diagrams have the characteristics as below, given mostly for future reference if needed. You can also verify the statements below from Figure - 13c:

- The slope of shear graph at any point is equal to the value of load graph at that point. *(The load value is equal to the derivative of shear function)*
- Moment graph's slope is equal to the shear graph's value. *(The shear value is equal to the derivative of moment function)*
- Shear value changes as much as the area of loading on the beam. *(Change in shear between two points is equal to the integral of load function between those two points)*. If a uniform load acts downwards, when you analyse starting from left end, it reduces the shear value over that span. In our example there was no uniform loading, so shear didn't vary over a distance.
- The moment will increase as much as the area under the positive shear diagram. *(Change in moment between two points is equal to the integral of shear function between those two points)*. For example calculate the area of the positive portion of shear diagram above. It is 9x4 =

36. As you see, the Moment is also 36 at point B, which increased by that value. This is when analyzing from left to right as we did.

- If there is a positive external Moment acting on the member, when we analyse from left to right, it will decrease the Moment value *(we did not have an external moment in the example)*.
- If a point load acts downwards, it reduces the shear value by that much (*in our example, points B and C*). If it acts upwards, it increases shear value. These are when analyzing from left to right.

The figure below shows the conventional way of taking positive values, which we also used.

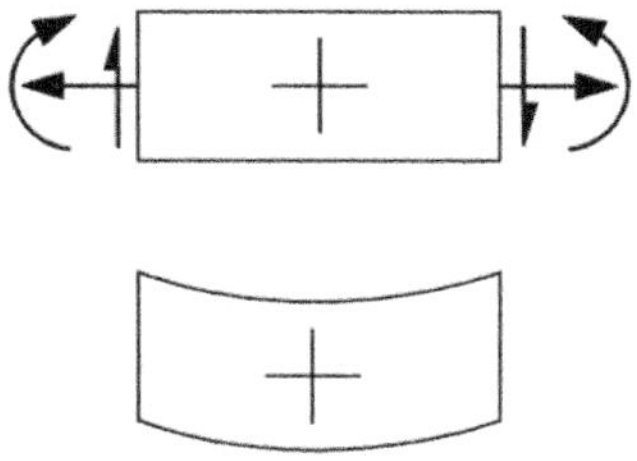

Figure - Statics 14

To verify if we also used this convention in our diagrams, check the values and directions in the results. You will see that when the forces we found on the right face of the beam segment was positive they were indeed downwards as in here. When the horizontal force on the right face of the beam pointed outwards it was positive. Also check the direction of the moment we showed when looking at beam segments. On the right face of the beam, they were in counterclockwise direction and positive as in the figure above. Positive moment causes the beam to bend as in the bottom part of Figure 14.

Knowing the internal forces and moments in members are very important for us to be able to design them to withstand these forces and moments. In a structural frame, for every member, these force and moment graphs are produced. They were being obtained manually in the past with indeterminate methods that we mention below and later discuss under structural engineering section. Now, since several decades, these are done by computer programs and by using more advanced methods, both of which, especially the former, have improved significantly since its emergence.

Determinate vs. Indeterminate Structures

In our beam example, we had only three reactions which satisfied the three equilibrium equations (in a 2D problem) which were also the bare minimum we needed for stability along vertical, horizontal directions and rotation. And since we had three equilibrium equations and three unknown reactions, would could solve all the reactions, which also meant that we had a determinate structure.

But what if we had more supports holding our beam? For example what if we replaced the roller support at point C with a fixed support? Then we would have more than just three reactions. This means, we would now have more unknowns to solve than the three equations provided to us by equilibrium conditions. As you may recall from basic algebra, we cannot solve it when we have more unknowns than the number of equations. We call such case an "indeterminate structure". More reactions on top of the three would be extra, or in other words redundant. Every extra reaction on top of the three means one more redundancy, which is also called degree of redundancy.

In statics we do not yet deal with indeterminate structures. To solve those structures, we need additional equations, which we obtain through the use of displacements of the structure. We will discuss methods to solve these types of structures under structural analysis section later, but first, we must study displacements, which are discussed in the strength of materials section next.

Indeterminacy (redundancy) is a good thing. Almost all structures are indeterminate to so many degrees. We want redundancies, so that when something underperforms or fails, loads can be redistributed in the structure, and the redundant reactions can take over or take more load and still ensure stability. The beam in our example did not have any redundancy, so, if one of its supports failed, it would be unstable.

Trusses

Trusses are formed by connections of straight members that form triangles which are connected to each other by hinges. The whole truss assembly which is made of many triangles takes advantage of geometry, and carries the same load with far less amount of material, in comparison to fully solid slabs or beams with the same overall outer dimensions.

Members in a truss can only be in tension or compression but not in bending or shear. This is because they are connected to each other with pins (hinges) and therefore no moment can be transferred to the members. In addition, in trusses the loads are applied only at these joints and not transversely on members. These are the main characteristics that differentiate a truss system from a frame system (such as the beam column system of a building). In a frame system, unlike truss system, bending and shear can also exist in members in addition to compression and tension, because we can have rigid or stiff connections between members such as beams and columns, and, the loads can also be applied on the frame members directly, such as applying a distributed load on a beam itself. In a truss system we must not apply transverse load on members because they are weak in bending. Truss members are designed to resist only axial loads, meaning compression or tension, but not bending or shear.

Trusses can behave like beams, columns or even domes. In the figure below you see a truss beam and a solid rectangular beam of the same height. Their moment of inertia (explained under next section), therefore moment capacity, are close to each other, although the truss uses far less material which means significantly lighter structure, and enables passing larger spans as a result.

The most basic component of a truss is a triangle, which is made of three members connected to each other in the form of a triangle, at their ends. Then, many triangles are connected to each other to make up the whole truss system, to obtain any geometry required. Trusses can be used in bridges, large roofs, domes, buildings, in heavy machinery such as cranes and so on...

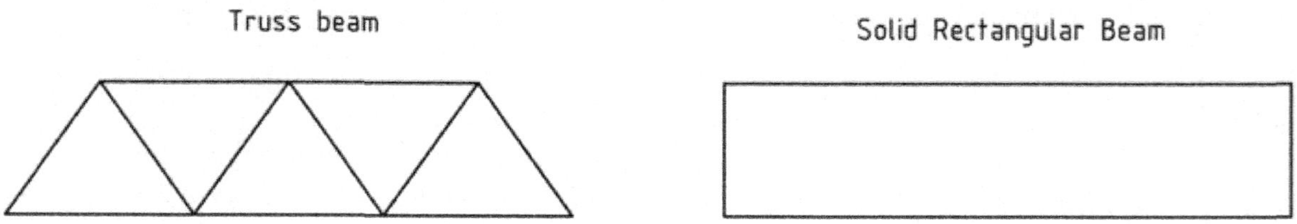

Figure - Statics 15

The first photo shows a large roof made of truss elements in the shape of a dome. The photo on the right shows a small pedestrian bridge made with steel angles and channels and W beams, arranged in the form of truss. Although all members are in compression or tension only, together the whole truss assembly is able to carry a considerable amount of bending moment. Also see Steel Construction section, and Chapter 3 > Bridge Classification > Truss.

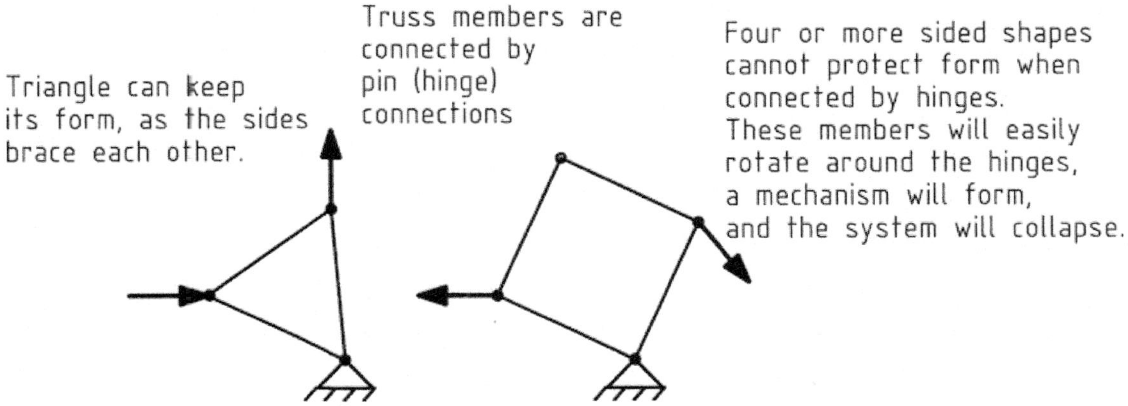

Figure Statics - 16

See figure above. Triangle is the only shape which, if made by connecting its members by hinges, that would still protect its form because each of the three sides braces the other two. A shape other than a triangle, if made by hinge connections, would easily collapse, because the overall geometry can change by rotation around hinges. Therefore, pinned assemblies with more than three sides must be reduced to triangles. This is why all the trusses you see are made of triangular components.

Compression members are generally governed (*meaning the most critical*) by their buckling resistance and tension members are generally governed by the tearing (rupture) strength of their connections, or the member's own yield strength. Especially buckling can be a very sudden failure.

We will not solve example in truss analysis but they are also ultimately solved with the same principles of force equilibrium, as we showed in statics, but with different methods. Still let's just list and describe the methods to analyse statically determinate trusses. There are two methods:

- Method of joints: In this method we look at force equilibrium of joints, which are hinges, by drawing their free body diagrams. Remember, no moment can exist on hinges so we can only have horizontal and vertical force equilibrium. The forces on the joint are applied by bars that connect into that joint.
- Method of sections: In this method, we remove a portion of the truss, and look at the equilibrium of one section.

When we solve trusses we assume that: no moment exist in the bars and they are perfectly straight, loads act on the joints only, and the hinges are entirely frictionless, thus resist zero moment.

Sometimes we can immediately understand that the force in a truss member must zero. We call these as zero force members. For example if only two members which are not on the same line connect to a joint, and there is not external force on that joint, both members AB and BC connecting to that joint must have zero force, otherwise the force equilibrium will not be satisfied, as you can see below in the left figure, because otherwise there would always be a net remaining force in x or y direction.

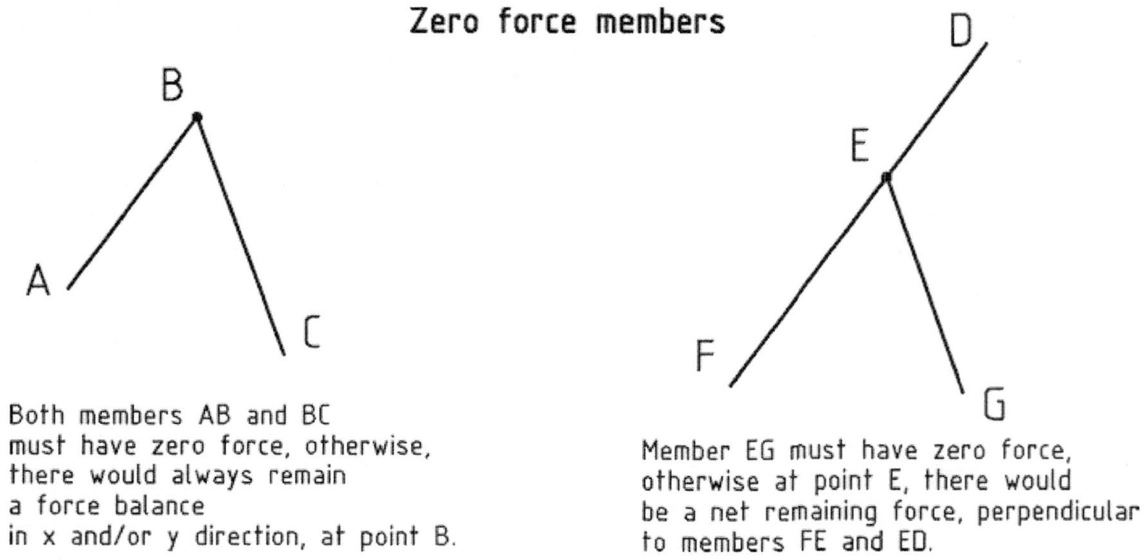

Figure - Statics 17

Another case when a truss member has zero force is when there are three members connected and two of them are colinear (members FE and ED) and one of them is not (member EG) as in the figure on the right. In that case, if the third member (member EG) had force, the force equilibrium would not be satisfied. Therefore, the force on the third member (member EG) must be zero.

STRUCTURAL BASICS: STRENGTH OF MATERIALS (STRESSES AND DEFORMATIONS)

In statics, we studied forces applied on objects but we did not consider what that object was made of, how it deformed, or when it failed. We assumed everything was rigid, which never deformed or failed. In this section, we will take these into account and study other very necessary concepts.

Stress

We can define stress as the force that falls (distributed) over an area. In other words, force per area:

Stress = Force / Area

Since we divided force by area here, then the units of stress must be pounds per inch square, $lbs/inch^2$ (psi) or Newton per millimeter square, N/mm^2 (Mega Pascal - MPa) or another stress unit that would conventionally be used in another industry.

To give an example, if we have a tensile force acting on the beam as below, we would have a tensile stress at the square cross section of that beam, trying to pull it apart. For example, let's say, the value of the tensile force was 20,000 Newtons (or 20 kN, read as "kilo" Newton), and the cross section area of the beam was, say, 40 cm^2, as below:

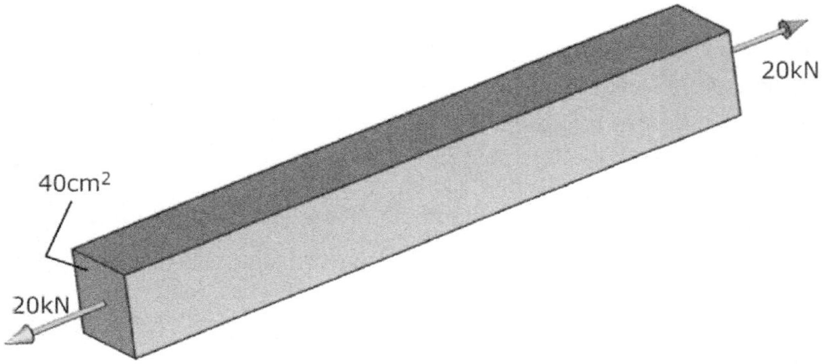

Figure - Strength 1

Then we would have a tensile stress value as:

$\sigma = 20000 \text{ N} / 4000 \text{ mm}^2 = 5 \text{ N}/\text{mm}^2 = 5 \text{ MPa}$ (Mega Pascal)

Note the symbol "σ" (sigma) we used, for stress. In civil engineering, sigma is used for axial stress, which will also be used as such, in the rest of this book. Axial means, as we saw in statics, something that acts in the direction of the main axis of a member.

Strength and pressure also have the same units as stress. Strength is the ability of a material to resist stress. Pressure is the effect of something external. Stress is produced in the material as a result of that pressure. Then we compare material's strength to the stress that is produced in the material.

For example, the beam above would not fail, if it had a tensile strength of more than the applied stress, which is 5 MPa. Strength has thresholds, such as yield, ultimate and rupture. We will show them soon.

The following are typical examples which are expressed in stress units:

- Strength of materials we use, such as concrete, reinforcing steel or structural steel strength
- Bearing and shear strength of soil
- The pressure that the soil applies horizontally to retaining structures (lateral earth pressure)
- Fluid (and in civil engineering, fluid almost always means water) hydrostatic pressure

Concrete used in general construction usually has compressive strength specified as anywhere from 3000 to 6000 psi in the US (although much higher strengths can also be obtained and used). For example for columns and foundations we may need the strong 5000 psi, while for beams and slabs 4000 psi may be ok, and for a slab on grade it can be 3000 psi. In countries with metric system concrete strengths can usually be specified as anywhere between 25 to 50 MPa for common construction projects. In some special projects, concrete with much higher strengths that may well exceed 10000 psi can be used. By concrete strength, we meant compression here, not tension. Concrete is useful because of its compressive strength. It is weak in shear and even weaker in tension, which is compensated by using reinforcing steel. Commonly used reinforcing steel typically has 60 ksi (60000 psi) strength. Steel is strong both in tension and compression.

As shown in statics, a force can act at any angle on an area, but, we can reduce it to its perpendicular and parallel components with respect to that area. Then, the stress obtained from the component that is perpendicular to the member's cross section surface will be called as "normal" or "axial" stress, and the component that acts parallel to that surface will be called as "shear" stress. So in the example above we had axial stress, which was tension in this case.

To calculate shear stress, we would divide shear force by the area where surfaces tend to slide against each other. For example take a look at Figures - Statics 4a and 4b under statics section. The contact area is the surface area that both beams touch each other - although we do not have such configurations in reality, and we deal with shear stresses within a member.

To give a numeric example, if the force we applied in Figure - Statics 4a and 4b was 10 kN, and the contact surface area where two sides tended to slide against each other was, say, $40cm^2$, we would obtain a shear stress value of:

$\tau = 10000 \text{ N} / 4000 \text{ mm}^2 = 2.5 \text{ MPa}$

Shear stress is shown by " τ " (Tao), <u>which will also be used in the remainder of the book.</u>

A final note about stress, which we will not mention again, is that stress is not a vector but it is something called "tensor", which carries more information than a vector because in addition to magnitude and direction, stress also carries information of a plane surface (perpendicular or parallel) on which the force (vector) acts.

Strength

The highest level of stress that a certain material is able to take is called the "strength" of that material. As we will discuss, there are different thresholds when yielding, such as plastic yield, ultimate yield, rupture / fracture yield, in all of which, the word strength can also be used in place of the word yield.

Strength also has the units of stress as: Force / Area

Stiffness and strength are often confused with each other, but they are different. Stiffness is the force per unit deformation as we will see. In general, stiffer materials are stronger too, but it may not always be the case. For example, two members may fail under the same stress, thus their strengths will be the same, but one may go through much more deformation or deflection than the other to reach to that point. The material that deforms less is stiffer, although the strength of both members may be the same. *(For example a rubber material may be very strong, but it has low stiffness, and even if it may take a lot of stress before yield, it will take little force to deform it.)* Stiffness is not only in terms of axial deformation. There are shear, bending or torsional stiffnesses as well. We will discuss stiffness too.

Deformation

Deformation is the amount that a member's body changes shape, which means deforms, under load. It is expressed in terms of length, angle or volume. For example a steel rod can elongate a certain amount when pulled, a concrete column will shorten a certain amount under heavy loads from above. Up to a certain stress level, deformation can fully recover when the load is removed. After that, some permanent deformation will remain. We show axial deformation with the symbol "δ".

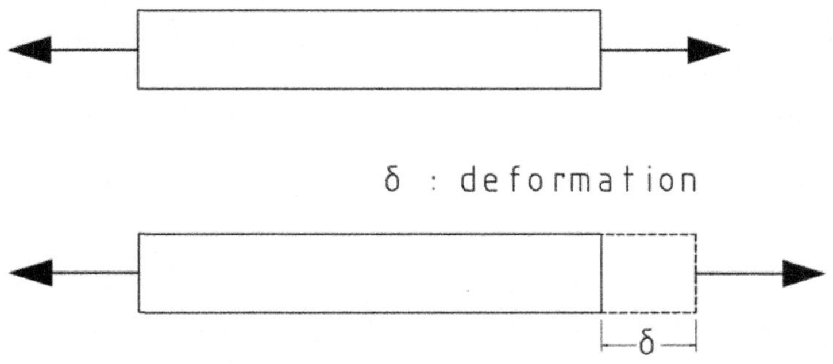

Figure - Strength 2

At this point, let's clarify: Deformation is often confused with deflection. They are not the same. Deflection is <u>not</u> the deformation of the body of material itself. Deflection is the movement of all or part of a certain structural member or structural assembly in a certain direction, under force. The body of the member of members do not necessarily deform in this case. For example a concrete slab may negligibly deform, but with the whole structure together, it may significantly deflect horizontally during an earthquake. Or when a beam bends, the middle point of the beam deflects with respect to its original horizontal position and there is bending rotation although the axial deformation is negligible.

<u>Important Note:</u> In the rest of the book, when we use the term <u>"displacement"</u> it will generally mean one or more of the <u>axial or lateral deformation, deflection, drift, bending rotation (or simply bending).</u>

swinging or twisting that may or may not happen simultaneously. Otherwise we will use the applicable term, if we mean one specifically. Also note that commonly, displacement tends to be more often used interchangeably with deflection but regardless of that we will use it as described in this note.

Strain

Expressing deformation may not be very informative for us always. It tells us something but what if two rods of different lengths deformed the same amount for example? We can not consider these two deformations equivalent to each other for our structural calculations because they deformed differently per their original lengths. We need another term that takes the original dimension into account as well. And that term is strain. So, strain is the deformation per original dimension of that material. Similar to deformation, strain can be axial, angular or volumetric.

So we can write strain as:

$$Strain = \frac{Deformation}{Original\, Dimension}$$

As you see, we divided length by length here, such as mm/mm or in/in. This means, strain has no units. We could do the same for angular or volumetric strain, and would be left without units. Axial strain is expressed by the symbol "ε". So it can be written as:

Axial strain = $\varepsilon = \delta/L$ = axial deformation / length

As an example, if we have a concrete test cylinder of 12 inches in length, and when we load it under compression it shortens by 0.03 inches, then it means the axial strain is: 0.03/12 = 0.0025 in this case. In addition to axial, there will also be lateral strain here, perpendicular to the cylinder's main axis, even if we did not apply load in that direction. We will see this under the term Poisson's ratio.

Moment of Inertia

Moment of inertia, shown by the letter "I", is a geometric property of a body's cross section, which increases when its mass accumulates further away outward from the cross section's center, with respect to axis of bending. The higher the moment of inertia, the more resistance to bending rotation. It means, for any member that may bend, such beams, columns, slabs, we must know this property.

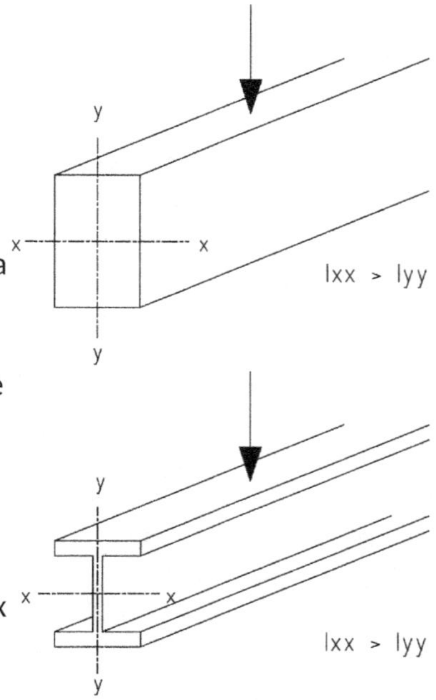

Figure - Strength 3

In the figure, for both beams, the moment of inertia with respect to x-x axis is higher. This is why it is better to orient these two beams as shown in the figure to take on this particular load from above. For a rectangular beam, $I=b.h^3/12$, where b is the beam width and h is the height, which also confirms why bending resistance is more when loaded as shown in the figure, because the height of beam has great influence on moment of inertia, by its cube, than its width. Similarly, for the I-beam, we can find the moment of inertia by dividing it into three rectangles, two for flanges and one for the web, finding individual moments of inertia for all rectangles by the formula $I=b.h^3/12$, and then adding them, where the height for all rectangles in y direction will again influence the result the most. Both b and h are distance units. So moment of inertia has the unit of length to the power of 4, such as mm^4 or in^4.

Also see that if the height of both beams in the figure above in y-y direction were equal, moment of inertia of the I-beam would not be much lower than of rectangular beam, *(because height is what matters the most)* which means, an I-beam is far more efficient section for bending, with respect to a solid rectangular beam of same height, because we use far less material but achieve almost the same moment of inertia.

Modulus of Elasticity (E)

Now we are at a point that we can combine the concept of stress with strain, which gives us one of the most important properties about a material. Stress and strain can be related to each other as:

$$\sigma = E \, . \, \varepsilon$$

In the formula above, we had already introduced stress, σ, and strain, ε. They are related to each other through "E", which is called "Elastic Modulus" or "Modulus of Elasticity". It is also known as "Young's Modulus", after the British scientist Thomas Young, who studied it in 19th century.

Above we said that the strain, ε, has no units. This means, Elasticity modulus must have the same units as stress, such as MPa (N/mm^2) or psi ($pounds/inch^2$).

Modulus of elasticity, E, is a <u>unique (inherent, characteristic) property</u> of a material, which means, the relationship between stress and strain for every material is <u>unique</u>. E is <u>one of the most important</u> material properties used in civil engineering and other fields such as material's science, metallurgy, mechanical engineering. Not only materials that we use in our structures, but also soil and rock have their own elasticity moduli. For low stresses E is usually very close to or can be assumed as a constant value. This makes our job much easier during design. For higher stresses, we may need to incorporate varying (nonlinear) E values into our design.

Bending (Flexure) of Beams

Note: This may seem as one of the relatively more technical topics if you are not an engineer, but as always, the intent was to explain everything as plainly as possible by also considering nonengineers. If something below seems too technical, remember that the info from here will only be needed later under structural analysis section and nowhere else, but it is still recommended to pay attention because this topic describes the most basics of how frames behave in its most fundamental level, on which ultimately a lot of things in structural engineering depend.

The trusses that we showed at the end of statics section, have no moment transfer between their members. For them, everything takes place through axial forces and the members do not experience any bending, and there are no beams.

On the other hand, a structural frame is a structural assembly made of beams and columns. The most fundamental characteristic of a structural frame is the transfer of moment between the beams and columns, and the bending of members.

Therefore trusses and frames are inherently two distinct major types of structural systems. (Another structural system is surface structures, such as tanks, silos, etc..., which is not covered in this book)

When analyzing structural frames, to understand first the bending of beams is essential. Beams are elements that resist forces through their bending. In some beams there can also be compression. In such cases, it is called a beam-column. In this section however, we will look at only beam bending, which is also called beam flexure. The bending formula is based on the Euler - Bernoulli beam theory after their study in 18th century.

To understand "bending of beams" means, we are able to mathematically plot the bent shape of the beam, and calculate slopes and deflections at any point of interest on the beam, find the maximum moment and its location, under the action of forces and moments and the support conditions. As we will show later under structural analysis section, slope-deflection equations are fundamental in solving indeterminate structural frames.

The bent shape of the beam is known as the "elastic curve". It is the shape of the line that passes from the center of the beam's cross section (geometric centroid), in other words, it is the shape of beam's centerline as shown below, which is a highly exaggerated figure *(the beam bending below is shown as not symmetrical on purpose, to emphasize that bending is not necessarily circular, symmetrical or regular in shape... but solely depends on applied loads, beam stiffness and geometry, and support conditions)*:

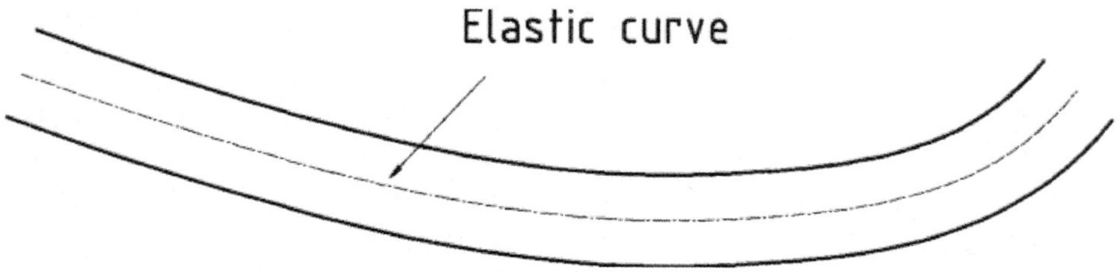

Figure - Strength 4

To determine the elastic curve means, we must find a relation between bending moment and beam geometry. To do that, we assume that we apply pure bending on a beam (i.e. no shear or axial stresses other than from bending itself). We assume that the deflections and slopes are very small, the beam behaves as linear elastic (its E doesn't vary), beam cross section is symmetrical, forces act towards the center of the beam (not eccentric), plane sections remain plane and each cross section of the beam remains at 90 degrees to the neutral axis (we will describe neutral axis in next section). Under such assumptions, and after some math which we will skip (which could be found in many structural focused resources under beam bending), it can be shown that the radius of curvature (ρ) of elastic curve is obtained in terms of bending moment (M), elasticity (E) and moment of inertia (I) as:

$1/\rho = M/EI$

Here ρ (Greek small letter "rho") is the radius of curvature of the elastic curve, which is the radius of an imaginary circle that would have the same radius of the curve at the particular point. And the reciprocal of the radius of curvature is the curvature itself, which is denoted by the Greek letter "kappa", κ. Thus, $\kappa=1/\rho = M/EI$.

The equation above can lead us to the equation of the elastic curve, because now in our hand we have a relation of how the beam bends (curvature), in terms of its elasticity, moment of inertia and the applied bending moment. To obtain equation of elastic curve, let's express it as a function. See below.

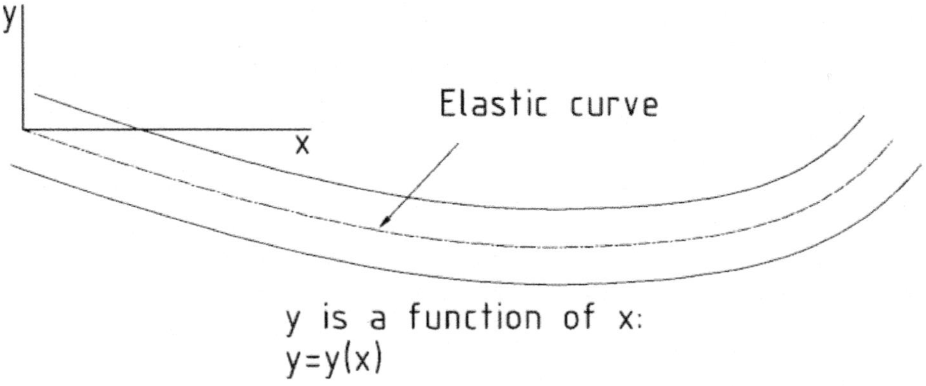

Figure - Strength 5

If we call the beam horizontal axis as x axis, and the vertical axis as y axis, then the deflected shape of the beam (elastic curve) can be written as a function of x, such as: y=y(x), where y is the position of the elastic curve line for any given x location (notice how the origin of x and y axes in the figure below starts from the beginning of elastic curve). After some math and by assuming that the deflections are very small so that we can ignore a certain complex but negligible term in the process, we can obtain the second derivative of this function y(x), written as y''(x), as equal to M/EI. So,

$$1/\rho = y''(x) = dy^2/dx^2 = M/EI$$

This is the beam equation based on Euler - Bernoulli beam theory, conveniently called as the differential equation of elastic curve, and one of the most critical contributions in structural engineering, which made possible to build far better structures after its development and application since couple hundred years ago. Ultimately, so much of structural engineering stems from right here.

Note: Derivative is the mathematical expression that shows the rate of change of a function, or in other words, slope of a function at a particular point. So then the first derivative, y'(x) is equal to the slope of the elastic curve for any given point. This is also used a lot in solution methods we will describe below, and also in the slope deflection equations we will see in Structural Engineering>Structural Analysis. y''(x), which is the second derivative, is therefore, rate of change of rate of change, in other words, the slope of the slope, or the rate of change of slope, often written as y'' for convenience. Above, the expression y''(x) means the same thing as dy^2/dx^2. They are just different ways of showing 2nd derivative.

So now we have just obtained a direct relation between the bending moment, modulus of elasticity, moment of inertia and the beam's elastic curve function in x-y plane, although this is still through its second derivative and not itself. Then, to obtain the y(x) itself from the second derivative y''(x), since integral is the opposite of derivative, of course the most straightforward thing to do would be to apply integral twice to y''(x), which is called the double integration method. There are also other methods which we list below. Again, to wrap things up here: Obtaining y(x) means obtaining the elastic curve itself, which means reaching the solution we look for, such as, having the actual equation of elastic curve, slopes and deflections anywhere we want, the maximum moment and its location, in short, solving the beam. The methods to solve the beam can be listed as:

- Double integration method
- Moment area method
- Conjugate beam method
- Virtual work method

Double Integration Method:

In this method, as we said above, the equation y''=M/EI is integrated twice in order to obtain y, which gives the deflected shape of the beam. Integration yields one constant at each step, thus we will have two constants. These constants can be solved by imposing movement conditions of supports (called boundary conditions), such as for example if at x=0 we had a pin support, the beam cannot deflect vertically there, so y must also be zero. So we can impose x=0, y=0 to the equation we find. The moment must also be zero at a pin so we can also impose x=0, M=0. Or for example if we had a fixed support at x=L, where L is length of the beam, we would have zero bending (thus slope, which means the first derivative of y' would be 0), and zero deflection there as y=0, so in other words into the elastic equation we would insert x=L, y'=0 and X=L, y=0 and so on... This way we can solve for the constants, which gives us the actual equation y(x). This method is appropriate when taking integral is easy and gets us to solution quick. In other words, when there is only one equation throughout the length of the beam which is easy to integrate and not divided into segments over the beam length.

Moment Area Method:

Double integration can be difficult sometimes, when performing calculations manually especially if the function y(x) is made of different segments or difficult to integrate. In such cases moment area method, which is a graphical method can be useful. Here first we obtain the moment diagram, as described in statics. Then we simply divide the values of this diagram by EI ("E" times "I"), to obtain an M/EI graph.

Below may seem complicated at first, but actually it is not, so just pay a little attention, because in the next few short paragraphs a really fundamental subject (moment area theorems and slope deflection) is described. <u>This is the connection link from what you learned so far in this book about the structural fundamentals to what you will learn in structural engineering > structural analysis section later.</u>

Remember, derivative means rate of change, in other words, slope. y' is the derivative of y which is the rate of change of y with respect to x. This simply describes <u>slope</u> of elastic curve. So if we call the slope as θ (Greek letter Theta), then we can write it as θ = y'. Then, if we take the derivative of both sides, we can write θ' = y''. And from before, we know that y''=M/EI. So, θ'=y''=M/EI. *Note: θ is more properly written as dy/dx, and θ' as dθ/dx, where "d" stands for derivative. So we have dθ/dx=M/EI.*

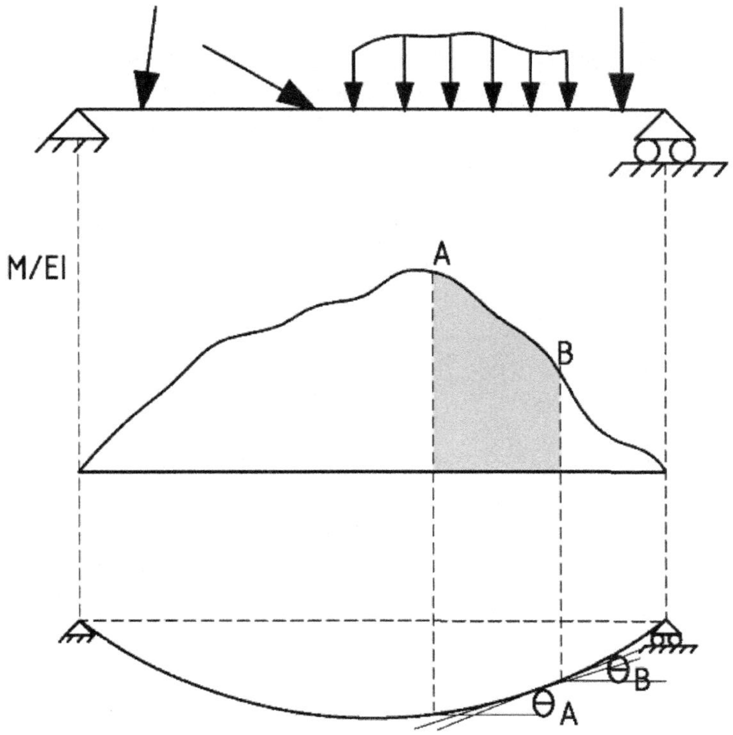

Figure - Strength 6

Now if we take integral of both sides, it can be shown that the slope of point B (θ_B) with respect to point A (θ_A) on the elastic curve is equal to the area under M/EI diagram between point A and B as shown above. This is called the first moment area theorem. So, in other words $\theta_{B/A} = \theta_B - \theta_A$ represents the bending rotation of point B, with respect to point A.

And from the integration it can also be shown that the deviation of tangents between two points is equal to the moment of the area under M/EI diagram with respect to the first point. (*Moment of an area is found by multiplying the area between A and B with the distance between the centroid of the area to the point of interest, such as point A or B*). This is called the second moment area theorem.

With these, we can find slopes and deflections anywhere. We can also obtain the relation of moment to rotation and swinging and also E and I. These form the essential inputs of the slope-deflection method that we will see under structural analysis section, to solve indeterminate frames, which becomes an ingredient to structural analysis methods even the computer programs use today for solving frames.

Other than the double integration method and the moment area method, there are also other methods such as the conjugate beam method and work-energy (unit load) method to solve elastic beam, which we will not discuss. Work and energy principles is a powerful concept that has many applications, including solving indeterminate structures, analyzing structural stability.

Bending Stress:

One very important use of the bending analysis of beams that we have just studied above, is determining the bending stresses in the beam. As a result of bending, a part of the beam experiences compression and its other part experiences tension in the form of axial stress. So bending stress is axial stress. The tension and compression zones are separated by a neutral axis, where no tension or compression stresses exist. Magnitude of axial compression and tension stresses due to bending is zero at neutral axis and increase towards the top and bottom fibers of the beam as seen below.

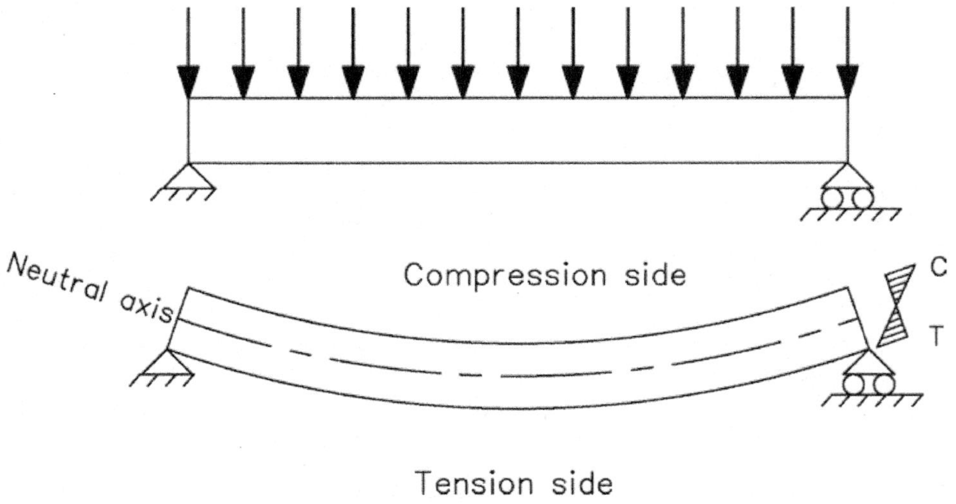

Figure - Strength 7

The axial stress at a certain point along the height of the beam can be derived from the formula we gave above for the elastic curve and it can be expressed as:

$$\sigma = \frac{-My}{I}$$

σ is the stress in axial (long) direction of the beam, M is the bending moment at that location (*at that location means, the value of M that we can see from a bending moment diagram*), y is the distance between that point on the cross section and the neutral axis. "I" is the moment of inertia of the beam cross section. *(Note: Negative sign is just because when we have positive moment, the stress will be compressive when y is positive).*

For example, you saw that an I-beam will have much more moment of inertia than a rectangular beam of the same value of cross section area. So from the formula above, the I-beam will experience far less bending stress under the same bending moment, in comparison to a beam of same amount of cross section but rectangular in cross section.

Again, the stress we expressed above is due to bending only, and it acts in the axial direction. If in addition to this axial (bending) stress, we had other axial stresses due to axial loading, then we can add them to each other, as they would all be axial. So for example, if a beam was already under compression from a compressive force, and if it also experienced bending, we add the compressive stress to the compressive portion of bending stress and subtract the compressive stress from the tension portion of bending stress (and the result will be in the direction of the larger stress). A beam which is subjected to both bending and axial stress is called a beam-column.

We still have shear stress in the beam above. That did not go away. But in this section we are talking about bending stress, which is in axial direction.

Also pay attention that the beam above is simply supported, similar to the beam example we showed in statics section. The simply supported beam can now hopefully be better understood, because here all the axial stress we have in the beam is purely from bending, as the beam end will move horizontally freely and no horizontal reaction will occur (*in the example in statics section if our externally applied force was not inclined, we would also have no axial stress there*).

Stiffness

Stiffness is the amount of force or moment that needs to be applied on a material for a certain amount of axial deformation or bending rotation (which is also a deformation). In other words, stiffness is the load per unit deformation. Here, by load, we will mainly mean axial force or bending moment, although there are also torsional stiffness, shear stiffness or rotational stiffness of a spring. So, if the load is axial force, then the axial stiffness is the axial force per unit axial deformation, and if the load is bending moment, then the bending stiffness is the bending moment per unit bending. Stiffness is denoted by the letter "k". So we can write as:

Axial stiffness:

k_{axial} = Axial force / axial deformation

Bending stiffness:

$k_{bending}$ = Bending moment / bending rotation

In other words, if it takes a lot of load to deform a material, it is stiff in that direction. These stiffness values are used during analysis of indeterminate structures to obtain additional equations we need from deformations.

Stiffness depends on elastic modulus of the material, E, but these are not the same thing *(which are often confused with each other)*. Stiffness, whether axial or bending, increases as E increases, by a directly proportional relationship which we will see below. So, for an infinitely stiff, in other words, rigid material, E would also be infinite, which does not exist in the universe.

Try to see why E would be infinite for a rigid material, also from the simple equation $\sigma = E.\ \varepsilon$ that was given before. This is because, a rigid material does not deform, so its strain would be zero, no matter how big is the stress. If you divide stress on the left side of the equation by zero strain, then E would be infinite.

In reality, if the stress exceeds the elastic limit of the material, the elasticity of the material will start to vary - not that it cannot vary below the elastic limit but for our materials such as steel and concrete we most commonly take it as linear below elastic limit. So when considering stiffness, we normally take the initial linear value, unless we know that the elastic limit will be exceeded and we also want to perform a plastic analysis. Now let's take a closer look at axial and bending stiffnesses...

Side Note: For example in structural analysis programs, sometimes we set the E value of certain members to extremely high numbers (even if we cannot set to infinity) in comparison to other members' elasticities that, they practically become rigid members, and the programs treat them as rigid members. This makes analysis much simpler, instead of having to deal with the negligibly small deformations of such members, had we not set their E values very high.

Axial Stiffness:

In the equation $\sigma = E.\ \varepsilon$, if we insert P/A for stress, and insert δ/L (axial deformation / length) for strain, we obtain the formula for axial deformation in terms of the force and Elasticity of a material:

$$P/A = E.\ \delta/L$$

which is commonly rewritten as:

$$\delta = \frac{P.L}{A.E}$$

where δ is axial deformation, L is length, P is axial force, E is elasticity, and A is the cross section area.

If you look at the equation above, it makes sense. Axial deformation increases as the force and original length go up, and it decreases as the cross section area of the member increases or the elasticity increases (which means it is stiffer, as shown in the relation right below).

As far as the relation between axial stiffness, k, and E, let's remember that we had said that stiffness = k = Axial Force / Axial deformation = P/δ

So, if we insert this into the previous equation we just wrote above, we get:

$$\text{Axial stiffness} = k = A.E / L$$

So, <u>axial</u> stiffness is equal to modulus of elasticity, E, multiplied by the cross section area, A, of the member that resists the force, divided by the length, L of the member. Thus, stiffness depends on both E and the geometry. But E doesn't depend on geometry. E is an inherent, characteristic property, which is unique to that material, as mentioned before. E is independent of a material's shape and dimensions.

Now let's take a look at the units. Since k=A.E/L, we can write the units of k as:

Area x (Force / Area) x (1 / Length)

This makes the units of axial stiffness as: Force / length. This confirms the force / deformation relation.

Bending Stiffness:

Above we were able to write the axial stiffness = axial force / axial deformation relationship in terms of material properties and geometry. Now we need to know the same type of relation for bending stiffness, where we can express it in terms of material property and geometry. The bending equation we obtained under bending of beams section, also included the stiffness in it. Let's rewrite it:

$\kappa = 1/\rho = M/EI$

Here κ was the curvature. If we divide moment by curvature, we get the bending stiffness:

$$\text{Bending stiffness} = k_{bending} = M/\kappa = EI$$

And the units are: (Force / Area) x length^4 *(since E is force / area, and I is length to the power of 4).* Thus, we will have unit of bending stiffness as: Force x area. This can also be verified from M/κ: (Force x length) / (1/length) = Force x area.

Look at a simple demonstration of bending stiffness below. We have the same beam, apply the same load the same way but the cross section is oriented differently in each case.

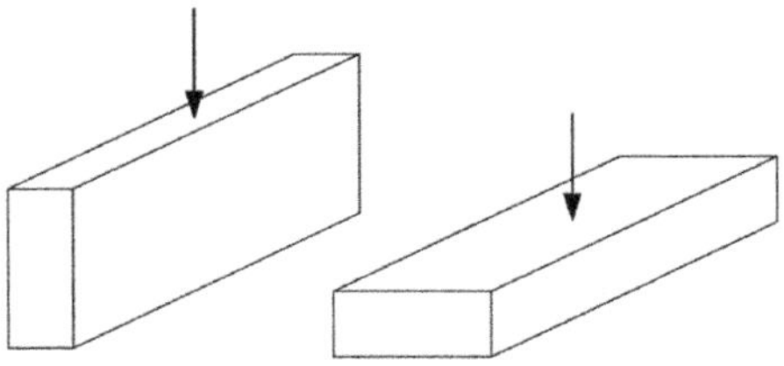

Figure - Strength 8

In the two cases above, the beam will bend different amounts, because although E is the same, but the moment of inertia "I" with respect to bending axis is different in both cases. So the bending stiffness EI is different in two cases, and as you can guess, the one on the left is stiffer, since its height, therefore, moment of inertia with respect to bending axis is higher. The example above is also good for remembering the fact that both beams have the same strength but different stiffnesses in such case.

Let's clarify an important point that may cause confusion. What if we had the situation above for beams of different lengths? Then a beam's stiffness, therefore bending per given moment would still be the same but the amount of bending thus its deflection would be different. For example the beam on the left is obviously stiffer in bending according to this loading axis here. But if it was longer than the beam on the right, both beams might have the same vertical deflection in the middle, although the bending per moment of the beam on the left would still be less. So the length, L, is important to determine the deflection amount. For example if we connected these beams to a column in opposite directions (in other words, similar to forming a capital letter "T"), and we applied a moment to the column head where beams connect, the moment would be distributed to these two beams according to the ratio of the beams' EI/L because the more EI, the more moment resisting capacity but the more L, the less moment resisting capacity. That is why it is also possible to see bending stiffness as EI/L in many places.

Finally, as discussed, a result of bending moment, a beam deflects. We must remember that there is also shear present in the beam but shear deflections are negligible, unless the beam is very deep in relation it length, and/or the beam's shear modulus is very low.

Poisson's Ratio

We introduced axial strain before. Under axial load, materials do not only deform axially, but also in the perpendicular (transverse) direction. Poisson's ratio shows the ratio of transverse to axial strain as:

$$\nu = \text{transverse strain} / \text{axial strain} = \varepsilon_{transverse} / \varepsilon_{axial}$$

Note that "v" here is the Greek small letter "nu" (not the letter "v" in English alphabet). Poisson's ratio is sometimes also shown with the Greek letter "μ" (mu). Instead of the word "transverse", sometimes the word "lateral' , and instead of the word "axial", sometimes the word "longitudinal" is also used.

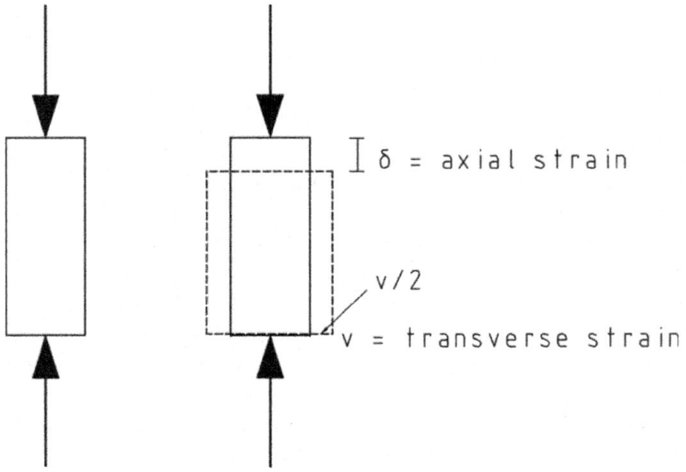

Figure - Strength 09

Under soil mechanics, you will also see similar parameters when we study lateral earth pressure.

Deformation of Springs

Studying the deformation of springs is important because we represent many things by springs, when modeling structures and soil.

An axial spring's deformation and force relationship can be written as:

$$F=k.x$$

Here F is the force that pushes or pulls the spring, k is the spring axial stiffness, and x is the amount of axial deformation of the spring. If you look at the units, k will have the unit of Axial Force / Axial deformation. Sometimes by convention a minus sign is written in front of the spring deformation formula above, just to remind that the resisting force is in opposite direction to acting force.

And a rotational spring's deformation and force relationship can be written as:

$$M=k.\theta$$

where, M is the moment, k is the rotational spring constant (rotational stiffness) and θ is the angle of rotation of the spring.

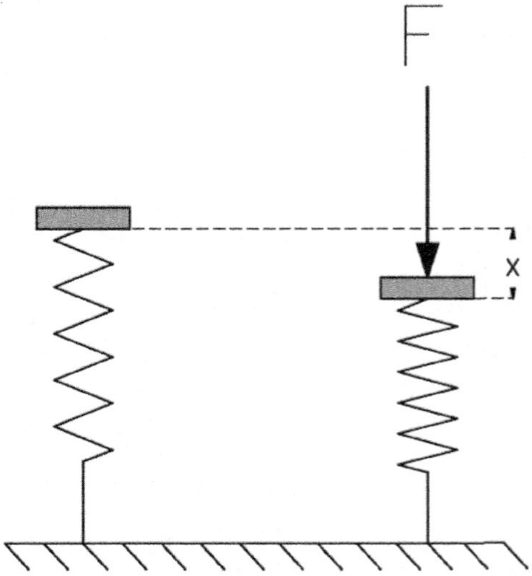

Figure - Strength 10

When the force is removed, spring returns to its original position, if it is elastic. This is based on Hooke's Law, after the 17th century British scientist Robert Hooke. Notice similarity of the formulas above with the stiffness formula shown a few pages back, where Force = (axial stiffness) x (axial deformation) or Moment = (bending stiffness) x (bending rotation). This is important in indeterminate structural analysis, including computer analysis, where stiffnesses of members and the structure are inserted in matrices and the forces and displacements are solved for the whole structure. That is why we use springs all the time, when modeling our structures. We use axial and rotational springs the most, and rarely torsional.

For example with an axial spring, we can represent reaction forces that the soil exerts to a foundation with many axial springs side by side. A rotational spring may be used when the beam at a fixed connection may yield and behave like a hinge, with its stiffness represented by the rotational spring. The important thing is, whenever we use springs in structural models, we set the spring stiffnesses properly as close as possible to members' real stiffnesses. These values can be constant (linear), but they don't have to. For example for nonlinear earthquake analysis, we can set the value of spring stiffness as linear upto a point, and then nonlinear after that as the earthquake damages progress (*i.e. yielding of the material, exiting elastic zone and entering plastic zone, as we usually assume elastic region as linear*).

Stress vs. Strain Curves for Concrete and Steel

Below, is a typical curve for the stress-strain relationship of concrete. The x mark you see at the end of the curve shows the point where the specimen suddenly gets crushed. This happens without much deformation for concrete. We call these types of materials as "brittle". When we build our structures, this is one of the main reasons that we must reinforce concrete with reinforcement steel, commonly called as "rebar", as in reinforcement / reinforcing bars.

Compressive strength of concrete is shown as f_c', which you can also see in the graph as the highest point of the curve, that corresponds to the maximum stress.

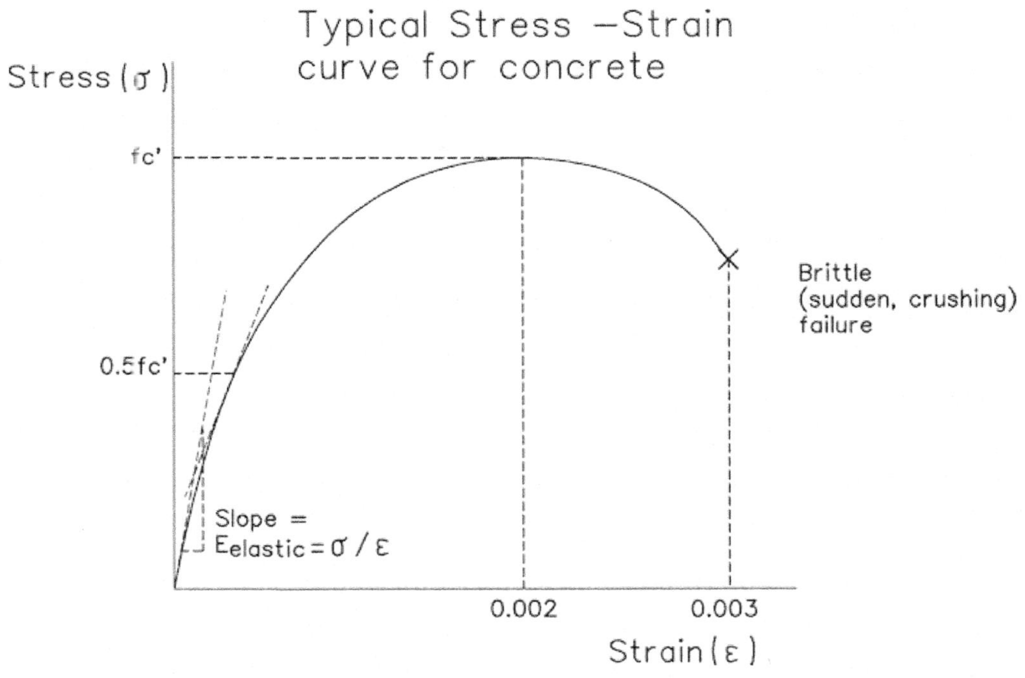

Figure - Strength 11

We have a curve here, which means its slope varies, and this means E will vary, since E = stress/strain. In most cases it is much more convenient to assume E constant. At low stresses, this assumption will not cause much error. So there are various geometrical methods used, such as the initial slope of the curve, secant modulus or tangent modulus, which are shown by dashed lines, to obtain a constant E.

The exact shape of this curve depends on concrete properties, rate of application of load, the shape, size and age of the specimen, effectiveness of concrete curing, admixtures and more. For example weaker mixes can have lower ultimate strength but higher ultimate strain, which means, they may be relatively more ductile than stronger mixes.

The strain of 0.003 is typically taken during design as the ultimate strain value that the concrete is able to handle. The strain value of approximately 0.002 corresponds to the maximum point of the curve (ultimate strength), which is about the same even for different types of concrete with different strengths and stress strain relations, which makes our job easier during concrete structure design.

This stress-strain curve is typical for an unconfined concrete test cylinder under compression. If we confined the concrete with reinforcing steel however, such as lateral reinforcing confinement (ties) of a concrete column, this would have a huge positive effect on this curve. Not only the ultimate strength, but also the strain absorbing capacity for that ultimate strength would considerably increase.

A typical stress vs. strain curve for steel is seen below. Although we did not show numerical values, the stress and strain values in this curve are much higher than the ones on concrete graph, which means, steel is much more stronger, and also able to tolerate far more deformation. As you see, the end point of this curve does not come suddenly as in concrete graph. Here, before we reach there, there is a lot of deformation first. This is good for us. It means, before failing, steel gives us plenty of warning.

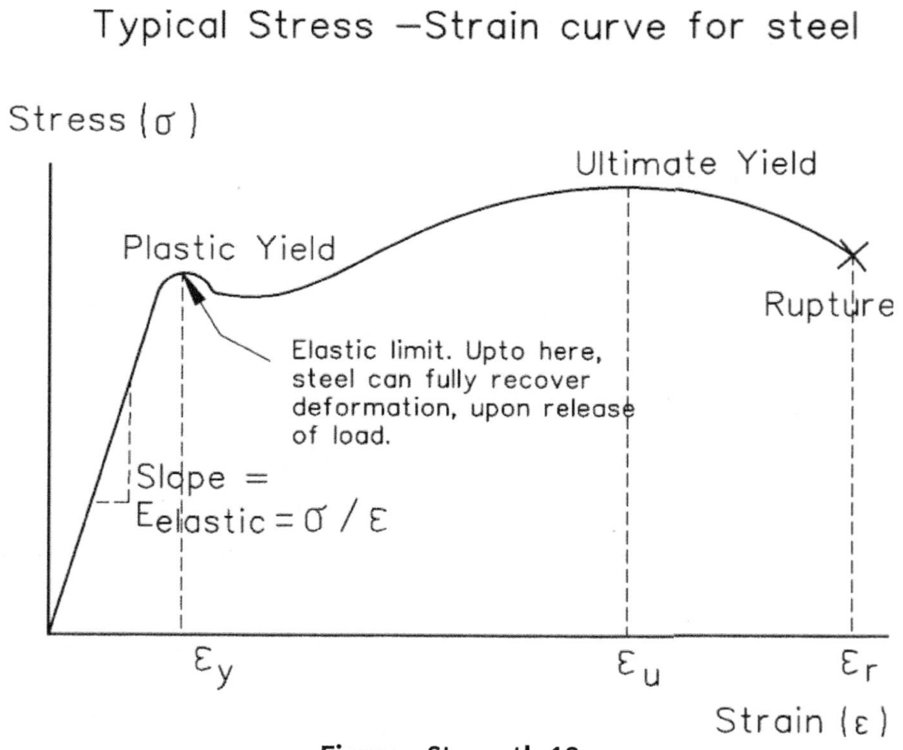

Figure - Strength 12

The area under these curves also represent the energy that the materials can absorb. Since we said that both its stress and strain values are higher, it means that steel can absorb far more energy than concrete before failing. This becomes vitally important during earthquakes. Because of high deformation and high energy absorption capacity, steel is a ductile material. Since steel gives us a lot of warning, we want steel to fail first in reinforced concrete structures, by limiting its amount, otherwise, concrete would fail first, which would be sudden, without warning, since as we said concrete is brittle.

In the graph above, see where it says "elastic limit". The graph up to this point is the elastic portion. In elastic portion, if we remove the loading, the material would return back to its original shape, by following the same loading curve backwards, with all deformation relieved. This is also true for concrete graph, where we would recover all deformation at low strains, below concrete's elastic limit (even if concrete does not have a well defined elastic limit as steel, as you can compare from both graphs). Here you see elastic portion as linear, which we usually assume so for steel and concrete, but in theory the elasticity modulus of a material in its elastic region does not necessarily need to be linear.

If we do not stop but continue loading after this elastic limit, the material yields, which is called the plastic yield point, and we enter into plastic zone. In this zone, if we removed the loading, some permanent deformation would remain, and some would be recovered. Plastic region can also be called as inelastic region. In this zone, as you can see, if we increase the load even a little bit, there will be a lot of deformation, in other words, E is lower here, moreover, it is not constant anymore. But regardless, the material still resists and takes on more load, even though with much more deformation now. This fact becomes useful during strong earthquakes. The highest stress point in plastic zone, which is the highest stress that the material can ever withstand is called the ultimate yield point, or ultimate strength of that material. In our designs often we do not want the materials to even approach here, and not even exceed elastic limit, but sometimes this will mean too much load needs to be resisted (uneconomical design), so, as relief, we can accept some plasticity in certain cases. The end of the curve, marked as rupture, is the strain where the steel member totally gets torn apart or breaks.

Soil and rock also have their own modulus of elasticity and they can also behave as elastic or plastic.

Strain Hardening

Strain hardening means, strengthening of a material while it undergoes plastic deformation. On the stress vs. strain graph for steel, it is the region on vertical axis that is between the plastic yield, and ultimate yield. Although strengthening sounds good at first, it means decreased ductility too. For example, bending a steel rod would be more difficult as we bend it more because it gets stronger, but it will also be more brittle at the bent location. Similarly if we bent a paper clip back and forth, the bent location will harden but it will break after a while in a brittle manner. When we tension a steel rod beyond elastic limit and it starts to deform plastically, and then release the load, and then load it again, we see that the elastic limit is now higher, so the material can take on more load elastically, but now its ductility is less, and it is prone to more sudden failure. Therefore, for example in moment frame buildings which we will discuss, as the reinforcing steel yields under earthquake shaking back and forth, the strain hardening can be be considered. Strain hardening is also known as cold working (as this happens when the metal is cold) or work hardening.

Stresses in Two Directions and Mohr's Circle

In our structures, we do not necessarily have forces that act from only one direction on members. Most often, we will have loads from multiple directions, which will cause stresses in materials also from multiple directions. This is also true for soil. The combined effect of the perpendicular stresses *(as we can resolve forces multiple directions into only two directions in perpendicular)* will be different than each one acting separately. We can not just add them up either. So we must have a way of studying their combined effect.

To do that, we can represent these stresses on a graph. This graph is called the Mohr's Circle, which takes its name from a German engineer, Otto Mohr, who studied strength of materials in late 19th century. Representing the perpendicular stresses with this graph enables us to draw a limit of failure, so that we can see what combinations of stresses will be below or above that limit for that material. This in turn enables us to design our materials accordingly and / or change the loads.

Now we will first describe how this graph is made, after that, show an example. It may seem complicated first but it is actually straightforward. To see the big picture now, take a look at Figure 13 below. This is the graph that we will obtain. The main things to learn from this topic are:

- the failure envelope,
- the angle of internal friction of the material, and,
- the failure plane that occurs within the material

You will see them all, especially the angle of internal friction more often, in soil mechanics.

This graph is made of two axes. The horizontal axis represents the stresses from both perpendicular directions. Let's repeat: the two perpendicular stresses are both represented on horizontal axis. Then what is the vertical axis for? The vertical axis is for representing shear stress, that would be produced in the material, as a result of those two perpendicular stresses. This is all for the axes. And the graph itself is drawn as follows:

1) In the lab, we test a material, by loading it from two directions (we think in 2 dimensions here). We record the loads, (thus stresses) at point of failure. Now we know a combination of two stresses, which are perpendicular to each other, which caused failure on this material.
2) We plot the values of these two perpendicular stresses on horizontal axis, as two dots.
3) We mark the mid-point of these two points as another dot, which will of course also fall on horizontal axis.
4) Then, using this mid-point as a center, we draw a circle so that it will pass from the two stress points we marked in step 2 (in other words, we draw a circle with a diameter equal to the difference in values of the two perpendicular stresses).
5) We draw a tangent line from the origin of axes (0,0), to the circle.
6) This tangent line is called the failure envelope. It was shown that, this line is the failure limit of our material, because it was drawn to a circle which was made using a pair of stresses that caused failure. So now, if we wanted to check whether another stress combination would cause

failure or not, without making another test *(in an ideal world)*, we can simply draw another circle with that new stress combination as described in the steps above, and see if that new circle remains entirely below this failure envelope, or crosses it. If it stays below the failure envelope, then that new stress combination is ok, and will not cause failure. If the new circle crosses the failure envelope, failure will occur and that new circle can not even happen for that material, because before getting there, the material will fail.

7) We measure the angle of the tangent line that we drew in step 5 with the horizontal. This is the angle of internal friction. It is a very important angle to know, especially in soil mechanics as we will see. It is shown with the symbol "Ø", which we will also use.

8) From the center of circle draw a radius to the tangent line (failure envelope). Measure the angle it makes with the horizontal. This angle will be used in a simple formula to determine the angle of failure plane within in the specimen itself, called the critical plane. This is the angle of the plane along which specimen fails in shear. The shear stress which we will obtain step 9 occurs along this plane.

9) Mark the point on original circle where it intersects the tangent line (failure envelope). Then from that point, draw a horizontal line, towards the vertical axis. The point on vertical axis where that horizontal line intersects, is the shear stress, that will occur in the material at the point of failure. So this is called critical shear stress. Note that this is lower than the shear value that would correspond to top of the circle.

Now let's give a numerical example. We will simply follow the steps above, nothing new. At any step, look at the graph below to verify values.

1) In the lab, we apply forces to a test cube, one force being vertical, and the other one equal from all its sides, horizontally. We will call the axial stress as $\sigma 1$, and the horizontal stresses from both orthogonal directions as $\sigma 2 = \sigma 3$, which means, we can only show just $\sigma 3$, for a 2 dimensional problem. We gradually increase the loads, and see that at a point of $\sigma 1$=600 kPa and $\sigma 3$ = 200 kPa, failure occurs.

2) We mark the points of 600 and 200 kPa on horizontal axis.

3) We mark the mid-point of these two points, which is 400 kPa, on horizontal axis.

4) Using this point (with coordinates of 400, 0) as a center, we draw a circle, with a radius of 200 kPa. (600-400 = 200, or similarly, 400-200 = 200)

5) We draw a tangent line from the origin, to the circle.

6) This tangent line is our failure envelope, which we explained above.

7) We measure the angle of this tangent line with horizontal, which is 30 degrees. This is the angle of internal friction, Ø, of our material.

8) From the center of circle, we draw the radius to the tangent line. We measure the angle it makes with the horizontal, which comes to be 120 degrees. The 120 and 30 are geometrically related as: 120 = 90 + 30. And the angle of failure plane is found as:

Angle of failure plane = 45+Ø/2 = 45+30/2 = 45+15 =60

9) We mark the point on the circle where it intersects the tangent line. Then from that point, draw a horizontal line to the vertical axis, see the point of intersection, which is the shear stress value at failure. We read it as: 173.26 kPa. Note that this value is lower than 200 kPa, which is where the top of the circle reaches (which is also the radius of the circle). This is lower as 173.26 because it occurred along failure (critical plane). 200 kPa shear stress, although higher, does not occur on failure plane, therefore it will not produce failure. So, in other words, in our specimen, for this particular loading, higher shear values than 173.26 kPa may exist, as much as up to 200 kPa, as long as they are not along the critical plane.

All these angle values, and critical shear stress value can also be found by certain trigonometric formulas which we are not showing but can easily be found in many relevant sources.

This circle is what we obtained as a result of using stresses which we knew that produced failure. If we made the test with any combination of higher stress values for example, and read their value at failure, we would obtain a larger circle, but that would still touch to the same tangent (failure envelope), as long as we test the exact same type of material (and in an ideal world).

Let's ask a question at this point. What will happen if for example we kept σ_1=600 kPa the same, but decreased $\sigma_2=\sigma_3$ to a value lower than 200 kPa? Intuitively you may think that since we keep one of the stresses the same but decreasing the other one, it will be ok, and failure will not occur. But this is incorrect. If you decrease the smaller stress, then for the same σ_1, we would be supporting the material with less support from its sides, thus it will fail. See it for yourself. Take a pencil and draw such a circle on the graph below yourself. Assume σ_1=600 kPa and $\sigma_2=\sigma_3$ = 100 kPa this time. Follow the steps above. The center of your new circle will be at 350 kPa on horizontal axis (*600-100 = 500, which is the diameter of your circle. So the radius is 250. Add 250 to 100 kPa, or subtract 250 from 600 kPa and the center of your circle is at the coordinate of 350, 0*). Now you have a bigger radius circle, and at a different center. Draw your circle. You will see that it will pass way above the failure envelope. It will easily cross it. It means, this stress combination cannot even happen for our material. As soon as you start to decrease $\sigma_2=\sigma_3$ to a lower value than 200 kPa, the new circle will cross failure envelope and the material will fail. And what would happen if you kept σ_1=600 kPa and increased $\sigma_2=\sigma_3$ to a value higher than 200 kPa? Nothing. The material will keep holding itself without failure, and even better. Because now you are increasing your lateral support for the material, and you will obtain even a smaller circle now, in comparison to the original circle for 600 & 200 pair. So this will be even safer loading condition. This is the logic behind retaining earth with retaining walls for instance. You can have a structure standing right near an excavation, but as long as you retain that foundation soil laterally with adequate force, you would be able to support that high surcharge load from the structure. Finally, try the same thing now by keeping 200 the same and increasing 600 to a higher value. Your circle will get bigger and immediately your material will fail, before your new stress pair can even exist.

Looking at the graph below, it looks like geometrically we can draw infinitely larger circles as long as we stay below the failure envelope. But this cannot happen. The material itself has a yield strength, and it will fail after a point and the failure envelope will start to flatten at higher stresses, until it becomes horizontal line, which means, we will not be able to draw a larger circle anymore. These are called Von Misses stresses for interested readers to look at, which we will not cover.

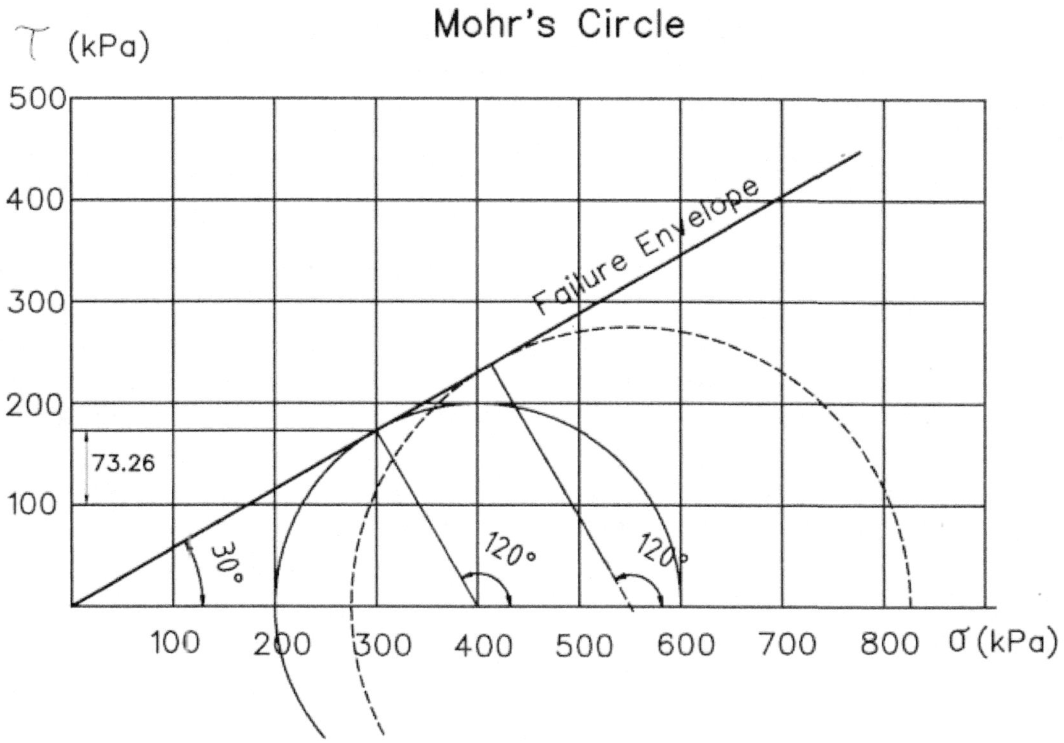

The two circles here are critical, as they touch the failure envelope. There can be infinitely many circles that touch or remain below failure envelope. As soon as a circle starts to cross failure envelope, the material fails, so that circle cannot exist.

Figure - Strength 13

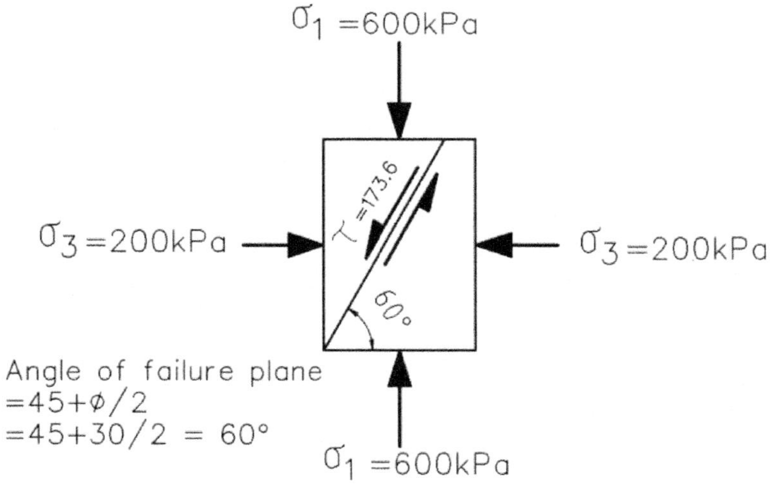

Figure - Strength 14

Sometimes, we need to draw the tangent line not from origin but from a higher point on vertical axis. It means, in addition to the vertical rise of the tangent line, which is because of friction, there is also something else to contribute to shear strength. You will see this in soil mechanics too, and in that case it is the cohesion of soil.

As you can see from all this discussion, shear strength is the key, to strength of a material. This is true for all materials you can think of, including concrete, steel and soil. What about water? Water cannot carry anything on it (we are not talking about uplift force of water that carries ships, that is different concept). For example you can step and walk on soil but not water. Because soil has some strength. But water has zero strength, which means it has zero shear strength and its angle of internal friction is zero.

For example, let's assume that by magic somehow you were able to mold liquid water into a cube shape without a container, and you (tried to) apply the different perpendicular stresses σ_1 and σ_3 to that cube. Before you can even give the two stress values σ_1 and σ_3 even the slightest difference, your cube would immediately collapse. There would not even be a Mohr's circle. All you will have will be some dots on the horizontal axis of Mohr's circle, where perpendicular stresses σ_1 and σ_3 can only be equal to each other. Water does not have the shear strength to sustain any difference of σ_1 and σ_3, in other words, a circle on Mohr's graph. You would not even be able to make the cube shaped water specimen in the first place, because water has no friction or cohesion between its molecules, so it cannot support any shear. If you freeze water though, now you not only obtain a cube which can stand on its own, but also can withstand perpendicular stresses σ_1 and σ_3 that are different from each other, because ice has shear strength. Now you would be able to obtain Mohr's circles, instead of single dots on horizontal axis. Therefore, a material's strength depends on its shear strength.

SOIL MECHANICS

We now start to study geotechnical engineering, which is another major branch of civil engineering. The concepts under geotechnical engineering must be well understood by any civil engineer, or anyone who wants to gain insight into civil engineering and how structures are built. We can now mention terms related to force, stress and deformation, without explaining them again, as we covered those earlier. If you are not an engineer and have not yet read the previous sections, it is recommended that you first go through those and come back to this point.

We will discuss geotechnical engineering under the following main sub-branches (we will discuss many other geotechnical subtopics under soil mechanics or foundation engineering):

- Soil mechanics
- Foundation engineering
- Stabilized slopes, deep excavations and retaining structures

In this section, we discuss the first one, soil mechanics. The goal of soil mechanics is to develop theory for the second and third items. One may also consider writing rock mechanics near soil mechanics in the list above, but it is quite separate subject than soil mechanics. Soil mechanics is studied first conventionally and applies in almost all situations whereas rock mechanics apply less frequently and mostly together with soil mechanics. Rock mechanics is not covered in the book.

We will cover soil mechanics, which is the longest section of this book, under the following subsections:

Phase Relationships

Phase relationships define the proportions of solid particles, water and air in a soil mass. These relationships are also used to determine compactness, degree of water saturation of the soil, which, then are used in earthwork volume calculations, as well as in determining some of strength properties.

The figure below shows how we typically represent the relative proportions of solids, water and air in a soil mass. At this point, it is reminded not to confuse soil and solids. Soil is the whole mass, including solids, water and air, while solids are the solid particles only, within that soil mass.

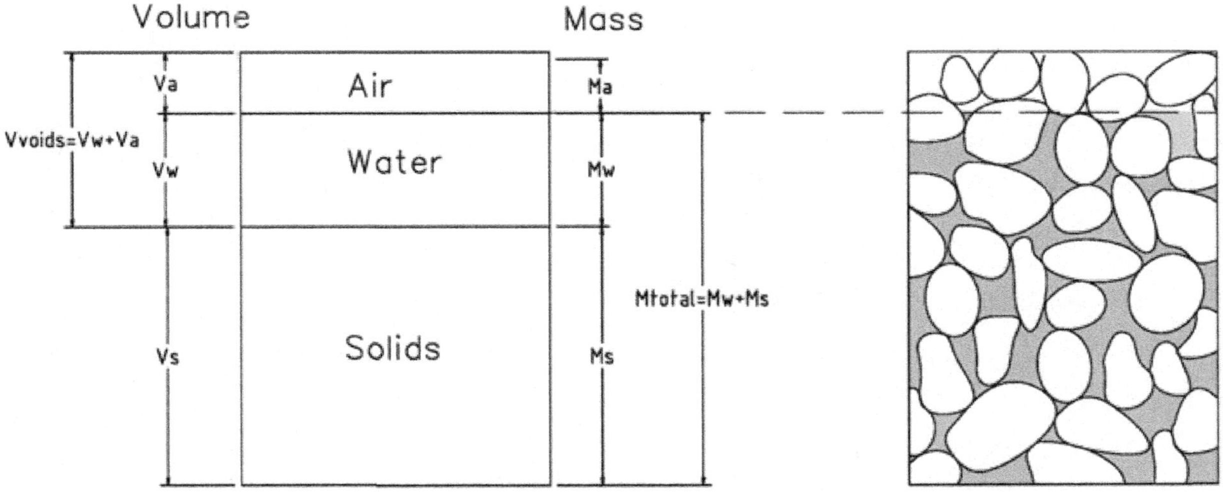

Figure - Soil Mechanics 1

We can write the relationships as below. We will not use them again anywhere in the book, but showing them only for learning and to serve as a reference if needed later:

- Density = Mass / Volume. In proper symbols, we write: $\rho = M/V$. "ρ" is the Greek letter "rho".
- Unit weight is not the same as density. It is found by multiplying density with Earth's gravity. Unit weight is shown by Greek letter γ (gamma). So: $\gamma = \rho.g$, where g is the gravity of Earth. We use unit weight instead of density, when dealing with the whole soil mass.
- Air's mass in the soil is neglected, as it is very small. It's volume however, can not be neglected. If the soil is fully saturated, we can take air volume as zero. In theory, a soil is never fully

saturated, because there will always be air bubbles, even if it seems fully saturated, but for our purposes we can neglect air volume, if we can assume that a soil mass is fully saturated.

So rewriting what we just said above:

$M_T= M_s+M_w$ (total mass of soil is the sum of mass of solids plus mass of water)

$V_T= M_s+M_w+M_a$ (total volume of soil is the sum of volume of solids plus volume of water plus volume of air)
Here T stands for total, S stands for solids, and W stands for water.

- Mass is a quantity of material. Weight is a force, which is obtained by multiplying mass with gravity of earth. So, W=M.g. Since we know from above that M=ρ.V, this means W=V.ρ.g. This also means W = γ.V. In other words, to find the weight of the entire soil mass or water or solids alone, we multiply their respective unit weights with their respective volume.

- Moisture content (m): It is important to know how much water there is in the soil. Moisture content, which we show by "m", is found as: Weight of Water/ Weight of solids (in percent).
So,
$m=W_w/W_s x100$
Again, pay attention that "s" is for solids, not the entire soil mass here.

- Degree of saturation, which we show with capital "S", shows how much of the void volume is filled with water, as percentage. So we can show it as: S = Vw/Vv

- Void ratio, which we will also see later, is a commonly encountered parameter. It is the ratio of the volume of voids in soil to volume of solids: Vv/Vs. Void ratio is used in calculating soil settlement, degree of compaction, critical state soil mechanics, which we will discuss.
Relative density (D_r), is a parameter that is used in earthwork volume calculations. It can be written in terms of void ratio as:

$$D_r=\frac{e_{max}-e}{e_{max}-e_{min}} x\,100$$

This is a parameter, which is expressed in percentage, that can always be seen in earthwork specifications of projects. The maximum void ratio e_{max} in the formula above, represents the loosest possible state and e_{min} represents the densest possible state. So in the formula above, if we had the loosest possible state of soil, the top part would be zero and the percentage would be 0. If we had the densest soil, then the nominator and denominator in the equation would be the same and we would have a 100% relative density. Neither of the values are possible and our values will lay somewhere in between, closer to 100%, and the specifications usually want at least 95% for structural fills and at least 90% for other fills.

- Porosity is the ratio of volume of voids to total volume. We can write it as: n=Vv/Vt, which also becomes equal to e/(1+e). The higher the porosity, the more permeability of the soil, which you will see when we study groundwater seepage.

- The dry unit weight, γ_{dry} of the soil is also important to know, which is the unit weight that contains no water. It can be obtained by drying the soil. So in this state, S, degree of saturation, will be equal to zero.

- $\gamma_{saturated}$, which we also write as γ_{sat}, is the unit weight of the soil when it is saturated, where all soil pores are completely filled with water (or so we can assume), so in this state S=1.

- Specific Gravity, which is shown by (G_s), applies to solid particles only. It is the ratio of the weight of solid particles to the weight of water of the same volume, in other words, how many times the solid particles are denser than water. Since it is a ratio of same type of units, the units cancel, which means specific gravity is unitless. So we can write it as:

  $G_s = W_{solids} / W_{water} = \gamma_{solid} / \gamma_{water} = \rho_{solid} / \rho_{water}$

  If we use SI units, the density of water is 1 g/cm^3, which is unity, so, G_s becomes equal to ρ of soil. But remember only numerically, since it has no units.

  Specific gravities of solid particles usually range from 2.6 to 2.8. It is better suited to use for fine grained soils. For coarse grained soils, there are lot of voids between particles so using Gs, when it is calculated this way, will have less accuracy as the grain sizes increase.

The formulas below will not be needed in any way or mentioned at all again in the book, so do not be confused. They are only given as reference for anyone who may need, because in different resources it is possible to find them arranged differently and or redundant with each other, which can be confusing. The ones below form a useful group which were derived by simple algebra from the parameters above.

$S.e = w.Gs$ $$e=\frac{G_s.V_t.\gamma_w}{W_s}-1 \qquad \gamma_{dry}=\frac{G_s.\gamma_w}{1+\frac{w.G_s}{S}} \qquad \gamma_{wet}=\gamma_w.\frac{G_s.(1+w)}{1+e}$$

$$\gamma_{saturated}=\frac{(G_s+e).\gamma_w}{1+e}$$

Soil Classification

Determining relative proportions of solid, water and air in a soil mass as we described above gives us some information about the soil, but this is not enough. We must also know what type and size of particles that solid part is made of, and the relative proportions with respect to each other if there are more than one type of solid particles. This is often the case, even if the proportion of one type may be far more dominant. We use this information to know the possible ranges of various soil properties such as strength, swelling / shrinkage potential, permeability... because all of these are well correlated to soil classes from past research *(and after that we can determine these more accurately with soil testing which will also be discussed later).*

The earth's crust was originally rock. Soils form as a result of decomposition and disintegration of rocks, through chemical and or mechanical processes. This process is called weathering of rock. As a result of weathering in different geographies and environmental conditions, there a high variety of soil types but for our purposes we can broadly classify them.

From civil engineering standpoint, from weaker to stronger, we can group earth materials as:

Organic soils, Clay, Silt, Sand, Gravel and Rock.

Organic soils, marshes, and swamps are not suitable for construction at all. For any type of project and for all our purposes, we completely ignore any support they might provide for a foundation, which is negligibly small to none.

If you paid attention we said earth materials above. We did not want to say soil there because we also included rock in the list. Rocks, which are usually far superior than soils in terms of being foundation materials, actually have their own mechanics, which is studied under rock mechanics subject. During learning civil engineering however, it is customary to first study and understand soil mechanics. After all, closer to the ground surface we almost exclusively have soils, and their behavior is more complex that needs to be studied and understood. We will not discuss rock mechanics in this book, but reminding that for some projects it becomes critical.

So leaving the two extremes, organic soils and rock out, in soil mechanics we deal with clay, silt, sand, gravel and their different proportions and types. The properties of the soil is greatly affected by the proportions of these types, moisture content, degree of compactness as well, which are not made by humans in controlled environments, therefore soil may be a highly variable material. *(This is why, geotechnical engineers may arguably be using engineering judgment more than civil engineers specialized in other disciplines. But this doesn't mean they are in the dark. It is actually the opposite. We have huge amount of past experience backed up with centuries of studies in strength of materials and well over a century in the field of soil mechanics specifically, which gives us powerful methods, statistical means of analyzing documented data, and a great deal of insight when making designs and solving problems. Add to it that now we are also equipped with computers and software which are increasingly powerful. So even though geotechnical field inherently cannot have detailed codes as in structural design, our decisions and design can still be rational, based on adequate backup literature)*

Out of the four groups we mentioned, such as clay, silt, sand, gravel, grain size of clays are the smallest. Then comes silt, then sand and finally the gravels have largest grain size among all. In any soil mixture one of these is dominant. In some cases, the percentage of the minor types can be so small that it can be neglected and we can classify the soil as if it is of one type. In other cases the mixture contains significant portion of each type. In addition, each of these four groups also have different types.

We can classify soils into two groups as fine grained and coarse grained according to particle size.

Clays and silts are fine grained, and,

sands and gravels are coarse grained.

The range of particle sizes of soils is the main criteria we use to classify them. This is determined by the sieve analysis for coarse grained soils. In the US, the standard sieve sizes range between #4 sieve, which has openings of size 4.75 mm (or 0.187 in), all the way down to #200 sieve which has openings of size 0.075 mm (or 0.0029 in).

We express grain size distribution as percentage in the mix. The particles small enough to pass even the #200 sieve are called fine grained particles and they cannot further be separated by sieve. For those, hydrometer test is needed, which is based on the settlement time of particles in water, where larger particles settle faster and smaller ones settle slower.

According to Unified Soil Classification System (USCS), which is one of the most widely used and oldest of systems, if 50% or more of the soil by weight is retained on a #200 sieve (0.075 mm), then that soil is mainly classified as a coarse grained soil. Furthermore, for that coarse portion, if 50 % or more by weight is retained by a #4 sieve (4.75 mm), the soil is mainly gravel, if not, it is mainly sand.

For the fine grained soils, the proportion of clay and silt with respect to each other is determined mainly by plasticity of soil as well as their particle sizes. Detailed criteria is given in USCS and in other soil classification systems. The size range of each type of soil, according to USCS are:

Gravels range between 75 mm to 4.75 mm

Sands range between 4.75 mm to 0.075 mm

Silts and Clays are smaller than 0.075 mm (and according to other classification systems, clays are even smaller than 0.002 mm.

Stones larger than gravels are called cobbles, which range between 75 mm to 300 mm in size. The rock pieces larger than 300 mm are called boulders.

There are other classification systems as well, such as by AASTHO, USDA, ASTM and the values they specify are close to each other and to USCS.

We can write the primary soil symbols for USCS as below:

- G: Gravel
- S: Sand
- M: Silt
- C: Clay
- O: Organic soil (clay and silt)
- Pt: Peat and other soils with high organic content

The secondary symbol is according to soil characteristics such as:

- W: Well graded
- P: Poorly Graded
- L: Low plasticity
- H: High plasticity

For example, clay of high plasticity is written as CH, or a poorly graded sand as SP and so on. The term plasticity will be discussed under soil tests. A lot of additional criteria is available in USCS itself, for the two letter classifications. The soil is highly variable material. In nature we do not always find just clean sands, silts or clays, but often encounter mixtures. The percentage of each is important to classify them, and as a result they show behavior that is not very easy to classify at once - although one of them will usually dominate a particular behavior - and they will have mixed properties such as stiffness, permeability, plasticity, strength and must be determined with tests that we will discuss.

In the list above, well graded means, the soil as a whole, contains enough particles of different sizes, which means they can fit in between each other well, which allows a more compact, denser, thus stronger soil mass. Poorly graded means soils that do not include sufficient range of particles, but mostly a narrow size range, so they can not fit well in between each other and cannot be compacted as effectively as well graded soils. For geologists, these two definitions are reversed.

Sands vs. Clays:

In geotechnical engineering, it is of utmost importance to understand the differences between clays and sands. We did not discuss most of the terms in the box below yet, but will show them all shortly.

CLAYS :

- **Fine grained**
- **Cohesive**
- **Plastic (expands/swells and shrinks)**
- **Weaker than sands in general**
- **Settles elastically and by consolidation**
- **Compacted by compression**
- **Very low permeability**

SANDS :

- **Coarse grained (also called as granular)**
- **Cohesionless**
- **Nonplastic (doesn't expand swell or shrink)**
- **Stronger than clays in general**
- **Settles elastically but not through consolidation**
- **Compacted by vibration**
- **High permeability**

Sands are generally stronger, settle, swell & shrink less, allow water to drain, so they are more preferable foundation materials in comparison to clays in general. That is why in many places it is possible to see clay replaced by sand. By saying strength, we mean shear and load bearing capacity. For example the strongest hard clay soil will be weaker than dense sand but it can be stronger than loose sand.

The term cohesion is so important that we sometimes call clays as cohesive and sands as cohesionless soils. As clays are very small particles, their surface attraction forces are significant and they can produce cohesion. Sands however are much larger particles and for them these surface attraction forces are very insignificant in comparison to gravity and frictional forces between them, which can generate considerably higher strengths. We will look at cohesion under strength of soil section below.

As for silts, they lay somewhere in between sands and clays as far as their particle sizes, as we saw in the previous page. Their origin is important to determine if they will behave closer to sands or clays.

Strength of Soil

Studying the strength of soil is the core of soil mechanics. We will study it under the following subtitles:

- Total and effective stress, pore water pressure: Stresses from weight of soil and water pressure.
- Shear strength: The source of strength of soil is its shear strength.
- Lateral earth pressure: The horizontal stresses in soil, and how they are retained.
- Bearing strength of soils: The ability of soil to support foundations.
- Stress increase in soils: As a result of new loads, the stresses in the soil will increase.
- Elastic settlement: Soils experience immediate elastic settlement, due to increased stresses.
- Consolidation settlement: Clays can go through considerable consolidation settlement.
- Slope stability: Types of slopes and the forces that stabilize or drive them are studied here.
- Critical state of soil: This concept combines few properties to obtain a unique failure state.

Total Stress, Effective Stress, Pore Water Pressure:

First let's make sure that one essential concept from basic physics is remembered. The pressure of a fluid at given depth is found by multiplying the fluid unit weight by depth in question, as:

Pressure = γ_{fluid} x d

Instead of depth, d often we may also write h for height. The formula above means, as the depth increases, the pressure increases in direct proportion. So, we will have a triangular pressure distribution as in the figure below. At a certain depth, the pressure of a liquid acts equally at all directions. So, as seen in the small figure on the right, if you assume a very small object of negligible height at a certain depth in a fluid, the pressure on all of its surfaces can considered to be equal. When we say fluid, we will mean liquid and specifically water, unless noted otherwise.

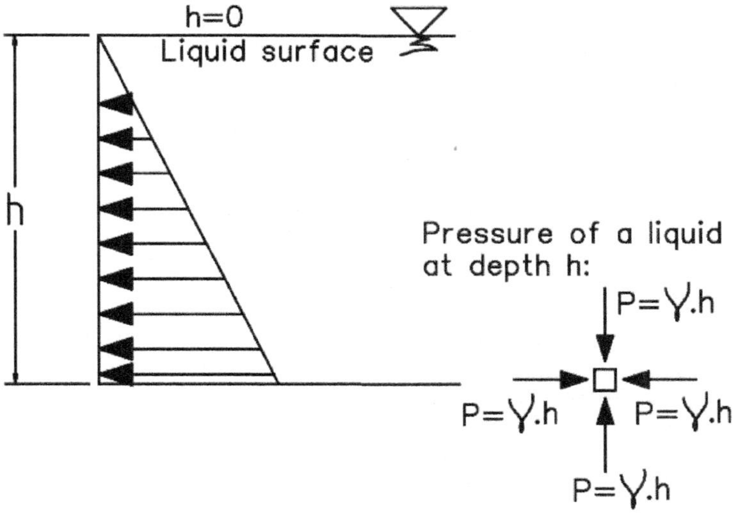

Figure - Soil Mechanics 2

Now we can continue with soil...

The total soil stress at a certain depth due to its own body weight is made from two components. The component that comes from the soil skeleton itself (called effective stress), which is a result of contact friction between solid particles, and the water pressure at that depth (also see Figure - 3b). All pressures are calculated as in the pressure formula given above. So, at a given depth:

Total soil stress = Effective stress + Pore water pressure

Effective stress is the portion that matters for strength of a soil. And we call water pressure as pore water pressure, or simply, "pore pressure". Rewriting the statement above in proper terms:

$$\sigma_T = \sigma' + u$$

where,

σ_T represents total stress

σ' represents effective stress. Pay attention that we used an apostrophe sign to identify it. This is always done so in soil mechanics to indicate effective parameters. Effective parameters are the ones resulting from the friction of solid particles in soil skeleton itself, which generates the soil strength. For example the effective angle of internal friction Ø' is the property of the soil itself, but not the undrained angle Ø, which also includes the effect of pore pressure (u).

u represents pore water pressure. It contributes nothing to strength of soil. On the contrary it makes the soil heavier and causes weaker contact within soil skeleton, by making the soil in submerged condition, through its uplift pressure on solid particles. So in geotechnical engineering, we prefer that the total stress is only made of effective stress, which means zero pore water pressure, although this cannot always be achieved and must be accounted for in the design if it cannot be avoided.

This formula above is one of the most important in soil mechanics, which was introduced by Karl Terzaghi about a hundred years ago, who is considered as the father of modern soil mechanics by many.

Figures 3a and 3b below shows nothing more than what we have just discussed in the previous few paragraphs. Pay attention to the following:

From ground down to the depth of h1:

Water starts at depth h_1. The upper layer of the soil above the depth of h_1 is not saturated, so water pressure is zero there, and the total stress is equal to effective stress in that part. So here,

$u=0$

$\sigma_T = \sigma' + u = \sigma' + 0 = \sigma'$

Where, $\sigma' = \gamma_{dry\ soil} \times h_1$

Below h_1:

Now the total stress is made both from the effective stress and water pressure. Note how the slope of effective and total stresses change at depth h_1 because of this.

At depth of h1 + h2:

Pore water pressure: $u = \gamma_{water} \times h_2$

Total stress: $\sigma_T = \gamma_{dry\ soil} \times h_1 + \gamma_{saturated\ soil} \cdot h_2$ (as you see, for total stress, we take into account the saturated unit weight anywhere below h_1)

Effective Stress: $\sigma' = \gamma_{dry\ soil} \times h_1 + \gamma' \cdot h_2$

above γ' is called the effective unit weight or buoyant unit weight or submerged unit weight and found as: $\gamma' = \gamma_{saturated\ soil} - \gamma_{water}$

Pay attention that we could also find the same effective stress (σ') value by subtracting pore pressure from total stress (since $\sigma_T = \sigma' + u$), at any point below the water table (surface of underground water).

If you pay attention to units, γ, the unit weight is in pounds per cubic feet (lbs/ft^3) or Newtons per cubic meter (N/m^3). If you multiply it by height, as in $\sigma = \gamma.h$, we again obtain units of stress, such as lbs/ft^2 or N/m^2.

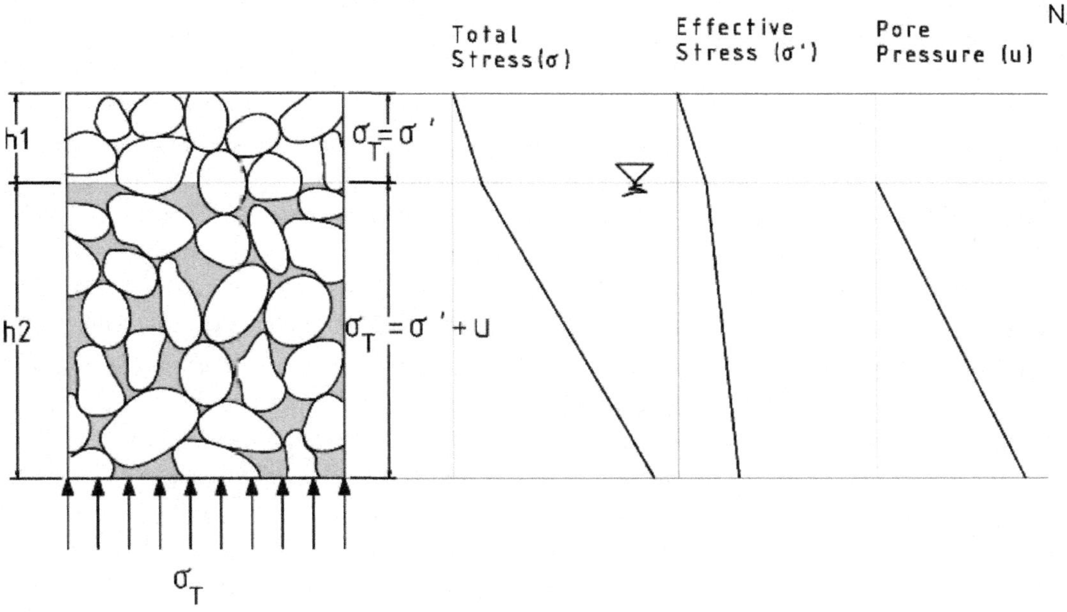

Figure - Soil Mechanics 3a

The Figure 3b below shows a zoomed in view, which shows the effective stress and pore pressure in greater detail, at depths below the water surface.

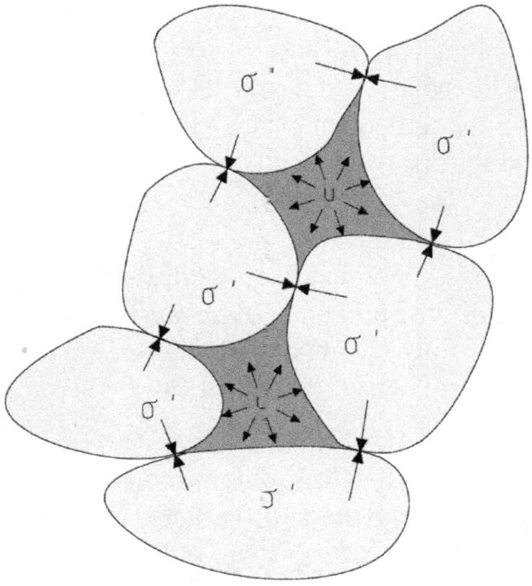

Figure - Soil Mechanics 3b

In some cases pore pressure can be negative. This happens when some water climbs above the water table just by sticking to soil particles, which is called capillary action. It means, now the water is just hanging to the soil, as if pulling it downwards. So here, we take water pressure as negative, which means, in that zone, the effective stress will increase because of it. This zone can be a few feet at most.

We may see negative water pressure also when we rapidly excavate clay soil. Due to removed overburden stress at the excavated level, the soil will heave a little, because it carried more stress in the past. This will cause decrease in pore pressure as in suction (imagine an expanding wet sponge), which is negative pore pressure.

Liquefaction:

In addition to natural stresses that exist in the soil which we have just studied, another important use of this knowledge is the liquefaction phenomenon, which deserves a side note at this point. One may argue that it is early to talk about liquefaction before studying other concepts that we will yet see, but the author placed it here on purpose to emphasize that even with the few sections we have studied so far under soil mechanics, we actually learned a great deal and can now discuss even an advanced concept such as liquefaction, when described even only in simple terms that we saw so far.

Now, visualize the following independent facts: 1) Liquid water can not support any load. 2) Loose sands are sensitive to vibration and they can easily be compacted through vibration, when they are also under pressure. 3) Earthquakes cause cyclic shaking motion in the ground...

When we combine these three facts, it means that if we had saturated, loose sand carrying a structure above, and if it is located in an earthquake zone, then there is a high risk that the sand may undergo compaction during earthquake. Here is how it works: When earthquake happens, it will apply sudden shear stress on the soil in a repeating and reversing, cyclic manner, where the soil will tend to go into compaction and its volume will suddenly tend to decrease, due to the gravity loads pressing from above. Compressed water (since we said the soil is saturated) will not have time to drain so quickly, and the pore water pressure (u) will suddenly increase. From the formula that you saw above, $\sigma_T = \sigma' + u$, this means, when water pressure increases, effective stress (σ') decreases, since the total stress from the structure and the soil above is still the same. Reduction in effective stress means, the friction between soil particles decreases (*more specifically, the shear strength generated by the soil decreases as we will see in the next section*). This is equivalent of saying that the soil's bearing strength decreases. If the strength loss exceeds a certain limit, soil settlement will also start to happen. This process is called liquefaction. The pore pressure can even increase up to a point of total stress, which means the effective stress becomes zero, which means the entire contacts between the solid particles disappear, and the soil basically turns into a liquid with floating solid particles in it. At that point the bearing strength of soil is zero.

In the process described above, the soil is not in a steady state. During each seismic shaking cycle, excess pore pressure builds up (accumulates), and consequently effective stress decreases further and further. Upon sufficient loss of effective stress, thus bearing capacity, settlement begins to happen, which also accumulates with each cycle beyond that point, if the shaking continues.

Liquefaction risk and its degree depends on the following factors:

- The type of soil: For liquefaction to happen, the soil must be sand or mainly sand type and behave as sand *(so if the sand contains clays, silts, gravels etc. the portion of those must be very small)*. As for clays, liquefaction is not a risk, because they are not sensitive to vibration. Nonplastic silts carry some liquefaction risk as well, as they also behave like sands. Gravels are too large and permeable for liquefaction pore pressures to develop.
- Degree of compaction of soil: The higher the void ratio (e) of the sand (the looser sand), the easier will be the liquefaction. Uniformly graded loose sands are the most susceptible.
- Depth of water table: Above water table liquefaction doesn't apply. For sands above water table, footings are designed for settlements from gravity loads, instead of liquefaction. Even if the sand is below water table, if it is dense enough there will be no liquefaction.

- Layering of the soil, sand depth and thickness: The soil on site may be layered, made of different types. At what depth the sand is located and its thickness is important. For example a layer of liquefiable sand may exist, but it may be located too deep and or its layer thickness may be too small, to cause any significant liquefaction.
- Earthquake intensity: The value of PGA (peak ground acceleration), the number of shaking cycles and the earthquake duration greatly affect the risk and extent of liquefaction. As the earthquake gets stronger, deeper layers will be affected as well. Note that earthquake intensity is also affected by the type of soil we have on site, as the soil can alter earthquake waves.
- Size and depth of footings: As footing sizes increase, the total stress will be less, but overall structural weight must also be considered. The heavier the structure, the more liquefaction settlement, which can be from small to entire sinking, including differential settlement. Heavy structures over water such as bridges on rivers, wharves etc... have more liquefaction risk.

As a result of liquefaction, there can also be other damages or hazards than just settlement, which are listed under Earthquakes > Geotechnical concerns subsection.

Liquefaction potential needs to be checked for any structure as required per codes. In many cases, we cannot avoid building on sites which carry liquefaction potential. The most common mitigation measures include the ones below, by applying either one or a combination of more:

- Making the soil denser, stiffer, by compaction or ground improvement methods. We will review ground improvement methods later.
- Installation of earthquake drains. These are perforated pipes wrapped by filters, which are inserted into the holes that are drilled into the ground, to provide drainage path for water to escape, which will prevent building up of pore water pressure. This also works as a ground improvement method too, because when the holes are drilled, the loose soil around the holes are forced outwards, which makes the soil denser at least at those locations.
- Lowering the groundwater table (a permanent solution must be in place, otherwise the water table will rise again)
- Simply removing and replacing the vulnerable soil (usually not practical, considering the thicknesses / depths involved).
- Creating artificial vibrations and settlements beforehand in the soil by inserting small explosives.
- Making the foundations deeper, to withstand sinking from any expected degree of liquefaction. Driven piles are preferred as they displace and densify soils around them.

Although the liquefaction phenomenon was known to engineers long before, after 1964 Nigata Earthquake in Japan, where large scale liquefaction caused entire buildings to sink although they were intact structurally, in depth studies started, such as the ones that were published in 1966 by Seed and Lee, and in 1971 by Seed and Idris, to understand its mechanism in detail and determining its potential by field tests, which were often used as references in many studies by others afterwards.

The liquefaction that we discussed above is due to earthquakes. Liquefaction can also occur because of <u>upwards</u> seepage of water, when it is so strong that it cancels the weight of the soil above and the effective stress in the soil becomes zero. This is also called "quick condition", or "boiling" of soil, where the soil essentially becomes like a boiling liquid as particles move upwards with water, upon exceeding a critical hydraulic gradient value. We will discuss hydraulic gradient in groundwater section.

Shear Strength of Soil:

A material's strength results from its shear strength. Soil is no exception. At this point, you may want to revisit Mohr's circle subject at the end of strength of materials section if you do not remember it.

To visualize better, let's give a few examples. Water has zero shear strength. It is not able to support any load *(uplift pressure that carries ships is entirely different concept)*. We explained this example in more detail at the end of Mohr's circle topic. For this same reason of zero shear strength, if we pour water on a sloped surface, it will flow. For the same reason you can walk on soil or concrete, because they have shear strength. *(Note: The strength of soil mostly is a result of lateral support, unlike concrete which can stand on its own)* To design foundations we must know how much shear strength of the soil. This is not only important for shallow foundations which simply bear on soil, but also for pile foundations, as their capacity is a result of skin friction of the vertical pile face and / or bottom end bearing, both of which also come from shear strength. Another case is retaining earth; it is much easier to retain stronger soil with higher shear strength. Another case are slopes, where stability increases with shear strength of soil.

Two sources make up shear strength of soil. First is the friction between solid soil particles, and second is cohesion between them. Friction is a better source *(which, for our purposes here, also includes bridging of particles due to interlocking)*. For example sands have high internal friction but low to zero cohesion. Clays have high cohesion but low to zero friction. That is why, sands generally support foundations better *(plus, clays also have the potential to shrink or expand, while sands do not)*.

We will now explore the subject of shear strength of soils under the following subtitles:

- Angle of Internal Friction
- Direct Shear Test
- Triaxial Test
- Angle of Repose
- Cohesion of Clay

Angle of Internal Friction:

We saw how angle of internal friction can be found, under strength of materials, Mohr's circle topic.

Let's look at it in a different way now. In the figure below, we push a solid block on a surface with friction, at the same time pressing from above. You can see the base reaction force that will form.

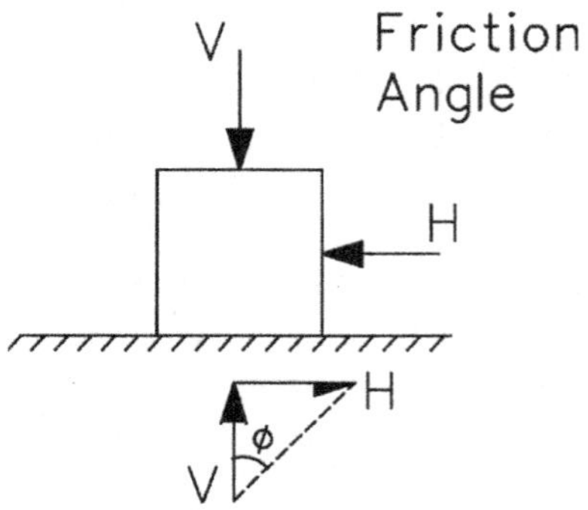

Figure - Soil Mechanics 4

If we drew the forces in scale as vectors, we can write the tangent of the angle as:

$$\tan \emptyset = \frac{H}{V}$$

Tan Ø is the coefficient of friction here. Ø is the friction angle. (*As a side note, the size of the contact area between the block and the table is irrelevant according to Amontons' Laws of friction - later Coulomb added to it that V is also independent of the pulling speed).* In the formula above, we have forces as H and V. If we divided both of them by the same contact area, A, the equality will not be affected. Since force divided by area is stress, where, τ=H/A (shear stress) and σ=V/A (normal stress):

$$\tan \emptyset = \frac{\tau}{\sigma}$$

And if we rearrange, we obtain:

$$\tau = \sigma . \tan \emptyset$$

The equation above is one of the most important relationships in soil mechanics. And as a reminder note, as we saw under Mohr's circle subject, when a block was stressed under perpendicular stresses, the plane of failure and the angle of internal friction were related as: θ = 45+ Ø/2, as in Figure 5 below. The equation $\tau = \sigma \tan Ø$ was first defined by Coulomb in 18th century by employing an early version of the direct shear test which we will show below.

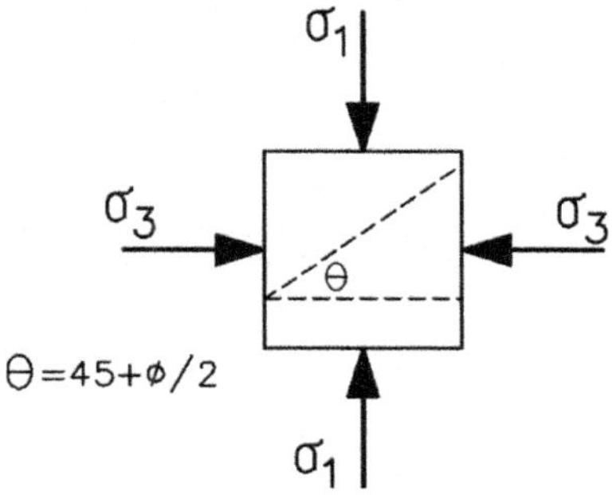

Figure - Soil Mechanics 5

Figure 6 shows how the graph of $\tau = \sigma \tan Ø$ will look like:

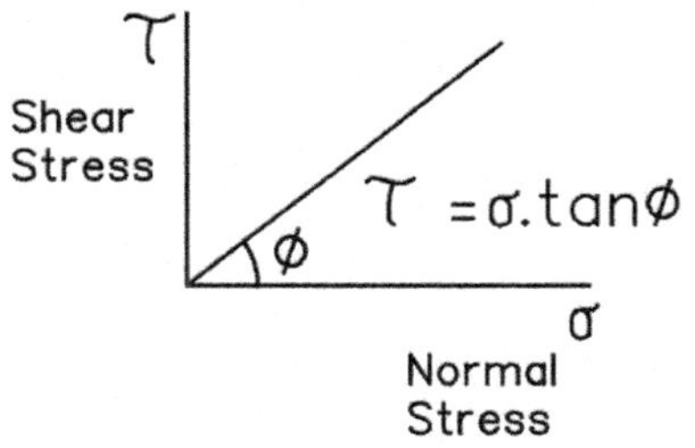

Figure - Soil Mechanics 6

This angle tells us how much friction is the soil able to produce inside. <u>The greater this angle Ø, the greater generation of friction, which means greater shear strength</u>, in other words, strength of soil. The value of Ø for water is zero, thus water has zero shear strength. So far we used total stress parameters (without the apostrophe sign) for conveniently describing the concept, but in reality we deal more with effective parameters (with the apostrophe), which is what matters for generating the strength of soil, therefore from here on we will use effective parameters. Dense sand can have Ø' of more than 35 degrees, which provides good strength. In the graph above, shear strength starts from zero. This is the case for cohesionless soils such as sands. If there was cohesion, then τ would start from a positive value as in Figure 12. That would mean, even if we had no σ1 and σ3 stresses pressing the clay cylinder or cube sample, thus no shear would yet be generated due to friction, inherently the clay would still have shear strength due to its cohesion. That is why, clays can hold themselves in the shape of a cylindrical mold, while sands cannot and will spill. But once we wrapped the sand in the shape of a cylinder and applied vertical and horizontal stresses, then due to its higher friction angle than clays, the sand would generate more shear strength than clay, even though initially the clay was able to hold itself without the help of a container and sand was not. The internal friction angle of sands generally vary between 30 to 35 degrees but they have zero to very low cohesion. Clays have low internal friction angle and in undrained condition this can even approach to zero (will be discussed), so the primary source of their shear strength is their cohesion.

Direct Shear test:

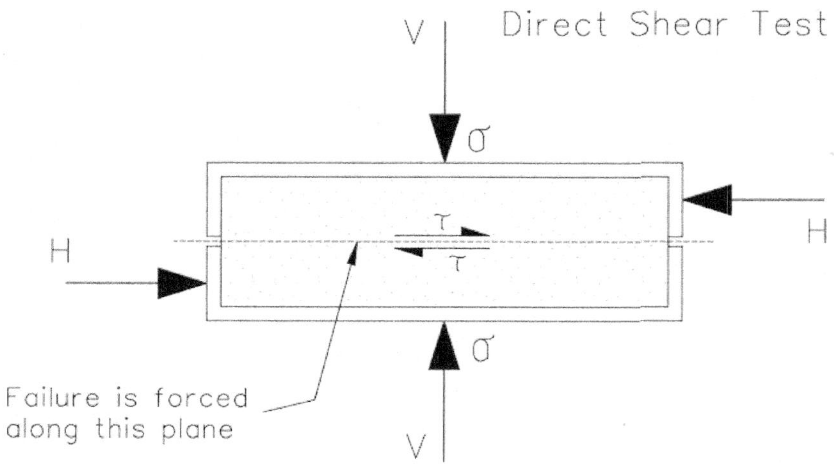

Figure - Soil Mechanics 7

Applying vertical and horizontal stress at the same time can be possible with a simple test apparatus as seen above. This test is called "Direct Shear Test", which is used to find the shear strength of soil, which is a simple and very old testing method that can be used for any kind of soil.

In this test, the failure plane is forced as horizontal, as seen above. As a result, it is not possible to draw Mohr's circles here, because to draw them, the failure must occur on the weakest plane naturally by itself. Not being able to obtain the critical plane is the greatest drawback of this test.

If we applied various pairs of H and V forces on dense sand, we could obtain a graph typically as below:

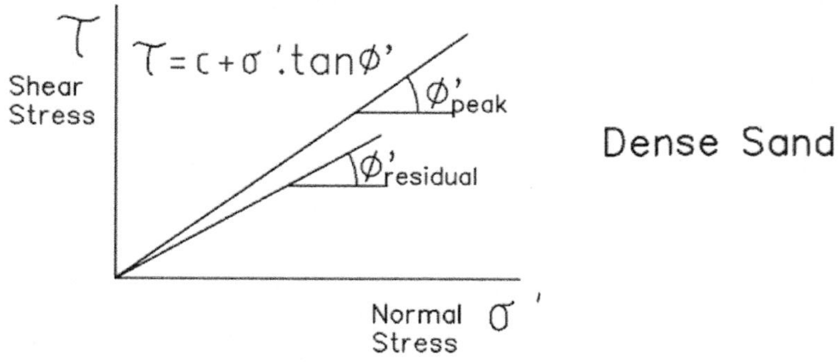

Figure - Soil Mechanics 8

In the figure above, you see the same graph that we showed in Figure 6 but this time there are two lines. The peak angle is there because, when we begin the test, it is harder to start initial deformation. That is why, that angle is called the peak angle. But then, once the soil starts deforming, the angle of friction Ø' will be lower and approach to a constant, which is now called the residual friction angle. The residual angle can also be called as the critical or ultimate angle. We will look at critical state concept after discussing several more topics. So for example if we had a hill slope and no deformation had occurred, we can use peak angle when analyzing slope stability. Peak angle is higher, which means higher strength. But if some deformation already took place, then we use the lower, residual angle.

We also wrote the cohesion intercept, c, in the formula in the figure above, but here it is zero. See the nonzero case in Figure 12. The line there starts from above zero shear stress, because of nonzero cohesion. So with cohesion, the Coulomb equation for failure envelope becomes:

$$\tau = c + \sigma' \tan \text{Ø}'$$

The τ vs. δ and ΔV vs. δ tables in Figure 9 below describe the behavior of sand during direct shear test. Here V is volume and ΔV means volume change. Top part of Figure 9 shows shear stress vs. deformation, which describes the same thing as in the previous paragraph, for the dense sand. The bottom part displays a typical volume change vs. deformation relationship. In both graphs, upper curves are for dense sand and lower curves are for loose sand. On both graphs, horizontal axes represent deformation due to shear, where two surfaces slide against each other.

Note: The curve for overconsolidated (OC) clay follows a similar type of route to that of dense sand, and the curve for normally consolidated (NC) clay follows a route similar to loose sand for both graphs. We discuss the OC and NC concepts under consolidation settlement topic.

The figure below shows that for dense sand first there is a slight decrease in volume (because we are also applying force from top and bottom) due to interlocking of particles, but once it is mobilized and the initial interlocking (plus frictional) resistance is overcome, it starts to loosen and the remainder of the resistance is due to friction only. Loose sands behave differently than dense sands. Try to visualize why *(it is because there was no initial interlocking and as we press from above they get denser)*.

After ultimate (critical) shear stress and void ratio is reached, critical state soil mechanics starts from that point. Ultimately they both reach to critical state, which will be the same for both dense and loose sand, given that they are the same sand. $\tau_{ultimate}$ is the same for both dense and loose sand. It is also called $\tau_{critical}$ or $\tau_{residual}$. $e_{ultimate}$ is also the same for both dense and loose sand. It is also called as $e_{critical}$. Therefore, critical state is a unique property of that sand, which is independent from how dense or loose the sand was initially. These τ vs. δ and ΔV vs. δ relationships described above are typically displayed on graph forms as below.

In design, in most situations we use the peak internal friction angle, which is the Ø' that corresponds to τ_{peak}. However, if shear deformations go beyond that level, then we must use lower friction angles. If the deformation is too much, we use the residual angle.

Note that in the figure below there is no undrained water pressure affecting shear strength vs. deformation, therefore it is a drained situation *(or a saturated soil can also behave as above as long as the excess pore pressure in it is zero)*. Also note that being dense or loose sand is a relative term that depends on stress. For example, the same sand which is below a very heavy structure can behave as loose, while it can behave as dense under a light, small structure.

Shear Stress (τ) - Deformation (δ) Relationship during Direct Shear Test

	Initial	Later	Ultimate
Dense Sand 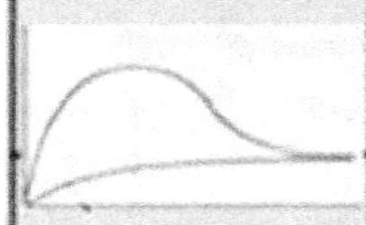	As deformation increases, shear stress quickly increases to a peak shear stress, τ_{peak}	Deformation increases further, but shear stress decreases from τ_{peak} to $\tau_{ultimate}$	Further increase in deformation does not affect shear stress. Shear stress steady as $\tau_{ultimate}$
Loose Sand	As deformation increases, shear stress slowly increases. There is no peak.	Deformation increases further, shear stress continues to increase towards $\tau_{ultimate}$	Further increase in deformation does not affect shear stress. Shear stress steady as $\tau_{ultimate}$

Note 1: Both shear stress and deformation start from zero.

Note 2: $\tau_{ultimate}$ is the <u>same</u> for both dense and loose sand. It is also called $\tau_{critical}$ or $\tau_{residual}$

Volume change (ΔV) - Deformation (δ) Relationship during Direct Shear Test

	Initial	Later	Ultimate
Dense Sand 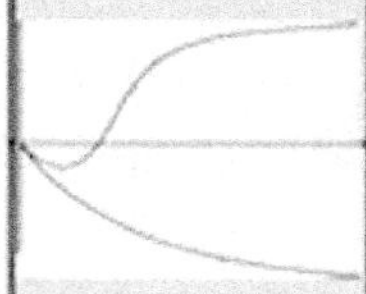	As deformation increases, volume slightly <u>decreases</u> at first. Then it recovers and starts to increase.	Deformation increases further, volume increases until the sand reaches ultimate void ratio, $e_{ultimate}$	Further increase in deformation does not change volume. Void ratio is steady at $e_{ultimate}$
Loose Sand	As deformation increases, volume decreases.	Deformation increases further, volume decreases until the sand reaches ultimate void ratio, $e_{ultimate}$	Further increase in deformation does not change volume. Void ratio is steady at $e_{ultimate}$

Note 1: Both volume change and deformation start from zero.

Note 2: $e_{ultimate}$ is the <u>same</u> for both dense and loose sand. It is also called as $e_{critical}$

Figure - Soil Mechanics 9

There is also the angle of dilatation ψ, which represents the slope of volume vs. deformation (ΔV vs. δ) curve, in other words, it is the rate of increase in volume versus shear strain. As described, first, dense sand gets compacted a little (remember there is also the vertical force pressing), then it starts to dilate during opening of interlocking of particles, which means increase in volume and the angle of dilatancy is the average angle that the particles slide during volume expansion. The angle of dilatation is maximum when the shear stress is at its peak. The angle of dilatation can also be calculated from the Mohr's circle, which can be constructed based on triaxial test.

Triaxial Test:

In direct shear test, plane of failure did not occur naturally but it was forced as horizontal. In addition, we could not control drainage conditions for water. Triaxial test can solve both of these problems.

The samples for testing must be obtained carefully from site and transported to the lab, without being disturbed. For sands this is very difficult to do, as they cannot keep their posture, but for clays it is easy, since they can keep their form, as long as proper care is taken. The sample is first wrapped in rubber membrane, placed in testing equipment and then compressed from above with the loading ram, while an all around pressure is applied with pressurized water, so that it will have both σ1 and σ3 pressures.

The cell pressure (σ3) should be adjusted to a value that matches as close as possible the natural stress that existed at the depth where the sample was taken.

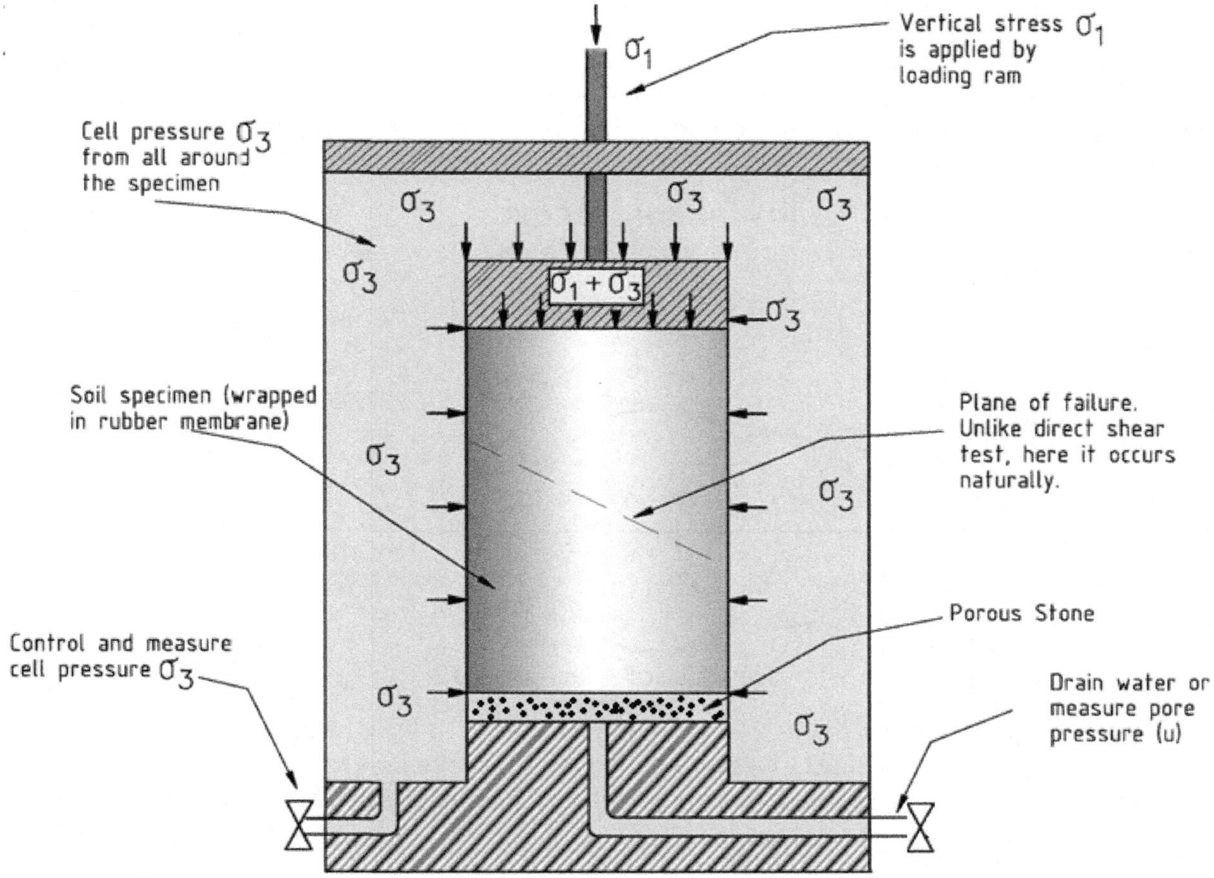

Figure - Soil Mechanics 10

We can perform the test for different loading pairs and obtain several Mohr's circles as shown below. Unlike the direct shear test, in triaxial test Mohr's circles are possible to obtain because the natural failure plane can form by itself. The friction angle Ø obtained through triaxial tests are lower than the ones obtained from direct shear test, because they represent the failure at a naturally occurring plane *(something that occurs naturally will be able to go through the weakest path, with lower resistance)*.

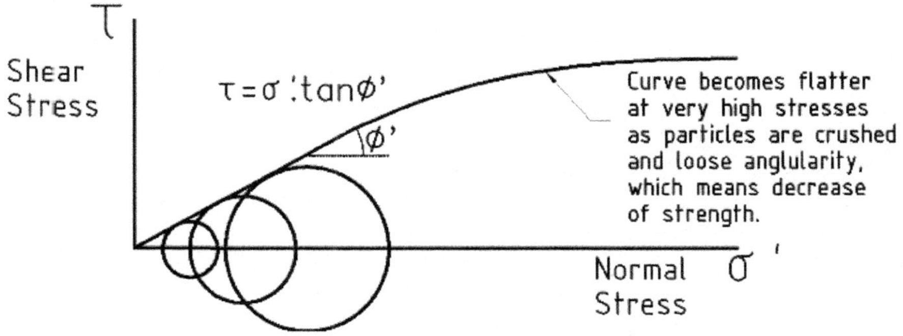

Figure - Soil Mechanics 11

As mentioned under Mohr's circle topic, no circle can cross the failure envelope and at very high stresses this line tends to get flatter. In sands it is due to particles getting crushed and becoming less angular which causes particles rearranging and in turn some more failure settlement due to this. The failure envelope is also called the Mohr-Coulomb failure envelope.

Above, the case for sand is shown, there is no cohesion so shear stress starts from zero. Clays have cohesion, which causes the tangent line to start at a positive value instead of zero as below. So, for clays, we can obtain a typical graph as:

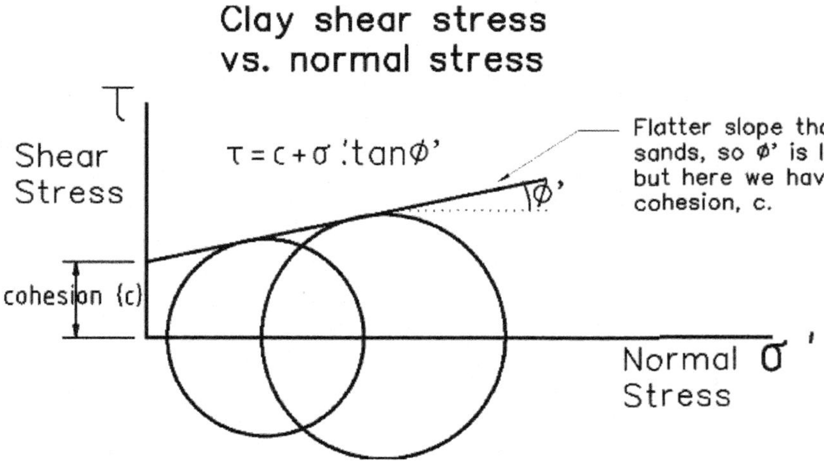

Figure - Soil Mechanics 12

The slope in the figure above is flatter than for sands, because Ø ' is lower, which means clay can generate less shear resistance due to friction, but here we have cohesion, c. Clays are weaker than sands, since friction is more important for strength, than cohesion. This failure envelope can even be flat in undrained conditions, which means zero Ø '.

As we said above, we can also control water drainage with the triaxial test. Try to see how, in Figure 10 above. If we do not let water to drain, it is called undrained test. If we allow it, then it becomes the drained test. Either or both can be used to better represent the natural conditions for a particular project. For example, to simulate a field condition when water did not have opportunity to escape, we can use undrained test. The drained test takes far longer time, which is its main disadvantage, due to very low permeability of clay, so they are used less. In the cases we need but don't have time do them, we can make approximations based on undrained test results and other factors. The rate of loading and in-situ stresses also affect test results.

So, the three types of triaxial tests are:

- Consolidated & Drained (CD Test)
- Consolidated & Undrained (CU Test)
- Unconsolidated & Undrained (UU Test)

The effective parameters, which are the true parameters of the soil (i.e. not the total stress parameters) can be found from the CD or CU tests. If we use CU test, we must also measure the pore pressure and draw the Mohr's circles and the failure envelope in terms of effective stress. Consolidation will be covered soon. It means the slow, time dependent settlement of clay.

From UU test, total stress parameters are obtained. In the UU test, for fully saturated samples, the failure envelope should be horizontal, as if Ø = 0, because no matter how much we increase the stress on the sample, all of that will be taken by the pore water (as it cannot escape anywhere), but not the soil anymore, which will be as if the soil's Ø = 0. Therefore, the Mohr's circles that we will obtain, will have the same diameter (i.e. same difference of $\sigma 1$ and $\sigma 3$, even if their values vary) enclosed by that horizontal failure envelope. And the shear strength becomes equal to cohesion of clay, where that horizontal envelope intersects the y axis, which will now be equal to the circle's radius, which is called the undrained shear strength, a parameter of great importance in actual undrained conditions, and written as c_u. If however, the sample is not properly saturated, then the failure envelope cannot be horizontal as the pore pressure cannot develop and the soil itself will carry the load increase, which means Ø will not be zero.

Angle of repose:

When sand is poured freely on a flat surface, it makes an angle with the horizontal. This is called the angle of repose. Higher angle of internal friction also means a higher angle of repose but they are not always the same.

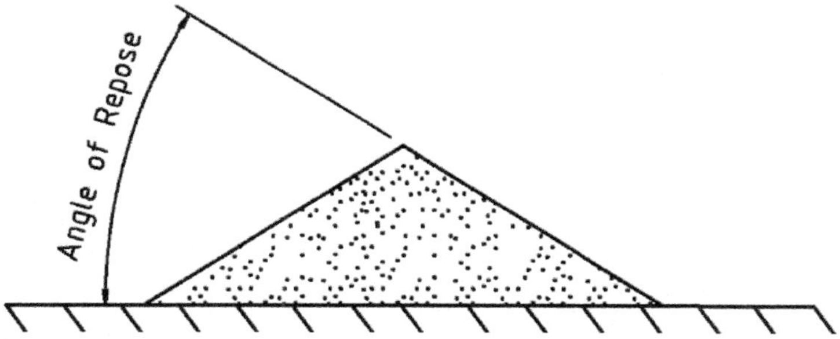

Figure - Soil Mechanics 13

Angle of repose is the same as angle of internal friction at critical state only. Otherwise, they are different, because the angle of internal friction can change according to the density of soil as we saw.

Clays cannot be spilled on a flat surface like this, so the angle of repose is for sands. The only exception would be when clays are submerged in water, as the cohesive bonds will be broken, they can then form a slope, but it will be flatter than sand, due to clays' lower angle of internal friction.

Cohesion of Clay:

As mentioned before, shear strength is a result of friction between solid particles of soil, and or cohesion between them. Cohesion is a force which holds the particles of fine grained soils together through cementing, sticking action. For shear stress, cohesion is the portion that we represent with c, in the shear stress formula $\tau = c + \sigma' \tan \emptyset$ that was shown. If you look at this formula, even when the angle of internal friction is zero, there can be cohesion, and in that case it will be equal to the entire shear strength, as τ=c. Figure Soil Mechanics-12 shows the cohesion intercept for the shear vs. normal stress graph. But how can we measure it, when we apply zero stress? The answer is that we apply the stresses, draw the Mohr's circles, and then make a best fit line for failure envelope. Where this line intersects the vertical axis (shear), is the cohesion value. Therefore, cohesion is an inferred value.

The sticking force is the result of surface attraction between soil particles, which can only apply to fine grained soils like clay. This force is too small to hold coarse grains together, so sands have little to no cohesion, but have considerable internal friction. Shear strength provided by cohesion is usually considerably lower than provided by friction, which is why sands are better foundation materials. Many soils contain both c and Ø components, even if one can be negligible or little. These are called c-Ø soils. Most of the information in these last two paragraphs were also mentioned earlier, but their importance cannot be overemphasized.

Lateral Earth Pressure:

Very often, we need to retain earth horizontally. To do this, we must be able to calculate the lateral (horizontal) stresses that the earth material will apply on retaining structures.

Imagine you pour sand on a flat surface. Try to put weight on it, and it will immediately fail. But if you laterally support the same sand by filling it into a container and load again, it will probably carry it, unless the load is too heavy or the container box is very weak. So, the sand's vertical load carrying ability increases dramatically when you restrain it from sides. If we want to explain this with Mohr's circles, you applied the same $\sigma 1$ in both cases, but when you put the sand in the box, there is now the lateral support of $\sigma 3$ that was not there before. So, now, the Mohr's circle would pass through $\sigma 1$ again but instead of zero for $\sigma 3$ (See Figure - Strength 13), it will pass from a positive value. This means, you will now have a much smaller circle, which will probably not cross the failure envelope for that sand.

To determine the horizontal stress in soil, we use something called "Lateral Earth Pressure Coefficient", which is shown by a capital "K". For a given vertical stress, we can find the horizontal stress as:

Horizontal stress = Vertical stress . K

So, K shows what portion of vertical stress is converted into lateral stress.

As shown before, the vertical stress at depth h = γ_{soil} . h. We can rewrite this to find horizontal stress as:

$$p_h = \gamma_{soil} \cdot h \cdot K$$

where p_h is the horizontal pressure.

K value in the formula above is not the reason of how much vertical stress will be converted to horizontal stress. K value is a result, which depends on how the soil and the retaining structure tend to move against each other, which can be represented in three cases of K values as below. We can list them from lower to higher values as:

- Active earth pressure coefficient (K_a)
- At rest earth pressure coefficient (K_o)
- Passive earth pressure coefficient (K_p)

Now let's describe each situation:

K_a: This is the case when the retaining structure tends to move away from the soil, so in other words, the soil is in active condition, as it pushes the wall, and during this, the wall moves away from it. This is the lowest K value because the soil pushes something that already tends to move away.

K_o: In this situation the retaining structure is either restrained, or too heavy to move at all. So, the soil is at-rest condition and nothing moves. In case of restraint, this can be provided by a slab, braced wall or a too rigid and heavy retaining structure. At rest K value is higher than the K value in active condition.

K_p: In this case, the retaining structure tries to push towards the earth, so the earth is in passive condition. This is the highest K value because here the soil can provide much more horizontal pressure when something pushes on it, and nothing is moving away from each other, but on the contrary, towards each other, such as the soil in front of the moving wall in the figure below.

In order for the full active or passive K values to be realized, the movement must already have taken place, in other words, for active, the wall must have already moved away, or for passive, the wall must have already pushed the soil. The deformation that must take place to fully mobilize the passive resistance is much more than the deformation for active state to fully develop. This is because in active state, after a short lateral deformation, the soil cannot push the wall further, as the wall already has moved away and the contact is lost, but for passive state, the wall can push further and further towards the soil, while the soil will keep developing passive pressure, while the full contact always exists.

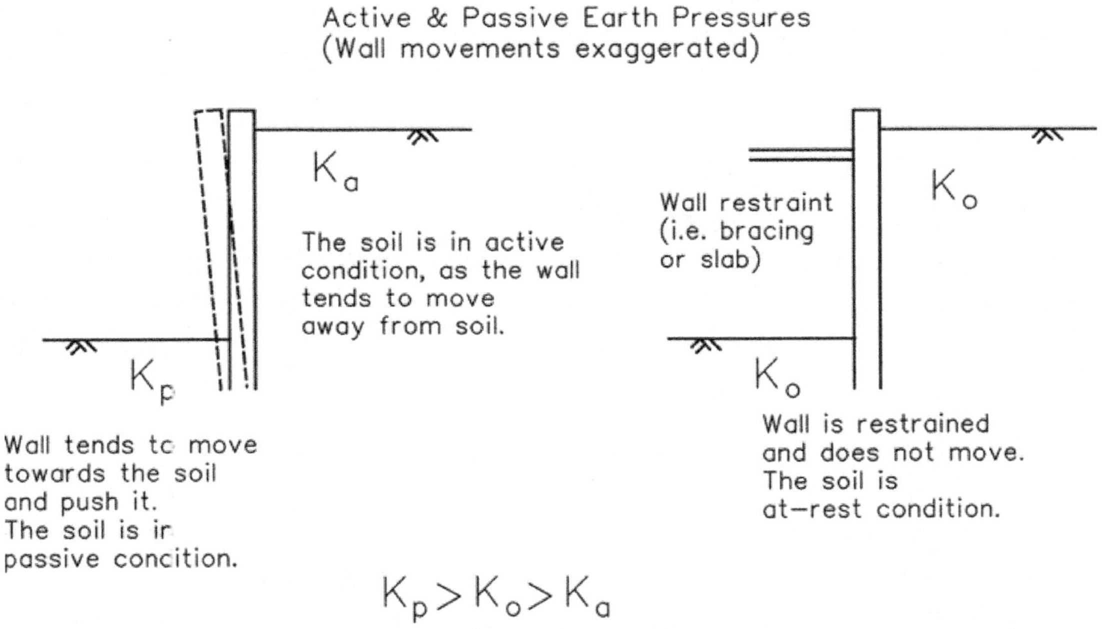

Figure - Soil Mechanics 14

To find K_a and K_p values, we need to know the angle of internal friction, Ø (but its effective value, not total, so, with the apostrophe). K values can be found as follows:

$$K_a = \frac{1-\sin\text{Ø}'}{1+\sin\text{Ø}'} \qquad K_p = \frac{1+\sin\text{Ø}'}{1-\sin\text{Ø}'}$$

These are called Rankine active and passive earth pressure coefficients, after the Scottish engineer who developed them in 19th century. It is possible to also often see them arranged differently with tangent of the angle, which is exactly equal to this formula. We showed only the basic equations that do not take into account the cohesion of soil or any friction between the soil and wall. It also assumes that the resultant active and passive forces are horizontal, wall surface is vertical and the soil behind the wall is not sloped, seismic effects were not considered and the soil is homogeneous and isotropic (the soil's composition is same everywhere and in all directions). The equations that also take some or all of these additional variables into account were developed by others, including by Coulomb who provided the shear stress equation that we saw. When surcharge load and groundwater is present, they are dealt additionally.

And the at-rest coefficient can be found as:

$$K_o = v/(1-v)$$

where, v is the Greek small letter "Nu", which is the Poisson's ratio of the soil. We had mentioned this ratio also under Strength of Materials topic. The value of K_o can also be approximated as: $K_o = 1\text{-sin}Ø'$ which is based on experiments, which was found in 1944 by Jacky, for normally consolidated soils. It is useful since it uses the angle of internal friction. To determine stresses in natural soil, such as prior to excavation work, we also use at-rest coefficient. σ_h/σ_v is the initial stress state of the soil, which means the ratio of horizontal to vertical stresses in the natural soil, at rest state, so K_o can also be written as σ_h/σ_v. The horizontal stresses in the soil can be increased with compacting. If we have to retain such soil after it is compacted, we also need to consider those locked-in horizontal stresses in the soil.

Although we have formulas as above, to determine the precise angle of internal friction, therefore K values, is difficult because it ultimately depends on the success of testing because soil is not a man-made material manufactured in controlled environments, and in addition there is the need to make preliminary calculations when testing might not be complete or not even performed yet. So engineering judgment is also often necessary to assume K values, but very close values can be estimated based on a significant amount of past data and experience.

Example for solving retaining wall under lateral earth pressure:

The simplified example below shows forces that act on a gravity retaining wall, which retains earth by the resistance provided by its own gravity. The figure on the left shows the retained earth and the wall. The middle figure shows the free body diagram. Here F_a represents the active earth pressure, F_p is the passive earth pressure, and W represents the self weight of the wall, which acts through the center of gravity. The reaction at the bottom is more on the left side, as the toe of the wall is pressed more, due to tendency of overturning (although it doesn't, if resisting forces are sufficient). The figure on the right shows the same thing, except that this time we represent the distributed forces as single forces, concentrated at the geometric center of their distribution, such as, F_a acting at 1/3 of h_1, which is at the geometric center of the triangle. The V force at the bottom is the shear resistance against sliding. N is the bearing reaction force provided by the soil to carry the wall, again simplified from the trapezoidal force shown in the middle figure.

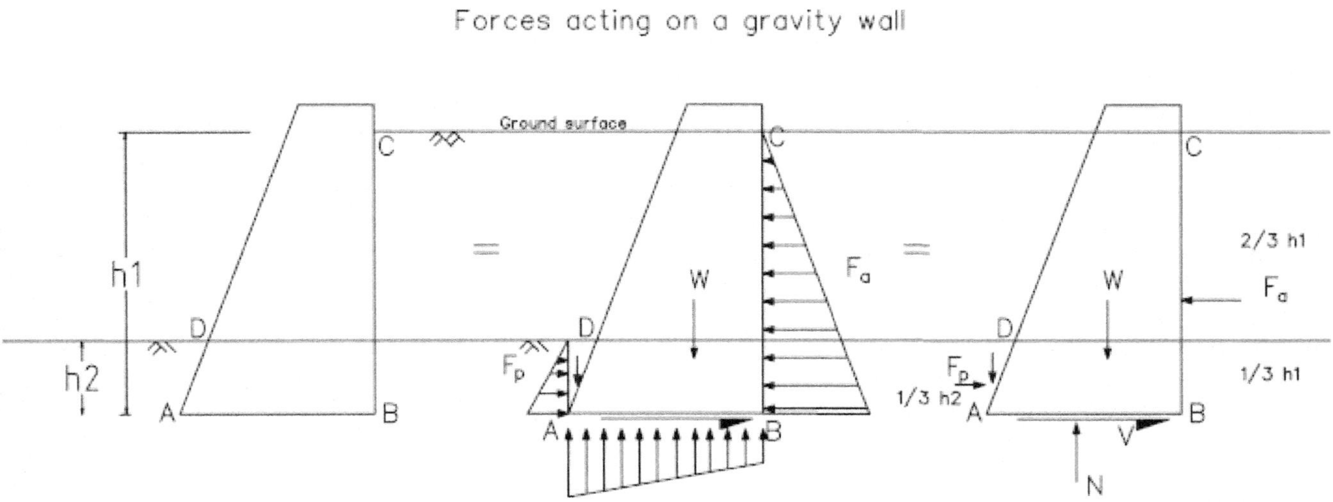

Figure - Soil Mechanics 15

Lateral pressure, P_h, at depth h is:

P_h =vertical pressure . K

Which can be rewritten as:

$P_h = \gamma_{soil}.h . K$

And since we know that Force = pressure . area,

F_a, the active soil force, is equal to active pressure multiplied by the area that it acts on. This area is height and the width, with height h. For width, we can take 1 unit, towards inside of plane of paper we are looking at. The lateral pressure is a triangular distribution so its average must be half of it, if we want to reduce it to single force, which we call F_a.

So F_a can be written as:

$F_a = 1/2 . h . (P_h) = 1/2 . h . (\gamma_{soil} . h . K_a)$

Rewriting this last one, we obtain,

$$F_a = \frac{1}{2} \gamma_{soil} . K_a . h_1^2$$

and with similar logic, for passive force we obtain,

$$F_p = \frac{1}{2} \gamma_{soil} . K_p . h_2^2$$

Remember that, these are not pressures, but forces per linear unit distance (1 unit) of wall into the paper plane since we multiplied pressure by area.

The weight of soil above the toe as in the figure, shown by the small downwards arrow right below point D on the passive side, can be ignored, which is a conservative assumption.

Again, all forces that are shown in the figure are per the unit length of wall, which is 1 unit (such as 1 foot or 1 meter), towards inside of the paper plane you are looking at. You can also verify this from units. First we have lateral pressure at depth h, which was: γ_{soil} . h . K, which has units of stress such as Force/Area. Then we multiply it with 1/2h, so now we have Force/length. So, the active and passive forces Fa and Fp that we calculated have units of Force / length, applied toward inside of the paper plane. Then if later you multiplied it with the wall length, towards this plane, you would get the full amount of force on the wall itself. But to design the typical section of wall in 2D, using force / length is sufficient.

After calculating these forces, we can then first check the overall stability of the wall against overturning, and against sliding. We do this similar to the analysis we made in statics section, where there must be equilibrium of all forces and moments for the structure to be stable, such as:

Sum Fx = 0, Sum Fy = 0, Sum Mz = 0

For example, if we take moments of all forces with respect to point A, we can simply check the overturning vs. resisting moments. And for sliding, we can add the forces acting in horizontal direction to see if there is enough sliding resistance. We can then divide the resisting forces with a safety factor for a safe design. In other words, we do not just take moments or compare forces to see if they are barely equal. Forces must also satisfy equilibrium even after we divided our stabilizing forces by a safety factor. For retaining walls, a factor of safety of at least 1.5 may be used, which varies according to applicable codes and standards and can be different for sliding and overturning.

Sliding and overturning are the basic overall checks for stability, but this is not enough. We must also check the settlement amount by comparing with bearing capacity of the soil. We must also design the reinforced concrete body of the wall to be able to withstand all these forces. There can also be other loads which we did not show, such as the surcharge load of structures or vehicles from the retained side, which then adds to the active pressure, there can be vibration forces, and in seismic areas earthquake must also be considered which makes the analysis more complex and may govern the design (be the most critical). If water accumulates behind the wall this will simply add to the force that the wall must retain therefore water must be drained, by providing drainage at the bottom of the wall and small holes on the wall called "weep holes". Check against uplift pressure of water must also be performed if we expect the possibility of water level to be high, even if rarely and short durations.

Active and passive zones of retained soil:

The active and passive zones we discussed above are also called as Rankine active and passive regions. They form angle with the horizontal that depends on the soil's internal friction angle. See below.

These regions are also called wedges. The angles shown here were calculated ignoring friction between the soil and wall. If not, they will be slightly different. Forming of wedges means, the soil is mobilized. If anchors are used, they must extend beyond the active wedge, to hold on to firm soil behind. The portion of anchors that stay within the wedge would contribute nothing to stability, as the wedge itself already needs support and tends to move. Again pay attention that effective friction angle Ø' is used.

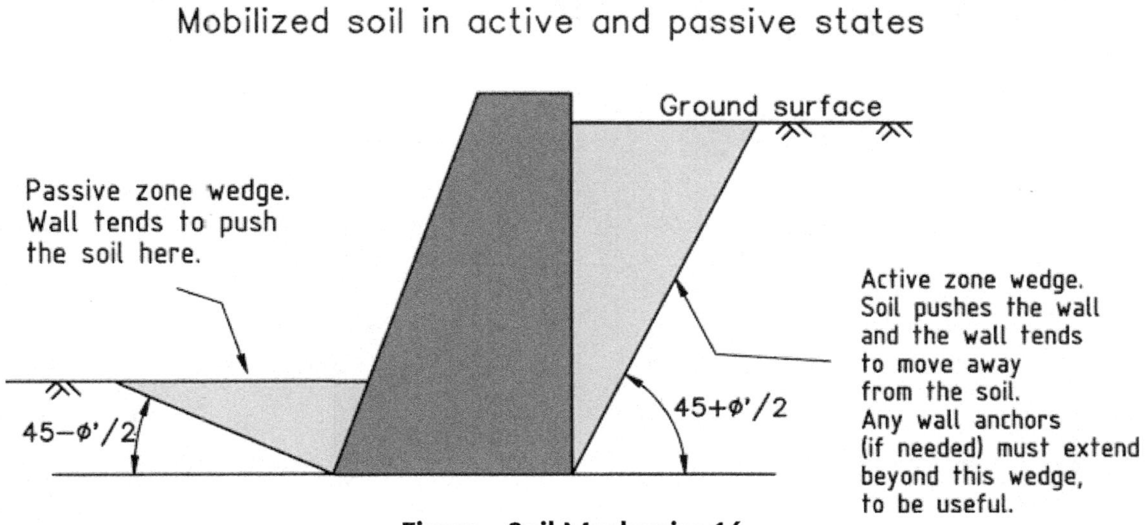

Figure - Soil Mechanics 16

Bearing Strength of Soil:

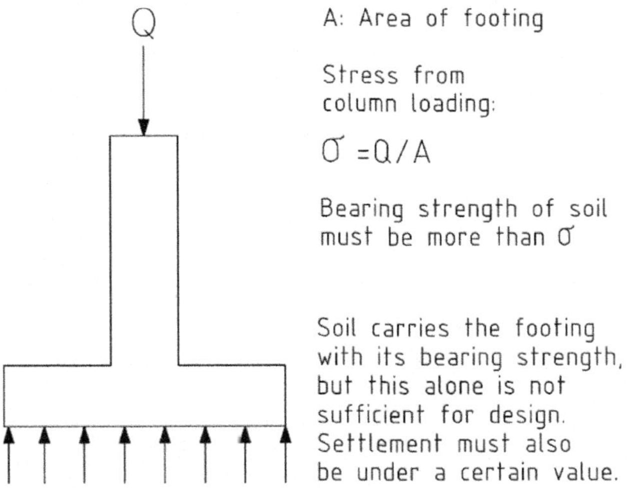

Figure - Soil Mechanics 17

Soils can carry the foundations up to the point of their bearing strength. Here "carrying" means that the settlements will be below a minimal acceptable tolerance, because in nature it is not possible to prevent deformations or settlements altogether. The stress from a foundation is simply found by dividing the load applied by the column by the footing area as $\sigma=Q/A$, where Q is the column load and A is the footing area. See Figure 17. Then we compare this stress, σ, with the bearing strength of the soil (after also applying a safety factor). If the stress σ is more than the safety factored strength, it means we need to either decrease the load from the column, or increase the bearing strength of the soil by one or more of the ground improvement methods that we will discuss, or increase the footing area, or apply more than one of these methods together.

The figure below shows the general shear failure mechanism, <u>when the soil is fully mobilized</u>. This can occur only in <u>strong</u> soils, because only strong soils are capable of forming well defined wedges like this. Wedge is the active zone which pushes. This is a sudden failure which happens catastrophically. The general equation to define this type of failure was first produced by Terzaghi.

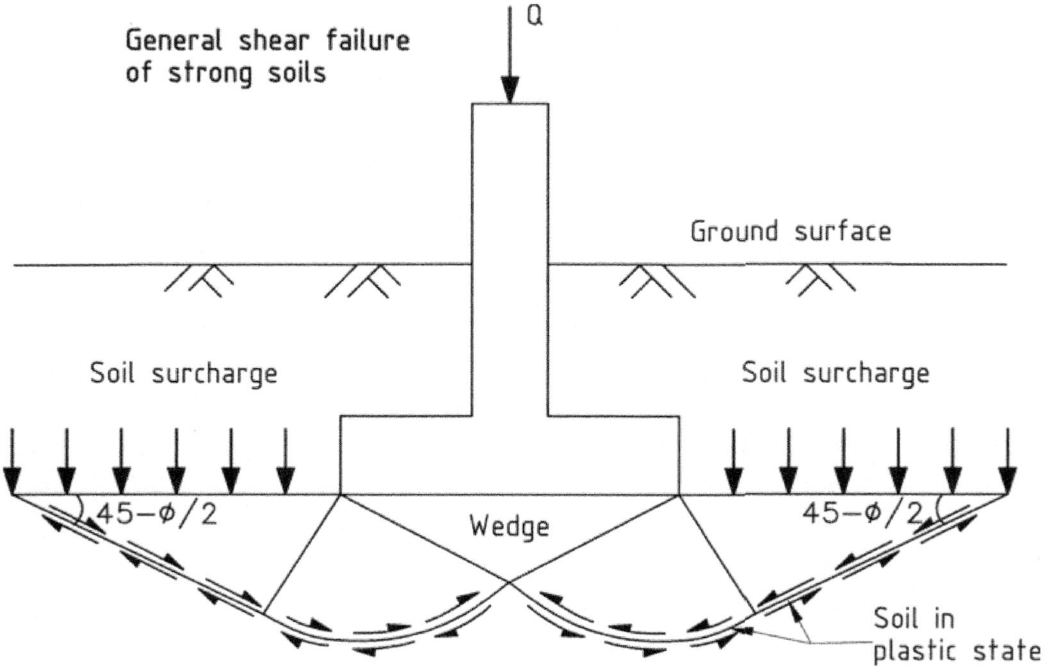

Figure - Soil Mechanics 18a

In Figure 18a above, you can see how the angle of internal friction affects forming of wedges. Also here the footing does not sit on the ground surface but buried at some depth. This is good, because the surcharge from the soil above, presses on the outer wedges, resists their heaving (upwards movement).

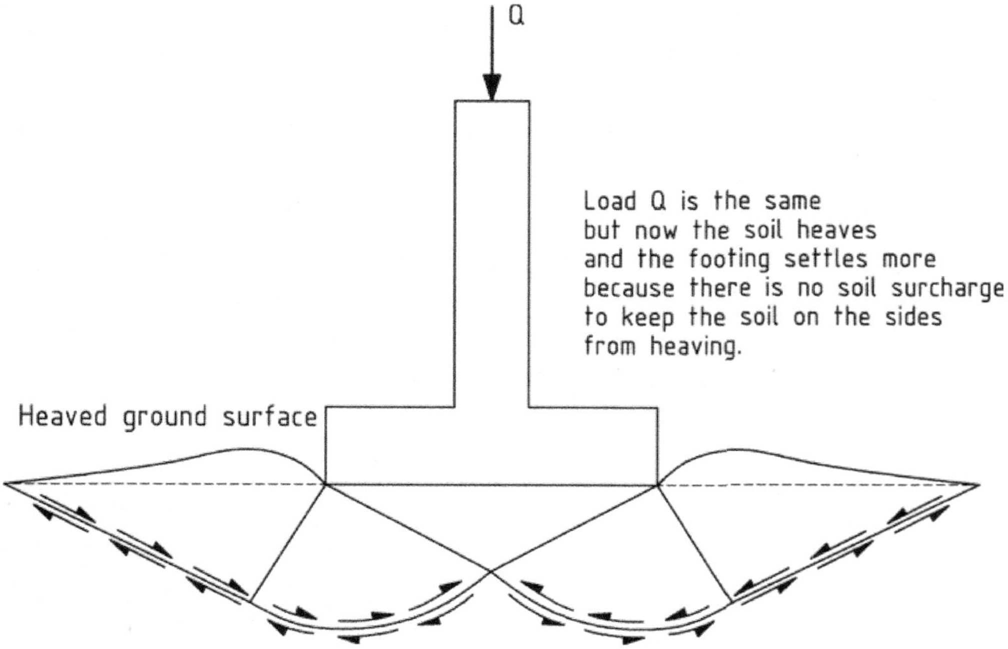

Figure - Soil Mechanics 18b

If the footing sat directly above ground (not buried) as in Figure 18b, this surcharge would not be present which means the bearing strength of the soil would be considerably lower as nothing would push the wedges down. In case of failure, the soil around the footing would heave.

If the soil is weak, it is not capable of forming well defined wedges as shown above and the failure happens in a punching manner as in the figure below, where the footing slowly sinks, and no heaving occurs around the footing. If soil had medium strength, the shape would look like somewhere in between Figures 18 and 19.

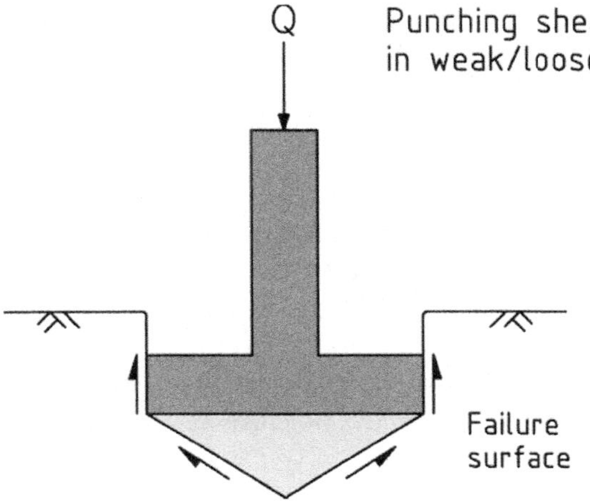

Figure - Soil Mechanics 19

The failures shown here represent the ultimate case, with a lot of deformation where total collapse occurs, which is far more than what we can tolerate. So this state is the ultimate bearing strength, which we write as q_{ult}. But for our design we need to divide this by a safety factor, which is the value we design to, which will mean tolerable levels of settlement. That is called the allowable bearing capacity, which we show as q_{all}. So,

$q_{all} = q_{ult} / FS$

where FS stands for factor of safety (or SF as safety factor).

Calculating Bearing Capacity:

Below is how we can calculate bearing capacity, for a square footing. For different shaped footings, such as circular or strip (continuous) footings, the constants below vary a little, which is the only difference. Keep Figure 18a in mind, when looking at the formula below. This formula is for dry condition, in other words, where water is deep far enough (such as, at least by 1 foundation width deeper than the bottom of footing) so that it doesn't affect anything. We can state the formula as:

Ultimate bearing capacity (q_u) = capacity from cohesion (q_c) + capacity from surcharge (q_q) + capacity from soil unit weight (q_γ).

Rewriting it, we have: $q_u = q_c + q_q + q_\gamma$

$$q_u = 1.2\, c\, N_c + q\, N_q + 0.4\, B\gamma\, N_\gamma$$

Now the equation above may seem complicated, but it isn't. Let's explain:

- All N values above depend on the angle of internal friction, Ø. The greater the friction angle, the higher N values.

- q_c, is the component of bearing strength as a result of cohesion. It is equal to $1.2cN_c$ as seen above. Here, c is the cohesion value of soil. It means, as the cohesion increases, bearing strength goes up.

- q_q, is the component resulting from the surcharge load, q, as shown in Figure 18a. It is equal to: qNq. The surcharge load can be from not only the weight of the soil but also from the load of a nearby structure. So, if the foundation was deeper, or if there was an existing structure near and above our footing, it means higher bearing capacity.

- q_γ is the component resulting from the unit weight of the soil, which is equal to $0.4\gamma BN_\gamma$. Here γ is the unit weight of the soil, and B is the foundation width. So the greater unit weight of soil, in other words, denser the soil, and also the wider the foundation, the higher bearing strength.

- For sands the cohesion component may be zero or very low, but through q_q and q_γ components, their bearing strength are still more than clays because as we said above N values depend on internal friction angle, which is higher for sands.

So to summarize, bearing capacity increases with angle of internal friction and cohesion of soil, nearby surcharge load, depth of foundation embedment, foundation width, and soil density.

The formula above is by Karl Terzaghi, who in 1943 came up with a general bearing capacity equation as above, for shallow foundations, for general shear failure as in Figure 18a. Different researchers also proposed different values of N values and different numerical coefficients too. For example Meyerhof and Vesic later made modifications and additions for other factors. These N values can be found in many soil mechanics references. For the formula above, shallow foundation means that its depth is less than or equal to its width, although values of more than its width are also specified in some places. For this formula, the soil is assumed to be homogeneous and isotropic. Also, friction and adhesion between the sides of the footing and the soil, inclined loading, or soil slope are neglected. The equation above needs to be adjusted for the case where water is present at shallow depth close to or even above the foundation, which determines if we use dry, saturated or effective unit weight. And finally, as stated before, we divide this ultimate bearing capacity with a safety factor to obtain allowable bearing capacity as $q_{all} = q_{ult} / FS$

When designing foundations, we primarily check two things: the bearing criteria as we introduced above, and the settlement, which is discussed below after several pages. Usually the settlement criteria governs the design but for smaller footings, bearing failure will tend to govern. For large foundations, settlement is more critical. We can never fully eliminate settlement, but keep it within tolerable levels.

Eccentric loading on a footing, and bearing of soil:

When a structure is subject to overturning effect, one side of the foundation will be pressed more. The retaining wall we looked at earlier, also had trapezoidal pressure distribution underneath. This can happen also for shallow foundations, through lateral forces from earthquakes, or from wind on a tall slender structure such as a tall chimney or a wind turbine. In such cases the pressure distribution below the footing will not be uniform anymore, but trapezoidal. See Figure 20a below. This trapezoidal distribution is what we want which happens when e<B/6. This means, everywhere at the bottom of footing, there is still contact with the soil. Here B is the width of footing. "e" is eccentricity, which is the distance that a single vertical force would act as on the right of the figure, to make up for the moment as on the left of the figure. In other words, the figures on the left and right sides are equal, since M=P.e. When e=B/6, we will now have a triangular distribution, although there will still be contact everywhere. Now look at Figure 20b. As soon as e > B/6, then that triangle will start to shift towards left and some of the footing area will lose contact with the soil. We do not want this. It means that suddenly we have a smaller footing now, which must still carry the same vertical load from above.

As e gets larger, the height of the triangle will grow, (*because force = pressure x area, and since the base of the triangle in horizontal is decreasing, its height in vertical must increase*) which means smaller area of soil carries the same load, and this means more stress on soil. This can now exceed the soil's bearing capacity and cause excessive settlement.

So, we must arrange the loads vs. the size of our footing such that this triangular distribution will <u>never</u> happen, and at all times the entire foundation bottom will contact the soil. Also pay attention that the pressure distribution's geometric center from centerline of the column corresponds to "e" value. To keep "e" below B/6, the most straightforward option is to increase footing width B if we have enough space. Or to keep e small, since e=M/P, we must either make M as small as possible, or P as large as possible, but too much increase of P may cause excessive settlement or bearing capacity failure.

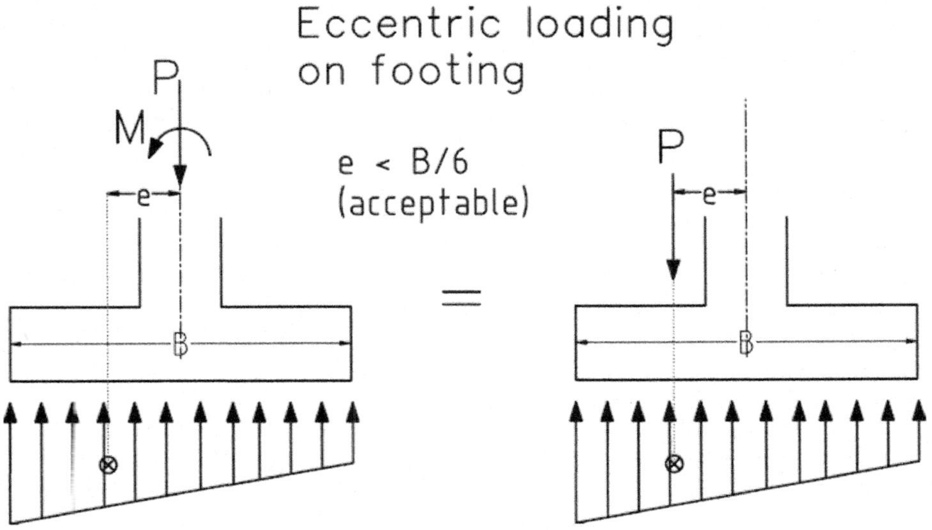

Figure - Soil Mechanics 20a

We showed two dimensions here but of course in reality it can happen in three dimensions, or the load can even be inclined, but the basic logic is the same. The eccentric loading also causes more bending stress in the footing body itself, which must be considered during design.

During foundation analysis and design, it is one of the primary objectives to satisfy this requirement below a footing that everywhere be under compression (or zero but never in tension as seen below) under all possible load combinations and worst case loadings, whether it is a single footing or a large mat foundation.

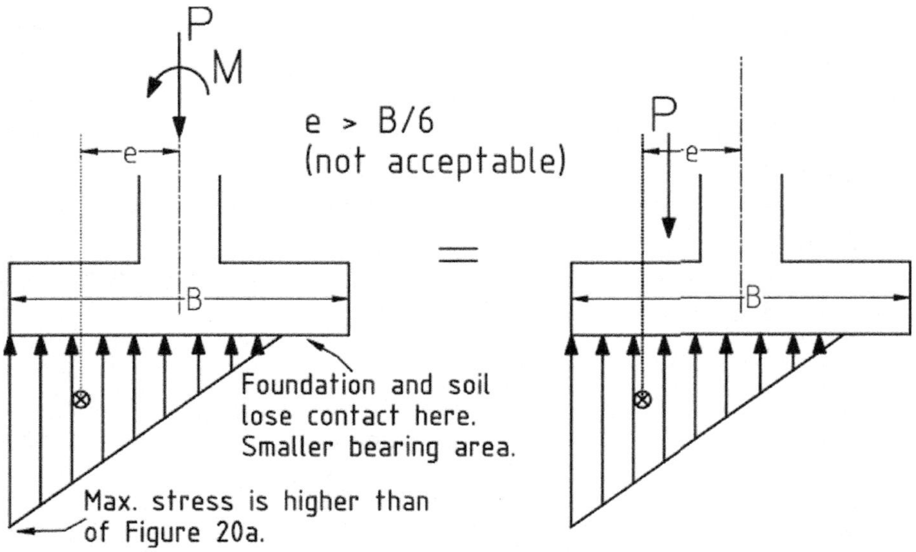

Figure - Soil Mechanics 20b

Stress Increase in Soils:

As mentioned, for checking adequacy of foundations, settlement and bearing criteria must be checked. We described bearing capacity above. Now let's look at settlement. To do that, we must first understand how stress increases in soils due to new loads, because this is what causes settlement. If you paid attention we underlined the word "new loads". This is to make sure that you do not confuse this with existing stresses in the soil, which we had covered in the beginning of soil mechanics section.

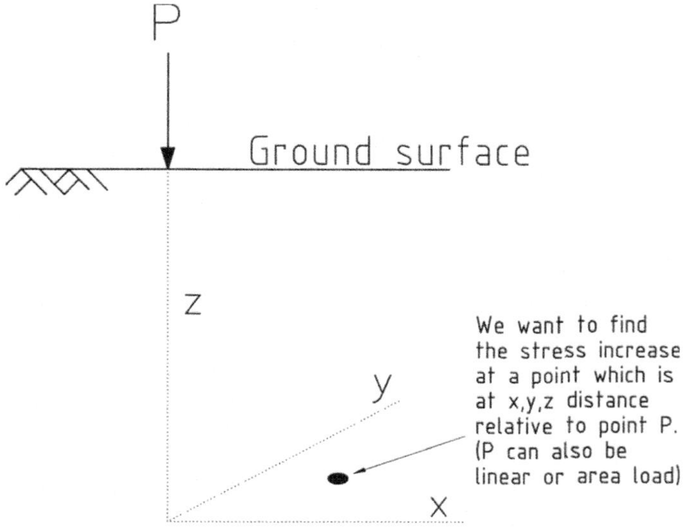

Figure - Soil Mechanics 21

We can apply loads in different ways on soil, which can be in the form of point loads, linear loads *(also called as strip loads)* and area loads. Due to these loads, the stress at a given point in soil will increase, unless that point is too far deep in the soil and/or simply far from the loading location horizontally. To find the stress increase there are a few commonly used methods. The most straightforward is to simply assume that the load gets distributed more and more in a linear fashion as we go deeper in the soil, at a geometric ratio of 1 horizontal to 2 vertical distance, which is called the 2 to 1 method (also shown as 2:1 method), as shown below. In this method, stress decreases uniformly with depth.

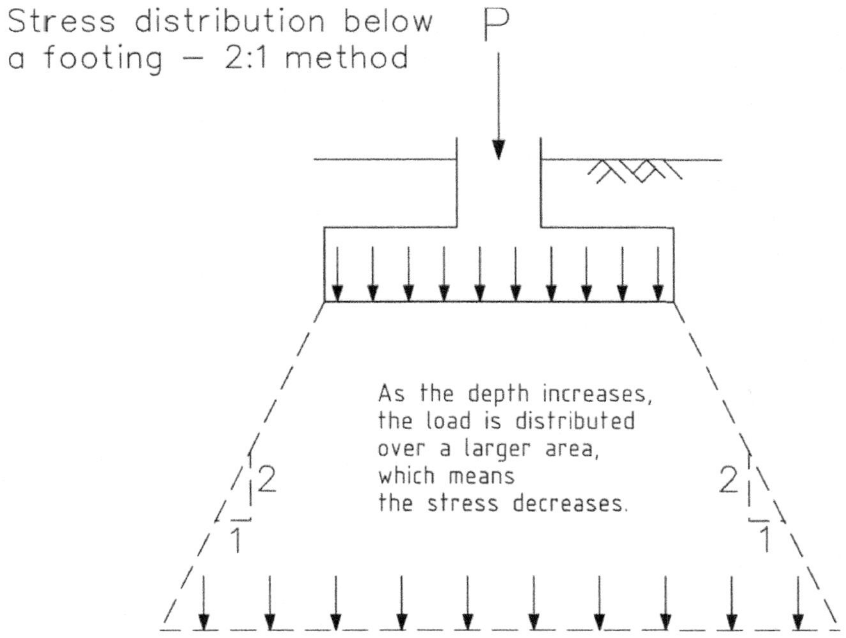

Figure - Soil Mechanics 22

Two of the other common methods are the ones by Boussinesq or Westergard. In Bousinessq method, based on a lot of data, stresses in the soil were obtained with respect to increased loads. As a result, stress points of equal value were represented by connecting them as lines, which are called stress contours, which look like the bulbs as seen in Figure 23 below. This is why these are also called as "stress bulbs" and less frequently also as "pressure isobars". Each contour corresponds to a certain stress value and its location with respect to the footing location and dimension can be compared by the square grid. The figure below doesn't show the contours exact layout with respect to the grid, nor their values, but just shows the concept. In this method, the soil is assumed as homogeneous, isotropic and elastic. The method by Westergard also considers varying soil layers. As can be seen below, as soon as a point gets further away from the foundation horizontally, even for shallow depths, the stress decreases at a very rapid rate and quickly approaches to zero. For different foundation shapes, the stress bulbs will also somewhat vary.

Each curve is expressed as a percentage of the stress q that the foundation applies on the soil. So for example where the soil touches the foundation, the q would be 100%. After that, as soon as the point gets deeper and further away, the percentage of q that the point will experience will decrease. The points that experience the same percentage of q are connected, which make up these contours.

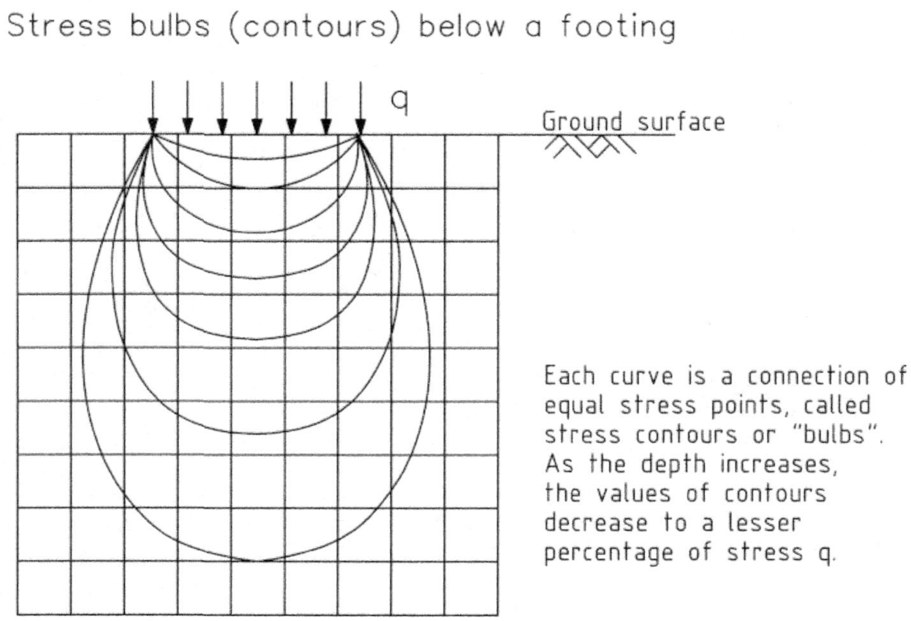

Figure - Soil Mechanics 23

Another common method is by Newmark, who provided charts to estimate stress increase in soils as well, which are also based on Bousinessq equations.

Elastic Settlement:

Soil settlement can occur in three ways:

- Elastic settlement, which is immediate, and applies to all types of soils.
- Consolidation settlement, which is due to water slowly escaping from the pores of clay under compression, which is also called as primary consolidation.
- Secondary consolidation settlement, which continues due to particle rearrangement even when the primary consolidation is complete. This is smaller in amount in comparison to primary consolidation. It is also called "soil creep".

For sands, elastic settlement is the only type of settlement that occurs, which is a result of immediate compression of soil skeleton. The more angular sand particles, the stronger the sand mass, due to interlocking of particles and higher internal friction. The more round the particles, such as river sand which was exposed to long time abrasion from water, the lower internal friction and more settlement. The soil density will also affect settlement. Vibration and being saturated also makes settlement in sands easier as discussed under liquefaction topic. This means, if we want to compact sand, we apply vibrating forces and add some water.

Different methods to estimate settlement exist, which will not be discussed here. They can be based on elastic theory or make use of results obtained from soil tests. For clays, obtaining undisturbed samples and transporting them to the lab is easy but for sands this is very difficult to impossible, therefore we perform in-situ (on site) tests. These field tests include most commonly, SPT, CPT, PMT, plate load tests as will be discussed under soil testing section.

For mat foundations, the important parameter to determine is the subgrade modulus of soil, also known as the modulus of subgrade reaction or coefficient of subgrade reaction. Its value depends on variables such as the type of soil, foundation stiffness, stress state of the soil, degree of compaction, depth, Poisson's ratio, foundation dimensions, moisture content and elastic modulus of the soil. It is easy to confuse subgrade modulus with the soil's Elastic modulus. Unlike the Elastic modulus of the soil, modulus of subgrade reaction is not a property of the soil. It is also not constant under a foundation, but changes according to what point we are looking at under the foundation, and whether the footing can be considered as flexible or rigid. Horizontal and vertical subgrade moduli are used as springs to represent force deformation relationships of soil when modeling the structure.

The pressure on the soil and deformation (settlement) are related to each other as:

$$k=q/\delta$$

where q is the bearing pressure (load/area), δ is settlement deformation (length) and k is the subgrade modulus, which means, k has units of force/length^{3}, or pressure/length, in other words, "foundation pressure per unit settlement".

Commonly at first, the relationship is linear elastic, but after a certain load is exceeded, soil will yield and there will be larger (plastic) deformations with the same rate of load increase *(in a structural model, these relationships are shown by springs)*. After the soil yields, the load-deformation graph is in plastic region, so if we remove the load, some permanent deformation will remain. There are k values for both horizontal and vertical directions (loads). *Compare what we have just described in this paragraph to the elasticity, stiffness, plasticity and yield concepts that we discussed under strength of materials section.*

Rigid vs. flexible footings require different settlement calculations. Under rigid footings, soil settles more at the edges, and under flexible footings, the soil settles more at the middle sections of the footing.

Small footings tend to be rigid and for them bearing criteria will be more important and will govern in design, rather than k. For their design, rigid methods can be used, which are more straightforward.

For a larger, flexible footing, from the formula above, since the pressure of footing on the soil is uniform and the footing settles more at the center, k will be the smallest at the center where the settlement is the most *(assuming it constant will not be appropriate, which would cause missing of the flexibility of the footing, which also affects the structure above)*. The larger the footing gets, the more flexible it can be considered and, settlement will govern where we use subgrade modulus shown above. The soil under a mat foundation can be represented with springs, using subgrade modulus, such as, by multiplying the area for each spring by subgrade modulus, which is in force / length^{3}, to obtain a force / length, which represent the spring constant for each area, and then we use these springs during computer analysis of the foundation *(or better yet, the soil can be divided into finite 3d elements and the true natural behavior can better be represented in 3D and analyzed in a finite element analysis, as the deformation also involves lateral stresses in the soil itself)*.

Modulus of subgrade reaction directly affects the results for foundation analysis such as the stresses on the soil, settlement, bending moment on the foundation, and even the stresses that may occur in the structure due to (differential) settlement. Therefore good judgment by the geotechnical engineer is essential in its determination, just as any other soil parameter. The values for dynamic (short term) and static (long term) loads can be different as upper and lower bound values.

Consolidation Settlement:

As we saw in previous section, elastic settlement applied to all soils. Now we will look at consolidation settlement, which is a time dependent settlement that is based on the speed that water exits from soil pores, and applies in clay soils only. It has primary and secondary phases. We will discuss primary phase only and will refer to it simply as consolidation in the remainder of this subtitle. For sands, consolidation doesn't apply as the water can easily escape from pores, so, all the settlement they experience is elastic. Consolidation process is very slow which may take months or years, because of clay's very low permeability. For the same reason, when we discussed triaxial test, we said that the consolidated-drained (CD) test is very slow and may not always be practical.

Consolidation occurs when new load is placed on clay, where water slowly exits pores and the soil settles. This is one dimensional, as the only way volume can change is through vertical settlement, since the soil will not be able to move laterally. Consolidation was first described by Terzaghi in the way we know today and as shown below, but actually the process was known since centuries. It is similar to pushing a piston which has a spring and water in it, which is an analogy that is commonly used to describe consolidation as seen below:

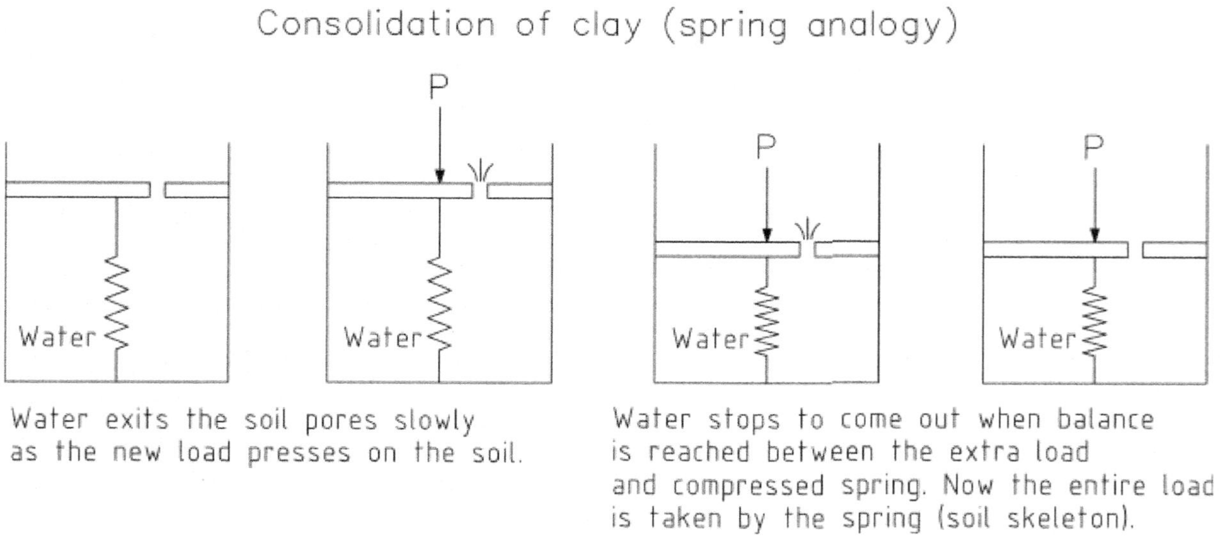

Figure - Soil Mechanics 24

In the figure above, water and spring represent pore water and soil skeleton respectively. The first piston represents existing ground. Then, as new load is applied, at first, water takes on the entire load. Then as water starts to come out, both water and spring carry the new load. As more water exit pores, spring gradually takes on more load and gets compressed more, which means settlement occurs, until enough settlement happens so that the entire new load is now taken by the spring (soil skeleton) again.

For faster consolidation rate, extra load can be placed over the construction area prior to construction (which is a very old technique), and/or gravel or other type of drains can be placed into the soil, to make the escape path of water shorter.

We can find the amount of settlement under given load for a given soil and also the time it takes for a certain amount of settlement to occur through soil consolidation equations. We show the equations below for information and will not need them later. The primary consolidation settlement is equal to:

$$S_c = \frac{\Delta e}{1+e_o} H$$

Where,

S_c: primary consolidation settlement amount (distance unit)

e_o: initial void ratio of soil

Δe: change in void ratio as a result of applied load

H: thickness of soil layer

The change in void ratio (Δe) depends on a factor called compression index, written as C_c, which is determined in the laboratory by performing consolidation test (oedometer test). After inserting C_c in place of Δe in the equation above, the equation becomes:

$$S_c = \frac{C_c H}{1+e_o} \log \frac{p'_o + \Delta p'}{p'_o}$$

Here p'_o represents the average effective stress at the center of clay layer before extra load is applied. And the $\Delta p'$ is the additional effective stress due to loading.

If the new load that we placed over the soil disappeared, the soil skeleton would push back upwards, just like the spring would do as in the figure above, which is called soil heave or swelling. This upwards push can not recover the entire settlement, because when we load the soil the first time, in addition to the process we described, some particles are rearranged in the soil, which is irrecoverable.

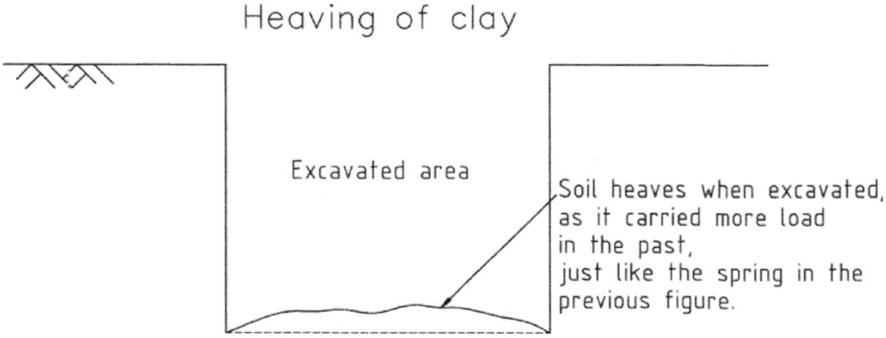

Figure - Soil Mechanics 25

The load removal in real life can happen by geological processes, such as water, wind, earthquake or glaciers eroding the soil, or man made activity such as excavation. This soil is now called "overconsolidated '. The degree to which this occurs is represented by overconsolidation ratio, OCR, which is an important parameter in soil strength calculations and effects the behavior of clays greatly, and in many cases reverses its behavior. Clays are soils with "memory" and OCR represents that.

Overconsolidation ratio is the ratio of maximum past effective stress, to current effective stress. So,

$$OCR = \frac{\sigma'_o}{\sigma'}$$

In the formula above,

σ'_0 : maximum effective stress that occurred in the past (also called preconsolidation pressure)

σ' : effective stress at present

As always, let's repeat because this cannot be overemphasized, effective stress is what matters for strength of soil, which is the reason we use them here, and not total stresses.

The higher the OCR, the better, because it means, in the past a heavy load was present, which would have caused good settlement and made the soil stronger. For example a soil with OCR of 3-4 or more is considered heavily overconsolidated, which means more compacted, stronger soil.

The void ratio (e) vs. effective stress (σ') is the fundamental relationship to know for consolidation. A typical graph of their relationship is as below. Logarithm is used conventionally, for easier reading.

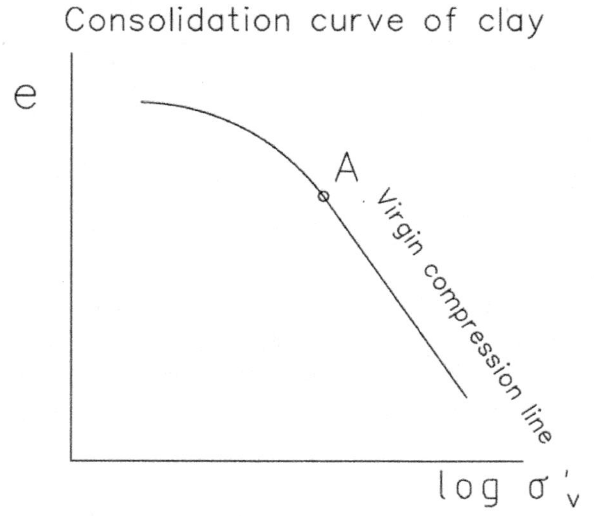

Figure - Soil Mechanics 26

The graph above, which shows the relationship between volume change and vertical stress on the soil, is the first thing we look at about consolidation. It describes that until the new stress reaches the level of past maximum stress, the settlement due to new load will be small (which will be good for our new structure for instance). This is the portion up to point A in the figure (when we go from left to right as we increase stress), which is the overconsolidated portion. Then, after reaching the maximum past load level and passing point A, then every load increase causes more settlement. This is called the virgin compression zone which starts at point A, which is now linear and most importantly, goes down at a steeper rate, which means more settlement for the same amount of load increase. It is called the virgin line because this clay has never experienced this much stress before. The point A on the graph can be determined geometrically by using tangent lines on the curve, such as using Casagrande's method.

If we release the new load, and then load it back it again, the curve seems like the one below in Figure 27. We can obtain these figures 26 and 27 in a lab, by performing oedometer test, which is also called as the consolidation test.

Figure - Soil Mechanics 27

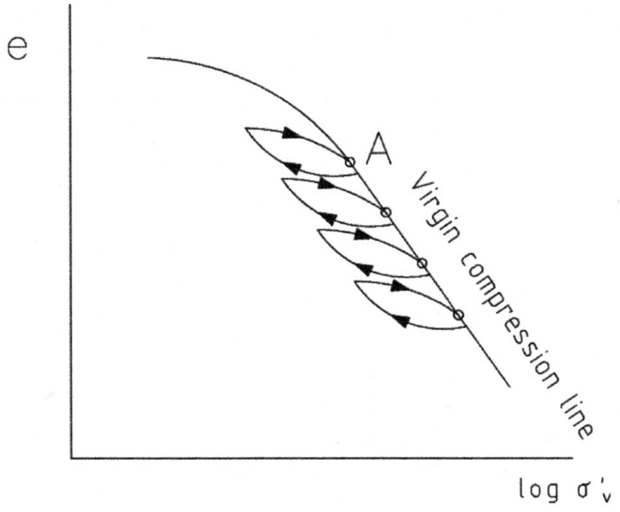

Figure 27 actually describes the same thing as Figure 26, but here we also show load and unload cycles. Again, when we load it first, up to the previous past maximum stress, the settlement is slow, then it becomes faster when we exceed the past maximum load. Then we release the load, the soil heaves, which are represented by upward arrows, but this heaving happens only partially. Then we load it again, now on a new curve, as the max past stress is now the previous cycle. Again it settles slow until reaching that previous stress, and when we exceed the past load, settlement is faster again, which falls on the virgin line. The cycles are repeated this way. Also pay attention that the virgin compression lines after each cycle fall practically on the same line.

Important note: Degree of saturation of clay also affects its settlement greatly. Above we assumed clay is saturated. Although we do not always find clay as saturated in the field, there is no guarantee that it will remain so (as it may rain heavily), and calculations are performed for saturated clay as they settle more. But it is important to know that unsaturated clays settle considerably less than saturated clays.

Slope Stability:

We build a lot of improvements near natural slopes, which may slide or fall, given enough time. The factors that affect the possibility, type and speed of a slope failure, in other words the stability of a slope include:

- Soil or rock type (sensitive clay type soils carry high risk)
- Slope angle
- Water content
- Loading condition such as intensity and duration
- Presence of streams or rivers eroding the toe of slope
- The type and depth of the firm layer below the slope in consideration
- Soil permeability
- Type and geometry of weaknesses in soil or rock. For rocks, the angle of rock layer vs. the slope angle, especially for weathered rock with joints, fractures, fissures
- Presence of plants and trees

Slope failures can occur in a variety of ways and speeds such as:

- Sudden fall or toppling of rock and or soil
- Lateral soil spread or creep, which occur very slowly
- Soil flowing continuously or in the form of an avalanche
- Sliding by rotation around a point, or straight motion along a very long slope
- Mixed pattern of more than one of the above

Slope failures can cause damage to property or even loss of lives. Sudden failures do not always mean catastrophe, or slow failures do not always mean no harm at all. For example, a slope may fail slowly, but if it is near an underground pipeline it can cause significant economical damage. We need to predict or take precautions in high risk areas. The prediction is not easy but some hints of failure are easier to detect, even without analysis, such as displacements that we can see, trees bowing, soil strata warping or tilting... We will discuss slope stabilization measures under a separate section later in this chapter. In this section, which is under soil mechanics, we will look at the mechanism of slope stability.

After heavy rains, the risk of slope failure is higher because the water content of the soil increases, which makes the soil heavier and not only that but it also increases the pore water pressure, which means decrease in solid particles' contact in soil skeleton (decrease in effective stress) which leads to decrease in shear strength. The similar condition can be simulated unwantedly, if for instance we have an earth embankment dam and the reservoir water is drained rapidly, which is called rapid drawdown, without the water having the chance to drain or evaporate, which leaves the slope saturated.

Loading the ground or excavating a slope, and the fact that the soil will be in drained or undrained conditions, in other words short or long term, also require different parameters to be used.

- For loading such as building an embankment, the critical phase is the short term undrained condition, where the pore pressure increases, thus effective stress is reduced, until drainage is complete. In the long term, when drainage has sufficiently progressed, soil settlement occurs and the soil gets stiffer. Unlike excavation, where we must deal with natural soil, embankment work has the advantage that we can control the materials.
- For unloading case such as excavation, the critical phase is the long term, drained condition. When we excavate, cohesive soil tends to heave as it carried more load before, which causes pore pressure to decrease (as if suction in an expanding sponge), which makes effective stress (thus shear strength) even more, which is good for us, but after the water drains, that suction effect will disappear, and the effective stress will be less, which means lower shear strength and higher chance of slope failure. Do not confuse this item with rapid drawdown we mentioned above, where there was no excavation but just rapid draining of dam reservoir water.

If there has been shear deformation (failure) on a slope already, then we might want to use the residual (critical state) parameters (See Figure 9).

By performing slope stability analysis, we understand the nature of the slope, potential of failure in the short and long term, evaluate it under seismic loading conditions. For a slope to be considered as stable and safe, the safety factor we obtain when we divide the resisting (stabilizing) forces by driving (destabilizing) forces must be higher than 1 (i.e. 1.5), which is given in codes. As always, the resisting forces are a result of shear strength of the soil.

So,

FS (factor of safety) = resisting shear strength / driving shear force.

These destabilizing forces will try to create shear failure along a failure surface. The resisting force is the shear strength which resists failure along that failure surface. We ignore any possible contribution to stabilizing force from the roots of plants and trees even if they are present down to the depth of failure plane, since they can not be relied on or predicted well and may just slide with the slope too.

The driving (destabilizing) forces include:

- Weight of the soil,
- Weight of water, if any
- Structures built on the slope, if any
- Water seepage forces, if any

The value of FS that we must use varies not only due to slope conditions but also the magnitude of harm it may cause (risk) in the event of failure. Different values are listed in different sources and codes although they are close. Generally though, a factor of safety of around 1.5 is applied for slopes where the failure pose the greatest risk to human life and economic property.

Slopes can be classified as finite and infinite slopes, for analysis.

Infinite Slopes:

Infinite slopes are very long in comparison to their depth, and we can assume them as infinitely long. This type of analysis is less frequently required, in comparison to the finite slopes we will discuss next. Below you see a free body diagram of a sliding block on an infinite slope, with its driving and stabilizing forces, which forms the basis of analysis for infinite slopes.

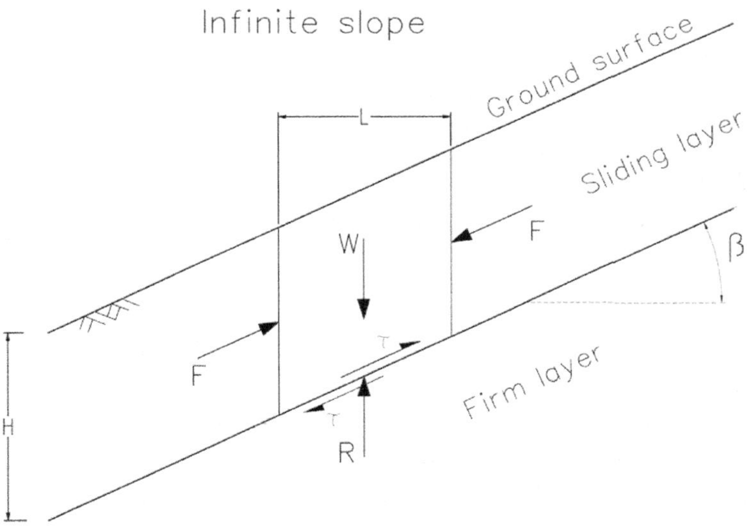

Figure - Soil Mechanics 28

Factors that affect the stability of a slope as seen above are:

- Slope angle from horizontal (β): This is the greatest factor that influences the driving force.
- Internal friction angle (Ø'): As discussed, angle of internal friction of soil is a main factor in determining shear strength, which is the resisting force. For coarse grained soils, the drained condition will apply, and effective values are used, unless there is liquefaction. For undrained clays, Ø can be taken as zero and the shear strength will be equal to undrained shear strength as was described under triaxial UU test.
- Cohesion of soil (c'): Sands can have little cohesion too, but if a mass of sand has zero cohesion, which means it is perfectly frictional without any contribution of cohesion, the failure surface will be an infinite slope. Cohesion is also a major factor that affects how cohesive soils fail.

- Thickness of sliding layer (H): The size of the block is determined by its height, which is the vertical distance between the ground surface and the firm layer. By firm layer we do not necessarily mean rock, but it can be dense soil as well, if it has considerably more stiffness than the weak, sliding layer above.
- Soil unit weight (γ): This affects weight of the soil block, shown as "W" in the figure above which affects vertical stress along the failure plane. We take the saturated unit weight to account for water as well. That is why slopes have higher risk of failure after heavy rains, in addition to decreased friction of solid particles due to pore pressure. Of course soil unit weight affects the effective stress, which is the source of shear strength, but for that we use effective unit weight, versus the saturated unit weight (which is more) mentioned in the previous sentence which can occur after heavy rains, even if usually the slope may be dry.
- Pore water pressure (u).
- Water seepage forces if present and not negligible.

And why didn't we include the width of the block, L, as shown above? Because it is redundant. Any change in L will of course affect the weight of the block, but it will at the same time increase the sliding area, the area on which the shear strength is developed. So, the larger L, the more W, but also the higher τ_r, which is the resisting shear force. The effect of L cancels out.

Also important to know that, when we divide driving forces by the resisting forces, we reach to an important conclusion for sands. Assuming zero cohesion, we see that an infinite slope made of sand is stable, IF, the angle of slope β is smaller than the angle of repose of sand (*which was introduced a few sections back*), which is equal to angle of internal friction Ø', in this case. This is a conclusion that makes sense, because remember that when we introduced angle of repose, we said that it is the angle for sand to keep its natural posture. This conclusion is independent of height, length or unit weight of soil.

Finite Slopes:

A finite slope, contrary to the infinite slope, has considerable height of weak layer, with respect to the length of the slope. In the figure below three typical cases are shown simultaneously to be able to compare them easier, but of course it will be only one of these arcs.

The analysis is made by assuming that the arc rotates around its center. The distance to the firm layer below, cohesion, angle of internal friction, height of the slope and the slope angle influence stability. Basically the weight of the arc tends to rotate the arc, while the shear resistance along the curved failure surface counters it, therefore we find a safety factor with similar logic as above, but by dividing the resisting moment to driving moment this time.

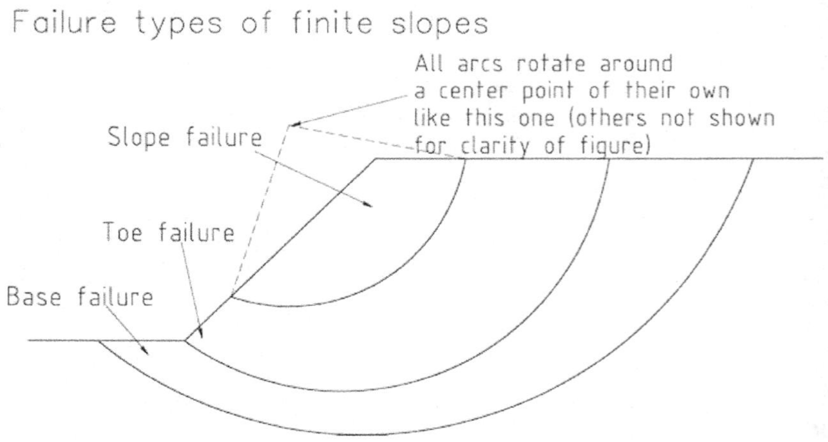

Figure - Soil Mechanics 29

In the figure above, you see no firm soil layer beneath the rotating arches, at a close distance. The figure below shows the firm layer at shallow depth, close to the slope. In this case, the arc can only extend down to the firm layer, and not more.

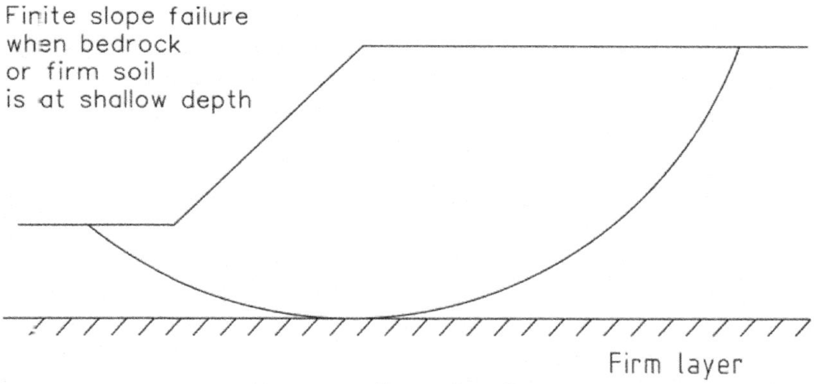

Figure - Soil Mechanics 30

Actually this smooth rotational surface is idealized for homogeneous soil which doesn't exist in nature, which should also be noted, but close enough failures can occur. When there are existing weaknesses such as fractures, faults etc... in soil mass, this can also greatly affect the soil failure surface.

We can use Taylor slope stability charts and coefficients based on soil geometry and strength, for analyzing total stresses of undrained case of cohesive soils or drained situation for cohesionless sand. These provide factors of safety, which can give us good approximations, and can then be applied to more complex real problems.

The figure below shows another common method to analyse rotating slopes, which is called the "method of slices". Here, we divide the arc into imaginary vertical segments, and for each piece the stability for forces and moments are analyzed with similar logic as above and then we take the sum of all stabilizing versus driving forces. Method of slices can produce more accurate results than simply writing resisting and driving moment equations for the whole slope as above. As long as we reasonably divide the arcs, their width don't need to be precisely equal to each other. So for each piece, we draw a FBD and show the weight, the resisting shear force and reaction force at the bottom and the forces from the adjacent pieces. This problem cannot be solved with just equilibrium equations and assumptions are needed. Various approaches to solving method of slices exist.

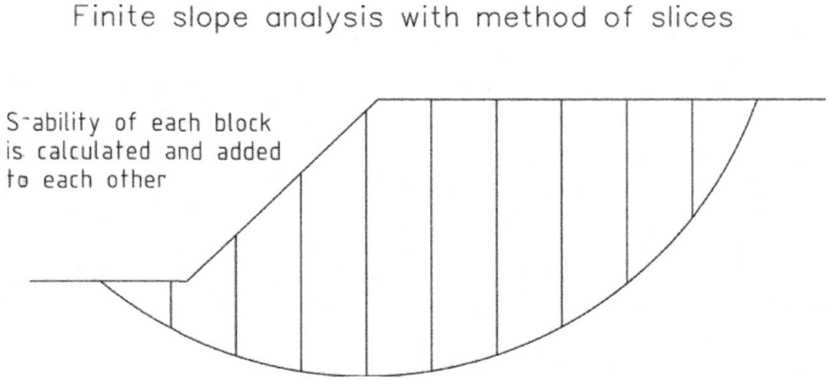

Figure - Soil Mechanics 31

Methods described above are the basic, common ones, which can be grouped into equivalent static limit equilibrium type of analysis, which are not suitable for complex (layered, irregular, sensitive) ground conditions. In this type of analysis, deformations are not calculated, only a factor of safety is found as seen above.

Other than the static analysis as described above, seismic analysis of earthquake triggered slope failures also needs to be performed in earthquake zones. Here liquefaction may also occur, where we have saturated sand slopes. Earthquake analysis of slopes is considerably more complex. Apart from analytical solutions which can serve as simple checks, numerical methods such as finite element methods (see Computer Aided Analysis section), or dynamic analysis are used.

Critical State of Soil:

Earlier we saw the τ - σ' graph under strength, and e - log σ' graph under consolidation sections. Critical state theory goes up one more level and combines these two relations. Those two graphs which are in two dimensions, are combined in a three dimensional graph, as seen below (try to visualize the three axes in your mind):

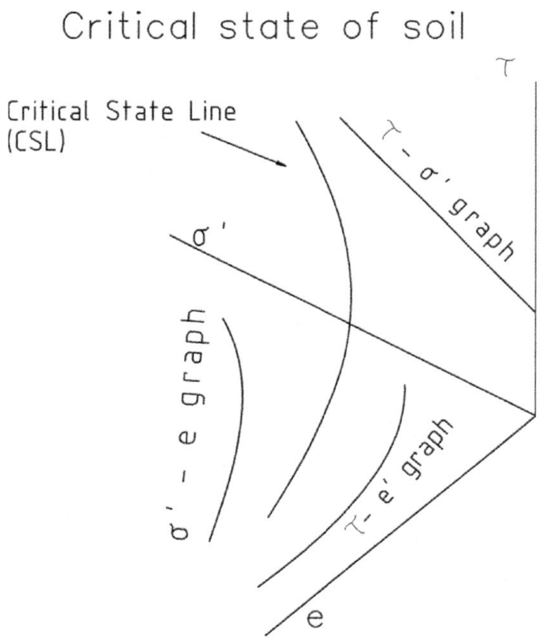

Figure - Soil Mechanics 32

This three dimensional line is called the critical state line. Its footprint on each of the three planes are also shown. The three axes of this graph are also called as p,q,v axes conventionally, which can be directly obtained from τ - σ' and e - log σ' relations. As was described in Figure 9, the void ratio, e, ultimately approaches to critical void ratio, which is a unique property of that soil, independent of the initial void ratio. Critical void ratio was first defined by Casagrande in 1938, and the concept was first published by Roscoe in 1958, after which it was increasingly studied. Critical state is relatively newer concept than the others.

The critical state (such as the CSL line is a unique property of that material. For each soil, there is a unique critical state graph. It is independent from the initial soil condition. All soils will ultimately tend to reach its critical state void ratio, which is their unique property. Depending on the consolidation state (i.e. normal vs. overconsolidated) of the soil, this graph may shift. Before reaching to critical state, a soil's density and drainage condition determines on which side of these lines the soil lies, and the path it follows to reach to critical state line. These subjects are studied in advanced soil mechanics courses.

And with this, we finished strength of soil subsection under soil mechanics, which is the longest subtopic in the book. Now let's move on to other subtopics of soil mechanics.

Subsurface Investigation

To build or even propose any structure on a site, first its ground conditions must be understood. For this, we use various methods, but before even performing a minute of ground investigation on site, a good amount of information can be obtained from various sources such as: Previous data about the site from previous projects, government and local agencies which can provide site development history, existing geological, topographical, agricultural maps, aerial photos, and even by talking to local people. Geotechnical engineer's visual site inspection is also important, to identify potential issues. After these, the engineer will have preliminary understanding about the site, even before testing.

The best information however, although the most expensive, is obtained by looking at the actual foundation soil on site, which is only possible by drilling boreholes, from which we obtain samples and test them in the field or lab to understand its properties, many of which were introduced earlier. Depth of boreholes and distance between them are determined based on the size and type of the proposed structure, our knowledge of ground conditions as obtained in the previous paragraph and also per the criteria required in codes or standards, such as for example FHWA for highways or bridges in USA.

Methods of drilling vary based on our purpose and ground conditions. Note: Anywhere in this book, ground conditions mean not only the soil/rock, but also underground water and its depth (water table). Methods of drilling borehole include bucket augers, hollow stem augers, rotating wash and hand augers. Other than the hand auger, all require machine power. Undisturbed recovery of samples, or at least with minimum possible disturbance is of highest importance. Samples can be obtained by continuous augers, split barrel, shelby tubes, pitcher tubes, each of which are better suited for different conditions. For rocks, samples are obtained by coring. Rock sample quality depends on the length of core that can can remain intact without breaking into segments, (which gives us RQD, Rock Quality Designation). After soil and or rock samples are obtained, they are tested to determine their properties. Rocks are most often found in larger depths than soil. So usually, as we go deeper underground, the strength of earth generally increases, although this is not a uniform increase and may even fluctuate.

The information obtained from boreholes are shown on boring logs. A generic example of a boring log prepared as an example is shown below.

Example Boring Log

Project Name:	Equipment:	Groundwater Depth (m):	Drilling Date:
Project Location:	Surface Elevation (m):	Sampling Method:	Borehole No:
Contractor:	Total borehole depth:	Hole Diameter (cm):	Caving Depth (m):
Engineer:	Drilling Method:	Borehole backfill:	
Reviewed by:	Groundwater measuring date:		

Depth (m)	Sample	USCS Soil Symbol	Graphic Log	% Recovery	Material Description	SPT Blow Count	Dry Density	Liquid Limit (%)	Plastic Limit (%)	Water Content	Comments
0											
1											
2											
3											
...											
...											

Figure - Soil Mechanics 33

These logs are the most essential documents when preparing geotechnical reports. In addition to ground conditions, the logs will also provide information about issues encountered during borehole drilling and obtaining samples, which is also valuable information to an experienced geotechnical engineer. Boring logs must be completed on site right when and where they were drilled - not in the lab later. The format and the items included will vary between projects. Some of the items that you see in the boring log above need no further clarification as they are obvious. Let's clarify the remaining ones:

- Caving Depth: During drilling, caving of hole may occur which needs to be noted. This is especially a problem for sandy soils. Caving depth also gives an idea of the soil itself.
- USCS Soil Symbol: We mentioned soil classification systems such as USCS before under soil classification section. The symbol of the soil in that system is indicated here.
- Recovery %: During sampling, not all of the material can be recovered all the time. This percentage of recovery can be shown here as an estimate.
- Graphic Log: The commonly used graphical symbols for the type of soils encountered on site are shown here, for immediate recognizing.
- SPT Blow Count: Standard Penetration Test is a common field test by advancing a rod into the soil with hammer blows. The more blows required to advance the rod, the stronger the soil. SPT results are correlated with important soil properties based on a great deal of past data, so the SPT blow count results are valuable to see.
- Material Description: In here, engineer writes his/her own description of the material.
- Dry density, liquid limit, plastic limit are measured in lab. We have not looked at liquid limit and plastic limit yet, which will be introduced in the next section, soil testing.

Visual observation during drilling of bore holes will also provide useful information to experienced personnel and engineers, such as the soil's appearance, the pushing effort and torque spent to advance the borehole, the depth of caving of soil, the required quantity and returning back of drilling mud (the liquid which is pumped into the hole to prevent caving) as the hole progresses. In some critical and rare situations, even cameras or electronic sensors can be lowered into a borehole.

Soil Testing and Measurement

To measure and know soil properties and parameters, we need to perform testing, which is a major part of soil mechanics. Tests can be grouped as the ones that we can perform in the field, and the ones we need to do in a testing lab. If we can perform a test in the field, we do so. It is just better to test the soil in its natural place, in undisturbed state. If we must do the testing in lab, the samples must be extracted carefully with minimum disturbance as possible *(which is called undisturbed sample, but in reality there is always some disturbance, which can be affected from so many factors including the skill of people who obtain the samples, the method applied, the condition and size of the equipment, the way the samples are stored and for how long etc...)* and then they must be carefully and quickly transported to the lab. If we cannot transport the samples quickly to the lab, we must employ standard procedures to make sure that the samples will remain in their natural state to the greatest possible extent, such as the water content, temperature, volume. The tests listed below are well defined in detail in standards, such as in the ASTM standards in the United States, DIN standards in Germany and other standards in other countries. The tests we perform in the lab are the ones that need more complex equipment or more precise measuring.

The skill of technicians and the state of equipment also play a large part in making field or lab tests. For maximum possible level of accuracy, the testing methods must not only be accurate and strictly follow standard procedures, but the ones in the lab must also be chosen and aligned to represent the actual soil conditions as close as possible especially because natural soil is a variable material, sensitive to many factors. Determination and use of properly judged soil parameters is often considered as a challenge by geotechnical engineers that requires engineering judgment. In determining parameters more than one method should be used for better accuracy.

Performing proper subsurface investigation and testing of course comes at a cost, but this is money well spent. Claims that arise later due to ground or foundation problems, which can be a result of inadequate and / or inaccurate subsurface exploration or soil testing and measurement, constitute a considerable part of construction claims, and the costs can be astronomical.

Before performing any test, the first thing to do is a visual inspection of the site. The material can be seen by drilling holes and obtaining samples, digging test pits or trenches, which provide a direct viewing of the soil right at its location. If digging pits or trenches is practical, (i.e. the open trench does not cause delay to other activities) it can be the best. From test pits or trenches, large undisturbed samples can be obtained easily, directly. Direct visual inspection makes it possible to directly see the soil, which will provide valuable insight. An experienced engineer can also simply take a soil sample in hand and visually inspect it for its various characteristics such as texture, plasticity and consistency. Also, if there is odor from soil, it is a sign of organic content, which means very weak soil. The more organic part, the more smell will be. If there were prior improvements on the site, test pits allow to see the existing foundation elements, so they are especially advisable for such sites, if there is space.

The tests listed below are not given just to have a list of soil tests. Reading them should also help wrap up / better understand the subjects in mind which we have so far studied under soil mechanics.

Field (in-situ) Soil Tests:

Standard Penetration Test (SPT):

This is a very common field test, and also one of the oldest type. SPT blow count, N value, is one of the first things a geotechnical engineer looks at, to have an idea about the soil. In this test, a standard hammer is dropped on standard equipment to drive a needle into the ground, and the number of blows (shown by N) to penetrate the needle by a certain distance, such as 6 inches is counted and recorded for every depth interval. As the soil gets denser and stronger, the N value grows. Therefore, low N values indicate weak soils. The N value for the entire relevant soil depth is valuable to know, because from past research they are extensively correlated with many soil properties such as bearing strength, relative density, shear strength, strata information, and more. SPT can be performed on both fine and coarse grained soils, but it is especially beneficial to use for sands because we cannot obtain undisturbed sand samples to take them to lab for testing.

Cone Penetration Test (CPT):

CPT can also be performed on all types of soil, except gravels, but in comparison to SPT, it performs better for clays. Information such as existing stresses and pore pressure, undrained *(since water has no time to escape)* shear strength of clay, and relatively more continuous data of soil stratigraphy with respect to SPT can be seen. CPT results are also correlated with many soil properties.

Vane Shear Test:

This is a very easy and popular method to measure undrained shear strength of soft to medium clay. In this test, a vane is penetrated into the soil and the blades are rotated as instructed in standards. The torque to cause shear in the soil is found and the results are correlated to shear strength. This test is not suitable for sands or gravels.

Plate Load Test (Plate Bearing Test):

This test involves directly applying load to a smaller size footing built specifically for testing and measuring settlement at the actual site. This way, we can have a good idea of load vs. settlement, subgrade modulus, bearing capacity at the actual site, since we know the size ratio of the test footing to an actual footing. It is a good way to check our theoretical foundation analysis results. The test footing must be placed at around the same depth as actual footings, as the strength of soil varies with depth.

Pressuremeter Test (PMT):

With this, hydraulic conductivity (which we will cover in groundwater section), elastic shear modulus, existing (in-situ) soil stresses, can be measured. This is more suitable for soft rock to dense sands, but not for clays, loose sands, silts. The results can be used in settlement and bearing capacity calculations.

Sand Cone Test:

Through this test, we can measure the in-situ density of the soil.

Nuclear Moisture and Density Test:

It is a very simple, practical test which is frequently performed on sites with soil compaction work, to determine moisture content and dry density, which are important parameters for phase relationship calculations. The reason of its common usage is because it doesn't require to take soil sample and disturb the soil, the possibility of technician errors are minimized and it can be done in several minutes which is far faster than the lab tests (if you also consider transportation of samples to the lab).

CBR (California Bearing Ratio) Test:

This test measures the shear strength of subgrade soil for roads and pavings.

Flat Plate Dilatometer Test:

Through this test we can determine overconsolidation ratio (OCR), type of soil, elastic modulus, undrained shear strength. Not suitable for gravels but can be performed for sands or clays.

Plasticity Field Tests:

The tests below are for roughly determining the plasticity of the soil in the field. These are best used for fine soil particles that pass the #40 sieve.

- Dry Strength (Breaking) test: Used to measure the soil's cohesive and plastic characteristics.

- Shine Test: Determines plasticity.
- Roll test: Determines plasticity.
- Ribbon test: Compliments roll test, to measure plasticity.
- Wet shaking test: Determines plasticity, and used to get an idea of silt and sand.

The primary test for plasticity is the consistency test, which is listed below under laboratory tests.

Piezometer Test:

It is performed by inserting tubes at various locations in the soil, to measure pore water pressure.

Groundwater Tests (observation wells):

By utilizing wells, water table depth can be determined. We can also measure permeability and seepage rates by pumping the water and determining flow rate by measuring the heights and radiuses from the well as shown below.

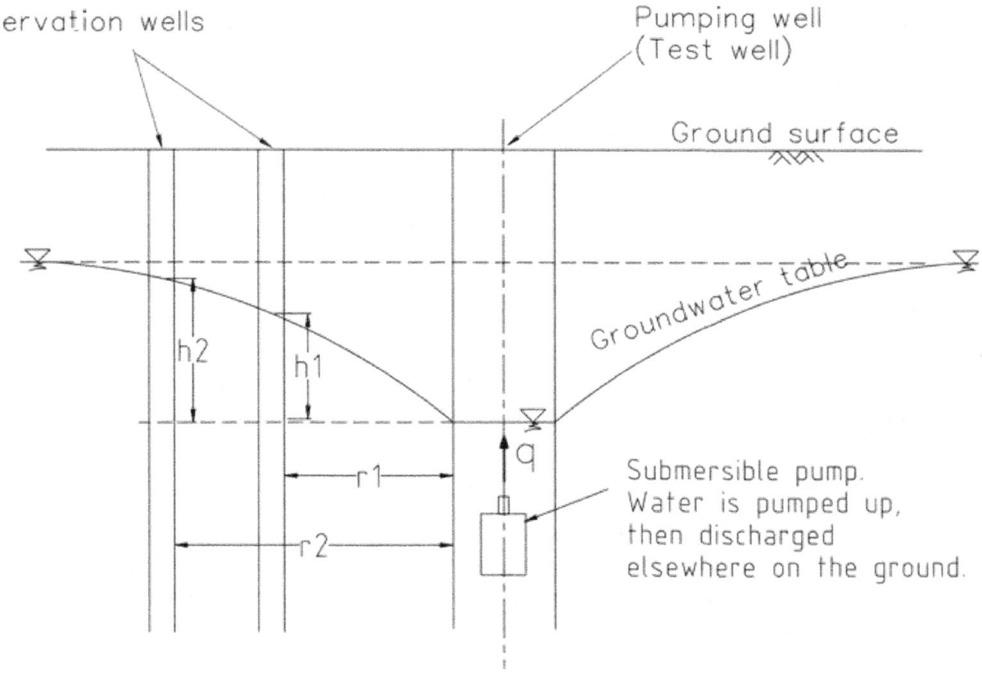

Figure - Soil Mechanics 34

Laboratory Soil Tests:

Now let's list the tests that are performed in the lab.

Sieve Analysis:

This test is used to find the particle size distribution of coarse grained soils, by passing them through sieves of successively decreasing sizes. The smallest size sieve is the #200 sieve with an opening of 0.075 mm as discussed under soil classification. If more than 50% of the soil by weight remains above the #200 sieve, it is considered as coarse grained, if not, fine grained soil. The sizes of smaller particles which also pass through #200 sieve can then be determined with hydrometer test as described below. Through sieve analysis, grain (particle) size distribution is plotted on a graph. Knowing this, enables us to tell the name of the soil in the first place, such as gravel, sand, silt or clay, so in a way this test can be regarded as the most fundamental one. Knowing the soil type has more use than just knowing its name. We can also use this information to determine design parameters correlated with that soil type.

Hydrometer Test (Sedimentation Test):

This is for determining particle size distribution of fine grained soils, because the sieve analysis listed above is not suitable for fine particles. The test involves separating different size particles by making use of their different settlement rates in water, as larger particles will settle faster, per Stoke's law. So at every certain interval, different size particles can be removed and particle size distribution is found.

Permeability Tests:

Permeability is an important soil parameter that shows how easily water can flow through soil. For coarse and fine grained soils, we have two different types of permeability tests. Permeability tests are based on Darcy's Law, which describes the flow of fluids through porous medium.

Falling head test is for low permeability, fine grained soils. Here water cannot flow easily through soil because of low permeability. So, we pour a small amount of water on the soil and wait for it to drain. The time that passes for draining of water to permeability of the soil is correlated through equations.

For coarse grained soils, constant head test is performed. In this test, water is continuously poured over the soil in a container which lets water out from below. The water flows easily since the soil is coarse. The amount of water that we must keep pouring is measured, to maintain water level the same. Then this water flow rate is related to the permeability of the soil.

Unconfined Compressive Test:

This test measures compressive strength of a self standing, cylindrical clay specimen. By definition it can not be applied on sands, as sands cannot stand on their own. As seen below, the force at failure and the load vs. deformation rate is measured. Water will not have time to drain, therefore the undrained shear strength (not the effective) is measured with this test. So in other words, in the shear stress equation τ=c + sinØ, everything will be in terms of total stress.

Figure - Soil Mechanics 35

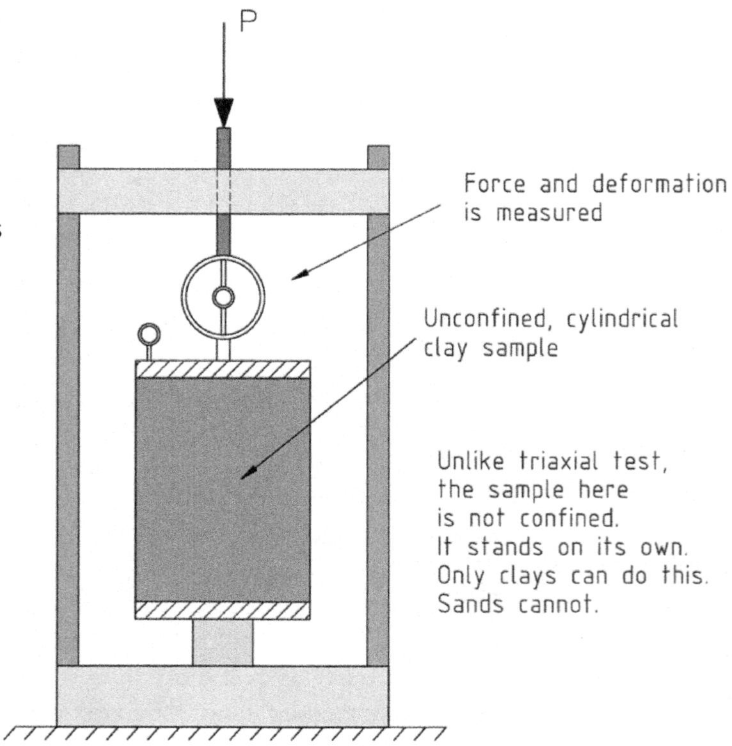

Consolidation Test (Oedometer Test):

We described this test before under consolidation subject in strength of soil section. Another name for this test is the confined compression test, because the clay sample is confined within a metal cylinder. At the top and bottom of cylinder are porous stones, for water to exit. So, the test is done at a slow enough rate (days or weeks) to allow consolidation and water drainage in clay. The void ratio vs. applied load, and load-unload cycles are plotted as discussed under consolidation subject.

Moisture Content Test:

We can easily determine the moisture content of a sample of soil by weighing it first, then drying it in oven and then weighing again, measuring the difference in weight. This is called oven drying method.

Consistency Test:

Plasticity is an indicator of how well the soil can deform without breaking apart or cracking. Plasticity applies to clay soils only, which is expressed in four states according to moisture content (%). From lower to higher moisture content, these are: Solid, Semi Solid, Plastic and Liquid. As the water is taken in or let out, the soil's plasticity transforms between these states, accompanied with volume change, as seen below. Consistency test is used to determine these limits.

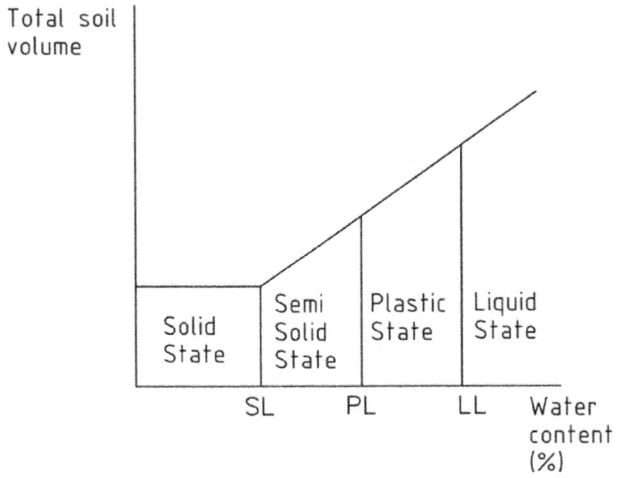

Figure - Soil Mechanics 36

As seen above,

SL: Shrinkage limit, between solid to semi solid state

PL: Plastic limit, between semi solid to plastic state

LL: Liquid limit, between plastic to liquid state

These are also called Atterberg's limits, after the engineer who defined it in 1911. Later, the test procedure and defining the limits were standardized by Casagrande in 1932.

Plasticity Index, PI, is one of the first things to know about fine grained soils. It is found as:

Plasticity Index = Liquid Limit - Plastic Limit

PI = LL - PL

Larger PI indicates more potential for expansion and shrinkage of the soil, which is undesirable below foundations, slab on grade or paving. Clays have high plasticity, therefore have high potential to expand and contract which is a why in many cases we need to replace them with sand before building slabs or foundations on them. The swelling or shrinking of clay can also cause differential settlement of the structure. Centers of slabs may rise or edges of slab on footings may rise or sink as a result. Roughly speaking and just to give a number here, a PI of 30 or more may mean highly expansive soil.

Compaction Test (Proctor Test):

Maximum dry density is an important soil parameter that is always specified in projects where soil compaction is involved. Through compaction test we can find this value and also the water content at which it was obtained. Compaction is performed by dropping a hammer of a given weight from a specific height at a specified number of times, onto the soil that is in a cylinder with known volume. For the same soil, there are different curves for different compactive efforts as seen below. For each curve, the point we obtain the maximum dry density is called the Optimum Moisture Content (OMC).

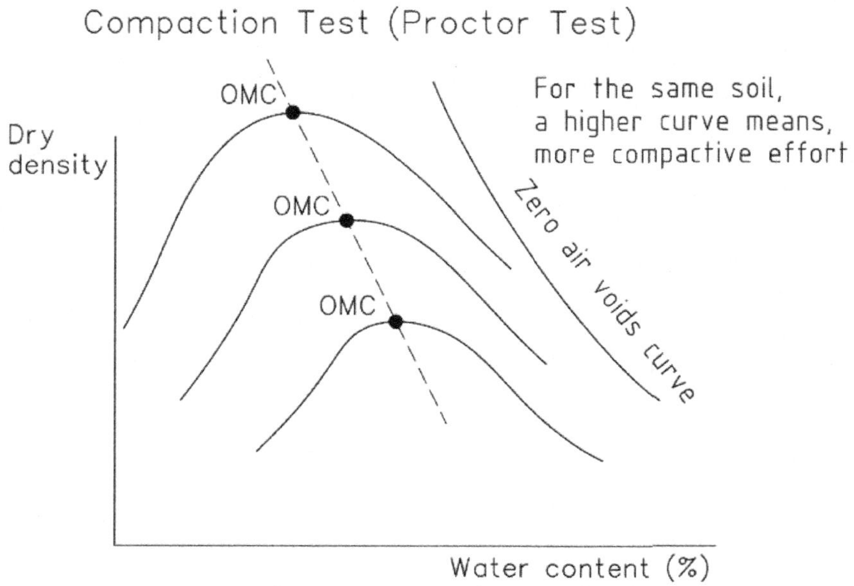

Figure - Soil Mechanics 37

As compactive effort increases, we can obtain higher maximum dry densities and corresponding lower Optimum Moisture Contents (OMC) as you can see from different curves as they shift towards left. And for the same compactive effort, different types of soils would produce higher or lower curves, where the curve shape can also vary a little.

As you can see from the curves, increasing water content enables better compaction at first. Better compaction increases dry density, as in the left part of each curve, until an optimum point (OMC) is reached, which is the highest point on each curve. After that, adding more water will only cause unnecessary volume build up (swelling) and will have the effect of lowering max. dry density, which is represented by the right part of each curve, where the curve descends. For the same soil, optimum points can be connected to draw an OMC curve as shown by the dashed line that connects the top points of curves. The zero air voids curve represents the ideal version of this line, which is only theoretical, since it cannot be achieved in reality because there will always be some minor pores remaining, no matter how much compactive effort we spend. This means, the maximum dry density we obtain here is only theoretical. When we need to perform compaction in the field on actual projects, 90 to 95 percent of this maximum possible dry density value is required by the specifications. For example, a backfill that will support load on it directly will usually be required to have 95% compaction, while the backfill which will just be placed around the structure or in utility trenches can remain at 90%.

Specific Gravity Test:

The value of specific gravity of solid particles of soil can be determined with this test.

Direct Shear Test:

This test was discussed under Strength of Soil>Angle of internal friction section.

Triaxial Test:

This test was discussed under Strength of Soil>Angle of internal friction section.

Geotechnical Reports

A geotechnical report (also commonly called as "soil / soils report") gives detailed information about the ground conditions of a site, including recommendations from geotechnical engineer. Geotechnical reports are prepared using various sources that include:

- Visual inspection of the field, to see existing structures, slopes, fills, geology, streams, utilities etc..., also called reconnaissance. From this point, how to proceed with next steps are decided.
- Data obtained from subsurface investigation such as by boring holes
- Data from nearby previous projects
- Results of soil tests
- USGS topographic and geologic maps, air photos
- Well reports that were logged before, which can be obtained from state governments

Geotechnical reports are almost always required, but for the most straightforward and simplest projects with no apparent problems or challenges from geotechnical point of view, the requirement may be waived. For example in International Building Code (IBC) 2018, which is adopted in all states in USA, per section 1803.2, the building official is allowed to waive the requirement for a geotechnical report as long as the items such as expansive or questionable soils, pile foundations, shallow groundwater table and a few more as listed in section 1803.5 are not present. In addition, the state, local codes or the project specifications may have additional criteria (i.e. the size of the project) or they may just simply dictate to have a geotechnical report in any case. Generally, regardless of how easy a project may seem, even if the criteria per codes is met, it is still prudent to have a geotechnical report.

Geotechnical reports include items below as applicable:

- Description of site, observations by the geotechnical engineer, and his/her recommendations.
- Photographs of site, that show typical features, important locations.
- Data from subsurface exploration, boring logs and subsurface profile, soil classification, a geologic map and cross section of the site with adequate scale to cover details, groundwater condition with dates of measurement.
- Discussion of the field and laboratory test results, soil materials on site, soil classification, soil parameters such as, allowable bearing pressures for static or dynamic loads *(often the first thing a structural engineer looks at in the report)*, horizontal and vertical subgrade moduli for static or dynamic loads which vary under a large mat foundation and will be different for each soil layer,

estimated settlements, internal friction angle (i.e. undrained, effective, residual), cohesion, drained, undrained and residual shear strength, shear modulus, elasticity modulus, frost depth, permeability, maximum slope angle for excavations, lateral earth pressure coefficients including dynamic lateral pressure during earthquake, sliding friction coefficient, minimum embedment distance and minimum diameter for piles, pile capacities and force deflection relationships for lateral and vertical loads, clay consolidation coefficient, plasticity, expansion and shrinkage potential, chloride content, applicable safety factors, and any other relevant soil mechanics parameters. It should be clear whether a value is an isolated, average or a conservative value or if it includes a factor of safety. There can also be recommendations in codes or standards for certain values based on the soil types on site, which can serve as limits.

- Recommendations and discussion about the type and design of foundations, retaining structures, tie back anchors, settlement and differential settlement potential from both elastic and consolidation settlement, heave potential, surface and subsurface drainage, ground improvement, corrosive soils, earthwork, slab on grade, slope stability, embankments and backfilling, erosion protection measures, landslides, existing structure foundations and their influence zone and surcharge, underpinning and shoring recommendations, backfill borrow locations, soil structure interaction potential, seismicity of the site and site class, fault lines, earthquake acceleration information, liquefaction and its settlement potential, surface rupture, landslides, lateral spreading, subsidence *(sinking of ground due to loss of material below such as groundwater)*, flood potential due to earthquakes, seismic analysis methods, design earthquake response spectrum (see earthquake section), possible geotechnical issues or problems on site, whether they are complex, unusual or not, noteworthy items that were encountered on site or in laboratory.

Structural loads (gravity, earthquake, snow, wind...), layout, size and elevation of footings, any basement, grading and all other relevant items about the structure must be adequately understood by the geotechnical engineer while preparing the report. Similarly, the items listed above must clearly be understood by the structural engineer. These can only be achieved through constant communication between the engineers in all stages from preparation of soil report until structural engineer modeling the structure, to avoid problems or inefficiencies later such as the parameters in the geotechnical report being overconservative, which may mean unnecessary foundation and or ground improvement costs, or the structural engineer's approach to modeling not being the most effective and so on... Generating geotechnical reports needs extensive experience and engineering judgment of a geotechnical engineer. If the structural engineer determines that there are missing information, he/she should refer back to geotechnical engineer, and not try to obtain data himself/herself.

Geotechnical reports are used mostly during design phase but during construction they can still be consulted. In case of a claim which may even be years after construction, this report will be critical.

Groundwater, Dewatering, Drainage, Seepage, Flownets

Groundwater is essential for our survival, which we use for various purposes, and it is the source of considerable part of our drinking water. There is more than thousand times more groundwater than the water on all lakes or rivers of the world, according to USGS website.

In civil engineering, the presence of groundwater close to the foundation of a structure often presents a challenge that needs to be dealt with such as when designing foundations, tunnels, retaining walls, slopes and so on, while still watching out for any negative environmental impacts.

Dewatering:

Dewatering, which is sometimes confused by the term drainage, is removing surface or groundwater during construction by most commonly pumping action, so that the construction activities such as excavation, utility works, foundations can continue. It can be done by opening a well or wells, called well points in the ground and lowering the water table through submersible pumps, or remove it from surface by sucking out with pumps, to an acceptable discharge location. Soil type, depth of existing water table vs. depth of required water table, size of area, seepage of water into and out of the area, environmental concerns, administrative / legal issues and as always the cost and construction schedule are the main factors that effect the amount of dewatering we need to do for how long and in what rate, which affect the number, depth and size of wells, number and capacity of pumps.

Note that dewatering can cause ground settlement, due to increase of effective stress in the soil when water is removed, which may be an issue if there are existing improvements nearby and especially if the excavation is deep or at least not shallow. Soil type also affects this. Ground improvement, which we will discuss under foundations section, may even be required in such cases.

Dewatering can be performed by the general contractor, or the contractor whose work is negatively affected from water, or ideally it can be subcontracted to a firm who specializes in dewatering. This is an item that may not be visible in the drawings, since it falls under contractor's means and methods, but must be paid attention to during budgeting and scheduling of the project. Dewatering during tunneling differs from dewatering for other structures. Tunnels will be discussed in Chapter 6.

Drainage:

Exposure to water has undesired effects to foundations or roads. Therefore, we must design permanent means for water to be diverted away. Therefore, unlike dewatering, which is a temporary operation, drainage is a permanent system to protect structures from water effects. As this is a permanent system, it needs to work naturally, by gravity action. It is rare that pumped systems are used for drainage of a structure, which are costly to maintain and operate.

Drainage can be on the surface, or underground. Surface drainage will be discussed again as applicable, such as for highways and airports. They can include the drains below ground surface at shallow depth to drain away surface water. Underground drainage systems are mainly two types, such as consolidation and earthquake drains. We will mention consolidation drains in ground improvement methods section. Earthquake drains were mentioned under liquefaction.

Seepage:

Seepage is the flow of underground water through soil pores or rock fractures. Seepage can be measured. Sometimes it can exert force on foundations, basement walls or retaining structures that we cannot neglect, affect the stability of a slope, apply force to a cofferdam base concrete from below...

Below we discuss important terms related to measuring seepage quantity. These are fundamental concepts in water resources engineering branch of civil engineering, that will be introduced later.

Water Head:

Water head is a vitally important concept in hydraulics, which is a major branch of water resources engineering. Although water resources engineering section is later in this chapter, we must introduce the water head concept now, to be able to explain the seepage subtopic that we are discussing.

The total water head of a fluid results from its pressure, velocity and height, per Bernoulli's principles. Water head has units of distance (such as feet or meter).

We can write the water head (also called as "total head" or "total water head") and its three components we have just mentioned as below:

Total Head (TH) = Pressure Head (PH) + Velocity Head (VH) + Elevation Head (EH)

So in other words, the total head of water increases if at least one of its components increases, such as an increase of pressure, or velocity, or elevation of water.

Again, the important thing to remember at this point is that each of the three components above are all in distance units (feet or meter). Elevation head is already in distance unit. And the terms in pressure head and velocity head were organized so that they produce distance unit. This is all that needs to be known at this point. We will come back to it later.

Hydraulic Gradient:

Now that we described what water head means, we can explain the hydraulic gradient.

When water flows in soil, it cannot continue with the same head, because of frictional resistance within the soil, and the water trying to find its way thorough the pores. Some head will be lost within the soil. Hydraulic gradient is the amount of total head loss of water, per the distance it flows.

Rewriting what we have just described in equation form, where water flows from point A to B:

$$i = \frac{\text{Total Head at Point A} - \text{Total Head at Point B}}{\text{Distance between AB}}$$

which means head loss between two points per the distance traveled, where, i is the hydraulic gradient. Pay attention that both upper and lower side of the division are in distance units. Since they cancel, it means, hydraulic gradient has no units.

The higher the i value, the more total head loss occurs between A and B. Or similarly, for the same amount of head loss, the smaller the distance between A and B, the higher the hydraulic gradient.

Permeability, Discharge Volume, Hydraulic Conductivity:

Now that we introduced water head and hydraulic gradient concept, we can discuss permeability.

The flow of water in soil largely depends on its permeability. The coarser the soil particles get, the quicker the permeability grows. Permeability of the soil is a property of that particular soil. It is also called the coefficient of permeability, and written as "k".

The velocity of water flow through soil (or any porous medium), is directly proportional to the medium's hydraulic gradient per Darcy's Law. This direct proportion is equal to coefficient of permeability of the soil. So we can write this relation as:

$$V=k.i$$

where, V is the velocity of flowing water, i is the hydraulic gradient and k is the permeability of soil.

It is often more valuable for us to know the amount of flowing water per given time, which means volume/time expressed as "Q", rather than its velocity, V, which is only in distance/time. To obtain volume/time from distance/time, we must simply multiply the distance/time, by the cross sectional area of flow, which is square of distance, so that we will get distance cube / time. Rewriting this:

Volume / time = (distance / time) . (distance squared)

Volume / time = Velocity . Area

Rewriting it,

$$Q = V.A$$

where, Q is the "discharge volume" of water.

Above we also said V=k.i, then we obtain,

$$Q = A.k.i$$

See figure below, that explains this:

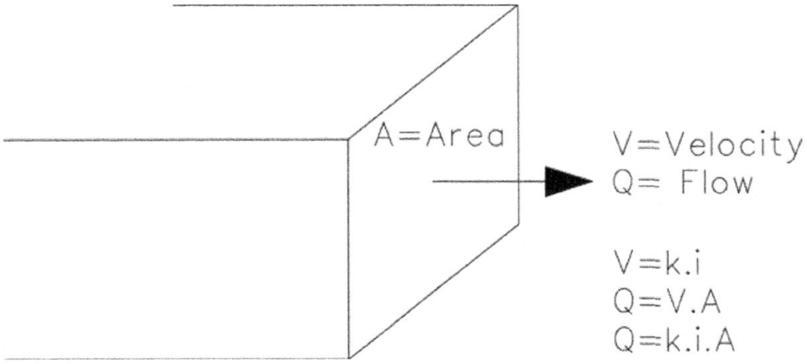

Figure - Soil Mechanics 38

Permeability is a unique property of a soil medium (or any porous medium). It greatly depends on the soil's porosity (we mentioned this term under phase relationships section) and how well pores are interconnected. If a soil medium has poorly connected or isolated pores, the permeability will be very low or practically none. Permeability is independent of the fluid that flows through it. All soil masses will in theory have different values of permeability, even if these values may differ very slightly between two particular soil masses. From fine grained soils such as clays to all the way up to gravels, there is millions of times of difference in permeability.

A term that is often confused with permeability is the hydraulic conductivity. Hydraulic conductivity is not a unique property of the porous medium. It represents the greater system of flowing fluid and porous medium together. It shows how easily a given fluid can flow in a given medium. Thus, we also need the fluid properties such as density and viscosity, in order to calculate hydraulic conductivity. So for the same medium, if the flowing fluid changes, the permeability will remain the same, but the hydraulic conductivity, and in turn amount of fluid flow will be different. *(Note: As far as the viscosity of fluids, there is the kinematic and dynamic viscosity, where the kinematic viscosity is obtained by dividing the dynamic viscosity by fluid density)*

Flow Net:

A flownet is a representation of water seepage through soil, by showing the flow in imaginary flow channels, which gradually lose total water head. As you can see in the figure below, the water flow was divided into approximately equal 4 channels that carry q amount of flow each.

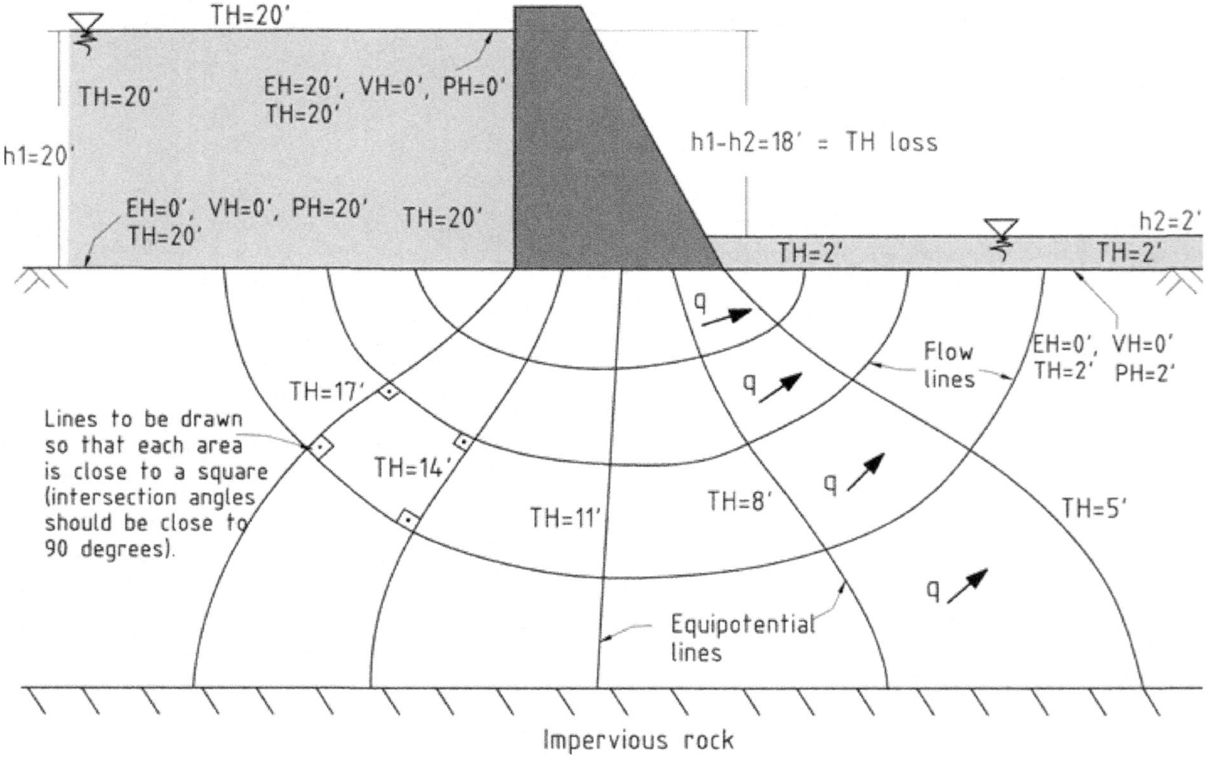

Figure - Soil Mechanics 39a

The lines that divide flow channels are called flow lines. So there are 3 flow lines. Total flow is Q=4q. We also drew perpendicular lines to them, which are made by connecting points of equal total head (TH), called equipotential lines, which successively drop from left to right. So there are 5 equipotential lines, and 4 flow channels seen in the figure below.

On the left side of the dam, above the soil, anywhere in the reservoir water volume, the total head, TH, of water is 20 feet. On the right side of the dam, you see that the TH dropped down to 2 feet. This loss occurred because of the soil medium through which the water traveled. Otherwise, if there was no loss, from simple communicating vessels principle, the water head on both sides of the dam would be the same and the standing water levels would also be equal (which would make the dam useless).

The reason that the TH is the same (20 feet) at any point on the left side, is because of the total head equation we wrote before:

Total Head (TH) = Elevation head (EH) + Velocity Head (VH) + Pressure Head (PH)

So within the reservoir, anywhere in the water body, if the elevation head decreases, then the pressure head goes up. Velocity head is zero in both cases. Same logic goes for the right side of the dam. Everywhere in the water body on the right side of the dam, TH is 2 feet, no matter where we look at.

The only place that the water head changes is where the water travels though soil. As it travels, it loses TH due to friction between soil and water. The total head dropped by 20-2=18 ft, when we compare the static water levels at both sides of the dam. After knowing that, along the path that the water traveled, we simply divided this 18 ft loss into appropriate equal intervals, each of which is one of the equipotential lines of successively dropping TH. So for instance, where it says TH=11, that equipotential line is a connection of points where the TH=11 ft and so on.

To know the TH is important because when we know it anywhere below the dam, from the TH equation we wrote above, we can also calculate the pressure head there. This will help us to determine the uplift force (as Force=pressure x area) that the water exerts on the dam for example. The reason that there are 3 flow lines and 5 equipotential lines is not random. It is because there are certain rules that needs to be observed when drawing these. For example, the lines should be intersecting at right angles or as close as possible to right angles. Also, each area between two flow lines and two equipotential lines must be as close as possible to a square, which are shown in the figure. If you observed these rules, you would also reasonably draw the lines as drawn in the figure above, because it is a self correcting process. *(Or for example you may take the time and draw twice the number of lines than what is shown here, but it would not provide you any extra benefit other than better resolution, because in the end, the total flow and the total TH drop are the same, and all you would get would be smaller gaps between TH drops and smaller flow channels).*

Now see figure 39b below, to make sure that one critical concept is well understood. Figure 39-b is the same as Figure 39-a, with only one difference, which shows a piezometer tube inserted to the equipotential line of TH=5. The water level in the tube will be as shown, 5 feet, and not 2 feet, because it represents the TH for the equipotential lines of TH=5 feet.

If we wanted to limit the flow of water even more, (*because after all, this is a dam and we want to retain water, and not let it flow under it*) thus achieve even more TH loss, we can for example insert a sheet pile wall below this dam. In that case the flow would be drastically reduced because now the path that water must travel will be much more. The flow net shape would also be very different, by flow channels now having to travel all the way below the wall which means more equipotential lines, which we will not display but it is a typical figure that can be seen in many sources in the subject.

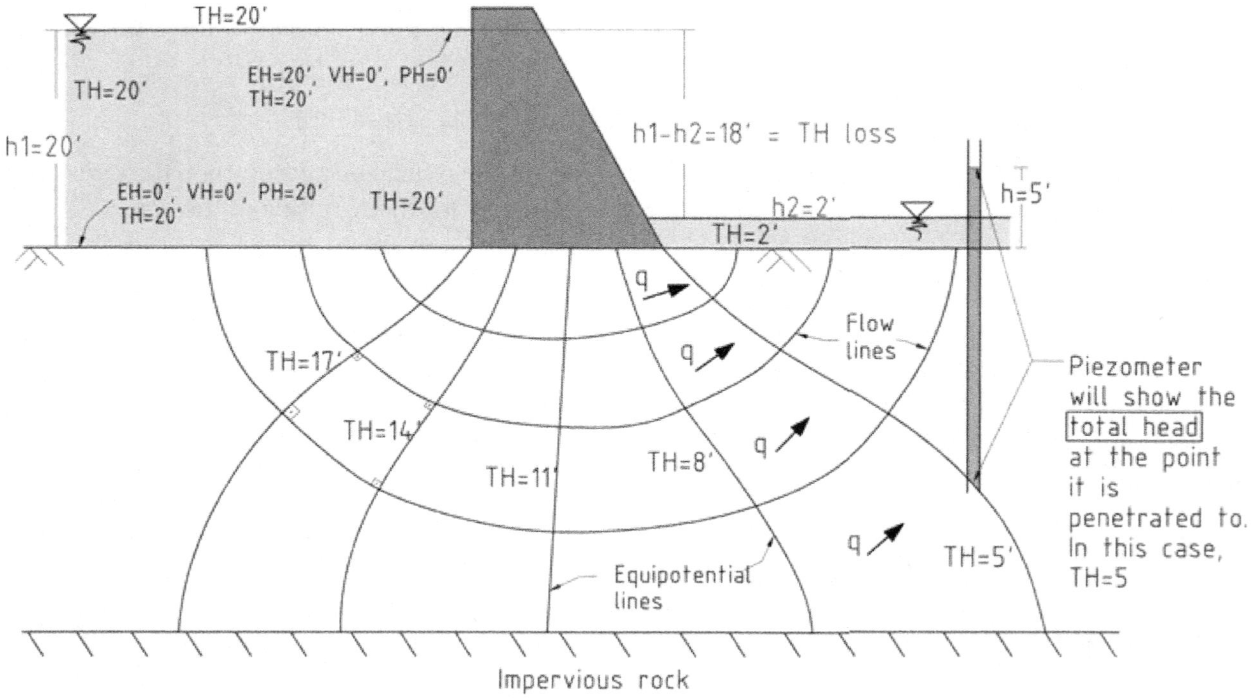

Figure - Soil Mechanics 39b

Freeze - Thaw Cycles (Frost - Heave):

Varying temperatures below and above freezing will cause the water in the soil to freeze and melt repeatedly. This cycle will damage the foundations, as water expands when it freezes. The frozen water will form "ice lenses" in the soil which will keep growing, which is also fed by the capillary rise of groundwater at lower levels. So, for this to happen in a soil, the soil must be porous enough to allow water to pass but these pores must not be so large to prevent water from climbing through capillary rise. Clay pores are too small and sand pores are too large for this. In between lies the silty soils, with such suitable pores, therefore they are the most susceptible to freeze-thaw. The swelling of the soil due to frozen water in it, is also called frost heave. Therefore, another reason that we bury the footings at a certain minimum depth below ground surface is to prevent the footing from being exposed to this cycle. The colder the climate, the deeper the effect of freeze and thaw cycle, and the deeper we must bury the footings *(note that with the use of a special insulator, shallower depths are possible)*. For example in the US, the frost depth varies between 0 to 8 feet per FHWA criteria, between the warmest and coldest climates of the country.

And with this, we finished soil mechanics, the longest section in this book *(although structural related sections are longer in total)*. So at this point, we have covered the basics from structural and geotechnical disciplines, which should provide a fairly good background for most of the remainder of the book. Now let's move on to other subjects of Chapter 1.

EARTHWORK

Construction earthwork activities are performed on a site for two main reasons:

- To bring the ground to desired elevation, by excavation (cut) and backfilling (fill), and grading,
- To remove and replace unsuitable natural soil with stronger soil, such as, replacing weak organic soils or expansive clay with sand and gravel mixes, which may also include treating them with different methods, to serve as foundation fill, or as road and pavement sub bases.

In this section we will discuss the first aspect, which is what we mainly mean by earthwork. The second one also falls under soil stabilization and will be discussed as a section under Chapter 4.

Cut and fill methods vary, depending on type of soil, its moisture content, the goal that we want to achieve and location. The backfill to fill a trench, behind a retaining wall or under foundations will be all be done differently. The excavation to prepare the grade for a building, to dig a trench, to remove massive amounts of soil for a dam foundation, or to open the route for a highway will also be performed differently. For example planning and execution of earthwork for a highway route will be far, far more complex than the one for a building project.

Earthwork is sensitive to weather. It is more difficult to lift and move wet soil than its dry state. It is far more difficult to excavate frozen soil than the soil in mild temperatures. The construction equipment travel with much greater difficulty on soft, wet ground, than dry, hard ground. This means the work schedule must consider the estimated days of rain, snow and freezing conditions.

Filling (Backfilling) and Compaction of Soil

Properly made fills will be stronger and stiffer than a majority of natural soils. This is because we select the material and compact it as well. Compaction is done in 6" or 12 layers, in other words, we do not just fill the entire soil at once and start compacting it. By backfilling and compaction, we usually seek to obtain 90% to 95% of maximum dry density of that particular soil. We discussed maximum dry density under proctor test (compaction test) before, but for backfilled soil which is already placed, we can quickly measure the density through nuclear dry density (and moisture) test as discussed under soil tests section. As the backfilled soil will be of known properties to us through prior tests, which is brought, placed and compacted by us under our control - unlike the naturally existing earth material on site - the backfill s density value that we measure in the field after its compaction will already be well correlated with other properties of that soil affecting the backfill's strength and permeability, in other words, we do not need to perform the whole set of other soil tests we listed before on backfilled soil.

Through compaction, we obtain a denser, stronger soil which will settle less, and have lower permeability as well. As we mentioned before, fine grained soils are compacted through pressing action, while coarse grained soils are much better compacted with vibration and pressing together.

In cases where the fill amounts are not massive, such as around foundations, behind retaining or basement walls, we try to avoid backfilling with clays because of their swelling potential and use sand

and gravel, even though a little costlier. When we cannot avoid using clay, we must take precautions for drainage and/or account for higher pressures due to swelling during foundation design.

Even when we need to raise the natural ground elevation by filling, we still need to scrape the topsoil first, which is the thin uppermost portion of the soil that generally ranges from several inches upto over a foot because it may contain a lot of organic materials and it is the weakest layer that can settle a lot.

Below as a simple example cross section view, we show the natural ground and a proposed road surface. As the proposed surface is higher, we need to place fill here. The figure below also shows how the surveyors mark the elevations and horizontal distances. For example, to set and mark the elevations, they must first have a point of known elevation nearby. If they don't, they need to carry it from the nearest official benchmark to the site, by performing successive measurements. After that, by knowing the line of sight of known elevation, with the help of vertically held measuring rods, they can determine the elevation of any point as required. For the horizontal layout, again they first have a benchmark with known location, from where they can start the road centerline. After that, they can layout the limits of the road on cross section from that centerline at every certain interval.

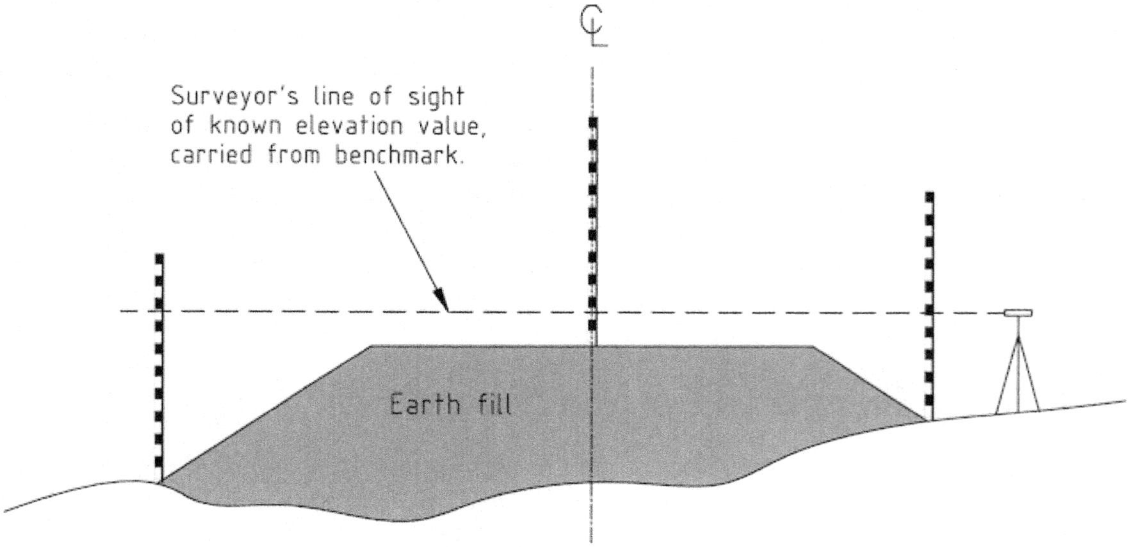

Figure Earthwork - 1

Cutting (Excavation) of Soil

To lower the elevation of the ground, or to remove the unsuitable material, we excavate the soil. The excavated soil may or may not be used as backfill, depending on project and the nature of materials. The excavation equipment and procedures may be somewhat different, according to the type and size of project. For example, narrow trenches are dug with hydraulic excavators, and in some cases, lateral temporary support will be necessary. The route for a highway can be excavated and graded with dozers and graders where massive amounts of materials are removed and in some cases blasting methods can be used in rocky areas. Also note that excavating tunnels is not considered earthwork as we mean here. It is tunneling, which is different, and will be discussed in Chapter 6.

Earthwork Quantity Calculations

We need to know the quantity of material that we will excavate or backfill, in order to plan our budget and schedule field operations. This gets increasingly more important as the size of the job grows.

The two main criteria to consider for earthwork quantities is the geotechnical requirements, and, keeping the earthwork to a minimum. For most structures with relatively small earthwork quantities, the geotechnical aspect is the only or the most important criteria. For those small projects, obtaining the required strength in the ground for a sufficient foundation is far more critical than keeping the earthwork to a minimum, which is already little. For other type of projects where there are massive amounts of earthwork, such as roads, highways, airports, keeping the quantities to a minimum is also important, while we still consider geotechnical aspects (*other criteria for layout for highways will be listed in Chapter 4*).

Earthwork is kept to a minimum by balancing the cut and fill amounts and making them adjacent to each other to the greatest extent possible, so that, not only the net amount of soil borrow or export, but also the hauling distances are kept to a minimum. This of course works only when the natural excavated material can be used for backfill as per civil engineer's approval.

Earthwork volume can simply be calculated by multiplying the total area of excavation and backfill with the average height of excavation and backfill respectively. For projects other than highways, this logic can be applied very easily, by determining the area from project plans, which can then be thought as if it was made of a suitable number of squares. For each square we determine the average height difference from the contour maps of existing vs. the proposed site elevations. After that the volume of earthwork for each square is calculated and we simply add them up, to come up with total cut and fill amounts. Computer programs can also do this very easily, once the existing and proposed contour maps and the project area limits are entered into the program. For highway projects, we employ different methods. Since we know the route, we obtain vertical cross section views at every certain interval along the route, such as 100 feet or meters. Then, we can multiply that interval length by the average of the two cross section areas at both ends of that interval, which is called as average end area method. We can also find the average area differently, if the area is irregular over that 100 meter or feet route length. Also see Chapter 4 > Geometrical Design of Highways > Earthwork Quantities.

During quantity calculations, different densities of the same soil material in different phases of earthwork must also be considered such as:

<u>Bank volume</u> means, the earth in its untouched natural state, before excavation.

<u>Loose volume</u> applies to excavated material in loose form. This volume is more than bank volume, for the same amount (mass) of material.

<u>Compacted volume</u> is after that soil is backfilled and compacted. Provided that an effective compaction operation was performed, this volume should be less than the bank volume.

Under phase relationships section in soil mechanics, we discussed the parameters to be used in earth volume calculations. The degree of compaction, as discussed under proctor test also affects these quantities which can then be used to plan the field operations. These volumes, as well as the difficulty and methods to excavate and backfill, will greatly depend on the type of soil, therefore, it is important to understand the soil types that will be handled on an earthwork project during planning.

Only for solid rocks, the bank vs. fill volume relationship will be the opposite, which means, their bank volume will be smaller than their volume after they are placed as embankment, because rocks are the most compact in their natural state.

Earthwork Equipment

There are different types of earthwork equipment for various purposes. Often they work together to complement each other, to achieve the most efficient operation. Based on the size and scope of the project, we can have an estimate for the number of equipment, as we will already know the rate of production for each. Equipment work performance (production rates) are affected by various factors including below. This is a general list and not all factors apply to every equipment:

- Weight of soil material to be handled
- Size of area, haul distances, equipment speed of operation and travel, area covered per cycle
- The time needed to load the vehicle
- Rolling resistance, which is the resisting force to equipment's wheels from the ground. Softer ground can have a lot more rolling resistance than a smooth firm ground. Loads on tires and tire contact area also affects this.
- Estimated actual working minutes per hour.
- For each equipment, there will be correction factors to be used in estimates, such as the factors for operator skill, soil type, bucket filling efficiency and others as applicable.
- Density and moisture content of soil

Now let's list commonly seen earthwork equipment:

Dozers:

Dozers are highly versatile, powerful equipment used for their strong traction ability that requires high pull or push force, for a variety of tasks in a construction site, such as earth hauling for short distances, spreading soil, assisting other earthwork equipment such as towing or loading, clearing land for structures or road path, ripping of existing pavings, scraping or loosening top soil layer, side cutting of hills, or other heavy duty tasks. For these variety of tasks, different types of blades are attached to their front and/or rippers to their back (in photo, below both are seen), which are also adjustable in height and angle, to provide even more flexibility, although the blades cannot be adjusted as much as grader blades. Dozers can travel on wheels or tracks. Tracked dozers can provide higher traction force than any other earthwork equipment in this list. Wheeled dozers can move faster than the tracked ones and can be used in longer distances, but this comes at the expense of lower traction power. For mass earthwork projects, dozers can haul material economically only short distances, up to around a thousand feet or less, which greatly varies depending on the equipment itself, soil type, density and moisture content. Longer haul distances require scrapers for economy.

Graders:

Graders are very useful for fine grading and shaping (finishing) of earth in many ways, such as grading, shaping banks, digging ditches, spreading soil at the precise elevations and at the highest possible range of angles, using their highly adjustable blades but they provide less traction power than dozers. A scarifier can also be attached in order to loosen the soil first, which makes grading easier afterwards.

Grader blade is highly adjustable in all directions, to cut soil in different ways. At the back side, also see the scarifier with blades like teeth, to loosen the soil first.

Loaders:

Similar to dozers, loaders also have a lot of uses. They are primarily designed to load soil to trucks and make excavations, rough grading, and backfilling, but with their powerful front bucket which can also be raised at a great distance from the ground, and high operating speeds and maneuverability, they can also be used to carry, lift, push, pull, tow things at the job site for a huge variety of secondary tasks and to help a lot of different trades, so they are probably the most widely used construction equipment. Bucket shapes vary according to the material to be handled. In some instances their front bucket can also be replaced with forks, to function as a forklift.

Like many other construction equipment, loaders especially, can be on tracks or wheels

Hydraulic Excavators:

Hydraulic excavators have much longer reach than loaders and the bucket orientation is the opposite which enables them to dig below the ground level. They are very useful to dig deep and narrow trenches, such as the ones for utility pipes, but also commonly used for excavating wider areas. Hydraulic excavators are also very versatile and can be used for a wide variety of secondary tasks.

On the left is a hydraulic excavator. On the right is a backhoe, which can serve both as a loader and a hydraulic excavator (but smaller) with buckets at its front and back.

Rollers:

Rollers are used to compact the soil, asphalt or RCC (Roller Compacted Concrete) with their heavy cylinders. For clay soils, pneumatic tire or sheepsfoot rollers are the most suitable. For sands, smooth drum vibratory rollers are used so that particles are rearranged by vibration to fill the voids between each other. For gravels, dual drum vibratory rollers will be more suitable, which can apply heavy blows. To facilitate better compaction, as we discussed under compaction / proctor test, the soil needs to be brought to its optimum water content, as determined by the tests. Therefore depending on the field moisture content of the soil, we might want to either add water or wait for the excessive water to dry.

Two rollers working simultaneously to compact a highway base material.

Scrapers:

Scrapers are specifically used to load, haul, spread, and dump soil, which must have been loosened beforehand with dozers or graders. After performing these tasks, a scraper returns to its original location and starts a new cycle for working on the adjacent section to be spread. Scrapers can haul soils economically longer distances than dozers, starting generally from around several hundred feet, up to a mile or more. Longer haul distances than that require the use of trucks, for efficiency.

Trucks:

Trucks are used for hauling earth for as long distances as necessary, and they can travel at much higher speeds. A loader is additionally needed to load the materials to the truck, but as the distances increase, this is certainly justified.

Trenchers:

These have the sole purpose of digging narrow trenches precisely and efficiently.

Draglines:

These are the oldest type of excavators. They are still used when needed, due to their longer reach than the hydraulic excavators. Draglines can also be positioned near a river bank etc... and dig underwater. It is made of a tower crane with an attached bucket with cables. The crane is brought to the proper position and then the bucket is moved with the help of cables to excavate earth.

Clamshell Excavators:

These are similar to draglines. They also operate with a tower crane and cables, but instead of a bucket, they have two pieces of buckets that can open and close, similar to opening and closing of a clam shell.

Construction Coordination of Earthwork

Below are some of the common construction coordination items to pay attention during earthwork:

- As mentioned, equipment like loaders, dozers, hydraulic excavators can be used for many more things than just earthwork. During planning of especially large projects, it must be made sure that the equipment and operator(s) are adequate and reliable, since other trades will also need them frequently, for many tasks that will need strong push, pull or lifting effort, such as loading, unloading and moving materials around, towing vehicles or other equipment, help with performing temporary construction items such as erosion control, fences, scaffolding. These should be coordinated with earthwork subcontractor when they are already mobilized on site.
- The earthwork subcontractor having reliable and adequate equipment and crew is important because during his work, access to many other activities can get interrupted and the entire job schedule afterwards will very often depend on how fast their work progresses.
- On jobsites with massive amount of fill, the surveying of volumes and dry density testing must follow backfilling work without lagging. This is also important for accounting purposes, to be able to submit accurate cost of the amount of fill placed in a timely manner.
- If water table is at shallow depth or we must excavate during rainy season, an effective plan for dewatering must be in place which must be communicated with all affected parties.
- If contaminated soil is encountered, its digging, storage and removal is a specialty task. The material needs to be piled up at a separate location and a specialty company must pick it up. During excavation, presence of water and rainy days makes this even more complex as water will spread the pollutants, so the work must be scheduled for days without rain. As for everything else, the requirements must be verified with the local jurisdiction.

FOUNDATIONS

We will discuss foundations under the following subtopics:

Earlier in this book, we gave some background in structural engineering and soil mechanics. We can now introduce foundations, under foundation engineering branch of geotechnical engineering. Any structure we build must bear on earth. Therefore, we must design the very bottom of our structures, which are called foundations, to transmit forces from structure to earth without too much settlement, and ensure stability against overturning, resist sliding, absorb vibrations. We can group foundations into two main categories, as shallow foundations and deep (pile) foundations. This section is intended to apply to foundations of any type of structure, even though we often give examples from buildings.

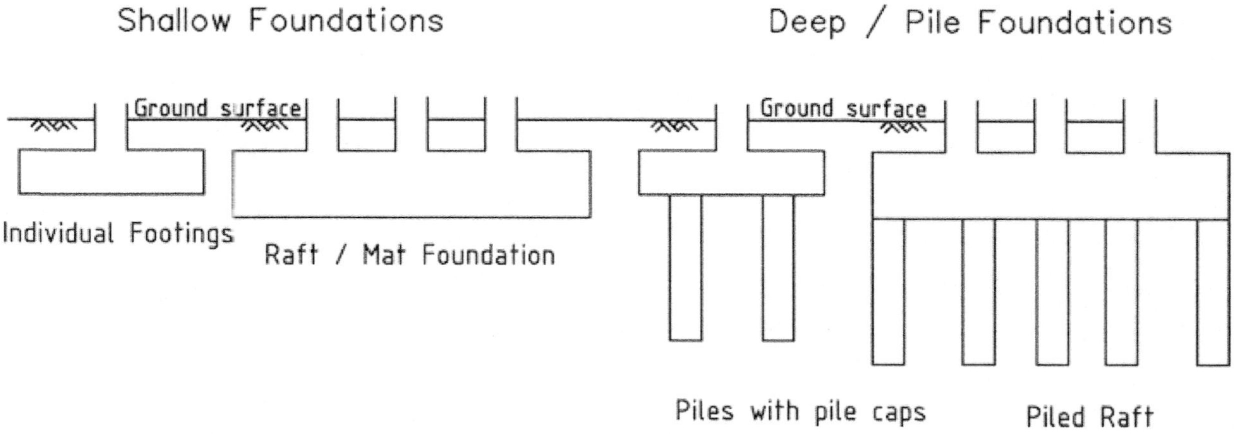

Figure - Foundations 1

Below, several of the important items when designing foundations are listed. These are general points for any foundation type. Most of the concepts mentioned below were introduced earlier in this book.

- The loads that come from the structure to the foundations, which are obtained from the structural engineer. In addition to regular gravity loads, eccentric loading, overturning effects, lateral loads, vibrations, uplift effects also need to be taken into account as applicable. The worst case load combinations must be considered as always. Depending on the type and size of the structure, different cases may govern the foundation design.
- Other than structural loads, information from the geotechnical report is the other main thing we need. This report includes the soil parameters to be used, liquefaction risk, groundwater conditions, frost depth, plasticity, shrinkage and expansion potential and all other items as applicable for the ground on site. See Soil Mechanics > Geotechnical Report section for more details on geotechnical reports where most items mentioned would apply to any project.
- The stress from foundation, vs. the bearing capacity of the soil and the settlement is of primary concern. The geotechnical report provides allowable bearing capacities. Differential settlement of foundations can cause large unexpected stresses in the structure itself, and have serious consequences. Preventing settlement altogether is not possible.
- The decision as to whether we can use shallow foundations or must use deep (pile) foundations will be made per the criteria in applicable codes, such as given in ASCE 7-16 Section 12.13.9.2. Structural loads vs. soil bearing capacities, liquefaction, lateral spreading potential are a few of the criteria to be checked. On some projects we use both shallow and pile foundations, depending on the regularity of load distributions from the structure and or the underlying soil/rock profile. Pile foundations are much more stronger than shallow ones but this comes at the expense of significantly higher costs and longer construction schedules.
- Ground improvement can enhance strength and stiffness of the soil. This affects the design.
- During earthquake, the behavior (response) of free ground without a structure on it, vs. the same ground supporting a structure will be different. This is called Soil Structure Interaction

which will be described in earthquakes section. This will sometimes need to considered for foundation design if the structure is heavy and stiff enough with respect to ground material.

- Foundation design ultimately depends on the success of soil testing and the accuracy of the parameters determined through them. Soil is a non-homogeneous material, therefore, we aim to place our soil response estimates within upper and lower boundary limits.

Shallow Foundations

Shallow foundations are all we need when the soil at a shallow depth has adequate strength for the loads that the structure will impose. Shallow foundations transmit the loads from the structure to the ground by spreading the load on to a larger area, therefore, they are also called as spread footings. What means "shallow" varies in different sources. Usually width to depth ratio of the foundation is a criteria, but what matters more is how they behave and how we analyse it, rather than this ratio.

As long as the project conditions allow, we prefer shallow foundations, which are much faster and economical to build than deep foundations. We do not need any special construction equipment to build shallow foundations, except the excavators or loaders for excavation. Shallow foundations are arranged to obtain uniform pressure distribution on the soil. This will also prevent differential settlements. It means, the sizes of single, combined and continuous footings from one column or wall to another may need to change, as each may have different amount of load. Shallow foundations can be designed as rigid vs. flexible. The pressure distribution on the soil will vary for each, and also according to soil type (fine or coarse grained soil). The larger a footing, the more flexible it behaves.

We cannot build the shallow foundations directly on the ground surface because the top soil usually contains organic material and it is not suitable, we need to stay below frost depth and also for higher bearing capacity, overburden pressure from soil above is needed. The codes provide a minimum depth. For example IBC 2015, section 1809.4 states that the minimum depth of a shallow foundation must be 12 inches below undisturbed ground surface. This 12 inches will still be too shallow in many places due to freeze thaw cycles and bearing capacity concerns.

Shallow foundations can usually be grouped as below:

- Individual (single) footings
- Combined footings
- Continuous (strip) footings
- Mat (raft) foundations

Single / Isolated / Individual Footings:

Single footings are made to transmit single column loads to the soil. Several important points about individual footings:

- Single footings are used in light, small structures.

- As the loads from the columns increase, or the bearing strength of soil decreases, the size of footings will also increase, to better distribute the loads on the soil. The different size of footings that may be present under the same structure is because of different loads from their columns, as the bearing strength of the soil below will assumed to be uniform.

- The shape of the footings will usually be square but when eccentric load / moment is expected on the footing by the column above in one direction, it may also be rectangular. Space limitations may also require the shape to be rectangular but in those cases we can also make combined footing in rectangular shape that will combine the two column footings together.
- If the individual footings are too close to each other, their zone of influence in the soil may interfere. This may cause too much stress on soil. In these cases, we can also prefer combined footings. If it comes to a situation that the number of individual footings are just too many, such as more than half of the building footprint, we can then prefer mat foundations and simply pour the entire foundation as one big raft.
- The column to footing connections can be made as fixed or pinned connections. Punching shear resistance *(see Structural Engineering>Various Structural Terms)* is also required. These will be determined during the analysis of the whole structure. The footing itself must be analyzed for bending as well. All of these, and whether we connect to a steel or RC (Reinforced Concrete) structure above will dictate footing thickness and its reinforcing details.
- It is not preferred to use single footings in zones with strong seismic activity, because single footings tend to move independently, even when they are connected to each other with grade beams. This independent movement may cause unexpected load distribution in the structure and/or differential settlements.
- If used in seismic zones, single footings must resist lateral forces too. This resistance can come from two sources. The sliding resistance at the bottom, and the passive lateral resistance on the side (vertical face) of footing in contact with the soil. The deformation amount to produce full passive resistance (*see lateral earth pressure under soil mechanics*) is much higher than the deformation to reach critical friction angle for sliding shear (see Figures - Soil Mechanics 8 and 9). Therefore, we may ignore the passive resistance, which will leave us at the conservative side, and count on only the sliding resistance for lateral resistance. In other words, we can not tolerate / wait for that large lateral movement to happen for passive reaction to develop fully.
- If we build shallow foundations on expansive soils, even if we compact the soil well, in the long term, the grade beams may be subject to horizontal and vertical push from the swell of soil, which can be very high. Therefore they must have sufficient reinforcing to withstand these effects but at first we must seek ways to separate them from such soil which can be done by flexible joint fillers or open blocks, to allow for swelling movement.
- Grade beams can be used to tie single footings or pile caps to each other. Grade beams help to:
    - distribute load more evenly between footings,
    - reduce size of the footings it is tied to,
    - reduce differential settlement,
    - resist loads from moment frames during seismic events,
    - increase seismic response predictability

Grade beams can also carry walls over them, but unlike continuous footings which simply rest on grade, they do this by behaving as beams (in other words beam action, by bending).

- We design a single footing primarily for three criteria: One way shear, two way shear, moment. One way shear effect tries to shear the footing at a vertical plane at a certain distance from face of column. A two way shear effect tries to shear the footing all around the column, again at a certain distance d from face of column, which means it is a punching (see Structural Engineering > Various Terms) shear stress. And the moment effect tries to bend the footing from its edge upto the face of the column. There may also be other criteria such as overturning, if applicable.

If interested, you may also want to view author's article about single footing design at: http://constructmagazine.com/geotechnical-engineering/foundation-engineering/shallow-foundations-foundation-engineering/single-footing-design/

Combined Footings:

Sometimes two or more individual footings supporting separate columns would be just too close to each other, and it makes more sense to build them together, which is called a combined footing. A combined footing can support upto several columns or shear walls (*which we will often call as "vertical elements"*) tied together with one footing, but it is still smaller than to tie the entire structure's vertical elements together. Combined footings can be square, rectangular or trapezoidal shape.

Combined footings are not only made from geometrical necessity and to save formwork, but also because the connection of a few columns into a single footing enables behavior like a single frame unit, which provides a capacity to resist much larger moments than the total of the individual columns', which will be a big contribution to lateral stiffness and seismic resistance of the building in that direction, although the effect of such increase of lateral stiffness will need to be analyzed so that it doesn't cause any other unwanted effects such as torsion or other irregularities in the building.

When designing the combined footings, the geometric center of application of loads from different vertical elements should be known. After that, we must proportion the geometry of the footing so that this resultant load is applied within acceptable limits of foundation geometric center to not cause too much eccentric loading on the footing.

Continuous (Strip) Footings:

These are the footings that have their lengths much larger than their widths. They are built below walls, a single line of columns, or anything that may continue as linear, to support linear loads along their length. These footings' structural design is similar to beam bending and shear. Continuous footings may also be curved, such as when supporting a large circular fluid storage tank (*for a small tank, a single circular footing may make more sense*).

Mat (Raft) Foundations:

This is a large foundation in one piece, which covers the whole footprint of a structure as a single element and spreads the entire structure's weight. The name "raft" is also used because this system resembles a raft that floats on soil. Mat foundations can accommodate heavier loads than combined footings, but still not as much as deep foundations. Even if making many individual footings would structurally and geotechnically be okay, mat foundation can still be preferred to save formwork and reinforcing detailing and work schedule, even if it means more concrete. Per NAVFAC DM 7.02, mat foundations should be preferred if the sum of individual footings would exceed half of the foundation area, if the soil contains cavities or compressible lenses (so that the raft will bridge over irregularities), or when resistance to water uplift is needed, or if the individual footings are governed by settlement and the foundation needs to equalize these individual settlements (since each of those settlements may be different from each other). Same situation is true where we must excavate and form a lot of grade beams as well, between individual footings. In short, the raft foundation makes everything uniform and act as one at the bottom of the structure, by distributing the load much more uniformly over the entire footprint. Mat foundations are made thick enough and with sufficient rebar, to serve as a single rigid base for the entire structure. Their top surface can also serve as the base floor level. They are also great at resisting overturning moments by providing a concentrated heavy mass at the bottom of building, and also resisting water uplift forces for structures which would not be heavy enough, had we not used mat.

Some of the important design considerations for mat foundations include:

- Please review Soil Mechanics > Strength of Soil > Elastic Settlement, subgrade modulus (modulus of subgrade reaction) again at this point, if you need to. One common approach to analyse mat foundation deflections for structural design is to look at it similar to a beam sitting on an elastic medium, such as a large number of springs that represent the soil, which is also called the Winkler modeling. The analysis also considers the flexibility (stiffness) of the mat, therefore, flexural (bending) stresses in the foundation. The flexibility of the foundation also causes redistribution of soil resistance from below nonuniformly. Considering the flexibility of a mat foundation is unlike a single footing which is much smaller and can be taken as rigid with uniform bearing pressure distribution from below. For convenience however, the engineer may first assume rigid mat with constant bearing pressure, to serve as a base point in mind and for establishing the approximate geometry of the mat, and then proceed with flexible design.

- In analyzing mat foundations, finite element methods are now most commonly used, by dividing the mat into a large number of elements, supported by springs. As a result we can estimate internal stress distributions and deformations in the mat and distribution of bearing resistance from below and then design the mat itself accordingly (which can also be performed by the program). Note that since a mat foundation is large enough, some local, small peak bearing pressure values which exceed the allowable values might be acceptable. It is important to represent the effects of basement walls, shear walls, columns, piles if any, soil subgrade modulus as accurately as possible to get good representation of mat stresses and deformations.

- The bearing pressure distributions obtained by the structural engineer should be checked in a settlement analysis by the geotechnical engineer, who may come up with revised subgrade reaction moduli under the mat area. Then these new values can be used by the structural engineer, and the process described in the steps above and this step are repeated.

- Since mat foundation has large area, the bearing capacity failure with the wedges we described before under bearing capacity subject is difficult to form, and the bearing capacity is very high. Therefore, on strong soils, such as sand and or gravel, most probably settlement governs the design. On softer soils such as clays, deep seated failure should be checked, as the settlement can occur not only in the shallower depths but also in greater depths too, due to the size of the area and load *(see Figure - Soil Mechanics 23, and visualize in your mind that as the footing and structure gets larger and heavier, the stress bulbs also get larger and can reach deeper layers and compress soils there, if they are weak)*.

- When we have basement walls around a mat foundation, during earthquake the lateral force is resisted by the sliding at the bottom of the mat and the passive earth resistance of the basement wall. But the deformation amount to mobilize these two resistances can be different, as explained under the similar item we mentioned for single footings. The structural and geotechnical engineer must coordinate this. Construction activities also can affect the soil properties at the perimeter and base of foundation too. *(Note that the mat working together with the basement walls and ground level slab diaphragm, creates a backstay affect for a tall building - a good thing for us, as if it is socketed into the ground as we describe in Chapter 2 when discussing basement slabs)*
- If the settlement values we get from analysis end up to be unacceptable, ground improvement and/or pile foundations may also be considered. When pile foundations are used, the system can be a composite system such as piled raft. In such case mat foundation also serves as one big pile cap for all piles. Analyzing the behavior of composite piled rafts is not straightforward and best be done by numerical computer methods such as finite element analysis. In this analysis each pile may be considered as a spring depending on configuration of piles.
- Soil structure interaction must be considered when applicable, which is discussed under Earthquakes > Geotechnical Concerns section of this chapter.
- For convenience and understanding design easier, it is also possible to imagine a mat foundation as if it is an upside down slab, as if the soil bearing pressure acts from above, while the columns carry that load and the slab itself (for this upside down visualization). The mat area can be divided into logical strips. Then we obtain internal demands, such as shear and moment diagrams for the strips, which are used in detailing concrete reinforcement to resist these loads.
- For mat foundation thickness, punching shear criteria from columns and shear walls, overall flexure or shear may be the governing criteria. Resisting high uplift forces with thick, heavy mats when there is high water table with respect to basement, can be another factor.
- Column and wall faces, interior corners, thickness changes are important for rebar detailing, as sudden internal force and moment changes, higher values can occur at these locations.

Construction of Shallow Foundations:

To start construction of shallow foundations, first the natural ground must be brought to grade as per project documents, by performing earthwork. This grade must either have adequate bearing capacity or its properties must be improved by one of ground improvement methods. A layer of low grade, plain concrete may also be poured on grade at this stage, to level the footprint area, isolate footings from soil below and provide a firm base, or we can just make neat cuts in the ground depending on project circumstances. After that the formwork starts. Shortly afterwards usually rebar can follow behind without waiting for finish of forms *(i.e. a start-to-start plus lag relation in schedule)*. A lot of measuring is done at this stage, to correctly locate and install formwork both in layout and elevation. While forms and rebar continue, cables and pipes for electrical and plumbing which run at the foundation level are also placed, so good coordination between these trades will need to be ensured by general contractor. Upon finishing these, the area will need to be inspected and accepted, before pouring concrete.

After the foundation is poured, waterproofing is applied as required. There are mainly two types of waterproofing, such as membrane waterproofing or bituminous dampproofing. The location of water table with respect to foundation and whether there is a basement level or not, affects the system to be used. In case of higher water table than the footings, the uplift pressure may also need to be relieved using subsurface drainage, which requires membrane waterproofing. If the water table is deep enough but water can still make the surrounding backfill wet through rains, then bituminous damp proofing will suffice. After the footings are poured and waterproofing applied, typically the backfill and then the sand layer is placed, and on top of that vapor barrier is laid, to isolate the slab on grade from moisture that may come from below. Finally the slab on grade reinforcing is placed and the slab is poured.

In case of mat foundations we have a large concrete volume. It may not be possible to make this pour all at once, although with proper coordination very long lasting pouring operations can be performed. If pouring at once is not possible, then we need to separate the pour into two or more pieces by dividing the foundation area, separated by construction joints (most commonly) in vertical plane. Additional rebar will be needed at such joints, detailed to make sure that the joint will not cause weakness but ensure pieces to work as one. Pouring in segments may not only be because of logistics but also to limit stresses caused by cement hydration heat *(A too large mass of concrete poured at once can mean too long of a path of heat escape, which can produce large thermal stresses inside the concrete. But to reduce hydration heat, other measures can also be taken as discussed under concrete section).*

Before any pour, all steel embed plates that will serve as base for any steel columns and or for various other purposes are also placed level with the top elevation of foundation which are carefully laid out by surveyor per coordination with other trades, and then firmly tied rebar to prevent moving from various vibrations and forces during concrete pour.

Deep / Pile Foundations

When the strength of soil at shallow depth is insufficient in comparison to expected loads, when we have highly expansive soils, we may need ground improvement or pile foundations or both. The heavier, larger and the more critical the structure, the more and deeper piles we need. For example for a high rise located in a highly seismic zone, or for a large bridge, we will need them. Pile foundations are as much as several times more expensive and take longer time to build than shallow foundations.

Each project will have its unique requirement for the optimum combination of pile material, number and depth of piles, installation method, pile caps or raft foundation. A logical way to compare the different options is to know the overall load carrying capacity obtained per unit cost, and not just to compare cost of linear feet of pile installation, which would miss some variables of each unique case.

Piles are often grouped under pile caps. In some cases, all piles support one single mat. If this is not done, still the pile caps need to be tied together with grade beams, so that their possibility to move and vibrate differently during earthquakes, thus cause unexpected behavior in the structure is restricted *(structures and structural assemblies each have their own unique natural frequencies of vibration).*

Since we can control and make the pile material itself as strong as we want, the soil's supporting ability is the limiting criteria for us, rather than failing of the pile material itself. We can however improve soil strength with ground improvement methods. In addition to theoretical methods, we often find the capacity of piles for all types of loads by installing extra actual piles in the field and testing them.

In addition to soil settlement, the axial shortening of pile, long term effects such as shrinkage and creep of the pile should also be considered when calculating overall settlement of the structure above. The heavier the structure, the longer the piles, the more pronounced these effects can be.

On the left, you see a cast in place pile during pour. The hole was first dug with the auger at the tip of the pile rig, then the casing was placed before pour. The piles will support a small retaining wall to retain the foot of the hill behind and separate it safely from the highway in front. A casing is used to retain the dug hole, which signifies sandy soil that would not be able to support itself. In the smaller photo on the right, you see a closer zoom of the same location which shows the rebar cage being lowered into the casing of the hole, immediately after the pour while the concrete is still wet. Here the pile rig functions as a crane to lift the rebar cage, which was already positioned and laid right near the hole beforehand.

Loads on Piles:

Shallow or pile foundations resist the following type of loads. Here we will look at piles:

- Vertical downwards (gravity) loads (such as dead and live loads) that puts pile in compression
- Vertical loads acting upwards, also called uplift, such as during earthquake or water uplift

- Lateral loads and moment as a result of earthquake, wind or even centrifugal forces on the structure above
- Lateral earth pressure acting on the pile body, when the piles are used to retain earth

Gravity (Compression) Loads:

Piles resist vertical downwards loads in two ways, as shown in Figure 2 below:

- The tip resistance at the bottom, also called as "end bearing resistance"
- The frictional resistance over their side surface area, which is also called "skin friction"

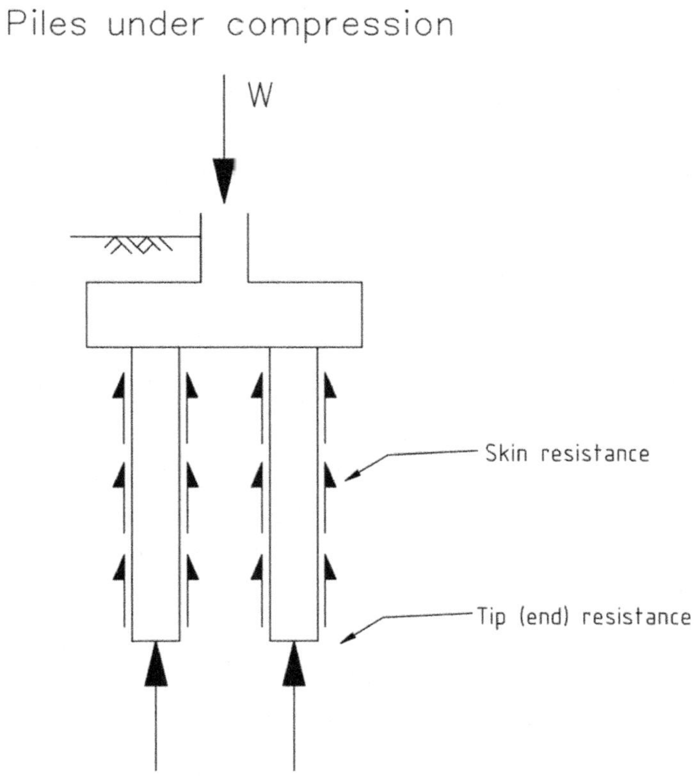

Figure - Foundations 2

A pile can have either or both, depending on the soil around and below it. For example, if there is weak soil around the pile face but it can bear on a firm soil or rock below, in our design we will only count on their end bearing resistance, or, if there is not much end bearing capacity below, but fairly firm soil around the pile, the only source of its strength will be from side (skin) friction. In some cases, both can be present, but generally they are designed mainly either as skin friction or end bearing piles. If the layer below on which the pile bears is rock or even weathered rock, it can provide very good end bearing support. Here when we say end bearing, the pile is not terminated as soon as it meets the firm soil or rock. It is still driven or drilled for some more distance, and socketed into the firm soil or rock.

In the figure here, the skin friction is shown as constant but in reality it will vary with depth, as the soil shear strength will vary with depth. As discussed in soil mechanics, for sands, this is mainly a result of friction, and for clays, it is mainly a result of cohesion, and friction generates more resistance. As the depth increases, generally the angle of internal friction will increase, which means greater skin resistance per length of pile if it is in sand, but this increase per length of pile cannot continue indefinitely. After a certain depth, further driving pile will only increase tota resistance but <u>not</u> the resistance <u>per feet</u>. This depth is specified in terms of pile diameters, such as 20 pile diameters in NAVFAC DM 7.02, which means it is dependent on the pile diameter. Presence of groundwater also affects skin resistance.

Load carrying capacity of piles can be derived from theoretical analysis, load tests performed on actual size test piles on site, various soil test result correlations to pile capacity, the resistance we measure and record during pile driving operations (applies to driven piles only) and computer analysis by finite element methods. Obtaining data from similar past projects in the area may also provide feedback. By saying load carrying capacity, we also refer to other type of loadings listed below, not just gravity loads.

The relationship between the skin friction of on the pile vs. its vertical displacement along the pile length is known as the pile t-z curve, where t is the skin friction shear stress and z is the vertical displacement. Therefore, it is obtained for multiple pile depth ranges. This should not be confused with Load vs. settlement curve, which only gives the total vertical load on top of pile vs. its total vertical settlement. These can be used to determine pile embedment length. The t-z curve and p-y curve (discussed below) are useful in modeling piles during computer analysis.

Side note: Soil Downdrag is another force that tends settle the pile down, by dragging the pile downwards. Downdrag can occur as a result of liquefaction *(see soil mechanics and earthquake sections)*, lowering of groundwater table, consolidating (settling) compressible soil around the pile.

Vertical Upwards (Tension) Loads:

In certain cases, upwards loads may act on piles, which is also called as uplift or pull out. Upwards forces can be a result of water uplift force, where all piles resist uplift, or from wind or earthquake where piles only on one side may have to resist pull out from overturning effect. The weight of the structure is generally sufficient to stabilize these uplift or overturning effects, but when it is not, we must rely on the piles' pull out resistances. As seen in Figure 3 below, the direction of skin resistance is reversed here, and also any possible end bearing resistance is irrelevant now and does not apply. In some cases, if we have rock at a reasonably reachable depth, we can make the pile tip socketed into the rock, which can greatly increase uplift resistance. This becomes relatively more important for structures such as bridges, large chimneys, wind turbines, marine decks... Buildings however, are usually inherently resistant to uplift due to their weight, although overturning is still one of the conditions we check for them with load combinations. If the skin friction as shown below is not enough and we do not have rock at shallow depth, then usage of batter (inclined) piles is another solution, which can significantly increase uplift resistance *(this will necessitate different structural details at foundation level because of different earthquake response and also increasing punching risk into the foundation or deck (i.e. marine deck), by the side in compression).*

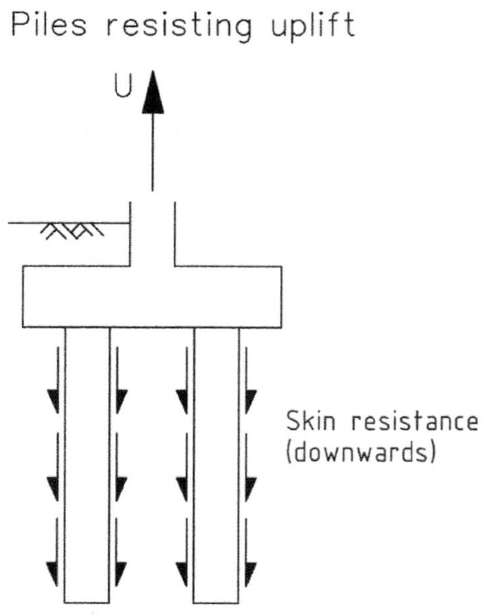

Figure - Foundations 3

Lateral Load and Moment top of Pile:

The forces such as wind and earthquake will not only tend to cause overturning and uplift, but also lateral forces on piles. Other sources of horizontal forces can be as vehicle or ship impacts on bridges, centrifugal force as a result of vehicles turning curves on bridges, lateral outward push forces from an arch above. These will create a complex pattern of shear force and moment within the pile, which will be generally more at the upper portions of the pile, which is why rebar ties are placed with smaller gaps (denser) there. The horizontal force will be resisted by the vertical faces of piles and the pile cap. In the figure below, to keep things simple, reactions are shown uniform although they vary around the pile perimeter and depth. Vertical forces and a typical laterally deflected pile shape are also not shown here for clarity of figure.

When analyzing piles under lateral loads (and for buckling under vertical loads), the pile will deflect in a complex shape, so it is important to estimate a point on pile with zero lateral deflection, which is commonly known as rotation point, or point of fixity. This helps us to come up with an effective pile length *(effective length may roughly be described as the working portion of a column or pile under loads).*

Side Note: Point of fixity is also useful when analyzing retaining walls embedded into soil, or columns under buckling. The zero deflection points on column determine effective column length (see structural engineering, structural stability section).

If a laterally loaded pile is considered flexible (i.e. length is large, with respect to diameter), the pile itself will have greater deflections, but the moment on it will be less (as some of it is relieved by flexible deformations). For a pile which can be classified as rigid, deflections will be less, but moments will be more.

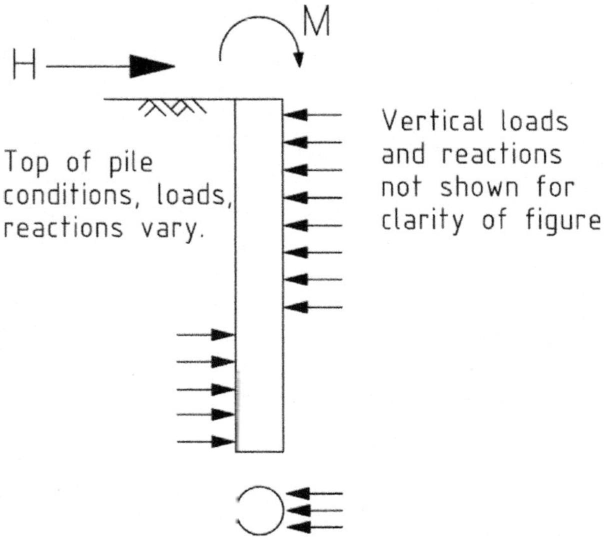

Figure - Foundations 4

The force per unit length of pile vs. deformation relationship for a laterally loaded pile at various depths is called a p-y curve when drawn on a graph, where p is the soil lateral resistance per unit length of pile and y is the deflection for that depth range. We say various depth ranges here, and not the entire pile length to avoid oversimplification because reactions and deflections vary along pile length. This means, when modeling in a computer analysis, we can represent these p-y relations as springs for each depth range, and in the end we will end up with multiple lateral springs attached to the pile, where the slopes of p-y curves for each spring will vary with varying depth. This procedure helps us to find the lateral resistance of the pile. The shape of the p-y curve and the shear and moment diagram along the pile depend on various factors such as the nature of loads such as static (slowly applied) or dynamic, soil type, friction between pile and the soil *(which also varies with depth, as the effective stress increases with depth)*, pile construction method *(driven or drilled, since it affects the soil around it)*, pile dimensions *(short piles behave rigid, which can make the soil fail first and slender piles behave as elastic, which means pile fails first)*, whether there is a group action if multiple piles, and the condition of top of pile and pile cap (if any). To determine the effect of these, the best way is to do a full scale load test, because these are a lot of variables that affect the pile response simultaneously.

When a group of piles *(piles combined under the same pile cap)* are laterally loaded, outermost piles on one side (the side that pushes the soil) will have the most load. The outer piles on both sides will also be under greater amount of compression and tension due to overturning forces.

Horizontal Loads on Pile Body:

Piles can also serve as retaining structures when installed adjacent to each other. In such cases, lateral forces act on the side of the piles.

Modeling of Loads on Piles:

In civil engineering, we represent structures in mathematically solvable formats, in order to analyse them. These are called structural models. The simply supported beam example you saw in statics section was a model for instance, although the simplest one. We were able to solve it. You will see other model examples later in the book. Foundations can be modeled as well. See an example pile model in Figure 5 below. When modeling piles, the reaction forces from the surrounding soil can be shown as if they were springs (which may even be springs with nonlinear force-deformation relations each). So for example to model a pile, we can draw the pile and many springs around it, to represent forces in different directions from soil, such as the skin friction force, axial force from end bearing, lateral support of soil. We use axial or rotational springs when modeling foundations and structures too, for simulating different behavior. We could also include damping component (similar to a piston) to each spring to be more accurate.

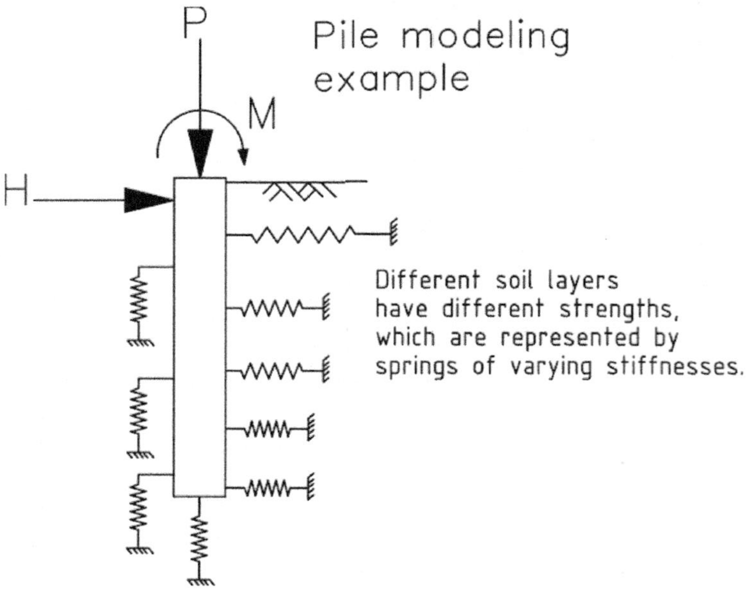

Figure - Foundations 5

Pile Groups:

Instead of making them work individually, piles are often grouped under a pile cap, to make them work together. This can enable the whole assembly generate better seismic performance.

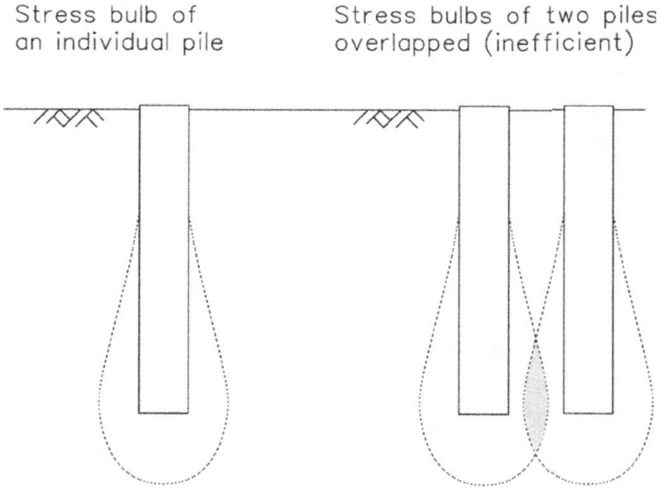

Figure - Foundations 6

An important criteria to pay attention here is that, since each pile is supported by the soil around it, the piles in the group must not be too close to each other in order to avoid needing support from the same soil volume as seen on the right side of Figure 6. This distance that we must leave between piles is influenced by the pile diameter and the soil type itself. As the pile diameter increases, the gap between piles also needs to increase.

Usually, the vertical load carrying capacity of a group of piles is less than the sum of individual piles that form the group. The factor that shows this proportion is called the group efficiency. It is found by dividing the capacity of the group by the sum of the capacities of individual piles, which means it is usually less than 1. In cohesionless soils however, sometimes, also depending on the pile spacing, the group efficiency can be more than 1, when the driven piles densify the soil around them. In those cases, the efficiency can be taken as 1 or in other words, 100%. In cohesive soils, the efficiency of group is less than 1. Regardless of soil type, if piles are too far apart from each other, then the overlap as in Figure 6 above will not happen and the group efficiency can be taken as 1. One must also remember to think in 3D when looking at the figure here, in other words, the group can also have a square pile cap with 4 piles or even bigger pile caps with shapes other than rectangles with more piles. When a pile group is subjected to lateral load and or moment from the structure above, the contribution of each pile to overall group capacity will vary depending on the pile's location within the group, therefore not all piles will contribute the same. Soil structure interaction, which will be described later, also affects the overall capacity of the group, which is a complex analysis.

Types of Piles by Installation Method:

Piles can be grouped as driven piles and cast-in-place piles according to their installation method.

Driven piles are first manufactured elsewhere, then brought to the site and installed by hammering or pushing them into the ground. They can be made of concrete, steel or timber. Composite piles that are made of more than one material, such as steel H or I sections encased in concrete or steel pipes which are filled with concrete are also used but they are not common. It is more difficult to build them, make them work as single unit and make the splices for long pile lengths.

Cast in place term applies to piles that are cast from concrete in drilled holes at their final location.

Each project has its unique requirement for the type of pile foundation influenced by factors such as the structure itself, soil and seismic conditions, available space, and available means, budget and time.

Although it is more challenging to do, if needed, with these methods we can also install piles as inclined (battered piles) upto different angles depending on the method and ground conditions, which greatly improves lateral and uplift load resistance of the structure although the gravity load carrying capacity is a little reduced.

The common types of pile foundations according to methods of installation are introduced below.

Driven Piles:

Pile driving is a very old construction technique, dating back to ancient Roman times. In this method, the piles are hammered, or pushed from top into the ground, with or without vibration. During this hammering or pushing, they displace and push the soil around them outwards, which causes the soil to get denser around the pile. This is desirable as it also serves as ground improvement and increases the skin friction capacity of pile. Because of the displacing of soil, they are also called displacement piles.

Driven piles can be made of precast concrete, steel or composite sections, pipes or timber. Steel sheet piles are also a type of driven pile since they are pushed into the ground, but they work differently than slender piles. Reinforced precast concrete piles are the most suitable in locations containing water, because they are more resistant to water effects than steel or timber.

The theoretical analysis that we will perform beforehand can show us pile locations, and preliminary lengths, but we can not determine the exact installation length of each pile this way. The actual pile lengths are finalized in the field, according to feedback we receive from the soil during pile driving. As we drive the piles, we will encounter increasingly more resistance, both because the total pile length in the soil is increasing, thus more skin resistance is provided, and also because generally the soil gets stronger as we reach deeper. Therefore, during driving, when we reach to a certain depth that will require more number of blows than a certain standard limit per length of pile advancement, it will be determined as the "refusal point" of the soil. That is the depth that the required pile capacity is determined to be satisfied where we stop driving that particular pile. This number of blows is stated per unit length of advancing the pile and determined beforehand through prior test and theoretical analysis that depends on soil type, pile type, and equipment itself. Potential energy to kinetic energy conversion equations are used for dropping a hammer of certain weight from a certain height.

Due to geometrical constraints during bringing the piles from factory to the jobsite, and / or maximum length of pile that the pile driving equipment can handle, the pile length we can install will have a limit, so it is not always possible to reach to the desired installation depth with just one pile segment. In these cases, the piles need to be made from more than one pieces by joining them (splicing) at their ends. Steel section piles or precast piles are suitable for this type of installation.

After the driven piles are installed, there will be a portion that remains on the ground surface. These are cut off to proper height to receive pile caps or footings over them.

Pile driving is one of the loudest activities in construction which also produces strong vibration and impact waves. This aspect must be considered in design and needs to be coordinated with the local jurisdiction and neighboring facilities.

Drilled Shafts (Piers) / Caissons / Cast in Place Concrete Piles:

The other pile type is the cast-in-place, which are also called caissons, drilled shafts, drilled piers. This time, instead of hammering or pushing a pile into the ground, we first dig a hole, remove the soil, lower reinforcing steel cage (that was prepared outside beforehand) into the hole, and then pour concrete inside the hole (or the concrete may also be poured first and then rebar cage lowered). We may use a removable steel casing during this pour to serve as formwork of concrete if the soil has the potential of collapsing (caving in), as in the case of sands. After the concrete is poured the steel casing is pulled out. Also see photos in the beginning of pile foundations section which is a cast in place concrete pile.

Drilled shafts have advantages over driven piles such as:

- Diameter flexibility: The driven piles that we discussed above can have diameters up to a certain limit only. This is because driving a large diameter pile would require so much impact force and so large machines that it would not be practical. For drilled shafts however, this problem doesn't apply. Because of the freedom to dig a hole with as large diameter as we want, these piles can have any diameter that we require, which means extremely heavy loads can be supported. In other words, drilled piers are the largest, strongest type of foundations that are used underneath the heaviest structures with the largest gravity and seismic load demands, such as large bridges, very tall buildings, towers etc... Having larger diameter also means, drilled shafts can be shorter than driven piles, for the same required pile capacity.

- Elimination of pile driving vibration and impact.
- Installation of cast in place piles require lower height clearance than driven piles. The most height needed here is when lowering the rebar cage into the hole in the ground which is less than the full pile height plus the hammer and its drop height in the case of driven piles. The drilling rig for drilled shafts also require considerably lower height clearance.
- Unlike driven piles, drilled shafts can be produced as tangent, secant walls that are right adjacent to each other, to form an underground wall, which can serve as a retaining wall after the soil is removed by excavation from one side later.

The disadvantages of drilled shafts in comparison to driven piles include:

- Since no soil is displaced outwards and densified as in the case of driven piles, we can not obtain the same improvement in skin friction capacity. In other words, ground improvement aspect of driven piles is absent here. So these piles are also called as non-displacement piles.
- The pile itself is produced in the field, which is more difficult to do than the factory made driven piles in controlled environment. If we use casing, the concrete is formed and produced much more regularly but if we don't, sometimes the concrete may just leak into the fractures underground. In all cases, more concrete is used in this case which means more waste.

The digging is done by special equipment called drilling rig. At its tip, an auger, bucket or core barrel can be attached, which can be of different sizes. In addition to the required pile size, the soil and groundwater conditions also determine the drilling method.

The steel cage is made on site and then lowered into the drilled hole, and it either stands by itself or with the help of slurry or steel casing. To make sure that the cage is centered in the hole and sufficient concrete cover will be obtained at vertical faces, small blocks of stone or concrete is attached to all around the cage at its outside. After the rebar cage is in place, concrete is poured.

The pour is performed usually with a long hose called tremie, which is lowered all the way to bottom, and then raised slowly as the concrete is cast. In some cases instead of using steel casing, slurry (a high consistency liquid) may be filled into the hole to prevent the hole from caving in, which gradually comes out, as the concrete is pumped at the bottom (concrete is denser than the slurry so it stays bottom). In such cases tremie should always remain in wet concrete, to prevent mixing of slurry with concrete. To achieve this without seeing inside the hole is possible by monitoring the amount of pumped concrete into the hole and estimating the concrete height. This also helps to monitor concrete losses towards surroundings. Pouring with tremie may not be necessary in firm and dry soil conditions, and in such cases concrete is simply filled into the hole with free fall (again starting from bottom sections).

The rebar at the top of the pile must be exposed by removing the top portion concrete, because rebar here will be embedded in the pile cap or mat foundation concrete that will be cast above it. Concrete at the top must be removed with care and using a proper method, in order not to damage and bend the rebar, as well as not causing too much vibration to rebar, which would cause loss of bond with the concrete pile body below. Not only this will weaken the pile but it can also cause rebar corrosion in the long term, if water can reach inside later through small cracks caused by demolition of concrete at top.

In addition to the theoretical design of piles, test piles are installed and loaded at the site, to determine pile capacity before proceeding with the actual piles.

Continuous Flight Auger (CFA) / Auger Cast In Place Piles:

This is a common method of installing cast in place piles. It is a suitable method for medium to soft soils. The name comes from the auger, which is called "continuous flight hollow stem auger". The auger has a hollow core, from where the concrete is poured, as the auger which drilled the hole is slowly raised. During the raising, the auger is still rotated in order to remove the drilled earth upwards. Again the concrete mix volume that is pumped into the hole is monitored. After the hole is completely filled with concrete, rebar cage is lowered into wet concrete.

CFA pile diameters is limited to the size of the auger which drills in one pass, so the diameter is limited. The drilled shafts do not have such limitation therefore CFA piles can be of upto a limited strength but this is enough for many applications, which makes them a very common method.

One significant plus with respect to drilled shafts is that CFA piles do not require slurry mix or casing during construction, because here the concrete is instantly poured into the drilled hole that eliminates risk of caving. Therefore they can be produced faster than drilled shafts in soils with caving potential. Similar to drilled shafts, CFA piles can also be made as a retaining structure as secant or tangent walls. They also do not produce impact noise and vibration as driven piles.

Mini Piles / Micro Piles:

These are small diameter piles used for a variety of purposes and soil types, at relatively lighter loads. They are easier to install in almost any location because of their small sizes, including areas with limited access, congested areas with a lot of existing features such as foundations and underground utilities and so on, and on sites where pile driving or auger-cast piles may not be possible due to for example large size boulders in the ground. Their tip can also be made larger with grout for improved load carrying capacity. They are frequently installed at an angle. During construction they cause less disturbance than all piles discussed earlier. They can be used under footings of new construction, in stabilizing slopes, as soldier piles when retaining earth (will be shown), and frequently as a foundation strengthening (underpinning) measure, to improve load carrying capacity of existing shallow foundations where we can core the existing footings or mat foundation from top, open hole, and install micropiles downwards. Micropiles are called as such because of their small diameter, which provides flexibility in their installation but this doesn't mean they are shallow. They can reach to bedrock at fairly large depths.

Pushed Piers (Jacked Piers):

In this system, which is also a foundation strengthening measure, the piles are attached to the foundation, and then advanced downwards by hydraulic jacking. So, the reaction force for jacking is the weight of the structure itself. The force for jacking is monitored and jacking continues until the desired pushing force is reached, which means the advancement length is enough for the desired pile capacity.

Helical (Screw) Piles (Piers, Anchors):

These are steel screws that are advanced into the ground by torque, which are often used to strengthen existing foundations. The torque to advance the screw is measured, and upon reaching a certain value, advancing can stop, which means the capacity is reached. These piles are for lighter loads in comparison to the previous ones we listed, but they are lower cost and easy to install. They can also be used as tiebacks to resist lateral loads, to strengthen an existing retaining system or be designed and installed on a new system.

Items to Pay Attention During Installation of Piles:

Below are a few example issues to pay attention during pile foundation construction:

- Pile installation is a critical item for the construction schedule, as everything else will follow it, therefore, it is critical to find a pile subcontractor with adequate experience, machinery and crew for the job, who will be always on board with project management. Pile installation also prevents surrounding sitework to proceed, due to pile equipment moving around which can sometimes create challenges for other sitework activities if the sitework has schedule critical items in that area. The most important thing in order to follow schedule is to have adequate number of pile rigs (and the crew for each) on site when needed. For large projects there may be hundreds or even thousands of piles overall, at a single or distributed locations.
- Installation of piles is more specialized work than building shallow foundations. The skill of crew and especially pile equipment operator play important role in the quality and speed of construction. The geotechnical engineer should also be present on site as often as possible.
- Driven piles may unwantedly tilt from vertical, or just slide sideways, especially in weak soil, due to heavy blows delivered on top of pile. The skill of crew is important here, because if some piles are too much off location or vertical alignment, new ones may need to be installed, so the contract terms must clearly account for such situation. Predrilled holes also called pilot holes, *(upto a certain portion of pile depth which are also smaller than the pile diameter so as not to cause end bearing and especially skin friction loss)* may help with this issue and enable easier pile driving, but adds to the costs and schedule. Predrilled holes can also help us confirm that the subgrade is as expected, before installing the actual pile.
- Pile equipment are large, have high hourly costs and difficult to transport, which may also need special permits while transporting. We must ensure that all piles are correctly installed, before letting the pile rigs leave the jobsite.
- Adequate area must be planned for the pile rigs to maneuver around. We must also ensure that the pile equipment or any other construction equipment will not travel over any installed piles, which may happen if the installation was done all the way to the ground level.
- Soft soils, such as clay, especially in rainy periods, may not have adequate strength to support the weight of the heavy piling equipment. In some difficult cases, it may be necessary to remove the top portion of the soil first, even may be upto few feet, and place some gravel instead, so that the pile rigs will not sink. This must be considered during the planning stage.

Ground Improvement Methods

Ground improvement is improving engineering properties of soil to support new or existing structures, to make it stronger, harder, denser, when it is considered inadequate for our proposed project requirements. *Note: Soil stabilization for highway and airport pavings, although also improves the engineering properties of the ground, is regarded separately than ground improvement as discussed here, and will be discussed in Chapter 4.*

The goals of ground improvement include mainly to:

- Increase the soil's shear strength and bearing capacity
- Increase the soil's density
- Achieve more uniform stress distribution in the soil
- Decrease settlements
- Increase liquefaction resistance (applies to sandy soils, n/a for clays)
- Decrease expansion and shrinkage potential (applies to clays, n/a for sands)
- Decrease permeability of the soil, thus control of seepage and groundwater
- Improve frost heave resistance (applies to silts)
- Improve stability of slopes
- Improve soil stability for the purposes of tunneling

Ground improvement may help us eliminate installing piles, which can usually be costlier and may take longer. In some difficult cases and/or heavy structures, even when we improve the ground, we must still install pile foundations. Both are costly, and usually critical path (*a branch in a schedule that delays the whole schedule if it delays*) activities that need to be carefully studied during planning stage.

Although each structure and the soil underneath is unique, at present there is considerable past experience in ground improvement applications, to significantly improve soils for any type of application we may require. It is therefore only about good planning to apply the suitable method and determine its extent for a particular job. To apply proper method to what extent along with foundation design will need to be well coordinated between the owner, construction manager, structural and geotechnical engineers as early as possible, to minimize cost and schedule impacts.

To measure the effectiveness of ground improvement, the performance of the soil both before and after ground improvement can be compared by applicable soil testing methods we covered earlier, such as SPT performed before and after ground improvement.

Below, several common ground improvement methods are listed.

Replacement of Unsuitable Soil:

The most straightforward method of ground improvement is just to remove the unsuitable top portion of the soil as necessary, and then replace it altogether with stronger fill, but this is not always possible. This method can be economically applied only for shallow depths of upto around 2-3 meters or so. As the depth increases, the issue is not only the larger quantities of materials, but also having to support a deeper excavation, and in urban settings, emergence of issues with foundations of adjacent structures.

Compaction:

Compaction is making the soil denser, stronger and more impervious by causing deliberate settlement and reducing its volume. Compaction methods produce vibrations and noise to varying degrees depending on the method, some of which which may be a disadvantage in densely populated areas. The most common methods of compaction include:

- Roller compaction: As mentioned before under earthwork equipment section, rollers with heavy cylindrical wheels are driven over soil to compact them. For sandy soil, vibrating type of rollers, and for clay soil, pneumatic tire or sheepsfoot rollers are suitable.
- Deep dynamic compaction: Dynamic compaction at deep layers can be achieved by dropping a heavy weight repeatedly from a certain height. This method produces strong vibrations and noise. This can be effective, on average upto compacting around 5-10 meters of depth.
- Rapid impact compaction: This is done with a hydraulic hammer, which repeatedly impacts the soil at high frequency. It is only suitable for sands at and shallow depths.
- Vibro compaction: This method does similar in soil to what a vibrator does in wet concrete. It is basically penetrating a large vibrator into sandy soil, thus causing it to get denser. Water jet can be used for making the process easier. Due to compaction there will be overall volume reduction, thus settlement at the site, so, where the vibrator is placed, new sand can be poured from above to make up for the lost volume, unless this elevation reduction is desired.
- Explosive compaction: This is artificially creating earthquake conditions in loose sands by small explosions underground, so that liquefaction will occur and the soil will densify before we build on it. Sand in deep layers can be liquefied and densified with this method.

Grouting Methods:

Grouting is a very common method which may also be used together with compaction. Grouting is introducing a chemical or particle based grout material into the ground using different methods, usually by injection. In addition to cement, additives such as fly ash, bitumen, lime are also used in the mix and with different combinations. Through grouting, all goals of ground improvement that we mentioned a couple pages earlier can be achieved.

Common grouting methods are listed below:

- Permeation grouting: This is probably the most common and oldest method. It is basically doing what we described in the paragraph above, which means injection of grout into the soil. It works better for coarse grained soils where the grout can permeate easily. Here the grout mainly "permeates" into the soil without causing any significant displacement in the soil.
- Compaction grouting: This means injecting stiff, low slump grout mix into the soil at given intervals, that will form grout columns in the soil. These grout columns will both carry the structure and also push the weak soil outwards which will densify the soil which is one of the main goals. This is a suitable method to remedy existing footing settlements because it is easily applied around existing structures and also the displaced soil can push existing foundation upwards, therefore the gain from this method increases when there is load from above.

- Jet grouting: High pressure jet grout is applied in a rotating manner, by raising the hose, to form grout columns in the ground. The columns serve the same purposes as in compaction grouting.
- Polyurethane grouting: This is an expanding material that is applied under existing slabs or footings, to fill voids, cracks and level the surface, while also creating a watertight barrier.
- Compensation grouting: This method is applied in order to fill (compensate) the voids, fractures in the soil or rock with pressurized grout.

Other Techniques for Ground Improvement:

Some of the other common ground improvement techniques include:

- Injection Systems: This is the injection of various liquid solutions into the soil, to improve it.
- Compaction Pile / Stone Column: We can make sand, aggregate or gravel columns in soil. The columns themselves and the displaced and densified soil around them, means stronger, stiffer soil. For compression, they work similar to piles *(but not for tension or shear because they do not join into footing like concrete piles)*. When applied in clay, the columns also work as vertical drains. In sands, it increases liquefaction resistance as well, due to densifying the soil and providing drainage, so that pressure buildup is reduced. They are made right below footing locations, as single or in groups of a few columns, and the eccentricity between the centroid of the stone column(s) and the footing must be kept to a minimum.
- Deep soil mixing: Soil is mixed with cement based materials, which can be as wet or dry mix. So it can be thought as a form of soil stabilization but at deeper layers and for the purposes of supporting structures, not pavings. We discuss soil stabilization for pavings in Chapter 4.
- Ground freezing: The soil in frozen state is much stronger than the soil in mild temperatures. In addition, it is totally impermeable because of ice. Thus we can freeze the ground for various purposes such as, to temporarily strengthen the soil under existing foundations, supporting open excavations, stabilizing slopes, supporting heavy construction equipment, digging tunnels in difficult ground conditions, controlling seepage, and so on... Freezing can be done by inserting metal pipes into the ground, in which a fluid is circulated through pumps and draws heat from surroundings. In cold climates, passive ground freezing systems can be used, which do not require external energy to work, in order to keep the ground frozen. Ground freezing also has the environmental advantage that no foreign materials are introduced into ground.
- Preloading and consolidation drains: This is simple and very old technique known for centuries, which is preloading the site to consolidate mainly clay soils. Settlement occurs beforehand, to reduce post construction settlements. As consolidation is slow, this method can only be used where there is enough time in schedule such as months even over a year. Large depth of compressible soil can be consolidated with this method, which is making the soil overconsolidated. The higher the degree of overconsolidation, the more compact and stronger the soil will be, as discussed in soil mechanics. The process can be faster by installing vertical consolidation drains, which provide a lot more draining path for water to drain faster. They can be made by drilling small, deep holes in the ground and filling with permeable, coarse grained soil, or permeable geosynthetic filter drains. Vertical drains are then connected with horizontal

drains on the ground, to remove the water away from the area. Upon installation of drains, the area will then be loaded, for consolidation to take place, prior to construction of the structure.

Foundation Repair & Strengthening (Underpinning)

Many of the ground improvement methods we discussed so far, also apply to existing foundations, but other than improving ground, sometimes we can also decide to strengthen the foundation itself by making repairs, modifications or additions. The reasons for foundation strengthening can be either too much settling or differential settlement of existing foundation, or any planned new construction or modification to the structure itself or adjacent to it. Sometimes we need to perform both ground improvement and foundation strengthening. During this, the possible new deformations and settlements in the ground must be closely watched. Foundation strengthening, which may also be called as underpinning, is usually performed in areas with difficult access, where there is already the existing structure itself and therefore the method must be chosen accordingly, and we usually cannot use large construction equipment.

As the depth of underpinning increases, such as more than a few feet, the lateral pressure of exposed soil also starts to become significant which may also affect the structure above and the method itself.

Several common foundation strengthening methods are listed below:

- Mass pour: This excavating foundation soil in small steps and replacing with concrete as we proceed, so we are basically gradually making a thicker foundation here to reach stronger soil below. So this is applicable for shallow foundations where strong soil is within shallow depth. For temporary support of undermined locations, hydraulic, mechanical jacks can also be used.
- Excavation support for adjacent structure: This is done to support the future excavation of a proposed structure near an existing structure. It is partially excavating below the existing structure's foundation, support it by filling with concrete and then proceeding with excavation further. Also may be called as pit underpinning.
- Temporary supporting by using jacking, shimming.
- Piles and beams: Small piles around footing (single or continuous) are installed, then connect the piles by beams on top, and finally support the wall on top of these beams.
- Various underpinning methods with piles which were listed under pile foundations section:
    - Helical (Screw) piers / piles
    - Pushed (Jacked) piers / piles
    - Micro (Mini) piers / piles

During foundation repair or additions, the modified or new foundation generally must not exert load on any existing foundations that may be nearby (remember Figures Soil Mechanics - 22 and 23). This usually means that the new foundation must be at least as deep as the existing ones, or if not, there should be some horizontal distance in between.

Differential Settlement of Foundations

When the foundation of a structure settles as a whole and fairly equally everywhere, this will not present a problem unless the settlement amount is "too much" for that project circumstances.

But when different parts of the structure settle in different amounts, even if these settlements are smaller than the ones in previous sentence, this can cause issues in the structure. A small level of differential settlement is tolerable but over that, we need to take precautions.

As mentioned above, differential settlement is a major reason for foundation repair and strengthening. Some indicators of too much or differential settlement of foundations of a building include:

- Cracks, warping, sagging on floor slabs, uneven stairways
- Gaps between exterior stairs and the structure,
- Doors or windows tilting, sticking due to the shape distortion of their opening,
- Diagonal cracks or deformation of walls such as bulging, damage to bricks veneer facade,
- Wall displacement, gaps between the walls and the interior or exterior slab on grade,
- Excessive buildup of stresses in structural members which can show cracks, yield signs
- For tall buildings, even if the base slightly rotated due to differential settlement, the effect along the height of building can accumulate which can even cause P-Delta effects *(see Structural Engineering>Various Terms)*
- Tilting of chimneys from vertical

These may be visible soon or even many years after construction completion. The more ductile the structure, the more its elements are able to accommodate rotation, the less severe will be the stresses caused by differential settlement.

If we can identify differential settlement potential during the design phase, this is of course the best and we can just take care of it by adjusting our foundation design, structural model or proposing ground improvement. For example when the soil properties under a large footprint area is different, we can either modify the foundation design and or propose ground improvement as necessary before the structure is built. Or if we have a structure that will impose considerably different weights to part of footprint, such as one side of the structure being taller than the other, we may consider piles under the taller section while the shorter side may still be okay with only the mat, or another option may be to perform construction in phases, such that the taller side will be built and settle (greater amount) first, and then the shorter side can be built, instead of building both sides simultaneously.

If we fail to expect differential settlement potential during the design phase but can only identify it after construction by seeing one or more of the signs listed above, this is of course far more costlier to solve. In such cases, we need to perform foundation strengthening and or ground improvement below the existing structure which may also present difficult access conditions.

RETAINING STRUCTURES AND STABILIZED SLOPES

Under soil mechanics section, we already covered topics that would serve as background to this section. Now we will discuss their construction applications. Elevation differences between two surfaces can be from natural terrain or excavations or fills we make. They can be overcome by retaining structures or slopes. If excavations are above a certain height, they will be classified as deep excavation.

Retaining Structures

To support elevation difference between two adjacent surfaces, retaining structures, commonly called as retaining walls are built. To chose type of retaining system, we consider factors such as height and length of elevation to be retained, available construction area and access, nearby structures that impose surcharge load, groundwater conditions, nearby underground structures or utilities, and other applicable general items as listed in the beginning of the book. In all elevation retaining projects, if there is available space, a sloped excavation should be considered as an alternative. *Note: The two "adjacent surfaces" mentioned above may also be at the coast. Coastal protection structures are discussed in Chapter 9. Designing and building retaining structures at the coast while dealing with sea water and dynamic waves is far more challenging than building on land. This section is for the ones on land but most things mentioned here also apply to coastal protection structures.* Common to all retaining structures, water must be avoided behind the wall within the retained soil. This can come from rain, leaks from nearby utilities or groundwater fluctuations especially on sloped sites. Water will reduce the strength of retained soil and add to unit weight which means more load. If the backfill is clay soil, it can cause consolidation settlement. In colder climates it can also cause freeze thaw problems with silty soils.

Gravity Retaining Walls:

The most straightforward retaining wall to design and build is a gravity retaining wall. As the name implies, it works simply by its weight. Gravity walls do not need to be made of mass concrete only. They can also be from masonry blocks or stone. Under soil mechanics section, when we studied lateral earth pressure, we showed a sketch of a concrete gravity wall and the forces acting on it. As also seen there, overturning and sliding resistance must be checked, plus, the soil must have enough bearing strength to carry the wall without excessive settlement. Also, surcharge loads, vibrations and earthquake forces must be considered. After these are analyzed for overall failure, we also design the wall body itself to withstand the internal forces caused by all these, such as its rebar details for a RC wall.

When designing gravity walls, cantilever walls, counterfort walls, we generally ignore the passive earth pressure in front of the wall, which is a conservative assumption, because as mentioned, it takes a lot of deformation for passive resistance to be fully mobilized, and we neither want to rely on, nor accept it.

To prevent water accumulation behind the wall, drainage holes are provided, which are also called "weep holes", in addition to drainage pipe at the bottom on the retained side. Precautions must be taken to prevent clogging of these. Accumulation of water is not only bad because it increases retained weight, but it also saturates sand fill, which increases liquefaction risk during earthquake.

Granular and permeable fill materials should be used behind the wall, even if it may be costlier and brought from far. We also prefer to make the ground surface on the higher side right behind the wall impervious, but sometimes it is not possible, if the higher side needs to have plants such as in a park.

Cantilever Retaining Walls:

A more efficient section (as far as the material used and the height of soil that can be retained) than a gravity wall is a cantilevered wall, because here the retained soil's own weight is utilized to resist overturning and sliding of the wall. The disadvantage in comparison to gravity wall is that it requires more labor to build these walls and the back of the wall must be excavated completely first, for forming and pouring the heel, and then it must be backfilled and compacted all the way up.

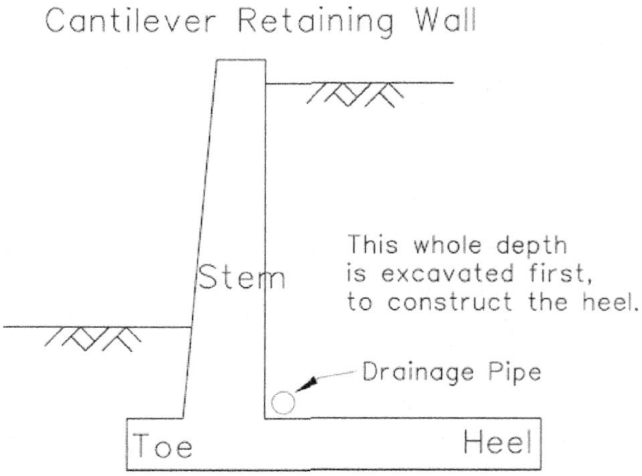

Figure - Retaining Elevations 01

Counterfort / Buttressed Walls:

Counterfort wall gets considerable support from its buttresses.

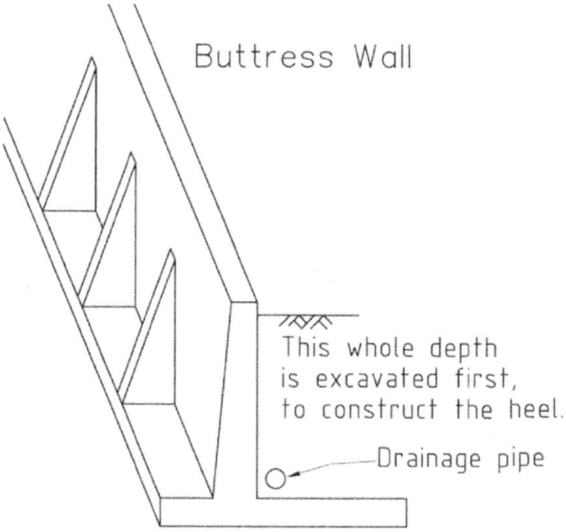

Figure - Retaining Elevations 02

Tangent / Secant Piles:

Piles can also be used as retaining structures. These are cast in place piles poured adjacent to each other, to make the walls as in the photo below. After the piles are installed, the excavation proceeds. Sequencing retaining wall vs. excavation this way opens up a lot of possibilities in construction, in addition to the previous retaining systems we discussed above.

A deep excavation is supported by tangent piles. Header beam and the beams below at every given interval make sure that the piles behave as a supporting wall unit. To further strengthen the support, header beams are anchored with some slope into the ground, with the anchoring machine as seen on the left. This means we are supporting the excavation by both vertically socketing piles into the ground and also through the lateral resistance provided by the piles.

The piles are installed by skipping a pile and then installing the skipped piles to reduce errors. As can be seen in the photo, the top of piles are connected together with a concrete beam called header beam, and then at every certain depth interval as the excavation proceeds, additional beams are made which are also held by anchors.

Diaphragm Walls:

A diaphragm wall is a thick reinforced concrete wall built underground to support deep excavations, prevent passage of water and usually to serve as permanent structural foundation walls later, which can support a high loads and help us achieve cost savings especially for foundation and even lateral force resisting system (LFRS) of the structure. They are made before the excavation takes place, to serve as a retaining structure for the upcoming excavation. They are made in sections (panels), which are usually at least several meters wide, by first leaving gap in between the panels and then closing the gaps. Near higher surcharge load locations from existing structures nearby, panel widths may be reduced.

Diaphragm walls can support very deep excavations and may get over 100 meters in depth and well over a meter in thickness. They are more efficient than the piled walls discussed above in terms of retaining elevations, because a piled wall has piles which are circular in cross section, which is not the ideal shape to resist lateral load. Diaphragm walls also have fewer joints (only where panels join each other) than piled walls, which is a positive for water resistance and structurally. Their verticality is ensured by first building temporary guide walls (also called guide beams) which are 1-2 meters in height and made as precisely vertical as possible on each side of the future diaphragm wall, and also inserting end pipes all the way down at both ends of the panel and then excavating in between these guide walls and end pipes / stop ends. There are special excavators used for this operation, which lowers its buckets with ropes into excavation in soft soils, and uses a hydraulic system in hard soils. As the excavation proceeds, slurry is temporarily filled into the dug wall volume, to prevent excavation from collapse by caving in. After the required depth is reached, the reinforcement mesh is sunk into the excavation, and then concrete is poured through tremie pipes that can reach all the way to the bottom. As the concrete fills the excavation, temporary slurry, which is lighter than concrete comes out *(remember we described a similar concreting process when discussing cast in place piles)*. Diaphragm walls can be braced form inside and/or also anchored into the soil. When the future structure, such as the basement level slabs of the building are erected, now the slab diaphragms brace the diaphragm wall from inside. In weak surrounding soil and/or tight site conditions, either alternative may be impractical, in which case, top-down construction may be an alternative at the basement levels.

Slurry Walls:

This is a temporary method used for supporting excavations during construction. To build slurry walls, first trenches are dug and then filled with slurry. Then the excavation can proceed in between the made walls or on its side. They are also used as a barrier to groundwater flow to control seepage. So they are in a way similar to diaphragm walls, but unreinforced, much shallower, lighter and they do not serve as a permanent structural wall.

Sheet Pile Walls:

This is a system of interlocking steel wall panel pieces, which are either driven (impact hammer) or pushed (vibratory hammer - more common) into the ground. They are usually made of steel sheets but precast concrete panels are also used. They can also be referred to as continuous walls. After they are driven one side can be excavated and the sheets can be supported by anchors as necessary. They can also carry vertical loads through end bearing at their tip and skin friction.

Anchoring Systems (Anchored Walls / Tiebacks):

Retaining walls can be made even more efficiently, if we also mechanically attach the wall itself to the stable soil behind the wall. Stable soil is the wedge as shown below, to which anchors must extend. By using anchors, we can obtain far thinner and flexible wall sections than the walls mentioned above, and can retain higher elevation differences for a given wall stiffness. Anchors can also be used as support for thick diaphragm concrete walls, or tangent / secant walls as well, all of which we mentioned above.

Anchor walls are relatively newer system, although it has been widely used since at least 40 years. Anchors are also called tiebacks or ground anchors. A common application is that they are grouted and prestressed, but other designs are also possible. The design must take the wall's flexibility into account especially for thinner, sheet pile walls. For example at the locations where the wall bends, moments can be reduced, or at the locations where the wall moves with active pressure, stresses will be reduced. There is also pulling forces from tiebacks. As a result of all these, there are force redistributions in the system. Taking all these into account requires more complicated analysis than the wall types mentioned above. As always for retaining structure design, the choice of the lateral earth pressure coefficient is important. After loads and moment distribution is obtained, the wall itself (such as its thickness, material and grade) is designed. The sloped line in Figure 3 below is the limit of wedge. Anchors must penetrate behind it, to be useful. When supporting deep excavations, in addition to anchors, internal corner braces and rakers can also be used, to brace walls from inside. During anchor installation, there can be slight inward or outward movements of wall and anchor pressures are monitored.

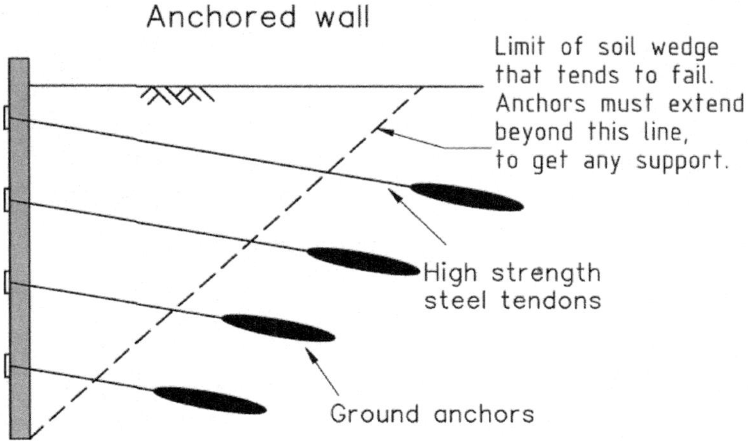

Figure - Retaining Elevations 03

The advantages of anchored walls with respect to gravity walls include:

- Relatively smaller access space is required during construction. This makes a difference especially in urban areas and can prevent issues with adjacent properties.
- Much higher lateral earth pressure can be retained with the help of anchors. If we wanted to retain the same elevation difference with retaining walls, they would require very heavy cross sections and pile foundations etc... which would make them extremely costly.
- In urban areas, there are nearby structures almost all the time, which means we will have limited space for construction and high surcharge loads on the retained elevation caused by these structures. This makes anchored wall systems often the only practical choice.
- Anchored walls usually take less time to build.
- For gravity walls, pile foundations may be needed. Anchored walls do not need such heavy foundation support as it is already supported by the anchors' tensile force from behind.
- We do not need to excavate, backfill and compact the soil behind anchored walls. Not doing this can save considerable time and money, eliminates the risk of damaging the wall during backfilling and compaction which involves construction equipment and vibration.
- Anchored walls eliminates the need for temporary support of excavation, because excavation is supported as soon as anchor are installed and the excavation work can proceed downwards.

- As they are thinner, anchored walls can be more easily fitted near existing structures.

An anchored wall that supports a hill slope near a highway. After its completion, its surface is covered with shotcrete.

Anchored walls also have some challenges in comparison to regular retaining walls:

- The design and construction of anchored walls is not as straightforward as traditional walls. During construction close coordination is needed between the excavation and the wall and installation of anchors, special equipment is needed, the speed of construction is important and the deflections must constantly be monitored.
- Soil properties must be known more accurately in anchored wall design than the traditional wall design. There is extensive past experience and research to determine soil properties, but in the end, soil is a natural material and the soil test results can be influenced by many factors, as discussed. Even so, in addition to theoretical knowledge, there is a huge amount of previous construction experience about anchored walls now that the engineers can make anything work and it is a matter of optimizing the costs and schedule only.
- The existing underground work such as existing utilities, foundations, and other features in the soil to be retained must be known precisely, to avoid conflicts during drilling of anchors. There can also be administrative problems as the anchors may extend to adjacent property.

As can be seen from above, the advantages of anchored walls outweigh their disadvantages in comparison to traditional walls, therefore, they are widely used especially where we have high elevations to support, limited work space and adjacent structures nearby.

In engineering, it is always good practice to visualize and understand all possible ways a structure may fail (failure modes). The list below will also remind many concepts that we discussed earlier in this book. Anchor walls may fail from a variety of reasons such as:

- Wall material failing from excessive bending
- Bond failure between tendon and grout, which causes pullout failure
- Bond failure between grout and soil, which causes pullout failure

- Failure of the lower front side of the wall (toe), as a result of insufficient passive earth pressure
- Localized compression failures on the wall, if anchor bearing plates press excessively on wall
- Rotation (swing) or bending of wall before installation of anchors
- The failure of tendon material itself from tension
- Downwards sinking of wall itself into the soil
- For slopes, overall rotation failure, which includes the wall and anchors altogether as a result of anchors not reaching far enough beyond the wedge, or bond failures as mentioned above.

Design steps for an anchored wall:

- As always, the first thing to do is to look at the overall structure and determine the external loads and the reaction forces, draw free body diagram and solve the reactions. The loads acting on the wall and anchor system include:
    - Lateral pressure from the retained soil
    - Lateral pressure caused by surcharge loads (i.e. existing structure, street and vehicles etc...)
    - Self weight of the wall
    - Vibration loads
    - Effects of water weight and seepage
    - Seismic effects

  Which are all resisted by the reaction forces provided by the anchors and the wall. These loads are determined and inserted into applicable load combinations.

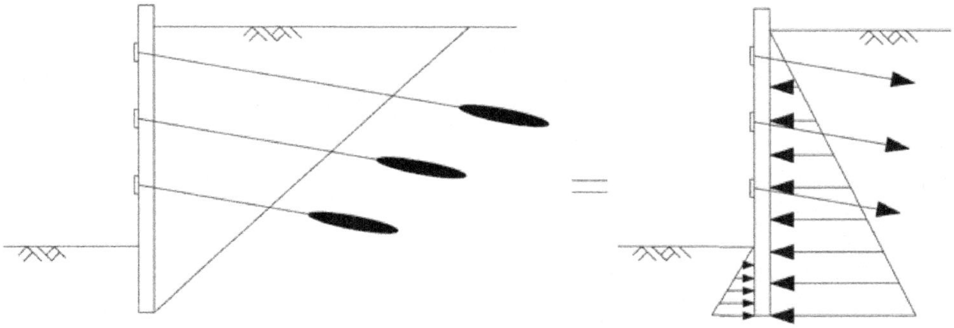

Figure - Retaining Elevations 04

- As we said, the anchors must penetrate all the way behind the active soil wedge. Therefore we must determine the angle of this wedge, which depends on the friction angle of the soil.
- We find the forces and moment diagram on the wall and the tension force for anchors. For the wall we look at the moment diagram to see at which sections we might have critical stresses. The deflected shape of the wall (not shown in the figure above) is a complex line based on the redistributed loads and moments due to bending and the force of anchors.
- We can now assign preliminary values and repeat the steps, until all results are satisfactory.

Ground anchors also have other uses than just for retaining soil. They can be used to stabilize slopes or to counter uplift forces from water in certain situations, such as slabs below water, dams.

Mechanically Stabilized Earth (MSE) Walls (Reinforced Earth Walls):

In this type of system, we directly reinforce the soil itself. These reinforcements are then connected to vertical face panels, which are connected to each other to make up the whole retaining wall. Soil is very weak in tension and the reinforcing strips take the tension in the soil through friction between the strips and the soil. So in this system, the wall face has to resist only little force, because the reinforced soil is now able to retain itself. In other words, the structural value of front panels are not much.

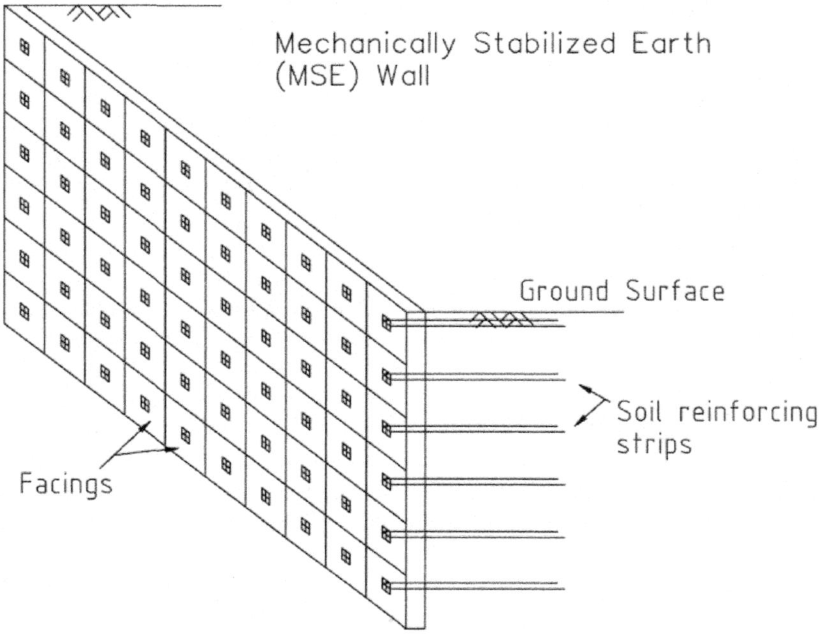

Figure - Retaining Elevations 05

The main disadvantage in comparison to anchor walls is that we need to excavate the soil to be retained first, for placing the strips, and then backfill and compact, which makes this method unsuitable in areas with limited available space. Still, MSE walls have various advantages including:

- As mentioned above, the soil supports itself here, which is the biggest advantage of this system. Because of this, several important benefits can be realized:
  - The wall face resists very little force.
  - Larger heights can be supported. To support the same height, the previous systems we mentioned would need to support enormous amount of lateral earth pressures.
  - Larger settlements or differential settlements can be tolerated.
  - Earthquakes are much better tolerated, and deformations can be very low, because here nothing is trying to restrain something else in the first place. The soil mass itself is reinforced and stands by itself (remember that the front panels resist negligible load).
- Where the retained earth is sloped, the reinforcing strips can also support bending that results from shear stresses along the failure plane (*see slope stability*), in addition to tension.
- Reinforced earth walls are easier to design and build, in comparison to the previous systems discussed. Using specialized construction equipment and crew is not needed as in anchor walls. Site preparation is straightforward and the construction can progress rapidly.

For reinforcing material, steel or geosynthetic materials are commonly used as grids or as strips. The facing panels can be of many different materials, such as welded wire mesh, shotcrete, precast concrete, wood panels and lagging, geosynthetic or metal sheets, metal plates...

If water is present behind the wall, in addition to the problems we mentioned before, such as reducing the strength of soil and increasing load on the wall and possible settlement of soil behind the wall, water will also reduce the friction between the soil and the reinforcement strips.

Bracing:

In many situations we can also brace retaining structures from the inside. Braced excavation as below is an example of at-rest condition of soil lateral pressure, as was discussed under soil mechanics. Bracing can be as horizontal, as seen in the photo below, for relatively narrow locations such as trenches, or they can be as corner braces or diagonal braces supported from the excavation floor.

Many braces were placed, due to lateral loads from closeby structures which impose surcharge loads and vehicles that cause vibration loads near the excavated area.

Gabions:

Gabions are made by filling stones and/or rocks into large boxes made of steel wire and work as gravity walls. They can be used for both earth retention or slope stability, as well as controlling erosion.

Gabion walls have the advantages of allowing easy and fast construction, being permeable and allowing water to drain out easily, tolerating considerable amount of soil movement, being stable due to their weight, having aesthetic looks with stones. To protect from corrosion, the wires can be coated with pvc or galvanized. Their exposed face may be stepped as seen in the photos or it can be flat on the outside and stepped on the inside. In any case, they are either leaned towards the slope by several degrees, and/or stepped, to ensure stability. A geotextile layer is placed between the gabions and the retained soil to prevent migration of fine particles into the cage, to maintain permeability of the gabion wall. Gabions with varying details can also be used at shores, where they protect against wave erosion, absorb wave energy or serve as foundation to breakwaters.

Crib Retaining Walls:

Interlocking pieces of timber or precast concrete can be used to make a gravity retaining wall which are called crib walls. Cohesionless soil is filled between the blocks to make it work as a gravity wall, at the same time draining water easily.

Soldier Piles & Lagging:

In this system, piles are placed into the ground at certain intervals and the gap in between is closed with horizontal beams called lagging.

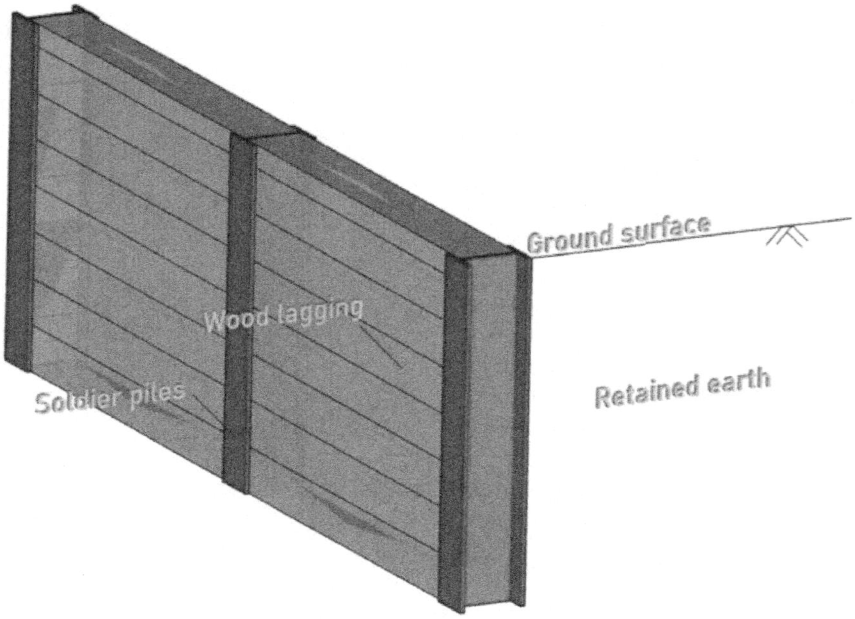

Figure - Retaining Elevations 06

The piles can be driven or placed in predrilled holes. The piles are usually H steel section and beams can be made of wood or precast concrete. See figure below. This system can also be used with ground anchors.

Interested readers may also want visit author's article about retaining wall design at: http://constructmagazine.com/geotechnical-engineering/retaining-structures/retaining-wall-design

Stabilized Slopes / Sloped Excavations

If we have adequate area around a proposed excavation, one option is to simply excavate the sides as sloped, instead of having to build retaining structures. Without any slope stabilization measure, the amount of steepest allowed slope depends on factors including the soil type and strength, maximum depth of excavation, whether a simple or benched slope will be made, groundwater table depth, the duration that the excavation will remain open. The allowed maximum slope values are provided in detail in OSHA standard 29 CFR 1926. We can of course make the slopes flatter if we want, which is obviously safer but costlier and requires more space.

We can also chose to stabilize slopes and obtain slopes which are not only steeper but more stable and stronger. We studied slope stability from soil mechanics standpoint before. Keeping that in mind, below we summarize slope stability measures. These slopes can be the existing ones, or, the ones obtained as a result of excavation or filling as embankments.

Slope Stabilization with Tiered Excavation:

If there is enough space available, the best and the most straightforward way for achieving slope stability is making a tiered excavation with low slopes.

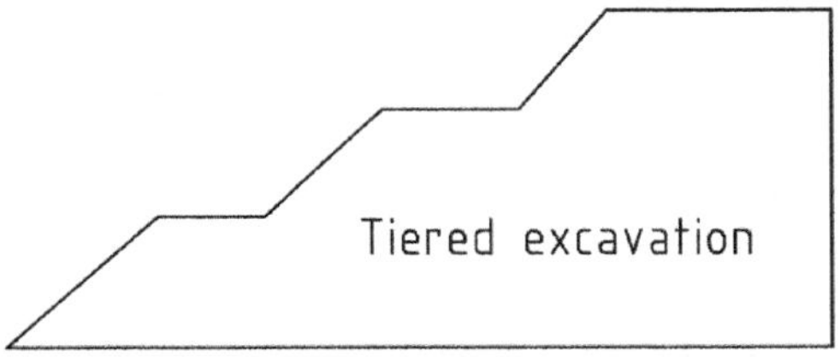

Figure - Retaining Elevations 07

For structures near slopes, the most straightforward option to protect them from slope failures is to provide setbacks on the side of the structure adjacent to slopes, or to provide low slopes. The slope can be on the downhill or uphill side or even both, with respect to structure's location. Due to land limitations however, providing setbacks or low degree of slopes is not always possible. Also see photo under Chapter 4 - Highways>Supplementary Components>Retaining Structures, Stabilized Slopes.

Keyed and Benched Slopes with Fill:

In this method, first we excavate the slope to provide keys and then backfill on top:

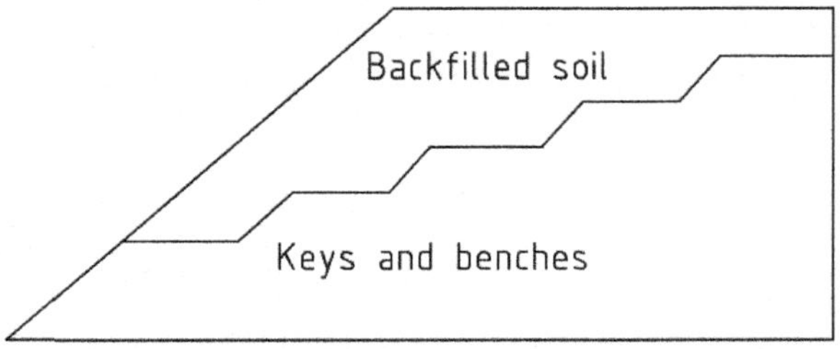

Figure - Retaining Elevations 08

Micropile Slope Stabilization:

Micropiles can also be used for stabilizing slopes, in a manner to provide stitches along the potential failure surface. The slope can also be an infinite slope, although only a finite slope is shown below.

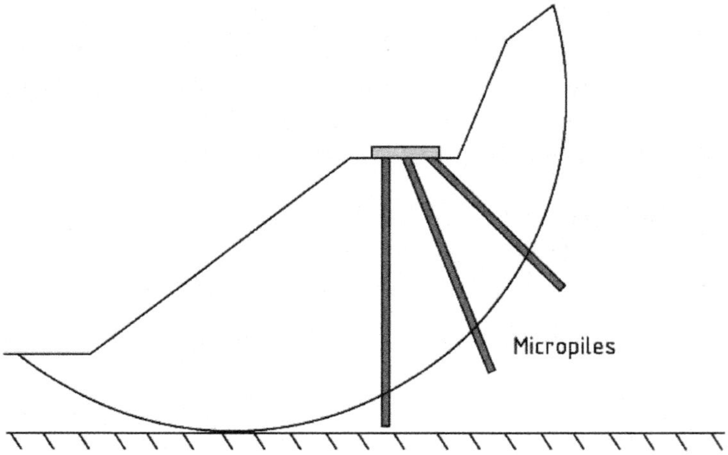

Figure - Retaining Elevations 09

Anchor Block Slope Stabilization:

Anchor bolts can also be used for slope stabilization. On the surface of slope, in order to provide the reaction force for the anchors, they are connected by beams.

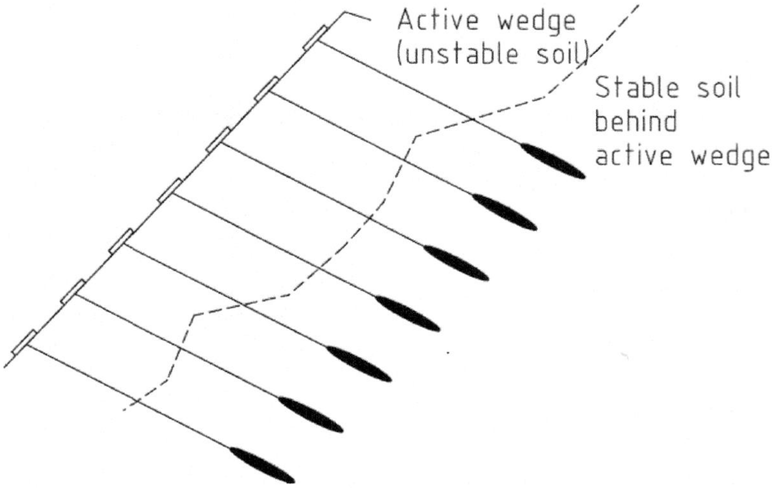

Figure - Retaining Elevations 10

Reinforced Soil Slopes:

This is the same thing as MSE walls we introduced above. The only difference is that this time the soil reinforcing are used for slope stability. The reinforcing stripes are laid such that they intercept the potential surface of slope failure. The stripes can be of steel or geosynthetics.

Geosynthetics:

Geotextiles, geogrids, geocomposites, which are all subgroups of geosynthetics, can also be used as reinforcing in highways and embankments. Geosynthetics, which are actually synthetic polymers, have wide variety of applications in geotechnical engineering, not just slopes. They can be used for reinforcing, drainage, filtering and separating a permeable coarse grained layer from fine grained layer (so that the fine particles will not migrate and block permeability of the coarse layer). See photo below, where geosynthetics is applied together with plants.

Covering of Slopes with Meshes:

To prevent rocks, stones, soil from falling, a mesh can be mechanically attached on a slope, especially if it is steep. The method and materials will differ based on the ground conditions and slope angle.

Slope Bioengineering:

Protecting slopes with natural methods by plants would be the most desirable for many reasons, although it is often not often possible to ensure adequate stability. This method is often applied in combination with other stabilization methods such as geosynthetics, as seen in the photo below.

Drainage of Water:

Drainage is also an important aspect of slope stability and often needs to be provided in addition to slope stability measures listed above. As shown in soil mechanics, presence of water affects the stability of slope negatively therefore we must design ways to remove it. Drainage is usually provided at the top and bottom (toe) parts of slopes at a minimum. Drainage can be achieved by either providing permeable drainage paths in the slope (passive drainage), or continuously pumping water away (active drainage). In both cases, the system needs to be maintained. Active systems are far less frequent due to the cost of continuously running and maintaining a pumping system, unless it is meant to be temporary.

STRUCTURAL ENGINEERING

In the beginning of the book, we covered the most basic structural concepts, which were also useful in soil mechanics and the following sections. Now we will return back to structural engineering discipline, and discuss many more essential concepts which builds on what we discussed earlier under structural basics - statics, and, strength of materials sections.

We will explore the structural engineering section under the following subtopics:

A good way to understand what structural engineering is all about may be to first look at the topics of classes learned during an undergraduate study. Structural engineering has the most number of classes in an undergraduate civil engineering study, which is typically 5 to 6 classes, in comparison to 1-3 classes for other four main branches of civil engineering. The list below includes the structural discipline classes that are typically taught as the core, probably in most places of the world as far as content, although the exact curriculum and titles of the courses may vary:

- Statics: This is the most fundamental structural and civil engineering class. This is why it made sense also to show this in the very beginning of the book, since we mentioned and will mention the terms and concepts from statics in many places afterwards. Statics discusses the most basic concepts about forces and their interaction with rigid bodies in order to calculate external reactions, force and moment diagrams. Material properties, stresses and deformations are not yet taken into account in this class, and all objects are assumed as rigid bodies.

- Strength of Materials: Taught after statics, this class takes into account the material properties and how the materials deform and fail when forces act on them, and introduces very essential concepts such as stress, strength, elasticity, plasticity, deformation, strain, stiffness... For example, although the book doesn't intend follow any course or exam topics, we had to cover the summary of this class right after statics, which was also not quite optional but the logical order, to be able to continue introducing further about civil engineering.

- Dynamics: This class studies objects that move and accelerate, so here things are not in static realm anymore but dynamic. It is the first class on the path to earthquake engineering, which constitutes a considerable part of structural design in most parts of the world. After this introductory class, further subjects in dynamics and earthquake engineering are typically studied at masters level. Although we did not give a summary of this, a few concepts from here will be mentioned or explained as applicable under earthquake section later.

- Materials Science: Not directly under structural engineering but this can be considered as complementary in nature to structural classes and deals with the properties of materials that are used in structures such as concrete, steel, masonry and wood. Materials science is also heavily studied by metallurgical engineers for metals, and it can be considered as a separate engineering/science branch by itself. We discuss three materials in this book, which are concrete, steel and masonry which contain many concepts that would be discussed in a materials science class from civil engineering angle.

- Structural Analysis: This is a class that can only be taken after statics and strength of materials are completed. It builds on top of those basic classes by starting to analyse simple structures. In this class, indeterminate structures which cannot be solved in statics are also solved, including solving not only the reactions but also finding their displacements. This is the core class where the essentials of structural analysis are learned. In this section we will give a short summary of this class, by describing indeterminate structure analysis methods, and solve a few examples.
- Reinforced Concrete Design: Having completed the previous classes listed above, in this class the principles of design of reinforced concrete structures is studied.
- Steel Design: Having completed the previous classes listed above, the students are introduced to principles of design of steel structures.

The classes mentioned above are the typical components of an undergraduate degree, but they can only serve as bare minimum, only a baseline and a starting point to an engineer, who wants to design real life structures. An engineer must also obtain considerable experience near other engineers before being able to design structures, plus knowledge in at least several of the topics listed in the following sentence, by either self study or studying within a masters program. The graduate (masters and beyond) level classes that can be taught under structural engineering can include but not limited to: Structural Stability, Bridge Design, Earthquake Engineering, Theory of Elasticity, Structural Dynamics, Advanced Steel Design, Finite Element Method, Theory of Plates, Advanced Concrete Design, Structural Optimization, Structural Reliability, Prestressed Concrete, Concrete Admixtures and more...Now let's start to explore the subtopics we listed in the previous page...

Structural Analysis

In the beginning of the book, we covered statics and strength of materials, which provide the background to study structural analysis methods that we will describe here. In statics we saw a determinate structure. Now we will look at "indeterminate" methods. The methods we will see here are the fundamentals of analyzing and solving structures, which were being used by engineers in the past, but they are still essential to learn. Today, almost all structural analysis are performed by computer software using advanced mathematical methods, once the engineers model the structure. Nonetheless, even the computer methods also ultimately depend on logic from fundamental methods. Now let's first see what are indeterminate structures, and then list the methods to solve them.

Indeterminate Structures:

In statics, we said that for a 2 dimensional analysis, there are three equilibrium equations such as:

$\sum Fx = 0 \quad \sum Fy = 0 \quad \sum Mz = 0$

These tell us that the sum of forces in all directions, and the sum of moments must be equal to zero, for a body to be in equilibrium. If we had a structure with supports that provided a total of three reaction forces countering the forces in x,y axes and moment around z, then we can solve it as we did in statics section. That was a "determinate" structure which we could easily solve, because we had three unknowns and three equations (again, in 2D).

But what if a structure had supports such that they provided more number of reactions than three?That is called an "indeterminate" structure. In that case we will have more unknowns than the number of equations and we cannot solve it with the simple method as in statics.

In Figure 1 below, we show a few random examples of simple indeterminate structures to several degrees at most. The structure itself is shown on the left, and its free body diagram on the right. For each structure, we count and add the reactions that the supports can provide, and then we subtract 3 from that sum, because we have 3 equilibrium equations. It means, now we found the number of extra reactions, beyond the three equilibrium equations. This extra number of reactions is called the "degree of redundancy" or the "degree of statical indeterminacy". *(Just as a note, there is also a kinematic indeterminacy that we will discuss)*. In the figure below:

Structure #1: We have 3+1 = 4 unknown reactions and 3 equilibrium equations. Therefore the structure is indeterminate to 4-3 = 1st degree.

Structure #2: We have 2+1+1 = 4 unknown reactions and 3 equilibrium equations. Therefore the structure is indeterminate to 4-3 = 1st degree. Note that the support at point A is a pin support too, although it may seem like a fixed support. We know it is a hinge because, first of all we show the circle that represents it, and more importantly, as discussed before, trusses only have pin supports. A fixed support can introduce bending moment to the members connecting to A, which is not proper for a truss, because truss members are only designed for axial forces, and they do not carry any bending moment. Also note that having a truss doesn't affect how we count the external reactions.

Structure #3: We have 1+3+2+1+2 = 9 unknown reactions and 3 equilibrium equations. Therefore the structure is indeterminate to 9-3 = 6th degree. Look at pin support at point E. Although we showed the support as inclined, we can still resolve its reactions to horizontal and vertical components for convenience.

Structure #4: We have the same structure as #3, with the only exception that there is an internal hinge to the right of point B now. Its location doesn't matter for our purpose. It is only shown because it provides an extra equation for us, because we know that at that location moment is zero. That is why we now have 4 equations here instead of 3, and it reduces the degree of determinacy to from 6 to 5.

The directions of reaction forces and moments in the figure below could also be drawn in the opposite direction as well. Upon solving, the true directions would emerge once we get positive or negative number for each - if we get positive, it means the direction we drew (or assumed positive) was correct.

So, to solve the reactions of these indeterminate structures, we need more equations than the three equilibrium conditions can provide to us. The additional equations we need, are obtained from the displacement versus force relations of the members in the structure. We learned about displacements in strength of materials section. This is why we said that this section is founded on the knowledge from both statics and strength of materials sections and takes them further by introducing new knowledge.

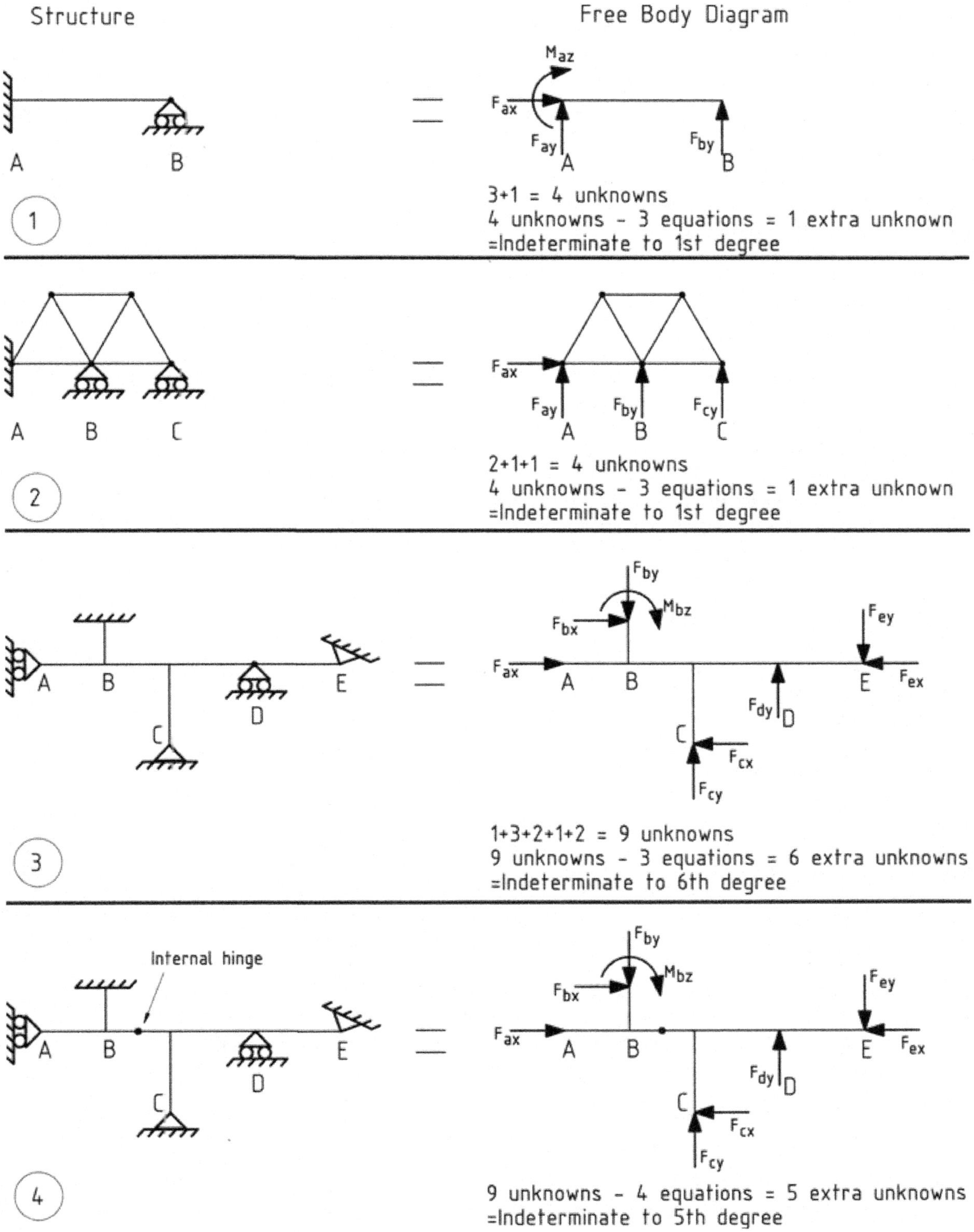

Figure - Structural 1

Indeterminate Structural Analysis Methods:

Note: Below, basic structural analysis methods are summarized. For engineers or students, it may serve as summary and or refresher. For nonengineers: For any part that may seem complicated, you can proceed without spending too much time, but still try to read attentively at least because these are the core of structural analysis, explained as plainly as the author could, and, while reading them, even if you do not fully study all details, you will still be seeing information about various fundamental behaviors of structures. Besides, if you understood Statics and Strength of Materials sections earlier in the book, those are all the foundation you need, in order to be able to follow here.

Reminder: As we mentioned in strength of materials section, when we say "displacement" in this book, we mean one or more of the deformation, deflection, drift, bending, swinging or twisting which may or may not happen at the same time.

The fundamental methods to analyse indeterminate structures can be listed and grouped as:

- Force (flexibility) methods: Forces are unknowns
  - Force method
  - Virtual work method
- Displacement (stiffness) methods: Displacements are unknowns
  - Slope-deflection method
  - Moment distribution method
  - Matrix displacement method

During learning these methods in a structural analysis class at undergraduate level, they are first studied under the following assumptions:

Assumptions for Basic Structural Analysis:

- All materials are linear elastic.
- The structural analysis is performed considering only the original shape of the structure, which is called the first order analysis. So, the new conditions that would emerge as a result of displacements, yield of material, buckling, cracking etc... are ignored, which would fall into the realm of second order analysis which is significantly more complex.
- For frames, axial deformations are negligibly small in comparison to bending, so that they are neglected and taken as zero, which means, all rotations and deflections are assumed to stem from bending. Again, this assumption is only for frames *(structures where members can transfer moments to each other, such as columns or beams)*, not trusses *(where there are no moments but only axial forces in members, and members are connected by hinges only)*.
- If more than one frame members meet at a rigid connection, they rotate or deflect the same amount, if there is a rotation or deflection there. So, the displacement of members are compatible with each other *(in other words no breaking, or deformation of connections)*
- Superposition principle applies, which means, we can separate a complex system into simpler systems and the sum of the actions (*external forces, internal forces and displacements*) in the simpler systems will add up to the actions in the complex system.

- All hinges (pins) are entirely frictionless, which means they transfer zero moment.

Most of the assumptions above are possible when the displacements are small enough. Solving structures under the above assumptions still gives us close enough results in many real situations. The analyses that do not necessarily include one or more of these assumptions are the subjects of a more advanced study, and will not be discuss here.

Solving indeterminate structures are more complicated than solving determined ones because we now have more unknowns than the three equilibrium equations as we mentioned above. So the bottom line is that the indeterminate methods are ultimately all about obtaining those extra equations needed, or doing something that would take place of them, and then solving the unknown reactions which would then provide us with reactions, force and moment diagrams and the final displacements everywhere.

To obtain the extra equations, in addition to the equilibrium conditions as in statics, indeterminate methods use compatibility conditions, which means geometrical conditions based on displacements.

More precisely, compatibility means two things:

- Displacements where adjacent members connect, need to be compatible with each other, since the members must keep their original undeformed connections, and,
- Displacements of the structure at the external supports need to obey, in other words, be compatible, to what those supports will allow.

These conditions provide us with extra knowledge, to solve the unknown reactions.

To apply these displacement conditions, we need to know the force-displacement relationships of members, as we studied in strength of materials section.

Let's remember them:

Trusses: Here the deformations are axial, and we saw the force deformation relationship as:

$$\delta = \frac{F.L}{A.E}$$

where, δ is axial deformation, P is the axial force, A is the area, L is the original length of the member and E is the modulus of elasticity of the member.

Frames: Frames are structures that are made of beams and columns that resist moments. Here we assume axial deformations as zero, and all displacements are due to bending. In strength of materials section, we expressed bending differential equation as:

$y'' = d\theta/dx = M/EI$

where, y'' or dθ/dx is the rate of change of bending rotation, M is the bending moment, E is the modulus of elasticity and I is the moment of inertia of the member.

Now, keeping in mind the introduction above at all times, let's describe the different indeterminate structural analysis methods below...

Force Method / Flexibility Method / Superposition Method / Method of Consistent Deformations:

The most straightforward method to learn when solving indeterminate structures is probably the force method because in this method we directly seek to solve the redundant reactions themselves. Both trusses and frames can be solved with force method.

The extra equations that we need beyond the three equilibrium equations, are obtained by looking at the displacement conditions of the structure, which are called compatibility conditions as was mentioned above, which are known to us. By using these, we can write equations that relate these known displacement conditions to unknown forces, which are called compatibility equations. We can then solve the unknown forces by simply solving all equations simultaneously. This is the bottom line of what we do in force method. The compatibility conditions can be determined just by looking at the geometry and supports of the structure. Take a look at the examples in the figure below.

For Structure #1: We know that at point A and B, any rotation or translation movement will be zero, since these are fixed supports.

Structure #2: We know that at points A, B, C and D there cannot be any vertical displacement. We could also say that at point A rotation or horizontal movement are also zero. What conditions to use are up to us for convenience as long as we obtain sufficient number of compatibility equations equal to the degree of indeterminacy (the number of extra unknowns).

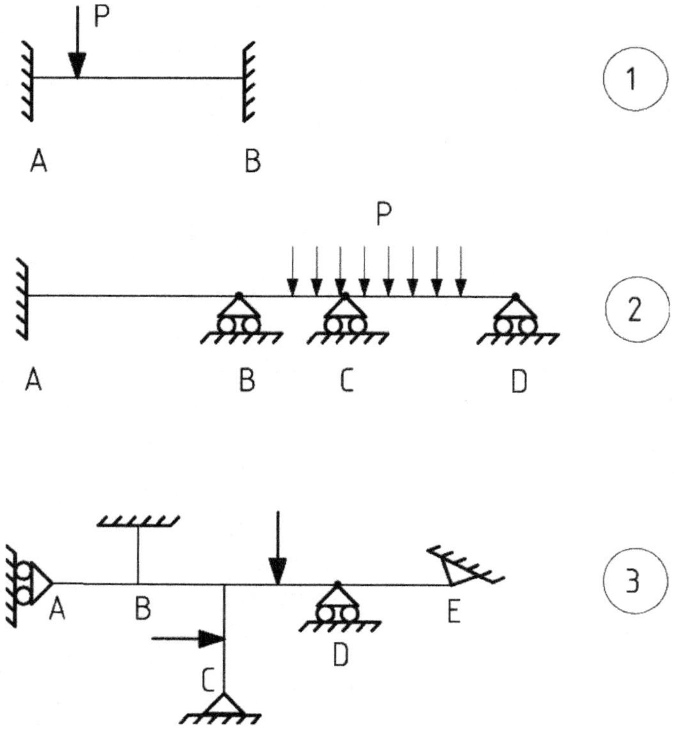

Figure - Structural 2

Structure #3: Similarly, depending on support conditions, we can write the compatibility equations for all support points. Which ones to chose is up to us, as long as we have adequate number of them. We can even make conditions at any other point than supports. For example we could even choose a point between A and B although it would be less convenient for us and very unnecessary. The part of the member to the right of that point must deflect the same amount as the part to the left of that point, because the member is not breaking.

Since we are trying to solve the forces here, but we know the displacement conditions from compatibilities, in force method, displacements are known, and the forces are unknown. Here each extra unknown reaction force requires one compatibility equation. Each additional equation increases the calculation effort exponentially. So this method is the most suitable when the degree of indeterminacy is low. For example a structure which is statically indeterminate to 5th degree requires far more effort to solve than a structure indeterminate to 2nd degree with force method.

Here are the steps to apply force method:

Step 1: Since we seek to directly solve the redundant reactions, we must first determine the degree of static indeterminacy in the first place, like we showed in Figure 1 a few pages back.

Step 2: We remove or reduce some of the supports to decrease extra number of reactions (degree of indeterminacy), to make the structure as if it is determinate. Which support reactions we reduce or remove are up to us, except the one that would make the structure unstable at a given direction (which happens if there is nothing else in that direction). Reduce means, for example if we had a fixed support, we can reduce it to a pin support, so that the moment resisting ability, therefore that reaction, is removed. To make up for the removed or reduced supports, we insert reaction forces or moments in their place. Now we draw the new "determinate" structure, with the original external loads and the newly introduced reactions in place of removed or reduced supports.

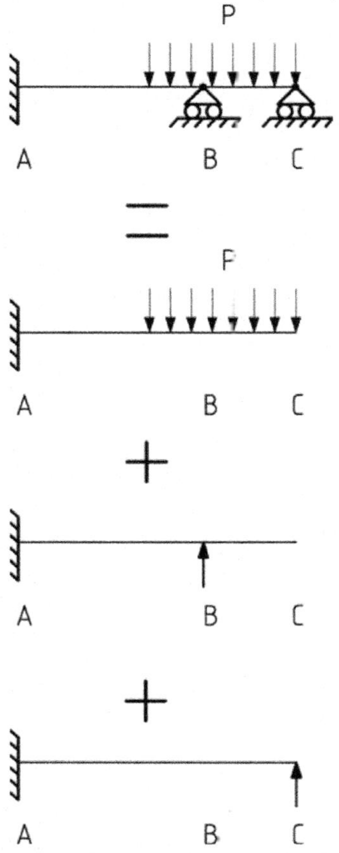

Note: When choosing the assumed reaction forces, we do not need to choose the correct direction. After our analysis, if we find a reaction as positive, it means that the direction we chose for it was correct. If negative, it means that the reaction acts in the opposite direction of what we assumed.

Figure - Structural 3

Step 3: Now we separate the structure we obtained in step 2 into substructures which all look like the same determinate structure. An example is shown in Figure 3, which is a structure indeterminate to 2nd degree, as you can see. The first determinate substructure will only have external loads from original problem, without any unknown reactions we introduced in step 2. Then for each unknown reaction force we introduced in step 2, we draw one substructure with that reaction load only, as shown in Figure 3. The sum of deflections at each substructure (also called sub assembly) at a given support point must be compatible with (*add up to*) the conditions imposed by the supports in the original structure, because remember, under assumptions, we said that, the principle of superposition applies. See Figure 3. Here the sum of vertical deflections at points B and C due to each individual force must add up to zero because the original structure did not allow vertical movements at those locations in reality.

Note: Similarly, as stated under compatibility assumptions, the displacements of members adjacent to each other must be compatible with each other so that they keep their undeformed connection. Here we did not use that condition but used only the conditions at B and C imposed by supports for convenience, which were enough to provide us the two extra equations we needed.

Step 4: For each substructure, we write reactions and deflections in terms of each other. For example as we saw in strength of materials section, for a frame or beam, we can use the M/EI graph to obtain slopes and deflections by moment area method, or for a truss or hinged frame system, we use the axial deformation to axial force relationship as δ=FL/AE. So now we have compatibility equations from each substructure, and the solved deflection from first substructure. Then, since superposition principle applies, the sum of all substructures must be equal to the condition in original structure.

The deflections can be calculated with moment area method, virtual work method, or for many typical situations, there exist tables that provide these deflections which are already calculated with these methods, which are available for our use.

Step 5: Now we solve all equations simultaneously to satisfy the displacement conditions of the original structure. The solution yields us the unknown reaction force(s). Solving all reactions means, we are done. We can now draw axial force, shear and moment diagrams anywhere.

Note that in step 4 we could also describe the process by saying that we assume unit value of 1 for unknown reactions, and find their deflections as a number. Let's call it δ. But then it will mean that that number, δ, will have to be multiplied by the reaction force itself, let's call it R, to yield the deflection from that reaction, let's call it Δ. So we will have Δ=δ.R, which would lead us to same result for reaction forces when we solve the compatibility equations. The meaning of δ here is that it is called the flexibility coefficient. Since it is the deflection which was produced as a result of 1 unit of load, its unit is displacement per unit force. This is why the force method is also called as the flexibility method.

In step 4, we could obtain the deflections also by virtual work method or simply by looking at tables prepared for all possible common loading conditions and then simply add them up using superposition.

In real life structures, there can be fabrication errors, temperature changes, settlements and so on. These can create forces in the structure. In force method, these can be accounted for, by including them (as given nonzero displacement amounts) in the compatibility equations.

The equations we generate from force method may also be solved with matrices. In that case, the forces and deformations are written in matrix form and related to each other through the flexibility matrices. *(The other method is the stiffness (displacement) method that we will discuss, which uses stiffness matrices, and that is the one used as a fundamental ingredient in computer analysis of structures.)*

Now, let's wrap up what we discussed under force method by solving an easy example, which is only indeterminate to the first degree.

Example - Force Method:

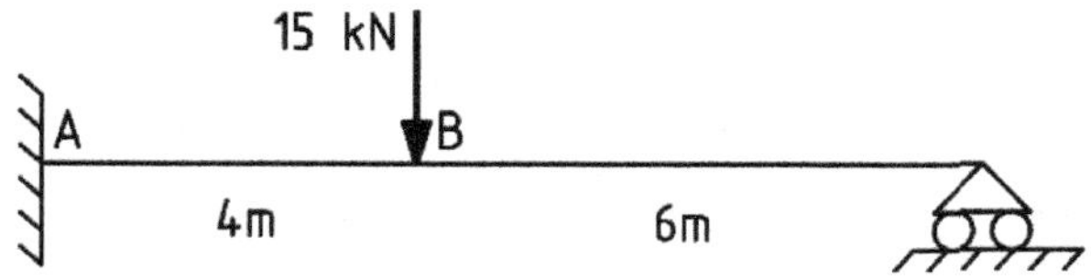

Figure - Structural 4a

The figure above is similar to the beam we looked at under statics section, with the only difference that we now have a fixed support, instead of the pinned support at point A. This makes our beam indeterminate now. Because now we have a total of 4 reaction forces, (3 from point A and 1 from point C), with respect to 3 equilibrium equations available to us. And for simplicity this time, we changed the loading from inclined to vertical (even if the load was inclined, *the horizontal force component would not affect other reactions*). Now let's go through the steps we listed above, to solve this problem.

Step 1: Our structure is indeterminate to the first degree, such as: 4 unknowns - 3 equations = 1

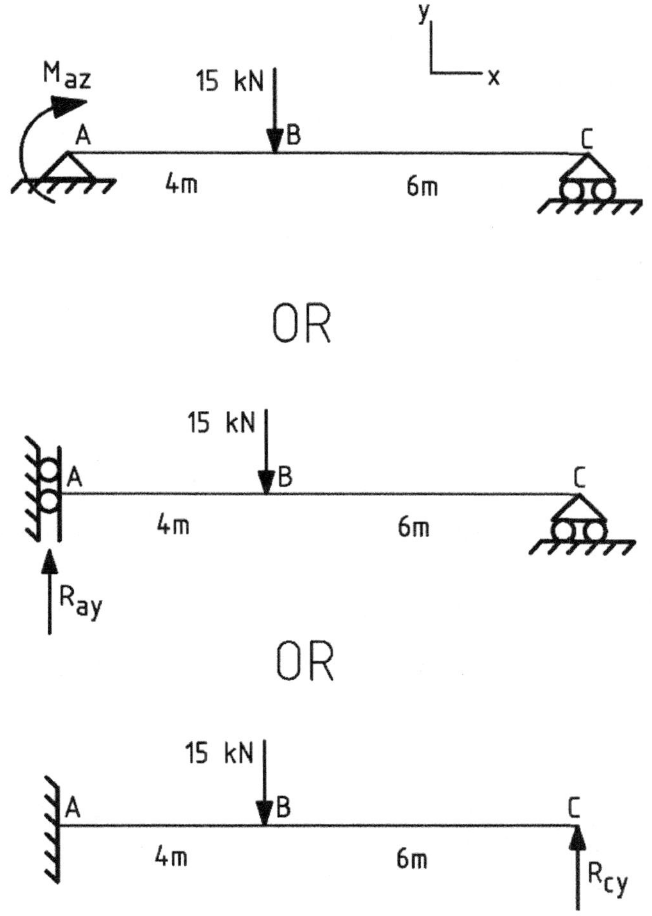

Figure - Structural 4b

Step 2: Since our structure is indeterminate to the first degree, we need to reduce or remove the support for a total of only 1 degree. We have different options for that, as shown in Figure 4b. Which one is up to us. For example as you can see above, we can reduce the three reactions provided by the fixed support at point A to 2, by either making it a pinned support (top figure), or by making it a fixed-roller support (middle figure) that is able to move up and down but fixed (not hinged) to the beam, or instead of these, we can totally remove roller support at point C (bottom figure) since it is not possible to further reduce it, as it could only provide 1 reaction. And for each case we provide the appropriate external load in place of the removed support. So, we may choose any of the alternatives above. Let's choose the last alternative at the bottom for simplicity, and remove the roller support at point C, so now we have a determinate sub-structure as below:

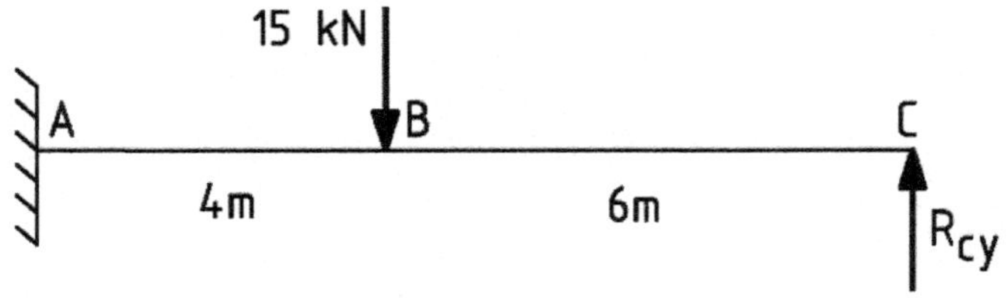

Figure - Structural 4c

Step 3: We separate this new (determinate) structure in Figure 4c into substructures as follows:

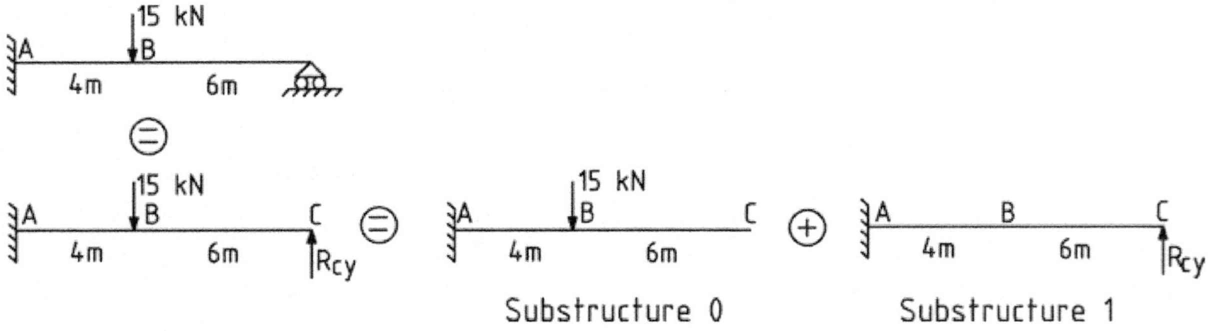

Figure - Structural 4d

This means, if we solve the two determinate structures (substructure 0 and substructure 1) above, and add their effects up (superposition principle), we obtain the solution for the original structure.

Step 4: See Figure 4e below which shows how we find deflections of substructures 0 and 1.

The two expressions below come from the moment area method that we discussed under strength of materials section, even if it seems complicated at first glance, actually nothing new here:

Δ_{CO}= (-60/EI.4.(1/2)) . (4.(2/3)+6) = -1039.92/EI (downwards)

Δ_{C1}= (10R/EI.10.(1/2)) . 10.(2/3) = 333.33.R/EI (upwards)

Here is how we calculated the vertical deflection for substructure 0, Δ_{CO} : According to the second moment area theorem, the vertical deflection at Point 2 with respect to Point 1 is equal to the moment of the area under M/EI diagram, with respect to the vertical axis that passes through point 2. So in the terms above, the first parenthesis is the area of the triangle (i.e. 4x15=60 is height of the triangle, 4 is length), and the second parenthesis is the length of moment arm of the (triangular) area *(do not confuse moment of an area with moment of a force)*. The Δ_{C1} is also calculated with similar logic.

So in both cases we calculated the vertical deflection of point C with respect to point A, by finding the area of M/EI diagram and multiplying that area by the moment arm of the area to point C.

Now although we calculated these deflections separately, in reality we know that the vertical deflection at point C is zero, because in reality there is the roller support there, which does not allow vertical movement. So we can write:

Δ_{CO}+ Δ_{C1}=0

This equation is our compatibility equation. This is a geometric condition that we know for sure. This is how we get the extra equation to solve our indeterminate structure. We needed only one of these here, because our structure was indeterminate to only first degree. If we had a more complex structure, we would have more degrees of indeterminacy and we would have needed more equations.

Let's write $\Delta_{CO}+\Delta_{C1}=0$ in terms of EI and R, and since the term (1/EI) is common to both:

(1/EI)(-1040+333R)=0

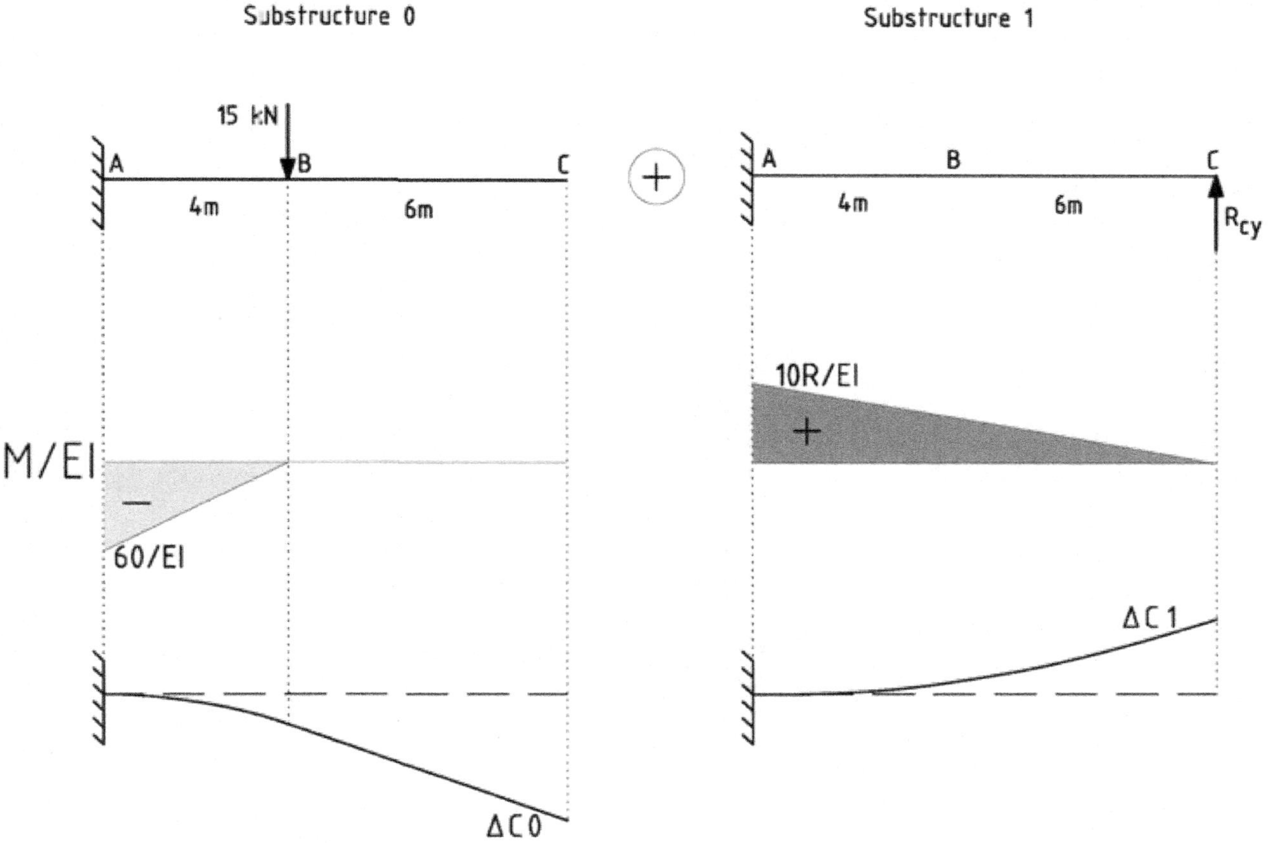

Figure - Structural 4e

As you see, the value of EI doesn't matter for finding the reactions. Inside of the second parenthesis must be zero to satisfy the equation, so we can solve this without knowing the value of EI (*we would need the value of EI only if we wanted to find the amount of deflection along the beam length*). So,

R=1040/333=3.12 kN (acts upwards)

Note: We could have found R also by assuming a unit force. We could have assumed R to be unity, as 1, and then find a known deflection per that unit force R=1. So now we would again have Δ_{CO}, which we know its value as -1039.92/EI, and instead of an unknown Δ_{C1}, which we had to write in terms of unknown R as 333.33.R/EI, we would now have a known δ_1 as we found it from a known unit force of 1, which would come to be 333.33/EI, since R is 1. So now to find the reaction R, we need to find how many times δ_1 must be increased, so that it becomes equal to Δ_{CO}, since we know that at point C we must *have zero deflection in reality. So we can simply find R as: R= Δ_{CO}/ δ_1 = 1040EI / 333EI = 3.12 kN acting upwards again. And δ_1 is called the flexibility coefficient as explained a couple pages back.*

We found the unknown reaction R, which means we solved the indeterminate problem. Now we can find the rest of the reactions simply with static equilibrium equations as we did under statics section.

For example, let's find the moment reaction M at point A. Total moment at A must be zero. So,

$M_A + 3.12*10 + 15*4 = 0$

$M_A = 28.8$ kN.m (acts counterclockwise)

Now let's find the vertical reaction at A. The sum of vertical forces on our structure must be zero. So,

$F_{Ay} - 15 + 3.12 = 0$

$F_{Ay} = 11.88$ (acts upwards)

Now that we know all reactions, we can draw the shear and moment diagrams:

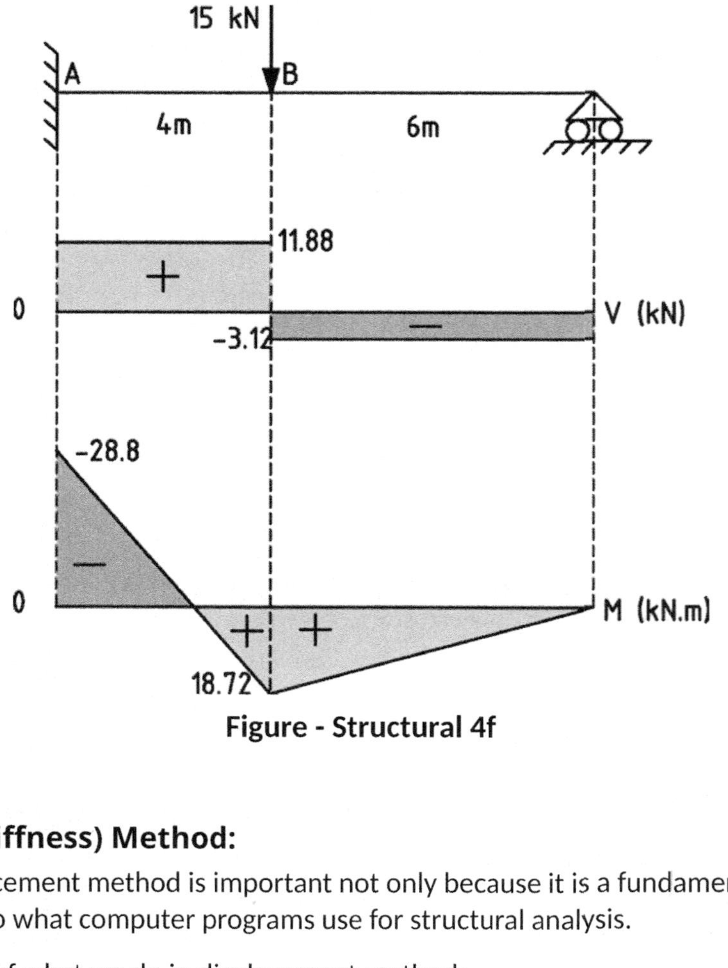

Figure - Structural 4f

Displacement (Stiffness) Method:

Understanding displacement method is important not only because it is a fundamental method but an essential ingredient to what computer programs use for structural analysis.

Here is the summary of what we do in displacement method:

1) We separate the structure by nodes and then find the degrees of freedom that the structure can move.
2) At these node points, we write force-displacement relations and equilibrium equations in terms of the degrees of freedom and forces.
3) We solve the equations, determine displacements and by using them, determine forces.

With displacement method we can also solve trusses, frames, plates, shells or anything else - but do not forget that the slope deflection and moment distribution methods (which we will discuss) that fall under displacement method only apply to frames.

Now let's expand what we wrote in the three steps above, by also inserting more steps:

Step 1 - Discretization: It means separating the structure into elements (members) by using nodal points, which are the points where the displacements are solved. Identifying the nodal points, in other words, discretization is very straightforward and can also be done visually for frames or trusses. For those, we can easily identify where a certain change occurs in the structure, such as a geometry or a material change of a member, member joints, supports, or pins in case of a truss. Nodes can be selected at any point actually.

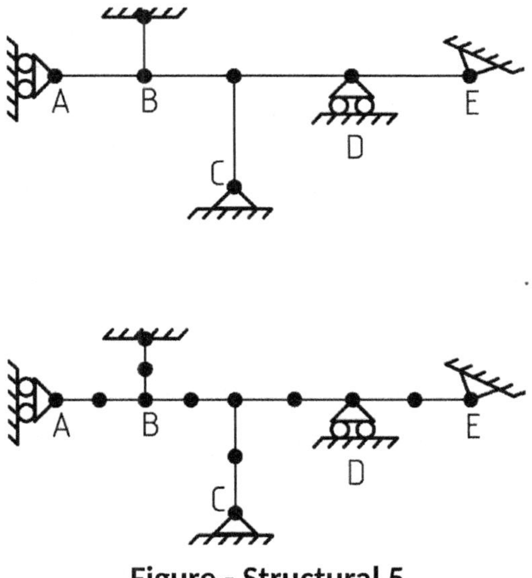

Figure - Structural 5

In the top part of Figure 5, we show nodes at obvious locations, such as joints or supports, which would be sufficient in practice, in order to solve this structure with displacement method. However, we may choose to be more precise, and assign the nodes more densely as seen in the bottom figure, although it would be trivial, and it would make our job exponentially more tedious (it is already a tedious task to solve the structure at the top by hand, which involves solving several simultaneous equations). Assigning nodes as in the bottom figure is trivial because with the top figure we can already represent the displacement response of the structure sufficiently for most purposes. Also see figure Figure - Bridges 9, in Chapter 3 as another example for discretization by nodes.

Step 2 - Kinematic indeterminacy: We identify and assign degrees of freedom, in other words, possible displacements at each node, which means identifying kinematic indeterminacy. These should be adequate to entirely show how the structure will move but at the same time be kept to a minimum to reduce calculation load. The sum of degrees of freedoms from joints is equal to kinematic indeterminacy of the structure, which is the number of unknowns. So now these are unknown displacements at nodes. In displacement method, displacements are unknown and forces are known (written in terms of displacements). Let's give a few examples about assigning degrees of freedom, in other words, determining kinematic indeterminacy:

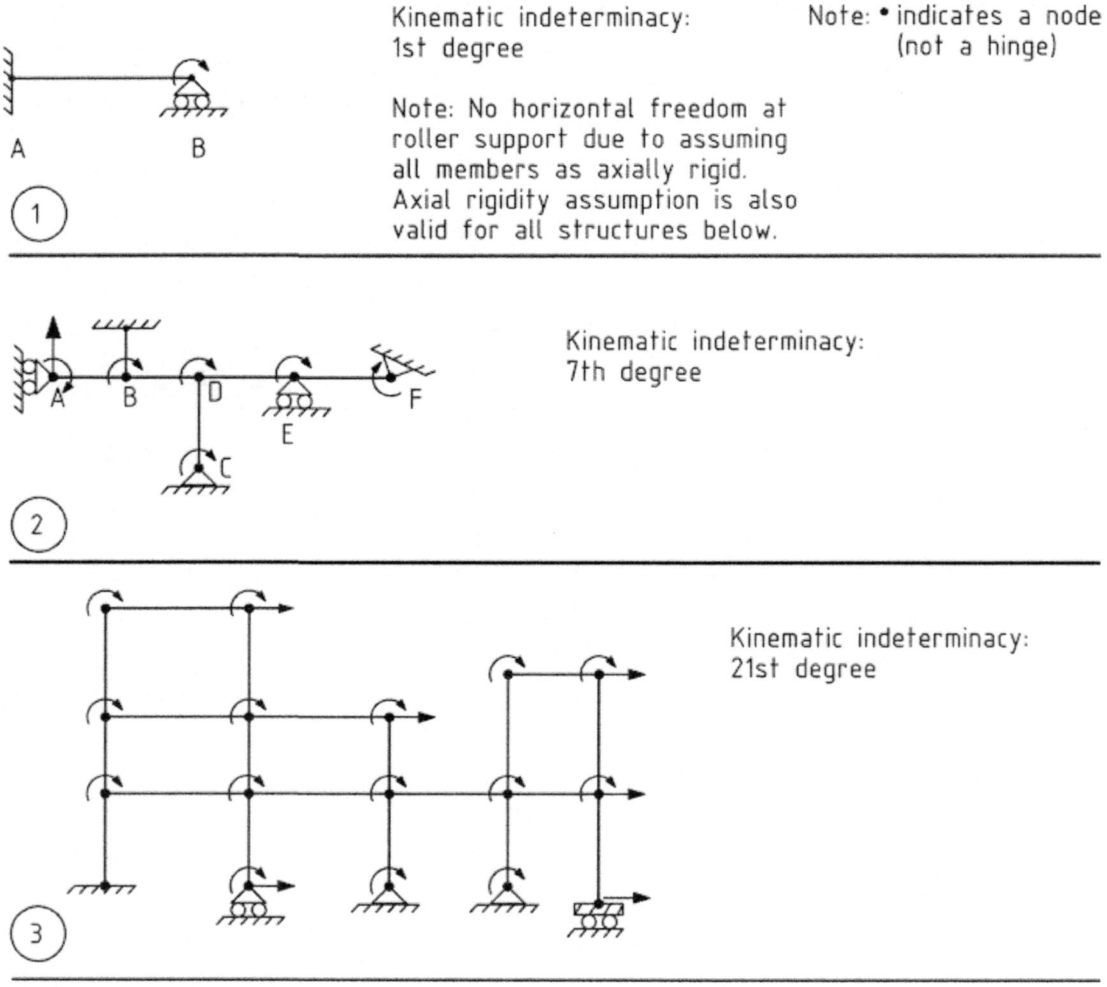

Figure - Structural 6

Again, the circles you see in the figure above are <u>not</u> hinges, but they are nodes assigned by us, to discretize the structure and then determine the degree of <u>kinematic</u> indeterminacy.

And let's explain the importance of axial rigidity assumption: As we listed under the assumptions of indeterminate structural analysis methods several pages back, the axial elongation or shortening of members are neglected, in other words, members are <u>assumed axially rigid</u>. This helps to reduce the number of unknown deflections (degrees of freedom), thus the calculation effort. So if you look at the three examples above, for structure 1, the fixed support allows no movement, so there is no degrees of freedom there. At the roller support on the right, moment cannot be resisted, so rotation is a degree of freedom, while translation is not, although the support itself would allow it, just because the members are assumed <u>axially rigid</u>. For structure 2, again, if you consider axial rigidity, you can understand why there is no horizontal translation at nodes B,D,E because the pin support at nodes F and A restrict horizontal movement, which reflects all the way to nodes B,D,E since the members are axially rigid. For structure 3, pay attention that we showed horizontal movement only at one node for each floor, again due to the fact that the other nodes at the same floor level move together due to axial rigidity of members and there is no extra degree of freedom for them.

And what would be the degrees of freedom if there was no axial rigidity of members? For structure #1, node B would be free to translate in horizontal direction in that case so the kinematic indeterminacy would be 2. For structure 2, similarly nodes B,D,E would be free to translate in horizontal direction and B and D would also be able to translate vertically, so the kinematic indeterminacy would rise to 12. For structure #3, with similar logic, the kinematic indeterminacy would then increase by 8 for extra horizontal translations and 12 for extra vertical translations for a total of 20 extra (under all possible loads, not necessarily only horizontal as in here). So now remember that we said even one more equation increases calculation load exponentially, now imagine adding 20 extra, if we didn't assume axial deformations as zero, which are already small...

Step 3 - Force - displacement relations: We write the force displacement relations. For frames, slope deflection relations (*which will be shown)* that relate moments to rotations and swinging are used which include bending stiffness (EI/L). For truss or pinned frame members, forces are related to axial deformations with axial stiffness (AE/L).

Step 4 - Equilibrium equations: The equilibrium equations are written in terms of force-displacement relations obtained in the previous step. There will be the same number of equilibrium equations as unknown displacements. From static equilibrium, at a frame joint where two or more members join, or at a hinge support, total moments must be zero. These are equilibrium conditions we can use. (Note: At a <u>fixed support</u>, moment is <u>not</u> zero, which is one of the reactions we are trying to find)

Step 5 - Now by using these equations we solve unknown displacements using simple algebra. When we have a lot of equations, matrices are better suited for solving.

Step 6 - Solved displacements are inserted back into force-displacement relations to find actual member forces (i.e. end moments for frames, and axial forces for trusses). This means we solved the problem, because, now we can just find the rest of the reactions easily using static equilibrium equations, by drawing free body diagrams of members and supports.

Step 7 - The shear and moment diagram of members can now be drawn.

Depending on the type of the problem, sometimes it is more convenient to choose force method and other times the displacement method. In displacement method, solutions of statically determinate or indeterminate structures are not different, and it can easily be applied in matrix form. If the degree of statical indeterminacy is low, force method will be easier. If the degree of kinematic indeterminacy is low, displacement method will be easier, even if the statical indeterminacy may be high.

Slope Deflection Method:

Slope deflection method falls under displacement (stiffness) methods, because we use unknown displacements of nodes to form equations, instead of unknown forces as was in the force method. Slope deflection method can analyse frames only, and trusses are out of scope. It was the first specific method of analyzing frames, which was started to be used in early 1900s and widely used until a more practical method called moment distribution method, which we will see below, appeared in 1930s (which was still based on slope deflection equations). Slope deflection equations are also an ingredient to matrix displacement methods, which are easily applied to the computer structural analysis programs (along with more advanced mathematical methods).

In slope deflection method it is assumed that all deformations in the structure happen because of bending of members, and, all axial and shear deformations are negligible compared to bending deformations. This is why trusses and pin connected frames can not be solved with this method, because they only involve axially loaded and deformed members which are not subject to bending, and their hinge connections cannot transfer moment.

In slope deflection method, we follow the same steps as described under displacement method above. The only thing specific to here is that the force displacement relations we mentioned in step 3 are the slope-deflection relations this time, and the rest of the steps follow using these relations.

So, the slope deflection equations are the key pieces of this method, because they allow us to write moments in terms of displacements as we stated in step 3 above. Slope deflection equations can be obtained by analyzing a beam segment in bending, through the use of moment area theorems as we saw before in strength of materials section. *Side note: This is done by superpositioning (adding) of the bending vs. deflection relations of the end moments and the span loading on to each other for a beam segment (as the one seen below) but we will not show here the steps to derive them, which can be found in many texts focused in basic structural analysis. We will directly display the equations.*

The terms in slope deflection equations and the positive sign convention is as shown below, which is:

- Counter clockwise moments and bending of beam are positive
- Counter clockwise chord rotation (swinging) is positive.

It is important to know that the moments in the figure below are external to the member. So do not confuse them with internal moments which are used to draw moment diagram. Per this sign convention below, when we find the external moment at A as in the figure above as positive, the internal moment at A will be negative, since you can see that the beam is bent concave down at A *(see sign convention for internal moments in statics section, if you do not remember it)*. But again per the sign convention below, when we find the external moment at B as positive, as you can see the beam is concave up, so internal moment will also be positive.

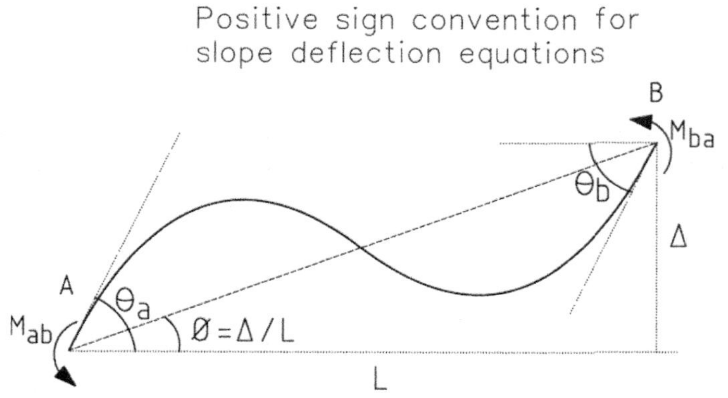

Figure - Structural 7

This point is important to remember when drawing moment diagram at the end of a problem, after finding all reactions because as was explained before in statics, in a moment diagram, we draw the values of internal moments in the member.

The slope deflection equations are:

$M_{ab} = 2EI/L\,(2\theta_a + \theta_b - 3Ø) + M_{ab(fixed)}$

$M_{ba} = 2EI/L\,(2\theta_b + \theta_a - 3Ø) + M_{ba(fixed)}$

$Ø = (\Delta_b - \Delta_a)/L = \Delta/L$

Now let's explain the equations. Consider the member AB as seen in the figure. It can have any arbitrary load. Remember, the principle of superposition applies. So we can think of this member as separate members, and assume that there will be all possible movements, which can be the bending at both ends a and b, and the swinging of the beam body. Point A is the near end, and Point B is the far end. Look at the first slope deflection equation. Here is what it is made of:

M_{ab} : the moment at node A of the member AB

EI : bending stiffness

L : length of member AB

θ_a : the rotation angle at point A

θ_b : the rotation angle at point B

Δ_b & Δ_a: These are the displacements of points B and A relative to a reference line, such as the horizontal line at the bottom. So here $\Delta_a = 0$ and $\Delta_b = \Delta$. Their difference is important, which is Δ. When we divide this by L, we obtain the amount of swinging of the beam, written as Ø (you may also see this as Ψ). This is the swinging amount, which is also called as "chord rotation". Do not confuse this with the bending rotation.

$M_{ab(fixed)}$: the fixed end moment at point A. This is not to be confused by the end moment, M_{ab}. We will explain the fixed end moment below.

Ø : the swinging angle, also called as chord rotation, if, there is any swinging (sway). In the formula to find M_{ab} or M_{ba}, note how the effect of Ø is subtracted from the effect of bending rotation angle θ when finding moment. This is because, the swinging, Ø, in the same direction of the bending, θ, will ease the bending burden on the beam, thus decreasing the moment produced at the member end.

Same logic also applies to M_{ba}, this time point B being the near end, and point A being the far end.

> *Note: For M_{ab}, and M_{ba}, it is also common to see one equation instead of two as above, which considers point A as the near end, and point B as the far end. This can be written as:*
>
> $M_{nf} = 2EI/L\,(2\theta_n + \theta_f - 3\varnothing) + M_{nf(fixed)}$
>
> *where n stands for near end and f stands for far end. This equation is the same as the two above.*

Now let's clarify what fixed end moments are...

Fixed end moments: These are the moments that we insert into the slope deflection equations, which were shown above, which are found by assuming the piece of beam segment as fixed at both ends as in the figure below, no matter what its actual joints might be.

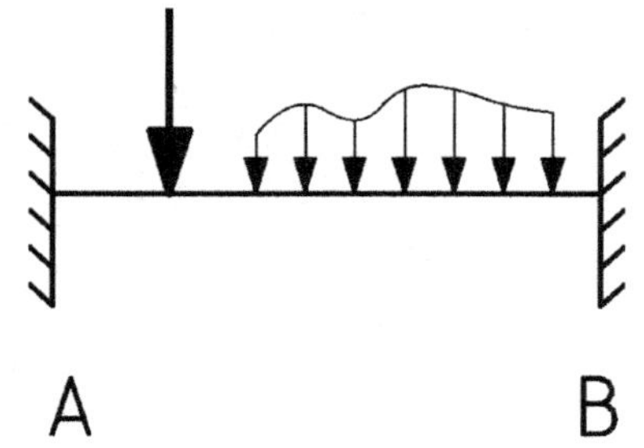

Figure - Structural 8

This beam segment might have any type of loading on it. Point loads, uniformly or arbitrarily distributed loads, which may act anywhere on the beam... The point is that, the beam is considered as fixed at both ends, and, there is a load acting between two nodes in transverse direction. In other words, if there is no transverse loading acting on the beam between points A and B then there is no fixed end moment.

The fixed end moments can most straightforwardly be calculated by the most fundamental method, the force method or also by virtual work method. However, this would be an exercise only for learning purposes but otherwise trivial, because they were already calculated for all possible common loading conditions and are displayed in tables which can easily be found in many structural analysis texts. The possible common situations of loadings include point loads, distributed loads of uniform or triangular shapes, applied at various positions on the beam and since the superposition principle applies we can also combine them, therefore from the fixed end moment tables we can obtain fixed end moments easily for practically any common loading situation we will encounter. So, the fixed end moment tables normally give us all we will ever need. Exceptions are loads that would act with an irregular distribution shape for some rare reason, or beams with an irregular geometry or nonuniform stiffness along its length or cross section, which can sometimes happen for architectural reasons or would mostly be for theoretical academic analysis purposes. And for unusually shaped load distributions, in real life we would conservatively simplify irregular loads to a regular distribution, plus the codes dictate us simplified, regular shaped loadings in the first place for most situations.

After finding the fixed end moment for a node of a frame segment, we insert the fixed end moment into the slope deflection equation for that node and while doing that we apply the same sign convention we showed above.

Now let's solve a few examples, that demonstrate how the slope-deflection method is applied. The first example will be the same one that we solved with force method.

Example 1 - Slope Deflection Method:

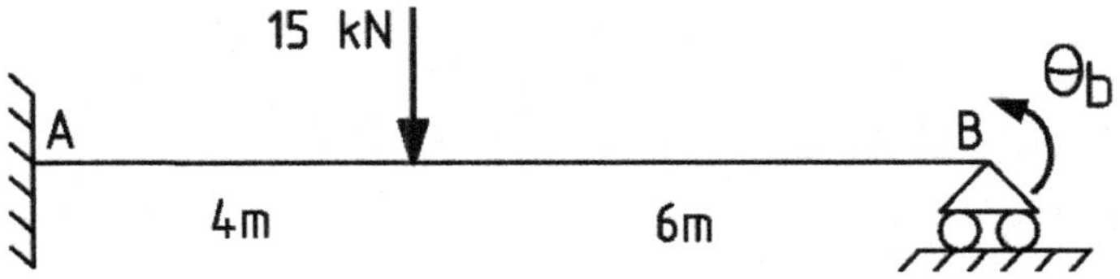

Figure - Structural 9

Let's rewrite the slope deflection equations for both point A and point B:

$M_{ab} = 2EI/L\ (2\theta_a + \theta_b - 3Ø) + M_{ab(fixed)}$

$M_{ba} = 2EI/L\ (2\theta_b + \theta_a - 3Ø) + M_{ba(fixed)}$

We know a few things here right from the start. We immediately know that:

- Since support A is fixed, there cannot be any rotation at point A, so θ_a must be zero.
- There is no swinging here (but only a rotation from bending at B), so Ø must be zero.
- We have a hinge at Point B (a roller hinge), which means, the moment at B, M_{ba}, is zero.

Now we can solve our equations. But first let's find our fixed end moments, $M_{ab(fixed)}$ and $M_{ba(fixed)}$.

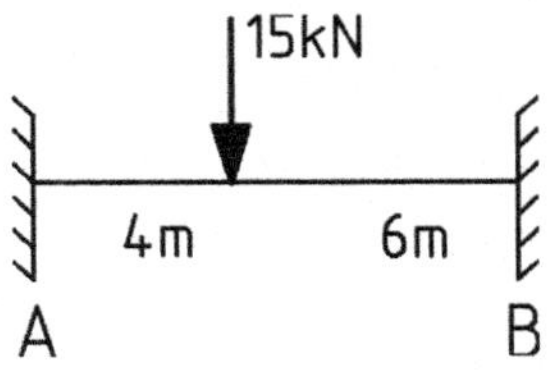

Figure - Structural 10

$M_{ab(fixed)} = Pab^2/L^2$ *(you can easily find this formula in a <u>fixed end moment table</u> for this loading condition, where a = 4m, b = 6m, P = 15kN and L = 10m)*

$M_{ab(fixed)} = 15*4*36/100 = 21.6$ kN.m

$M_{ba(fixed)} = -15*16*6/100 = -14.4$ kN.m *(this time see that a and b are reversed)*

Above said that $M_{ba} = 0$, Ø =0 and θ_a=0. So,

$M_{ba} = 2EI/L\ (2\theta_b + \theta_a - 3Ø) + M_{ba(fixed)} = 0$

$2EI/10\ (2\theta_b + 0 - 3*0) - 14.4 = 0$

$\theta_b = 144/4EI = 36/EI$

Let's insert θ_b into the equation that we wrote in the beginning for M_{ab}.

Again, $M_{ab} = 2EI/L\ (2\theta_a + \theta_b - 3\varnothing) + M_{ab(fixed)}$

$M_{ab} = 2EI/10\ (0 + 36/EI - 0) + 21.6$ kN.m (see that EI values cancel)

$M_{ab} = +7.2 + 21.6 = 28.8$ kN.m

Now you can see that this is the same answer that we found when we solved the same problem with force method in the previous section.

From this point forward, the rest is the same as was shown in force method example, where we find the same reaction at B by simply using static equilibrium equations and draw the shear and moment diagrams with the same values, which we will not repeat. Note that the moment reaction is +28.8 at A, (so its direction will be the same as the moment at point A as in Figure 7) but the internal moment starts from minus 28.8 as we explained a few pages back under sign convention. As you can see, to the right of support A, the beam is concave down, so it indeed starts with negative internal moment.

Note: We did not need to know the value of EI to solve the reactions or draw force and moment diagrams. But if we wanted to calculate the rotation amount of B, which is θ_b, we already know it in terms of EI as: $\theta_b = 36/EI$. So if EI values were given to us, we can easily find the value of rotation at B. This will produce an angle in radians. See that the higher the EI value, the less rotation will be per the equation above, which makes sense because higher EI means a stiffer beam.

Also see how we already applied the 7 steps that we listed when describing the slope deflection method. For step 1, we did not need to separate it to nodes as it was just a beam but regardless of that there are two nodes, such as A and B. For step 2, we only had rotation at point B, θ_b, which we showed. For step 3, we wrote the slope deflection equations for M_{ab} and M_{ba}. For step 4, we wrote the equilibrium equation at point B, where we knew the total moment had to be zero as there is a hinge there. For step 5, we solved by the unknown displacement θ_b. For step 6, we inserted and solved θ_b back into the slope deflection equation for M_{ab}, and found its value, which was what we were looking for. It means we solved the problem and can find the rest of reactions and draw shear and moment diagrams, as mentioned in step 7. In the two examples below we will not list the steps again but if you pay attention you will see that they are followed.

Example 2 - Slope Deflection Method:

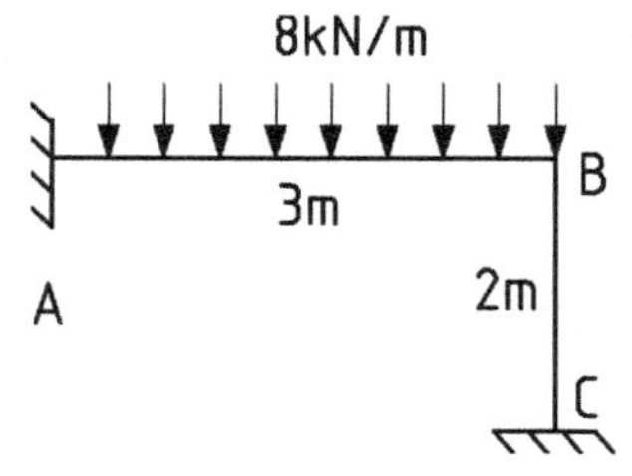

Figure - Structural 11

Our structure is indeterminate to the 3rd degree statically, such as 6 reactions from two fixed supports, minus the 3 equilibrium equations of statics. *(Note: If we used force method, we would need to write three equations and solve them simultaneously. That is why force method is less convenient as the number of statical indeterminacy goes up.)*

To apply the slope deflection method, we look at our structure and see that there are 3 distinct points, as A, B and C, which are our nodes.

Then we look at the kinematic indeterminacy, or in other words, degrees of freedom, which means, how our structure can move. We can see that the only possible movement for our structure is the rotation at node B. Remember, fixed supports cannot have rotation or translation. So the rotation at B, which we can write as θ_b, is the only unknown here.

As you know, the slope-deflection equations that we will write, will include fixed end moments, so let's find the fixed and moments first. We have a fixed end moment only for member AB. We do not have any fixed end moment for member BC, because there is no transverse loading acting on it.

$M_{ab(fixed)} = wL^2/12$ (see this in a fixed end moment table)

$M_{ab(fixed)} = 8*3^2/12 = 6$ kN.m

$M_{ba(fixed)} = -6$ kN.m

$M_{bc(fixed)} = M_{cb(fixed)} = 0$

Now let's write the reaction moments we are looking for:

Moment reaction point A:

$M_{ab} = 2EI/L\ (2\theta_a + \theta_b - 3\emptyset) + M_{ab(fixed)}$

We know that rotation at A (θ_a) is zero. We also know that this structure cannot sway (swing) as it is restrained from both directions, so chord rotation (Ø) is also zero. We already found fixed end moment. So we can write the equation as:

$M_{ab} = 2EI/3\ (0 + \theta_b - 0) + 6 = 2EI\theta_b/3 + 6$

Moment reaction point C:

$M_{cb} = 2EI/L\ (2\theta_c + \theta_b - 3\emptyset) + M_{cb(fixed)}$

We know that rotation at C (θ_c)is zero. We also know that this structure cannot sway (swing) as it is restrained from both directions, so chord rotation (Ø) is also zero. There is no fixed end moment here because there is no transverse loading to the member BC. So we can write the equation as:

$M_{cb} = 2EI/L\ (0 + \theta_b - 0) + 0 = (2EI/2)\theta_b = EI\theta_b$

So now we know both M_{ab} and M_{cb} , in terms of the unknown displacement, θ_b. To find θ_b, we can use the equilibrium at point B, by writing it in terms of moments using slope deflection equations. In other words, at point B, the total moment is zero and we can use this knowledge.

So,

$M_{ba} + M_{bc} = 0$

Then we write these two moments by slope deflection equations, with θ_b in them, and can solve θ_b.

$M_{ba} = 2EI/L\ (2\theta_b + \theta_a - 3\varnothing) + M_{ba(fixed)} = 2EI/3(2\theta_b) - 6 = 4EI\theta_b/3 - 6$

$M_{bc} = 2EI/L\ (2\theta_b + \theta_c - 3\varnothing) + M_{bc(fixed)} = 2EI/2(2\theta_b) + 0 = 2EI\theta_b$

Since $M_{ba} + M_{bc} = 0$, we can write,

$4EI\theta_b/3 - 6 + 2EI\theta_b = 0$

From here we get:

$\theta_b = 9/5EI$

Now that we found θ_b, we can insert it back to M_{ab} and M_{cb}.

As we found above, $M_{ab} = 2EI\theta_b/3 + 6$ and $M_{cb} = EI\theta_b$

Then, if we insert $\theta_b = 9/5EI$ into these equations, we get:

$M_{ab} = (2EI/3)(9/5EI) + 6 = 7.2$ kN.m (Answer)

$M_{cb} = EI(9/5EI) = 1.8$ kN.m (Answer)

These were the reactions we were looking for. The rest is simple statics solution. Since we found the reaction moments and already know the external forces on each member, we can calculate the horizontal and vertical reactions on each as well. Then we can draw the force and moment diagrams for both members. So for member AB and BC we will have the reactions as in the figure below.

We found the reactions in the figure below using statics as:

M_{ba} and M_{bc} can be found by inserting the value of θ_b which we found above, into the equations of M_{ba} and M_{bc}, from which we get:

$M_{ba} = -3.6$ kN.m and,

$M_{bc} = 3.6$ kN.m

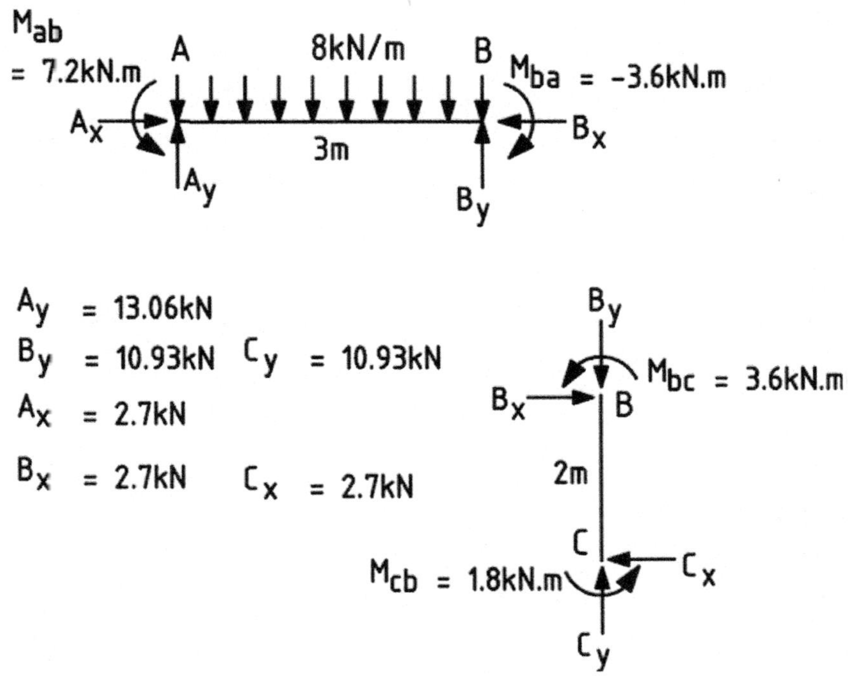

Figure - Structural 12

From moment equilibrium of member AB, if we take moment with respect to point A(and knowing that M_{ab} = 7.2 kN.m as we found before), we get:

7.2+3By-3x8x3/2-3.6 = 0. From here we find

By=10.93 kN

And from vertical force equilibrium Ay+By = 3x8. So we find

Ay=13.06 kN

From vertical force equilibrium of member BC, By = Cy , so,

Cy=10.93 kN

From moment equilibrium of member BC, if we take moment with respect to point C (and knowing that M_{cb} = 1.8 kN.m as we found before), we get:

1.8+3.6-2Bx=0. From here we find

Bx=2.7 kN

And from horizontal force equilibrium of member AB, we find

Ax=2.7kN

And from horizontal force equilibrium of member BC, we find

Cx=2.7kN

All values above are per the directions as shown in the figure above.

And, the axial force, shear force and moment diagrams for members AB and BC will look like in Figure 13 below (for convenience we drew member BC as horizontal).

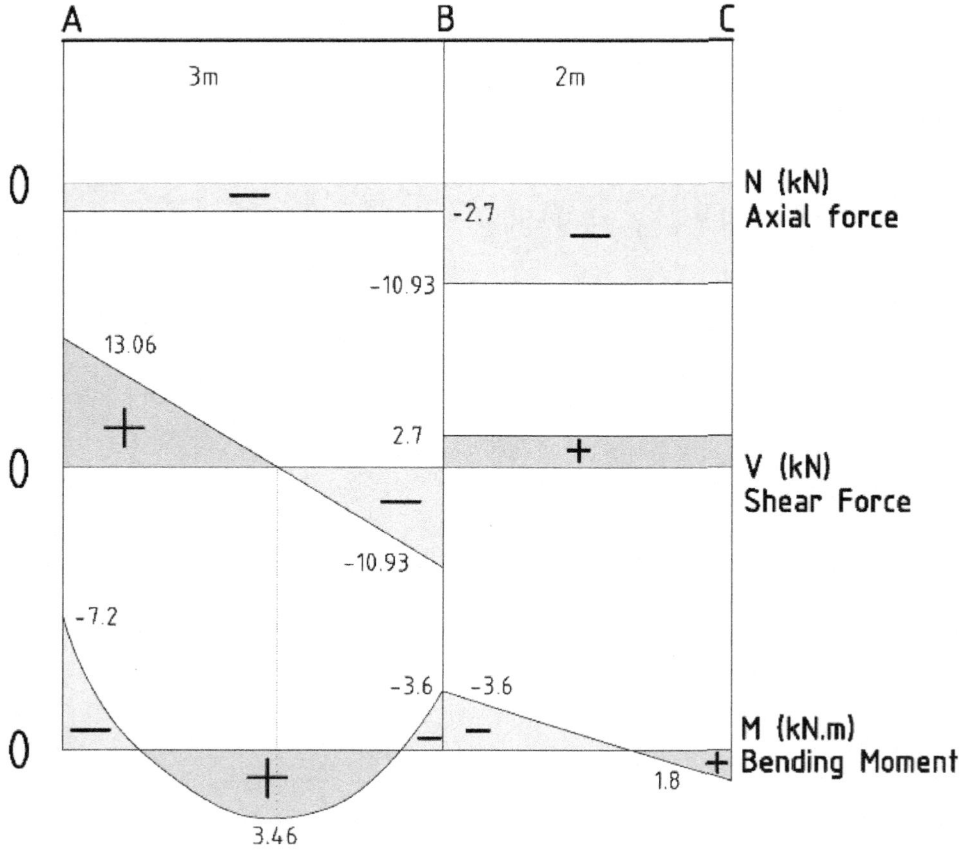

Figure - Structural 13

Notice a few things in the graph above:

Notice that at point B, there is the continuity of moment, with same values across members.

Also pay attention that the maximum moment occurs where shear is zero, just like we showed in statics section. And how did we calculate the maximum moment? Again as stated in statics section, the change in moment is equal to the area under shear diagram. After knowing this, it is easy to calculate the maximum moment. First from simple geometry we find the location of zero shear, which is 3x13.06/24=1.63 m from point A (in other words, we know that a change of 3x8=24 kN occurs over a distance of 3m, so how much distance it takes to reduce 13.06 to 0 is the location of zero shear, so we apply he proportion of 13.06 to 24). Then we can calculate the triangular area under shear diagram up to this point, which is: 13.06x1.63/2=10.66 kN.m. It means moment changed by this amount towards positive. So it brings the negative moment of -7.2 to: -7.2+10.66 = 3.46 kN.m which is the maximum moment, which occurs at 1.63 m from point A.

Note that in this problem the statical indeterminacy is 3, but kinematic indeterminacy is 1, which means we would need more effort if we wanted to solve this with force method, instead of a displacement method such as slope deflection method.

Hopefully the example above further clarified the slope deflection method... To see the deformed shape of the frame, see Figure - Structural 26. Let's solve one last example...

Example 3 - Slope Deflection Method:

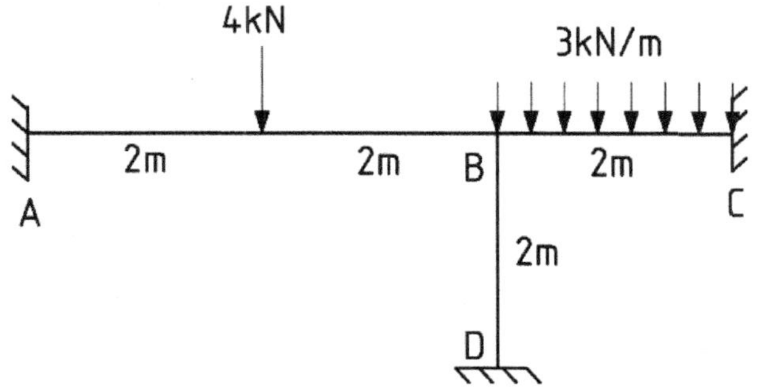

Figure - Structural 14

This time our frame is made of three members. To apply the slope deflection method, we look at the structure and see that there are 4 distinct points, such as A, B, C and D, which are the nodes.

Then we look at the degrees of freedom and see that the only possible movement for our structure is the rotation at node B. Remember, fixed supports cannot have rotation. So the rotation at B, which we can write as θ_b, is our only unknown here.

Again let's find our fixed end moments first, because we will use them in our equations.

$M_{ab(fixed)}$ = PL/8 (again, the formula can be found in a fixed end moment table for this loading)

$M_{ab(fixed)} = 4*4/8 = 2$ kN.m

$M_{ba(fixed)} = -2$ kN.m

$M_{bc(fixed)} = wL^2/12 = 3*2^2/12 = 1$ kN.m

$M_{cb(fixed)} = -1$ kN.m

$M_{bd(fixed)} = M_{db(fixed)} = 0$

Now let's write the reaction moments we are looking for:

Moment reaction point A:

$M_{ab} = 2EI/L\ (2\theta_a - \theta_b - 3\varnothing) + M_{ab(fixed)}$

We know that rotation at A (θ_a) is zero. We also know that this structure cannot sway (swing) as it is restrained from both directions, so chord rotation ($\varnothing$) is also zero. We already found fixed end moment. So we can write M_{ab} as:

$M_{ab} = 2EI/4\ (0 + \theta_b - 0) + 2 = EI\theta_b/2 + 2$

Moment reaction point C:

M_{cb} = 2EI/L ($2\theta_c + \theta_b - 3Ø$) + $M_{cb(fixed)}$

We know that rotation at C (θ_c) is zero. We also know that this structure cannot sway (swing) as it is restrained from both directions, so chord rotation (Ø) is also zero. We already found fixed end moment. So we can write the equation as:

$M_{cb} = 2EI/2\ (0 + \theta_b - 0) - 1 = EI\theta_b - 1$

Moment reaction point D:

M_{db} = 2EI/L ($2\theta_d + \theta_b - 3Ø$) + $M_{db(fixed)}$

We know that rotation at D (θ_d)is zero. We also know that this structure cannot sway (swing) as it is restrained from both directions, so chord rotation (Ø) is also zero. There is no fixed end moment on member BD. So we can write M_{db} as:

$M_{db} = 2EI/L\ (0 + \theta_b - 0) + 0 = (2/2)EI\theta_b = EI\theta_b$

Now all these are in terms of the only one unknown, θ_b

To find θ_b, we can write the slope deflection equations for all three member moments at node B, and add them up to zero, as the total moment at node B must be zero, as anywhere else on the structure.

So,

$M_{ba} + M_{bc} + M_{bd} = 0$

And as we found above:

$M_{ba} = 2EI/4(2\theta_b) - 2 = EI\theta_b - 2$

$M_{bc} = 2EI/2(2\theta_b) + 1 = 2EI\theta_b + 1$

$M_{bd} = 2EI/2(2\theta_b) + 0 = 2EI\theta_b$

Since $M_{ba} + M_{bc} + M_{bd} = 0$, we can write,

$EI\theta_b - 2 + 2EI\theta_b + 1 + 2EI\theta_b = 0$

$5EI\theta_b - 1 = 0$

$\theta_b = 1/5EI$

Now that we found θ_b, we can insert it back to M_{ab}, M_{cb} and M_{db}. So we get:

$M_{ab} = EI\theta_b/2 + 2 = EI(1/5EI)/2+2= 1/10+2 = 2.1$ kN.m (Answer)

$M_{cb} = EI\theta_b - 1 = EI(1/5EI) - 1 = -0.8$ kN.m (Answer)

$M_{db} = EI(1/5EI) = 0.2$ kN.m (Answer)

These were the reactions we were looking for. After this, we can insert θ_b into the equations for M_{ba} , M_{bc} and M_{bd} and can find those as: -1.8, 1.4 and 0.4 kN.m respectively.

Now that we know the reactions and external forces on each member, we can easily calculate the horizontal and vertical forces on each member from equations of statics, which we will not explain again. You should be able to find them the same way we showed in the previous exercise. To see the deformed shape of the frame, see Figure - Structural 26 and compare it with the moment diagram in Figure 15 below. The final moment diaphragm will look like the one in Figure 15 below.

In displacement methods, we have the same number of unknown displacements and equilibrium equations all the time. For example if we had a pin support at A, C or D, then there would be another rotation at that point, which means, now we would have 2 displacement quantities that we would need to solve. And we could get an extra equilibrium equation at pin support by knowing that the moment there must be zero. So we would have 2 unknowns and 2 equilibrium equations this time.

Also note that if we used force method, this problem would have needed solving of 6 unknowns (as it is statically indeterminate to the 6th degree), in comparison to solving only 1 unknown when we used displacement method. So as you see, the displacement method was far more efficient here.

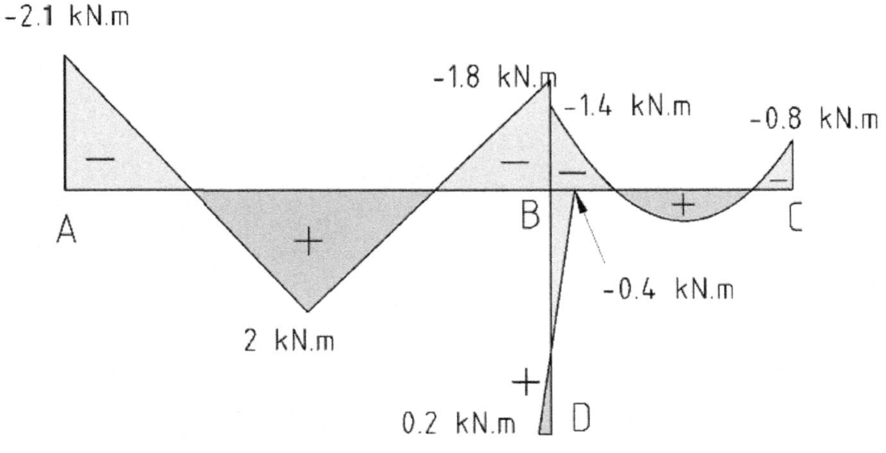

Figure - Structural 15

To summarize, let's list various useful points to remember when applying slope deflection method:

- As we saw in the examples, if a member (such as CD in example 3) has no transverse loading acting on it, both of its slope deflection equations M_{cd} and M_{dc} will have no fixed end moments.
- If the swing, rotation or translation amount(s) at a joint or joints in a problem is/are already given, it (they) can still be treated as degree(s) of freedom and inserted into the slope deflection equation(s), but this time as known quantit(ies), which will be useful. This can happen for

example when a support settlement, a member's fabrication error or its thermal expansion/contraction amount, or the amount of creep (see concrete section) is known.

- Right where the member enters into the fixed support, the rotation θ is zero, but there is bending moment. As you know from statics, fixed support is the only support that can resist moment and generate moment reaction.
- If the structure geometry and supports and the loading with respect to structure are symmetrical, and the EI of members are also symmetrical, then we can reduce the number of unknown displacements by taking certain displacements equal. See Figure 16 below as an example. Here the rotations at B and C are equal. Because as you see, the structure geometry is symmetrical, the load is applied symmetrically to the structure, and all members have the same EI as given. Therefore we can conclude that the rotations at points B and C are equal to each other. Even if the EI of member BC was different than AB and BD in the figure above, as long as AB and BD had the same EI, the symmetry would still be valid. Note: We could also have one vertical point load acting in the middle of BC, or a uniform vertical load between BC and again there would still be symmetry. Also see the preserving of 90 degree angle at joints B and C, since the members rotate the same amount (they do not rotate relative to each other, which was one of our main assumptions we listed before). And finally in the figure below, can you instantly identify the point of zero moment? It is where the curve of members AB and BC change direction, towards the middle of members. Also see Figure 26, under inflection point.

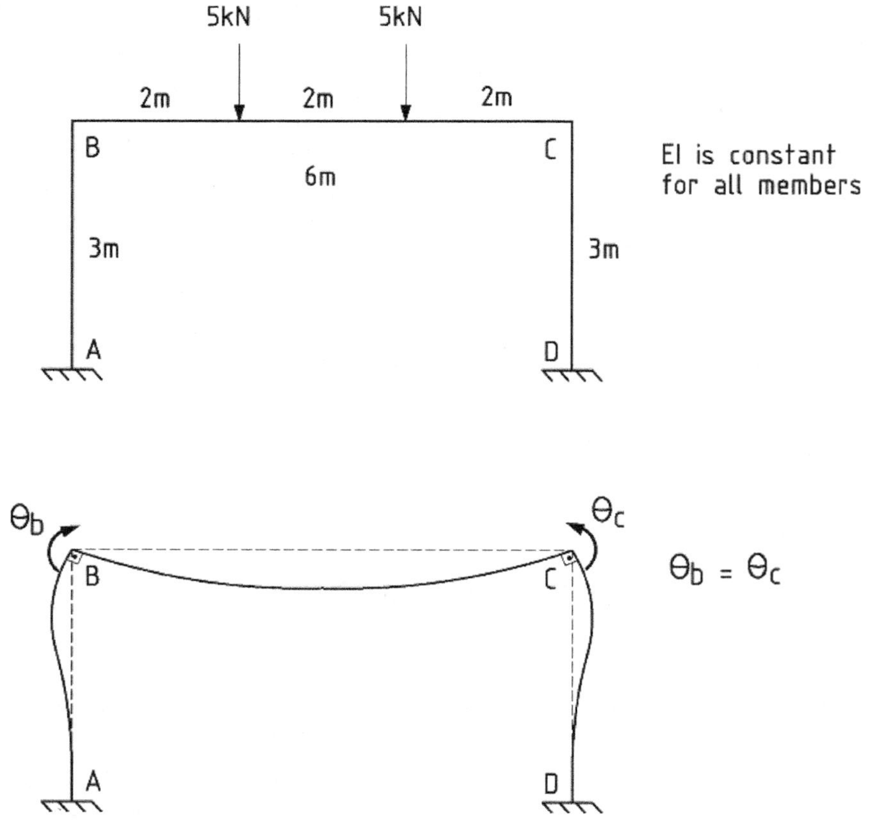

Figure - Structural 16

Moment Distribution Method (Hardy Cross Method):

It is a practical approximation method introduced in 1930s by Hardy Cross in USA. It is also a displacement method, which is applicable to frames only but not trusses. The method was developed by using slope deflection equations.

When solving frames, this method is more practical than the force method or the slope deflection method because we do not need to solve simultaneous equations here. When invented, moment distribution method gained worldwide popularity due to its practicality, until the advance of computers in 1970s. Regardless of that, it is still taught in structural analysis classes at undergraduate level as one of the traditional ways of solving indeterminate frames and beams.

Here are the steps for this procedure, where side sway is prevented, in other words braced frame: (we will also use these steps in an example to reinforce understanding):

1) This method works by distributing moments between members as in the steps below. To do that we need to draw a table. The table is made such that for each member there is a column. It will be shown in the example. And to start filling this table, we need to find two things first: Fixed end moments and distribution factors for every member at each node.

2) We calculate fixed end moments at member ends, no matter what type of supports we have, and imagine as if all supports were fixed type, as explained before under slope deflection method. Here this process is called clamping the joint. Now we have fixed end moments at every node but they are not necessarily balancing each other yet (add up to zero), although they must, to satisfy equilibrium. These unbalanced moments must be distributed to members joining at the nodes, which is what this method is all about and its name comes from.

3) Stiffness attracts load (whether force or moment). The stiffer a member (higher EI/L), the higher its distribution factor and the more portion of unbalanced moment it attracts. The distribution factor from a joint to a member is the ratio of that member's bending (flexural) stiffness to the total of bending stiffnesses of all members meeting at that common joint. Therefore, the distribution factors of members meeting at a joint, must sum up to 100%, in other words, 1. *(Side note: The distributed moments are related to the rotation of that joint by slope deflection equations.)* Also as you may notice, as L, the length of member increases, its distribution factor will decrease, which only makes sense because a longer member will be more flexible (less stiff) against bending and therefore can take on a lower portion of the moment in the joint.

   Distribution factor of a member = $(EI/L)_{member} / \sum(EI/L)$all members at that joint

   In the formula above, if a member's far end is hinged, we multiply the d.f. of that member by 3/4 *(although we will not show how it is derived, it is obvious that a member with far end hinged can rotate more, as some of its moment burden is relieved, so it would be able to attract less moment from the common joint)*

   So now we have a table with distribution factors and fixed end moments filled.

   Notes about distribution factors:

   If a member is framing into a fixed support, its distribution factor is 0 at the fixed support, because distribution moment in a member can only develop if there is rotation, but a fixed support does not allow any rotation (or think this way: the fixed support would absorb the entire moment, and nothing would be left to the member).

If a member connects to a hinge, its distribution factor is 1 there (or think this way: since the hinge cannot take any moment, all moment is taken by the member).

If a cantilever member is framing into a joint, that member's distribution factor is zero. In other words, it cannot receive any unbalanced moment as it cannot generate any force or moment reaction with its free end.

4) Look at the table you prepared which has fixed end moments and distribution factors now. As the joint moments are yet unbalanced, we must first balance the moments at each joint to satisfy equilibrium, which means, we add the negative of the unbalanced moment to that joint to make the total moment at that joint zero, and then, we must distribute this newly introduced moment among the members according to distribution factors of members.

   The members receive this distributed moment to their near ends which frame into that joint. And to their far ends, there will be carryover moments as a result of this newly added moment. This carryover moment has the same sign and is equal to half of the distributed moment received, for a regular shaped member. Note: If the far end is hinged, no carryover moment can go there, as a hinge cannot resist and take on any moment.

   We do this for every node. When we distribute moment at a joint it is called unclamping the joint. So we unclamp and distribute moments at each joint one after another.

5) After we distribute moments and carryover moments at each node, there may still be unbalanced moments, but this time considerably less. We do another cycle, starting at step 4 this time, and repeat distribution. We do cycles until all unbalanced moments become zero or negligible *(usually at most 3-4 cycles will provide adequate precision)*. Then proceed to step 6.

6) For every node, we sum up distributed moments from above. The sum is the result moment we wanted. This is the moment that is actually acting on the end of that member. For nodes where only members join (without support) and hinge supports these must add up to zero. Only at fixed supports there will remain a balance, as they are the only place that can resist moment.

7) The rest is just statically solving other reactions and drawing axial force, shear force and bending moment diagrams from equations of static equilibrium.

Now let's solve an example to demonstrate the steps above. The structure will be the same as example #2 that we solved under slope deflection method.

Example:

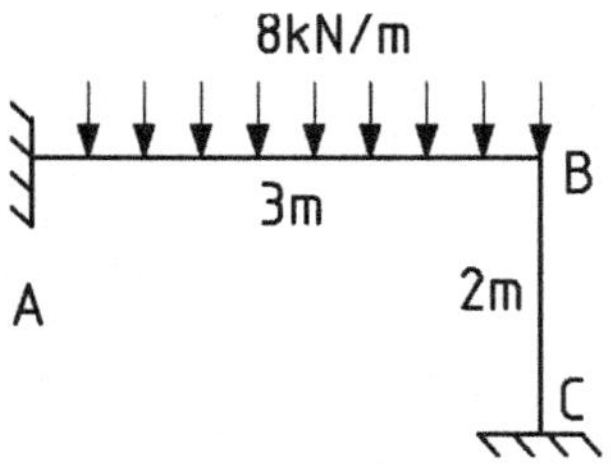

Figure - Structural 17 (same as Figure - Structural 11)

Step 1: We draw a table, with a column for every member:

	A	B	B	C
DF				
FEM				

We draw our table as above. Members AB and BC are shown. The distribution factors (DF) and fixed end moments (FEM) will need to be filled here first.

Step 2: Now we find fixed end moments for each member. We are working on the same problem that we solved when we studied slope deflection method. We had calculated fixed end moments there as:

$M_{ab(fixed)} = 8*3^2/12 = 6$ kN.m

$M_{ba(fixed)} = -6$ kN.m

$M_{bc(fixed)} = M_{cb(fixed)} = 0$

Step 3: We now calculate distribution factors. As noted in step 3, the distribution factor of a member is the ratio of that member's bending stiffness to the total bending stiffnesses of members meeting at the same joint. So the distribution factors can be calculated as:

$DF_{BA} = (EI/3) / (EI/3+EI/2) = 0.4$

$DF_{BC} = (EI/2) / (EI/2+EI/3) = 0.6$ (we could have obtained this by 1-0.4=0.6 also because the total of member DF's must be 1 at a joint). Now we have all fixed end moments and distribution factors. We can now insert them into our table:

	A	B	B	C
DF	0	0.4	0.6	0
FEM	6	-6	0	0

Step 4: As seen above, we have an unbalanced moment of -6 kN.m at joint B. The +6 kN.m moment at A is not an unbalanced because the fixed support there absorbs all of it as a reaction moment, and, we cannot distribute it back to the member AB (in other words, the DF of member AB at fixed support is 0).

So now there is only one unbalanced moment that we need to deal with, which is at joint B. Since we have a net moment of -6, we need to add +6 there, to make the total moment at B zero.

Now this newly added +6 moment must be distributed to the members. We already have the distribution factors. So, member BA takes 0.4x6=2.4 kN.m and member BC takes 0.6x6 = 3.6 kN.m as distributed moment as in the figure below.

Then we send half of these distributed moments as carryover moments to the far ends of the members, as 2.4/2= 1.2 to point A and 3.6/2= 1.8 to point C. *(Important to remember, as we noted in Step 4 that, if the far end was a hinge we could not send a carryover moment there.)*

Carryover moments are also entirely absorbed by fixed supports. None of it goes back to the member. Now our table looks like below:

	A	B	B	C
DF	0	0.4	0.6	0
FEM	6	-6	0	0
DM	0	2.4	3.6	0
COM	1.2	0	0	1.8
DM	0	0	0	0

Step 5: We see that there is not an unbalanced moment anymore (although it could have been the case, and we would go back to step 4 and keep distributing moments and sending carryover moments). Since there is no unbalanced moment anywhere, we can proceed to step 6.

Step 6: We add up all the moments. This was what we were looking for. Finally the table looks like:

	A	B	B	C
DF	0	0.4	0.6	0
FEM	6	-6	0	0
DM	0	2.4	3.6	0
COM	1.2	0	0	1.8
DM	0	0	0	0
Total	7.2	-3.6	3.6	1.8

As you see, we obtained 7.2 kN.m at joint A, and 1.8kN.m at point C, which is the same answer as we found in example 2 of slope deflection method. Then we obtain the same reaction values from statics equations, just like we showed before as in Figure 12 and same shear, moment diagrams as in Figure 13.

See that at joint B, the moments of -3.6 and +3.6 (which can also be seen in Figures 12 and 13) cancel each other and add up to zero, which satisfies moment equilibrium.

The steps we described and the example we solved above was for a frame where side sway was prevented. If we have a frame where side sway is not prevented (unbraced frame), we cannot determine the fixed end moments at first, because the frame is free to sway, and there will be unknown swing (chord swing amounts) that we do not know. To solve a frame where the frame is free to sway, we follow the steps below (we will not solve an example for this, but just describe steps):

1) We insert an imaginary brace to our structure at a location that would prevent sway. This brace will be in the form of a roller or pin support, perpendicular to the sway direction, so that it will only be able to provide a reaction force to brace the frame against sway, but not resist moment.

2) We analyse the structure we obtained in step 1, by following the same steps described earlier for a braced frame and solve it. Here we also obtain a reaction force from imaginary brace, that braces the structure. Let's call it R.

3) Then to make up for the sway, we draw a copy of the same structure, without any of the external loads *(since those were already included in the structure we analyzed in step 2)*. This time on this structure, we impose an imaginary amount of sway (deflection) only, at the sway location. By using that deflection, with the help of slope deflection equation (by using the swing amount as chord swing, Δ), we can determine the fixed end moment. Then we solve the imagined frame by following the usual steps for distributing the moment, because now we have the fixed end moments. As a result of this analysis, we obtain a total horizontal reaction. Let's call this H. The proportion of R from step 2 to H, R/H=k is the factor that we need to multiply our assumed deflection amount to reach to actual value. Because when we do that, H will be equal to R, thus cancel it. We want R to disappear because in the original structure there was no brace, thus no R in reality in the first place. So in other words, we simply multiply all results we obtained in our analysis at this step, including moments, forces and sway amount that we had assumed, by the factor k. This will be the result of this step.

4) Then we add both solved frames from steps 2 and 3 by superposition to reach the final result. So then, the final result will be a total of the moments and forces and the deflections, from steps 2 and 3, including the side sway. For multistory frames, there will be more than one side sway and for each of them, we add another condition with imaginary sway amount to solve.

And in step 3 why did we have to assume a value for the sway amount? Because if we didn't, we could not have obtained any fixed end moments to start our analysis in step 3. The value we chose for the sway amount was not important because in the end, we obtained the H, and when we divided R by H, we obtained a k factor, which we then used to multiply everything that we obtained in step 3 to reach to correct values. In other words, if we had assumed another value for sway, then we would find a different R/H=k value, but then upon multiplying that different k value with the amount of sway we now assumed, we would have still arrived to the same actual sway result form this step.

Matrix Displacement Method / Matrix Stiffness Method / Direct Stiffness Method:

In real structures there needs to be a lot of nodes, which means a lot of equations, far, far more than we can solve manually, which is why moment distribution method was widely used before advance of computers, instead of solving many nodes. After computers reached to a certain computing power in 1970s, engineers were able to implement direct stiffness methods even if the structural model contained a lot of nodes and involved solving large matrices. Since then the computer analysis methods always improved, which also involved advanced numerical methods and integration.

We described discretization under displacement method and said that for skeletal frames or trusses it was very easy. In such cases, we can just assign the nodes at obvious points and direct stiffness method can be applied. Complex bodies however, such as members with complex shapes, foundation soil, structural shells can be solved by dividing them into greater number of nodes, which were not distinguishable, which creates a mesh *(in both cases we must keep the number of nodes at a minimum to reduce calculation effort, without sacrificing accuracy)*. The mesh is constructed automatically by computer by assigning nodes in proper places, which creates finite elements connected by nodes. Finite Element Method (FEM) uses more complex methods than direct stiffness method to calculate interaction between nodes, and to solve the force, displacement and stiffness matrices for equilibrium. Today practically all structural analysis are performed using computers. We will look at this in more detail later in this chapter, under computer aided analysis section. Now let's return back to direct stiffness method, to see the basics.

After discretization of the structures, similar to what we said under displacement method, we need to obtain the force displacement relationships and the equations of equilibrium and solve displacements. To do that, we first generate member stiffness matrices and then combine them into a single global stiffness matrix. We obtain the global matrix by transforming the members' x-y-z coordinates into global x-y-z coordinates. Now let's discuss individual element matrices first and then see how they are converted to global matrix.

Stiffness, as discussed under strength of materials, is the force or moment required to give a member unit displacement, so, it relates loads to displacements. Stiffnesses are calculated by giving a unit displacement to each degree of freedom for that element one by one while keeping all others restrained, and finding out the corresponding forces that are required along each degree of freedom to produce displacement only in that degree of freedom. For this, for frame members, slope deflection relations that relate moments to rotations are used (EI/L). For truss or pinned frame members, axial stiffness are used (AE/L). These loads per unit displacements are now the stiffness values. So, this process of finding stiffnesses is repeated for every degree of freedom of every element.

The terms in the matrix are arranged such that for example, the term k_{42} means, the stiffness value at row 4 and column 2. This means, the force or moment needed along the degree of freedom #4 to have the unit displacement along the degree of freedom (DOF) #2, while all other displacements (DOFs) are kept as zero (restrained). So again, stiffness coefficients in the stiffness matrix are all load per unit displacement. A term can be positive, negative or zero. The stiffness matrix is a square matrix, where the size is equal to the number of degrees of freedom (i.e. if DOF is 3, then it is a 3x3 matrix). It is also a symmetric matrix, in other words, for example the term k_{42} will be equal to k_{24}, which holds true when we have linear elastic materials or there is not some damage in the structure to cause plastic behavior. So, the displacement produced at DOF 1 as a result of a unit load at DOF 2 is equal to the displacement produced at DOF 2 as a result of the unit force at DOF 1, which is called as Maxwell's Reciprocal Theorem. It holds true as long as the member and the whole structure behaves linear elastic.

So then for example for an element with two degrees of freedom, the matrices [k][d] = [f] will be formed by writing equations in the form of:

$k_{11}d_1 + k_{12}d_2 = f_1$

$k_{21}d_1 + k_{22}d_2 = f_2$

And then

$$\begin{bmatrix} k_{11} & k_{12} \\ k_{21} & k_{22} \end{bmatrix} \begin{bmatrix} d_1 \\ d_2 \end{bmatrix} = \begin{bmatrix} f_1 \\ f_2 \end{bmatrix}$$

where f is force, k is stiffness and d is displacement. These are vector matrices, which means, they contain both magnitude and direction information.

If a stiffness term k_{ij} is zero, it means that the force on node i cannot produce any displacement whatsoever on node j, no matter how large that force is. For example on a truss member, if a force acted transversely on a node (90 degrees to axial), it can not produce any deformation in that member *(in other words, in that member's node i or j)* at all, because truss members can only carry axial loads. It means, that load cannot be transferred between nodes i and j, for that particular load and member.

After obtaining the individual element matrices, we then combine them in one matrix by placing them such that their common nodes intersect, as in the figure below. So, at the common nodes, we simply add the stiffness terms from individual matrices. After obtaining this matrix, we multiply it by a rotation matrix to bring the global matrix coordinates to global x y z coordinates.

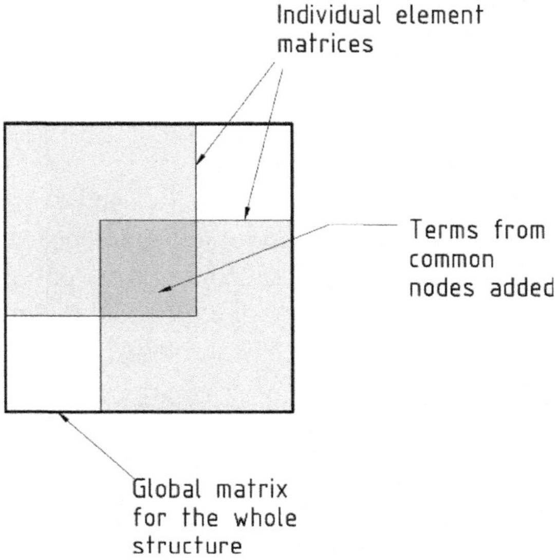

Figure - Structural 18

Converting to global stiffness matrix is to relate member forces and displacements to global force, equilibrium and displacements for the entire structure. Through these matrices, the structure is mathematically represented as an assembly of discrete members that meet at common nodes. Similarly, load and displacement matrices are represented globally. Global matrices relate as:

[K][D] = [F]

where,

[K]: stiffness matrix *(which considers geometry (length, area, moment of inertia) and material properties (modulus of elasticity), and how members are connected within the structure)*

[D]: unknown displacements *(axial deformations, deflections, rotations as applicable, so displacement vector includes all degrees of freedom)* vector matrix

[F]: matrix that contains vector of loads (forces and moments)

Note: For element matrices lower case letters, and for global structural matrix capital letters are used.

The displacement matrix [D] contains nonzero terms for all nodes and their DOF that are free to move. The only zero values in displacement matrix will be for the DOF along which there are reactions because of external supports.

The force matrix [F] will have nonzero forces only along the DOFs where an external or reaction force or moment exist and everywhere else in the matrix it will be zero.

Then, upon application of external forces and external support conditions (boundary conditions), we solve the displacements of all nodes, in order to find forces, strains and stresses everywhere.

The method described above is very suitable to adapt to computers. In our structural model every node in 2D will have 3 degrees of freedom (two translations and one rotation), and in 3D, 6 degrees of freedom (three translations and three rotations), unless we assign external support at a node. After that, stiffness matrices are formed for the members and then the structure, to establish the equations of equilibrium and solve them to determine the forces and displacements at each node and support, and finally generate the force and moment diagrams everywhere.

Work and Energy Method:

If you may remember from physics, work is equal to force multiplied by distance in force's direction (or moment times rotation). Since we deal with deformations and loads in structural engineering, we can apply the work and energy principles also to solving structures. Work-energy methods are powerful means that have applications in structural engineering and other fields. For example we can use them to solve frames, trusses, plates or shells, structural stability problems.

When external forces are applied on a structure, they perform work. This work is considered as being exactly equal to the energy that is stored in the structure as a result of deformations, which is called as strain energy, assuming that there is no energy loss. This is called the conservation of energy principle. For this to be valid, the assumption is that no heat energy or kinetic energy is produced, in other words, there is no friction and the loads are applied slow enough. To apply the work and energy principles, the virtual work concept is used, which means, applying imaginary forces and moments (called as dummy load and chosen as having unit value) to anywhere we want, and find their corresponding deformations. For each applied load, we can solve one deformation. This way we can obtain the additional equations we need when solving indeterminate structures.

Approximate Methods of Indeterminate Analysis:

In some cases, making an approximate analysis is useful to quickly verify results of a computer analysis, or a manual analysis done by one of the methods described above. It can also be used to make a quick preliminary design of a structural assembly.

With this, we conclude discussing the methods for indeterminate structural analysis.

General Steps of Structural Analysis and Design Process:

So far we discussed many important concepts in structural engineering and will do more. If you are not an engineer, it is easy to lose track among many different concepts, although the author sought to organize and present them as clearly and in logical order as possible. So, here is a road map to tie the concepts together in mind, which summarizes the steps in the most general sense, to analyse and design a structure. We did not cover some of the terms or concepts below but we will.

1. We construct structural model based on project requirements and past experience.

2. We set the items to apply to our model such as, the design parameters, material properties, (earthquake) analysis methods, settings, design approach to take such as ASD or LRFD, loads and load combinations (gravity, earthquake, wind, snow, water... as applicable) per codes.

3. Structural analysis is performed on our structural model, based on all the settings, analysis method and loads we apply. We check all applicable load combinations and take the worst case (most unfavorable) loading conditions into account. The analysis was being done manually in the past, before several decades ago, relying on the methods we covered. In our age, it is performed by computers by using more advanced methods but still the fundamental concepts that we covered in statics, strength of materials and structural engineering sections apply.

4. We obtain force and moment diagrams, and displacements everywhere in the structure. We modify the structural model and make refinements as necessary, such as the type and size of members, their configurations, supports, connections, loads as applicable, and repeat analysis until all results are acceptable, which means, the capacity of the structure is more than the demand from it (capacity design), and not only that but also it will structurally function in an acceptable manner (serviceability design / design for deformations), based on the performance level we expect from it. Based on the loads as shown in force and moment diagrams, concrete reinforcement / structural steel details are produced per codes. Structural analysis programs can also produce these details automatically for engineer's review and input.

Structural Design Approaches

We perform structural engineering to design structures that are not only strong enough and can stand without collapsing which means member and structural strength must be more than load demands, also known as capacity design / capacity level design / ultimate limit state, but also serve their intended purpose, and without too much displacements, such as deflections, deformations, cracks, vibrations, story drifts, settlement... and causing any discomfort to people who use it, also known as service design / service level design / design for deformations / serviceability limit state. In capacity design we limit ductile behavior (thus, detailing) to certain members while others reliably carry the loads with adequate capacity in their elastic range (i.e. strong column - weak beam), therefore capacity design means higher certainty in predicting performance and ensures overall reliability. *Side Note: Design for deformations is as equally important as design for capacity, but it has especially greater importance for structures such as very tall / large, seismically isolated therefore flexible structures, evaluating existing buildings, and other relatively more deformation sensitive structures as applicable.*

In addition to elastic capacity of load carrying system members (with negligible plastic deformations), other members can show plastic, ductile, permanent deformations and dissipate energy (note that a design may also be entirely elastic). In the first one (elastic design) we stay in the assumed linear zone and in the second one (plastic / ductile design) we extend into nonlinear zone of members force - deformation relationships. Therefore, linear and nonlinear analysis are the two main approaches to structural analysis (which are further grouped into static or dynamic). We will discuss structural analysis methods under earthquakes section (because for most of the world, structural analysis means analyzing structures for earthquakes).

In connection with the previous paragraph, if we want to look at things at member level, within a structural model, the components of a structure are designed as either force controlled, which are designed according to strength criteria and expected to remain elastic and modeled elastically, (even if in reality they may show small plastic behavior), or deformation controlled, which can show ductile behavior with nonnegligible irreversible plastic deformations after yield point, which must be modeled inelastically. For example during a seismic event, force controlled members must behave essentially as elastic, which ensure that the structure safely stands, and the deformation controlled members dissipate energy that enters the structure while undergoing predictable, controlled, gradual inelastic deformations and also contributing to overall strength. *(i.e. A steel brace under compression will be force controlled because after a point it may suddenly buckle. We cannot predict or rely on its deformation capacity. But a beam can show gradual deformation due to bending. It can be a deformation controlled member.)*

There are also other criteria to satisfy as applicable to that particular structure, which include structural stability, structural irregularities, disproportionate (progressive) collapse, fire, blast resistance...

The two main approaches when designing structures in the US are:

Allowable Stress Design (ASD).

and,

Load and Resistance Factor Design (LRFD)

Overall, both methods compare the available strength of members with the strength that is demanded of them, but do so in different ways.

ASD method is also called Working Stress Design (WSD) method.

LRFD is also called the strength design method as stated in ASCE 7.

LRFD is newer than ASD method. It gradually replaced ASD in the last few decades for the most part but ASD still did not disappear, and can still be preferable in certain situations.

A design cannot use partially LRFD and partially ASD, which can result in unsafe design. This doesn't mean that a structure can not be separately designed using each, using one as a check method.

Allowable Stress Design (ASD):

In ASD method, stresses in structural members cannot exceed a percentage of nominal strength.

The stresses are found as a result of service loads on member, without applying any load factors to them. Nominal strength is the strength of a material in a particular situation, as determined by applicable methods given in codes and specifications, before application of a safety factor.

In ASD, to obtain allowable strength, we divide the nominal strength by a "Factor of Safety", also called as "Safety Factor", shown by the Greek letter Ω (Omega), which is used to account for unpredictabilities for both material strengths and loads.

So,

Allowable Strength of the Material = Nominal Strength the Material / Factor of Safety (Ω)

And, finally the following must be satisfied:

Allowable strength > Demanded strength

For example, for a given situation in the structure, if the nominal strength of a certain member was determined as 100 MPa, and the code gave us a factor of safety of 2 for that, it means that per ASD criteria we can allow a maximum stress of 100/2 = 50 MPa for that member in such situation. Now, after we perform a structural analysis, if we find out that the stress in this member will be 60 MPa, this would be unacceptable, since it is more than its allowable strength, and we would either need to select a stronger material, or a larger cross section, or arrange the configuration of the member or structure itself differently, or revise the design to reduce the loads and so on... Then we analyse again, until we confirm that the stress in that member does not exceed allowable strength that is available per ASD. The safety factors, method to determine nominal strengths and the load combinations to use are provided in applicable codes.

In ASD we remain in the linear elastic range (see Figures 11 and 12 under strength of materials section). Therefore, for very sensitive structures, such as water tanks, nuclear reactors, chimneys, where even small cracks or deflections are not tolerable or less tolerable and we need more strict deflection control, or structures with higher stresses such as prestressed structures, ASD can be more suitable.

Load and Resistance Factor Design (LRFD):

Similar to ASD, LRFD also seeks to ensure that the available strength of components and the structure will be greater than the strength demanded from them under load effects, which are determined by load factors and load combinations given for LRFD this time.

For calculating the available strength, in this method again we have the nominal strengths of members, as calculated by the methods given in codes. Then these nominal strength values are again reduced, but this time by multiplying with the factor "Ø", called as the strength reduction factor, design factor or resistance factor, which is smaller than 1, to obtain a design strength. The nominal strengths and the resistance factors are again for a given situation as per in codes and specifications. They depend on the degree of confidence in accurately predicting the nominal strength of the member, the nature of applied loads such as axial, or shear force or moment, the way that the member can fail and the consequences in the structure if that member failed. A typical value is 0.85 for certain steel members for instance (remember, they can be 1 at most). For concrete it can be lower values (because it is harder to predict concrete strength, than steel) in different situations such as in Table 21.2.1 of ACI 318-19.

To find the strength demand in LRFD, unlike in ASD, the loads are increased by certain factors and combined in load combinations for LRFD which are given in applicable codes. Then with these loads, structural analysis is performed, which shows us the demanded strength values from members. More predictable loads are increased less, and less predictable loads are increased more in the combinations, which also consider their possible simultaneous occurrences.

Again in the end, the resistance produced by the members and the structure must be greater than the demand produced by the load combinations.

So,

Factored Loads < Factored Resistance

must be satisfied.

LRFD, which is relatively newer and more complex, has now greatly replaced ASD and used more commonly since several decades. It is a more finely tuned approach than ASD and can consider uncertainties better. It can produce more uniform level of safety everywhere, and at the same time may enable more economical results, such as reduced cross sections of members.

For an engineer, it is not a cycle of repeated choosing between two methods from one project to the next however. Typically he/she uses one approach always or most of the time. The code may enforce one method in the first place, depending on the location or the type of structure.

Loads and Load Combinations

The loads that will act on our structure are combined in various ways per code criteria and per the design approach we chose, and applied to our structural model in order to obtain the internal loads and displacements within the structure for the load combinations that would produce the worst case conditions. Then these force and displacement demands are compared with acceptance criteria.

Different types of loads are combined under load combinations by using load factors. This is true for both ASD and LRFD approaches, but of course their load factors and combinations are different and vary for different structures. So at this point let's list the type of loads, and then show some load combination examples.

The main types of loads are:

Dead Load (D): Results from the weight (earth's gravitational acceleration times mass) of the structure itself and the components that are fixed on to it. Always acts downwards and in static manner.

Live Loads: (L): These are the loads from people, vehicles, furniture etc... in other words, the loads that result from the usage of structure. Live loads do not include construction loads or environmental loads such as wind, rain, snow, ice, earthquake etc...

Earthquake Load (E): It is a dynamic, cyclic load that can act in any direction, which is a result of the acceleration of the ground below the structure. So similar to gravity acceleration, earthquake acceleration also acts internally on the entire mass of the structure, but here instead of producing a static weight load as in dead load which acted always downwards, earthquake acceleration produces inertial earthquake load, which can act also laterally and in a dynamic manner. This is unlike the wind load, which simply pushes or pulls the structure from the outside.

Wind Load (W): Wind acts as compressive and / or suction forces externally on outer surfaces of a structure. The suction force is the reason why roofs of buildings fly away during strong hurricanes.

Snow Load (S): The load from weight of snow. The minimum snow loads are provided in codes for every geographical area in each country.

...and several more including loads from rain, ice, vibrations, impact load from trucks or ships, waves, tsunami, floodwater, fluids, groundwater pressure, material pressure, lateral earth pressure, blast and any other applicable load for a particular structure, such as in ASCE 7.

In addition, there are loads that arise from structure's own behavior, such as the ones resulting from differential settlement of foundations, the structure working against dimensional restraints, loads due to the effects of concrete creep, shrinkage, expansion or contraction due to temperature changes, change of humidity, steel manufacturing errors, fatigue and so on.

Dynamic loads are the loads that are applied fast enough that they are considered to change with time, and able to cause dynamic effects, unlike static loads. Dynamic loads include earthquake loads, vibration loads, traffic loads (that can also result in fatigue), impact loads, such as from a vehicle impact or blast and more... Often (but not always) they are simplified and treated as static, which greatly simplifies calculations while still allowing us to remain on the conservative side.

The new loads that will emerge as a result of the displacements of the structure due to the loads we counted above are called secondary or second order effects. In certain situations these must also be accounted for. *(Remember that in structural analysis section, one major assumption that we listed in the beginning was that second order effects were ignored. That is for basic structural analysis for which we presented a summary. Taking the second order effects into consideration will add considerable level of complexity into structural design, which will be the subject of advanced structural analysis, also combined with structural stability concerns. We will describe structural stability in the next section.)*

The codes clarify the design approaches to use (may allow only one or both ASD or LRFD), show the load combinations and the minimum loads to be used in any situation, in addition to many essential criteria to ensure public safety and comfort. These are the loads and their criteria that every engineer must obey when designing structures. Structural analysis programs are now automatically able to help the engineers in applying code required loads and load combinations along with comparing the analysis with the desired code criteria. In each country, codes provide load combinations for different types of structures. For example, the few combinations we show below are for buildings in the US, but for bridges for example, these are not applicable *(the assumptions while producing those combinations would not be applicable for bridges)*. In another country different combinations and approaches can be used, such as the Eurocode in European countries. Also, various countries simply adopted or base their codes on US codes and made modifications per their needs.

A few of the load combinations from ASCE 7-16 are listed below as examples, which we will not use anywhere but showing to explain the concept.

LRFD example combinations:

1.4D

$1.2D + E_v + E_h + L + 0.2S$

Here the first load combination tells us to multiply the dead load by 1.4. Then we apply this on our structural model, perform structural analysis and look at the results. The second combination tells us to take the dead, live, snow and earthquake load together, multiplied by the factors as shown. Again, we take this combination and perform another analysis from scratch and again look at the results. In the second combination, the factor for dead load is smaller in this case, since we also consider other loads at the same time. There are also load combinations that we didn't show here, with different loads and or factors. We apply them on our structural model, perform structural analysis to find the worst one for anywhere in the structure (but from experience we can usually guess which one will be the worst case for a particular structure even before trying them all).

When we described LRFD a couple pages back, we said loads are increased by load factors. This is generally true, but in some load combinations, the load factors can be less than 1 for some of the loads, when they act simultaneously with others. We may still select these combinations for the structure or for some members, if, after checking we see that that combination produces worse effect than other combinations. For example another load combination for LRFD is 0.9D + 1.0W. Now here D, dead load is not increased but reduced. But we can still choose this combination for certain members or the whole structure, and design accordingly, if, as a result of structural analysis, we find that this load combination for that member or the structure produces the worst effect (similar logic is also true for ASD). You may be asking, why did we reduce the dead load in 0.9D + 1.0W. This is done to deliberately allow the overturning effect of wind to the whole structure become more significant, by reducing the weight, which is the stabilizing force, therefore creating a worse condition.

For ASD, a few examples again from ASCE 7-16 are:

D + L

0.6D + 0.6W

$1.0D + 0.7E_v + 0.7E_h$

For example here, for the first combination, we will take dead and live load together without modification of any factor and add them up, but in the second combination, we apply the dead and wind loads together, and reduce the dead load to its 60%, and take 60% of wind, add them up and check if it will produce the worst effect somewhere in the structure and so on... So, again we check and compare the different load combinations to find the worst one that applies to any member or the structure. For example, for a large but lightweight structure with a large surface area, the second combination which includes the wind load may govern. But that same load combination would most probably will not be the critical one for a concrete building of a few stories. The third one listed above, in addition to dead load, includes both horizontal and vertical component of earthquake load.

The codes give us not only the load combinations and load factors but also the minimum actual load values to be used where and in what situation. We can never take load values less than the minimum loads in codes for a particular situation, even if we are sure the actual load will be less for our project. For example in the US, we cannot take for instance the live load for a hospital corridor above first floor to be less than 80 psf (3.83 kPa), or a patient room less than 40 psf, as per the values in ASCE 7-16, Table 4.3-1, even if we know that the values in our particular project for such corridor or patient room will be less *(there would not be a lot of difference anyway)*. However, if the engineer has a specific reason to believe that it could be more, he/she must use the higher value.

Structural Stability

A structural assembly may fail because of insufficient material strength, such as when a member's material yields from too much stress. That type of failure is not dependent on the geometry. For example, when we compress a short concrete cylinder from above, it will fail when the stress in the member exceeds its strength. Or when we pull a steel rod with increasing force, it will first yield and eventually experience ultimate failure. These failures do not depend on the geometrical and support conditions. They simply result when stress exceeds strength, as discussed under strength of materials.

Structures may also fail due to geometrical configuration of members and supports, even when the stress they experience may be well below their yield strength. In fact, this type of failure occurs more frequently than material yielding failures and can be very catastrophic. In the previous paragraph, we said "short" cylinder on purpose, let's say, a foot length. Now imagine in your mind, another concrete cylinder with the same exact concrete mix, same strength and same diameter, but this time assume that the cylinder is much, much taller and looks very slender (let's assume its diameter is 6 inches but height is let's say, 10 feet now). Again we put it on a flat surface and apply a compressive force from top, in the same way we applied to the short cylinder before. Now can you visualize that, it may first tend to buckle, even before the material yields? This failure is an example of a geometric failure, which is also called failure due to structural instability. So the keyword in the previous paragraph was yield, but in this paragraph it is geometry. Therefore, in addition to making the structural members' materials strong, we must also ensure structural stability, which depends on geometrical conditions, how the members and their supports are arranged, member sizes, stiffnesses... Ensuring structural stability is a major part of any structural design. Failures from structural instability are usually sudden, catastrophic.

Structural stability is studied in masters level classes of civil engineering (although some of its concepts such as buckling is introduced in steel design classes at undergraduate level). It is important to understand structural stability, to be able to understand many things about structures better. Therefore, we will mention some of the important concepts about it in this section.

Structural stability can be defined in various ways. Here let's define it as the resistance of a structural assembly or structural member to failure which would result from geometrical and support conditions, stiffness(es) of member(s), upon application of compressive loads to it or the structure.

If a structural member or system tends to return to its original state after a disturbing effect, the system is called stable, or, in stable equilibrium. If a structural system or member can not return to its original position after a disturbance, it is in unstable equilibrium.

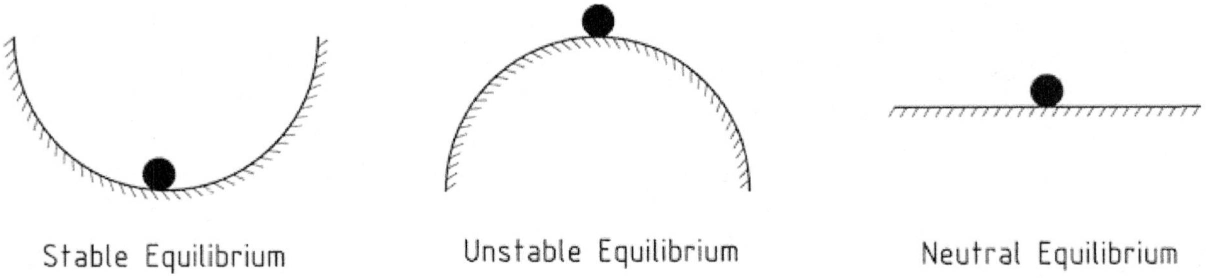

Figure - Structural 19

For example, even if we applied a large disturbing force to the ball on the left, it would want to return to its original configuration. For example, a steel column in tension, a roof made of a tension system, is in stable equilibrium. If we increased the tensile force, the material would keep elongating until the material itself yields or its support gets broken or torn, but these are not stability failures. A column in compression however, or for example an arch, which works by compression, is in an unstable equilibrium. If the force compressing them becomes "too large", their equilibrium will be destroyed and a sudden failure would occur - in case of a column it will be relatively slower in bending manner (although still not so slow) but in case of an arch a sudden and catastrophically snapping. Both the column in compression and arch will not return to their original configuration upon failure. That "too large" force is called the critical force for structural stability, and shown as P_{cr}. For structural assemblies, P_{cr} is a compressive force, which would turn a stable system into an unstable system, which means, the system will not return to its original geometry anymore and not only that but also suddenly, catastrophically fail. Here we say "compressive", because structural stability is about the compressive forces. See the Figures 20a through 20d below and imagine applying the forces in the opposite direction to the columns, in other words, in tension, instead of compression as shown. Would we need to worry about reaching a P_{cr}? No, we would keep pulling it as much as we want, and the columns would neither rotate, nor buckle. We would pull without worrying about a critical force, until the column material itself yields in tension or until the connection below is simply torn apart, which would make the problem a material yield or rupture problem, but not a stability (geometric) type of failure.

Buckling:

Buckling is a common cause of stability failure, which means it is also a sudden failure (the degree of being sudden or slower also vary according to different types of buckling for which we will not go into details) therefore we want members to yield first, before they buckle. The figures below are basic possible configurations for buckling, in other words they are not arbitrary examples, and you would see these also in steel design or structural stability classes for instance.

In Figure 20a, we have a simply supported column. In an ideal world, as long as we apply the P force right at the geometric center or column cross section, nothing will happen to column, unless we increase the stress up to the column material's yield strength. But in real world, deviation will occur from the ideal geometric center sooner or later, even if it is extremely small. This is not only because the application of load will not be perfectly at the center and completely downwards, but also because no material in the real world is made perfectly homogeneous. So we start applying the load, and once it reaches a critical load, P_{cr}, the system will no longer be stable and the column will buckle, *(provided that we did not exceed the yield strength first, before we even get to P_{cr})*. This critical load given in the formula below was found by Euler almost three hundred years ago. In the figures below on the left you see the columns, supports and the applied load, and on the right, you see how the columns deform based on the support conditions. In the formula below, "E" is the modulus of Elasticity, "L" is the unbraced length of the column, "I" is the moment of inertia.

Figure 20a is the simply supported case. The critical load for this case is:

$$P_{cr} = \frac{\pi^2 EI}{L^2}$$

The formula above is the basic buckling formula. Figures 20b - 20d represent different end support conditions where in each figure, the column on the left is shown with its supports, and the corresponding column on the right shows that same column's deflected shape, under that particular support condition. See the different critical loads and deformed shapes in each case. Try to animate in your mind, why the deformed shape is as shown below in each case.

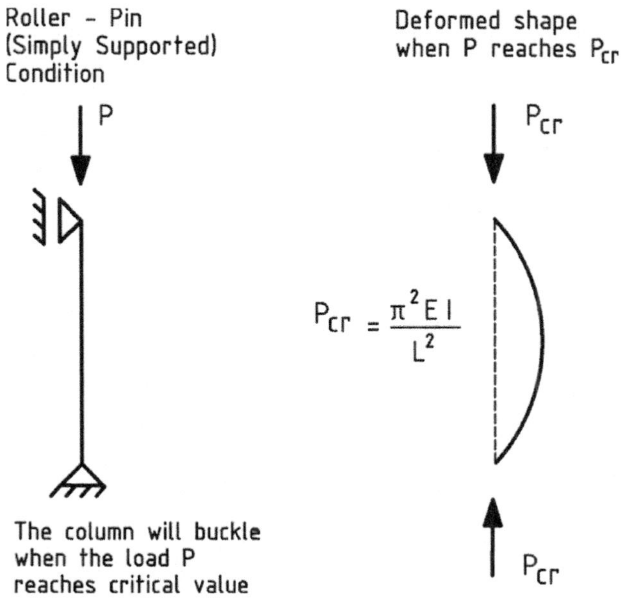

Figure - Structural 20a

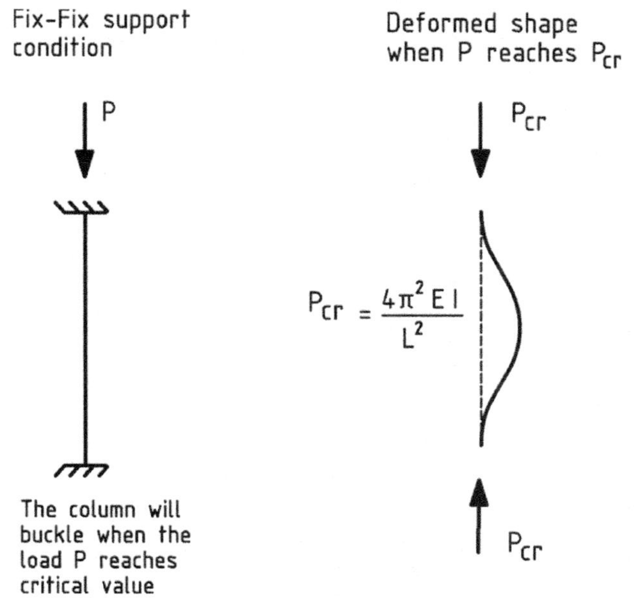

Figure - Structural 20b

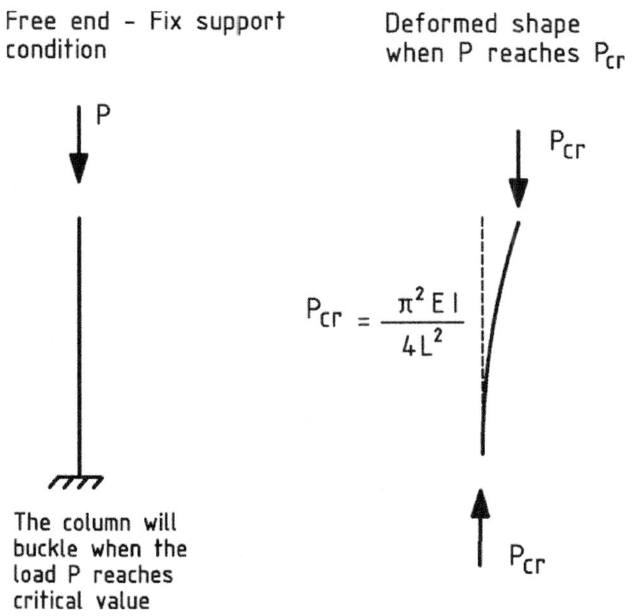

Figure-- Structural 20c

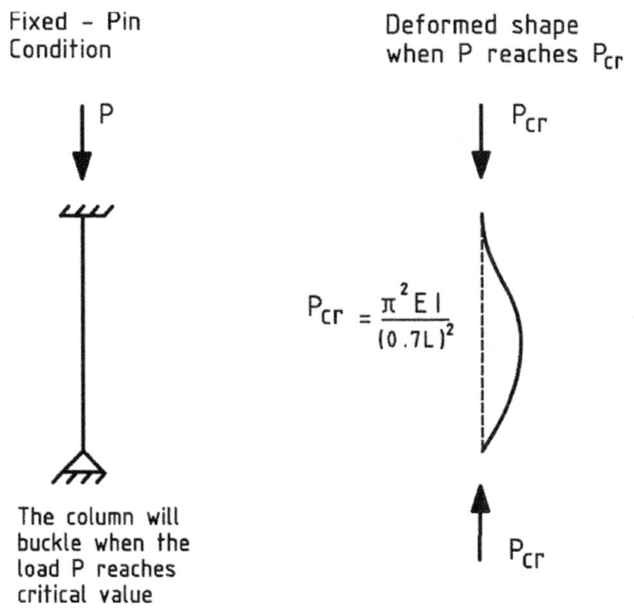

Figure - Structural 20d

As you can see in these figures:

The P_{cr} value in Figure 20b is 4 times more than the one in Figure 20a (both ends fixed provides better support for ends, which causes much higher P_{cr})

The P_{cr} value in Figure 20c is only a quarter of the one in Figure 20a (since one free end causes that end to deflect very easily, hence the much lower P_{cr}) and so on....

P_{cr} values are found by imposing different end conditions for each case in the differential equations that define the deflected shape of the columns, which is not shown here but can be found in almost any structural stability focused reference.

To increase critical load, and therefore to increase the buckling resistance, at least one or more of the following can be done. Keep the P_{cr} formula above in mind, when reading the items below:

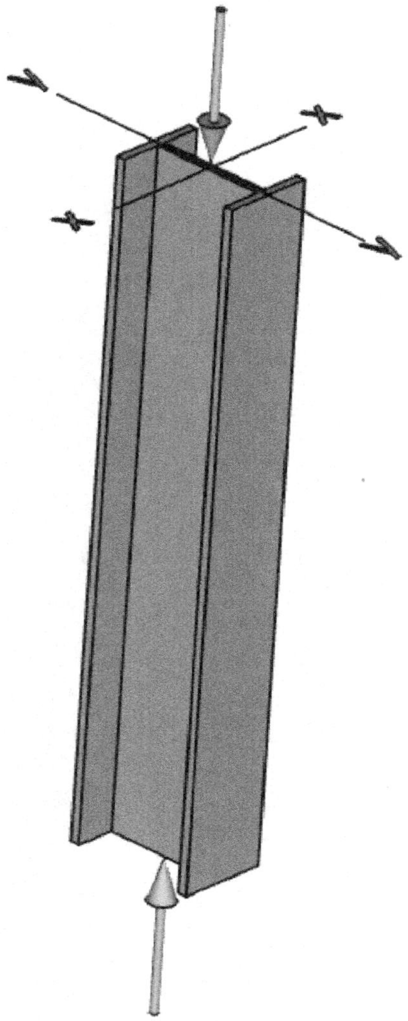

Figure - Structural 21

- The member can be braced (which reduces the free length "L" available to buckle). This means reducing the "effective length", which is the length that contributes to buckling, after the effects of supports are taken into account. As in the figures above, although the column lengths are the same, the buckling portion lengths are different in each case, which is the effective length for buckling. Also see the bracing topic under temporary construction section.
- The length of the member can be decreased (bracing has the similar effect, as it divides length into segments). The members with large length to cross section ratio are considered slender. Reducing the length means, making them less slender. Slenderness ratio is a number that is found by dividing the effective length of member, by radius of gyration of the member. Slenderness ratio is an important parameter used in the design of compression members. The smaller this ratio, the higher buckling resistance.
- The member must be aligned so that its strong axis takes on the critical load. See Figure 21. Bending or buckling around x-x is harder than around y-y because the moment of inertia $I_{x\text{-}x}$ is much higher than $I_{y\text{-}y}$. This means the column would buckle about y-y when P reaches P_{cr}. But if we wanted to make y-y stronger and placed bracing to column to support buckling with respect to y-y, the effective length for buckling with respect to y-y would significantly decrease, while it would remain the same with respect to x-x, so now x-x may become critical for buckling.
- From manufacturing process, any initial crookedness (eccentricity) in the column can be minimized. Theoretically this can never be fully prevented, but for practical purposes it can easily be kept below tolerable limits, such as at the levels of L/1000 or L/1500, where L is the length of the member.

- Again from manufacturing, we can prevent any initial (residual) stresses in the member, even before it is loaded. Residual stress can result from uneven cooling of member, uneven cooling of welded shapes, cold straightening of members, which fall into the realm of metallurgical engineering.
- We can use a material with a higher elasticity modulus, (E).
- We can use cross section with higher moment of inertia (I).
- One or more ends can be modified to a more restraining condition. See the different end conditions we showed above in Figures - 20a, 20b, 20c, 20d. So for example changing from a free end to a pinned end, or changing from a pinned end to a fixed end would all increase P_{cr}.

A reminder just in case: Do not confuse buckling with bending. They are different terms. Buckling is as described in this section, which is the sudden failure of a member under compression, due to geometry. Bending on the other hand, is what we saw many times before since the beginning of the book, which occurs when a transverse load or moment is applied on a member, such as the load that a bridge deck or a building slab exerts on a supporting beam, or the moment that a swaying frame applies at beam-column connections during an earthquake.

Above we looked at buckling of columns, which occurs in the form of bending. There are also other types of buckling. For example, the buckling of arches or the buckling of shell structures happen differently than buckling of columns. When columns buckle, at the critical load, the buckling starts with a shape only slightly different than the original straight shape of column. But arches for example, experience snap-through buckling, where, a significant change in shape occurs very suddenly, which can be more catastrophic. There is also lateral torsional buckling which can happen in axially loaded channels, angles, which do not have symmetrical section, or simply (especially light) beams under bending when their compression side tends to buckle laterally, if unbraced. These different types of buckling will not be discussed further here. For columns buckling can occur as local buckling at a relatively small location or as a bending buckling which occurs over the whole effective length of the member (as in Figures 20a-d). Buckling can also occur in beams under compression, similar to columns.

Figure 22 below shows some random structural assemblies, with compressive loads applied. When the critical load is reached in each case, the structure becomes unstable and buckling occurs. At this point we say a "mechanism" is formed, as it resembles a moving mechanism, which is not acceptable and considered as a total (and catastrophic) failure. In each random case below, the critical load, where the buckling starts first and of course the buckled shape will be different. How to find the P_{cr} and the buckled shape each case below is out of this book's scope and studied in detail in a structural stability course. Still, the P_{cr} and the buckled shapes all depend on the same following variables:

- Geometrical configuration of the structural assembly
- The elasticity modulus, E of each member,
- Moment of inertia "I" of each member (with respect to its weak axis),
- Length, L, of each member,
- Support and connection types and their locations,
- Location, type and stiffness of springs

Of course, the we do not just look at these for individual members but for the entire structure and how the members will interact. In the random example figures below, notice that we randomly used axial or rotational springs, different types of supports, some members are drawn with thicker lines to indicate that their bending stiffness EI is relatively higher (or in some problems we may even assume rigid members). Although it is not an easy exercise and we will not cover more details on it, give it a minute and try your best to visualize where and how each assembly in the figure below would buckle first.

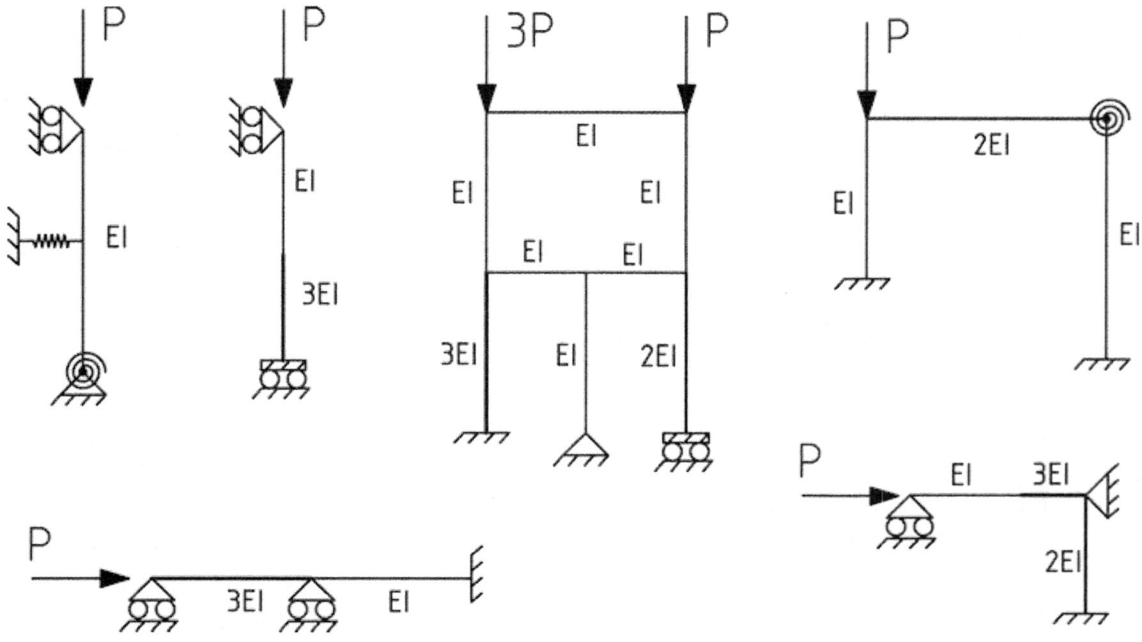

Figure - Structural 22

Various Terms & Concepts in Structural Engineering

Below is a list of some of the frequently encountered terms and concepts in structural engineering in alphabetical order. The list below was not made with the intent to cover each and every structural term like a dictionary. Therefore, other common structural terms may not be here but mentioned in different parts of the book - if covered in this book. The list below just emerged and expanded naturally while the author wrote the book, to either include the terms that were not covered anywhere else or to describe some terms that were mentioned in other places which also deserved separate explanation in the author's opinion. It ended up including many useful terms, which may come handy when looking for information.

Structural engineering and structural earthquake engineering are inseparable, so inevitably some earthquake engineering related terms are also seen here, but more earthquake engineering specific terms can be found under "Earthquakes and seismic structural analysis" section later in this chapter.

Arch Action:

Arch action is resisting transverse loads by making use of the arch shape, so that within the arch only compressive loads would exist, which can be resisted by the material or pieces that makes up the arch, just by being under compression, without the need for taking any tension. For example through arch action, certain length of span can be passed by arranging construction materials, such as stone blocks, in a way that they form an arch and naturally carry themselves and loads from above just by pressing against each other, without needing any attachment or reinforcement in between, because within the arch there will be no tension but only compression. If we made a complete circle and uniformly loaded it transversely from outside, again only compression would occur within the circle (also called ring effect) and the material that makes up the circle can carry very high outer transverse loads, just by being under compression itself.

Before arch action got discovered more than 2000 years ago, only straight, horizontal beams were used in order to cover freespans, which were not efficient and could only cover a small distance and needed columns often. Through an arched system, far larger spans could be passed and larger loads could be carried from above efficiently and naturally, just because of arrangement of the materials close to arch shape. This at the same time meant saving body weight (dead load) of the structure, *(so arches were used even if we did not need to pass a freespan)*, reducing burden on the foundations and increasing earthquake resistance, because lower mass means lower earthquake force. Overall, it was nothing less than a revolution in construction technology. Ancient Romans were the first to widely implement it, although the discovery might have been made much earlier.

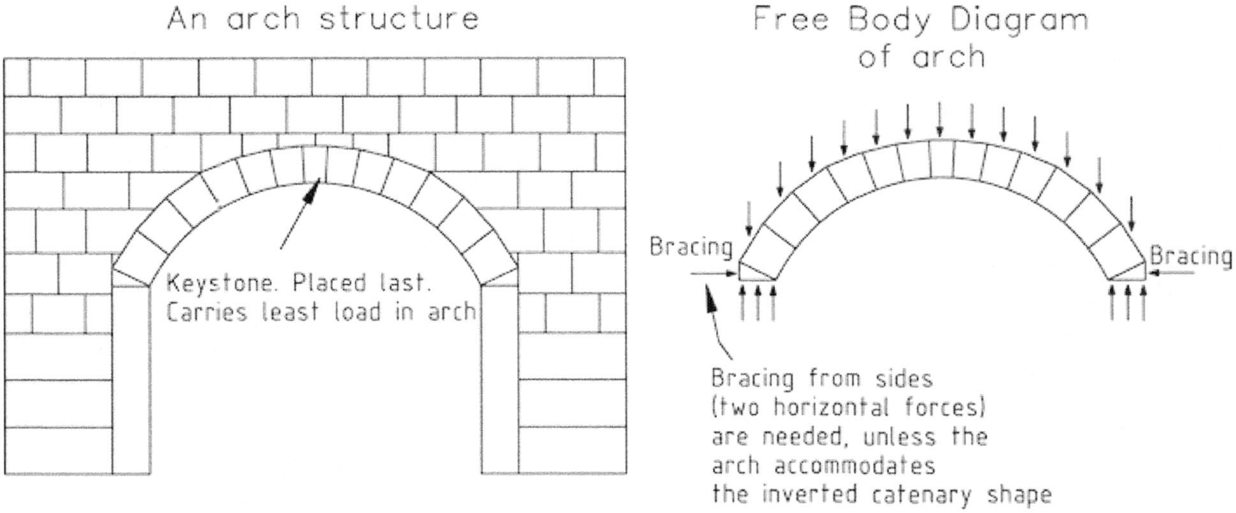

Figure - Structural 23

In the figure above, on the left you can see an arch wall. On the right, you see the FBD of the arch. As the weight of the structure presses on the arch, the stones that make up the arch naturally compress against each other without needing any attachments or reinforcing. The force in the arch also creates outward push towards both sides, unless the arch is shaped sufficiently close to an inverted catenary, which we will show below. This outward push is countered by horizontal reactions, which are shown as bracing forces in the FBD above, is provided by the brick columns at both sides of the arch.

Stone arches are constructed by using an arch shaped formwork. All stones are placed on the form from both sides of the arch starting from the bottom ends of arch, and finally the top stone called the "keystone" is placed, which is the key element of the arch, as shown in the figure. After the keystone is placed, the arch now has the connection in itself to transmit the compressive force between the stones and the formwork can be removed. Now the arch not only naturally carries itself, but also a considerable amount of load from above, if any.

A free standing arch, in other words, an arch that doesn't need lateral bracing, must be able to accommodate an inverted (turned upside down) catenary shape. A catenary shape is the natural shape that a rope or chain, such as a necklace *(or for example the cables of a suspension bridge after suspension from two towers at both ends)*, naturally takes by itself under the action of gravity, when simply held from its both ends, as shown approximately in Figure 24 below. The rope always takes the catenary shape naturally. It never fails to achieve the catenary shape no matter how far or close, how high or low we hold the ends with respect to each other. It always remains in tension, because that is the only thing it can support (a rope cannot support shear or compressive stress). Similarly, if we want the stone blocks to support only compression so that they will not have to support any tension and/or be braced from sides, the way to make them do this is to bring them together in the form of the inverted catenary shape. The blocks should at least be able to keep this shape within the cross section of the stones that make up the arch (ideally the middle third of the cross section), in order to be in compression everywhere, in which case, no tension will exist in the arch and no side bracing will be needed as was the case in Figure 23. *Note: Although catenary shape line looks like a parabola, it is not a parabola (for interested readers, it is represented by a "hyperbolic cosine" function).*

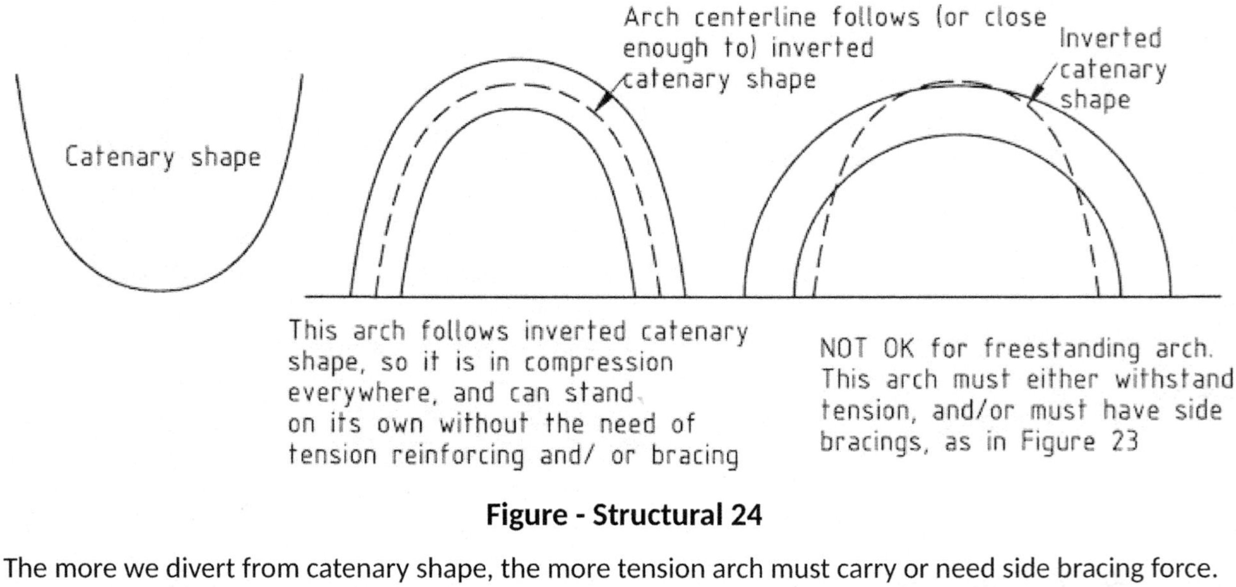

Figure - Structural 24

The more we divert from catenary shape, the more tension arch must carry or need side bracing force. Most arches don't follow catenary shape, thus need bracing but in almost all cases we already provide this anyway, such as in Figure 23. That is why we see circular or close to circular arches more often than an inverted catenary shaped arches.

Arches are still used today, although it is not as critical to use them as in the past, because now we can accommodate large amounts of tension forces and can have horizontal free spans too. For this we use reinforcing in concrete, or use structural steel, long prestressed beams or even very long suspension cables. If we didn't have these technologies, which were invented within the last couple hundred years, the arch would still be the best way for passing spans. Since they enable natural and efficient way of passing spans and carrying loads, arches are still used today and most of the time in combination with newer technologies, to build bridges or wide span roofs etc... but sometimes this is also due to aesthetical reasons. For arch dams, the arch effect is still utilized as the primary load resisting system, to have compression everywhere and avoid having to use reinforcing in the dam body and also to save concrete volume. As the water pressure increases with depth, the arch radius is decreased and the thickness of the dam is increased, which also gives the dam an aesthetical appearance. For tunnels, arch or ring effect is relied on, depending on the tunnel shape. Arch effect also have uses in other engineering disciplines. For example submarines can resist large amount of water pressure this way.

A dome can be constructed if we revolve an arch around its central vertical axis. But arches and domes differ in some ways. For example, a freestanding triangular conical dome can be stable, as it can brace itself in 3D, while a triangular arch cannot, as it needs to follow catenary line as discussed. When a dome shape is stable, the arch we obtain from its cross section might not. But the opposite is not true: if an arch shape is stable, the dome that we construct by rotating that arch will always be stable. This is why, when studying domes, we can divide the dome into arches, study it, and if we find out that the arch is stable, the dome will also be stable. A dome needs to have a bracing ring around its base if it is made from an arch that needed horizontal bracing. This is something we can see around large domes.

An arch load transfer path also forms naturally by itself in a deep rectangular beam which is loaded from above, to transfer the load by forming compression zone from the load towards the supports on both sides. The underside of the beam will be in tension. The compression and tension zones in the beam can be represented by truss analogy, which is called the strut and tie method, which is a term introduced later under this section.

Buckling:

It is a failure mode of a material under compression. For more, see structural stability section.

Composite Members:

Members made of two different materials working together, such as structural steel and reinforced concrete, so that the resulting member will take advantage from both materials, are called composite members. Slabs, beams, columns, shear walls, braces, piles... can all be made as composite.

For example a composite metal deck is a steel deck filled with concrete that contains reinforcing mesh in it. A composite column can be made by pouring concrete inside a steel hollow tube section, or encasing a structural steel "I" shaped column in concrete.

Other than members alone, there are also composite structural systems. For example a building can be designed with both reinforced concrete and structural steel lateral force resisting system to work together during an earthquake. This can provide redundancy in the structure, and make use of both systems' advantages.

To avoid a possible confusion, reinforced concrete also takes advantage of both concrete and steel but it is not regarded as composite member. It is simply regarded as concrete which is reinforced with steel.

Coupling Beam:

See subsections shear walls and also beams, in Chapter 2 > Building Frame and Structural Members.

Cracked Section Properties:

As a result of yield of reinforcing in concrete, there will be large strains, which will also mean the concrete cracks. This means the cross section of the member suddenly decreases, with lower moment of inertia, which means less resistance to bending. For earthquake analysis, this must be considered, per the criteria given by codes, if the section is likely to crack under design circumstances. For only gravity loads however, considering this will not be as critical but may still be used.

Cyclic Behavior:

During earthquake, ground movement causes cyclic behavior in the structure by shaking it back and forth. In such cases, an idealized load-deformation curve can resemble the figure below:

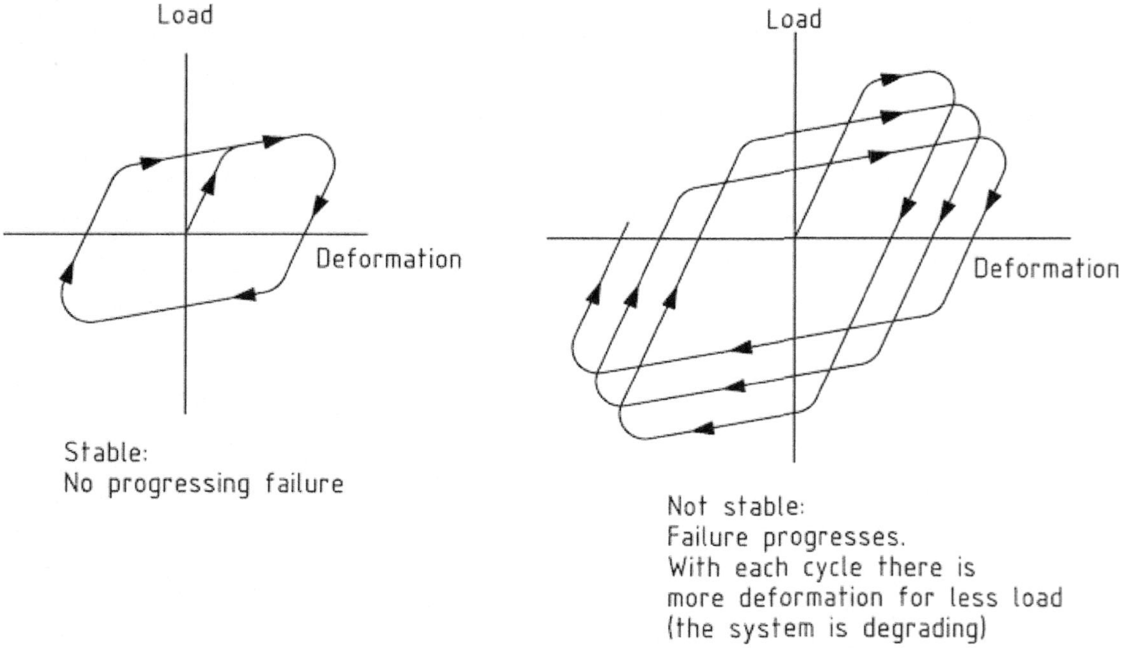

Figure - Structural 25

The larger the interior area of this shape, the more energy the structure can dissipate during earthquake. If the shape of this curve is somewhat stable after each cycle, it indicates good performance. If the deformation side of the shape gets longer and load side shortens with each cycle, it means an unstable and unpredictable system, such as the one on the right figure.

Degrees of Freedom:

This term was introduced under statics and also used various times in structural analysis section. It means the number of ways a system is free to vary (move in our case).

Diaphragm:

It is a structural horizontal area such as a floor slab, deck or roof, which holds together the other members that surrounds it, such as beams, columns, shear walls, tilt-up panels etc... by laterally bracing them. The ability to provide this bracing effect is called the diaphragm action. So the building slabs are not only there to carry our and the furniture weight. Their other, equally important duty is to connect and brace all other members that surround them. During wind or earthquake, diaphragm transfers and distributes lateral loads to and from vertical members around it, for everything to act together as part of structural frame, which enables far greater lateral load resisting capacity than a frame without diaphragms. *(Would a table be more resistant to side sway with or without the table top?)*

Diaphragms are grouped as rigid or flexible. A rigid diaphragm deforms negligibly in relation to the drift of the story it is on during earthquake. To better understand what we mean, imagine your kitchen table's legs are fixed to the floor. The table top will be the diaphragm. Now if you try to push the table horizontally from its side, the table top does not deform or negligibly deforms, but the legs of the table will sway, (unless there were bracings between the legs), since we said the legs are fixed to the floor. They will sway so much that you would probably be able to see it. If we tried to twist the table in horizontal plane, again the table top (diaphragm) would not deform but the legs would sway to accommodate torsion. So, rigid diaphragms can transfer both lateral loads and also torsional moment to members that surround them. Rigid diaphragms can be made of cast-in-place concrete, precast concrete or composite steel deck. Also important to remind that, we said "push" only to give example. In case of an earthquake, the floor would accelerate, which would cause acceleration in the table mass, so nothing would "push" it. Pushing (or even pulling) would happen from wind action.

Rigid diaphragms distribute horizontal force to the surrounding vertical members in proportion to those members' stiffnesses. So if the table legs had the same lateral stiffness, they would take on force equally. Do not forget: Stiffness attracts load. It is not the first time you see this in the book. You also saw it under structural analysis section, under moment distribution method *(in that case we talked about moment load and rotational stiffness)*. We will say "stiffness attracts load" again at least several times more in this book. To understand why stiffness attracts load, imagine you put two boxes on the floor, one weighing 200 pounds and the other only 20 pounds. Now push both equally with same force, with each one of your hands on one box. The 20 pound box will tend move easily, while the 200 pound box will not, and you will tend to try to push the heavier one with greater force if you want to move it the same amount as the lighter box. As was discussed under strength of materials, stiffness is load per deformation. If we take the movement of the box as deformation here, the heavier box would need 10 times more force, just to move equal amount as the 20 pound box. *(In other words, in a way it is stiffer against pushing although here this stiffness is a result of friction)* So: If two members displace the same amount, it will take more force to displace the stiffer one than the weaker one. This is why diaphragms distribute force to members in proportion to their lateral stiffnesses. This is also why, when studying moment distribution method in structural analysis, stiffer members took more of the unbalanced moment in the joint, because the rotations for all members were assumed to be the same (which was one of our initial assumptions in structural analysis section), as they connected to the same joint.

The other type of diaphragm is flexible diaphragm. According to ASCE 7-16, if the maximum diaphragm deformation along its length (laterally during earthquake) is more than twice of the average lateral drift of vertical elements immediately below that story, then that diaphragm is considered as flexible.

This large deformation means, flexible diaphragms are not as capable to transfer lateral movement or building torsion to and from vertical members as rigid diaphragms. They can be made of wood, or metal decks without concrete topping. In case of flexible diaphragms, for lateral force distribution tributary area method is used, which we will not discuss. Diaphragms that are in between flexible and rigid are considered as "stiff".

Diaphragms at two adjacent floors of a building, connecting to shear walls on the sides, form an assembly similar to a box, which is called the box effect.

A structural truss in horizontal plane can also serve as a diaphragm.

Ductility:

Ductile materials can undergo large deformations before failure. This means they can absorb a lot of energy, and give warning before failure, which is desired. See Figure Strength - 12 under Strength of Materials. Ductility is the portion that falls on to horizontal axis that would correspond to between the plastic yield point, up to rupture yield point, from ε_y to ε_r. For steel, this distance on horizontal axis is long. But for concrete, it is not. The longer this distance, the more ductile is our material and the more it can deform at yield point and energy it can absorb, such as during earthquakes.

We must achieve a certain level of overall structural ductility. For example AASHTO LRFD Bridge Design specifications Section 1.3.3 requires that a bridge structure must be able to show large deformations before failing, which means a ductile structure. This will not only depend on the ductility of members alone, but also how they are connected and their reinforcing or structural steel details. For example ductility in a reinforced concrete building frame is ensured by proper detailing of reinforcing steel especially at expected plastic hinge locations, beam column junctions. The transverse reinforcing in members, such as beam stirrups, column ties and hoops are the main reinforcing that ensure ductility by wrapping the concrete core that remains inside the area the enclose.

To understand easier, designing for ductility can be thought as inserting fuses into a structure, so that during an earthquake after a point the fuses will yield and be able to keep deforming and absorb energy with their ductile deforming ability, but while doing so, they will also ensure that the forces on the rest of the structure will remain below a certain limit, which we design to withstand those forces without yielding (force controlled members). For example although concrete itself is brittle, in our reinforced concrete structures we want ductile failures to occur first, by detailing such that the rebar yields first, before any brittle concrete failure occurs in a sudden crushing manner.

Ductility should not be confused with flexibility. <u>The opposite of ductile is brittle. The opposite of flexible is stiff.</u> See the term stiffness under strength of materials section.

Elastic Analysis:

See this term later in this chapter under Earthquakes > Earthquake Analysis Categories. Materials and structures behave elastically in the beginning (with none or negligible permanent deformation), before the stresses get considerably higher and they enter into plastic zone. We may require certain members or the entire structure to remain in elastic range, which would increase the forces they must withstand.

Equilibrium:

This is the most fundamental concept of all, in civil engineering. The net forces and moments acting on an object must be zero, for that object to be in equilibrium. It is the starting point of all our structural and geotechnical analysis. See statics section in the beginning of the book.

Factor of Safety (FS) / Safety Factor (SF):

In all engineering design, factors of safety (FS) are applied at various points, in various forms. For example the ASD or LRFD incorporate FS in various forms and places because FS is also a general name of the concept that covers all the factors/coefficients etc... which accounts for wrong assumptions, construction errors, unexpected circumstances during structure's lifetime, unforeseen material defects, loads that are greater than expected and so on... Even if in a theoretical world everything was foreseen perfectly, it just does not make sense to build things exactly to expected loads without any safety factor or redundancy, and nobody would feel comfortable using such structures. Given enough time, one or another negative thing will happen. Application of factor of safety are shown or integrated into the codes and standards, in countless ways and forms but on top of that engineers can still exercise judgment, as long as they meet the code requirement. Having said all that, engineers never forget about costs and schedule either. The goal of a design is to make the most economical structure, without sacrificing safety, functionality and comfort, by also meeting code.

Fatigue:

Fatigue is the weakening of a material due to <u>long term, repeated, large number of load cycles</u> that can <u>regularly</u> occur, <u>even if these loads caused low amount of stress</u> in each cycle. It is especially pronounced for steel structures where bolted and welded connection locations are affected most. Fatigue can apply to concrete structures as well, which can cause cracks, fractures and rebar corrosion and can also cause fatigue in the rebar itself. Structures exposed to repeated dynamic loads such as bridges (especially cablestay or suspension bridges), coastal structures, wind turbines, buildings with high number of cycles of live loads (i.e. buildings that are used by a lot of people) are most exposed to fatigue. The members must be designed with such adequate cross section that the fatigue stresses will not be able to progress beyond an acceptable limit. Traffic loads, vibrations and oscillations from machinery or other sources are most typical causes of fatigue. The number of repeated load cycles that will require fatigue consideration are specified in codes for different situations as applicable. Even if a structure complied with all serviceability and strength requirements, fatigue consideration may still be critical. When we say fatigue stresses we mean only the repeated long term <u>varying</u> stresses, not the permanent stresses resulting from static dead or live loads. The permanent static loads do not contribute to fatigue and including them may cause very overconservative design for fatigue criteria. And for varying stresses, we need to consider the difference of both extremes in opposite directions. For example if a cyclic stress exists as 5MPa in one direction and 5 MPa in the other direction, our fatigue stress should be taken not as 5 but as 10 MPa.

Fixed End Moment:

See slope deflection method that was discussed under Structural Engineering > Structural Analysis.

Flat Slab:

See Chapter 2 > Building Frame and Structural Members > Slabs

Frame:

A frame is a structural assembly where beams and columns join by being fixed into each other, therefore transferring moments. Beam bending and consequently the slope deflection equations form the fundamentals of frame analysis, where the end moments of members are calculated depending on flexural stiffness EI, length L, and the rotations (or even swinging) of members. Beam bending was discussed in strength of materials. The methods shown under structural analysis section for frames, such as slope deflection and moment distribution, also used beam bending knowledge as discussed.

Hysteretic Behavior:

Hysteretic behavior occurs under high stresses, after the material yields and enters into plastic zone, which causes inelastic force deformation relationship, during which, load vs. deformation curve that we showed (Figure 25) under the term cyclic behavior does not follow the same path anymore and its shape changes (degrades) similar to the right side of the figure. These loops are called hysteretic loops (cycles). During first stages of hysteretic loops, there may be some strength gain as a result of strain hardening *(see Strength of Materials section)* but later strength and stiffness both decrease.

Indeterminate Structures:

A structure where the number of reactions is more than the three equilibrium equations (in 2D) is statically indeterminate. Under slope deflection and displacement methods, we also showed kinematic indeterminacy. For kinematic indeterminacy we specifically say kinematic. Otherwise an indeterminate structure is understood as statically indeterminate. Determinate structure such as a simple beam was solved in statics section. Indeterminate methods was studied under structural analysis section.

Inflection Point:

The point on a beam or slab or column where the moment diagram changes from positive to negative or vice versa, which means, a point with zero moment. This means, the member's bending direction will change at that point.

In the figure below, the bottom two figures are from Examples 2 and 3 that we solved under slope deflection method before. Compare the deflected shape of the frames, the inflection points and the moment diagrams of those exercises, which were shown in Figures Structural 13 and 15.

Knowing the fact that we have zero moment at an inflection point is useful, which provides us with an extra equation if we are making an approximate indeterminate analysis during preliminary design for instance. The approximate locations of inflection points can be determined from the approximate deflected shape of a beam or frame which can generally be drawn reasonably well even without analysis by simply observing the constraint conditions visually by the engineer. Note: We did not cover approximate analysis method but only mentioned it under structural analysis section.

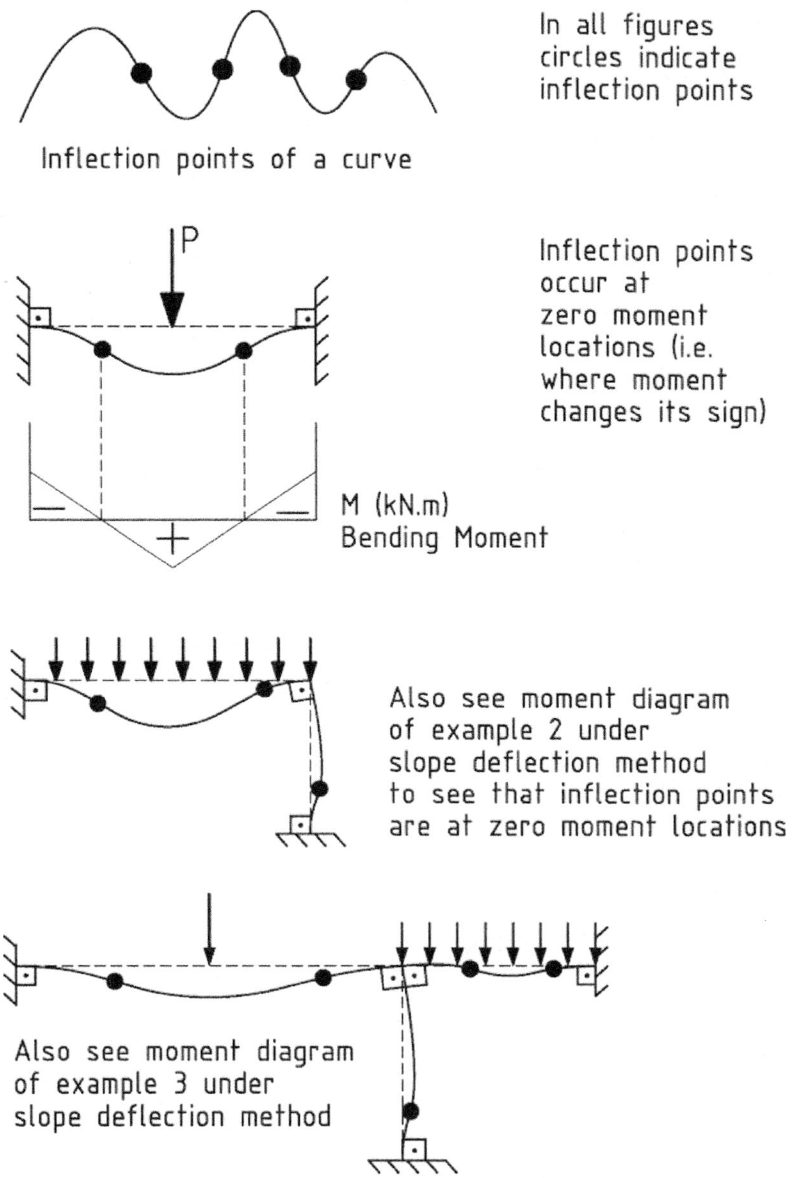

Figure - Structural 26

Also notice that at fixed supports, since rotations are zero, frame members leave supports at 90 degree angles (but immediately after leaving the support, this angle starts to change). Also notice in bottom two structures that where members join to each other, the right angle of the joint remains the same, which means, members do not rotate relative to each other (*we also took the rotation of members at joints equal when solving the exercises*). If members rotated unequally, it would mean some plastic deformation there, which goes against the main assumptions we made in the beginning of structural analysis methods (so it can happen but only at high stresses and requires more advanced solutions).

Knowing the inflection points is also very useful in that we can place the weaker spots of a structural frame member at these locations. For example structural steel column splices, reinforced concrete column, beam or slab rebar lap splices should be located at or near these inflection points, where the moment demand is already zero or close.

Influence Line:

An influence line is a graph that shows how the shear force, axial force, moment, or deformation varies at a given point, as a result of a moving unit load over a path on a member. So in other words, all values we draw on an influence line graph, are for a single point of interest which doesn't move, as a result of a unit load that moves along the entire segment of the structure that is relevant. After we draw it, we multiply the unit load by the actual load value, which also multiplies the values on influence line to give us the real values.

When multiple loads exist, we can add their influence lines to each other. Influence lines are useful to see effects of moving or different positions of loads for a desired point of interest within a structure. As you might have guessed, this becomes important especially during bridge design, when we analyse the effects of moving vehicle loads to see at a glance the worst case, in other words, the maximum shear and moment value that will occur on a given point, according to the loads acting on different points.

Again, the graphs are not the usual shear or moment diagrams. These are influence lines for points C and D for shear and moment, with respect to a vertical moving unit load form A to B. For example for point C, as the load moves from A to B, shear increases at first *(minus sign simply indicates the direction, not a minus load)*, then when the load reaches and passes point C, it suddenly reverses the direction and as the load gets further towards B, it decreases in magnitude until it becomes zero, when the moving load reaches point B. Same logic goes for point D and also for moment graphs too, if we calculated moment.

As a shortcut to determine the location of the critical load that produces the worst effect at the point of interest, the shape of the influence line, and the locations where worst case (maximum or minimum) values of influence line occurs, Muller-Breaslau principle is used (which we will not discuss).

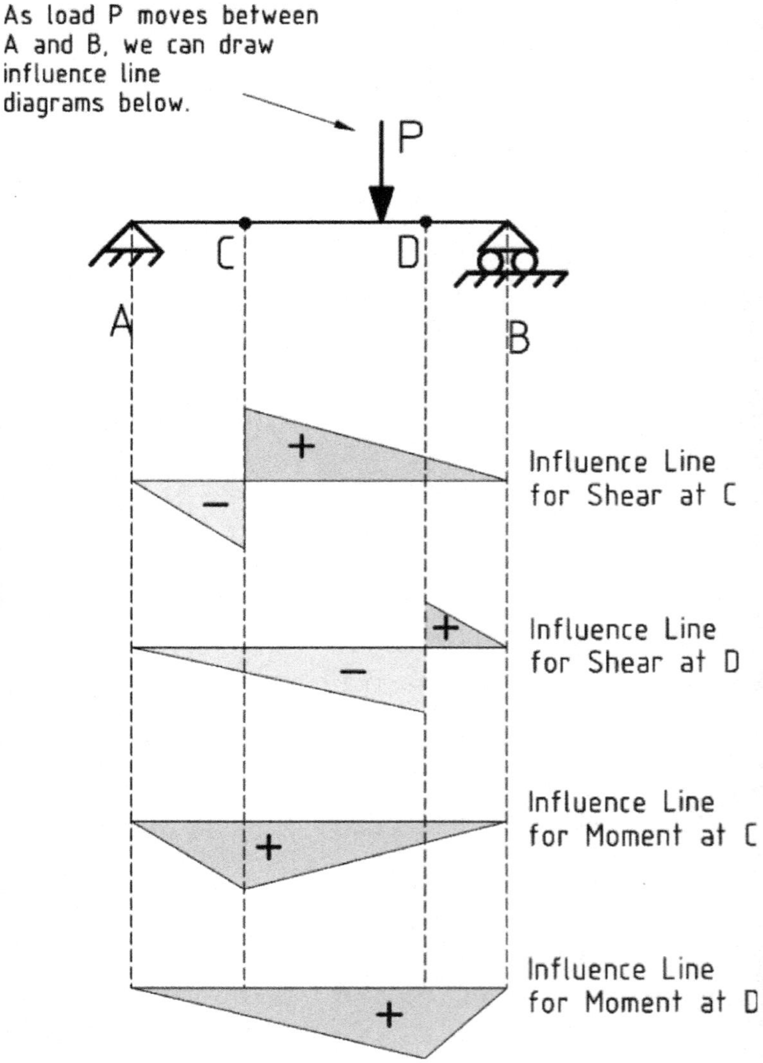

Figure - Structural 27

Joints: Construction, Expansion, Isolation, Control Joints:

Joints are used to separate parts of structure from each other in different ways, for various reasons.

Construction Joint: These are used in concrete structures to make construction easier or to allow for the next placement of concrete. So the two faces of concrete touch and adhere to each other but structurally they are separate pieces as they are not poured monolithically but separately, after the first section hardens. Construction joints can be specified during design, or, a contractor may prefer to use them for his convenience (upon engineer's approval) such as when the next delivery of concrete delays during a concrete pour, or for certain members such as large footings, large area of slab on grade, or long retaining walls which would be difficult to pour and finish in one day. If that is the case, a checkerboard pattern should be followed between successive pours, meaning, pouring one section, leaving the next empty, pouring the following... and then coming back and filling empty sections.

The proposed layout and connection rebar detail of construction joints should be submitted ahead of time by the contractor to the structural engineer for approval and recommendations.

Construction joints must be well prepared before pour to bond between old and new sections. For this, surface of hardened concrete is cleaned and any loose concrete must be removed, such as by water jets or wet sandblasting. Similarly, when we pour directly on rock (such as dams), the rock surfaces must be free of loose material, and be saturated with water to prevent it from absorbing concrete water necessary for concrete hydration, but should be surface dry.

Expansion (Movement / Flexible) Joint: These are for allowing relative movement of two sections of a structure, so that they can work separately and will not cause each other any undue stress and displacement. The movements can be because of loads, thermal expansion or contraction, wind, earthquake, overall structural deformations as a result of concrete creep and shrinkage, settlement and differential settlement and so on. Through expansion joints, a structure is basically divided into segments and to achieve this complete division, the joint can continue through anywhere within the structure as necessary, such as slabs, walls, roof deck, paving (if there is a basement underneath), to make this "slice" of the structure. Expansion joints can also be used within parts of the structure such as separating a masonry block wall from a concrete column.

Examples of usage include *(by keeping in mind the causes of movements listed above)*:

- In buildings we can use them for distinct parts of the structure which will tend to or better work separately.
- At bridges, they can be placed at the border of the bridge deck and the continuing road at grade, or even between sections of decks on different sides of a pier.
- For retaining walls, when there is a change in grade, section or alignment, where differential movements can occur.

Expansion joints are made by filling the gap between two structural segments with flexible material and then closing with proper surfacing material, depending on where it is used. The location(s) and spacing (if needed) of expansion joint(s) is/are determined based on factors including the size, geometry, materials and the type of structure, the type of its load resisting system, regional temperature variances, seismicity of the area, long term creep and shrinkage, traffic loading if any, and of course structural analysis. And the width of a joint is determined according to the anticipated movement at that location based on the factors we have just mentioned. For example, when joining two segments of a building with expansion joints, the joints at higher floors need to be wider because during earthquake higher floors will move more laterally.

Expansion joints can sometimes also be called as movement joints and it can also mean movement due to contraction, in addition to expansion, but the key thing is that pieces are working separately here, in other words, we are not talking about contraction as we will describe below under control / contraction here. Rebar do not extend through these expansion joints, or, in certain cases dowels wrapped by plastic sleeves can be used that allow relative movement in their direction. Loads, gap width, surrounding materials, whether the joint is located indoors or outdoors, future maintenance considerations, cost and other inputs as applicable, will be variables in determining the specific detail to use. For example where we need watertightness waterstops are used.

Another usefulness of expansion joints is that they eliminate thermal stress build-up between two sections of a structure. In a large structure without expansion joints, thermal loads may cause significant stresses.

Expansion joints should only be used when necessary because it is an extra cost item and it also means more effort and coordination during design and construction. They may also need maintenance later, of which, the frequency will depend on the success of design and installation.

<u>Control / Contraction Joints:</u> No matter how well we design and pour concrete, after a certain width and or length, it will crack. Control joints are deliberately made to prevent random cracking, in other words, to "control" where the concrete will crack, due to temperature shrinkage or concrete shrinkage *(see Concrete>Unfavorable effects)* effects. By making control joints, we make weaker sections at the locations we want, so the cracking will occur only at those lines, which will be much better than random cracking in terms of aesthetics and user satisfaction, and it will appear as if it did not crack, although the cracking still happens along those weakened lines we introduce and we cannot see them. On slabs and concrete retaining walls or concrete exterior walls, this type of joint is used. The rebar can still continue through the smaller cross section here and they ensure that the cracks that form at that smaller cross section are of small width.

On a slab on grade, we provide joints at every certain distance, such as every 10-15 feet or so, a distance which varies and depends on several factors including the thickness of the slab, concrete mix *(i.e. generally the lower water cement ratio, the wider gap we can have between two joints)*, amount of reinforcement in the slab, expected temperature effects, effectiveness of curing and so on. The cut sections should be kept as square as possible. Depth of cut should generally be around 20-25% of the slab thickness and the width around 1/8 inch. Although making the joints during hardening of concrete is also possible, control joints are usually made after the concrete is hardened and gains little strength to withstand the effects of cutting, but the wait should not be too long to cause cutting operation to be difficult and also to cause cracks to start establishing themselves at random locations before cutting. This means cutting should be done at least several hours after but no later than about a day of pouring the concrete. Transverse and longitudinal joints should be cut immediately one after another. Because the timing of sawcutting is so critical, the operation must continue as long as it is necessary, regardless of day or night shifts of work, and proper coordination is needed ahead of time for large projects. The cut faces must be kept moist during curing period. If a slab on grade is post tensioned, the gap between joints can be wider. This joint can also commonly be called as contraction joint.

The joints we use in dams are also called contraction joints, but that is a very different situation, although the goal is similar, which is separating very large 3D concrete blocks from each other against the negative effects of thermal contraction and concrete shrinkage.

Isolation Joint: These joints are for separating fixed objects from their surroundings, such as around manholes, foundations, utility boxes, two sections of slab from each other, columns or foundation walls from slab on grade, or, as defined in ACI 318, the adjoining parts of a concrete structure, usually a vertical plane. This separation is needed to tolerate relative movements and vibrations in all directions that may happen due to live or dynamic loads. Rebar does not continue through isolation joints.

Cold Joint: This is not a deliberately made joint, but an unwanted occurrence when casting concrete, which must be remedied. Cold joint occurs when the concrete cannot be poured monolithically, due to inadvertent delaying of pouring the next concrete batch while the previously poured batch hardens. Also see this term under concrete section, later in this chapter.

Kinematic Indeterminacy:

The number of ways a structure can move is its kinematic indeterminacy. Also see this term under Structural Analysis > Indeterminate Structural Analysis> slope deflection and displacement methods.

Linear Analysis:

See this term later in this chapter under Earthquakes > Earthquake Analysis Categories.

Moment-Curvature Relationship:

The moment-curvature curve graphically represents the variation of moment with respect to curvature (how much it bends) of a member's cross section. Moment curvature relation can be calculated for both columns and beams.

As shown under strength of materials section, moment causes bending stress as σ=My/I. Then, corresponding strain at the extreme fiber (*because extreme fiber shortens or elongates the most*) can be calculated from from ε=σ/E, as we now know the stress, σ, and we already know the Elasticity modulus E of our material.

Now, when we know strain, from simple geometry we can calculate the curvature value, which shows the amount of bending geometrically, for a given moment value. So these steps are repeated for all four points A, B, C, D (explained below) shown in the figure at a minimum, to be able to construct the curve.

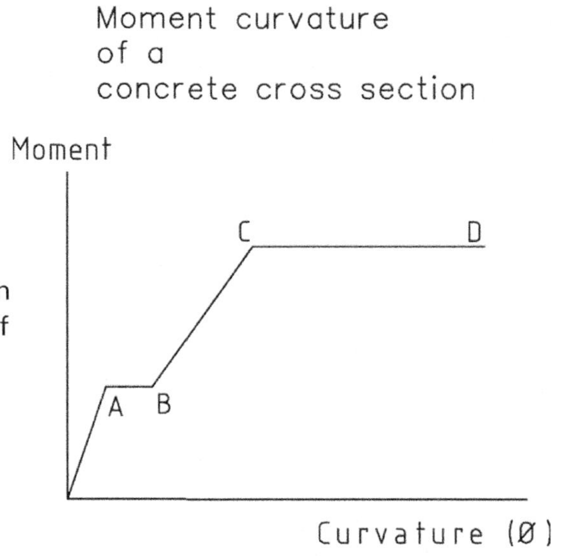

Figure - Structural 28

Moment curvature relationship is a useful tool to understand the ductility of a cross section, such as even after the material becomes plastic and cannot carry higher moment, there will still be available bending capacity, thus energy dissipation potential. In other words, from this relationship we can know how much the beam or column can keep bending without failing. This is important to know during nonlinear analysis in seismic design, to prevent sudden collapse mechanisms to form. Moment - Curvature curve is also useful for comparing stiffnesses of different members because it considers the member geometry and locations of reinforcing steel.

Important points when plotting the moment vs. curvature are as follows:

- the moment value right before concrete cracks first time, (point A on the graph below)
- the moment value right after concrete cracks first time, where rebar starts to take on stress and at this point curvature increases considerably without increase in moment, (line A-B)
- the moment value where rebar starts to yield (point C). After this point the graph is not linear, although for simplicity this graph is shown as made of straight lines here
- finally the point where the section completely fails and can not bend any further (point D).

M-N Interaction Diagram:

Moment (M) vs. axial load (N) of a column can be represented on a graph, which is one of the first things to look at for column and shear wall design because it tells us capacity of that vertical member based on critical moment and axial load combinations. An M-N combination point must fall inside of the M-N curve, for a column or shear wall to be able to withstand it. As a result of applying loads to our structure as discussed under load combinations section, we obtain different combinations of M-N demands from a vertical member and compare it with the members' limiting M-N curve.

The nominal axial load capacity is the sum of the axial loads that can be carried by rebar and concrete, given that we accept an ultimate strain limit on concrete which is 0.003. The nominal moment capacity is found by multiplying rebar forces and concrete forces by their distances to the neutral axis of the cross section. Then the relationship of this axial load vs. moment is drawn. Nominal axial load and moment can be reduced by strength reduction factors too.

Monolithic:

A monolithic structure acts essentially as one piece. When pouring concrete, monolithic pour is wanted as possible, which means all relevant components are poured in one continuous pour, without waiting concrete to harden between different areas.

The word monolithic is also used in other contexts. For example there are old structures that are carved out of rock as one piece, which are monolithic structures. In the context of underground utilities, it means a manhole made of a single piece.

Nodal Points / Nodes:

Nodes are used to divide a larger body into smaller elements. See Structural Analysis > Indeterminate Structural Analysis Methods > Displacement, Slope deflection and Matrix displacement methods, and also Computer Aided Analysis section.

Nonlinear Analysis:

See this term later in this chapter under Earthquakes > Earthquake Analysis Categories.

One way / Two way slab:

See this term under Concrete > Reinforcing Steel

Over-reinforcement:

See this term under Concrete > Reinforcing Steel

Overturning Moment:

It is one of the criteria to check for a structure, to confirm that it will not be overturned. For example lateral earthquake forces or wind will try to overturn a structure. Since moment is force times distance, the taller the structure, the more moment will be experienced at the base. For a building, overturning moment will be calculated by the lateral earthquake force on each floor, times that floors height, which are all added up, to find the moment at the base. Overturning moment is often countered adequately by the structure's own weight. If not, the foundation's contribution is needed to resist it.

P-Delta (P-Δ) Effect:

P-Delta effect is a second order effect (the effects produced based on the deformed body of the structure). P-Δ effect is caused by gravity loads acting (compressing) on a laterally deflecting structure, so it is a matter of structural (geometrical) stability. Here we analyse the structure based on the deformed geometry, and must satisfy equilibrium for it, then do the same for the next new geometry and so on, and see if the result diverges (failure), or converges (stable). As we mentioned under structural stability section, failures as a result of geometric instability (structural instability) are usually sudden and catastrophic. It is difficult to analyse P-Δ in inelastic range and under dynamically varying loads, but we can make simplifications to elastic and static loads.

Let's say lateral force is acting on a structure, which causes lateral movement. Now the weight of the structure, which is a vertical force (P), will act at a distance from its previous line of action, which is the eccentricity (Delta). This means a new moment is created, just because of this new eccentricity. So now we must examine this new situation, and come up with a new eccentricity amount and repeat this cycle at least a few times. We must also take into account any possible reduction of the lateral stiffness of the structure. Then we will see that the result may either converge, which is good, because in each new calculation, incremental eccentricity will be less, or, it may diverge, which means each incremental eccentricity will get larger which will ultimately mean collapse.

During earthquakes, the lateral force acting on the structure must be resisted by systems such as shear walls, bracings or moment frames (see Chapter 2) or their combinations. For shear walls or braced frames, P-Delta effects are usually not a problem, but moment frames can only resist lateral force by creating moment at joints so they sway to sides considerably more. Swaying means more eccentricity which means higher P-Delta effect. The more slender columns, the more swaying. Since P-Delta effects arise from the lateral drift of structure, drift control plays an important part in controlling these effects. In most structures, where we can make linear analysis and displacements are small, the original geometry of the structure can be assumed to remain the same and P-Δ effects will be negligible.

There can also be local deformation of members, which is called P-Deformation effect, shown as P-⌋, such as when the member itself starts to buckle with respect to its own center axis (chord) because of compression. It is generally less of a concern than P-Δ effect, although it may be included if the analysis is very detailed. So again, P-Δ is for overall structure or part of it, and P-⌋ is for a certain member.

Pier:

A pier is a very wide column, such as pier of a bridge or archway. For a different meaning of "pier", also see Chapter 9, Coastal and Marine structures.

Plastic Hinge:

As we saw before under strength of materials section, steel behaves as elastic at first and then yields and enters into plastic region. A point around which a structural member is purposefully allowed to, or upon analysis expected to yield and rotate, is a plastic hinge. We may want hinges to develop at the locations we want, during an earthquake. For this, the members that are supposed to develop hinges are designed such that at certain locations after a point they will yield such as end of beams, and act as fuses by spending energy *(deformation controlled members)*, so that the other members that are supposed to withstand loads without yielding and carry the structure *(force controlled members)* can do their job without developing hinges, such as the column ends.

Portal Frame:

A portal frame is a structure that resembles a portal, where a beam is supported on columns. The connection of beams to columns are fixed type, which enables moment transfer. So it is basically a two dimensional structure, and by making multiple of these tied and braced to each other in third dimension, any length of structure can be obtained. Although this type of structure is inherently single story, large interior spaces can be obtained such as for hangars, warehouses and similar structures, where unobstructed wide and high interior spaces is desired.

A portal framed building structure, which will be used as a storage building.

Progressive Collapse:

Also called disproportionate collapse, it happens when a primary structural element such as a column fails, which triggers failure of adjacent elements, which creates even more failure. It can cause failure of an entire structure or a disproportionately large part of it, with respect to what that original failing member supported. Precast concrete buildings are relatively more vulnerable to progressive collapse but it can happen in any structure without proper design, or it can also occur as a result of extreme events such as a nearby blast or a very heavy impact.

To prevent progressive collapse, alternate routes for loading must be provided with redundancies in the design. For example, say, if a column suddenly failed as a result of a blast, the beams over that column can be designed to work in tension, as in catenary (cable) action, and still carry the structure above without total collapse and save lives, even though there will be very large displacements and structure may or may not be repaired.

Punching Shear:

Punching shear failure occurs when a structural member such as a column pointing toward a flat member, such as a slab surface or foundation just "punches" that flat member and causes it to fail under high shear stress, typically similar to as in the figure below, along the broken surface.

Here to avoid the shear failure, we could use shear caps or sufficient reinforcing detail. We want to avoid shear failure since it is brittle, and try to make bending failure critical (happen before), which we can tolerate, (depending on the performance level we aim for), since it is ductile.

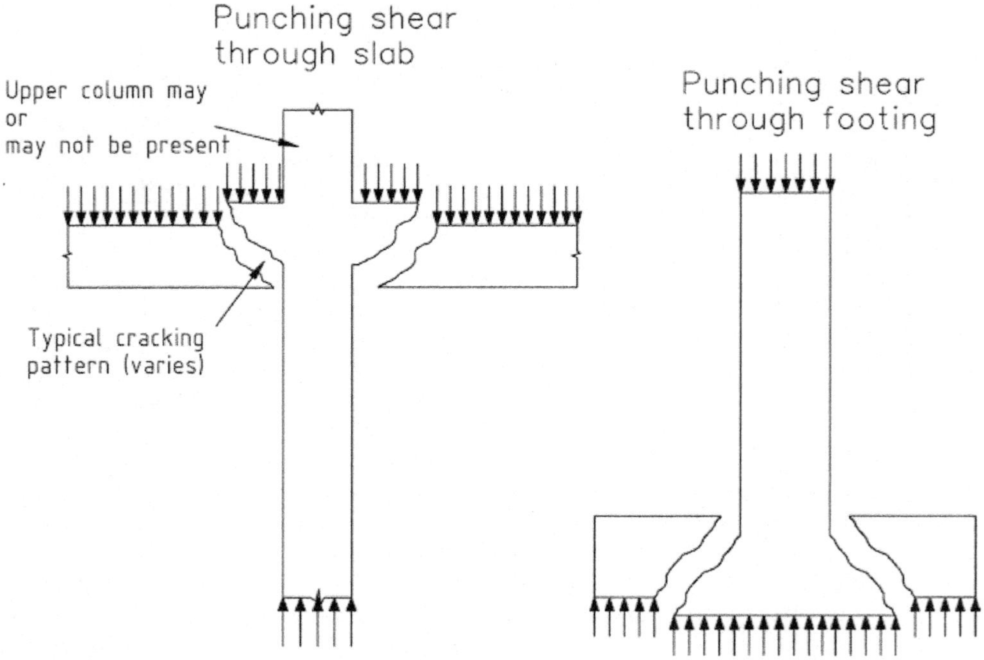

Figure - Structural 29

Redundancy (Degree of redundancy):

The number of reactions that the supports are able to provide beyond the three (in 2D) equilibrium equations. See Statics > Determinate vs. Indeterminate structures.

Rupture:

Rupture type of failure can happen in different ways for steel. It can happen after the ultimate yield point, when the member simply breaks or gets torn apart. See Figure - Strength 12 also. Rupture may happen at low stresses too, even lower than yield stresses due to nature of loads, connections, and member alignments, for example when certain members are torn as if a sheet of paper.

Second Order Analysis:

In basic structural analysis that we discussed, one assumption was that we did not consider second order effects, which means the new situation arising from the displacements of the structure. See also the term P-Delta effect and, Earthquakes > Earthquake Analysis Categories > Second order analysis.

Shear Caps:

To avoid punching shear failures, shear caps are provided. They are made by thickening the slab at column-slab joints, so that the same shear force is met by a larger area, which means reduced stress.

Shear Failure:

Shear failures are brittle (which means sudden), which we do not want. We must design our members so that they will fail in bending (flexure), which is ductile, before they fail in shear.

Short Column Effect:

A vertical member's stiffness for lateral loads depends also on its length. As length increases, lateral stiffness decreases. For example if we had columns of equal cross section and material, but different lengths, the shorter one would be stiffer, which means it could take on more of the lateral load during earthquake because stiffness attracts load. This will make the shorter column more vulnerable, so short columns must be avoided, unless we deliberately want them to take on more lateral load.

Short columns can occur because of situations such as:

- A partial height wall built near a column and not separated by an expansion joint. This can happen especially when a building owner decides to add such a wall later during remodeling for instance, thinking that it is a nonstructural item which requires no engineer input, but the wall may create the short column there, although the original design did not intend it.
- A stair landing, parking ramp or a beam joins the column at intermediate height.
- One side of the ground is higher as in the figure below:

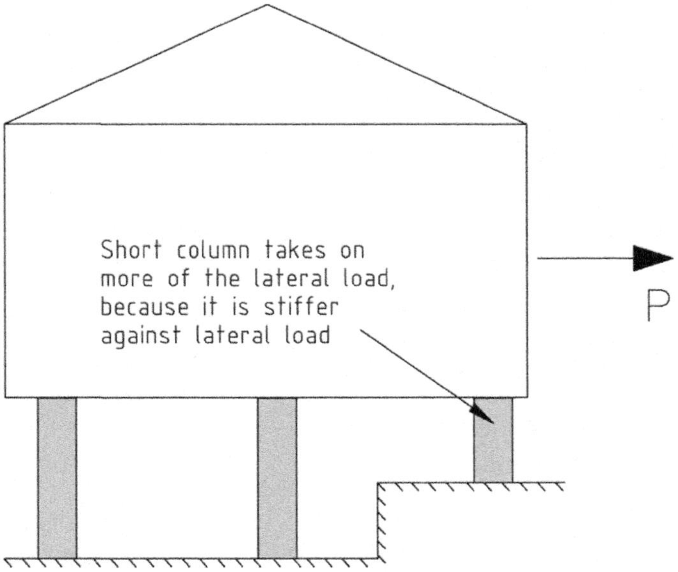

Figure - Structural 30

The short column situation can be remedied by applying one of the measures such as:

- make other columns laterally stiffer so that they are in range of lateral stiffness of short column,
- remove the obstacle that cause the column to work as short column, or provide expansion joint between it and the column, to accommodate column movement and effectively separate the obstacle from column

- reduce the short column cross section to the point possible, in order to make it less stiff
- join the column and the adjacent wall to eliminate the short column, which creates a shear wall. If we do this, we must check its effect on stiffness distribution of floor, to not cause building torsion
- reinforce the short column per special details as in codes

Soft Story:

It is a floor in a building with considerably less stiffness than the others. Such floors will make the most lateral drifts during earthquakes. The shear resisting ability of floors should not be different from each other too much, which would create a soft story that can lead to failure. The term "weak story" is often used interchangeably with "soft story", which indeed generally is the same floor, but technically the term soft refers to a low stiffness, while the term weak refers to low strength (something may be strong but not stiff or vice versa). Often, the soft or weak stories happen at the first floor, where the floor height can be higher, with more slender columns, due to having a lobby etc... at such floors.

Stiffness Matrix:

A stiffness matrix shows force deformation relationships of a structure in matrix form. Also see matrix structural analysis, and computer aided analysis sections.

Story Drift:

Story drift is one of the primary criteria set by codes, to be satisfied for all buildings, due to earthquake and wind effects, no matter what structural system or material we use. Story drift criteria, which falls under deformations / service design, is so important in buildings that, it could be considered as equally important in comparison to the buildings' design for load carrying capacity.

This criteria basically tells us that a building's floors must not laterally move beyond a certain limit as a result of earthquake or wind. It is usually measured as a ratio by taking into account the drifts vs. floor height between two adjacent stories. Too much drift can create secondary moments and p-delta effects, too much plastic strains and damage in structural members. It will also cause damage to nonstructural components and discomfort to occupants.

Story drifts are influenced by the stiffness of a building. Story drift is a useful indicator of the overall effect of all types of displacements in the structure. Column rotations, overall torsion of the building, slabs' nonrigid deformations, partial or intermediate floors if any, non-vertical columns if any, all influence story drift and make evaluation more complex, which also interact with other building responses such as bending of beams. For example too much overall building torsion may cause using story drift criteria less meaningful.

Not only for structural but also for nonstructural components, such as facade, partition walls and other attached components to the structure, story drift is a good overall limiting measure, as far as the displacements these components can tolerate. Therefore, limiting story drifts will mean lower displacement effects for overall and member level. Lower deformations can also be better predicted and analyzed. Story drift ratios can also change from floor to floor, which allow us to compare displacements, stiffnesses throughout the height of the building.

Structural Systems:

In a general sense, one way to group / list structural systems might be as below. Some structures are made from more than one of the systems combined. A reminder here is that there is no one / best / correct way of grouping / listing and this will vary in almost every source because it can be made by primarily considering different criteria, in different ways and going into different level of details according to how one is making the list.

- Frames (columns, beams, diaphragms, shear walls working together)
- Trusses
- Braced systems
- Surface / Membrane structures
- Cable structures
- Beams (i.e. beam bridges)
- Tower structures
- Arches / Domes
- Gravity structures
- Retaining systems
- Tube / Pipe structures
- Stone / Bearing wall structures
- Embankment / Road structures
- Tunnels

Strong Column - Weak Beam:

Where columns and beams join, we want columns and column-beam connections to be stronger than beams, so that the beams will fail first but not the columns. Column failure would be much more catastrophic than beam failure, such as causing progressive collapse. Another reason to want beam failures is greater energy dissipation, as it will be in much more locations than column failures.

Strong column - weak beam condition can be obtained by allowing plastic hinges to form at the ends of the beams, which would behave like hinges, dissipate the energy, while columns remain functioning. In other words, we want to make the moment capacity of the columns and the column-beam joints, stronger than the moment capacity of the beams. In concrete structures, special reinforcement detail, and in steel structures special connection details are needed for these locations. Also see Figure - Buildings 3, in Chapter 2. This failure must be from tension, not shear, because tension failure is ductile and can consume lot of energy during elongation of rebar, whereas shear failure is sudden and brittle.

If for a reason certain columns cannot satisfy strong column weak beam requirement, they cannot be a part of lateral force resisting system and carry gravity loads only *(which means the lateral forces for the entire structure must still adequately be resisted but by other columns and or walls)*. But this doesn't mean those columns do not sway with the building. They must still be able to withstand deflections from swaying, while supporting the gravity loads they are supposed to carry at the same time.

Strong column - weak beam concept is one of the most critical items to make a building resist earthquake. Another important point to be satisfied at beam-column joints is to prevent shear failures, which is brittle, and instead allow flexural moment failures, which is ductile. These two criteria ensure high level of ductility at these locations.

Strut and Tie Method:

It is a useful method of structural analysis of concrete structures (regularly reinforced or prestressed), where the stress flow paths in concrete members are represented by truss members, such as by compression members (struts) and tension members (ties), all of which are connected by nodes. The path / trajectory / flow of stresses depends on the geometry of the structure and where the loads act. As always the main objective is to satisfy force equilibrium through logical force paths with these struts and ties. Where there is a sudden change or discontinuity in a member's geometry, irregular shaped geometries (such as a shear wall with openings, corbels, pier caps, frame corners, slabs with openings etc...), where there are concentrated loads, these cause irregular stress distribution in the members. Strut and tie method is useful in representing the stress flows in such situations, and determining reinforcement layout, which means a large sphere of application in structural analysis. It is also used in members where shear is critical such as deep beams *(a beam with low span (length) to depth ratio - i.e. per Eurocode 2, section 5.3.1, if length to depth ratio is below 3, it is considered a deep beam)*, nonslender shear walls *(a wall with low height to width ratio)*. Shear critical means, flexure is not critical. Flexure critical members are the ones such as slender walls or long shallow beams, where a straightforward beam analysis would be sufficient to represent stresses.

When a concrete member is loaded, there will be tensile and compressive stress paths, which naturally form, in the simplest way possible as everything else in nature *(for example it is the natural flow that a proper finite element analysis program would generate)*. In strut and tie method, we must approximately, reasonably represent those natural stress paths in the simplest way possible with truss members, therefore, using strut and tie method requires good understanding of stress flow. There is not only one way to do this, but we can still follow certain rules to make the model *(for example, struts generally being parallel to direction of expected cracking, ties being perpendicular so that they take tension, nodes placed in front of supports, external loads and around corners, a minimum angle between members...)* Using a 2D or 3D model depends on the complexity of the situation. After a preliminary model is drawn, it can iteratively be refined.

Where we have ties, we need to provide reinforcing steel with proper details, to take on that tension. The nodes also must be checked. The detailing of a node point will depend on the angle and geometry of connecting members, whether compression or tension or both type of members are connected, and whether equal or unequal stresses are pressing / pulling on the node faces, all of which have effect on the node strength. Nodes must be in equilibrium, as anywhere else, which means, the compression and tensile forces acting on nodes must be in equilibrium, as also done when solving trusses.

Codes and standards such as Eurocode 2, ACI 318 and AASHTO LRFD Bridge Design Specifications allow using strut and tie method, and describe how it can be applied. The truss analogy was known since over a hundred years ago (Ritter, and then Mörsch) and then studied by various researchers. Building on top of that knowledge, strut and tie method were first mainly described to apply to entire structures by Schlaich et al. in 1987, and others have also contributed in separate studies.

Superposition Principle:

It means that the total effect on a structure caused by actions is equal to the algebraic sum of effects that would be caused by individual actions acting separately. Here action means, applied forces, moments and the displacements they cause. For each of the actions, we can draw a substructure, analyse the effects, and then simply add the results produced in the substructures to obtain the results that would be obtained in the original structure. This is a very useful principle in structural analysis. It can be used when solving indeterminate structures, as was shown under structural analysis section.

For the principle of superposition to be valid, the structure must behave as linear elastic and the displacements should not create any second order effects, as we assumed in structural analysis section.

Superstructure:

It means the structure itself that are above the foundations.

Surface Structure:

A surface structure is a distinct type of structure, where the forces are carried with surface tension, such as liquid tanks. Discussion of these are out of this book's scope.

Tributary Area:

Tributary area of a beam or column is the portion of slab that is carried by that beam or column. It is found geometrically by simply drawing a line between two beams, which is equidistant from both beams, in other words, a line that is located at half the distance between two beams. When we draw lines for all beams in this manner, we can see the tributary areas for all beams and columns.

Truss:

A structural assembly made of members joined by hinges, that make triangles. Truss members only carry axial loads in tension or compression but <u>not</u> moment. A truss can be made in the form of anything, such as a beam, column or even slab where it can also resist bending moment as a whole, although the individual members do not carry moment. Considerable structural efficiencies can be achieved with trusses. Trusses were discussed in more detail at the end of statics section.

Waffle / Ribbed Slab:

See Chapter 2 > Building Frame and Structural Members > Slabs.

Repair of Structural Damages

Structural repair is mainly bringing parts of a structure to comply with current codes, if deficiencies are found by tests, visual inspection or structural surveys.

Structural damage or deterioration can be caused by many factors including:

- Poor design and/or construction practices
- Using inadequate quality or quantity of materials during construction
- Settlements (especially differential), due to inadequate foundations, which affect the structure
- Weather, water effects
- Inadequate preventive maintenance
- Inadequate efforts to remove or reduce the source of damage (i.e. not removing or repairing the source of leaking water near a structural element)
- Earthquakes, impacts, blasts. We will discuss earthquakes later in this chapter.

The repair cost of structures are usually many times the cost of properly doing or preventing the items above. Therefore, before things get to the point of structural repair, the first thing to do is to prevent damage from happening. Also see "Repair of concrete" under concrete section.

CODES, STANDARDS, GUIDELINES, SPECIFICATIONS

When designing structures, first of all we must know what codes, standards and guidelines will apply. In addition, the project will have its own specifications as well. Each country has its own codes to follow and each state or jurisdiction within a country may also have more specific requirements.

Code: They are written by governments or entities that are approved by government. Codes are ultimately about public safety and comfort, in other words, they set minimum limits for structural safety and serviceability for everything we want to design and build and they are enforced by law. They provide the minimum or maximum values, details, parameters, equations to use, which assumptions to make in what situation, which alternatives to choose from, and so on. When government adopts a standard or guide and it becomes legally enforceable, it becomes a code. Codes are enforceable by the applicable authority for where a project is located, regardless of the terms of a contract between an owner and a contractor, *(which may bring even more stringent requirements on top of the code)*. Codes can refer standards or guidelines.

Standard: It is a set of detailed technical definitions, instructions, processes, criteria for providing products or services, and sets common language to standardize things. They are written by governments or approved organizations. For example, ASTM International (USA), DIN (Germany), BS (UK) and more standardization organizations in the world provide standards for essentially everything. If a standard is referred to in contract, its use becomes mandatory for that particular project. Standards are also referred to in codes, guidelines, manuals, specifications. Standards provide huge savings and added value into economy, by ensuring that everyone follows the same path and speaks in same, equal terms for a particular product and service, which prevents a lot of confusion and inefficiencies, and makes everything more effective, in terms of planning, costs, execution, usage and so on. Note: Codes may commonly be called as standards too in daily language, although this is not technically correct.

Guideline: These are similar to standards or codes, except their language is less strict. They are also reliable and very useful documents and can include a lot of fine details similar to standards. In many projects and type of work they are also followed very commonly and referred to in contracts. When included in a business contract, guidelines or standards or manuals are also enforceable. About a certain subject, a guideline may appear first, and then over time, after a lot of experience and feedback from the industry and academic sphere, they can evolve into standards and codes.

Manual: It has more or less an equivalent meaning to guideline but generally more directed towards actual application or usage, such as instructions of doing something (in design or construction).

Specification: Specfications can mean either the ones developed by approved organizations to serve as manual, guide and standard, or they can also mean the specifications developed by the consultants of a project, for that project specifically, although they do not differ much for same types of projects, other than project specific requirements. Project specifications provide additional requirements for materials or services on top of codes, standards, guidelines or manuals, such as project specific product characteristics, installation requirements, performance criteria, dimensions and so on, specifically for a particular project. They are part of the contract between the owner and contractor, which must be followed by the contractor. Specifications refer to codes, standards, manuals or guidelines often, but even if it doesn't, the most stringent code requirement applicable in that jurisdiction would still govern.

All five items listed above are ultimately based on proven and well established engineering theories, long time prior experience and experiments, statistical studies and calibrations, and expert judgments and agreement of highly competent teams of engineers. On top of the established engineering background they are based on, they introduce assumptions to guide engineers to be on the safer side, to ensure that everything is more standardized, simplified, uniform and easier to understand, therefore avoiding confusion and make everyone's job easier, such as engineers, architects, manufacturers, owners, installers and contractors. As a result, they create safe, functional and more cost effective results, which save a lot of time and money in an industry and reflects to overall economy of a country.

Several of the most commonly used codes and standards in the US in civil engineering and construction are listed below. Many states and local jurisdictions have their own requirements on top of them. Other countries have their own, and many also has adopted US codes and standards by modifying to their own needs, or some published their own. For example in Europe, there is the Eurocode, which are followed by many countries or in India there is the Indian Standard, IS, and so on... The first numbers in the list below indicates the ID number of the document and any numbers after the dash indicates the current version number as of the date of last update of this section, in February 2023:

- ASCE 7-22: Minimum Design Loads and Associated Criteria for Buildings and Other Structures - American Society of Civil Engineers - www.asce.org
- ACI 318-19(22): Building Code Requirements for Structural Concrete - American Concrete Institute - www.concrete.org

- Steel Construction Manual - 15th Edition - American Institute of Steel Construction - www.aisc.org
- Specifications for Structural Steel Buildings - ANSI/AISC 360-16 - American Institute of Steel Construction - www.aisc.org
- Seismic Provisions for Structural Steel Buildings - ANSI/AISC 341-16 - American Institute of Steel Construction - www.aisc.org
- International Building Code (IBC) for buildings-2021 - International Code Council Inc. - www.iccsafe.org *(In 1994, organizations that published the former widely used codes UBC, BOCA and SBC merged into one organization as ICC and since 2000, IBC replaced those codes.)*
- A Policy on Geometric Design of Highways and Streets, 7th Edition 2018 - American Association of State Highway and Transportation Officials (AASHTO) - www.transportation.org *(This reference is also commonly called as "The Green Book")*
- Standard Specifications for Highway Bridges, 17th Edition - AASHTO - www.transportation.org
- AASHTO LRFD Bridge Design Specifications, 9th Edition - AASHTO - www.transportation.org
- TMS 402/602-22: Building Code Requirements and Specification for Masonry Structures - The Masonry Society - masonrysociety.org
- SEI/ASCE 37-14: Design Loads on Structures During Construction - American Society of Civil Engineers - www.asce.org
- NEHRP documents: National Earthquake Hazards Reduction Program (NEHRP) provides guides and recommendations on earthquake related design, construction. Federal Emergency Management Agency, FEMA, which is one of the four agencies that contribute to the NEHRP, has a lot of useful guides, for earthquakes and other hazards - www.nehrp.gov

There are also other codes or standards from the organizations listed above and other organizations such as ASTM, American Wood Council, American Iron and Steel Institute, American Welding Society, Department of Defense (as far as construction too), Metal Building Manufacturers Association, Steel Deck Institute, Steel Joist Institute. Construction also involves trades other than civil engineering related disciplines, such as for electrical, mechanical, fire protection and so on, for all of which there also exist a lot of codes, standards, guides and manuals. In addition to adopting the ones above, each state and jurisdiction can also have their own codes and standards, such as the California Building Standards Code - Title 24 in California, for buildings.

COMPUTER AIDED ANALYSIS

Note: Chapter 2 > Structural Modeling and Analysis of buildings section introduces concepts about computer modeling of a structure, which is a continuation of, and complementary to this section.

Even before computers, engineers were able to build structures such as the Empire States Building or the Brooklyn Bridge. They used structural and geotechnical engineering concepts developed in the last few centuries, many of which are either mentioned in this book. Today engineers use computer software to analyze structures. In comparison to the past, these programs use more advanced methods, do the job much faster, thus the whole process require fewer engineers, can provide better accuracy in representing behavior, so bigger and more efficient structures can be designed and built.

Computer aided analysis and design of structures is a topic that can be approached in various ways. Our approach in this section will be to summarize main ways of modeling. Then, as a continuation, in Chapter 2 we discuss more aspects in modeling and analysis, by also using buildings as focus / example. Note that the structural analysis itself is discussed from different angles in structural engineering and earthquake sections on this chapter, and in Chapter 2.

Modeling of structures can be done mainly as:

- modeling small number of distinct elements to represent distinct members of the structure which are connected by nodes or hinges which is commonly called as discrete modeling,

or,

- a continuum of a large number of small elements connected by a large number of nodes, which divide structures / members into far smaller pieces that work together under applied forces, which is commonly called as Finite Element Method (FEM).

We can use these methods alone *(more often the first approach)*, or, apply both together for parts of the structure, to get the best of both, for efficiency. In short, both methods essentially do the same thing, which is representing how a structure will deform and experience stress under given set of forces, based on its (the structure's) stiffness by applying principles of structural analysis.

In discrete modeling we are confident that we know how members will generally behave and we will not miss any critical response even if we represented them with relatively small number of distinct elements, to behave the way we expect. On the other hand, we use FEM when we need to understand complex force deformation behavior by dividing everything in question into a large number of very small elements that work together under the action of forces, to naturally come up with stress and deformation outcome by itself, which we might not have predicted entirely beforehand. With FEM the complexity, solving difficulty and the number of equations increases in comparison to discrete models.

Although discrete models are simpler, it doesn't mean FEM is better in every way. They have different usefulness. This is why, in many cases we can use both in a model. The decision to modeling approach is affected by the level of complexity we need, the expected failure locations and modes, degree and location of expected nonlinearities. But do not forget that in addition to advanced mathematical methods suited for computer solutions, both also use the same fundamental structural engineering principles as essential ingredients that were discussed earlier in this book under statics, strength of materials and structural engineering sections (and of course soil mechanics for geotechnical analyses).

Computer analysis approximates force-displacement relationships as linear equations which are combined as [f]=[k][d] matrices. Then, as discussed under matrix stiffness method in Structural Engineering > Structural Analysis section, these matrices are combined and their coordinate systems are adjusted by multiplying with rotation matrix in order to obtain one global matrix as:

[F]=[K][D]

where,

[F] is global external loads vector, [K] is the global stiffness matrix, and [D] is the global displacement vector that we are after at nodal points.

We want to solve this system upon applying external loads and support conditions, to find displacements at each node, which means solving the structure, because then we can also know the external reactions, strains and stresses, force and moment diagrams everywhere.

Whether we use discrete modeling or FEM, if a structure is symmetrical (and the loads also act uniformly), we can solve in 2D. Similarly in geotechnical analysis, if the problem has axial symmetry, the loads and deformations are expected in a 2D plane, the geometry is constant in the third dimension, and the material can be assumed as homogeneous, a 2D analysis is sufficient. For example if we are analyzing retaining wall, or a slope, which continues for a long distance (and without any other intersecting features along the way) and under the same loads into the paper plane (3rd dimension), we can draw its cross section in 2D, and solve in 2D, or solve multiple 2D sections and impose them to a 3D solution. This was more critical before 2000s but since about 1-2 decades software and computers are powerful enough that we are also able to routinely perform complex 3D analyses for any situation.

The bottom line about the usefulness of computer aided analysis is that, although in theory we can manually solve structures and geotechnical problems with traditional, analytical solutions and principles, many of which were mentioned or introduced earlier in this book under structural and geotechnical related sections, we utilize enormous calculating power of computers to apply advanced mathematical numerical solutions which can enable solving large and complex situations far more faster and efficiently (including increased accuracy in most cases), for better designs. Now let's take a closer look at the two types of modeling we mentioned above.

Discrete Models

In discrete models, which are generally adequate for most projects or most parts of a structure, we represent the elements of a structure with elastic 1D line and 2D surface elements elements connected by a limited number of nodes and we are confident that the right behavior can be represented adequately at the right place this way. In most cases a discrete model of a structure with a few hundred elements and nodes can adequately represent the overall behavior of a structure in a far more simpler and efficient way than modeling the whole structure as a continuum of small finite elements with many thousands of nodes. This statement applies to even for determining seismic response of the structure.

For example, go back to the beginning of the book, where we solved a simply supported beam in statics. You see only 1 line as a beam and a two nodes that represent the supports. That was all we needed to see the internal forces and moments; we could still draw the axial force, shear and moment diagrams of the beam. For that we did not need to represent the beam as a continuum of thousands of small elements connected by thousands of nodes which would require computer solving many nodes with a large matrix. Yes we could still obtain force and moment diagrams, but that would be totally inefficient and unnecessary. Even to see the deformed shape, we could simply employ methods as described under strength of materials section. The examples that we solved in structural analysis section were also discrete models. If we tried to analyze the same examples by telling the program to mesh the entire frame in 3D, thousands of nodes might have been generated, with little extra benefits (and far heavier calculation load), because after all, the location of maximum moment or shear will not change, nor the value of forces and moments at intermediate locations, nor the overall deflected shape, nor the reaction forces, as long as our few nodes represented it with adequate precision (did not cause us to miss important behavior). Therefore, for analyzing skeletal frames such as column and beam frame elements, or trusses, 3D mesh of a continuum of elements as in FEM is generally not needed unless we expect complex behavior. Those can be simply solved by discrete models by making nodes at obvious member junctions, supports, hinges etc...

Let's list some of the important points about discrete modeling:

- In discrete modeling, members can be modeled as force or deformation controlled - remember we discussed this concept in the beginning of Structural Engineering > Structural Design Approaches section. Force controlled members are modeled as elastic.
- 1D line element represents the entire cross section of that member as if concentrated at its geometric center, but care must be taken if the analysis is nonlinear and many other things are modeled in more detail.
- A beam for example would be represented by line element, (to which the real cross sectional geometry is assigned), with connecting nodes at its two ends to other beams or columns. The nodes can be made as rigid links. A slab can be represented as an assembly of smaller 2D rectangular surface elements connected by nodes at its four corners to other surface elements or beams or columns. All these nodes will have 6 degrees of freedom in 3D. The only nodes that will have less degrees of freedom would be the external supports.
- Discrete modeling is not just for 2D but also for 3D models. We can very well represent structures in 3D with discrete modeling, only by using line or surface elements. In fact, most of the time using discrete modeling we make models in 3D, by placing elements and their connection nodes in x y z space.
- With discrete modeling we can represent shear force vs. shear deformation, axial force vs axial deformation and moment vs. rotation with springs at certain locations (such as zero length springs concentrated at certain locations), when we want to model behavior more specifically such as nonlinear modeling, by assigning nonlinear force-deformation properties to springs. So we force (because we can predict) to the model where nonlinearities occur as plastic hinges. This way, between the hinges, we do not need to make the members nonlinear and can define them as linear elastic. Yielding of these springs will cause the stiffness of the entire structure to be recalculated and decrease. So if we know what we are doing, and the nonlinearities will indeed happen at those locations, this is the most computationally efficient way. For example at columns and beams, these hinges may not be right on the connection point, but near the beam ends, to represent beam yield there, to satisfy strong column-weak beam criteria. For a deep and long beam carrying a large gravity load, we may also want to add another hinge in the middle. For a shear wall, we may define a hinge at mid height of a floor. Other than knowing the right location, it is also important to define stiffness as accurately as possible. The yield point of axial, shear or rotational springs will represent the force, shear or moment resisting

capacity of those members where springs are located. We may assign the yield points as upper limits in order to avoid large forces and moments for those members. We can also make hinges as not concentrated but having a certain length, which allows us to represent distribution more accurately than concentrated hinges if needed, but concentrated hinge models are used most commonly. Even though we can account for nonlinearities in hinges, it is hard to represent the exact post yield behavior. When we want more details, we use fiber type or finite element modeling, although it is less common.

Fiber Type Models:

Apart from concentrated hinges, a special subgroup of discrete models is fiber type modeling. In this modeling approach, the member cross section is divided into at least several fibers that run axially along the entire or a part of member length. Any inelasticity can also be distributed within these fibers as needed. Fiber type models, as far as their complexity, lie between discrete representation of elements and continuum finite element method. In comparison to concentrated hinge models, fiber type models allow defining materials themselves more precisely because here each fiber can be assigned its own stress strain relationship, therefore fiber type modeling allows more complex load effects and interactions and inelastic deformations within members more precisely *(which comes at the expense of increased calculation load because of integrations along fibers).*

Fiber type modeling can be applied to any member but it is most appropriate for slender shear walls, beam-columns where we have both axial loads and bending moments, by modeling it as made of fibers, which allows more precise defining of axial load vs. moment relations. For reinforced concrete we can represent concrete with a few fibers and rebar with another fiber where each have their own stress strain curve. Fiber type models are used less commonly than concentrated hinge models and the best way to use them is in combination with concentrated hinge models, where we can benefit from efficiencies of both, to represent the structure's behavior.

Continuous Meshed Finite Element Method (FEM)

With the advance of computers after mid 20th century, matrix methods, which, engineers had already started to develop in theory, could then be used to solve large matrices, along with advanced mathematical methods which facilitated the development of FEM. By 1970s, FEM started to gain widespread acceptance. It enabled solving complex physical shapes under loads, which is not only useful in analyzing structures but also machines and parts. FEM is also used in other contexts such as aerospace, automotive, chemistry, fluid dynamics... and it was one of the biggest engineering advances in 20th century because it allows us to closely approximate material behavior in its most fundamental level in 2D or 3D in sufficiently high resolution, therefore accurately predict the deformations and stresses of even objects of complex geometry under the influence of complex set of loads in 3D.

With FEM we can see for example how the stresses and deformations will be distributed on a connection, on a large mat foundation, a member under buckling, and many more complex cases which we will list, where accurate representation of stresses and deformations, load paths must be understood as part of a complex behavior, which we would not be able to easily predict ourselves and model with discrete line or surface elements and hinges, and when analytical solutions would be too difficult. For example if we represented a steel connection with a single node, or the large mat as a single plate or shell, we would understand nothing about the real distribution of stresses and deformations within the connection or the mat. FEM would bring the realistic deformation and stress distributions, load paths, in other words materials' behavior as real as possible, which would naturally occur instead of us forcing them, based on applied forces and our model's geometry and stiffness. This will often lead to not only safer, but also in most cases more efficient and economical designs that meet strength, serviceability and stability and other criteria as applicable.

So, as long as modeled correctly and the processing software has adequate capabilities, FEM is almost the equivalent of physical testing, simulated in virtual world. But representing everything in a structure as such is not needed. Typically we use FEM to represent only the locations with relatively complex situations in a structure, and can take care of the rest in adequate detail with discrete models as discussed above. Representing a structure or a part of it with FEM when we could also do it with a discrete model is not needed and would unnecessarily increase calculation load.

In addition to basic structural analysis principles, FEM also depends on numerical methods and advanced mathematical solutions (i.e. variational or Galerkin methods). A numerical method reaches a solution by approximating, instead of trying to solve equations exactly (as in analytical method, also called closed form). Numerical methods were started to be used with the advance of computers. FEM is a numerical method. There are also other numerical methods than FEM, such as the finite differences method (which is also easily applicable to computers) or others. If we tried to solve the force deformation interactions of a 3D object exactly, we need analytical equations to solve infinite degrees of freedom with infinitely difficult equations, which is not possible. So instead of that, we try to get an approximate solution which is still close enough to the real solution, by dividing the object into smaller pieces and simulate how they work together which is far more easier to solve. In mathematics, for a differential equation, boundary conditions are the constraints to be fulfilled, which enable us to solve them. When performing FEA, a boundary condition is how we set the environment for a model, such as, external supports, external loads or pressure, imposed displacements, symmetry conditions (if applicable) that make the solution easier. Imposing boundary conditions less or more than necessary can produce errors in displacement, force and stress results that we obtain.

FEM works by dividing objects, such as structural members and their connections into a large but still <u>finite</u> number of, <u>smaller</u> <u>elements</u> with assigned stiffness and strength properties and connected by <u>nodes</u>, which form a <u>continuum</u>, (in other words a <u>continuous mesh</u>) and allowed to <u>interact naturally</u> under given loads. Separating the entire assembly into <u>discrete</u> elements by nodes is called discretization, as mentioned in displacement method and similar to discrete method, but here this is in much larger number and in smaller scale and form a mesh in 2D or 3D *(i.e. shell or plate elements for 2D and brick elements for 3D)*. The elements interact with each other through equilibrium equations and mathematical relations. This description also essentially fits to discrete method, and as explained above both ultimately depend on the same structural analysis principles, therefore sometimes for the sake of simplicity both discrete and continuum methods conveniently called as finite element method. The definite distinction is that in continuum finite element modeling we do not need to use hinges, and there are no single line elements which are as large as beams or columns for which discrete modeling would often suffice in the first place. We need continuum finite element modeling wherever we cannot reliably represent by discrete modeling. This is why in many cases we use both in combination, to get the best of both. The smaller the elements, in other words, the more number of nodes, with better precision the structure is represented, but of course it takes longer time to calculate, since each node is a point of calculation. Therefore, after a point where more precision would not bring extra benefits, the reduction of element sizes, in other words, increasing the number of nodes in the mesh should stop. Number of nodes can be in the amount of even millions, for 3D analysis of large, complex structures.

Meshing:

We can divide a linear skeletal object into 1D line elements. 2D elements can be quadrilateral (4 sided but not necessarily square or rectangle) or triangular, where each node can have 3 degrees of freedom (x, y translation and z rotation). 3D elements can most commonly be tetrahedron (4 faces), pyramid (which can have a base of usually 4 or 5 sides but can be more) triangular prism, or hexahedron (with 6 faces as on a cube but not necessarily a cube or prism with right angles always) where each node can have up to 6 degrees of freedom (translation and rotation along x,y,z). A 2D or 3D mesh will almost always have more than one type of these shapes combined. The 2D and 3D elements can also be divided further using intermediate nodes. When a node is restrained from moving along or rotating around a certain axis, it means an external support where reaction force or moment will be generated. All nodes other than supports, will have 3 degrees of freedom in a 2D, and 6 DOF in a 3D problem.

We can assign the nodes ourselves, in other words construct the mesh manually, or simply tell the program divide a certain line or area member into equal size segments as in discrete method, although dividing equally will not often be possible. But these two ways are only for simple situations. Other than these two, when we say meshing, what we most often mean and understand is the auto meshing. An auto generated mesh may include many unequal pieces, in the order of hundreds of thousands of small elements working together, which is why doing it manually would be far beyond practical. To generate an auto mesh, the programs follow some general rules such as the ones below, whether for a 2D or 3D system *(but of course the 3D would be exponentially more complicated)*:

- For straight, flat members, triangular or rectangular elements can be obtained. For curved elements or boundaries, irregular shapes, changes in geometry or material, more nodes are used to better represent the actual condition. Of course this will increase computational load.
- The nodes must be arranged such that the elements will have as regular shapes as possible. As the element sizes decrease (resolution of mesh increases), the likelihood of them being more regular shape increases, but of course finer mesh means higher computation burden.
- When adjacent members of a structure are meshed separately, these meshes must be compatible and be able to work together. For example a building floor slab has strong rebar connections to a shear wall, to work together not only for gravity but also seismic loads. When these are meshed in our model, they need to have common boundary which enables the slab and wall to work compatibly, not independently, as if not connected and torn from each other.
- For irregular bodies, it is not possible to make elements of equal size, but still nodes should not cause sudden and significant element size changes. The size ratio from element to element should generally be less than 2. The elements themselves must not be too irregular. Ideally the ratio of shortest to longest side of an element should be 1 but this can rarely be achieved. Adjacent elements that are too disproportionate with each other can cause unrealistic stress

peaks and fluctuations, which leads to unrealistic yield or convergence *(error in analysis when results do not converge to a stable state)* at those locations.

- Nodes should also take real life behavior into consideration and not just the geometry. For example, at the edges and especially the corners of a rigid footing, there are high stress and strain gradients (concentrations). Same is true where a pipeline crosses a fault line. The mesh needs to be finer in such locations in order not to miss stress peaks which can govern the design, at least for those locations. Or for example the deformation throughout the thickness *(don't confuse with width)* of a shear wall will not vary much, so dividing the thickness of the wall into many finer elements would also be inefficient. So even if we applied all the criteria above, but skipped these, our mesh will be inefficient. These are where engineers must especially pay attention to, even though the programs can now commonly detect such things.

Example Uses of FEM:

Some examples where using FEM is necessary <u>or</u> can provide better representation are listed below:

- Large, complex, valuable, irregular shaped, unsymmetrical structures of any type and their foundations, such as tall or irregular shaped buildings, almost any bridge or dam structure, towers, coastal structures, and more. Buildings which are not symmetrical from bay to bay.
- Composite piled mat foundations, or rafts without piles, which can show complex 3D behavior.
- Foundations of large structures, such as dams, bridges, tall buildings, heavy structures and the behavior of soil below in 3D including soil structure interaction (see earthquake section) under dynamic, cyclic earthquake effects.
- Analysis of weak to medium soils, especially when effects such as soil structure interaction cannot be neglected, or when underlying rock's variable properties such as faults, joints and shears are considerable.
- Representing buckling, lateral torsional buckling, cyclic deformations, shear deformations, effects of strength decrease at large deformations, complex connections, stress concentrations.
- Analyzing the 3D behavior of frames, trusses, plates and shells, such as slabs, shear walls, retaining walls, surface structures. For example, slabs are traditionally analyzed as a group of beams, but when we analyse with FEM, we can better understand and see the overall behavior and identify deformations and stresses everywhere including the effect on supports and how it works together with beams, columns and shear walls.
- Underground construction that involve interaction of various items such as existing foundations, tunnels, pipes, layers of soil, diaphragm walls etc... For example, tunnels in urban locations can be analyzed together with the ground and existing underground features.
- Transmission pipelines and simulation of seismic ground movements over long distances. The pipeline can be modeled as shell element, and the surrounding soil in 3D elements. The contact interface between the pipe and surrounding soil must be represented with proper friction.
- Effects of ramps, stairways, shafts to the lateral stiffness and vertical force transfer of structure.

- Slope stability problems for natural slopes or embankments where we need more detailed analysis for deformations, stress and pore pressure distributions.
- Individual or simultaneous analysis of geotechnical problems such as consolidation, slope stability, earthquakes, elastic settlement, seepage, retaining structures, shallow or deep foundations, drained or undrained conditions, overburden pressures, underground features, unisotropic or complex shaped slopes and ground conditions...
- Analysis of marine and coastal structures and their foundations where the presence of water, waves, highly variable loads and the necessity of using piles creates a complex problem.
- Deformational effects such as creep and shrinkage (which are also time dependent), temperature, differential column shortening, relaxation of steel... as applicable, and their cumulative effects on a structure.
- Detailed analysis of behavior of steel beam column connections (panel zones).
- Analysis of staged construction, such as bridges, wide span roofs, or any other structure where intermediate stages may create critical situations for structural capacity and stability.
- Analysis of masonry walls, especially when the masonry walls work together with the structural frame, create masonry boxed sections, walls with openings, coupled walls, irregular and complex geometries where the behavior of masonry wall elements can be mixed and complex. By lateral analysis of masonry wall systems with FEM, we may determine true load paths, understand load interactions between masonry and the structural frame and diaphragms during seismic events. Of course also for RC walls FEM is useful where there are a lot of openings, coupled walls, complex flow of forces.
- Analysis of deep excavations including its different stages, to find out anchor and or bracing loads, wall deflections, adjacent existing structure settlements/lateral deflections as applicable.
- Help finding *(or confirming what is found with other methods)* the natural period (see earthquake section) of any structure, as long as the mass and stiffness distribution of the structure is represented adequately. Analysis of effect of vibration modes on the structure.
- Modeling basement walls, slender shear walls, shear walls with openings (i.e. with shell elements).
- Analysis of unbraced slender members with difficult to predict behavior.
- Analysis of wind turbines, which must constantly withstand dynamic loads and fatigue cycles in addition to gravity and seismic load analysis.
- Analyzing diaphragms with large and or a lot of openings, irregularities, footprint changes such as podium level slabs, basement level slabs that work with basement walls.
- Some situations are challenging even for FEM. For example, simulating fracture, cracking of concrete or situations where we want to simulate exactly how two materials work together such as rebar or anchor and concrete, bond slippages. Fortunately for these we can make simplifying assumptions with adequate confidence and make FEM closely represent behavior.

The items above are just a sample of many situations where analytical solutions would either be too difficult or cannot provide adequate details or both and also using discrete method will also not cover sufficient details. FEM greatly enhances our ability to estimate stresses and deformations and overall system behavior that occur naturally in 3D without us forcing any behavior, as long as we model realistically enough. FEM or other numerical methods are also used for simpler problems that could be fairly easily solved by analytical equations, to understand those problems better or serve as a check.

The example list we gave above for use of FEM doesn't mean we can never use discrete method in some of these cases. For example instead of modeling a shear wall with a shell element and analyse with FEM, we could also use discrete modeling and represent it by a column element which has perpendicular rigid elements on both sides. Modeling as such will not be as accurate as FEM, but can still give us idea about stresses and deformations and will be much simpler. Bottom line is that, the best modeling uses both approaches as applicable to take benefit from their respective advantages by considering the structure's complexity, expected failure modes and locations, any nonlinearities...

Applied Element Method (AEM)

A relatively newer method than FEM, AEM is another numerical computer analysis method which is particularly better suited and used for collapse or near-collapse analysis, extreme loading (such as blast or a large vehicle impact) analysis of structures or demolition simulations. In this method, the structure is represented by discrete elements, which are connected by springs, therefore this method does not use mesh and is not bound by the meshing rules of FEM where elements form a continuous mesh that do not break away from each other. This makes AEM far more efficient for collapse analysis where the discrete elements are allowed to move relative to each other.

EARTHQUAKES and SEISMIC STRUCTURAL ANALYSIS

Earthquakes influence almost all types of design in civil engineering. In this section, we first introduce earthquakes, and then look at the concept from structural earthquake engineering point of view.

Note: All sections under Chapter 1 are intended to be general knowledge to apply to any type of structure, not specific to a certain structure type, and this section is no exception. However, in this section to tell concepts easier and for convenience, buildings are mentioned much more than any other type of structure. Even in those places, it is reminded that the concepts and terms presented in this section still apply to other types of structures, except the ones apparently specific to buildings only.

We will cover earthquake section under the following subtopics:

Earthquake Analysis Categories - Page 262
Earthquake Analysis Methods - Page 267

Mechanism of Earthquakes

Earthquakes happen due to large, distinct land masses at earth's crust, called tectonic plates, moving relative to each other. These movements can range anywhere from less than an inch to over a foot per year. The locations where plates move relative to each other are called faults or fault lines, along which a great majority of earthquakes happen. The plates can move away from, towards or slide along each other. When plates move towards or away from each other, this will cause the plates to move up and down with respect to each other, which is called a dip-slip fault. The sliding of plates horizontally along each other is called a strike-slip fault, which produces more number of earthquakes. For instance, the San Andreas fault in California is a strike-slip fault. Not all faults are visible on the ground. There are many fault lines that do not reach all the way up to earth's surface but can still cause earthquakes that we feel. These are called "Blind Thrust Faults". Extensive geological and seismic studies have been performed to determine the locations of major and minor faults all over the world, which provided a great deal of information. As a result of the movements of plates, stresses build up over the years in the brittle rock of earth crust. After a point, this stress can not be tolerated and it is suddenly released. This sudden release of energy is the earthquake. Earthquakes happen deep in the ground at a minimum of a few miles, and a great majority of them are too weak for us to feel, however, if they are strong enough, they cause shaking at the surface and can damage structures. So, although significant earthquakes that can cause major damage are rare, structures must be designed to resist them.

We still do not have a way of predicting earthquakes beforehand in a sufficiently reliable way, although there were many attempts and considerable research about it, which still continue. We are however making better and better educated guesses, as our knowledge about previous earthquakes, faults and the mechanism of earthquakes improve. For example, we have very detailed fault line maps now for anywhere in the world and historical data to varying details. So then, although for a particular city we can not exactly tell when an earthquake will happen at what magnitude, we can say for example *"an earthquake of magnitude 7 or higher is expected to happen in this city with %50 possibility within the next 20 years, or with %70 possibility within the next 50 years"* and so on... We can make such guesses because of our knowledge about the specific fault lines and also it is not uncommon for earthquakes for a given area also to follow a somewhat regular pattern, in terms of the approximate magnitude and frequency. The word "often" in previous sentence can vary greatly from one geographical location to another, from a dozen to hundreds of years, which creates a challenge when we try to focus our attention in much shorter time such as within our lifetimes.

Epicenter is the point on the surface of earth, directly above focus point (which is deep in the ground) where the first breakage (stress release) occurs, which is the source of seismic waves. The vertical distance between epicenter and the focus point is called the focal depth, which is one of the important parameters to know for the damage potential of an earthquake. The shallower focal depth, the more damage potential of earthquake for our structures. The focus point is directly on the fault plane deep in the ground, but on the surface, there might be a distance between the visible fault line and the epicenter. This is because the fault plane is not necessarily a vertical plane, but may be an inclined one.

An earthquake ground motion may not be the same between two points on the ground, even if they are at the same distance from the earthquake epicenter. This is because, the amount of ground motion is much more severe in the direction of rupture propagation of fault. This is called "directivity". Earthquake acceleration decreases rapidly in the perpendicular direction to the fault.

When a section of earth surface is under tension as a result of an earthquake, some sections can also sink, while others remain in place. This is called "Horst and Graben".

Elastic rebound means, the earth's crust is bent under stress until the main energy release, after which deformation occurs, and the crust returns to its original orientation, but not location.

Earthquake Waves (Seismic Waves):

We can group earthquake waves as:

- P-Waves (Primary waves): These are compressional in nature, which cause push and pull action to the ground, that cause movement in the same direction of wave propagation. They are longitudinal type waves similar to sound waves, and also called pressure waves, as they compress or dilate the medium they are in, with the push and pull action. This is the fastest earthquake wave type and can travel through any medium such as solid, liquid, or gas, and reach to the structures first.
- S-Waves (Secondary waves): The second type of waves we feel during earthquake are S-waves. They can only move through solid rock but not liquids. They are transverse waves; cause the up and down motion perpendicular to wave propagation direction, which means they are shear waves, so the energy transmission is slower than of P waves. As seen before in soil mechanics, shear strength of water is zero. This means, S-waves can not travel through water (by studying the trajectories of these waves, it was proven that earth had a liquid outer core for instance). They need a solid medium with sufficient shear strength, to be able to transfer shear. S-waves are slower, larger and also cause up and down motion which can move the earth with greater force, hence be more damaging than P-waves' one directional effect.
- L-Waves (Love waves): It is similar to S wave. It causes transverse motion on horizontal plane with high amplitude, perpendicular to wave propagation.
- R-Waves (Rayleigh waves): They cause elliptic motion in vertical plane only, which means, both in the direction of propagation and perpendicular. They are high period, long duration waves.

P and S waves we mentioned above are body waves. L and R waves are surface waves. These reach to the structure at different instances, but may also join each other, similar to many waves echoing in an empty room or reflecting and joining each other on a pool surface. When P and S waves reach to the surface, the energy is reflected back into the earth's crust, which causes more shaking at the surface than deeper in the ground. For example, miners underground may feel less shaking than the people on the surface. Surface waves can also cause as much damage as body waves, depending on the ground characteristics, such as how much the ground reflect, amplify or absorb earthquake waves.

In harder soil, the waves travel faster. Knowing the shear wave velocity of a particular ground is important for us to distinguish the soil type, as we will see later. The soil type and its parameters below a structure affect the structure's response to a given earthquake.

Seismic waves are like all waves; they transfer energy from one place to another without moving material. Strong earthquakes can generate waves that can travel as much as several times around the earth, as they gradually decrease in size (called attenuation of wave) and then finally disappear.

The location of an earthquake can be determined by measuring the time it takes for seismic waves to arrive at the observatories. Then a circle is drawn around the observatory for that radius. Then this is intersected with a circle from another observatory, and a third one, to determine exact location, which is called triangulation. This is the same logic of how maps work on our cellphones as they receive signals from at least three satellites.

Aftershocks can follow the main shock of an earthquake. They can continue days, weeks, months or even years. The stronger the earthquake, the larger and more aftershocks there will be, continuing longer time. There can also be foreshocks before the largest shock, but this is not always the case, because earthquakes happen by sudden release of energy, without smaller warning shocks, similar to trying to bend a ruler, which withstands bending up to a point without any failure but at one point breaks suddenly. So the main shock is usually much larger than the aftershocks or foreshocks if any.

Magnitude of Earthquakes:

Earthquakes are detected and recorded by seismographs. A seismograph consists of a freely hanging weight over a base on the ground. When earthquake shakes the ground, the base moves together with the ground but the weight does not, and this way it draws lines on the base. The size and number of the lines and the time it takes for the weight to draw them shows the size and duration of earthquake.

Magnitude of an earthquake tells us how much energy it releases. There are methods to measure it such as the Richter scale, Moment Magnitude Scale, as the widely accepted measuring standards. Each whole number increase in Richter scale means 10 times more amplitude of waves and 32 times more energy release because the scale is constructed in logarithmic scale.

Earthquake magnitude scale has no lower or upper limit. Scale 2.5 or less earthquakes occur actually hundreds of thousands of times each year, but they are not strong enough to be felt by us in any way, even if they occur close to where we live. Magnitude 4 earthquakes, for instance, happen considerably less, such as few thousand times a year, but they can be felt if people are close enough to the earthquake region, but still with virtually no structural damage. Minor structural damage usually starts after magnitudes of 4.5 to 5. Earthquakes larger than 7 for example, can occur a few times per year totally, but when they happen, they are not only felt but can also cause severe damage to structures. In other words, as the magnitude increases, the number of occurrences geometrically decreases but destruction potential also geometrically increases. At most few earthquakes over magnitude of 9 can be seen in the world in approximately a century. The largest recorded earthquake in history happened in Valdivia, Chile, in 1960, with a magnitude of 9.5 according to US Geological Survey (USGS).

Magnitude is defined in different ways, such as the magnitude felt at the surface because of earthquake waves (surface magnitude) and the overall magnitude of the event of earth crust movement (moment magnitude). What matters more for civil engineers are the Peak Ground Acceleration (PGA), Peak Ground Velocity (PGV), Peak Ground Displacement (PGD) at the base of structures, and the duration of earthquake, which ultimately depend on the magnitude of earthquake.

The duration of an earthquake also greatly affects the level of damage it causes to structures. The damage generally increases as the magnitude increases, although the relationship cannot clearly be defined in the form of an equation. Earthquakes start at a point and progress very fast as the fault ruptures. The magnitude and duration of the earthquake increase as the ruptured fault length and the affected area increase. The duration also depends on the reflections and resonances in the ground, similar to echoes heard after a strong sound, or ripples and reflections on a pool surface after an object is thrown into the pool. Especially over an area with soft soil at the surface, the surface ground shaking can continue for relatively longer time with longer periods, which is another reason why softer grounds are less suitable materials to build on especially for higher, longer period structures as will be discussed.

Geographical areas are classified according to probability and severity of expected earthquakes to happen, and earthquake maps were produced for every country. In the US for example, Pacific coast, southern coasts of Alaska, and a small area between the states of Missouri, Tennessee, Illinois, Arkansas and Kentucky carry the highest risk for frequent and severe earthquakes, while several northern and southern states have the minimum risk. All these maps were prepared according to historical earthquake data and the location of faults. The levels of earthquake risk are expressed as:

%50 probability of occurrence in 50 years - 72 year return period (can happen every 72 years, in other words, in a year, it has 1/72 chance to happen), or

%20 probability of occurrence in 50 years - 225 year return period (can happen every 225 years, in other words, in a year, it has 1/225 chance to happen), or

%10 probability of occurrence in 50 years - 475 year return period (can happen every 475 years, in other words, in a year, it has 1/475 chance to happen), or

%2 probability of occurrence in 50 years - 2500 year return period (can happen every 2500 years, in other words, in a year, it has 1/2500 chance to happen).

And then we can round the return periods above to 75, 225, 500, 2500 years respectively. Above, as the return year period increases, it means a bigger earthquake. Different codes and countries may require different percentage and return periods for each type, complexity, size and location of structure.

After we know the probability, we look at the map that shows the spectral acceleration, "a", in terms of gravitational acceleration "g" for the structure that we want to build. These maps are provided by government agencies and in codes. This will enable us to get the peak ground acceleration "a" that will affect the structure in terms of gravitational acceleration "g". So for example the less frequent, stronger earthquakes will mean higher acceleration values to use in the response spectrum graph (will be introduced below). In other words, the response spectrum graph we use, depends first of all on the return period we chose here. Again remember that we said "probability" here. It is never possible to predict what exact magnitude of earthquake will happen when and where. We can only predict a probability percentage over a certain period (in years) and geographic area. Of course, as our historical databases and methods improve, the accuracy of these predictions also improve.

Therefore, for us, there is not a design earthquake "magnitude". An earthquake of magnitude 7 may happen 100 miles away, and not do much damage, while a magnitude 6 earthquake with an epicenter of 10 miles away can do much more damage to our structure. Type of ground, distance from fault line and direction of earthquake wave progress affect the ground shaking intensity for a certain structure. So, the shaking can show variances even in a relatively small geographic region of the size of a city or even town. For structural engineers, what matters is the earthquake acceleration value which will influence their structure at its base (and the duration of shaking). This acceleration value is used to calculate the overall earthquake force that affects the structure, called the base shear. It inherently depends on the logic that derives from Newtons' law, F=ma *(see the comment in section C12.7.2, ASCE 7-16 for instance, which mentions mass, acceleration and inertial forces)*, although it (the base shear) is not calculated directly as in Newton's law but per the relations given in relevant codes (such as ASCE 7 in the US for buildings and other structures), which also considers other variables. This force is then adjusted as necessary to obtain the actual design values. After that we apply a proper analysis method to find how the forces and displacements will be distributed in our structure. The mass, height, overall geometry and member configurations, materials, structural system, connections, foundations, ground conditions will determine the earthquake loads and how they will be distributed.

For different periods of earthquake occurrences, different performance levels are targeted. For example, we want the structure to remain totally elastic (essentially no permanent damage, and availability for immediate continuation of its use) for a certain lower level (lower period) earthquake intensity. For a higher earthquake intensity, we can accept some structural damage. For even higher level of earthquake that can be expected for that area, our only goal will be to prevent total collapse and save lives, although the structure will not be usable at all again.

Seismic Structural Analysis Terms and Concepts

In structural engineering section, we listed some common structural terms and concepts, but excluded terms which were specific to earthquakes. We will list them here. This time the terms do not appear in alphabetical order, but presented in an order that will hopefully help easier understanding of the concepts by starting from more fundamental knowledge to more specific terms.

Natural Periods of Vibration:

Denoted by f, frequency is the number of times a repeating event occurs over certain duration. Period, which is denoted by T, is just the opposite of frequency, which means, how long it takes for one cycle of a repeating event to happen. Therefore, T = 1/f.

A mode of vibration is the pattern that an object vibrates and the shape that it takes (i.e. displacements that it displays) repeatingly at each vibration cycle, during its free (natural, not forced) vibration. As examples of mode shapes, see the typical shape of the first (fundamental) mode for a building in Figure 1 below, and the typical shapes of first and higher modes in Figure 2. Natural period of vibration of a structure is the time it takes for that structure under a certain mode, to complete one vibration cycle by itself, when subjected to shaking action, such as an earthquake or wind. So each mode has its own natural period, and, as the order of mode increases, its period decreases (frequency increases). For short and or inflexible, stiff structures, usually we can observe only the first mode, and the higher modes they generate are too small for us to feel or to contribute to motion significantly, and this decrease is more as the mode increases. For tall, long and or

Side Note: For any physical object, there are as many as three times of vibration modes as the number of atoms that object contains (as each atom can vibrate in x,y,z directions), which fall into the realm of physics and chemistry.

flexible structures however, there will likely be more than just one nonnegligible modes up to several modes, that may have effect on the vibrating behavior of the structure. For our purposes, when analyzing structures, we only deal with the first several modes that a structure can display overall, and often only the first mode, unless that structure is tall and or flexible one as mentioned. That is why, since we see it all the time, in civil engineering, we call the first mode as the the fundamental mode, and its period as the fundamental period of vibration, which is the mode with the longest period. Commonly but incorrectly, the first mode and its period is also called as the natural mode and natural period, but as explained above this is technically incorrect because all vibration modes of a physical object are natural, not just the first one. That is why, this section has the title "natural periods" (plural), but not "natural period" (singular). Also, there can be not only translational, but also torsional modes.

No matter how strong or weak is the shaking action, the natural periods of a structure do not change. They are unique for that structure in its entirety, like its fingerprint. They are the result of the structure's own body itself, irrelevant of any external effects, which is why they are called "natural". No other structure in the world has the exact same mass, height, geometry, structural system, materials, arrangement of members, connections, supports and foundations which all affect the natural periods and how the structure will respond to an earthquake. Response means, the structure's acceleration, velocity and displacements as a result of contribution from all modes that apply, which are mostly just the first few modes. Therefore, all structures have their own unique periods and responses to earthquakes. This also means, as soon as we modify the structure or its foundations (or even the soil below it) in any way, its fundamental period and all higher order mode periods will be different. Height (or length in case of bridge for instance), mass and stiffness of a structure are the three most important factors that affect its natural periods of vibration.

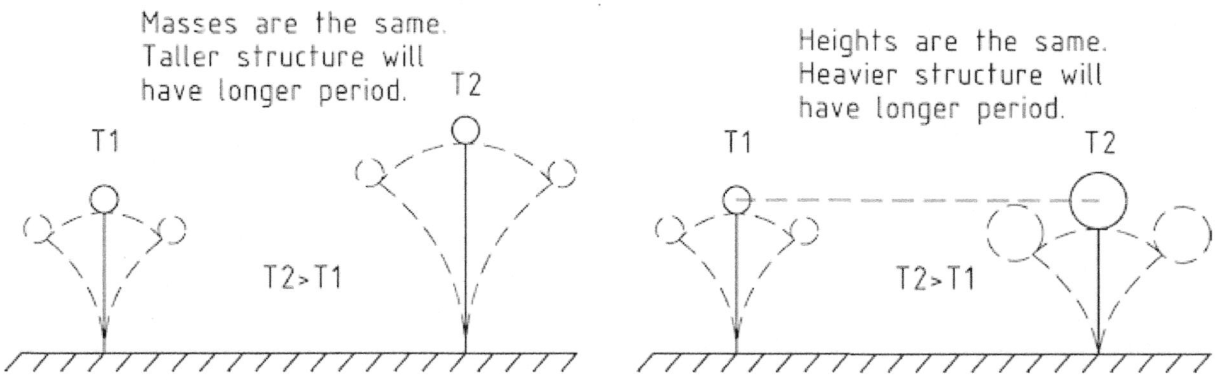

Figure - Earthquakes 1

Figure 1 above illustrates the effect of height and mass on natural period (for convenience only first mode is shown). In the figure, on the left, there are two objects with same amount of masses attached to rods of different lengths. The taller structure will have longer period in this case. On the right, two different amounts of masses are attached to rods such that their center of masses are at the same height. In this case the heavier structure will have the longer period. It will not matter if the ground in the figure shook more or less in the beginning and with what duration. The periods of all rod+ball assemblies will not change, as discussed above. In fact, even the internal friction within the rod+ball assemblies (called damping of the structure), will not affect their natural frequencies. But as soon as we make modifications to a structure, such as modifying its mass or height, or modify the stiffness of the rod, the natural period of that structure will change.

So, the taller the structure or the more mass it has, the more flexible it is, the longer its natural period, T. For example, if we wanted to give some typical and approximate values, for a single story building, T may be 0.05 second, for a midrise building of 12 floors, 1 second, for a very high industrial chimney may be 2 seconds, for a very tall skyscraper or large suspension bridge, it can be as high as 5-6 seconds *(try to animate in your mind the shaking with a period of T=6 seconds, and why it needs to be very a large structure but cannot be a single story house for instance)*. A rough rule of thumb for buildings is that the fundamental period in seconds is roughly equal to the number of floors divided by 10. Presence of basement floors, which are much more rigid than upper floors and basically sockets the building into the ground, also have effect on the natural period.

In addition to height and mass, the arrangement of mass within the structure is also important. For example for an elevated water tank, if we kept the height of the tower the same, but located the tank itself at mid height instead of on top, the period would be lower. Or if we put the structure carrying the tank on a firmer base with a better fixed connection, or if we made the structure stiffer, in other words less flexible, the period would be less. Short period structures experience more seismic force (also see the earthquake response spectrum graph in Figure 5, where shorter periods have higher acceleration). Now, again try to visualize. As the structure gets more flexible, it will take longer time to vibrate back and forth, which means longer period, but less force demand from its own structural system. Higher flexibility and ductility both increase the period of a structure, and limit the forces that the structure must resist at the expense of more deformation, some of which are also permanent. Flexibility should not be confused with ductility, as we explained in Structural Engineering and Strength of Materials sections. The opposite of ductile is brittle. The opposite of flexible is stiff.

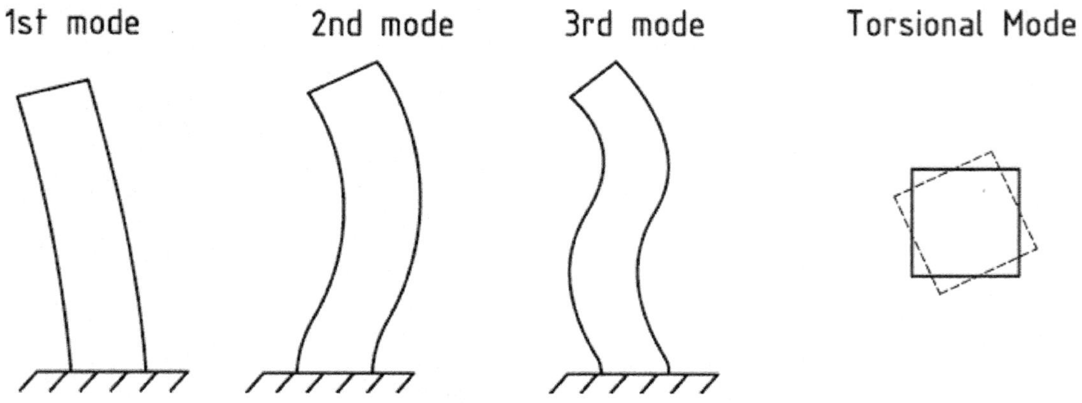

Figure - Earthquakes 2

Figure 2 above shows higher modes of vibration. For short buildings, there is just not enough height to produce higher modes of vibration, but as the height increases, the more significant the higher modes become, but still, the modes' significance decrease as their order increase *(2nd mode is less significant than the 1st, and 3rd mode is less significant than the 2nd and so on)*. This is because in higher modes there are more number of zero displacement points (nodes) to produce *(more strict criteria to meet for a vibration)*, which can be produced with more difficulty unless the exciting force from outside matched or got close to the frequency of that mode, which is called resonance that we will discuss below. Natural modes shapes and their periods are determined by eigenvalue analysis. From each vibration mode, the amount of mass that is contributed to the movement at a particular location is important.

Note that the vibration modes shown above are separately for each east-west and north-south directions. In other words, both EW and NS directions will have their own first second third and higher modes of vibrations, unless the structure is exactly symmetrical in both perpendicular directions. In addition there is also the torsional mode of vibration shown on right, as if looking down from top. Overall, our structural analysis will reveal which mode will be the most critical and appear first, such as first mode of EW, or first mode of NS or first mode of torsion, and similarly for higher modes as well. The modal analysis that we will describe in the following pages, also called Response Spectrum analysis, take these modes into account and determines how much mass contributes to each mode, which makes it possible to determine the inertial earthquake forces.

It is fundamentally important to understand that free vibration starts after an initial force disturbs an object, and then leaves. The modes we mentioned above can only happen in a freely vibrating body, that is why they are natural modes. If the initial excitation force had continued repeatedly, then that would not be free vibration, but a forced vibration. In a forced vibration, the frequency of vibration can only be equal to the frequency of applied force and nothing else, no higher or lower modes, just one mode, with the same frequency (period) of the applied force. The earthquake acceleration shakes the ground and as a result produces an inertial force in the structure, which ultimately depends on Newton's F=ma, to shake the building with forced vibration (earthquake does not happen only at an instant and disappear, it continues shaking for a while, but after the earthquake stops then the structure can switch to free vibration).

What we seek to achieve during earthquake is to prevent that the period of earthquake shaking matches or gets very close with the period of a natural vibration mode *(of course this means only the first mode for a short/stiff structure because for them 2nd and higher modes are too insignificant)* of that structure. If that happens, it would mean resonance and may cause damage and even failure, because resonance amplifies vibration, in other words, the closer the earthquake period to a structure's period, the more earthquake energy can enter into the structure which causes more damage. Imagine a swing. There is only one possible period for it to swing back and forth naturally (if it swung freely). Someone who pushes the swing (forcing the vibration) must match the natural period of that swing with her hand when pushing it. In that case, she would be able to keep the swing moving with very little push effort each time. When we say resonance amplifies vibration, it means the acceleration, velocity and displacement is amplified. Another example of resonance can be a small boat and a large ship on the sea. The small short period waves with small wavelength will rock the small boat, but for the large ship they will not matter at all. On the other hand, if the waves were large, with longer period, then they will rock the large ship, but the small boat will only smoothly go up and down with the wave surface without experiencing any rocking motion. Yet another example is, not allowing soldiers to walk on a bridge with the same pace, which avoids forced vibration which could match or be close enough to one of the natural vibration modes of the bridge. That is why, the natural mode frequencies are also called as resonant frequencies, because a forced vibration will create resonance only if, it matched with one of the natural modes frequencies.

Natural period of firm soil is shorter. Natural period of soft soil is longer. The depth of soil to the firm rock layer below also affects its period. The larger the depth of relatively softer soil layer above the bedrock, the longer its natural periods. So for example it may be possible during an earthquake that on average a certain height range of structures gets damaged more, if the period of the ground matches the period of such buildings (height affects the period of buildings the most generally). Of course results will vary for each building but we are thinking average here.

Soft soils amplify the ground acceleration, like a ripple effect *(imagine gel in a bowl)*, while hard soils and rocks dampen it. Also add the fact that soft soils' bearing strength is less and they will settle more. Therefore, it makes more sense to only build short, stiff structures with short periods on soft soils. The 1985 Mexico City earthquake actually happened quite far from the city, but the ripple effect under the city was strong, because Mexico City is built on soft soil, which was an old, dried lake bed.

Most soils have natural periods from 0.4 for firm soil to 2 seconds for very soft soil, which match with periods of many short to midsize structures. Buildings taller than 20 stories usually have periods of more than 2 seconds, but this doesn't mean they never match the ground period and less vulnerable.

As mentioned, flexible structures will have larger periods. So when an earthquake starts to cause yields and cracks in a structure, the structure becomes "softer" and more flexible, and its natural period will start to increase. If the ground also had a longer period, the structure's period will start to match the ground period, which will make things even worse. Note that in case of a steel structure, the opposite can also happen, if the cyclic shaking action causes strain hardening of steel which makes the structure stiffer, also shortening the period *(but now the building's ductility is reduced, which means reduced energy dissipation capacity - also see ductility under structural terms)*.

Given what we discussed so far about vibrations, let's list some earthquake challenges for tall buildings:

- As mentioned, tall buildings can produce significant second or even third modes of vibration. This means, tall buildings can produce additional deformations which would require us to provide higher shear strength and ductility limit for them.
- Another risk about second and third modes of vibration is that although the first mode of vibration is the most critical, second or higher modes of vibration have shorter periods, and therefore the probability of matching the ground period is higher.
- Tall buildings dissipate energy slower than short buildings. This is because the energy is spent during each shaking cycle but the periods of tall buildings are longer, which means their shaking is slower. This means energy stays in for longer time. The frequency of short buildings are higher, which means they can dissipate more energy per second than tall buildings.
- Short buildings behave in first mode only. For them a pushover study, which is a type of static analysis that we will describe, can be applied to find earthquake resisting capacity. For tall buildings we will need dynamic analysis which is more complex.
- Another reason for faster energy consumption of shorter buildings is that their damping coefficient is higher, such as around %5, which means in each cycle they spend 5% of the energy, while for taller buildings the damping coefficient tends to be lower.
- For wind, unlike seismic forces where short period structures experience more force, longer period structures will see higher lateral forces and base shear force - imagine a long period tall building or long suspension bridge keep swinging back and forth under prolonged wind loads. In other words, for a tall building where wind loads can also be significant, if we design for a shorter period for earthquake, we might be missing the longer period effects of wind.

These disadvantages do not mean tall buildings are more risky during earthquakes. As long as we model the building well, interpret the results from the analysis of the model correctly, design per codes and build by following standard construction practices, a tall building will also be able to provide life safety, for its design earthquake level.

We can calculate the fundamental period of a structure using different methods. For buildings, the fundamental period can approximately be found as shown in ASCE 7-16, section 12.8.2.1. This would not be valid for nonbuilding structures, as their stiffness and masses are distributed differently than buildings. Natural periods can be more precisely calculated with an applicable method such as by taking into account the mode shapes of the structure, obtained from methods such as an eigenvalue analysis, Rayleigh method. The fundamental period has great effect on the earthquake forces we design to, within a structure. Based on these forces the structural system and members can be adjusted (which means stiffness and in turn the period is adjusted), to find optimum between lower forces (but more drifts) or lower drifts (but more strength). Story drift limit is also a major design criteria for buildings.

Eigenvalue Analysis:

Eigenvector is a term used to describe natural modes of vibration. An eigenvalue analysis shows the basic dynamic behavior of a structure, such as the natural periods and mode shapes and how each mode participates in overall behavior. This is done before performing the actual earthquake analysis.

Damping:

Damping is the reduction in vibration amplitude, due to energy losses. When we push a swing, if there was no damping, we would be able to add increased motion (response) to the swing indefinitely with each small push. In the real world, in our structures, fortunately there is always damping, which limits the amount of maximum response during an earthquake, and after the earthquake stops, it is because of damping that the vibration decreases gradually and finally stops.

During earthquake, seismic energy enters into the structure, which must be safely dissipated at a sufficiently high rate by the structure. If the structure's rate of energy dissipation is slower than this energy input, failures will start to happen. The higher the damping, the faster this energy dissipation. Damping is a force. It increases as the velocity of motion increases and can be found as:

$$\text{Damping Force} = c.V = \text{viscous damping coefficient} \times \text{velocity}$$

which is part of a general equation of motion that we will show. Therefore the faster the earthquake shakes the building (more velocity), and the larger the viscous damping coefficient, the higher the damping force will be. As the damping slows the structure down, its own value will also decrease due to decreasing velocity. The damping we described above is due to <u>friction</u>, not yield. This happens even when the structure shakes a little, just by friction, even before any damage such as yields or cracks.

During major earthquakes, when the yielding and cracking starts, some energy is also lost due to these permanent, plastic deformations. These also consume energy, which is something that engineers may still rely on, depending on the targeted performance level, such as collapse prevention. The damping at this level is the hysteretic damping (see Structural Engineering > Structural terms). Due to all yields and cracks, the original viscous damping coefficient of the building will also change in this stage, which should be accounted for in nonlinear analysis (which we will discuss).

Damping ratio (coefficient) shows how quickly the structure dissipates energy as a percentage of the previous cycle. A heavy concrete structure for instance, will provide more damping ratio (such as 5%-10%) than a light steel frame (such as 2%-5%), because the concrete building will have more internal friction within. The higher this percentage, the further decrease in response spectrum curve, which we will discuss, which means lower accelerations affecting the building.

Normally, the nonstructural components will also have some small contribution to the stiffness and damping of the building, such as interior partition walls, exterior facade, some of the installed components, in other words, they will also absorb some of the earthquake energy in order to move, and contribute to damping through their friction with the structure, but of course they also increase the mass, which is the downside, and they cannot be relied on in analysis. For example old buildings with heavy nonstructural brick partition walls can absorb a lot of energy, which means high damping ratio, but their mass are also high which means more burden on the structural system and foundations.

Now while we are discussing damping and energy loss, let's also mention seismic isolators and seismic dampers. These can significantly improve the performance of structures under lateral loads, (capacity, deflections, floor accelerations) and make more efficient structural designs possible. Other less frequent energy dissipating systems can also be used, which we will not cover.

Seismic Isolators / Base Isolators:

A very effective means of energy dissipation is achieved by using seismic / base isolators. These are flexible pieces which separate the foundations from the structure above (so they are placed in between) and absorb a great deal of energy during earthquakes even before the energy enters into the structure in the first place. Seismic isolators may absorb energy though friction, viscosity of fluid, rubber deformation, or the yield of metals that they contain, and they are the most effective means of consuming a great deal of earthquake energy, greatly reducing or totally eliminating damage to the structure. This way the structure can move nearly as a rigid body during an earthquake. With this system, we should generally be able to keep the structure to behave as elastic, or reasonably close, and the nonlinearities should mostly or entirely be limited to the isolators themselves.

Seismic isolators can be used for almost any type of structure, with various applications and cost impacts, but their effectiveness will vary according to the stiffness and height of the structure as explained below. Often their cost is compensated by being able to use a less expensive structural system. When seismic isolators are used, the utilities that enter to the building must be able to accommodate the expected movements of the building at the base.

Using seismic isolators will increase the natural period of the structure, which results in lower accelerations affecting the structure, in other words, it results in lower accelerations matching with the structure's natural period, when considering the response spectrum curve (which we will discuss), because as the building period increases, the acceleration values on the response spectrum curve decrease. In addition, the isolators add a certain amount of damping, which brings down the response spectrum curve, which also results in lower accelerations affecting the building. Due to increased period effect, base isolators are more effective when used for structures with high mass and stiffness of low to mid height on medium to hard soil, irregular buildings, but their benefit drops when used in light and flexible structures, tall buildings, structures on soft ground. This is mainly because the isolators already have long periods, and when used with long period structures, their periods start to match and interact with each other, which decrease the effectiveness of the isolators. For too high or flexible buildings, instead of base isolators, seismic dampers, may make more sense.

The isolators need to be placed so that their center of rigidity is as close as possible to the center of mass at the level of base isolator application. This is to avoid overall torsion of the structure. When we use isolators we can model them for nonlinear behavior, while the structure above can be modeled linearly. Each project has unique requirement of type and capacity dampers. As the movement distance toleration capacity of dampers increases, the complexity and cost of dampers increase exponentially.

Seismic Dampers / Energy Dissipators:

These are energy absorbing devices placed within the structure itself. They are more suitable for relatively flexible buildings. Energy dissipators can consume energy when deformations start during earthquake. So they cannot prevent the energy entering into the building like seismic isolators, but similar to shock absorbers, they consume the energy that already entered into the system. They are placed at locations such as on the cross bracings of a steel building or a liquid tank system on high floor of a tall building. Dampers take various shapes and forms. The types include friction, viscous, yielding, liquid tuned dampers. They not only serve as energy dissipators for earthquake safety but also as control mechanisms for deflections and accelerations during both seismic and wind events.

Peak Ground Acceleration (PGA) and Earthquake Forces on a Structure:

Knowing the magnitude of an earthquake is important but what matters in the end for civil engineers when designing structures is the actual ground motion that will shake the structure, as mentioned.

Peak ground velocity, PGV is the maximum velocity of a point on the ground and PGA is the peak acceleration of a point on the ground as a result of the earthquake. For both PGV and PGA, we are interested to know the values that occur at the actual project site, right beneath our structure. PGA can be expressed as a percentage of gravitational acceleration.

According to Newton's Law, if an object has acceleration, then it means there is a force acting on it, which is found by the formula F=ma. The force here is called the inertial force, which is a result of the inertia of the mass of the structure, when the ground shakes with a certain acceleration. So, the earthquake is similar to pulling the carpet beneath your feet with a certain acceleration, but not pushing you on your shoulder. Pushing or pulling on your shoulder would be similar to what the wind does. Wind force is an external force, not an inertial force.

So, if we know the peak ground acceleration that will affect our building, *(underlined because it may be smaller than the PGA on site)* we can know the maximum force that that acceleration will produce, since we already know the mass of structure. This becomes the primary ingredient to seismic analysis of our structure, which is shown as the base shear in Figure 3 below. And the F=ma is shown just to emphasize the origin of the concept, otherwise the actual acceleration on the building is found differently, as shown by the formulas in applicable codes, although it still ultimately depends on F=ma.

This explains why lighter frame buildings usually perform better in earthquakes, such as wood frame or light steel frame buildings, because mass, m, for lighter buildings is less, and it means the inertial force for a given ground acceleration will be less too. Less inertial force means less demand on the structure and foundations.

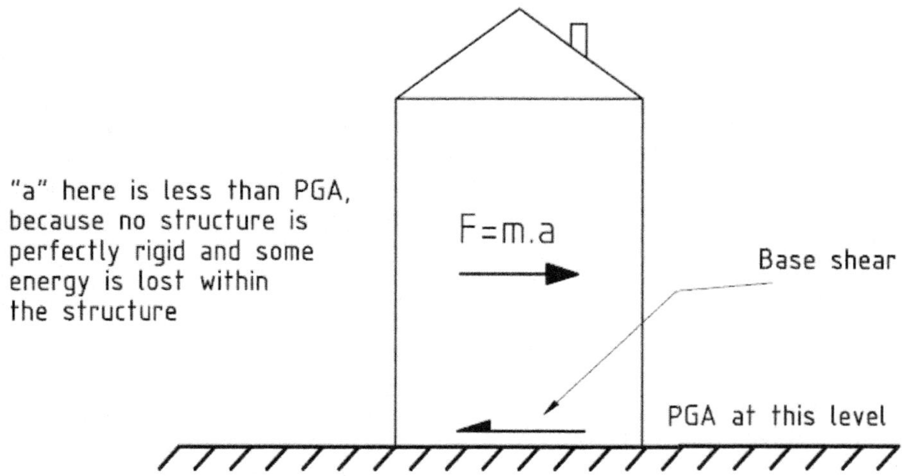

Figure - Earthquakes 3

After finding base force, it must be reduced by a certain factor, which is called Response Reduction Factor (R), to account for energy losses. This is because no structure is perfectly rigid, and some of the force resulted from PGA will be lost due to energy absorption within the structure (good thing).

> **Side Note:** Actual elevation of "base" as shown in Figure-3 may not be straightforward to determine at once. Existence of basement levels, slope of grade, soil stiffness also affect it, which varies between grade level and top of shallow footings or pile caps. Depth of piles are normally not considered to bring the base level down because where the ground introduces earthquake force to superstructure is important but piles mainly move as part of the ground.

PGA is not constant but varies by vibrating back and forth. At a particular moment when PGA is at its peak, let's say 1g for strong earthquakes for example, it is as if, our structure has its full weight in horizontal direction too. The whole lateral force resisting system therefore must resist this enormous lateral force for that moment, plus the cyclic shaking even if with lower accelerations, for the duration of the earthquake. Poorly built structures can start suffering damage at a very low PGA values.

Only knowing the PGA on site for calculating the total earthquake force on a structure is not enough. That PGA does not always create the worst case for our structure because that acceleration corresponds to a period (T), which may be considerably different than our structure's natural period of vibration (T). In that case, that acceleration will not mean the maximum acceleration as far as our structure is concerned. We should look for such an acceleration value that, its ground motion period matches or close to our structure's natural period of vibration *(remember resonance concept, and the example we gave about a big ship vs. small waves or the boat vs. large waves)*. In other words, the response of the structure depends on the period of ground motion and the structure's own natural vibration period too, not just the peak ground acceleration. This is where the earthquake response spectrum comes in, which we will look at next.

Duration of earthquake is also important. Earthquakes can last anywhere from several seconds up to a minute or even more. Lower acceleration that shakes for 20 seconds may cause more damage than a high peak acceleration that lasts only one second or less. As long as the earthquake continues, it will keep sending energy into the structure. The structure must be able to consume this incoming energy rapidly enough. As the number of cycles (thus duration) increases, if inelastic deformations started, the strength and stiffness will keep decreasing. For example in a concrete structure if the rebar started to yield, the plastic regions can continue to spread further. As far as the soil, if it is prone to liquefaction, the increasing duration will also cause the soil pore water pressures to build up more and more.

In the figure below, you see a simplified, static (slow enough, not dynamic) distribution of earthquake forces on a building. The higher the floor, the greater the force. This is called the equivalent static force. Later under earthquake analysis methods, we will see equivalent lateral force method which is the most straightforward method and uses this distribution, which is the simplest way of assuming earthquake forces on a structure. We say equivalent static because in reality the force is dynamic but we assume its static equivalent for simplicity. We must consider earthquake forces in both orthogonal directions. Also see the base shear at the bottom, which is the source force as a result of the ground acceleration. The distribution in the figure is triangular because we assumed equal mass of floors. The triangular force distribution below would not be accurate for an irregular shaped structure. A modal response analysis that would capture vibration modes *(remember mode shapes we showed)* and participating mass for each mode would be more appropriate in that case, which would distribute forces more realistically in each floor, and can also capture torsional effects.

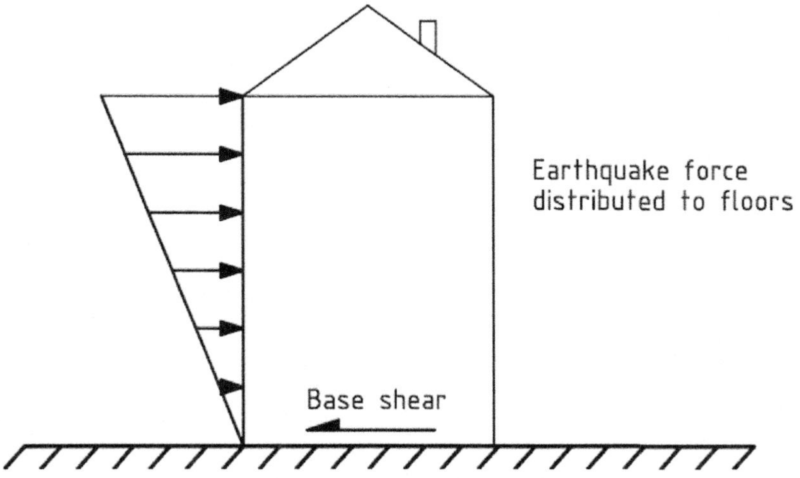

Figure - Earthquakes 4

Earthquake Response Spectrum:

Response means, something's "answer" to an external effect. Spectrum means, something that shows a range of values. In statics section remember we saw action and reaction forces. Now we are in the realm of dynamics, and instead of static reactions, here we have dynamic responses.

An earthquake response spectrum is a graph that shows the maximum responses of different structures with a range of periods under a given earthquake vibration. Each of the structures here can be considered as SDOF - Single Degree of Freedom linear oscillators, which are also damped (otherwise oscillation would last forever). The responses here are acceleration (a), velocity (v) and displacement (x) and they are plotted against period, T. In other words, we have a-T, v-T, and x-T graphs. Response spectrum is also called as "site response spectrum".

Side Note: The a-T, v-T, and x-T graphs are mathematically related to each other as velocity is the derivative of displacement with respect to time, and acceleration is the derivative of velocity w.r.t. time, from basic physics.

The importance of this graph is that, the PGA discussed in the previous section, which is the maximum acceleration response value for our structure, can be obtained from this response spectrum graph.

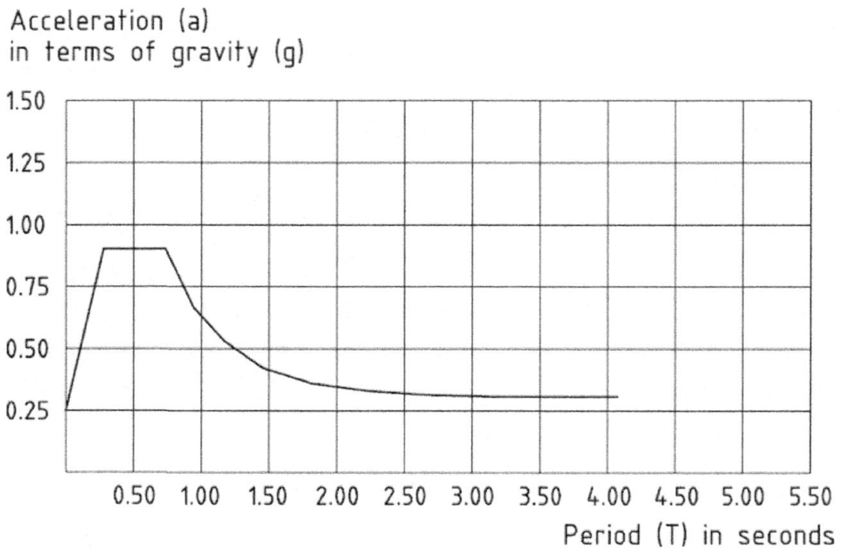

Figure - Earthquakes 5

To build this graph, first we apply the same earthquake motion to a lot of oscillators with different periods. For each oscillator, we plot the a-t (not period T, but time t), v-t, and x-t graphs. Then, from each graph we obtain only the maximum response values. Finally we plot these maximum values on to a graph of a-T, v-T, or x-T, such as the one above, where each point on x axis (for period T) represents the period of a different oscillator, and the point on y axis (for either a or v or x) represents the maximum value of that oscillator taken from the (a-t or v-t or x-t) graph of that oscillator.

Since we only took the maximum values from each oscillator's graph when that maximum value occurred, it means, time-history doesn't matter here. Only the maximum responses versus their corresponding period for that given earthquake on that given site is important.

After obtaining the response spectrum graph, we can come up with a more smoothened graph by statistical methods for each of the acceleration, velocity and displacement graphs, which are called the design response spectrum graphs. These are the ones we use. For example the a-T graph typically looks like the figure above. The numbers and the shape of the graph in the figure above are arbitrarily given for example purposes, but possibly still within reasonably close ranges to commonly seen values, and the general shape of the graph typically looks similar to the one above.

Here is how we use this graph: As we know the natural period T of our building, we can just simply locate this period value on the graph's horizontal axis, and then read the corresponding maximum acceleration value *(or displacement or velocity, depending on which graph we look at)* on the vertical axis. This maximum acceleration value determines the main base shear force that affects our building after applying formulas per codes.

Since we used the period of our structure to obtain the maximum response as described above, it means, buildings with different periods on the same site will respond differently to same earthquake, and this also means, a building will respond differently to different earthquakes on the same site. This is a user friendly graph, unlike the ones before the smoothing out (simplifying), which we didn't show.

For example, in the graph above, if our structure had a period of 1 second, the corresponding acceleration value could be read as mid way between 0.5 and 0.75g, let's say 0.63g *(which is where T=1 second intersects the curve)*. This also means, for this particular earthquake and structure combination, we avoided the highest site acceleration values in our structure, such as around 0.9g as in the graph. The movements due to the higher acceleration values than 0.63g in this case just cancelled out within the structure and could not affect it, since they happened with too low period (high frequency) for our structure. If we had another stiffer building with a period of 0.5 seconds, then its peak acceleration can be read from the graph above as 0.9g, a <u>higher value</u>. From the general shape of the graph above we can infer that shorter, stiffer buildings will generally experience higher acceleration values than taller buildings, since they have lower periods. *(This also means shorter/stiffer structures will experience higher forces within themselves, in other words they will "pull" more force to their own skeleton - remember we said "stiffness attracts load" several times before and will keep saying)*. By using the site's response spectrum graph, we can make sure in our design that the period of the structure will <u>not</u> match the site period that would produce maximum values. In other words, we want the acceleration to our structure to come from <u>not</u> the maximum acceleration value of the graph, but from <u>lower</u> values.

Although the shorter, stiffer building in the example above experienced more acceleration, it would undergo <u>less displacement (motion)</u> than the taller one. This is why, the nonstructural components in a taller or less stiff building will have to tolerate larger motions for instance. *(Try to visualize. Shaking of a short stick, with high acceleration, higher frequency (smaller period) but smaller displacement, vs. shaking of a much longer stick which will be more flexible, with lower acceleration, lower frequency (longer period), which comes at the expense of greater motion. In addition, your hand would need to hold the stick more firmly in the first case of shaking the short stick with greater frequency - similarly, buildings with shorter periods see higher base shear and forces in their structural body)*. During design, the drift of the structure and parts of it must be kept under certain values, to limit damages not only to the structure but also to shafts, partition walls, exterior walls, glass and so on, which gets especially more critical for taller, more flexible buildings.

Seismic codes give us very straightforward and simple response spectrum graphs to follow, looking similar to the one above, at least for buildings which are not tall or irregular. For more complex and taller structures, using basic response spectrum method is not adequate, but the codes also provide procedures to construct project specific response spectrum graphs.

As mentioned, there are two other response spectrum graphs, with the same horizontal axis as period (T), but for the vertical axis one has the displacement and the other has velocity. The velocity vs. period graph follows a similar pattern to the acceleration-period graph. The displacement vs. period graph shows in general that, as the period increases, displacement increases.

An important variable that affects the scale of the response spectrum graph is the <u>damping ratio</u>, which we introduced. As damping increases, which is desirable, it will change the scale of response spectrum graph, by lowering acceleration values for the same periods. So, horizontal axis will remain the same, but the values on vertical axis will decrease, which is good. Lower acceleration means lower inertia force demand. Response spectrum graphs are often shown as drawing several plots on the same graph for different damping ratio values, such as for 0% damping, 5% damping etc... (typically 5% is used most). Even a very small (such as 2%) increase in damping, can lower peak acceleration considerably.

As mentioned before, soft soils amplify earthquake acceleration. When the soil changes from hard to soft, the graph above is shifted upwards, and towards right too, meaning, for the same period (T), the acceleration will be higher. For stronger soil (or if we make the soil stronger with ground improvement), the flat portion at the top will shift leftwards and its length will decrease. This means, the building will experience less than maximum acceleration, which is desirable. So for example, if we made the soil stiffer, for the structure example which had a period of 1 second, if the graph shifted leftwards, it would mean, the corresponding acceleration that intersects our structure's 1 second period would be lower than 0.63g, which would mean lower base shear, which is better. So, the more we improve the ground, the better will be for a structure that has a period over a certain value such as a building over 5-10 floors or so. If our structure had a period of 0.5 second in the graph above, which means, a shorter, smaller structure such as around 5 floors, improving the ground and shifting the flat portion leftwards and narrowing it, may not mean a difference, since the structure will still experience the maximum acceleration anyway as its period of 0.5 second will still touch the upper flat portion *(but the ground improvement will of course improve the stiffness and strength of the soil and may still be needed).*

There is also a response spectrum for vertical accelerations. It is made by the similar logic but has different period and acceleration values. What matters more for us is the horizontal accelerations.

Other than the earthquake force, permanent ground damage can also cause severe damage to structures, even more than the earthquake force would in certain cases. Ground movement is larger at locations closer to epicenter of the earthquake.

The earthquake analysis we perform using the earthquake response spectrum as above is called the "Response Spectrum Analysis". A more detailed option is "Time History Analysis". We will come back to both, under earthquake analysis methods section.

Time History Graph:

Instead of a response spectrum graph as above, a time history graph can also be used to perform time history analysis, if there is sufficient earthquake data about the region, which can provide values based on real accelerogram recordings. By earthquake data, we mean the time vs. acceleration and time vs. displacement data for a particular region, which can later be used when analyzing structures in that region. With time history analysis, more detailed study is possible. We will discuss both response spectrum and time history analyses, along with other earthquake analysis methods.

Vertical Force Component of Earthquakes:

Earthquakes also cause additional vertical loads on structures but these are usually not critical to overload the vertical load resisting system of a structure, as the structures are already designed for gravity loads with a lot of safety margins. In addition, the vertical acceleration of earthquakes are usually lower than their horizontal acceleration. However, this doesn't mean we can totally disregard seismic vertical acceleration which can be both upwards and downwards. Per ASCE 7-16, we take a part of horizontal earthquake acceleration multiplied by the effect of dead load to come up with a vertical earthquake load component in certain loads cases and if the spectral acceleration is not very small.

Vertical earthquake forces may be more critical for items such as:

- Heavy concrete structures with high dead load in comparison to their live load, including the loads from nonstructural components

- Long cantilever beams or slabs, signs
- Domes, especially the wider ones
- Concentrated mass at the upper parts of a structure such as large water tanks
- Bridges

The ratio of vertical component to horizontal component of acceleration is also important but usually this is not more than half of PGA in horizontal direction.

Earthquake Analysis and Design Variables:

Below are some of the most important variables used when performing earthquake analysis for structures. The list below can further help learning the concepts and serve as a useful summary list. All these variables are defined in detail, their values or calculations methods shown in codes such as ASCE 7-16 in the US, under seismic design criteria, Chapter 11, as well as many other country's codes, structural guides and manuals.

Base Shear (V):

As mentioned before, this is the total lateral force that we take at the base of the structure as a result of peak ground acceleration which matches the period of the structure that we use in our design. The base shear equals to the total of lateral inertial forces distributed throughout the height of the structure. These inertial seismic forces are found by one of the earthquake analysis methods.

Damping Ratio (%):

We described damping ratio under damping and response spectrum subtitles.

Deflection Amplification Factor (C_d):

To adjust the elastic base shear from PGA and bring it to design level, we divide it by R factor as mentioned. However, the lateral deflection of the structure due to earthquake at this force level doesn't necessarily represent the full deflection, because we must also consider certain yielding effects afterwards. This yielding behavior means, there will be additional deflection of the structure such as story drift or yielding of certain members, <u>until it reaches ultimate strength</u>. Therefore, the deflection at this design level is multiplied by the deflection amplification factor C_d, to obtain the expected drift of the structure, which also accounts for inelastic (plastic) deflections at the ultimate force level. This means, C_d is the <u>ratio of inelastic drift to elastic drift</u>. Low C_d in general means a less ductile, more brittle structure. Also see Figure - 7. In this way, C_d is similar to overstrength factor, Ω_0, that is listed below, because similar to C_d, overstrength factor also brings the values from the same design level to ultimate level, but in the case of Ω_0, we deal with force. Values of C_d, R factor and Ω_0 for different categories of building frames are provided in ASCE 7-16, Table 12.2-1.

So the R factor, Ω_C and the C_d together can be called as seismic response coefficients or seismic performance factors. They are the main representing factors of a seismic load resisting system. We first set a base lateral force for the building, and then determine force and deflection performance criteria by using these coefficients. After that, the design lateral forces are distributed to the floors and then to the component members of the structure using various methods.

The reason that we use these three factors, R, Ω_0 and the C_d, and do not design to elastic strength but consider yielding, is that we can obtain much more economical designs with this approach, and still maintain the life safety performance. In other words, if we wanted nothing to yield, and everything to remain elastic, it would require much stronger construction details which would increase costs - of course we are talking about the design level, in other words, strong earthquake here.

Demand Capacity Ratio (DCR):

Per ASCE 41-17, DCRs are used to show us irregularity of a structure and the magnitude and distribution of inelastic demands in a structure by calculating them for its members. DCR is a ratio of the demand to capacity. In other words, when the DCR of a member is high, it means that the member is taking relatively more load with respect to its capacity, and have more potential for nonlinear behavior. We can determine if linear procedures are applicable or not, by looking at these ratios. For each story, the primary component with the highest DCR becomes critical. And for each component, its DCR is taken as the highest one that results from the action of shear, axial force, bending moment or torsion. Linear analysis methods are much more straightforward and simpler than nonlinear methods, and they are preferred if possible, if our structure does not have a lot of irregularity. The lower the DCRs, the more likely that the responses will be elastic and the more suitable to apply linear procedures.

Importance Factor (I):

Earthquake importance factors are used to increase demand earthquake load on critical, essential structures that need to remain operational after an earthquake and or the structures that will pose the greatest risk to human life in case of failure. Per ASCE 7-16, importance factors range between 1.0 - 1.5. So the more essential and risky a structure, the higher importance factor it receives, such as, bridges, hospitals, firefighting stations which are among the most critical structures. The importance factor fundamentally affects the design forces, so it affects all analysis that follows. The higher the importance factor, the more elastic strength (capacity) the structure has, without undergoing large plastic deformations. In other words, with the increase of importance factor, the ductile detailing requirements are reduced (lower R value), but strength design requirements increase.

Overstrength Factor (Ω_0):

Overstrength factor increases the design force on the system. Using overstrength factor, the design base shear force that we obtain after dividing the unadjusted base shear with R value (described below), raises up to the ultimate yield strength of the structure, because after the design base shear is passed although some members start to deform, the force demand continues to increase for other members that are supposed to remain elastic, until the ultimate yield strength of the structure is reached so by using overstrength factor we increase the demand forces on them and then compare this with the capacity required from them per the targeted performance level *(this also means for members that are supposed to show ductile yielding behavior, overstrength factor would not be applicable)*. Remember the stress strain curve we saw under strength of materials section although it was for a single member but main concept is similar. When yielding starts, it doesn't mean the maximum capacity yet. Although now there are deformations, the system still has capacity to take on more force. Also see Figure 7. Overstrength factor becomes important for members that are supposed to remain elastic for the highest forces that they will experience (force critical), while other ductile members can deform in their inelastic ranges (deformation critical) after passing design base shear point, up to ultimate yield of the structure.

In other words, if we used only the internal forces in members obtained in an analysis by using design base shear, we would not be accounting for the range between design base shear upto ultimate yield where there is still demand increase for members that are supposed to remain elastic while other members start deforming but the structure can still attract more demand. Values of C_d, R factor and Ω_0 for different categories of building frames are provided in ASCE 7-16, Table 12.2-1.

Response Modification Coefficient (R):

It is also known as "R factor", "R value", "Response Modification Factor", or "Response Reduction Factor". As mentioned when discussing PGA, it is used to reduce the unadjusted, elastic base shear force, V_e, which would occur if the structure was linearly elastic, to a value called design base shear, which is a lower value as a result of yielding and dissipation of the energy. So,

R = Maximum Elastic Force / Design Force = Elastic Base shear / Design Base Shear = $V_{elastic} / V_{design}$

Or if we rewrite it, $V_{design} = V_{elastic}/R$. This means, whatever the base shear we find as a result of mass of the structure and the acceleration we use, we divide this base shear by R factor, and the value we obtain, which is much smaller, is the design value that our structure must withstand elastically which also means, inelastic behavior starts there. And what happens to the energy from the rest of the base shear? That is dissipated by the ductile behavior of our structure. So after elastic zone finishes and yielding starts, the structure can still take on the remainder of the earthquake load but this time instead of elastic behavior like a spring, it will yield and consume energy that way, by its ductile behavior. So basically R factor allows us to design to a lower earthquake force, at the expense of yielding, and therefore it is one of the most important factors in earthquake design. A higher R value does not necessarily mean always better earthquake performance though. Also see Figure 7. So by using R, we bring the response spectrum graph, which is the elastic spectrum, down to a lower, design spectrum.

We generally want our structures to be ductile. The higher R value, the higher the expected inelastic behavior, the more ductile the structure, the higher the period T, the lower the design lateral loads we must design to, the more criteria the ductile detailing needs to satisfy. For example, a structure with very high levels of expected plastic behavior can be assigned an R value of 7-8, which would reduce the unadjusted base lateral load obtained from linear elastic analysis to design lateral load by that factor, which is indeed a lot. This means, only a small portion of the base shear is resisted by the lateral strength of the structure, and the majority of it will be resisted by the ductile behavior, and we must design and detail the structure to show such behavior. For example a value of R = 1 would mean no modification to linear elastic behavior and more demand on the foundations and structural frame.

The lowest R values start from 1.25 which are listed in ASCE 7-16. The R, Ω_0, and C_d values are shown for different lateral load resisting systems in ASCE 7-16.

Structures with high R values such as moment frame buildings, have lower design base shear which means they are more flexible, will demand less lateral support from foundation, but their drift control is more difficult as they deflect more laterally, so they will have higher C_d values.

Detailing a structure seismically means, giving it more capacity for ductile deformation and energy absorption. Generally, the higher R value means a higher level of seismic detailing as mentioned above.

A rule of thumb to remember may be that, in areas with a medium or less seismic activity, it doesn't make much sense to use R values of higher than 3 or may be 4 which means we can save detailing costs. Similarly, code values also take the seismicity of the site into consideration, and require systems with higher R values as the seismicity of the site increases. In low to medium seismic areas, wind forces may govern the lateral forces instead of earthquake.

Shear Wave Velocity (Vs):

This is the velocity at which the earthquake waves travel in the earth's crust. There are different speeds based on the type of material. For example, for hard rock, the average shear wave velocity is higher than 5000 feet/sec, while for soft clay it can be lower than 600 feet/sec as per ASCE 7-16, Table 20.3-1, site classification. Other soils and rocks lie somewhere in between. Therefore, the harder to soil, the faster the waves propagate. Shear wave velocity is also correlated to void ratio of the soil, based on different confining pressures within the soil. As the void ratio decreases and the confining pressure increases, which means denser soil, the shear wave velocity increases. Shear modulus of soil, G, an important soil parameter that is used in dynamic load calculations, can easily be obtained from V_S.

Shear wave velocity is one of the main factors to determine the site class, as described below. For example V_S 30 means, average shear wave velocity between 0 and 30 meters depth. Basically, with shear wave velocity we can determine how stiff or soft the soil on a particular site is. For example on sites with soft soil, SSI - Soil Structure Interaction (will be discussed) is more critical. *Note: Do not confuse the V_S here, with the base shear V. They are entirely different things.*

Site Class (A,B,C,D,E,F):

Sites can be classified as Classes A, B, C, D, E, F, from harder to softer soil as shown in ASCE 7-16, section 20.3. For example Class A soil is hard rock with the highest shear wave velocity, where Class F is a weak soil that requires special attention. We can use the shear wave velocity value of the soil, and also other soil parameters such as the SPT blow count (N), undrained shear strength s_u, or plasticity index PI, *(see soil testing subsection under soil mechanics)* to determine the site class.

Site Factors (Fa, Fv):

F_a: Short period site coefficient. It depends on site class and short period spectral acceleration, S_S, as in ASCE 7-16, table 11.4-1.

F_V: Long period site coefficient. It depends on site class and long period spectral acceleration, S_1, as in ASCE 7-16, table 11.4-2.

Seismic Design Category (A,B,C,D,E,F):

This is classification of how strict the seismic design must be, based on the importance of the structure and the severity of ground motion expected at the project site, per ASCE 7-16. The groups are the consecutive letters from A through F, where A is the least severe and F is the most. For example a structure with seismic design category of F means a critically important structure which is also not far from an active fault, and its seismic detailing and design requirements are the strictest.

Seismic Response Coefficient (C_S):

It enables us to come up with a base shear force, based on the spectral acceleration and the mass of the structure. It depends on the design spectral response acceleration, response modification factor and the importance factor.

Spectral Response Acceleration Parameters:

These are used to find spectral response accelerations, as shown in detail in ASCE 7-16, Chapter 11.

S_S: Mapped short period spectral response acceleration parameter, from which we get the design value as S_{DS} by using F_a as shown in ASCE 7-16 section 11.4.4 and 11.4.5

S_1: Mapped spectral response acceleration parameter at 1 second, from which we get the design value as S_{D1} by using F_v as shown in ASCE 7-16 section 11.4.4 and 11.4.5

S_{DS} and S_{D1} are called the "design spectral response acceleration parameters", which are the actual ones we use in determining the spectral response acceleration, S_a., thus the response spectrum graph, by also using period (T) values.

S_a: Design Spectral Response Acceleration at any period. This represents the acceleration (vertical axis) values of the response spectrum curve for each period T (horizontal axis). It is calculated by using the S_{DS} and S_{D1} and period, T.

Geotechnical Concerns About Earthquake Effects

Earthquakes will cause not only the structure itself but also the foundation and the ground beneath to behave differently and these effects must also be considered, such as:

- Shallow or pile foundations, retaining walls, slopes all need much more complex analysis for earthquakes. Dynamic seismic forces, ground motion, soil dynamic properties, the rocking and energy dissipation of the structure, soil structure interaction must be considered.
- Whole or differential settlement as a result of loss of soil bearing can occur.
- Resonance (matching of period) of the structure and the soil below must be avoided.
- Inundation and flooding may occur at low areas, near tanks, reservoirs. Scour effects must also be considered during such floods.
- Underground structures are generally safe from earthquakes, as they move together with the ground. Exceptions include locations along fault lines, inside of a slope, where slope stability or lateral spreading problem may arise during an earthquake, or if the underground structure is in loose sand with high water table where the soil can liquefy. Especially pipelines are vulnerable to these problems, since they continue long distances and cover various types of terrain including fault lines and weak and or liquefiable soils.
- Soil rupture due to faulting or lateral spreading must be checked.

- Soils behave in a considerably nonlinear fashion during strong earthquakes. It is a challenge to predict accurate soil stress-strain relationships under strong, dynamic, cyclic shaking.
- Soft soils amplify earthquake waves, increase their period, when their layer thickness is large enough, which have higher chance of matching with fundamental (first) periods of tall / long / flexible structures. In other words, if we have rock everywhere, all the way up to the surface, it will transmit earthquake waves as they are, but if the rock terminates some depth below the surface, and we are dealing with soil instead of rock, which is often the case, the earthquake waves will most likely be amplified. The softer the soil and the deeper the rock-soil interface, the more amplification (including lengthening of duration). Therefore, parts of cities developed on areas such as on soft alluvium and sediment basins will experience greater earthquake accelerations and durations, due to the ripple effect, as in Mexico City example we gave.
- Foundation rocking and uplift can greatly increase earthquake response of the structure but some energy will also be dissipated and relieved from the structure in the process.
- Liquefaction: see below.
- Soil Structure Interaction (SSI): see below.

Liquefaction:

We discussed liquefaction under Soil mechanics>Strength of Soil>Total and effective stress section. Liquefaction can cause damage in the following forms, some of which may happen simultaneously:

- Due to pore water pressure increase and therefore decrease of effective stress, soil's load carrying ability decreases. After sufficient loss of strength, with each shaking cycle, settlement increases, as if sinking into a liquid, either as a whole or as in differential settlement, which can vary from very small amount to many feet.
- Lateral spreading of soil can occur in slightly sloped areas, when an entire soil mass spreads almost horizontally, which causes settlement of structure as it compresses the soil from above. The amount spread can vary from a few inches to hundreds of feet.
- Stability of slopes may decrease.
- Reduced soil support for piles, laterally and axially, soil downdrag (soil pulling pile downwards).
- During liquefaction the soil is weakened and weaker soils amplify earthquake waves.
- Some underground features which can be hollow inside, such as underground tanks, manholes, small light buildings with basement, may be uplifted, due extra uplift pressure of water plus soil mixture, which has much greater unit weight than water alone.
- As pressured water comes up, it will also drag sand with it, and there can appear piles of sand on the ground surface. Note: these may not always reach to the ground surface and be visible but if they do, it is clear sign of liquefaction.

- Liquefaction also increases pressure on building basement walls, retaining walls, sea walls, because now in addition to the pressure from the saturated soil, hydrodynamic force of water during liquefaction needs to be considered too.
- Although liquefaction is an unwanted phenomenon, in some cases, it can also serve as a natural energy absorber during earthquakes, which is a positive side effect. For example, in the past earthquakes when the buildings with shallow foundations sank as a result of liquefaction, in some cases they sank as whole rigid bodies with their structure intact, even though they might have considerably tilted and or moved towards one side during sinking (a quick search on the internet would show such photos). Other than proper structural design and construction, another reason that might have contributed to keeping the structural integrity (and causing far less causalities with respect to structurally collapsed buildings) might have been the tilting and or moving of the foundations relieving significant amount of earthquake load from the columns at foundation level and the structural frame, and also dissipating energy during this motion. Liquefaction can also help relieve lateral loads on pile foundations (and the p-y curves can shift to somewhat lower load values). Of course these examples are not given to say that we should allow or rely on liquefaction, but only to describe liquefaction better. We must always work to prevent or mitigate liquefaction if it has the potential to cause any foundation sinking or other types of damages as listed here, and, design and build our structural details and foundations adequately, to prevent structural failure as well.

Even if we have liquefaction risk at a particular site, the extent of this may be too small that it may not be sufficient to cause any of the harms listed above. Therefore, not just the possibility of occurrence, but also the expected degree of liquefaction must be evaluated during analysis in the first place.

Soil Structure Interaction (SSI):

The soil, foundation and the structure interact and affect each others' responses during a seismic event, which is called Soil Structure Interaction (SSI). SSI analysis looks at their combined responses.

Although SSI always happens, it is usually not significant and not taken into account, although its use is increasing. In an analysis that neglects SSI, which is the more straightforward and common case, the ground is just considered as a rigid base below the structure, which is called the fixed base situation. In this case, the foundation will have the same, unmodified ground motion as the source earthquake, which is called the free field ground motion, that means the free site response before construction, which the structural engineer usually inputs to the structural analysis as it is generally more conservative for the structure. When SSI is likely, we need to check it. SSI often ends up causing more favorable effects on the structure, in which case we ignore it and use the more conservative (worse) case of free field motion - but to come to that point we do the check anyway, just in case it would have caused a relatively worse case.

What basically happens during SSI is that, when the earthquake ground motion propagating in the surrounding soil medium reaches to the foundation, some gets reflected back into the soil, while the rest gets transmitted to the foundation and to the structure, causes vibration for them, and then returns back to the soil.

When SSI effects cannot be neglected, the foundation translates and rotates in 3D, thus increases the structure's period and causes it to behave more flexible overall. These will add to the responses of the structure, but in doing so, energy will also be dissipated. Therefore, SSI is likely to cause lower demands on the structure by relieving a part of earthquake load, but at the same time it can also increase the responses and the period of the structure. This means more displacements and also may cause resonance with the ground, if the ground period was also higher.

When we have stiff, heavy structures over soft ground, SSI effects can be significant. In addition, during SSI, since the soil is also affected by the structure, the soil deformations will also be more, which can cause higher level of structural responses. Therefore, having a soft ground is the primary criteria we look at *(i.e. for that we can look at the shear wave velocity for the site)*, and more specifically the stiffness of the structure vs. the stiffness of the soil to see if there is the potential for SSI. The stiffer the structure with respect to soil, the higher SSI effects and increasingly differing responses than the fixed base (no SSI) condition. Depth of foundation is also a factor, where deeper mats surrounded by basement walls acting together have more interaction potential than mats sitting on or close to ground surface. Liquefaction, lateral spreading of soil, highly variable soil layers also increase SSI potential.

SSI analysis is complex. It would take a specialized, experienced geotechnical engineer, and, structural engineer together to perform this type of analysis. It should only be performed if it is required and or we are confident that there is non negligible SSI and it can either save us money, due to less demand from the structure or prevent unsafe design. ASCE 7-16 allows SSI to reduce the base shear amount, if the structural model of the building considered a fixed base foundation (i.e. no springs to represent it).

Let's illustrate SSI with examples:

Example 1: Imagine we have a light, 2 story house over a very firm soil or even rock. During earthquake, would that light/small house be able to cause any change to how the rock would want to move? No, it would just go along with the ground motion. This house is just too light and flexible to cause any change in how the firm soil or rock moves. In this case, there is none or negligible SSI.

Example 2: Again assume rock or firm soil, but now instead of the small house, a 10 story building but with a small footprint. Will this alter the free field motion of the site? More than the house, yes, but not much, because although it is a tall building with more mass (thus inertial force), we said small footprint, which means, it is a flexible, slender structure, which would likely go along with the stiff rock or firm soil motion below. In other words, in these two examples so far, the rock or firm soil is too stiff to care about something too small or flexible over it.

Example 3: Now imagine soft soil, and over it we want to build a concrete dam. When the earthquake happens, would the heavy and stiff structure above cause changes in the way that the soil would want to move if there was no construction there? Certainly. Here the structure changes the response of the soil and in turn the soil also changes the response of the structure, which means there is nonnegligible SSI. The soft soil's motion is easily affected by the dam, which also affects the dam's response.

When SSI happens, it increases the natural period of the structure, and generally increases damping ratio. When the period increases, now it will correspond to a different part of the response spectrum curve. This means, modified base shear value with respect to the fixed base situation, which affects the whole analysis results.

SSI effects are also more pronounced and more complex for structures with pile foundations, instead of shallow foundations in addition to weak soils. Earthquake waves prefer harder mediums to propagate, where they advance faster, as we saw under the term shear wave velocity. Therefore, when we have pile foundations, in comparison to the free site, more of the earthquake waves will be modified. As a result, the piles will be under significant stress. There may also emerge gaps between the soil and the pile supported structure, which means more interaction and makes things even more complex. Whether there is a basement level or not would also affect the SSI. A building basement below ground level will cause reduction in foundation motion, due to lower ground motion below the ground surface.

Soil and structure interact with each other in two ways. The first is called the kinematic interaction, and the second one is called the inertial interaction.

- Kinematic Interaction: Because of SSI as described above when the foundation and soil interacts, the ground shaking that the structure experiences will be different than the free field motion. It is this modified ground shaking that we input into our structural analysis, which means kinematic interaction. So kinematic interaction is between the soil and the (usually deep) foundation system. Here the mass (thus inertia) of the structure and the foundation are not considered.

- Inertial Interaction: As a result of earthquake, the structure will vibrate. During this vibration, because of the inertia from mass of the structure, inertial loads (lateral force (base shear) and moment) will act at the base (foundation plus soil system), which means inertial interaction. So here the mass of the structure and foundation is considered. Also, the soil to foundation interface is flexible and not rigid in reality. Due to these two causes, motions will be generated at the foundation, which dissipate energy, and increase flexibility of the building behavior. This means the damping and the period of the structure increases.

  Here as a result of that vibration, we look at how the mass of the structure and its foundation transmit their inertial force to the back to the ground, which also causes additional deformations in the ground.

SSI analysis is performed in two ways:

- Indirect / Substructure method: Here, the foundation analysis and structural analysis are done separately. First the free field ground motions become input motions to the foundation. At this step, it is as if the structure above has no mass. After that the structure is analyzed based on foundation motions previously obtained. The stiffnesses for the foundation are represented by springs and dampers. This method is simpler to perform than the direct method, and more common. For both methods we need powerful FEA software.

- Direct method: In this method, the soil, the foundation and the structure are analyzed altogether. Therefore although its modeling requirements and analysis is more complex than the indirect method, it may be more straightforward to understand at first. In this method the soil has an outer boundary with the outside world, at an adequate distance from the foundation, and the whole soil mass inside that boundary (and below and around the structure's foundation) is represented by a continuum of mass automatically meshed in 3D by the FEA software, with proper resolution. To this outer boundary, we apply the free field earthquake ground motion from outside, and the soil transmits these motions to the border interface between the soil and the foundation, where the kinematic interaction happens. To correctly reflect the interactions at the border between the soil and the foundation is critical, to properly calculate the relative movements of both with respect to each other. The responses at this border interface are the result (sum) of both the structure and the soil responses while they affect each other at the same time under dynamic loads while considering all possible nonlinearities of the soil and the structure. When we consider the fact that SSI analysis is done under dynamic loads, in a nonlinear way for both the soil and structure, this is the most complicated analysis possible, where structural and geotechnical engineering are parts of the same analysis, therefore specialized, experienced engineers from both fields are needed.

Earthquake Analysis Categories

Structural analysis methods vary according to what we want to achieve. For example, we can make analysis for only gravity loads, progressive collapse, stability, fire safety or earthquakes. The analysis methods for these purposes also vary based on the type of structure, magnitude of loads and the level of accuracy we want. Progressive collapse and fire safety analyses are more specific types, which are performed by assuming certain members fail and then analyzing the effects of it for the whole structure. This section focuses on earthquake analysis, which considers cyclic earthquake loads, and already accounts for the gravity loads as seismic weight. By performing earthquake analysis, we determine force and displacement demands on a structure as a result of seismic effects, and find how it will respond. The effective seismic weight includes the weight of the structure itself plus all architectural and other components that are permanent part of the structure.

Before listing earthquake analysis methods, let's first understand how can we categorize them in the first place. These categories can be listed as:

- First order vs. Second order
- Elastic vs. Inelastic
- Linear vs. Nonlinear
- Static vs. Dynamic

Now take a moment and go back to Structural Engineering > Structural Analysis > Indeterminate Structural Analysis Methods, and read the assumptions we made in the beginning of introducing the analysis methods there, to relate the concepts in mind. As you can see, we had assumed only first order, elastic, linear analysis, and were not dealing with dynamic forces anyway so things were also static. Pay attention that, the assumptions there, are the same as the first parts of the four items we listed above. Structural analysis is learned that way, under those assumptions first, at undergraduate level. And those assumptions still work for many analyses, even in real life. But to have complete knowledge about structural analysis, one must also know about the second parts, which are significantly more complex. Below we will also describe the second parts, without diving into any math.

The terms in the fourth pair in the list above, static and dynamic, can be combined with any of the terms in the three lines above that (so we can come up with for example linear-static method or nonlinear-dynamic method and so on... as we will see under methods).

Another important thing to know is that the codes clearly guide the engineer to use which analysis category and method in what specific situation, so it is not a matter of preference by the engineer, but of course the engineer is free to perform any additional methods and/or approximate manual methods for checking the accuracy of the results, which in fact, he/she should. Now let's describe each category.

First Order Analysis:

This is the general name given to analyses that are based on the initial, undeformed shape and configuration of a structure. If a structure is straightforward and not large, and the loads and deformations are not high, this will be sufficient.

Second Order Analysis:

This type of analysis goes beyond first order analysis, and here we write the equilibrium equations by also taking into account the displaced state of the structure, such as due to drift, deflection, rotational or axial deformation yield, cracking, stability, rupture. Therefore it is more accurate than a first order analysis, but it takes much more knowledge, time and effort to setup the model and solve. In addition, the person who interprets the results must be experienced. Setting up the structural model and interpreting the results of this type of analysis incorrectly is worse than not doing this analysis at all.

The following can produce second order effects such as P-Delta and P-Deformation effects which can lead to structural instability:

- Displacements of members from all types of effects (compression, tension, shear, moment)
- Decrease in stiffness of members, which are considered during a nonlinear analysis,

This type of analysis will be needed when loads and deformations are not small, and the structure is large, flexible and or not regular.

As an example, Eurocode 3, Part 1.1, Section 5.2.1 allows a first order analysis when the critical global instability buckling load is at least 10 times of the design loading on a building structure *(same for bridges per Eurocode 3, Part 2, Section 5.2.1)*, during an elastic analysis. In other words, if the critical global buckling instability load is smaller than that *(which means it can happen earlier)*, then a second order analysis is required. For a plastic analysis, it increases the factor to 15.

Second order effects may be important for only individual members, part of a structure or globally. The criteria for analysis is clarified in codes. In nonlinear analysis we take second order effects into account.

Elastic Analysis:

In this type of analysis, the response essentially remains in elastic region. This is often confused with linear analysis, since elastic behavior is often considered as linear but it may not be the case always. As we said before, elastic means, upon removal of load <u>no permanent deformation remains</u>, where the member or structure returns back to original <u>by following the same load-deformation (stress-strain) path when it was being loaded</u>, even if that path was <u>not necessarily</u> linear. Elastic analysis will generally coincide with first order and linear analysis. Often it would suffice to perform elastic analysis which is much simpler, where we can still account for certain plasticities by using adjustment factors.

Inelastic Analysis:

Here the plastic behavior, which means irrecoverable deformation is also considered. This analysis will be needed when loads and deformations can be high. Again, modeling and interpretation is more difficult and critical here. Inelastic analysis also generally means second order and or nonlinear analysis.

Linear Analysis:

In linear methods, the relationships such as force and deformation are linearly proportional to each other, such as when the yield points of materials are not exceeded. It is a straightforward analysis method where the loads are small and structures are regular. Linear methods are not suitable for structures with high degree of irregularities, or for structures which cannot behave sufficiently elastic. If the structure complies with Demand Capacity Ratio (DCR) requirements (we described DCR before), linear analysis are permitted, which are far easier than nonlinear methods.

Even for complex structures and when we expect yielding at various locations, we can start with a linear analysis and then make adjustments or refine it by introducing nonlinearities for a nonlinear analysis after that. Linear, elastic methods are used more often than nonlinear methods.

Nonlinear Analysis:

Nonlinear methods are more complex, and take inelastic effects into account. Nonlinear behavior leads to changing stiffness locally or overall. What to include to which degree also depends on the configuration of the structure itself. For example, if expansion / movement joints are provided adequately, some of the items might be neglected depending on the particular project requirements. The factors that can be included in a nonlinear analysis include a long list of items, a few of which can be: structural or reinforcing steel yielding, cracking, anchor slippage, concrete shrinkage or creep, temperature effects, foundation rocking, SSI effects. A more detailed list can be viewed at Chapter 2 > Structural Modeling and Analysis > Modeling the Structure > Modeling nonlinearities. Note that most of those apply also to other structures, even though they are listed under the chapter for buildings.

Structural analysis programs perform linear analysis by default which will be sufficient in most cases. To do nonlinear analysis, considerably more information must be entered into the program and the model. In nonlinear methods, wrong assumptions may lead to large errors even if all others are correct, so, nonlinear modeling requires more time and expertise. It is also not easy to capture all nonlinearities. The challenge grows under dynamic, cyclic loads. Computer calculation load is also much heavier here. Nonlinear analysis results involve a lot of data, which takes an experienced team to set up and interpret otherwise it will not only be useless but also misleading. In addition, peer review may also be mandatory. The capabilities of the software itself is also a major factor in effectiveness of the analysis. Not all software packages can perform nonlinear analysis, although the widely used ones often do.

Nonlinear analysis starts similar to a linear analysis at first, as the analysis program keeps increasing the load. At the end of each load step, equilibrium and compatibility (force deformation relationships) must be satisfied. When the elastic yield of even one member is reached, the program updates strains, stiffnesses and deflections and the new cycle begins. The results (updated stiffness matrix and displacements) of the previous cycle become the starting input for the next cycle. So the overall relationship we showed under finite element method, [F]=[K][D], which is also valid for linear analysis, becomes [ΔF]=[K][ΔD] here, where Δ represents change for the nonlinear portion. (Note: Deformation is also often denoted as "U"). See figure below, for linear and non linear zones.

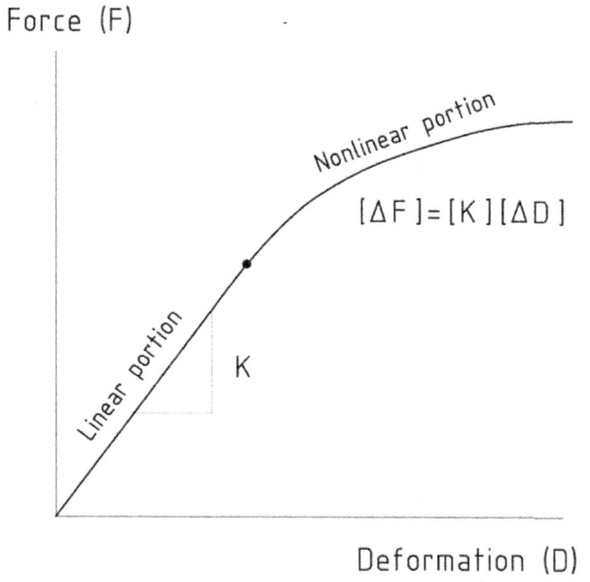

Figure - Earthquakes 6

Decreasing stiffness greatly affects dynamic response. For example, it causes increase in the period of vibration (T) and makes structure more flexible, thus deform more (*something softer will deform more and shake slower than something stiffer).* This also means that the structure will now have left with less capacity to dissipate energy - but remember that we come to that point only after consuming a lot of energy. Decrease in stiffness will redistribute forces in structure which may cause larger deformations and or nonuniform load distributions in certain areas or the whole structure. Larger deformations, when combined with the gravity load, can lead to even more (secondary order) deformations under P-Delta effects. The plastic regions in members can also proceed further with each new load cycle.

After the deformation of a member exceeds a certain limit, its strength and stiffness may not easily be predicted. Same goes for overall structure, therefore there are measures to limit deformations under a certain level and within certain regions.

Nonlinear analyses can be used relatively more often to analyze existing structures. In fact, the common use started probably more for existing structures than new ones. Use of nonlinear methods was only possible after sufficient computer and software capabilities were achieved. They were first introduced in 1990s and their common use started only after around 2010 or so, to encompass a wide variety of structures, existing or new. Tall, complex, critical structures are analyzed nonlinearly.

Static Analysis:

In static analysis, loads are applied very slowly, such that their rate of change is negligible with respect to the natural frequency of the structure. This means, the effect of velocity (which affects damping force), and the effect of acceleration (which affects inertia) are negligible. Therefore there is only static force deformation as in the form of a spring, such as F=kx, where F is the force, k is the spring constant, and x is the displacement. In static analysis, only the first mode of vibration is significant. As we are dealing with earthquake forces which are dynamic, it is immediately understood that static method simplifies real forces. Static methods are suitable for short, regular buildings, where higher vibration mode effects can not form or they are negligible.

Dynamic Analysis:

When forces and displacements vary quickly, including rapidly reversing, cyclic earthquake loads, which are close to the range of the structure's natural frequency, the dynamic effects need to be accounted for, such as for flexible and/or irregular structures including tall buildings, nonorthogonal structures, structures with reduced stiffness, torsional irregularities, where higher vibration modes are significant. Taking higher vibration modes into account is one of the most important benefits of dynamic analysis. In other words, we can obtain knowledge about the behavior of the structure under possible vibrations. Better knowledge of earthquake ground motion is required here, in comparison to static methods.

In this analysis, the force effects of velocity (F=cV, where c is damping, V is velocity) and acceleration (F=ma, which means inertial force) are also considered, in addition to the elastic spring force (F=kx) that we have just mentioned above under static method. This is because the loads are applied much faster so they become time dependent here. Now we add these two (F=ma and F=cV) new forces on top of the spring force F=kx we mentioned in static analysis. So now, the sum of these three forces (spring + inertia + damping) will be equal to the earthquake force F, which is time dependent (dynamic) in this case. Since the force is time dependent here, it must be calculated at every reasonable time interval.

Below is the general equation of motion in structural dynamics. A structure must obey this equation during its movement:

$f_{inertia} + f_{damping} + f_{spring} = f(t)$

$m.a \quad + \quad c.v \quad + \quad k.x = f(t)$

where,

f(t): time dependent earthquake force

m: mass of the structure

a: acceleration (this is also the derivative of v, velocity)

c: viscous damping coefficient

v: velocity (this is also the derivative of x, displacement)

k: stiffness to represent elastic behavior of building

x: amount of displacement

Considering derivatives above, the more proper way of writing of the equation above is as follows:

$m.q(t)'' + c.q(t)' + k.q(t) = F(t)$

Here q(t) is a time dependent displacement, and F(t) is the time dependent force. q(t)' is the velocity, which is the derivative of displacement q(t). Similarly, q(t)'' is the acceleration, which is the derivative of velocity, q(t)', and second derivative of displacement q(t). Therefore, we have two unknowns here, which is the time dependent force F(t) and time dependent displacement q(t). The structural analysis software solves this equation of motion in dynamic analysis in small time interval increments, such as in the Time History Analysis (THA) that we will discuss.

From the equation above, which is properly called as "*General equation of motion of a single degree of freedom system*", the whole topic of advanced structural dynamics emerges. For example, some of the important terms mentioned before, such as natural vibration period, modes of vibration and many related topics, emerge from this equation. We can then generalize this equation to the whole structure, if we represent the mass, damping coefficient, stiffness and force in matrix forms, and write them as:

$$[M].q(t)'' + [C].q(t)' + [K].q(t) = [F(t)]$$

This equation is then solved for the whole structure.

We can now proceed to earthquake analysis methods that fall under these categories we described.

Earthquake Analysis Methods

The category groups we mentioned in the previous section, are commonly grouped into linear or nonlinear, and, static or dynamic categories. So from now, we will group the methods as:

- Linear Static
- Linear Dynamic
- Nonlinear Static
- Nonlinear Dynamic

Now we will discuss the most frequently used analysis methods that fall under these, which are:

- Equivalent Lateral Force procedure (Linear Static)
- Response Spectrum Analysis (Linear Dynamic)
- Time History Analysis (Linear Dynamic version)
- Pushover Analysis (Nonlinear Static)
- Time History Analysis (Nonlinear Dynamic version)

Linear Static Methods:

Linear and static categories of analysis were described in the previous section. Those definitions can be combined, to come up with a "linear-static" category. Below we describe Equivalent Lateral Force Method, which falls under this category.

Equivalent Static Force / Equivalent Lateral Force (ELF) Method:

This is the simplest, most straightforward and most widely used method that is applicable to structures that are not tall and not complex and provides adequate results for such structures. It is sometimes also called the elastic or linear elastic method. It is not the most accurate, but gives reasonably accurate results for a lot of structures. Equivalent static lateral force means, converting something which moves (dynamic) into its equivalent that doesn't move (static), which allows much simpler analysis. Here the structure is modeled by considering linear and elastic stiffness everywhere. Being a static method, only the first mode of vibration is significant. See Figure 2 again, where we showed typical mode shapes. Similar to the shape of the first mode, this method resembles analyzing an elastic cantilever beam subjected to static forces from transverse direction. In other words, think of this method, as if the building was attached to a wall from its foundations as an elastic cantilever, and keep this in mind the whole time, when reading below.

In ELF method, first we obtain a maximum acceleration from the response spectrum. Then this is used together with the seismic weight to give a base shear force. This force is then reduced by a response reduction factor R (as explained before), due to expected energy losses in the structure. After that we multiply it with the importance factor. *(Note: The equation for base shear as also shown in ASCE 7-16, Section 12.8.1 is* $V=C_S.W$*, and* $C_S=S_{DS}/(R/I_e)$ *which makes the equation as:* $V=S_{DS}.I_e.W/R$*. Here V is base shear,* S_{DS} *is design spectral response acceleration parameter,* I_e *is the importance factor, W is the effective seismic weight of the building, R is the response modification coefficient. These parameters were listed earlier.)*

So we basically take the weight of the building and modify it with certain factors to come up with a base shear value to represent the earthquake force, and then distribute it along the height of the structure, as static lateral inertial forces, which looks like the triangular load distribution, as in Figure 4. The higher the story, the more lateral force that story generates. We say "generates" because these are inertial forces generated by acceleration and mass, and not external forces as wind.

Since we said maximum acceleration above, what we find is a maximum force that the building will experience. It means, from this procedure we can find the maximum possible stresses and deflections, instead of the time history of different values while the earthquake continues.

For structures with horizontal and vertical geometric irregularities, stiffness, mass or lateral load resisting system irregularities, tall and or flexible, displaying nonlinear behavior, ELF is not suitable. This method is also less suitable as the seismicity of the area increases.

We make the following assumptions when using this method:

- Every point on the structure will experience the same acceleration.
- The structure does not deform itself, it only deflects like an elastic cantilever, and because of this, base shear can be distributed along the building height.
- The structure is perfectly fixed to the foundation.
- The structure responds primarily to its first (fundamental) mode, which is only possible if it is not too tall or irregular. This is because, only in the first vibration mode dynamic force can be equivalent to static force and distributed at each floor level similar to Figure 4. If higher modes become significant for a structure, such as a flexible and or irregular structure, ELF will not be suitable and may not calculate responses adequately. In such case, modal analysis method, as we will introduce below should be used.

The use of ELF can also be extended with adjustments to account for:

- taller buildings up to a certain height where higher modes of vibration can take place
- buildings with low levels of irregularity and torsion,

- yielding of materials up to a limited point, to reduce forces, since yielding means energy loss.

As we assume the structure is elastic, the forces found from this analysis can be more than actual. This is because if we considered yielding at various locations, which may be the case in reality, some energy would be lost and the structure would not generate as much force, which would also be easier on foundations. So the results obtained with ELF are usually conservative. In cases where a nonlinear analysis is needed, ELF can still be used as a check, to identify unreasonable mistakes, as it is simple to do anyway. ELF will still help to see many things including the load path, and the preliminary member sizes, and would serve as comparison.

So to wrap up, the main steps to perform this method are:

1-Calculate design base shear for the entire building

2-Distribute this along the height of building to each floor level

3-Distribute these loads to individual lateral load resisting elements of that floor

4-We can now determine the lateral drift of the floors and the structure, and the forces in the members and see if the members and their connections will be adequate. Any risk of overturning of the structure can also be seen here, although it is rare.

ELF can be applied for a great majority of structures in seismic design categories D, E, F and for all structures in categories B and C per ASCE 7-16 table 12.6-1. Even if for some structures it is not allowed, in addition to the other more complex required method such as RSA we discuss below, the engineer may still perform ELF which is a relatively simple analysis method for his / her own understanding anyway, to help establish approximate, imaginary boundaries in mind.

Linear Dynamic Methods:

We described linear and dynamic categories of analysis in the previous section. We can combine those descriptions to come up with a "linear-dynamic" category. As discussed, dynamic category is suitable for taller, irregular, flexible structures. All these characteristics mean that the influence of higher modes of vibration can be significant, unlike static methods. What percent of the building mass participates in which mode can be represented with dynamic methods.

Two main methods, Response Spectrum Analysis (RSA) Method, and Time History Analysis (THA) Method that fall under this category, are introduced below.

Modal Analysis / Modal Response Spectrum Analysis:

Response spectrum itself was discussed earlier. In Modal Response Spectrum Analysis (MRSA) method, similar to the Equivalent Lateral Force (ELF) method, the maximum acceleration is obtained from the same response spectrum. Both ELF and RSA are linear elastic analyses, but ELF is static, while RSA is dynamic. When performing RSA, we can now consider multiple vibration modes and how the mass of the structure participates under each of those modes, where in ELF only first mode was considered. Considering higher mode shapes allows better representation of real behavior. As a result, for the structures that we cannot apply ELF due to complexities, irregularities and or height, we can apply RSA.

The approximate mode shapes shown in Figure 2 represent the typical shapes which a tall building structure will naturally take as a result of vibrating. The peak responses, such as lateral forces, displacements and the member forces in each story from each mode are obtained from the same response spectrum. Then we combine these responses from each mode, to come up with total responses in all x, y and z directions. So in RSA, we obtain a single, maximum response value, and time history of motion is not in consideration here, similar to ELF (which also saves calculation effort).

Although there may be many modes, usually only the first several are taken into account, unless we deal with very tall buildings or long bridges etc. since the effect decreases as the mode is higher. The first mode, which is the mode with the lowest frequency (highest period), has the greatest impact. Typically in codes, it requires us to take sufficient modes into account, such as the first few, so that at least 90% or 95% of the mass of the structure can participate to produce the peak modal responses. This is because, each mode shape contributes a different amount of mass at a certain location. Of course, all these few modes mean both orthogonal directions and torsion for a building, but for other structures different directions can also exist. Peak modal responses that we obtain from each mode are combined with modal combination methods such as the absolute method, which means simply adding them by one of SRSS, CQC, CQC-4 or an equivalent method. Forces are distributed at floors by using participation factors.

In response spectrum method, which is also a linear elastic method similar to ELF, using maximum values causes us to obtain the maximum displacement values and higher member forces internally than a yielding building. To account for possible yielding, modification factors are used. Being a linear method also allows to add displacements from different modes to each other.

This method can be used on more complex structures than ELF. It can be applied to structures with irregularities, discontinuities, multiple or nonuniform lateral systems. In ELF, we said it is as if we are analyzing a simple cantilever elastic beam vibrating under first mode, which makes the load demands relatively larger than other methods, such as the design base shear and moment at the base. In RSA however, considering multiple modes (which indeed exist) can allow us to come up with less demand lateral forces and moments, which means more efficient designs, for especially taller, more complex structures. RSA can be applied to wider range of structures than ELF and indeed it is very widely used, also because it requires not too much extra effort from the engineer in comparison to ELF. In fact, it should be performed in any case, even just for comparing the force, moment and displacement results for the members and the whole structure. As shown in table ASCE 7-16 table 12.6-1, this method is permitted for any type of structure in any seismic design category. So, the ELF and RSA methods are commonly performed for the great majority of structures together. Most commercially available software now have the ability to perform RSA.

Linear Response History Analysis (LRH) / Linear Time History Analysis (THA):

As mentioned earlier, time history is the plot of acceleration, velocity and displacement vs. time of an actual earthquake, obtained from accelerogram. Time History Analysis (THA) method studies how a structure behaves under an actual past or artificially generated earthquake ground motion data *(there should be at least a few of them for higher confidence in results)* as acceleration motion input at the base of the structure. This means more complex earthquake motion input, in comparison to RSA. Being a dynamic method, similar to RSA, THA also considers responses to higher vibration modes.

Considering time history allows us to know the behavioral history of the structure, as far as loads and displacements, for the entire earthquake duration, which is unlike RSA, which only gave the maximum responses. Time is introduced from a past or artificial earthquake data in increments of such as every 0.05 to 0.1 seconds, and for each step the response is calculated by the program by numerical integration (while the equation of motion that we showed earlier is also satisfied), that finds forces and displacements at every node of our model. The next step starts with the results of the previous, such as the displacement, velocity and acceleration. This means, with THA, we can know the sequence of responses of different parts of the structure as earthquake progresses. Another advantage of THA over RSA is that since we are going step by step here, for any step we can easily introduce non-linear properties and convert the linear THA into a nonlinear THA. THA is a more complicated analysis however, with few extra benefits, therefore performing THA over RSA is not justified always, and as a result, THA is used less. Also, we need the required number of earthquake data each of which needs separate analysis, which takes considerable time. The earthquake motion records for each perpendicular axis must be consistent with each other, to analyse in both directions. Nevertheless, it is recommended to compare RSA and THA.

As shown in table ASCE 7-16 table 12.6-1, this method is permitted for any type of structure in any seismic design category.

Non-Linear Static Methods:

We described nonlinear and static categories of analysis before. We can combine those definitions to come up with a "nonlinear-static" category. Pushover Analysis Method falls under this category.

Pushover Analysis:

In this method, which is common, the structural model is "pushed" at its nodes laterally by inertial forces up to its capacity. This pushing is done in a slowly increasing, non-dynamic, non-cyclic, but static manner, while inelasticities and hinges sequentially emerge and progress in the structural model, until a target displacement is reached at a given control point on the building, such as the center point on the roof, or until collapse, to determine the ultimate capacity of the structure. Target displacement is the maximum expected displacement (in other words, the displacement demand) under given earthquake. It is important to note that simplifying dynamic loads into static loads will mean bigger loss of accuracy when the structure gets taller and more complex. Pushover analysis can be applied to most buildings and structures and it is especially frequently used in seismic rehabilitation of structures as well.

In this method, for the overall structure, we look at the forces and displacements when we reach target displacement and compare them to allowed values. We can also compare the displacement values we obtain here with the ones obtained from response spectrum method, to see if they are close enough. Story drift and base shear values can also be checked. At the member level, as in any type of analysis method, the adequacy of both force and deformation controlled members can be checked.

Through pushover analysis, we can get a good feel of inelastic response of the structure, how it is distributed and will progress, potential weak mechanisms / any concentrations of locations with inelastic behavior, overall structural capacity and stability, easier than nonlinear dynamic methods, even in situations where a nonlinear dynamic analysis must still be used. Because of its ability to show useful information fairly easily, pushover analysis is often utilized as a supplementary checking method, as a means to refine the designs, rather than the determining method for final performance.

Since this is a nonlinear method, the inelastic yielding and all other nonlinearities that we listed before under nonlinear category and will list in Chapter 2, under modeling section can be entered into the model for every member or component. Remember we said "imagine an elastic cantilever attached to the wall", when we described ELF method. Now again imagine this cantilever, and we still apply the loads statically (slowly), but this time nonlinearities will occur. This means improved accuracy, but the modeling, analysis and result interpretation is more complex. As pushover method is a nonlinear method, in order to define the nonlinear strength and stiffness of members in the model, their force-deformation curves that accounts for nonlinearities must be entered.

The similarity of pushover analysis to EFL is that both are static methods, which means, the pushover method will not be as effective as dynamic methods when analyzing irregular, flexible, tall structures, with long vibration periods where dynamic forces and higher modes are nonnegligible. So, to use EFL or pushover analysis, the mass participation in first mode must be sufficiently high, which means a building which is fairly symmetrical, and not tall. To understand better, imagine we have two buildings, one is short to medium height, and the other is tall. Now we start to apply lateral inertial force to building models but slowly, since this is static procedure. This slow pushing would make sense for the short structure as it would only take that one mode shape anyway as we push it. But for the tall structure, when we push it slowly, although it will now only take the shape of the first mode, it also could have taken shapes of second, third or even higher modes, if we had applied the force fast enough (dynamically), which we didn't. It means we will have missed higher modes if we used pushover analysis, for that tall structure which could have produced higher modes. As a way of check, we can also compare the deflected shapes we obtain from ELF to pushover analysis. If they are close to each other, it may be likely that the first mode governs.

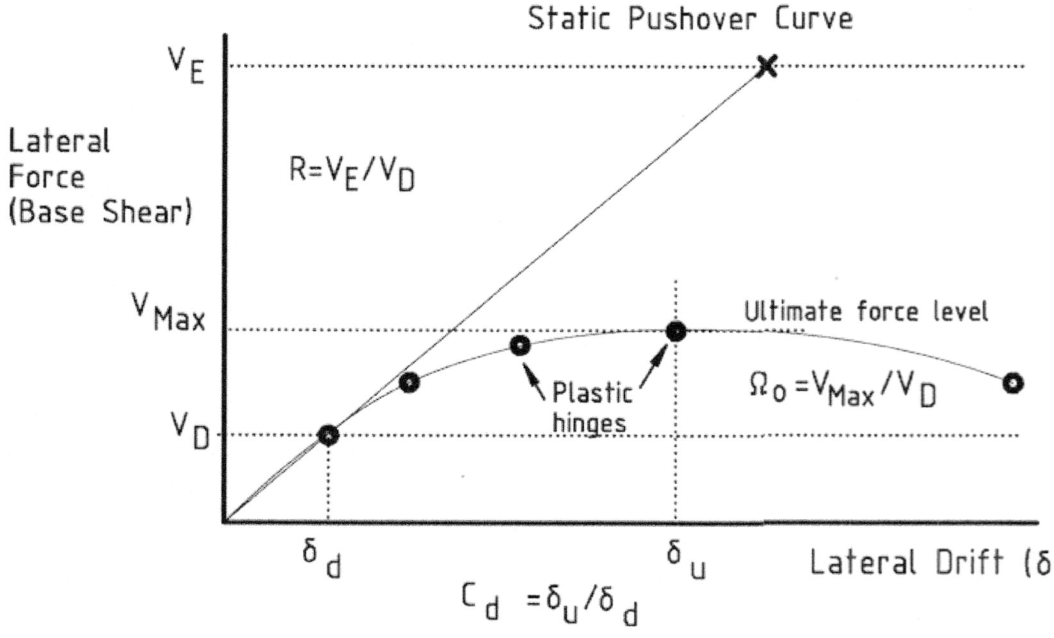

Figure - Earthquakes 7

See a typical pushover curve in the figure above which shows increasing lateral force versus increasing deformation response of the structure at the target point such as the roof. As we increase the load, the first yielding occurs (wherever was the most critical in the structure), which is the first (leftmost) plastic hinge in the figure, which means the end of the elastic behavior for the overall structure. By plastic hinges, we mean yielding or cracking of members here. As we keep increasing the lateral load, deformations increase and successive plastic hinges occur in other locations (wherever critical next) of the structure and because of this the load path may also change.

After a point when the building reaches its full yield strength, it cannot take on more load and will start to deform considerably more, even with the slightest or no increase in load, until collapse. So the force which corresponds to the top of the curve, is the full strength of the structure, which is shown as the ultimate force level in the figure. During this process, we can observe the failure sequence of members, sequence of displacements, and the effect on overall stability and identify weak spots. Then those weak spots in the model can be corrected, and the analysis repeated, where a new load path and new weak spots may emerge, then the analysis is repeated and so on, until the results are satisfactory. This process is also called the plastic hinge analysis. In the figure above you also see the response modification factor (R), overstrength factor (Ω), and the deflection amplification factor (C_d) that were discussed earlier. These definitions were given earlier.

In the pushover curve above, which may also be commonly called as a "backbone curve", the straight line from the origin that follows the initial elastic slope of the curve that goes all the way up to V_E level, would be true if everything were elastic. Therefore the point marked with an "x" at the top at V_E level, shows the response of structure if everything were elastic.

The curve above is also called as the capacity curve because it shows the force vs. displacement capacity of the structure, which also shows how ductile the structure is. The area below the curve shows the energy absorption (dissipation) capacity of the structure. As deformations increase, the building will be able to continue absorbing energy, the overall stiffness of the structure will decrease, the period of the structure will increase, in other words, the structure gets softer, but it will keep standing. This also means that the lateral force demand on the building will decrease. (*Imagine two identical buildings, on same site, with the only difference that one is stiffer. When the ground shakes them equally, the stiffer building will protect its form, but this means more force demand on its own lateral system, while the softer one will keep deforming with each shaking but in doing so it will spend energy and therefore demand less from its lateral resisting system. As another example, imagine you shake two same size cubes, one made of wood, and the other made of gel. Same logic applies.*) The larger the area under the pushover curve, the more ductile the structure. Higher structural ductility also means less lateral load on the foundations but more deformation of the structure (*the demand is the most at foundations, where they essentially carry the whole structure like horizontal cantilever beam*). We may want to limit these deformations, depending on the desired performance level of the structure (*the required performance of many buildings are not just collapse prevention but higher levels, such as immediate occupancy*). It means the structure must be able to resist higher force demands, which means the height of the curve, force, will increase, while deformation decreases. The ultimate force level in the curve above is the ultimate strength of the structure, the ultimate limit state.

After obtaining the curve above, we can compare it with the demand curve, which is obtained from the response spectra and then repeat the analysis. We also compare deformations with code limits. Other plots we can obtain from pushover analysis include: base shear vs. displacement, lateral force vs. hinge deformation, bending moment vs. plastic rotation, P-Delta effect. As always, to not miss the P-Delta effects and other relevant responses such as columns axial force - moment interaction, gravity loads must also be present. This process described here is valid for structures with satisfactory degree of redundancy, which is why hinges can form one after another. For structures which are not sufficiently redundant, there is no possibility of hinges to form in sequence as shown above, *(because there is one load path which cannot alternate to another)* and in such case the hinges form almost simultaneously, which can lead to structural instability and rapid, catastrophic collapse, as in a rapidly diverging P-Delta effect. When making redundant systems that can produce sequential plastic hinges, we must also not forget the symmetry because cracking or yielding in a nonsymmetrical way can cause torsion.

Non-Linear Dynamic Methods:

We described nonlinear and dynamic categories of analysis before. We can combine those descriptions to come up with a "Nonlinear-Dynamic" category. This is the most complex analysis category, which takes the longest time to construct the model and to calculate and interpret the results. The structural models include nonlinearities in all aspects, and the dynamic analysis enables us to capture the strong dynamic earthquake loads, higher vibration modes, without turning them into static loads.

Nonlinear dynamic method can be applied to any structure, as it is the most complex method which will give the most precise results if done correctly, but for a great majority of structures it will be too detailed, and will cost more than its benefit, so it is less common. Nonlinear dynamic methods are used on relatively smaller number of large, highly complex structures and for academical research. Its use increased gradually in the last couple decades or so, but still lags and will likely to continue to lag behind ELF, RHA and pushover methods. When there is soil structure interaction analysis get even more complex. In other words, we could perform all analysis methods described above by assuming fixed base for foundations, but when the soil and foundation affect each other's responses, things become significantly more complex, as discussed.

Non Linear Response History Analysis (Nonlinear Time History Analysis):

We introduced the THA method before under linear dynamic category and said that since this is a step by step incremental procedure, we could also introduce nonlinearities at each step and can transform it into a nonlinear dynamic procedure. Therefore, THA can be applied to both linear and nonlinear analysis. To what we described about THA in linear category, the same applies here with the addition of nonlinear aspects. Or, to what we described under pushover analysis, here in addition we add the very important aspect of accounting for dynamic, cyclic earthquake time history accelerations, instead of just simply pushing the structure slowly and laterally.

In linear THA method, we were assuming that the stiffness of the structure was constant during earthquake. In the nonlinear approach however, the stiffness of the structure is constant only within an incremental time period, such as 0.1 second, as long as the time increment is small enough for accuracy but not too small to cause too much calculation. So with each increment, the dynamic equation of motion that we showed under dynamic category is integrated. Within a given interval the acceleration is also assumed to vary linearly between each incremental step, as these are small time intervals. So the program starts with the initial, unyielded, uncracked stiffness, to analyze time steps. The next step takes the results of the previous and so on, as described before. The program considers the nonlinearities *(to see examples, see Chapter 2 > Structural Modeling and Analysis > Modeling the Structure > Modeling Nonlinearities)* in each step, as they should already be in the model. Based on the values found in each step, and based on the most current deformation state of the structure, the stiffnesses are updated. The equation of motion is solved for each modal coordinate. The mode results are also added to each other by superposition and the global result for the entire structure is found. The repeating of these steps produces the response of the entire structure based on time history.

WIND LOAD

Earthquake force is inertial as discussed, which results from acceleration of the structure's own body. Wind however, is an external force. Although wind and earthquake forces act inherently different, they both have lateral and vertical effects. Wind loads are critical for very tall buildings, suspension bridges, structures like high chimneys, towers, large freestanding signs (especially cantilevered), light structures with large cross section area in general.

The first thing to check for wind design is to determine the wind speed. In ASCE 7-16, wind speed maps are provided for the US, where the speeds generally decrease from the shore as we go inland. For example for the coast of Florida, wind speed as much as 170 mph (76 m/s) is given, for Texas coast 67 m/s, and for inland states, it ranges in the 40s, in m/s. These are based on measurements 10m (33ft) over the ground and 3 second gusts. In Eurocode however the speeds are based on 10 minute average speeds, which of course would be lower than 3 second gusts but the method of calculation is different than in ASCE formulas. Many countries in the world use one of these approaches. There is also the return period. It is the period which the strongest wind to design to is statistically expected to repeat itself. For example in ASCE 7-16, Figure 26.5-1B, this is 700 years for risk category 2 buildings and so on. The wind speed measurements must be made at open locations sufficiently far from structures of the city, as this can affect speed. As the elevation increases, the wind speed also increases. The roughness of the terrain influences the wind speed. Coastal areas receive the strongest winds in general.

Wind is generally less critical in design of buildings in comparison to earthquakes, unless we are dealing with very light and small or very tall buildings. In other words, the structure's weight and resistance to earthquake loads are often already sufficient to cover for the effects of wind. The load combinations given in codes also include the ones with wind, which must be checked too, regardless. No matter what, the exterior facade system and roof must still be analyzed for wind effects. Wind will act as suction force on the roof and since it is the highest point of the structure, it will experience the most wind force and turbulence. Especially corners, roof edges, overhangs are more exposed to wind effects. In codes, for certain roof geometries calculations are provided but not all roofs can have a standardized geometry which may require 3D analysis for unusual shapes. When analyzing roof, both the structure and the cladding of the roof must be analyzed. For the structure, as long as the roof is not very small, often the average wind speed will govern, but for the cladding, short local gusts at higher speeds.

Wind acts at the external face of the structure. During design for static analysis we can take it as a triangular distributed force (which increases by height) acting externally, which can be represented by equivalent external point loads on each floor, to simplify analysis. Wind force is proportional to the surface area it acts on and the square of wind velocity. The wind velocity will be affected by the neighboring obstacles in the area. As for the surface area, the solid vs. gross area also makes a difference, for surfaces with hollow sections or truss structures. Wind should not cause any part of the structure to yield and enter into plastic region, in other words, all resistance must take place elastically.

A wind tunnel test is sometimes not only useful but necessary. During a wind tunnel test, the structure and its surroundings are modeled fully and as realistically as possible, to a smaller scale. The wind is blown laterally, and at various angles. The calculated design strength in the structure must be higher than the failure point found by these tests. After wind tunnel testing, adjustment of overall mass and stiffness of design can be made (and in addition, solutions such as dampers, outriggers (in case of buildings) may be introduced) and the new design is sent to testing again. Several such cycles can occur. The benefits of performing wind tunnel test include:

- Slender, light, tall, irregular structures are often wind critical. The static loading given in a code or standard may not be sufficient to accurately predict the wind loads for these. A wind tunnel test is often required.
- In areas with adjacent obstacles such as hills, adjacent buildings, trees, wind velocity is affected, which may make the code based calculations to cause too high or too low wind force estimates.
- Wind tunnel test means more realistic forces, which means safer and more economical designs.

For very tall buildings with long periods, wind force and vibration effects are more significant. Depending on the site's wind and seismic conditions and the building itself, wind load may even be more critical than earthquake loads. It can also produce second order effects (See Structural Engineering>Various Terms). When the wind meets the exterior of a tall building, it will cause turbulence in an oscillating manner on both perpendicular directions with a certain period, swirling and forming vortices, given constant wind velocity *(This oscillating motion is called "vortex shedding" in fluid dynamics, where the frequency of oscillation increases as the fluid velocity increases, and decreases as the building width increases, which are related by a constant called the Strouhal number)*. In other words, the wind alternatingly applies suction forces on the building from two perpendicular directions to wind flow direction, with a certain period. If you remember the natural period of vibration topic under earthquakes section, you may immediately realize that this may shake the building in resonance, if it matches with the building's natural period, which is a situation we must avoid. Again, this effect is only for tall buildings over a certain height and natural period and the effect gets stronger as the height, flexibility and therefore the period of the structure increases. Also, for vortex shedding effect to emerge, the wind must be originally steady - a turbulent wind cannot cause vortex shedding. Therefore, the irregular exterior shapes of some iconic very tall buildings, such as Burj Khalifa, the world's current tallest building in UAE, are not only for aesthetics, but also to better distribute and make the wind more irregular, and prevent it from being steady. Even slight adjustments on exterior geometry of building can make a difference. Two same height buildings on the same site may experience quite different wind forces, even if they had slightly different cross sections.

Even when wind force or vortex shedding may never be enough to cause collapse, the amount of shaking can be above the comfort limit of occupants as far as motion perception, which can be measured in terms of acceleration caused by wind, i.e 2% of g, gravitational acceleration, in other words, 20 milli-g. This is also an important variable to consider when adjusting a very tall building's overall stiffness and mass and deciding whether to use systems such as tuned dampers *(tuned with building period)*, or outrigger systems or not during design. Of course, stiffer and heavier structure will have better wind vibration resistance, but this will also mean more load on the foundation and the structural frame, higher base shear. Also if you remember from earthquake section, we mentioned a typical damping ratio of 5%. For wind, we can use lower ratios because here there is no inelastic deformation and the building can swing back and forth elastically under wind much larger number of times than it would shake easltically or inelastically during earthquake.

This same alternating force effect between two sides of a structure also happens at long bridges such as suspension bridges. When the wind laterally blows parallel to the bridge deck, the suction force will oscillate between the upper and lower sides, which will create up and down motion of deck. In addition, for long bridges another effect called aeroelastic flutter can cause bending and also torsional (twisting) motion in higher modes which may bring the deck to an unstable state. Aeroelastic flutter happens as a result of the aerodynamics of the structure's geometry, which determines how it interacts with the flowing fluid around it, where forces and displacements are created as a result (imagine a flag in strong wind), which can ultimately destroy the structure if it gets sufficiently strong. During the well known collapse of Tacoma Narrows Bridge, only a 42 mph steady wind caused up and down motion due to vortex shedding and also put the bridge in repeated torsional motion due to aeroelastic flutter (which happened that way as a result of aerodynamics of bridge's geometry) that caused total collapse on November 7th, 1940, only a few months after its opening. Therefore, we want to keep these motions below acceptable limits by making the bridge deck geometry to have more favorable aerodynamic properties. In addition, the longer the bridge, the higher number of periods it can support and the more important higher modes of vibrations (both in torsion and bending) become. This means a higher chance to match with the wind's turbulence period and also the more significant aeroelastic effects can become. To determine the forces and responses as a result of vortex shedding and aeroelastic effects for large, critical structures such as cable stayed or suspension bridges, wind tunnel tests <u>must</u> be performed.

WATER RESOURCES ENGINEERING

As water is essential for all life, the construction of water related structures date back to very old ages. For example, one of the first known written law in human history, by the Babylonian King Hammurabi, deals with the irrigation sequence and duration for the hanging gardens. Today one of the main indicators of a developed nation is having the desired quality and quantity of water when and where necessary. Not only the distribution and collection but also controlling the water that exist on earth's surface in various ways is required for the needs of our civilization.

Water resources engineering, which is one of the five primary branches of civil engineering, is about structures that deal with the usage, control and management of water.

Water resources engineering can further be grouped into three sub branches as:

- Hydrology
- Hydromechanics
- Coastal and marine structures

Hydrology

Before planning for and designing any water related structure or system at all, first we must know how much water will that system handle. Determining the amount of water to be handled for water related structures and systems on land falls under hydrology engineering. Hydrology deals with distribution, movement, properties, quality, and environmental interaction of water on, below or above the surface of earth. The losses we experience due to events such as flood or water shortage at various places on earth is mostly because of inadequate or wrong water management practices we perform with respect to our needs or what we must expect from nature, for a region.

A hydrologic cycle is the full cycle of water on earth. Water stored in clouds fall on earth in the form of rain or snow, then flows on the ground or underground back to the sea, and evaporates back to be cloud again. During this cycle water can go as deep as a kilometer underground or can climb as high as 16 kilometers in the air. So, this cycle happens in a very large volume and in a very complex manner. Hydrologic cycle is also affected by what that geographical area includes, such as sea, lake, mountain, forest, desert etc... which is why it is possible to find an arid region adjacent to a region that receives sufficient amount of rain. The amount of evaporation and rainfall on the surface of earth is in balance given long enough time, because after all, the earth has a given amount of water.

Determining Storm Water Quantity:

A drainage basin, also called catchment or tributary area, is an area of land where the rainfall accumulates and drains into a common point. The figure below shows a river basin, with the main river and the sub branches contributing from sub basins. Catch basin area for large rivers can be very large. For example, the catch basin for Mississippi in the US is an area that is almost half of the continental US, if we also consider all sub branches contributing to it.

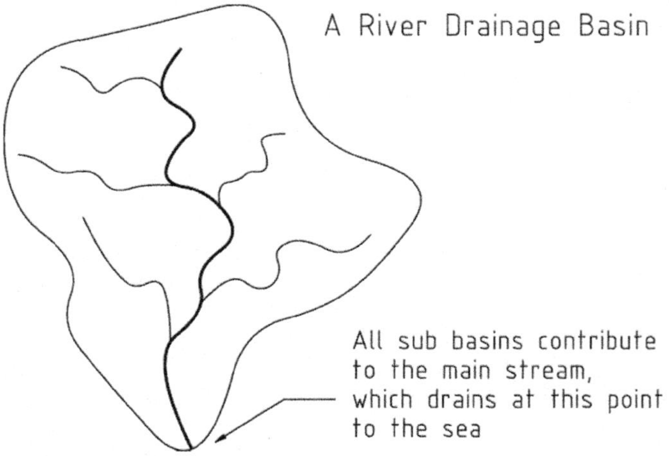

Figure - Water Resources 1

In hydrology, one of the main problems to deal with is finding out the amount of water that falls on a basin accurately. To do this, one of the most essential things to know is the rainfall intensity. When water falls on the ground, after the initial absorption by soil, it starts to flow on the surface.

A hydrograph shows rate of water flow vs. time, through a given point or cross section. It is used to calculate surface storm water runoff, which is then used when designing structures that handle storm water such as storm sewer systems. A simple hydrograph can be seen below. In this figure, instead of rainfall, the vertical axis can also be represented by precipitation or discharge too, which would only affect the vertical scale and the values of the graph, but the horizontal axis, time, will remain as time. From this hydrograph, a unit hydrograph can also be developed, by dividing the rainfall values with the average precipitation. A unit hydrograph is a hydrograph that results from a unit depth of rain falling uniformly on a catch basin at constant rate for unit duration. So it is the most basic type of hydrograph, specific for an area. It is an important tool used in design as it is used for developing hydrographs for similar storms but with somewhat different durations in same area, which simplifies things, by eliminating the need for developing hydrographs each time, which requires a lot of rainfall measurement effort. The more past measurements, the more reliable data we have. As shown in the figure below, for a river, a hydrograph represents both the regular flow of the river which is the base flow, and on top of it, the excess (storm) flow which is caused by the storm.

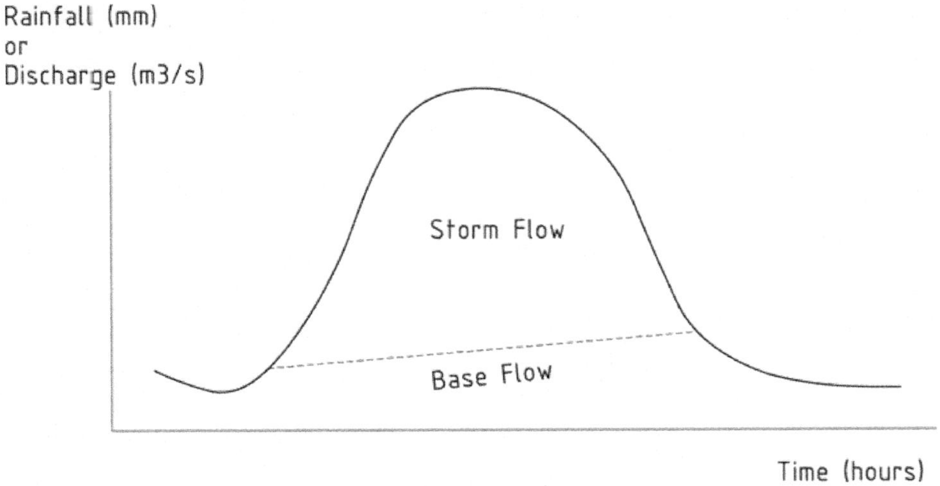

Figure - Water Resources 2

During rainfall part of water enters into soil, and some held by plants. After the soil is saturated, the rest just flows on ground surface (minus the water that continues to seep into soil and evaporate back to air), which is called surface runoff water. That is what causes floods, if measures are not taken. That is the water that our storm drain systems must handle in the cities, and the riverbed must continue to drain in rural or urban areas. This will mean higher river water level during storm, thus a greater width for the riverbed, where nothing must be built, unless precautions are taken such as flood walls, levees.

The amount of surface runoff water depends on:

- Rainfall intensity (amount of precipitation per hour)
- Rainfall duration
- Area of rainfall
- Slope of the area of rainfall
- Permeability of the surface
- Water holding capacity of the surface including plants
- Air temperature (as it affects evaporation rate)

Surface runoff increases with rainfall intensity, duration of rain, size and slope of rainfall area, and decreases with higher permeability, water absorbing capacity of the soil and air temperature. This is just one of the reasons to protect our natural ground, by not allowing a lot of impervious surfaces such as asphalt or concrete, and have lots of plants and trees, in order to reduce the amount of surface runoff water and to also to protect our groundwater levels by allowing water to seep into the ground. For example, the worst type of surface area for contributing to surface runoff is a totally impervious surface, which does not hold water at all, such as the watertight roof of a building or an asphalt parking lot. And the best case would be a natural sandy soil with heavy vegetation, which not only slows down and holds a great deal of water because of plants, but also allows water enter into the ground as sand is permeable. In very populated urban areas with heavy burden on storm drainage systems, more public parks, incentives for green roofs will provide relief. Large open areas such as public parks and grass sports fields can be built on lower parts of a city, which can receive excess amount of surface runoff in the worst case, and then gradually let it seep into the ground.

Among the items counted above, determining the rainfall intensity and then choosing the design rainfall intensity for a storm drain system or a river bed can prove to be not so easy. Sometimes, sufficient data is not available and engineers can use what is called a synthetic unit hydrograph, which is a hydrograph that is produced through mathematical approximations.

When finding the design storm rainfall value, an important parameter to know is the time of concentration, which is the time it takes for water to travel on the surface of the basin from the most remote part of the rainfall area, to the watershed outlet or other design outlet point. In other words, it is the longest possible time that passes between falling of a water drop on a basin until it exits the basin, which is only possible if that water drop falls on the most remote part of that basin. It is not a straightforward task to measure this but there are available methods. Time of concentration is independent of the rainfall intensity. So in other words, whether it rains a lot or not, as long as the flow is enough, time of concentration is the same for a given area with given properties. So in a way, it is a characteristic of that area, until something changes on that area either naturally or artificially.

Another important parameter to know is the peak runoff volume, Q, found by the equation written as below, which is probably the best known formula in hydrology:

$$Q = A.C.i$$

where,

A: is the size of the surface area under consideration

C: is a dimensionless coefficient, which is called the runoff coefficient, that relates the precipitation amount to runoff amount. It is based on the characteristics of the area, which includes the soil type, soil moisture content, type of soil cover (plants), recurrence period of storm, slope and shape of the area, topography including depressions and how much water pockets are formed, imperviousness. C values for every drainage basin for various conditions were extensively developed and available

i: is the rainfall intensity

These values are used for storm sewer, reservoir, river bed improvements etc... design.

Determining Quality and Quantity of Water Demand:

Although the quantity of water on earth remains the same, our usage and pollution always increase, and this means, the problems relating to water keeps getting more critical. Other than determining the amount of water that falls onto a basin as we have seen above, hydrology also deals with the quantity and quality of water to be supplied to an area.

Quality of drinking water will not be the same as quality of water for general use. To achieve this, different treatment processes are needed at different stages.

To calculate the quantity of water that is needed in an area is a complex study, which considers many variables from different areas of life, including considerations for future, and in the end only estimates can be obtained. The findings determine the projects that will be built, such as dams and water distribution systems, which require considerable sums of money to be spent by nations.

Hydromechanics

The hydrology branch of water resources engineering that was just discussed above mainly is about how water moves in nature and how much water we are dealing with in the first place. We then take that information and put it into use by actually designing structures that will control water. To do that we use hydromechanics principles.

Fluid is the matter that flows and deforms without a fixed shape. Another good definition would be in terms of shear stress. As we saw in soil mechanics, a fluid cannot support any shear stress, such as gas or liquid, such as water. As stated in soil mechanics, for our purposes in this book, fluid will mean water unless noted otherwise.

In hydromechanics, one of the main tasks is to calculate the flow of fluids in pressurized or open conduits for various purposes, such as the design of storm drains, flood channels, municipal water distribution systems, dam spillways and so on...

The most fundamental concept to know in hydromechanics is the Energy Grade Line (EGL) and Hydraulic Grade Line (HGL) concept, which are based on Bernoulli energy equations. To understand them, we must know what "head" in hydromechanics means, which was already introduced under Soil Mechanics > Groundwater > Seepage section. If you need, please first <u>read</u> that section at this point before proceeding further in this section... As was discussed there, head has distance unit such as in meters or feet, which represents quantities related to elevation, pressure or velocity of a fluid which are all organized in distance units. Hydraulic Grade Line (HGL) is an imaginary line which is the sum of elevation head and pressure heads. Rewriting,

$$\text{Hydraulic Grade Line} = \text{HGL} = \text{Elevation Head} + \text{Pressure Head}$$

Energy Grade Line (EGL) is an imaginary line that represents the <u>total</u> energy of fluid running from a conduit or open channel. It it equal to HGL plus velocity head:

$$\text{EGL} = \text{Total Head} = \text{HGL} + \text{Velocity Head.}$$

It is also called as Total Head (TH). So we can rewrite it as:

Energy Grade Line (EGL) = Total Head = Elevation Head + Pressure Head + Velocity Head

The figure below summarizes what we have just described. Here a fluid runs in a pressurized pipe. As the flow in the pipe continues from left to right, the EGL and HGL decrease due to frictional head losses. When the flow meets the pump, suddenly there is a big upwards push, equal to head of the pump, h_p. EGL and HGL run parallel, with their difference always equaling the velocity head.

See Figure - Soil mechanics 39a where we showed water seeping below a dam, and the equipotential lines. Each of those represented the elevation of hydraulic grade line there, equaling the elevation head plus pressure head, as we considered velocity was negligible and assumed as zero. The HGL is where the water would rise, if a tube was inserted into a pipe flowing under pressure. So that water level would represent the elevation of water in the pipe (elevation head), plus the pressure head that would cause that water to rise in that tube with respect to the pipe, like in the equation above and in Figure - Soil mechanics 39b.

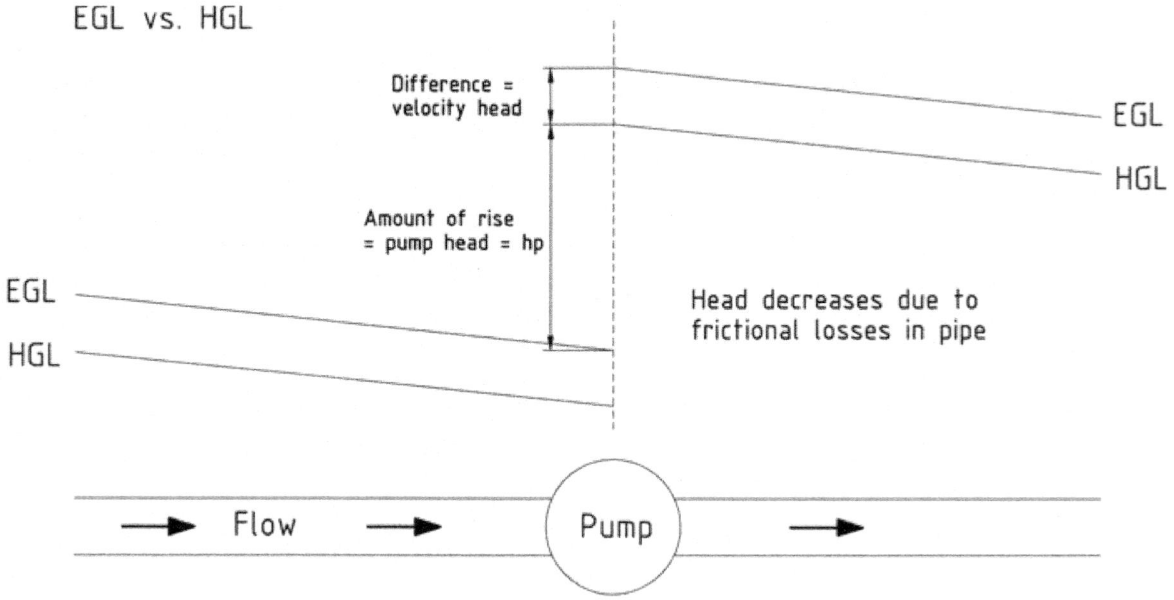

Figure - Water Resources 3

In case of a free flow, such as open channel flow, as there is no pressure, the pressure head will be zero and the hydraulic grade line will be equal to elevation head only, in other words the elevation of water surface. So, whether the flow is pressurized or not makes a big difference, therefore, the flow of a fluid can be grouped into two:

- Pressurized Flow
- Open Channel Flow

The energy concept presented above is applicable to both of these flow types.

Pressurized Flow:

Pressurized flow of a fluid happens in a closed medium such as a pipe or conduit. Here because of the presence of pressure, the hydraulic grade line is above the surface of water. The fluid must fully fill the cross section of the pipe, plus have some pressure, for a flow to be considered as pressurized. If the water in a pipe is not fully occupying the whole pipe cross section, and some air is left on top, then the flow essentially becomes an open channel flow, even if the water flew in a closed channel.

Open Channel Flow:

Open channel flow occurs where the fluid flows by gravity in an open channel with a free surface open to atmosphere (or at least atmospheric pressure), which means, this type of flow is not pressurized. To calculate the flow, we must first know the hydraulic radius. It is found as in the figure below:

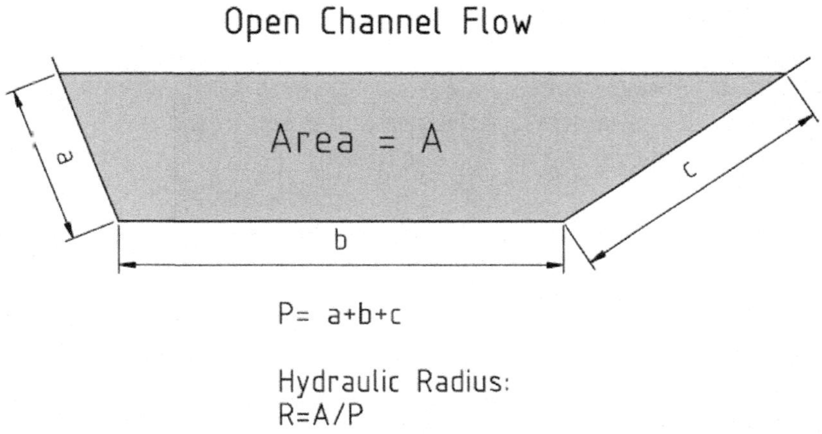

Figure - Water Resources 4

In the figure above,

A: Cross sectional area of flow (Units: ft^2 or m^2)

P: Wetted perimeter of cross section by the fluid, which is: a+b+c (Units: ft or m)

R: Hydraulic radius = A / P (Units: ft or m)

Then we use the hydraulic radius R to find the discharge volume, Q, from Manning's equation as:

$$Q = V.A = \frac{K}{n} A R^{\frac{2}{3}} \sqrt{S}$$

where,

Q: Discharge Rate (also called flow rate or amount of flow) (ft^3/sec or m^3/sec)

V: Velocity of flow (ft/sec or m/sec)

A: Area of flow (ft^2 or m^2)

S_o: Energy grade line's slope (Units: ft/ft or m/m so, unitless)

n: Manning's roughness coefficient (Units: unitless)

K: a constant based on the unit system used such as imperial or metric units.

The discharge rate Q, is directly proportional to A, R and S_o. It means, as the flow area, hydraulic radius and slope of energy grade line increase, the discharge rate, which means the flow, increases.

Q is inversely proportional to n. It means, as the roughness of the channel increases, Q will decrease. Different surface characteristics such as different materials, surface textures, vegetation cover etc... all have effect on this "n" value which is extensively tabulated and provided in charts.

Coastal and Marine Engineering

Another main branch of water resources engineering is the coastal and marine engineering. We discuss coastal and marine structures in Chapter 9. At this point, let's only introduce ocean waves, which is an important subject to know in coastal and marine engineering.

Ocean Waves:

Although a physicist would define it more accurately and completely, we can just say here that waves are propagating energy as a result of disturbance. After a disturbing energy such as wind creates the wave energy, which is a kinetic energy, it travels through all possible directions. We discussed seismic waves in earthquakes section. Although some basic characteristics of all waves are the same, ocean waves have their own characteristics as well, much different than geological seismic waves, and that is why we call them specifically ocean waves and not just waves. In this subsection though, from here forward, wave will mean ocean wave.

Ocean waves are formed by various effects and can carry a lot of energy. It is estimated that only a few percent of the energy of all waves in the world are enough to supply the entire energy needs of our civilization. Similarly for the US, according to US Energy Information Administration, if all wave energy at the US coasts were to be harnessed, it could generate up to 2.64 trillion kilowatt hours, which is 66% of the total electrical energy generation of the country as of 2020.

Waves play the greatest role in design of coastal or offshore structures. They can also cause coastal flooding or beach erosion. Tsunamis can be far more damaging.

Ocean waves can be classified as:

- Wind driven: This is the most common type, where wind transfers its energy to water. These are called surface waves too, or simply wind waves. They range between ripples to waves of as much as about a few dozens feet height. The faster the wind, the greater distance and the longer duration it blows, and the greater distance the wave can travel without obstruction, the larger the waves become until equilibrium is reached between wind and the wave. After that, the same wind will just drive the wave with the same height and speed, (thus energy), and the constant added energy by a constant wind is consumed for the most part by the small constant breaking of the wave, plus the very small energy losses from capillarity and friction from air and viscosity. When the wind slows down or stops, the wave breaking stops and the wave can continue hundreds or even thousands of miles by itself, losing its energy only very slowly, since there is some energy loss due to viscosity, capillarity and air friction which are very small. These continuing waves, after they moved out of the area they were generated, are called free waves or swells. To have random periods and heights for wind driven waves are typical, which are called as irregular waves. If the wind was weak, only small capillary waves (tiny ripples on surface) form, which quickly disappear due to surface tension effects being significant.

- Tidal waves: These are caused by gravitational pull as earth and moon rotate around a center of mass point between each other, and therefore they are the most consistent and predictable. For a given point on earth, between two high tides, there is 12 hours and 25 minutes. *(If you multiply 12 hours 25 minutes by two, you get 24 hours and 50 minutes. Earth needs 24 hours to make one rotation around itself, and an extra 50 minutes to catch up with the moon's rotation around the earth with respect to previous day. There are two highest and two lowest tides on earth at any given time. The high tide towards the moon happens as the moon pulls water on earth towards itself. And the other high tide on the other side of the earth opposite to the moon is formed by the centrifugal force resulting from earth and moon orbiting around one another. So for a certain location, between two high tides there is 12 hours and 25 minutes, which means, 6 hours and 12.5 minutes between a high and low tide)*
- Internal waves: These are created by water of different densities moving against each other deep in the sea, and they can be much larger than surface waves. They can affect safety of submarines by pulling them down to greater depths which exert more pressure than they were designed for. Internal waves can also break underwater as they approach mainlands.
- Tsunami: They can be caused by earthquakes, volcano activities, or massive landslides (splash waves). They are the fastest waves, which can travel several hundred miles/hour, in all directions from the source. Tsunamis contain far more energy than wind waves because, their energy comes from the entire sea floor moving vertically during an earthquake, which means a huge amount of volume. This also means that an earthquake where the seabed moves only horizontally will not produce tsunami. Tsunamis do not necessarily have to be large to cause damage. Even if a small tsunami wave reached a coast, it can still be larger than wind or tide waves, which may cause some damage, although less than the larger tsunamis.

The waves listed above are also called as gravity waves, because during their progress, gravity is the restoring force. If a wave is very small, such as less than several cm in length however, capillary forces of water also begin to dominate. As the length gets smaller, then it becomes a mainly capillary wave. With waves, water molecules do not travel, at least when the waves travel in deep parts of the sea. They only make circular motion as seen below. It is the energy that travels, until the wave breaks. Waves can travel by making different movements as discussed in earthquake section, which can be longitudinal, transverse, or circular. Ocean waves are the circular type, which is a combination of longitudinal and transverse movement. The figure below, shows the basic terms about wave geometry. This figure applies to the waves in open sea, and it is an assumed ideal shape. As seen below, in greater depths, circles get smaller which is explained by wave equations that are beyond this book's scope. Note that the figure shows exaggeration of height to length ratio of the wave. The wave height can be maximum of around 1/7 of wavelength. If the height is more than that, the wave will break. Here for the sake of showing things clearly, we showed height at a higher ratio to wavelength than it can be in reality. Below, you can also see the velocities and accelerations of water particles at the top, neutral and bottom points of the wave, indicated by "V" and "a" respectively. If you see how the velocity (or acceleration) arrows progress from left to right, you should be able to follow the circular motion for a particle from there too.

- Crest: The highest point of wave.
- Trough: The lowest point of wave.
- Period: The time it takes for one full cycle of a wave to pass a certain point, shown by T.
- Amplitude: Distance between crest and the natural surface before waves were present.

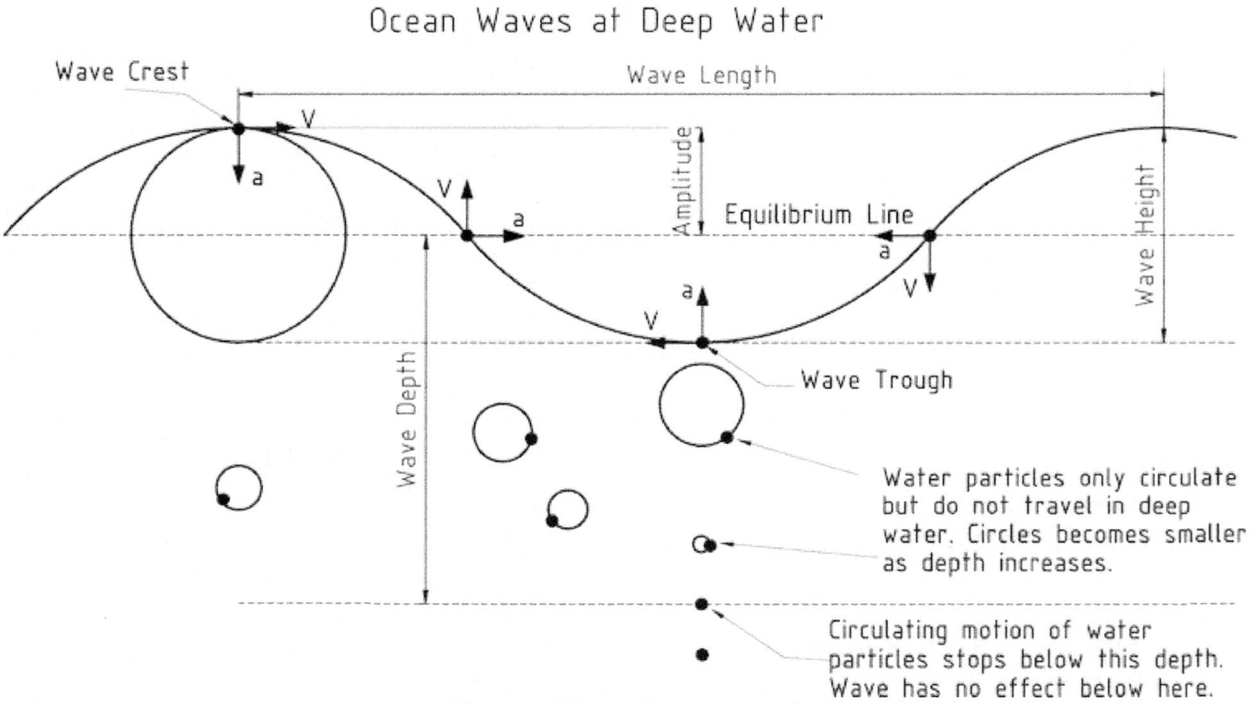

Figure - Water Resources 5

- Wavelength: The distance between two identical spots on two consecutive waves, such as between two crests. Longer waves are faster, as can be confirmed by wave equations.
- Wave height: The vertical distance between crest and trough.
- Equilibrium line: It is the undisturbed, zero energy water level if there were no waves.
- Depth: The maximum depth that a surface wave can cause motion in water body down below. Therefore, for example the portions of an offshore structure below the wave depth at open sea will not experience any wave force. This was found to be about half of wavelength.

The figure below shows the situation when the wave approaches shore. As soon as sea depth gets shallower than the wave depth, the circles become elliptical, because now from the bottom there is friction, which slows down the wave bottom, but the speed of wave at its crest still tends to remain the same. The amplitude increases, and the wavelength decreases, which means that the waves are getting closer to each other. The period however is not affected (*remember from earthquake section that the natural period of a structure was not affected from external effects*). After a while, the wave crest becomes too steep and it falls forward, which means the wave breaks. So for example, the breaking wave curling into the cool ellipse shape in which the surfers ride is because of this. And indeed this zone is also called the "surfer zone", in wave mechanics. That type of curling surfer waves form over a steep sloping beach. If the beach slope is mild, such as the one seen below, the waves will just break in a spilling/flooding form, without those curls we see on surfer waves. Or, the beach slope may be very steep, which will then cause a sudden increase in amplitude, with very little curling at the top as if splashing to a wall. Because of the wave breaking process, near shore currents are also created.

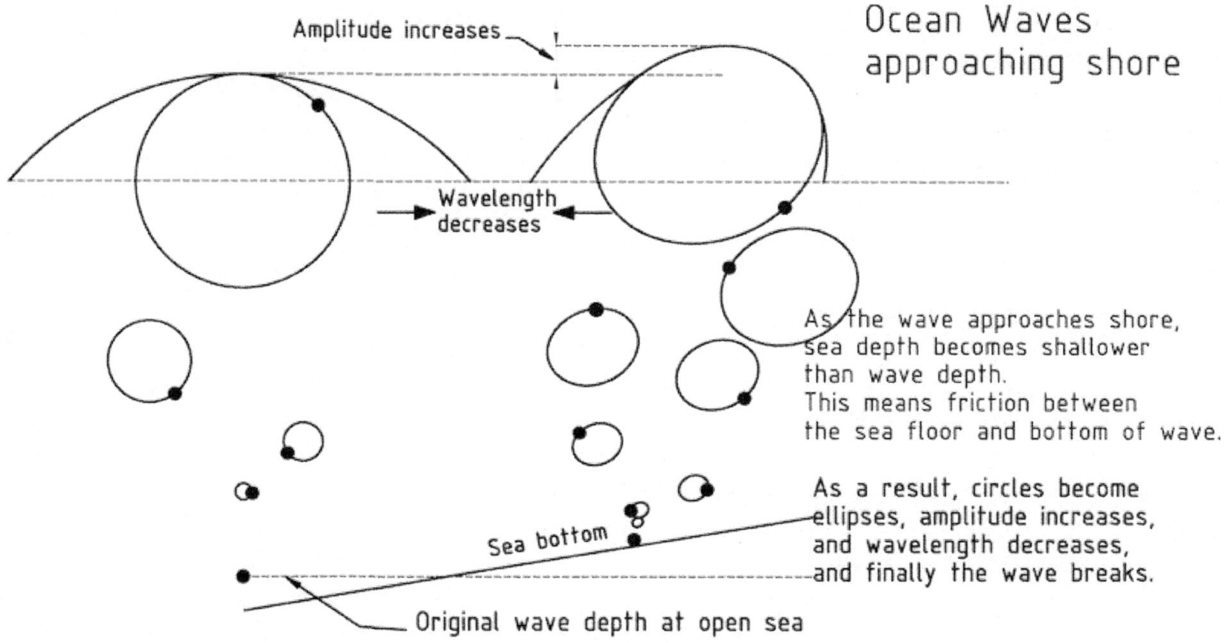

Figure - Water Resources 6

The amplitude of waves getting larger as they approach the coast can be disastrous especially in case of Tsunamis. For example, when earthquake happens out in the open sea and a Tsunami forms thousands of miles away from coast, the amplitude of the wave in the beginning may be very low, such as a meter or less (which is the amount the ocean floor moved vertically) and it seems harmless. But as explained, the energy of the Tsunami comes from the movement of an entire tectonic plate, which is huge. So when that wave which was only around a meter high out in the open sea approaches to shoreline and the sea depth decreases, suddenly this huge amount of energy greatly increases height of wave, to anywhere between several, up to a around 30 meters, and have very destructive effects. The speed of wave can be around several hundred km/hr in open sea, although it slows down when reaching the coast. A large tsunami wave, if it reaches flat land, river beds, can go inland for hundreds of km because that length of sea bed had moved and pushed that huge amount of water volume in the first place. That is also why the average wavelength of a Tsunami wave is very long, which can be as much as few hundred km. When it reaches the coast, it looks like flooding water of great height. If the trough of the wave arrives first, it looks like water is receding from the area for a short amount of time, before the crest of the wave hits. Contrary to the common knowledge, sometimes we do not see any recession first because the crest can also arrive first, which means no warning. The whole set of Tsunami waves are usually made of several waves following each other. The 2004 Tsunami in South East Asia caused more than 220,000 deaths in a few hours on different coasts. According to earthobservatory.nasa.gov website, the earthquake of magnitude 9.0 in Indian Ocean caused the Burma plate sea floor to move vertically up to 5 meters, and horizontally 11 meters at some locations *(deep 18 km underground, right where the earthquake occurred, the movement was up to 20 meters)*, and the coast of Sumatra island moved by as much as 3 meters. Wave heights over 10 meters were seen in Sumatra.

Figure 7 below shows the wave slowing down when it reaches the shore. This means, the part of the wave that is closer to the shore slows down first, then the rest of the wave that still travels at its normal speed catches up, until all of the wave that reached the shore travels at the same speed, which means the wave crest bends to align with underwater contours and the wave becomes parallel or almost parallel to the shoreline. This is called refraction as shown above. The refraction of waves can cause more energy focused on a particular area of shore which sticks out with respect to the rest of the shore. This is important when designing coastal structures. The short arrows in the figure are called wave rays, which represent the wave direction at those points, which are also perpendicular to the wave crest lines at those points.

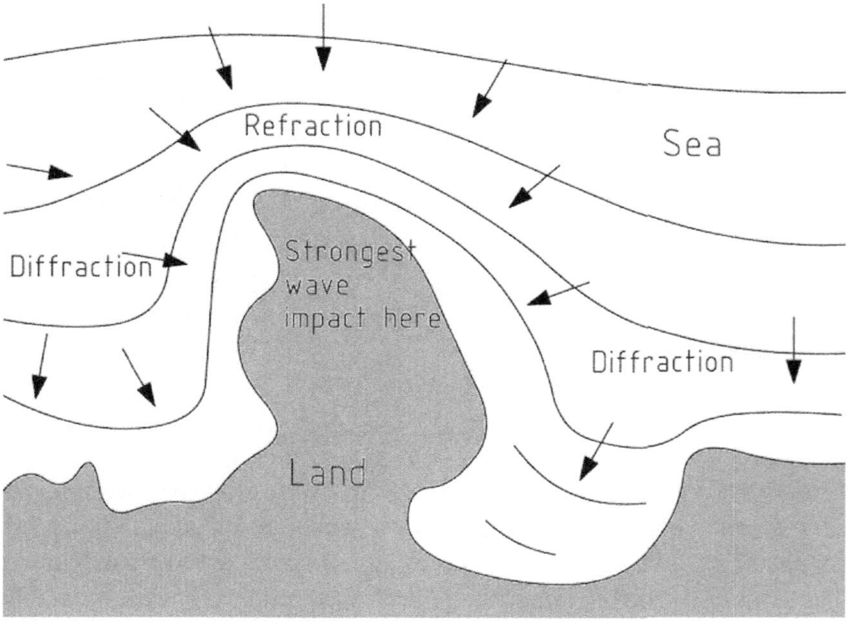

Figure - Water Resources 7

Waves can also reflect from coast by bouncing back if they meet smooth and rigid surfaces, which is called wave reflection. This can cause erosion at sea bottom, by taking away sediments such as in front of a sea wall. That is why, when building sea walls, the bottom is protected by rocks to reduce reflection by consuming energy and also preventing sediments in front of wall from moving away and cause scour.

When waves interfere, the effect can be either to amplify or to smooth out each other. If two waves add up, it can cause rogue waves, also called killer waves or freak waves, which can be unexpected and much larger than the usual wave sizes and can be dangerous for ships, even in calm looking sea. These can form from swell waves that were created before at far locations, traveling in different directions.

Kinetic energy of an ocean wave is the sum of kinetic energy of all water particles moving with oscillatory motion within the wave in that instant - but here keep in mind that it is not the particles themselves that travel, but the energy, since their motion is only oscillatory, as also shown in Figure 5. This energy depends on the amplitude, wavelength and wave speed. These in turn depend on the wind speed, the length of time and the distance that wind blew (called fetch), which makes the wave gradually larger, carrying more energy. The amplitude of the wave carries the wave's energy, and that is the most important parameter for us when designing structures that are to withstand waves. When the waves reach these structures, the energy is transferred to the structure, or sand or rock. Some of the transferred energy is lost due to friction in the form of heat, which is too small for us to feel, but the remaining transferred energy must be resisted by the structures, therefore, knowing wave dynamics is important. This is also important to know for wave energy generators. For example the power that the wave delivers to the generator is directly proportional to wave period and wavelength but it is proportional to the square of the wave height (height is twice of amplitude as in Figure 5).

Currents:

Ocean or sea currents are a concern for coastal and marine structures. Three major sources of sea / ocean currents are: Tides, seasonal winds, and general circulation of water within the ocean.

CONCRETE

Concrete is the most common construction material of our civilization, which is made by mixing cement, sand, aggregates, and often also admixtures, by water and letting it to harden. Concrete is strong in compression but weaker in shear and especially the weakest tension, therefore reinforcing steel is also used in most cases. In this section many important terms and concepts about concrete are introduced. When we say concrete in this book, we mean the first one that comes to mind, which is the cement concrete used in our structures. Asphalt concrete to make pavements is out of book's scope.

We will discuss concrete under the following subsections:

Concrete Mix

To understand concrete, let's first look at the ingredients that make up a concrete mix. If we had to describe what happens in one sentence: The sand and aggregate are bound with cement, which undergoes hydration reaction and hardens in the presence of water. These are the basic ingredients since Roman times. Even with the variations in just these basic ingredients, it is possible to obtain mixes that make so many types of concrete with different strengths, durabilities and other properties. In addition to these, we now also very commonly use admixtures, which makes it possible to achieve virtually unlimited variations and quality improvements. As far as the weight of a typical concrete mix, 150 lbs/ft^3 or 2500 kg/m^3 would be good round numbers to remember. When specifying materials for a design mix, it is also important to realize the availability of materials not far from the concrete plant and jobsite. Below the basic ingredients are introduced, and a few sections later, admixtures.

Cement:

Cement is the material that binds everything together in concrete. In other words, cement is the material that makes concrete, concrete, and it has the greatest effect on its properties. There are mainly 5 types of cement, per ASTM classification, listed from types 1 to 5. Type 1 cement is the general purpose cement, which is used most commonly in general construction. The other cement types have different purposes. For example Type 2 cement has better resistance to alkali reaction, and, produces lower heat of hydration, which becomes important when pouring mass concrete such as dams or very large foundations, where excessive heat generated by hydration may have adverse effects on concrete properties. Type 3 cement has finer particles, which allows it to react faster, and develop high early strength at the expense of higher hydration heat. It is used when we need rapid strength gain. It can be used at locations with tight construction schedules, such as multistory buildings, precast member manufacturing, quick repair of pavings, in cold weather construction and more. Its ultimate strength may not be the highest. Type 4 cement produces even lower hydration heat than Type 2 cement and is very suitable for mass concrete. Type 5 cement is especially used for the highest alkali resistance, such as for the concrete that will be in contact with (corrosive) soil or high sulfate content water, but its strength gain rate is slower than Type 1. Similar to Type 2, which also has high resistance to sulphate attack, it can be used for slabs on grade, underground concrete pipes. Different cement types combined with different admixtures, sand and aggregates with different characteristics in different ratios are used to achieve an unlimited variety of results for various types of structures, structural members, geographies, climates, construction schedules, underground or above ground conditions, while always keeping in mind the durability, strength and economy of the design.

Hydraulic Cement:

Some structures are built in or around water, subjected to constant water action including seepage or leakage, waves, contaminating chemicals, or severe climatic conditions. In these cases, hydraulic cement can be more useful. Foundations, manholes, fountains, pools, basement walls, elevator pits, chimneys, coastal and marine components are places where hydraulic cement is often used.

Water:

Water is needed in concrete for:

- Making the concrete workable by serving as a means to mix the sand, cement and aggregate (and admixtures if any), and,
- Allowing for cement to hydrate, which is the process that enables cement to bind everything.

Some of the water in concrete is only the excess portion, it is just there to make the concrete mix easier to work with and evaporates over time. Remaining water enters into hydration reaction with cement.

A primary variable to look at in a concrete mix is its water cement ratio, written as W/C, which is the ratio of weight of water to weight of cement. This value generally ranges anywhere from around 0.4 to 0.6. If we want to give an example, FHWA FP-14 specifications (section 552.03) limits this ratio to a maximum of 0.45 for Class A concrete on federal highway bridge projects. This is a fairly typical number to use also for different types of projects and or locations, which may be specified as the same or close in other applicable standards.

By keeping W/C ratio as low as possible, we achieve:

- higher strength,
- better crack control,
- higher adhesion between concrete and rebar,
- lower porosity, which means less permeable, stronger, more durable concrete.

A disadvantage of low W/C is less workable, more stiff concrete mix while placement, which can create difficulty in handling while pour, but this problem can be solved by adding superplasticisers, as explained in admixtures section. An optimum level for W/C must be found for each case.

The higher W/C ratio, which means the more water per cement, the weaker and less durable the concrete will be in general, because excessive water will remain unhydrated and increase porosity - but do not forget that although W/C is a great influence, it is not the only factor. Higher W/C ratio requires us to perform longer duration of curing, which may still not be enough to cover the defect caused by a high W/C ratio in the first place, because W/C has much more fundamental effect on the final product.

So, only just enough water should be added to the mix and nothing more, to allow for complete hydration of cement, plus some amount to make the concrete workable enough so that we can move it around and compact it properly into the forms.

Water content affects concrete's slump value. A slump test is performed very commonly to see how liquid is the wet concrete mix (also see concrete tests section). Concrete with low slump will be too difficult to pump and handle and concrete with a high slump value will be too liquid and probably weak.

Water should ideally be potable. It should be reasonably free of silt, alkali, salts, organic compounds and other impurities. Water containing chlorides or sulphates are especially unwanted, because they contain salt. This is also called chloride contamination. Eventually it also causes rebar corrosion.

After many decades of improvement, engineers were able to reduce the average W/C ratio of concrete mixes, at the same time still obtaining same workability and achieving even greater strengths and durability. For example, while in 1950s a common W/C ratio could be 0.60 or more, today we are able to produce much better concrete in highly controlled environments with W/C down to about 0.40s.

Aggregates and Sand:

Aggregates and sand *(can also be called as coarse and fine aggregates)* together make up most of the concrete volume by roughly around 70%, and weight by up to 75% (for asphalt concrete this goes up to 95%). Being cheaper than cement, aggregates reduce the cost of the mix, provide the main skeleton of concrete, and greatly reduce shrinkage. Aggregates should be angular shape rather than round, in order to provide better bond with cement and more friction within the aggregate itself and with sand. Cement paste fills the gap between sand particles and sand particles fill the gaps between aggregates, to provide denser and stronger structure, and other than that sands have basically the same purposes as aggregates. Good adherence between cement paste and aggregates (and sand) is very important. Based on size, angularity, chemical composition and texture, aggregates vary greatly and influence properties of concrete. Although aggregates are cheaper than cement, using too much or using too large size aggregates decreases concrete strength.

Roughly speaking, in a typical concrete mix, for every 1 part of cement there may roughly be anywhere from 1.5 to 4 parts sand and 2 to 8 parts of aggregate, mixed ideally with "just enough" water which can produce a reasonably workable concrete that can be handled, suffice for complete cement hydration but not too much water to reduce strength and durability of concrete, which may correspond to an amount of roughly 0.5 to 0.8 parts of water (if by volume) with respect to 1 part of cement, or about 0.4 to 0.6 w/c ratio by weight, as was defined in the previous page. An easy to remember rule of thumb may be 1:2:3 rule for cement, sand and aggregate respectively, all by volume. Here if we didn't have aggregates, it would be just mortar, not concrete, which would have lower strength than concrete.

Side Note: Cement, sand aggregate ratios can be expressed by weight or volume, but usually volume such as buckets etc... is understood in practice when we talk about concrete mix ratios.

As the aggregate and sand particle sizes decrease, they will have more overall surface area, therefore more water needs to be added for their contact with cement. On the contrary, if larger aggregates are used, less water will be needed, which not only means lower water cement ratio, but also decreases heat of hydration. But of course there is a limit to maximum aggregate size that can be used, because for constructability, we do not want aggregates blocking flow of concrete by getting stuck between the rebar, within the form or between the rebar and forms, and from design point of view, too large aggregate sizes will affect overall strength and durability negatively. The smaller and more intricate a structure, the smaller aggregate sizes will be. For most concrete structures you see in a city, the maximum allowed aggregate size would be around 1.5 inches. For mass concrete pours such as dams, aggregate sizes up to several inches is possible, but that is for the most bulky parts of a dam. The aggregate size is measured as the longest distance possible (diameter) between two outside corners of an aggregate. It is easily controlled by using sieves. The maximum dimension of each aggregate (width length depth) should not be more than 3 times of its minimum dimension.

Hydration, Setting, Curing

Hydration reaction takes place when cement particles contact water. When cement hydrates, the cement particles grow hydration products which form bond with other cement particles, sand and aggregate, which

Side Note: As for the water used for concrete hydration, a small portion of it is actually used in hydration. The rest evaporates. This is a major reason for concrete shrinkage.

form the body structure of concrete. When we pour concrete, at first the concrete paste still behaves almost like a liquid and it is able to fill any form shape we want. Then as hydration continues, the concrete reaches its initial setting (hardening) point, where the plasticity and workability of concrete starts to disappear. Usually this point is reached within the first 30 minutes or so of starting the pour. After that point, the concrete quickly starts to gain its strength and stiffness. A little after the initial setting, the concrete reaches its final setting point. This makes the concrete appear as a completely hard solid mass, but at this stage it is still far from its design strength. After the final setting, hydration reaction and strength gain still continues. An important aspect of hydration reaction is that it produces heat. This is positive in cold weather pours, but it can also be something to be remedied during hot weather or mass concrete pours, to prevent negative effects to concrete properties.

Concrete curing requires cement particles have enough moisture around them to hydrate, which means gain of strength and stiffness. The most important property of concrete, other than durability, is its compressive strength. The curing period to achieve design strength is taken as 28 days but in general shorter periods are possible with certain precautions. The strength gain is faster at first and it gradually slows down. For example, in general, approximately about half of the 28 day design strength is reached within the first several days of curing, and the other half takes about at least 3 more weeks. In reality, curing of concrete and strength gain never stops, and still some small strength gain *(usually around 10% or more)* beyond the 28 day design strength is achieved after 28 days in the very long term, but this portion is conservatively ignored. So for example a 4000 psi concrete means, with proper curing the concrete is designed to reach 4000 psi in 28 days, which is the value we use for strength of concrete in our design calculations. Poorly cured concrete can remain far below its design strength and durability. Curing period must be accounted for in the construction schedule.

Curing of a concrete slab may be performed by watering the surface of concrete after pour and letting it pond there to provide constant moisture, or by wet blankets / tarps covering surface of the concrete which allows less often watering frequency by holding moisture, thus saving labor, or by sealing the concrete surface to trap water inside to save even more labor, which is the reason that it is the most commonly used one. If one of the first two methods is used, it must be ensured that the surface of the concrete is constantly wet, without letting any drying-wetting cycles, so that the water inside the concrete, which is adequate for curing, cannot escape by evaporation. If the curing compound method is used, it must be applied a little after the finishing of surface is complete, to prevent water evaporation from concrete. The compound is applied as soon as the concrete hardens enough to carry the worker who will apply the compound, which may be around 60 minutes after trowel finish, which will vary with air temperature, concrete mixture, and other project conditions. Waiting too long will cause some moisture inside the concrete to escape, and may affect strength negatively. While a curing compound will wear off in time, a curing and sealing compound can also be used, which will be durable.

For formed surfaces, such as columns and beams, the forms also prevent water from escaping, and by the time they are removed, already considerable curing will have taken place, unless the forms are removed early. Therefore spraying curing compound to such surfaces after removing the forms is not always necessary and depends on the project requirements but recommended.

Using wet aggregates will help uniform curing and also counter the effect of water consumption inside the concrete due to hydration and water loss due to drying *(also see shrinkage under Harmful effects to concrete subsection)*.

An elevated (structural) concrete slab generally needs at least about 15 days of formwork support in practice, if it is determined by the engineer that the concrete has cured adequately and can support itself without sagging because sagging in early stages has negative effects on long term performance. Rapid curing may also be achieved if high early strength type of cement is used, which becomes important where the schedule is critical, but this is costlier and causes more hydration heat.

In Figure 1, approximate development rate for the compressive strength of concrete is shown with a typical curve shape, which will vary for each project. The mix design itself, skill of batch plant operator, skill of pouring crew, ambient temperature *(higher air temperature helps faster curing)*, effectiveness of curing and other project conditions will all affect this graph's values.

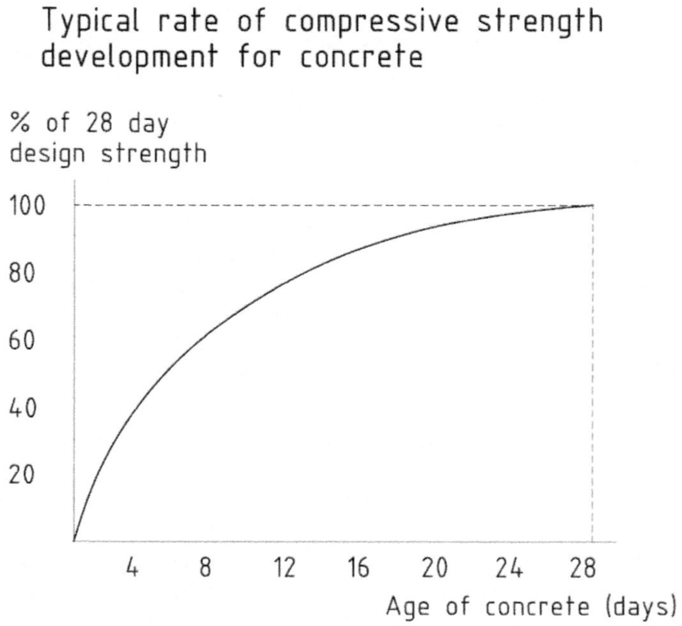

Figure - Concrete 1

Batching and Mixing

First of all let's not confuse batching and mixing with designing of the mix. The engineer of record designs or usually selects a particular mix. Only after that the concrete is produced by batching and mixing, when needed at the jobsite.

Through batching and mixing, we seek to obtain a homogeneous mix of desired properties with uniform moisture content. Water should not be added suddenly but gradually, uniformly to have the same amount of water with each batch and for proper mixing. During adding ingredients, entry of foreign particles into the mix, which may delay setting, prevent adherence, cause voids and cause corrosion of rebar in the long term, should be prevented.

Mixing should be adequate for enabling proper contact of sand, aggregates, cement and water, but excessive mixing is not good because the concrete will start to harden too early before being poured.

Handling & Placement

After the concrete is mixed, it must be transported to the jobsite as quick as possible. Because of this, batch plants not far from the jobsites must be chosen, or for large enough jobs, a plant may be erected at the jobsite. Then the concrete is brought to the final location where it will be poured. After this point, concrete handling and placement will begin. This phase is as important as designing, batching and mixing the concrete, as far as obtaining adequate concrete strength and durability. Improper handling and placement is often the reason for not obtaining proper results in the final product because this phase is inherently less controllable and open to more variables. Here some common aspects of placement and handling of concrete pours are introduced.

Above, we said that excessive mixing is not allowed because the concrete will start to harden early. The longer overmixing, the worse it will get. Because of this, concrete pouring operation must be very well coordinated. If a concrete batch remains too long in a concrete truck so that it is overmixed and starts to harden, that mix should not be used anymore. In such case, adding water afterwards to make the concrete more workable again should never be permitted because the concrete strength and durability will be negatively affected. Therefore, especially in jobs requiring many trucks with tight access to the jobsite, and in urban areas with a lot of traffic, coordination is essential. The time that passes between introduction of cement or water to the mix, and placement of that concrete to its final location should not be more than 90 minutes per ASTM C-94, Standard Specifications for Ready-Mixed Concrete.

Every separate concrete pouring operation means an additional mobilization cost, therefore, as long as there is nothing justifiable to limit pour amounts, it makes sense for a contractor to try to cast as much concrete as possible in every pour. This applies to any item, whether a large area of slab on grade, a massive footing, a long retaining wall, dam blocks and so on. In fact, some pours should be completed in one time, such as the beams and slab of a building floor, to make them work together.

After pouring concrete to its final location, while it is still wet, it should be vibrated by penetrating vibrators in it to let air bubbles escape. Insufficient vibration may cause voids (called honeycombs) in concrete, which have negative affect on strength and durability. Touching of vibrator to forms should be prevented, although it should still penetrate to sufficient depth, as full depth as possible, very close to the bottom of fresh concrete. The vibrator should be fairly quickly *(not too slow to cause segregation but not too quick to cause insufficient vibration)* and vertically immersed and removed.

Workability of concrete is an overall term used for how easily the concrete is pumped, placed and finished. To achieve better workability, we can use admixtures. Again, this should never be attempted by just adding more water into an already prepared mix brought to the jobsite.

For each project, many properties of the mix we will need is different, not only for its final, hardened properties, but also during when the concrete is placed. For example a highway concrete mix will have different workability, slump, aggregate sizes than a mat foundation concrete, where there is also a lot of rebar that the concrete must flow through and where different equipment will be used for placement.

Segregation should be prevented. It means aggregates and sand separating from each other and gather in different parts of concrete, which reduces strength and durability and results in surface defects, rock pockets, surface scaling and crazing, sand streaks. Segregation can happen if the concrete is vibrated too much, or poured long distances through hoses or pipes, or made to fall from too high, or not placed directly in its final position and pushed on horizontal or sloped surfaces for excessive distances.

When the section to be poured is too tall, such as high concrete walls, all of that height can not be poured at once, but in "lifts" (layers). The reason for not pouring at once is to prevent too much sudden liquid pressure at the bottom of the form, by allowing the first layer to start hardening. Then, without waiting too much, the next lift is poured so that both lifts will still mix. This can be done while pouring the upper lift, by inserting vibrator all the way to the bottom of upper lift and the upper side of the lower lift, so that they mix.

There are also inspections and quality control required before and during concrete placement, such as inspecting rebar and forms before pour, comparing the poured batch with mix design and performing slump tests, making sure placement is made correctly, such as avoiding segregation and doing vibration not too much or little, obtaining test cylinders properly and in adequate numbers.

The concrete trucks are lined up in front of a building to dump their concrete into the pump, so that the pour will not have to stop while waiting for the next truck, which may cause cold joints otherwise. The pump elevates the concrete all the way up through the tip of the hose through the temporary tower in the middle of the deck, which makes pouring easier and faster than lifting buckets by crane. This tower will be raised up, as the floors are poured.

Self Consolidating Concrete (SCC):

One very useful option about handling and placement is to use self consolidating concrete (SCC), also may be called as self compacting concrete. Although less common and may be a little more expensive, the use of SCC seems to be increasing. SCC essentially flows like a liquid, but still enables us to obtain concrete with adequate strength and (may be even better) durability. This is a specially designed high slump mix that can consolidate itself under its own weight and fills the forms without the need for vibration. The advantages include:

- As mentioned above, vibration is not needed to remove air bubbles because the concrete self consolidates / compacts.
- SCC can flow between the rebar that would be too congested for normal concrete flow.
- It can easily flow into irregular or hard to reach forms.
- SCC does not undergo segregation.
- On slabs, SCC self levels (troweling is still needed). With smooth forms at columns, beams and walls, it can create very smooth surfaces without or very minor need to finish further.
- SCC also enables us to pour larger, taller and also more complex shaped areas in a single pour.

These aspects bring a whole new range of construction possibilities that we could not make possible with traditional casting method. They also mean improved efficiency, speed, safety and cost savings for concrete placement and a little higher freedom in rebar detailing. The disadvantages are that it is more of a specialized pour for which not every contractor has the experience, and the formwork must be made stronger to support the wet weight and the mix itself is more expensive. SCC may be used for any size of project after a cost benefit analysis.

Cold Joint:

During concrete pour, if too much time passes between pouring two batches which were supposed to be poured continuously to form a monolithic concrete, the first batch will start to harden. When the next batch is poured near or on top of the already poured batch, a cold joint will form, which is not acceptable. This happens when there is poor planning and coordination of production-delivery-pouring operations. We say poor planning and not unexpected circumstances, because before a concrete pour all sources of possible delays that should be reasonably expected, should be accounted for. If, during pouring of concrete, a cold joint seems likely or certain, the measure to take depends. If the hardening of first batch did not progress too much, its surface can be roughened and the next batch may be allowed to continue. In some cases, we can place or drill dowels into the already poured section and continue with the next. Any action to take at this point will be subject to approval of the structural engineer of the project. Documentation of cold joints as detailed as possible is also important to accurately evaluate its structural impacts. The worst case scenario would be to demolish the poured section and repour everything, which might be a cost and schedule disaster for a project.

Admixtures

Admixtures serve various purposes for obtaining or improving various concrete properties either during pour or after it hardens. Their purpose is <u>not</u> to make up for a poor quality concrete mix, but to improve properties of a properly made mix. They usually increase cost, but cost reduction is also possible. Their usage and benefits are dependent on all factors that impact the concrete quality itself, such as the concrete ingredients used, environmental factors and the way concrete is placed. Admixtures can be cementitious or chemical. The cementitious admixtures should not be confused with the actual cement. For example Type 2 or Type 4 cement can be used for reducing hydration heat, which becomes important when placing mass concrete but they are cement types, not admixtures.

Cementitious admixture materials are mainly fly ash, silica fume, blast furnace slag. These have pozzolanic or hydraulic (behave similar to cement) properties to varying degrees. For example, silica fume, shows the highest degree of pozzolanic behavior, while fly ash and blast furnace slag can show mixed behavior. Through the use of these, cost savings, reduction of hydration heat, and increase in strength can be achieved to varying degrees, and they all especially increase durability such as by decreasing permeability *(see the term "Durability of concrete" in the following pages to see what properties are meant)*. The disadvantages may include increased curing duration *(except silica fume, which may decrease curing duration depending on the mix)*, decreased freeze thaw resistance due to having less air in concrete, concrete mix that may require more water which increases w/c ratio... There are actually a huge amount of variations in bottom line effects, due to varying conditions and mixes. Fortunately there is a lot of data available from past research and experience. Disadvantages may be remedied by using other admixtures, such as chemicals. Using industrial waste products such as blast furnace slag, silica fume, fly ash is also environmentally friendly, and can count towards points of getting LEED certification for a project.

For all admixture types listed below, there are different types and their varieties, proprietary brands. Uses of admixtures include:

- Accelerators are used when high early strength of concrete is desired by increasing hydration rate. This, in turn means, increase of hydration heat generation rate, although the total heat generated may not vary much. Temperature differences caused by this faster heat generation must be controlled to avoid thermal cracks. In other cases this may be useful, such as in cold weather pours. Accelerators are also used in shotcrete, as it is sprayed on nonhorizontal surfaces and fast setting rate is important. Accelerators that include calcium chlorides may have the negative effects of increasing drying shrinkage and rebar corrosion possibility.
- On the opposite side, setting rate of concrete can also be slowed down by set retarders, to obtain longer placement durations in situations where pour can not be completed quickly.
- It is possible to obtain higher strengths by using certain pozzolanic admixtures, such as fly ash, silica fume which also improve durability. Silica fume is highly pozzolanic with very fine particles, even smaller than those of cement. This increases surface area for reaction, which accelerates setting but lowers workability and increases water needed for hydration.
- Normally concrete can contain air voids anywhere between %1-3 by volume. Air entraining agents are used when more air bubbles in concrete is desired, such as thermally insulating concrete. These admixtures also enhance workability during pour and durability in the final product. This is beneficial for concrete which will be exposed to a lot of freeze-thaw or wetting and drying cycles, and against scaling of surface of pavings. They have the negative effect of causing some reduction in compressive strength. Having smaller air bubbles distributed uniformly throughout the concrete volume is better than smaller number of larger bubbles.
- Water reducing admixtures, such as chemical admixtures which are also called as plasticisers or superplasticisers, make it possible to use less water for the same workability of concrete, which

decrease water cement ratio, which means higher strength and durability for the same amount of cement. Because of this, they are likely the most commonly used admixtures. In other words, they enable higher slump concrete for a given water content. Fly ash, which is not a chemical but a cementitious admixture, can also improve workability due to its spherical particles, and can slightly reduce water demand during hydration.

- A pozzolan is not a cementitious material, but in the presence of moisture, it reacts with cement. Properties of pozzolans greatly vary. There are both artificial and natural pozzolans. They are less expensive than cement, but can be used near cement, and increase durability, although workability can be negatively affected. Pozzolans provide high resistance to salty water and sulphate action. Pozzolans are used since the ancient Romans, which is the reason for the durability of ancient Roman concrete and mortar.
- Some aggregates are highly reactive with the alkali content of cement (also see alkali silica reaction later). This is a harmful effect, but sometimes that aggregate can still be preferred for cost reasons, as long as we also use fly ash or pozzolan as admixtures.
- Admixtures can be used to reduce segregation.
- Admixtures can be used to reduce or prevent shrinkage. We will list types of shrinkage.
- Corrosion preventing admixtures can be used to protect reinforcing steel, in locations with high risk of corrosion in the long term such as structures in wet / coastal environments. Fly ash can be used to increase durability and corrosion resistance.
- Waterproofing admixtures also prevent corrosion by protecting concrete against water infiltration. For example silica fume reduces permeability and decrease risk of corrosion.
- Color changing admixtures can be used for aesthetical reasons.
- Concrete finishing operations and / or the inside face of forms have greatest affect on the look and texture of concrete surface, but this can also be influenced by using admixtures.
- We can obtain self consolidating concrete, which was mentioned in the previous subsection, handling and placement. To achieve this, superplasticisers (water reducing admixtures), viscosity increasing admixtures (to counter the tendency to segregate) are used.

Strength of Concrete

Compressive Strength:

The most the important property of concrete, and mainly what it is used for is its high compressive strength. It is of such importance that we name a concrete batch by its compressive strength. This value is expressed in terms of stress, such as pounds per square inch (psi), or MegaPascal (MPa). In many commercial construction projects, typically a value of at least 3000 psi (about 20 MPa) concrete is commonly specified by engineers for structural slabs, and strength values up to 5000-6000 psi (about 30-40 MPa) may be specified for footings, shear walls or columns that take higher compressive loads.

For many projects, stronger concrete are now also often used. For example, for lower columns of high rise buildings, it is possible to see concrete strengths over 10000 psi, which must be produced in highly controlled ways but many concrete plants are now able to produce this. Even much higher strengths were reported by researchers, through the use of different materials such as steel fibers. Usage of high strength concrete can also allow us to use higher strength rebar. Per ACI 318-19, Section 19.2.1.1, the minimum strength we can use anywhere is 2500 psi for structural concrete.

Concrete compressive strength is denoted as f_c. We use this also when expressing tension and shear strength, with addition of coefficients. The compressive strength of concrete will keep increasing in time, even after 28 day design strength period, but this rate of increase will slow down and approach to zero. In the long term concrete also becomes more brittle, undergoes creep and shrinkage.

The failure of concrete is caused by emergence and progressing of cracks, when it is loaded. For compression, this crack propagation rate is slow, but for tensile forces it is very quick. As we keep loading concrete beyond its elastic limit, its stiffness will decrease due to crack propagation, which means more deformation per stress increase. The cyclic action of earthquake also contributes to decreasing stiffness of concrete, which must be included in the structural model when making nonlinear earthquake analysis *(of course reinforcing steel will be working here too but here we are talking about concrete itself)*.

Usage of higher strength concrete can lead to cost savings by enabling less rebar and smaller members, for the same goal we want to achieve. This becomes especially true for columns and shear walls. Higher compressive strength also means higher modulus of elasticity, which means stiffer concrete, which in turn contributes to a stiffer structure. For structures where flexibility can be critical such as tall buildings under seismic or wind effects, this can be helpful.

Tensile Strength:

Concrete is weak in tension. In design we usually assume tensile strength of concrete as zero, although it is not. To make up for the defect of tensile (and shear) strength, reinforcing steel (rebar) is used in concrete. Tensile strength of concrete is important for plain concrete structures such as dams or some structural components where we may use plain concrete such as for example as allowed by ACI 318-19, Chapter 14. Tensile strength of most commonly used concrete vary between 300 to 700 psi (2 to 5 MPa), which roughly correspond to anywhere between 5 to 15 % of their corresponding compressive strength values. So there is some strength, but not enough for most applications, for us to include it in our design. Tensile strength of concrete, although low, is actually important because it is the determining factor also for cracking and deflections in concrete itself and shear strength (which is higher than tensile strength) which in turn becomes important when considering torsional strength, bonding strength of rebar and anchors. For example when a crack in concrete happens, it is because of tension *(of course that crack would happen much earlier if we do not use rebar)*. It is common to calculate the tensile and shear strength of concrete in terms of its compressive strength.

For example in AASHTO LRFD design manual section C 5.4.2.7 and FHWA Publication NHI-15-047 - LRFD for Highway Bridge Superstructures section 5.2.1.1.2, for normal weight concrete up to 10 ksi, the tensile strength can be estimated in terms of compressive strength as $0.23\sqrt{f'c}$ where f'c is the compressive strength in <u>ksi</u>, which brings tensile strength to around 10% of compressive strength.

In ACI 318-19, section 19.2.3.1, modulus of rupture is given as $7.5\lambda\sqrt{f'c}$, where f'c is the compressive strength in psi, which brings the tensile strength to around 7% - 9% for the commonly used compressive strength values of 3000 to 5000 psi and for normal weight concrete, where λ=1 per 19.2.4.3 and for lightweight concrete, λ=0.75, per 19.2.4.2 .

Durability of Concrete

Other than its compressive strength, the other equally important property of concrete is its durability. Durability is better resistance against:

- Weather effects
- Alkali-Silica reactions
- Water, including sea water, contaminated water
- Carbonation, Chemical effects, Sulphate action, Acids
- Erosion
- Wetting and drying cycles
- Freezing and thawing cycles
- High rate of vehicular traffic and abrasion if pavement concrete
- Loads from heavy equipment in industrial slabs
- High heat exposure

Low permeability is the main indicator of high durability, which affects most of the items listed above. To increase durability, we mainly use lower w/c ratio and certain admixtures. Durability of a concrete structure also depends on how well it is protected and maintained, but the initial design and construction is still of highest importance *(many ancient Roman structures made of concrete, mortar and stone, still stand or even continue to serve).*

Permeability:

Permeability is a result of voids in concrete and the micro cracks between the cement paste and the aggregates. The pores may be interconnected and may reach all the way from the surface of concrete to inside. Denser, higher quality concrete with lower water cement ratio reduces the amount of pores (porosity) and therefore permeability, which means more durable concrete. Lower porosity also means stronger concrete.

Especially for underwater concrete, low permeability is of utmost importance. For structures above ground, high permeability causes carbonation reaction which we will describe.

Permeability can be reduced by using low W/C ratio, selecting smaller aggregates, properly compacting concrete during placement, properly curing concrete so that the hydration process is more complete, using fine particles in concrete as much as possible such as fly ash, stone powder, silica fume etc..., using certain chemical admixtures.

Although not common, sometimes we may want permeable concrete. For example, some paving concrete are made very permeable on purpose to allow the rain water to penetrate and reach to the soil below, therefore reducing the amount of surface runoff and the burden on storm sewers.

There are also types of special purpose concrete to especially withstand various effects, which we can also group under the durability subject such as:

High Heat Resistant Concrete:

Concrete already has high heat resistance but for certain reasons we may want even higher heat ratings. In those cases, we must obtain more durable concrete mix.

Air Entrained Concrete:

It is a specially designed concrete mix to have small air bubbles inside, in order to have lighter concrete with better thermal insulation, and freeze-thaw resistance properties. All concrete mixes, not just the air entrained concrete, also contain a certain amount of air up to a few percent by volume.

Unfavorable / Harmful Effects to Concrete

Many exterior effects may harm concrete. We can broadly group these as:

- Physical / Mechanical effects: Overloading, temperature effects, fire, deicing solutions, freeze-thaw cycles, drying-wetting cycles, impact loads, creep, abrasion, shrinkage
- Chemical / Biological effects: Sulphate attack, biological processes, carbonation, alkali-silica reaction, chloride attack

As a result of these, concrete may lose strength and durability, crack, its surface can spall, rebar can corrode. Several of these effects are described below. Freeze thaw action was described before under soil mechanics, groundwater section, as it also applies to soils and foundations.

Alkali Silica Reaction:

Although most concrete aggregates are inert (not reactive with other materials), some aggregates containing reactive silica can easily enter into reaction with alkali in cement. Silica is the name given to a group of minerals of silicon and oxygen, which are the two most abundant elements in earth's crust. If there is also high enough moisture and high pH, and given enough time (at least several years or much more), reaction of reactive silica in aggregate and alkali in cement gradually form expansive gel around aggregates and cause expansion in concrete, and eventually cracks and spalling. The cracks invite other harmful effects, such as corrosion of rebar, freeze thaw damage, and they affect strength and durability negatively. As mentioned under admixtures, fly ash, silica fume, blast furnace slag can be used to counter this effect because they replace a part of the cement, which means less alkali and lower pH of concrete. If the aggregate is very reactive, then in addition to admixtures, low alkali cement may also be needed. Simply using nonreactive aggregate, making less permeable concrete (to reduce moisture inside) can also be other solutions. The cost effects must also be considered because aggregate makes up a considerable portion of the concrete volume. ASR is a form of alkali aggregate reaction.

Carbonation:

Carbon dioxide can penetrate into concrete through pores. The more porous (permeable) the concrete, the faster carbonation reaction will be. High air moisture and high CO_2 content makes this process faster, which means in moist urban areas with high CO_2 content, carbonation will be faster. During carbonation, carbon dioxide from air causes acid to form in the concrete pores that contain moisture, which reacts with the concrete. This reaction increases the pores in the concrete and also lowers its pH. Reduction of pH may also cause corrosion of rebar if the pH value goes below a critical value. More pores means permeability is increased and durability is reduced. The moisture levels must be within a certain range, for carbonation to happen. Too low or high moisture contents will prevent carbonation.

Creep:

Creep is the very slow (many years) time dependent deformation of concrete under constant, very long duration stress. The larger the stress, the longer the duration, the more amount of creep deformation can take place. The stronger the design mix, the more aggregates used, the stiffer aggregates used, the less will be creep. Creep also affects the long term stiffness of concrete. Creep can increase if concrete was loaded too early in the days following its casting, before fully gaining its design strength, even if that load is within the design limit and would not have caused any significant creep otherwise.

Creep and shrinkage may damage members, and in turn the structure in the long term, therefore it should be accounted for in structural design, such as with special rebar details (i.e by closure strips) or expansion joints. The larger the structure, the more effects can be. The systems attached to the structure, such as facade, elevators and other systems will also be affected more and more from these long term deformations, as the size of the structure grows. For complex shaped unsymmetrical, irregular structures, the effects can get more complex such as load distributions and deformations. For high rise buildings, their effect will add on top of long term / differential column shortening. Although creep and shrinkage cannot be avoided in the long term, their effects can be reduced. Relaxation, which means decrease of stress under constant strain, should not be confused with creep.

Shrinkage:

Shrinkage of concrete is an unwanted effect, which is the reduction of concrete volume due to various causes. Keeping the W/C ratio low, using certain admixtures or even fibers, slowing down hydration rate, reducing moisture transfer rate, reducing gaps in rebar placement will lower shrinkage. The change of concrete volume by shrinkage will cause cracks and unexpected stresses in structures, even before any external load is applied.

Shrinkage can be thermal shrinkage, plastic shrinkage, drying shrinkage and autogenous shrinkage.

Side Note: When we say crack, it must be remembered that cracking in concrete is ultimately a result of concrete's poor resistance to tension. Shrinkage and loads are the two major sources of tension in concrete.

Plastic shrinkage: During concrete pour, water can accumulate at the surface because it has less density than the aggregates, sand or cement. Plastic shrinkage occurs when the concrete is yet in plastic stage while hardening, with rapid loss of water at the surface of concrete by evaporation due to causes such as hot weather, low humidity, wind, sun exposure or their combinations. Before pour we can take precautions to reduce the effect of these. Set retarders delay setting time thus there will be more chance of water evaporation, and plastic shrinkage. The cracks that occur during plastic shrinkage when the concrete still sets and in plastic state can be much deeper than drying shrinkage cracks. The remedy is easier however, and in many cases the cracks can simply be closed while finishing the hardening / just hardened concrete surface, as long as we do not add any additional water. Plastic shrinkage can also happen due to wet concrete settling over rebar, or cement paste sinking among aggregates.

Drying shrinkage: It is the loss of water from concrete after hardening, by water escaping from the pores of concrete. It happens because of the difference in humidity between the inside and outside of the concrete. Therefore it develops slowly. In dry climates, drying shrinkage will be more. Water is lost by evaporation to air, or an underlying dry soil layer which absorbs moisture from concrete. Drying shrinkage produces stress in the concrete, and can produce a lot of cracks of varying widths on concrete surface which negatively affects durability. Drying shrinkage can be reduced by:

- Keeping w/c ratio low which reduces porosity of concrete.
- Curing concrete properly, which reduces porosity.
- Increasing stiffness of concrete, such as with lower w/c, using more aggregates.
- Keeping member surface area low, with respect to its volume. The longer distance water needs to travel to reach to the surface, the lower the shrinkage will be.
- Properly compacting concrete during pour.
- Applying sealants to concrete surface.

Thermal shrinkage: It can occur when there is excessive heat during hydration of concrete and during rapid cooling afterwards. This effect is especially present in mass concrete pours such as large foundations or dams. Cracks due to thermal shrinkage can also occur during the life of the structure later, due to temperature changes of weather, if contraction joints are not adequate or improperly made. Thermal shrinkage cracks can continue throughout the entire depth of concrete.

Autogenous shrinkage: It occurs as a result of the water loss because of the hydration reaction of cement especially during the first few weeks. So it is not because of exchange of moisture with the surroundings or temperature differences, and not significant as the ones described above but it can be more significant at low water cement ratios *(this is not to say that we should make W/C high)*. Autogenous shrinkage can greatly be reduced by keeping the concrete wet during the curing period, using water saturated aggregate, because otherwise the water needed for the hydration reaction must come from the pores inside the concrete, which will cause autogenous shrinkage.

Corrosion of Rebar:

Rebar in concrete will corrode if exposed to water and chloride. These can reach to the rebar if there are cracks in the concrete for various reasons, and or the concrete has high permeability. Concrete has a pH value of 12-13, which is a result of silica in portland cement. This high pH, in other words high alkalinity protects the rebar surface from corrosion by forming a protecting layer around the bars. If chloride, a corrosive chemical, is able to penetrate through the concrete cover, it can damage this protection at the rebar surface by lowering the pH and the corrosion process begins as soon as the pH drops to around 9-10. Chloride can be found in de-icing salts, sea water, soil. Carbonation, which we discussed, can also lead to corrosion since it decreases pH of the concrete. Corrosion is an electrochemical process, which uses oxygen. Rebar serves as anode and cathode as electricity flows between them, where water becomes the electrolyte that transmits the flow.

Corrosion of rebar reduces its cross section, so its strength is reduced. There are also corrosion products, as iron oxidizes (rusting), which expand as the Iron (Fe) atoms of rebar now also get bonded with the Oxygen (O) and Hydrogen (H) in water, and the total volume that the rebar occupied before can go up significantly, as the rusted layer increases. This means the rebar will try to expand and push the concrete from inside, creating tensile stresses in concrete and can even crack it, which is also why you can very often see the spalling of concrete at corroded rebar locations *(this in turn accelerates the rate of entry of moisture from outside, which makes the corrosion process even faster)*. From outside it can be understood by an experienced person with a hammer blow by hand, which would bring a dull sound, instead of a clear ringing sound from concrete. When cracks in concrete are visible from outside due to corrosion, these can visually be identified by their characteristic that they follow the rebar inside and the rust may also start to come out. For example, a vertical crack close to the corner of a column indicates corroded rebar at the column corner expanding, thus forcing the concrete cover from inside. The most serious negative effect of corrosion is the considerable reduction or disappearance of the bond between concrete and rebar. This is the most advanced stage of corrosion and practically turns a reinforced concrete member into a plain concrete member because of the loss of bond and rebar.

Corrosion process is very long and <u>takes many years.</u> The duration is hard to estimate and can vary a lot. Even after the chemicals reach to the rebar, there may be significant time before the corrosion effects start to happen. In earlier stages of corrosion, there is just rust over the rebar surface, without expansion yet. When we see cracking, spalling, delamination of concrete from corrosion, that is the last stage after corrosion already took place for a long time, but cracks can go worse if nothing is done.

Corrosion of rebar must be prevented by providing <u>sufficient concrete cover</u>, having impervious and <u>durable concrete</u>, <u>preventing concrete cracks</u>, using <u>cementitious admixtures</u>, using <u>coated rebar</u> such as special epoxy coating or galvanizing, and most of all, <u>preventing water or humidity</u> to get to the concrete in the first place, or at least not letting the water to stand on concrete for long periods, such as by installing proper plumbing or water distribution systems and maintaining them, and keeping the concrete as dry as possible, which can also mean providing slopes, and or waterproofing. Another measure is a process called <u>cathodic protection</u>, which is providing a small amount of current to rebar to counteract the flow of current from corrosion process. Also see concrete repair subsection later.

Reinforcing Steel

Concrete is strong in resisting compression but weak in resisting tension and shear stresses. This deficiency is overcome by placing steel bars - also called reinforcing steel, reinforcing bars, or simply rebar - in concrete to take tension and shear stresses, to contribute to compression strength and also to wrap the concrete members as a cage, which increases that members' strength, deformation and energy absorbing capacity and ductility considerably. Rebar is also great for reducing temperature and shrinkage cracks in concrete. As discussed, concrete is brittle, but steel is ductile material. Ductile structures absorb a lot of energy during earthquakes and do not fail suddenly but only after a lot of deformation. Rebar is sensitive to corrosion and fire, but concrete is not. So the rebar and concrete partnership is a good one, where one makes up for the deficiency of the other.

After structural analysis, stress and displacement demands everywhere is known. Per these, rebar details are finalized per the minimum criteria set in codes, or even more stringently as required.

Let's now discuss various general concepts about reinforcing. Starting this discussion with beams seems easier to introduce many concepts. And then we will also list some points about other members.

See the beam in the figure below. The bottom is in tension and upper side is under compression, as in the bending stress formula, $\sigma = -My/I$, as was discussed under Strength of Materials section.

Middle portion of the beam (between points B-C): Tension here is purely as a result of bending stress, which produces vertical tension cracks. Shear is zero here, as seen in the shear force diagram.

Both sides of beam (between points A-B and C-D): In addition to axial tension (at the bottom half) from bending, there is also shear, which is why the cracks are diagonal. Also remember the Mohr's Circle topic, under Strength of Materials section, where we looked at stresses from both directions. In these end sections, the cracks become diagonal because of the changing magnitude of tensile force (horizontal) from bending and the shear force (vertical), that does not exist in the middle section, which cause a different angle of plane of failure, just like the cube we looked at in Mohr's circle subject.

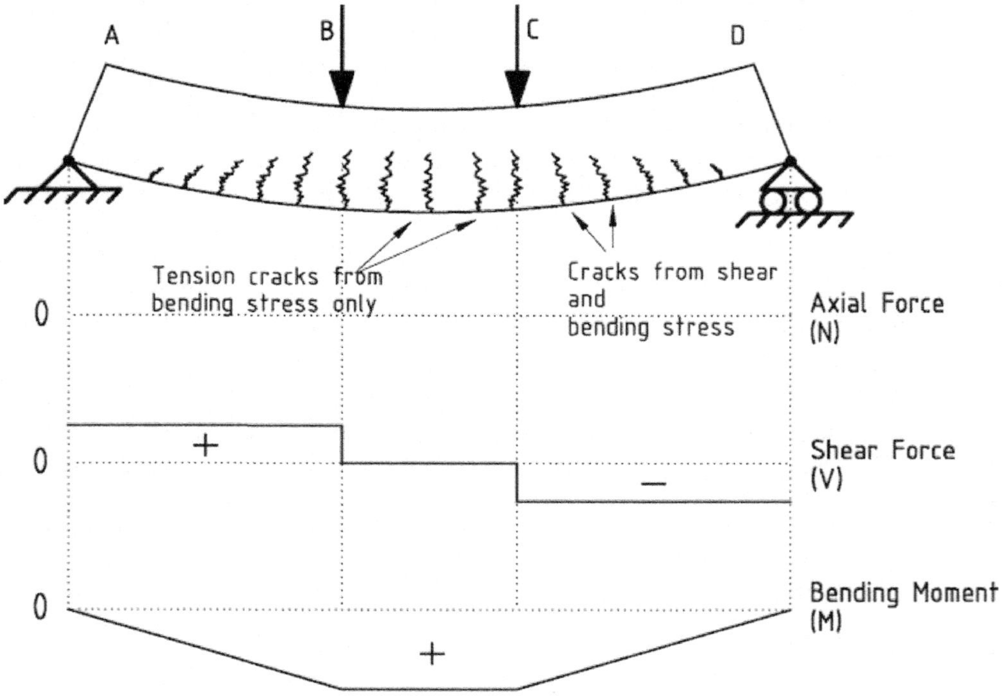

Figure - Concrete 2

Also pay attention to how the beam is supported. It is a simply supported beam. Supporting this way prevents any axial stress buildup in the beam, because with the help of roller support on the right, it is free to move horizontally so there is no axial stress except from bending itself.

Shear forces occur in all beams, and depending on the support and loading conditions, they are the highest adjacent to supports (*in our example its values are constant and the same on both sides*).

One important note here is that, the beam above cracks typically this way as in Figure 2 only when there is reinforcing bars in it. If there were no reinforcing, the brittle concrete can not sustain so much cracks like in the figure below and it would just suddenly split into two by having only one big vertical crack in the middle, between B-C only.

From shear and moment diagrams, we know the values of shear and moment everywhere on the beam. To resist tension, rebar must be at the bottom in the middle of the beam. At both ends we can bend it upwards, called bent up bar as in Figure 3 below, or we can just leave it at the bottom without bending and only use stirrups, which is done much more often, as in Figure 5. If we were to bend the bars to make bent up bars, extra labor is required, thus bent up bars are now less frequently used. Stirrups / hoops also serve as shear reinforcement. When the loads are reversed due to earthquake shaking or near columns (then the moment graph will be as in Figure 4), bent up bars would not be useful at both ends but the use of perpendicular stirrups will not be affected as they can serve the same whether it is a sagging or hogging beam. Note that, beams do not often fail from shear.

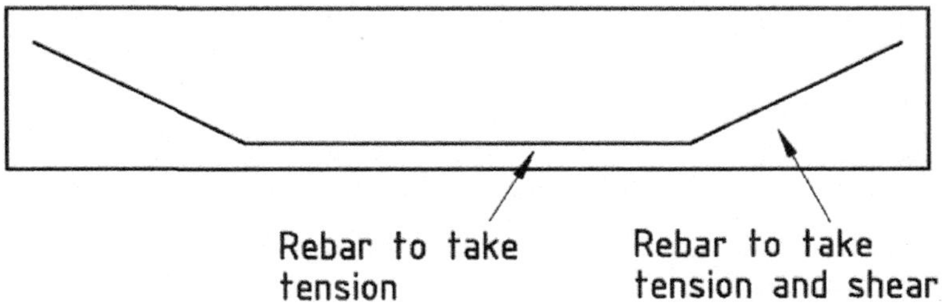

Figure - Concrete 3

Stirrups / hoops are for:

- shear resistance (in additional to diagonal bars if any),
- enabling better bond between longitudinal bars and concrete, hold them in place during pour,
- torsional resistance,
- buckling resistance by wrapping the beam around, (beams can also be in compression, such as during earthquakes when there are lateral forces)
- increasing ductility,
- providing support for lap splice locations,
- preventing movement of main bars during concrete placement, and to hold them in place.

As we get closer to the columns on both sides of a beam, we place stirrups more densely, to provide strong moment connection between column and beam. Also note that commonly the word "hoop" or "tie" may also be used interchangeably with stirrup, although there are differences. Hoops provide full closure and can be made of stirrups and ties and can perform the duties listed above but stirrups may not be closed and their primary purpose is shear resistance. In common usage, hoops and ties are more commonly meant for columns and stirrups for beams.

The tension bar you see at the bottom of Figure 3 above, will need to be on the top on both ends of the beam, because near and top of the columns, the moment direction in the beam will usually change, in other words, normally there will be sagging in the middle of the beam and hogging at both ends closer to columns. So for instance, after we perform structural analysis, for a beam that runs over several columns, bending moment diagram typically looks like the one below:

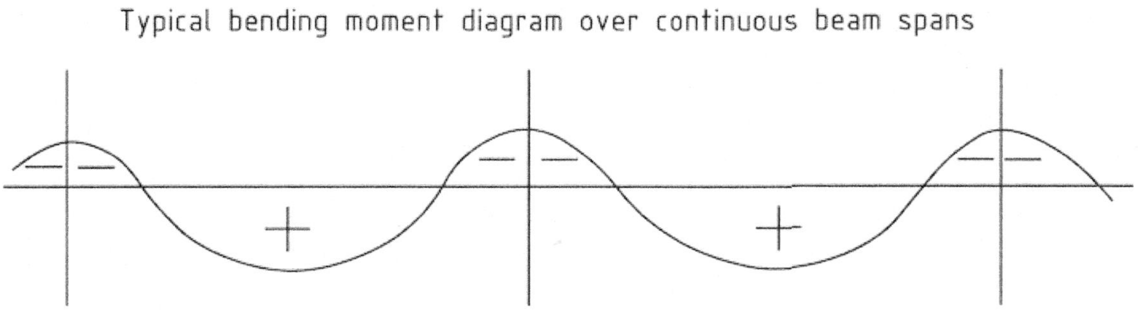

Figure - Concrete 4

The bending moment diagram shown above also looks similar to the typical deflected shape of the beam *(if exaggerated of course)*, sagging in the middle and hogging on the sides. So, the basic beam rebar per discussions above will typically look like in Figure 5 as below.

To design the tension and compression steel, lets assume we have the beam cross section as below. We place three reinforcement bars at the bottom and two at the top (stirrup is not shown for clarity). Then for the cross section, from equilibrium equations as F_x=0 and M=0, we check the tension steel force at the bottom against the compression force of concrete plus compression force of upper steel, when the concrete strain on top is: ε = 0.003, which is the ultimate strain of concrete, as in Figure - Strength 11 (and the maximum strain that corresponds to the maximum compressive strength is around 0.002, even with different concrete mixes).

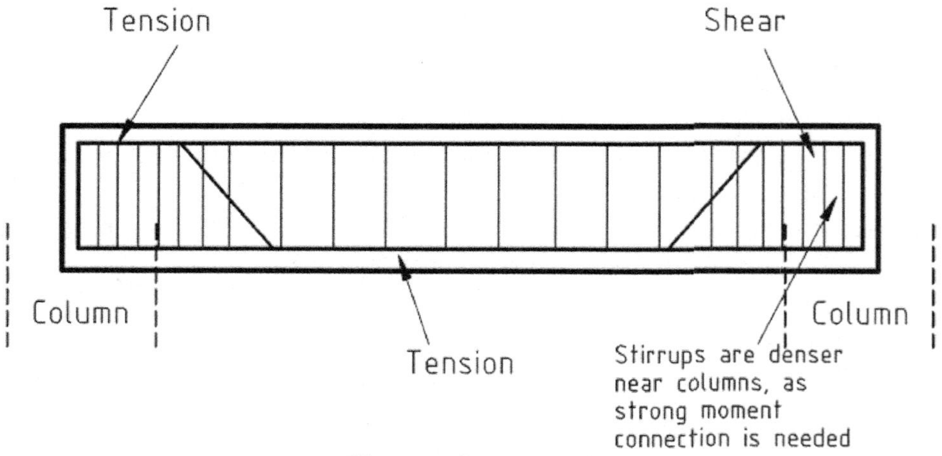

Figure - Concrete 5

Codes, such as ACI 318 provide values to use and the equations. We will mention Figure 6 again in the following pages. Here we can place more rows of rebar than what is shown here that will enable better crack and creep control and ensure that stresses are distributed more evenly to the beam cross section.

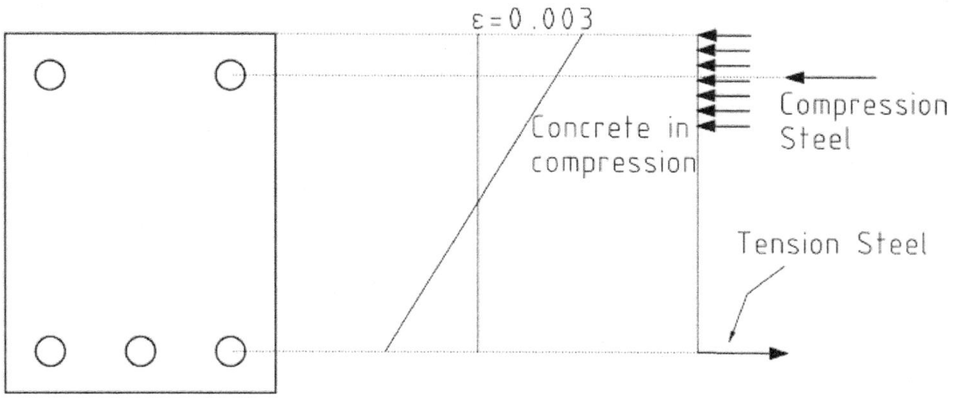

Figure - Concrete 6

Even at the parts of a beam where we expect only compression due to bending, we still need rebar, not only for ease of construction and to hold the stirrups in place but also because compression steel in beams mainly increase the beam's ductility and in turn energy absorption capacity, and to some degree its moment carrying capacity. The compression steel is also beneficial when the beam continues long spans where direction of moment changes at columns, and another benefit is that it helps control long term deformations of shrinkage and creep.

During construction, for interior beams, the beam rebar is first made right above the form, and then lowered straight down to the form. For exterior beams, the form for the interior and bottom face can be made first, then beam rebar is installed, and after that the exterior form face can be added.

Above, we focused our attention on beams, to introduce many concepts. Now let's also discuss some critical points about reinforcement of other members:

Slabs:

Slabs are designed somewhat similar to beams by separating into sections and reinforcing against bending for positive and negative moment sections but this is not all. There must also be sufficient rebar connection between the slabs and the beams and shear walls, which can be made by extending slab rebar all the way to the outer surface of the beam or wall and then bending it to form an L shape. For wide slabs, sagging control becomes critical. Collectors and chords (see Chapter 2, slab subsection) are made within the slabs by placing extra reinforcing to make up stiff, longitudinal "sticks" contained within the slab. The slab to structural wall connections must be heavily reinforced by also distributing rebar well into each other. For slabs shear is not critical as they are very wide, but in case of narrower widths and high loads, it can be countered by forming beams inside of the slab by tying top and bottom slab rebar by stirrups. Any possible punching shear stresses are resisted by shear studs. During earthquakes, we assume the mass of each floor is concentrated in each slab, and then transmitted to vertical members. To be able to make this transfer, there will be compression, shear and tension forces within the slab plane, and the slab section must be designed accordingly.

A slab can geometrically be classified as one way slab or two way slab. If the ratio of long side of the slab to its short side is greater than 2, then it is a one way slab. If this ratio is smaller than 2, such as a square slab or a rectangle which is close to a square, it is called two way slab. In two way slabs, the main reinforcement is provided in both directions, because the slab can bend almost equally or equally in both directions. In case of one way slabs, main rebar is placed parallel to short side of the slab. This is because, bending of the slab itself mainly occurs in short direction. This is true regardless of having beams only on long sides or on all four sides. Slabs made of other systems such as post tensioned slabs, composite concrete steel decks, even horizontal truss systems can also work as one way or two way.

At openings in a concrete member such as a duct opening on a slab, we must place additional rebar around the opening to better distribute the concentrated stresses and to ensure better force transfer along the edges and corners of the opening.

Shear Walls:

See Chapter 2 > Building Frame > Shear Walls. Also note that transverse ties for columns as explained below, have the same duties in case of a shear wall.

Columns:

Columns are mainly compression members, but during earthquake they may also take tension. Their primary reinforcing is the longitudinal rebars.

The transverse rebar in columns, called column ties and hoops, are wrapped around the longitudinal bars and they have very critical duties. Even if we placed a small amount of transverse rebar to provide confining pressure to the concrete core, this increases the compressive carrying capacity significantly. When we confine something on the side, it can carry much more vertical load, as we also saw under lateral earth pressure in soil mechanics and also the Mohr's Circle under strength of materials sections. Ties and hoops confine the concrete core and considerably increase ductility, energy absorption capacity, compressive strength, shear strength, moment carrying capacity and buckling resistance of vertical bars, effective cross section of a column. The increases are greater as their spacing decreases. They also hold the vertical column bars in place during rebar installation and concrete pour, confine lap splices. The transverse rebar should continue from the face of beam column joint upto a certain distance where yielding is not expected anymore. Closer to upper and lower ends of the column, tie/hoop spacing is reduced to increase moment capacity for better column-beam joint support *(for faster installation they may just be evenly spaced, therefore denser than needed in the middle)*. Remember the stress strain curve for concrete in strength of materials section. By using column ties, therefore confining concrete, both the strength and strain capacity of the concrete member are improved, in other words, all values in a load deformation graph increase, and the slope of the descending part of the curve becomes flatter, which means more ductility. The transverse rebar requirement is also based on compressive strength of the column and the design axial force.

Design of Reinforcing:

Things shown on a rebar design shop drawing include the information below at a minimum:

- Location, count, length, size (diameter) of bars and their bending details. The bar diameters in the US are commonly designated between #3 to #18. To find the actual diameter in inches, these are all divided by 8. For example, #5 bar means, a bar with a diameter of 5/8". To save labor costs, using fewer and stronger/thicker bars are preferable, which affects bar spacing, as long as crack control is still maintained. Making bar sizes as uniform as possible and allowing repeated installation patterns will also achieve labor cost savings.

- Grade of bars, which shows yield strength. Most commonly, grade 60, which means bars with yield strength of 60 ksi is used in the US for general construction purposes. Other grades also exist which usually range anywhere between grades 40 to 120. Usage of higher grades than the usual 60 grade, such as 75, 80 and 100 are increasing. With these we can use lower number of or diameter of rebar to achieve same purpose. This also makes installation easier with less congestion, which makes concrete placement also easier. As the grade of bar increases, we may also need to use stronger concrete, for proper bond between the rebar and the concrete and to protect balance of the stresses taken by the concrete and rebar. The spacing of transverse rebar (to wrap high grade longitudinal rebar) must also be lower and lap splice distances greater. For transverse rebar, such as ties, hoops or stirrups, usage of higher strength steel can also be helpful as it may allow slightly higher spacing for easier construction and lower congestion. Higher grade bars cost more, but this may be offset considering the advantages obtained.

- Type of bars (deformed bar or mild steel/plain bar). Almost always, deformed bars are used.

- Spacing of bars. There must be a certain gap distance between parallel bars, not only for cost or structural reasons but also because during concrete pour, wet concrete must be able to flow in between the bars. The largest size aggregate must be able to easily go through the gap between the bars, and not only that, but the vibrator must also be able to penetrate wet concrete through these bars. Reinforcing bars being too close to each other is called rebar congestion. The rebar spacing amounts depend on the bar diameters, and where the bars are placed in the structure. These criteria are set by applicable codes such as Eurocode 2, ACI 318. The engineer must not forget the actual geometry when designating the number and size of bars, hooks, splices, layers and bends of bars, to see if they fit with adequate space inside the member. For example for columns, if we have more than about 3% (to give a ballpark number) of column cross section area as total longitudinal rebar cross section area, things start to get congested, with little additional benefit as we raise the percentage.

- Splices (lap splices). Often, one piece of rebar can not continue full length of a member, such as the lower floor column longitudinal rebar extending to upper floor and terminating somewhere at midheight of upper floor. In these cases, the rebar from lower floor must be spliced with that of upper floor, which means, the upper floor rebar must start somewhere along the lower floor bar. This distance where two bars overlap is the splice distance, which must be longer as the bar diameters increase in order to provide adequate support. Codes show minimum splice distances and their criteria. Sometimes splices may need to be staggered to avoid congestion in a limited space. Lap splices must be located outside of the yielding zones of members, and be confined with transverse reinforcements such as stirrups, ties, hoops.

- Concrete cover. Rebar must remain inside of the concrete surface, so, there must be a gap distance between the rebar and inside face of formwork, which is called the concrete cover. It should usually be 2 inches minimum and mentioned on the drawings. This gap is not only needed for proper strength development but also for fire and corrosion protection of rebar. To achieve this gap, stone/concrete blocks, or, small plastic pieces called "chairs" are positioned on the interior face of the forms by tying rebar to these blocks.

- Development length. Rebar must be able to continue in concrete for a certain distance, to be able to develop its full design tensile strength. If due to geometric restrictions there is not such distance available, then headed bars *(also called as headed anchor / headed bar anchor)* or hooks *(bending end of bar into a hook)* can be considered. Usage of headed bars can help reduce congestion and enable simpler bar layout and easier installation. In comparison to hooks, headed bars can structurally work more effectively with the concrete and they also make larger bar diameter possible by eliminating the requirement for minimum hook radius.

Many structural analysis programs, in addition to complete analysis of structural loads, stresses, displacements, can also automatically detail rebar (and formwork) per code criteria very quickly after an analysis is done, which took days or weeks for engineers / drafters to do manually before. Of course engineers must still check the computer generated drawings. As long as the code requirements are met, the final design is upto the engineer. Another important goal to achieve is the constructability and avoid rebar congestion. The engineer must always keep in mind the actual field dimensions, available installation space, installation sequence and relative difficulty when considering alternatives. If necessary, the dimensions must be drawn with scale that will include the actual number of rebar inside the architectural dimension. After everything seems to be coordinated, the engineer must clearly communicate the design intent to rebar detailer.

Now let's list several of the most important concepts about rebar and its design that may apply to any member in general (we listed several member specific points in the previous pages):

- Generally, reinforcing steel can be classified into two, such as the main reinforcement, and the secondary (also called distribution) reinforcement. Main reinforcement steel takes the main load, such as tension and shear in the member. Secondary (distribution) rebars enable distribution of loads uniformly, preventing cracking in concrete (such as temperature or shrinkage cracks), hold the main reinforcing bars in place.

- For better bonding to concrete, the surface of rebars are deformed on purpose by providing small diagonal ridges that run through the entire bar length. These bars are called as deformed or corrugated rebar, which have considerably better bonding capacity to concrete in comparison to plain rebars without corrugation. Plain rebars, without deformed surface (also called mild steel bars) also exist but they now have little usage in the US, except where they are expected to slide, such as between segments of highway pavements or segmental bridges. In other parts of the world plain rebar may have more usage depending on the country, although always less than corrugated rebar in our time. Sometimes both can be used, deformed bar being the main reinforcement which takes the main load and mild rebar as secondary reinforcement, for cost efficiency. Note: Actually with sufficient detailing, in theory, plain rebar may as well be used in many situations, as long as their smaller bonding capacity is taken into account and they are detailed accordingly, with less gaps and more hooks etc... but this increases labor costs therefore in practice their use is now limited as described above.

- In many cases, hooks are provided at the ends of rebar for better bonding but this adds to labor cost. When deformed bars are used, in some cases codes may allow not to use hooks. The hook radius also depends on the bar diameter. Too small hook radius in comparison to bar diameter may not only cause too much stress and plastic yield of the bar, but it may also cause too much stress in the part of concrete that remains inside the bent area.

- We must continue the rebar for a certain distance in concrete. This length is called the development length. It is through this length that the rebar develops its strength to do its duty. The longer this length, the more strength is developed. The distance of the rebar to the face of concrete member, strength of rebar and concrete, the degree of deformations and friction on the rebar surface, the rebar diameter, whether the bars are bundled together, whether hooks or headed bars are used and a few more factors also all affect how much of development length we need. We need more development length for larger rebar diameters, lower friction on rebar surface, and for weaker concrete.

- We need a minimum amount of reinforcing in concrete members, for rebar to properly do its duty to reinforce the concrete, so that the capacity of the member is increased, brittle failures are prevented and cracks are controlled. This minimum amount of rebar is specified in codes as a minimum percentage of the concrete member's cross section area. However, specifying a minimum rebar area definitely doesn't mean that we can increase the rebar as much as we want, if we satisfied the minimum. We must also limit the maximum amount of total rebar for any member, because we want reinforcing steel to yield first, for a ductile failure. If we overreinforce and concrete fails first, this will be a sudden failure without warning (because concrete is brittle), which we don't want.

  Let's expand what we have just mentioned here a little more. Look at again to Figure - Concrete 6 a few pages back. Because of the reason we have just explained, we must make sure that the maximum possible strain in concrete, 0.003 will not be exceeded, in other words, concrete will not fail first. So the steel must be placed to such extent that the corresponding concrete strain value of 0.003 *(from the two opposite triangles for strain as seen in the figure, so it is simple proportioning the bases of triangles)*, will not be exceeded and steel will yield first.

  For flexural members such as beams, even if we add even a little reinforcing steel at the top in compression zone, which is called a "doubly reinforced" beam, this makes a big difference, and now we can place more tension reinforcement at the bottom, because we have just increased our capacity in compression zone at the top. To be able to place more reinforcement at the bottom means not only higher bending capacity, but it also means more steel deformation available to us (before breaking of concrete), which means much more ductile capacity for beam. In other words, here we still want the bottom reinforcement in tension to fail first, but now this can happen at a higher strength. Adding the reinforcing steel at the top is necessary anyway, so to hold the stirrups in place. Double reinforcement also reduces long term deformations such as shrinkage or creep, so it is also used at the underside of cantilevered slabs, to take on that long term concrete deformation.

- At member connections, such as beam to column or pile cap to pile connections, the ties must be placed with smaller gaps, to take on the larger moments at those locations.

- Crack control is another major duty of rebar. Placement of reinforcing in a staggering manner, using smaller bars spaced closer to each other rather than larger bars spaced with larger gaps even if both systems provided the same total rebar area, help better control of cracks.

- Welding of rebar is generally either not preferred or forbidden practice, as it may affect the properties of steel at the welded location, and may make it brittle. Another reason of not preferring welding is because of the relative difficulty to perform it within the forms. Instead, rebar couplers are used or the rebar are tied to each other with wires. In case of tying with wires, it is mostly the surrounding concrete that creates the bond between bars, rather than the wires. Therefore the bars must overlap for a certain distance, which is called the splicing distance as we mentioned, to have sufficient support from surrounding concrete.

- Several bars may be bundled together sometimes. In this case, they should not be very different than each other. For example, Eurocode 2 1992.1.1 section 8.9.1 requires that the bundled bars should be of same type and grade, and their diameter ratio is maximum 1.7.
- Rebar cages can be assembled offsite, which is more efficient than installing in the field. After the cages are pre-assembled, they are brought to the site and inserted and fixed to their final positions. This however, makes transportation less efficient. Construction of cages on site but still outside of forms is also possible. Options must be weighed.
- One of the very important reasons of widespread usage of reinforced concrete is that contrary to intuition but fortunately, concrete and steel have very close thermal expansion coefficients with respect to each other. If this was not the case, designing and using reinforced concrete structures would be more difficult and less effective.
- The items above pertain to reinforcing steel placed in forms prior to casting concrete. In some instances and for various reasons however, such as to add a new concrete section to an existing section, reinforcing steel is installed by drilling existing concrete, which can be called as post-installed rebar / anchors / dowels. Once installed, failure may occur in the form of breaking of concrete, failure of bond between dowel/anchor and concrete or the yield of the dowel/anchor. The adhesion between the existing concrete and the anchors are achieved by applying high strength grout into the drilled hole, which also should not shrink. The size of the bar and its development length in concrete (in other words the length of the hole we must drill) are the primary factors to design, assuming strong enough grout. The existing rebar in concrete may not necessarily serve as a reinforcement to post-installed rebar.

Installation of Rebar:

Design and placement of rebar is a major cost and schedule item in overall concrete work, along with the formwork and the pour of concrete itself. In construction schedule, once formwork of a section is complete, rebar can start there, and it doesn't have to wait finishing of formwork for the entire structure, if we have a large enough area. Similarly, once rebar is complete at a large enough section, concrete can be poured there, without waiting the finish of rebar in whole structure, if the structure is large enough, the project conditions allow it and we have approved construction joint details between the two pours. These three activities usually follow each other in this manner to save time, except for small work. Once an area is ready for concrete pour, the rebar and forms must be inspected and accepted by authorities for compliance with project documents. *Side Note: In construction, upon completion of some major/critical task, an official inspection is needed for continuation with the next task which will close that previous work. Examples include inspection of electrical conduits before closing the walls, inspection of underground utilities before backfilling, inspection of steel frame before closing with drywall and so on. For each project, different inspections are needed, depending on project scope and jurisdiction, but generally the same inspections are needed for the same type of projects. Concrete pour for example, is an important activity which needs inspection before and during pour because how the forms and rebar were installed and the concrete is produced on site has great effect on the structure's performance and consequently public safety. Coordinating timely inspections versus the dynamic construction schedule is often a challenge in construction project management.*

Before installing, rebar must be clean of foreign materials, to ensure good bond with concrete. Bars must be firmly tied to each other, plus, strongly fixed to their final position, so that they will not move during concrete pour when wet concrete impacts and then gets vibrated, or when workmen walk on them to place concrete. Installing rebar in tight form locations can prove to be difficult. If the rebar and formwork crew are from different companies, we must coordinate the sequencing of formwork installation versus the installing of rebar to make sure that at tight rebar locations, the portions of the formwork will be installed after the rebar is in place (of course the rebar cages can be made outside of the form and then be inserted into the formwork, but sometimes this will not work).

Concrete Tests

There are various tests for measuring properties of concrete. Several common tests are listed here. The tests listed below are well defined in detail in standards, such as in the ASTM standards in the United States and in other standards in other countries.

Compression Test:

Along with its durability, the compressive strength of concrete is its most significant property. Compression test is regarded as the main concrete test, since it directly measures the compression strength of the actual concrete being used, under an actual compressive load. The theoretical work of designing a mix must be supported by the actual test results. Compression test gives the best feedback for a new design mix. As a result, the design may be revised if needed. The compression test is performed in the lab. A cylindrical specimen is crushed under compression, while recording the compressive force and deformation. The size of cylinders used in the US is usually 6" in diameter and 12" in height. Different size specimens for the same concrete and same curing conditions may give different results so a standard procedure was established and should be followed such as in ASTM C39. The larger the specimen gets, the lower the average measured strength values may get, because larger members will inevitably have more number of weak spots in them. The 6"x12" is big enough specimen that the concrete can be considered as homogeneous. This 12" height is also sufficient that the middle portion of the concrete will be in pure compression. The rate of loading is also important, which must be applied at a standard rate. Loading too rapidly will yield higher strength values.

During pour, a small amount of wet concrete is taken from concrete trucks to fill cylinder shaped sampling forms. For each project and location, the number of samples to take and procedure to be followed are specified in standards, which clearly outlines every step for all situations. After the samples are obtained, they are taken to the lab and left to curing or left at the jobsite on purpose, to experience the same atmospheric conditions as the actual poured concrete in place. Tests can be performed on 7th, 14th and 28th days. Minimum of a few (usually 3) specimens should be tested on each day. The 28th day average compressive strength must exceed the required strength for that particular structural member *(tests before 28th day are also done to see if we can strip the forms earlier, by seeing the rate of progress of strength gain).*

A disadvantage of lab compression tests is that the curing history of cylinders can not exactly be the same as the curing history of the actual concrete of the structure, due to different shape and size of the cylinder versus the concrete in the actual structure, and the possible temperature differences, even if the cylinders are left in the field. Another disadvantage is, many cylinders must be obtained, because for each testing day as mentioned above, three samples are needed so that their average can be taken.

If the tested concrete specimens fail to reach the desired values, further testing will be needed such as by obtaining cores from the actual structure, or by one of the nondestructive methods. Sometimes a small amount of noncompliance may be acceptable, as long as the location is not the most critical and/or the degree of noncompliance is low, provided that the structural engineer approves it. As we saw earlier in structural design approaches, there are already considerable safety measures in design. If, with respect to mix design's intended strength, unusually high number of tests consistently show lower strengths, there can be other issues such as the accuracy of the test lab's equipment and testing procedures, batching and mixing, or even concrete pouring operation itself. The lower than intended compressive strength will also influence the durability of the concrete negatively.

Maturity Test:

Maturity test measures the degree of progress of the cement hydration. It is performed by measuring the temperature history of concrete which enables predicting the stresses. This method has the advantage of being a nondestructive test. The temperature history is a valuable information because heat is released off as cement hydrates and concrete gains strength. The hydration amount depends on how long time has passed under what temperature. There is also a certain temperature below which further cement hydration is negligible, which depends on the ingredients of that particular design mix. This test is especially helpful to know when to strip forms and shoring, when to stop cold weather protection, when to perform posttensioning, by only measuring concrete temperature, instead of waiting for longer periods. To perform maturity test, the strength of cylinders taken during pour must also be known from lab compression test and these must be correlated to temperature.

Other nondestructive test methods include penetration resistance, pulse velocity measuring, rebound hammer tests. Performing cylinder compression tests together with one of the nondestructive methods is desirable to estimate early strength during construction, for possibly earlier formwork and shoring removal. Especially for tall buildings, completion speed of floors is critical. Like maturity test, other nondestructive tests must also be calibrated with cylinder compression tests to improve their reliability.

Slump Test:

This is performed to see consistency (degree of liquidity) of wet concrete. This is the most common field test, done practically during every pour as described in standards such as in ASTM C143. Slump value is important because it tells us how easily the concrete will be placed. High slump may mean excessive water and possibly - <u>but not necessarily</u> - weak concrete. We said not necessarily, because slump is not a direct indicator of concrete strength. It is just a rough result of how much water do we have in unit volume of concrete. This is not to be confused by water / cement ratio, which is a direct (but not only) influence on concrete strength and durability (lower w/c leads to higher strength and durability in general, as discussed).

Project specifications mention the maximum allowed slump values. Slump test involves filling a bucket with wet concrete, then turning the bucket upside down, lifting it away, and measuring how many inches vertically the top of the wet concrete slumps (which usually varies between 1 to few inches). Higher slump indicates more liquid concrete, which needs less vibration. For beams, columns and walls, where concrete is vertically placed, higher slump values can be acceptable for concrete to flow easier through the gaps between rebar to reach lower sections and fill the bottom of forms completely.

Tensile Strength Test:

The two main methods are the split cylinder test and the beam flexural test. In split cylinder test, a concrete cylinder is laterally placed on a base, and pressed along its length, to cause fracture by tensile force. For beam flexural test, see Figure - Concrete 2. As can be seen there, in the middle of the beam, only bending moment exists, from which we can obtain the tensile stress from the equation $\sigma = -My/I$.

Concrete Pouring in Adverse Weather

Hot Weather Concreting:

As discussed, concrete produces hydration heat during curing. Too much heat has negative effects on long term concrete properties and it may even lead to early cracking. Therefore, during hot weather pours, precautions must be taken to reduce temperature. These include using small ice pieces when mixing concrete, cooling down aggregate and sand before mixing, using cold water, using admixtures to reduce hydration heat, pouring at night or morning hours when temperatures are lower. The more massive the concrete pour, the stricter precautions must be taken, as the heat will have longer path to escape and must stay in concrete longer, therefore in mass concrete situations slowing down the pour slightly without causing any emergence of cold joints can also help for gradual hydration of concrete and release of heat on shorter paths. Cooling is also necessary for making concrete to stay workable for a longer time, as higher temperature causes faster hardening.

Cold Weather Concreting:

If the water in fresh concrete freezes, the cement can not hydrate. In addition, freezing water expands and will create unwanted pores in the concrete. Freeze thaw cycles may also occur. All these will cause loss of strength and durability of concrete. Even in low temperatures above freezing, setting and hydration process and early strength gain period will slow down, and concrete may not even gain its intended 28 day design strength. Therefore concrete placed in cold weather must be protected as soon as it is placed, until it will gain a certain portion of its design strength. The thinner the concrete element to be placed, more heat losses and therefore the warmer the concrete mix must be.

The protection methods include:

- Maintaining a heated environment, such as using tents

- Covering the concrete with insulator blanket after it is placed, so that the heat of hydration remains in concrete much longer time
- Heating of aggregates before mixing (but this method can only be effective for a limited time)
- Using heat isolated forms (inside of forms and rebar must be free of ice before pour)
- Sealer curing compounds should be preferred instead of water curing

The following will also help:

- Using low W/C ratio, low slump concrete mix
- Using high early strength cement
- Using admixtures for (also see admixtures topic):
    - increasing heat of hydration, accelerate setting
    - allowing curing in lower temperatures
    - reducing water freezing point
    - obtaining air entrained mixture to produce controlled pores in the concrete to increase resistance against freezing

When casting concrete floors of a multistory building one after another in cold climates, due to longer curing times, the shoring and reshoring may need to stay in place longer time than a building in warmer climate, which must be taken into account when preparing construction schedule.

Concrete Pouring in Rainy Weather:

Usually in construction, rainy weather is a cause for not pouring concrete, which should also be incorporated into construction schedules ahead of time, in terms of expected weather delays. So we normally do not pour concrete in rainy weather. But even if it is rare, due to various reasons whole or part of a concrete pour might have to be performed under rain occasionally. If this is a slab, because of its wide surface area, this becomes critical and a source of poor quality concrete.

We discussed water cement ratio (W/C) before. When we pour under rain and as a consequence W/C increases, the strength and durability of concrete will decrease. This is the primary reason that we do not pour concrete under rain. Here only the upper parts will be affected, because it is not likely that the rain will be so strong that it will penetrate deeper than an inch into the slab, but even that it is enough for poor quality. Also, the finishing of concrete surface under extra water can cause surface problems of concrete later. The strong rain can also be sufficient to wash away the cement slurry and even sand at the top parts of the slab and leave aggregates exposed. In highly polluted industrial or urban environments, the rain could be acidic, which can also affect the chemical composition of concrete. Another adverse effect of rain is that it will slow down the pour. If the concrete was already poured, its surface finished and started to harden, the rain not be a problem.

Concrete Repair

Deterioration, crumbling, spalling, cracking of concrete can happen due to many reasons including alkali aggregate reaction, carbonation, water leaks, sulphate

Side Note: Cracks don't include very small cracks which are called "hairline cracks" which are stable, can be expected in concrete and do not necessarily create any structural issues. However if the cracks are growing, that can may mean a risk of failure.

action from soil or water, abrasion, erosion, rebar corrosion, cavitation caused by water flow, poor quality materials, improper concrete mix design, poor quality mixing (especially using too much water), poor quality workmanship during pour, poor quality finishing and curing, freeze-thaw cycles, insufficient concrete cover over rebar, acid exposure, excessive temperature variations, excessive shrinkage, poorly planned or made joints of any type, poor design of waterproofing / drainage / deck slope details or their improper installation, and of course errors in structural design or poor quality construction which may cause overloaded members among other effects... Especially for concrete structures that must withstand water, providing effective maintenance in the long term is more cost effective (and safer) than not doing maintenance and then having to perform costly repairs later. Good quality concrete in the right environment can last even thousands of years. Before repair, the causes of the problem must be found and solved, so that the deterioration or cracks will not occur later again, or simply happen at other locations in the future.

Repair of concrete means first removal of the damaged concrete <u>entirely</u>, until the undamaged concrete is exposed. This step is critical because even if the rest was perfectly done, failure to remove all damaged concrete will prevent new concrete patch to bond with the existing undamaged concrete. Removal methods of concrete most commonly include:

- Cutting, such as by saw or water jet,
- Breaking by hand or machine operated hammers,
- Sandblasting, shotblasting
- Scarifying

The removed and exposed shape should be as simple as possible, preferably square or rectangle. After the undamaged existing concrete is properly exposed, its surface must be cleaned, washed and dried. The existing surface is saturated, so that moisture necessary for cement hydration of repair material will not be absorbed by existing concrete. Materials that can be used for repair include:

- Concrete: Used in areas that needs extensive repair where the depth to repair is the largest, such as more than about 6 inches. Temporary shoring may also be needed.
- Epoxy bonded grout: Can be used for shallow surface repairs and sharp edges. Very sharp edges should be avoided. The repair grouts can be at least as strong as the existing concrete. It is suitable method to repair and fill cracks by injecting grout.
- Portland cement mortar
- Dry pack mortar
- Injected non-shrink grout to fill cracks, small gaps, voids, joints
- Shotcrete

After the repair, proper curing is necessary. This is not only for achieving durability and strength but also to minimize drying shrinkage of the repair material.

When the existing rebar is exposed, any rust and poor quality concrete around the bars must be entirely removed, such as by high pressure water, sand blasting or wire brush. If there is corrosion of bars, or a nonnegligible portion of rebar circumference is exposed during concrete removal, the concrete around the bars are entirely removed, even if the concrete itself was still of good quality. After the rebar is cleaned, it is coated and then mortar patch is placed back. If there is advanced corrosion damage in the rebar, which reduced the rebar cross section over a certain limit, a structural engineer's review is needed and rebar can be replaced or added if necessary before patching back the concrete, or a jacketing layer may be installed over the existing concrete.

In many cases, dowels can also be placed between the existing and new concrete to significantly improve force transfer between them. For this, the existing unharmed concrete is exposed, surface is roughened, dowel holes are drilled and dowels installed, and then new concrete is poured.

Sometimes even if the concrete is undamaged, the member's cross section may be inadequate, which we must increase which is called jacketing. In such cases dowels are also used. It is done by roughening the surface of existing concrete by chipping, for better bond between new and old concrete but the real bond to work together is achieved by anchors / dowels we drill into the exiting concrete, which remain in new concrete when poured. These anchors work mainly in shear. This similar logic applies for anywhere we add new concrete over existing concrete, including increasing size of foundations, columns, shear walls, adding new shear walls... Jacketing is not only done to remedy corrosion but more often also to strengthen members with inadequate cross section, seismic retrofitting. Self consolidating concrete is often used for jacketing, to be able to fill the forms completely and eliminate the need for vibration, where the space to pour and vibrate may not be available or limited.

The extent of the damage can usually easily be determined by sound testing with hammer blows, which can provide good info to experienced personnel. The good quality concrete will cause clear ringing sound with a strong hammer bounce, while the deteriorated concrete will be softer with dull sound and the rebound of the hammer will be weak. Other methods include ultrasonic and acoustic devices that measure the speed of sound in concrete, taking concrete core samples to measure properties in a lab.

Also see corrosion of rebar topic, which was discussed under harmful effects to concrete subsection.

Mass Concrete

Pouring large amount of concrete at once, such as dams or large mat foundations is a challenge. One of the concerns is the hydration heat because the escape path of heat is long. Measures such as slightly slowing the rate of pouring without causing cold joints can help shorten paths of heat escape.

Another challenge is the need for continuous supply of concrete trucks. This involves planning and coordination all the way from the batch plant to the exact pour location, such as the route and number of concrete trucks, crew, the site access plan and coordinating with all parties involved. Pouring large amount of concrete is especially challenging in remote / underdeveloped locations. For example, the capacity of the batch plant may not be enough for a complete pour at once, the contractor may not continue pouring nonstop for too many consecutive hours, lighting at night can be another challenge.

To solve or at least reduce the magnitude of problems above, concrete can be poured in segments. For example for a dam, the whole body of the dam is divided into blocks both horizontally and vertically. They are separated by joints to avoid thermal cracking later. These blocks are clearly outlined in design documents. A mass foundation can also be poured in segments, which are separated by construction joints, with proper rebar and dowel detailing in between the segments because we still want all segments to act as one monolithic piece. The segments are not poured in consecutive adjacent sections, but by skipping every other section as a checkerboard pattern, and then coming back to pour the missing ones to avoid uneven settlements, stresses or displacements.

Precast Concrete

Concrete is not poured in its final place always. Structural or non structural concrete members can also be poured in manufacturing plants, and then transported and installed at the jobsite. Other than the common shapes shown below, a wide variety of other shapes of varying complexities are also possible.

The advantages of precast concrete are that the members can be made in much more controlled manner in factories, which means the strength and other properties of concrete can better be predicted and controlled, and also the final dimensions in the finished structure can be more precisely obtained. *(Of course with good workmanship we also obtain accurate dimensions with cast in place concrete, but ultimately this is done in the field with heavy materials and equipment, vibrators on field installed formwork, therefore staying within tolerances are more challenging.)* We save time from forming, placing rebar, pouring and then waiting for concrete to cure in the field, as long as we take care of the design, approval, fabrication, delivery and installation processes of precast members timely. With proper design and installation, considerable resistance against earthquakes can also be achieved, including even multistory buildings made of precast members. It is also possible to use CIP concrete with precast members in combination, such as making the vertical members as CIP and horizontal members from precast.

Several of typical precast concrete sections

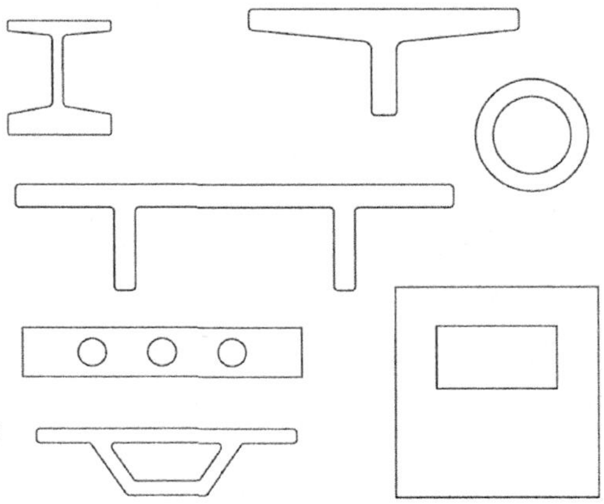

Figure - Concrete 7

Precast concrete member connections can be designed to transfer moment or not. Moment transferring connections must resist cyclic earthquake loads and satisfy capacity and ductility criteria.

The connections can be per the ones prescribed by codes or new type of connections can also be proposed as long as they are tested and reviewed (which is a far more expensive and lengthy route and not done unless there are special circumstances).

Among the disadvantages are that the precast members can only be made up to a certain dimension due to limitations during transportation or site access, and limiting the member sizes means more connections between members, which is structurally less preferred than a continuous jobsite pour. Precast members are also not cheap to fabricate and transport. Very large and heavy precast members require big cranes, which are expensive to rent. There is less room for error with precast; the structural design is more complex especially due to connections, versus the much more common and straightforward cast in place concrete structure design. The installation also requires more specialized crews but considering the experience of these crews, this should not be a disadvantage.

Based on these, to use cast in place or precast concrete depends on factors such as construction schedule and costs, traffic and site access limitations, type, size and complexity of project and load requirements, design and installation capabilities, or simply the preference of the owner and architect. Precast members are often used for buildings, bridges, tunnels, water distribution systems, barriers and more. Contrary to common conception, precast concrete buildings are not only for industrial buildings. Any type of building including multi story buildings can be built using precast members.

In addition to the more familiar jointed design, precast concrete structures can also be designed as if they will be connected to work as a cast in place concrete structure, which is a monolithic structure in principle. ACI PRC-550.1-09: *Guide to Emulating Cast-in-Place Detailing for Seismic Design of Precast Concrete Structures*, covers such topic. In such structures, between adjacent precast pieces, rebar from each piece can be connected such as by couplers or lapping and then pouring concrete or grout.

Tilt-Up Construction:

A common example of precast application is "tilt-up" construction. Tilt-up panels are usually poured on site and on the ground, since making larger pieces are structurally preferred and producing on site eliminates transportation of large pieces on highways and streets. After pour and curing, they are brought to final position and "tilted up" by crane, to immediately make up the exterior shell of the building. The panels include openings for windows too, which means faster construction schedules and generally lower overall costs for the project. Tilt-up construction can usually mean cost savings, due to elimination of elevated formwork.

Load bearing tilt-up panels eliminate the need for structural framing at the perimeter, and in addition to carrying vertical loads, they also resist lateral loads from earthquakes and wind. If non load bearing tilt up panels are used, we must have a separate frame to carry vertical and lateral loads. If the non load bearing panels must carry lateral earthquake loads by acting as shear walls, then the deflection of the frame and the panels must be coordinated. Non load bearing tilt up panels are less common.

One full panel can span up to three stories in height, which is usually the most number of floors built with this system, but it is possible to add more than one tilt up panel on top of each other, in order to make higher than two or three story buildings, upto around 6 stories, although it is not common. Tilt-up system is used most commonly for offices, warehouses and factories.

During structural design of a tilt-up structure, factors to consider include:

- Panel thickness and reinforcing. Upto a certain thickness, one layer of rebar will be adequate. Increasing panel thickness can mean two layers of rebar but thicker panels are stronger too. Panel thicknesses affect self weight, load carrying capacity, deflections and connection designs.
- Connections between panels and other structural members such as floors and roof must adequately transfer gravity, seismic, wind and all applicable forces and have acceptable deflections. The behavior of connections affect the behavior of the whole structure.
- Creep, shrinkage, temperature effects for the entire structure should be accounted for, which becomes especially critical during connection design as far as displacements and stresses.
- Strength of panels during tilt up (lifting) operation by crane can be critical, because the panels will be in diagonal position and supported from only few points, which can create a worst case scenario for the panel, even if it will not be subjected to that load ever again.
- After the panels are lifted, temporary braces must be installed to hold the panels in position and to provide resistance against wind. This must be done before the crane releases the panels.
- The contractor who will do the actual lifting of the panels should be consulted during the design phase, for achieving better progress during construction.

- Wind and thermal loads have more influence with respect to cast in place structures.
- Tilt up panel structures are prone to progressive collapse when erected. Adequate redundancies and alternate load paths must be provided in the system to avoid this. P-Delta effects must be studied during design.
- Design of solid panels without openings are much simpler than of panels with openings, but for most structures we will need to have many panels with openings for windows, doors ducts... These must be well coordinated timely, in order to avoid coring later, which will have cost and schedule impacts not only for the core itself but also due to possible revised structural details at the affected locations.
- The finished structure is designed to brace itself and work together as the slabs and walls brace each other. But during construction, the incomplete states of the structure must be considered and the critical stages must be analyzed, as they can be more critical.
- Panels, reinforcements, fixings, inserts should be standardized as much as possible, both for design and construction convenience.
- As a general rule, the panels should be designed as simple as possible when considering casting, lifting, connections and structural analysis.
- The panels need to go below a certain distance below the slab, but having foundation walls can eliminate this requirement, enabling shorter panels, as long as the panel to foundation wall connections are properly made. Having shorter panels in turn, can mean less burden on other connections, easier construction with schedule savings.
- The design should account for deviations from geometry and uncertainties as much as possible such as casting or construction tolerances, shrinkage and creep.
- Tilt up systems can resist lateral loads by itself, or they can also be used together with other lateral load resisting systems, such as the ones listed in Chapter 2. If combined systems are used, which can be good for creating alternate load paths, the lateral stiffnesses of each system relative to each other is important. During lateral deflections, overturning moment as a result of panel's own weight should also be considered.
- The connection locations should be designed so that:
  - the fire resistance of the structure is not compromised
  - the connections are protected from any outside chemical and physical effects
  - spalling or splitting of concrete does not take place
  - no extra difficulty is caused regarding installation and inspection
  - geometrical deviations and relative movement of elements are sufficiently tolerated

Before construction, the sequence of casting and erection of panels must be coordinated. During construction, the panels should be erected starting from where they touch each other towards the openings, since the openings can absorb tolerances. The crane position must be optimally arranged so that maximum number of panels can be reached and placed at each crane position. Blind lifting should be avoided *(the situation where the crane operator entirely depends on another person and does not see the final position of the panel during lifting)* if possible.

Some of the items above also apply to other type of precast concrete structures, such as structures with precast beams.

Precast Concrete Pavers:

Another very common application of precast concrete is the pavers. As always the case for precast concrete, these pavers are manufactured in controlled environments and specific desired properties are easily and cost effectively obtained. Precast concrete pavers offer a lot of advantages:

- Unlimited variations in geometry, color, durability, density and strength are easily obtained.
- Aesthetically appealing appearances can be obtained precisely, which is not always possible with poured concrete or natural stone.
- Natural stone can be expensive and difficult to obtain in desired qualities. With precast concrete pavers, appearance of natural stone can be imitated in a cost effective way, with specifically desired properties.
- They can also be made in the form of interlocking pavers, which can work together. Cracking both from loading or temperature (expansion and contraction) is almost entirely eliminated by using smaller units as they form a lot of joints, but can still work together.
- Very high compressive strength and durability, which can be much more than rigid or asphalt pavements can be achieved, which can reduce the need for long term maintenance to very low to none. Many decades of useful lifespan is possible.
- Wide range of applications and simple installation.
- High durability also means high resistance to freeze-thaw cycles, and extreme heat.
- Any damaged pavers can easily and quickly be removed and replaced, without unpleasant looking demolition and patches as in the case of rigid or asphalt paving. In case of natural stone, it may not even be possible to obtain the exact same looking stone after some time.
- Need for reinforcing such as welded rebar mat as in the case of rigid pavements is eliminated.
- The joints between pavers allow rain water to pass through therefore increase the amount of water absorbed by the ground, which helps to protect the groundwater level, and reduces burden on storm drain system.

Precast stone pavers are placed above compacted grade, sand and rebar mesh. Usage of mesh greatly improves load carrying capacity in a uniform manner, where point tire loads from heavy vehicles might cause local settlements otherwise. The joints between pavers can be filled with sand or grout and compacted by using a hand operated vibrating compactor and then the excess sand at the surface, if any, is swept off. The tool seen at the bottom of the photo is called "Angle Grinder", which is a very versatile tool heavily used in construction sites from start to finish, for cutting, sanding and polishing materials like stone, concrete, metal or wood.

Prestressed Concrete

We can choose to apply stress to concrete before it starts to serve its duty under actual loads. This is called prestressing. It can be done as pre-tensioning, or post-tensioning. So, prestressing is a general term that includes both.

In pre-tensioning, we stress the tendons before pouring, and, in post-tensioning we stress the tendons after pouring concrete and waiting for it to gain its strength. This "before" and "after" mean a big difference as far as how and where we make and use them. Ultimately they both allow more efficient structures where members can carry higher loads because in both methods we use ultra high strength steel, much stronger than regular rebar.

For prestressed members, there will be immediate and long term (time dependent) losses from prestressing force. These can be listed as:

- Frictional losses (immediate)
- Elastic shortening of concrete (immediate)
- Slippage of anchor at stressing (jacking) point (immediate)

- Relaxation of prestressed steel (immediate and long term)

Relaxation of steel is the reduction in stress when the length is constant. It starts immediately and continues indefinitely but at a decreasing rate. As a result of it, the reduction of prestress forces in the long term may not be negligible and this may need to be included as a variable in a nonlinear analysis.

Pre-Tensioned Concrete:

Pretensioned members are usually beams or slab segments, which are poured in plants, and then transported to their final location, or they can be poured on the ground on site, if there is available space and equipment. Here are the main steps for pretensioning (also see the figure below):

- The formwork (casting bed) is placed and prepared on the ground.
- High strength reinforcement tendons are placed in casting beds. They are vertically positioned with certain eccentricity below mid height, so that when we apply load later, the beam or slab can bend upwards.
- The tendons are tensioned with hydraulic jacks before the concrete is cast.
- Concrete is poured in the casting bed. Wet concrete fully encloses the tensioned tendons.
- Concrete hardens, cures for a certain period and gains adequate strength.
- Tension in tendons is released by releasing hydraulic jack, and now tendons transfer this stress to the surrounding concrete, which puts the concrete around the tendons in compression.
- After the forms are stripped, this makes the concrete beam (or slab segment) bend upwards.

As a result, the concrete member has initial stress even before being placed in its final location in the field. Inducing stress in a structural member even before it is in place might seem illogical at first but actually it makes sense, because when the beam (or slab segment) is installed at its final location and the loads are applied on it, that load must first overcome the effect of upwards bending, only to make it straight again, and only then it can start bending the beam or slab segment downwards.

The level up to which the actual applied load in the structure must first overcome upwards bending is the net strength gain for bending. This obviously allows more efficient geometry, such as, the same free spans can be passed with shallower beams. The only condition in this whole thing is that the steel tendons inside the concrete must be very strong. Other advantages of using pretensioned concrete are mostly similar to those of precast.

Cracks in pretensioned members are also better controlled, and crack thicknesses are smaller, since the members are prestressed in compression already.

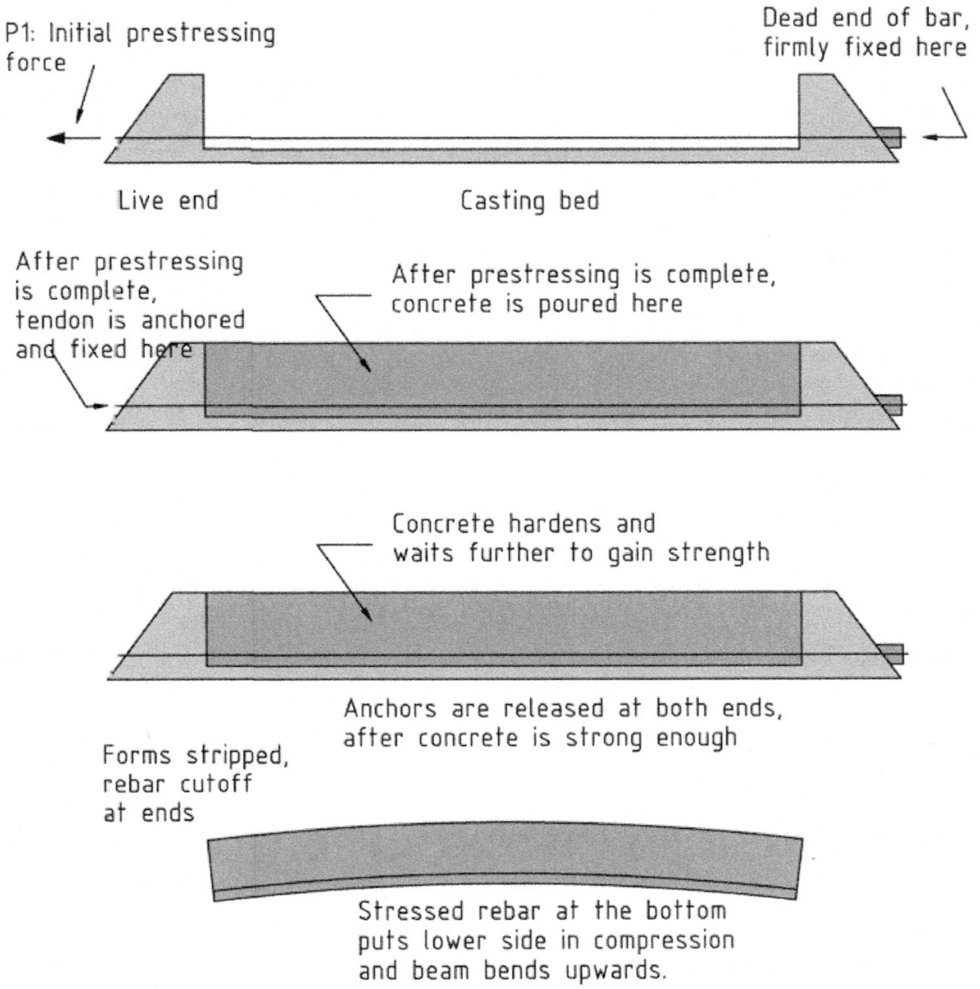

Figure - Concrete 8

Post-Tensioned Concrete:

Post-tensioning of concrete beams or slabs means, tensioning the tendons <u>after</u> pouring concrete. Here are the main steps:

- High strength steel tendons are inserted in protective conduits, in which they can slide.
- The tendons are loosely laid in place, but still by following the exact horizontal and vertical layout in the drawings.
- Tendons are anchored to the edge of slab / beam, (whichever member we are posttensioning).
- Concrete is poured and we wait for it to cure and gain strength.
- After sufficient strength gain of concrete, high strength steel tendons are tensioned. During tensioning the tendons can easily slide inside concrete because they are enclosed in protective plastic conduits in which they can slide as mentioned above.
- Grout is injected into the conduit, to fill it completely. This ensures stresses between tendons and concrete are transferred, and the wholeness of the slab or beam. Grout also serves as corrosion protection for the tendons inside.

- Tension from hydraulic jack is released and tendons transfer this stress to surrounding concrete.

This way, similar to pre-tensioning, we give a certain compression to concrete beam or slab first, at intended layout. This also means that the entire self weight of the slab or beam is carried by tendons, plus a lot of reserve capacity. So when the actual load is placed on the beam or slab later, it must first overcome the compression at the lower side of slab/beam to make it flat again, which is the net strength gain, also called balance load, and only then it can start to place the lower side in tension, which means stronger, more efficient members and passing longer freespans.

Sometimes, post-tensioning may be the only logical choice, when for example we have to pass long spans, build long cantilevered sections, need thinner slabs or shallower beams, slabs, or we need beams that are supposed to take high loads for a given slab or beam depth... Post tensioned slabs can also be made on grade, which is a great way to resist expansion of soils, carry high loads, crack control.

PT slabs can also help better minor crack control, due to introducing compressive forces in the plane of the slab. At the wall locations however, slabs are restrained which can lead to even more severe cracking than a regular slab, as we have lower amount of regular reinforcing in a PT slab. This is why we can make slip joints at wall locations, allowing tendons to slide through the wall. Placing pour strips or expansion joints are other solutions to prevent cracking in PT slabs. Pour strips are sections of slab where do not pour at first and postpone, to allow the adjacent poured slab sections to first deform under post tensioning stresses while still incomplete, instead of the whole slab trying to deform as a whole, which could cause severe cracks. If the distance between pour strips are too much, separating the structure altogether by expansion joints can be considered. While waiting for pour strips to be poured, the poured slab edges adjacent to the unpoured strip must be supported by shoring, since these slab sections will be in cantilevered form without linking to each other.

Post tensioned slab has considerably less regular rebar than a conventional cast in place slab. The purpose of the regular reinforcing here is to serve as secondary reinforcing and ensure uniform distribution of stresses in the slab and prevent localized failures, take on shear forces, better crack control from shrinkage, creep, and thermal stresses. Wire mesh, also called as Welded Wire Fabric (WWF) can also be used for regular reinforcing.

At anchorage locations, we must make sure that the edge of the slab has enough strength and stiffness to resist great local compression forces which then spread toward the slab inwards.

The advantages of post tensioning over pretensioning include:

- We can adjust the exact shape of conduit as we want before tensioning the tendons, so the tendons can follow any shape we want, such as the path where we will have tensile stresses *(but still we must make sure that the tendons run as smoothly as possible, without sharp curves, to avoid localized higher stresses)*. This is a big advantage for slabs or for beams that span over multiple spans, because the moment diagram sign, therefore the location of tension will keep changing (see Figure 4). We can not do this with pretensioned tendons easily because in that case we pull the tendons with great force first, which makes it difficult to make it follow a desired route with curves, although it can be done with specialized equipment.

- Pre-tensioned concrete members are brought to final position on site and placed to their final location in the structure after they are poured, which means more difficulty in making slabs or continuous spans. That is why pre-tensioning is suitable for simply supported members where the moment diagram is simpler, beams with a single span only, such as precast beam bridges.

- If we use pre-tensioned members, we need to leave support keys on columns, and make special connections where the pre-tensioned members will be placed. When using post-tensioned tendons however, the tendons are simply placed inside the actual formwork, and the whole floor and beams are simply poured in place, without needing complex connection details.
- Post tensioning can also be performed externally, which can especially be valuable during structural repairs. This can also be applied for new structures.

Some of the points that may be regarded as disadvantage or need more careful coordination include:

- Repairs can be difficult since the cables are not bonded to concrete, and the repair needs to be done very carefully because cutting of these stressed tendons can cause great damage. Future modifications such as making opening in a slab are also more difficult for the same reasons and in some cases may not even be possible.
- Analysis of post-tensioned slabs is more complex than regular slabs not only for the slab itself and the structure, but also for shoring calculations, due to load redistributions.
- Conduit and other embedded item layouts will have less freedom to not only interfere but even influence tendons. The structural engineer must approve the layout before pour of each slab which need coordination between a number of parties.
- In comparison to layout of ordinary cast in place reinforcement, more care must be given to tendon layout. If location of a tendon does not closely match its intended location in structural drawings, undue stresses, unaccounted load paths may emerge in structure upon tensioning.
- Post tension forces can affect perimeter beams (and in turn columns). High post tension forces can cause problems to facade in the long term as well, by excessively pulling perimeter beams inward and cause shrinkage and creep in slab.
- Post tensioning may cause nonnegligible secondary moments in the structure which must be accounted for during the design phase.

While post tensioning, the elongations are recorded and compared with desired values. If some individual tendons come to be out of tolerance this may still be acceptable depending on the type, size and location of the tendon and the local codes, subject to engineer's approval, because the overall intended effect on the slab might still have been achieved when we think of multiple tendons.

Shotcrete

Shotcrete (also called gunite) is concrete that is shot from a spraying gun, instead of being poured in forms. Its aggregate sizes are smaller than concrete, such as 1/2" is approximately the maximum used. Spraying of concrete opens a whole new level of possible uses, because vertical surfaces can be concreted without using forms. This is a big deal considering saving of formwork cost and time, the labor to pour and the flexible geometries that can be obtained. As shotcrete is sprayed on vertical surfaces, its fast setting is important, for which, certain concrete admixtures can be used.

Shotcrete can be dry or wet mix, depending on when the water is added. In dry process, water is introduced at the very end, whereas in wet mix, water is mixed in the beginning. Both has advantages and disadvantages and therefore different applications. For example, dry mix process can produce higher strength shotcrete, and the shotcrete can be transported over greater distances, but wet mix can be more easily produced, and generates less waste. Shotcrete is commonly used at:

- tunnels,
- slope stability projects,
- pools, tanks, large aquariums,
- renovations,
- domes,
- various aesthetical and architectural applications,
- repair of various of structures unless thick layers of concrete is needed.

Shotcrete is brittle, and also weak in tension, just like concrete. Therefore it is often reinforced with steel, which means, reinforcing bars or welded rebar mesh are placed first and then the shotcrete is sprayed. Reinforcing can be achieved mostly by steel, but use of synthetic fibers is also possible.

Lightweight Concrete

Lightweight concrete contain porous coarse aggregates, (and can still contain normal fine aggregates) which means lighter concrete that may weigh about 25% less on average than normal weight concrete. Porous aggregates can be natural such as volcanic pumice, or obtained through processing (usually heating) of certain other materials such as slate, fly ash, tuff, slag, shale, crushed bricks, clay.

As a result of the advantages below, lightweight concrete can be used not only in buildings but also bridges, pavements and anywhere else, including precast members, allowing us to benefit from this light, strong and insulating material.

Side Note: Lightweight concrete can also be achieved by using mixtures which contain no fine aggregates or aerated mixtures. For the other two, the terms "no fines concrete" and "aerated concrete" are also used.

Let's list important advantages and disadvantages and also compare to normal weight concrete:

- Lightweight concrete can mean overall savings for the structure due to reduced weight, although the lightweight concrete itself is more expensive (mainly due to extra cost of obtaining lightweight aggregate).

- Lighter concrete also means easier pumping and placement in the field, less burden on forms and shoring. It also means easier transportation, lifting and installation for precast members.

- High porosity aggregates also means better thermal (and sound) insulation. LW concrete is common for roof slabs or metal decking where fire rating is important, which enables thinner slabs *(so not only the concrete itself is lighter but the slabs can also be thinner for the same required metal deck slab fire rating, which means lower load on structure and foundations).*

- Even though aggregates are porous, typically used strength ranges for structural concrete are still commonly obtained with lightweight concrete. In other words, lightweight concrete doesn't necessarily mean concrete that we cannot structurally use, at least upto around medium strength levels (such as upto around 5000 psi = 35 MPa). Research is always underway to make concrete even lighter and stronger (and more durable of course).
- Higher amount of pores means better freeze and thaw resistance.
- High amount of pores in lightweight aggregates can absorb a lot more water than the crushed stone aggregate used in normal weight concrete. This can reduce available water necessary for concrete slump which provides workability, and the water needed during hydration of cement, therefore it is common to presaturate these aggregates before adding into the concrete mix.
- The absorbed water in aggregate pores can contribute to better cement hydration (curing) internally in long term. But a negative side here is that some excess water will eventually need to evaporate, and this takes longer time than normal weight concrete, which may cause moisture related problems in the long term such as carbonation (which can then lead to corrosion of rebar) or where lightweight concrete slab is used under flooring.

Finishing of Concrete

Concrete surfaces can be made to have various features, depending on their intended use. Some can be rough, with varying evenness and texture, others can be uniform and smooth or even glossy. If the concrete surface will be exposed, architects choose the pattern will be used.

There can be reasons other than architectural, for finishing concrete surface in different ways. For example, some must be very smooth with tight tolerances, to the degree that laser screeds are applied when finishing during final setting of concrete. For example, in structures conveying high velocity water flow, the surface must be also very smooth in order to prevent destructive effects of water and also to minimize energy losses due to friction. Some surfaces must be rough, or at least not very smooth, such as sidewalks, highway pavements, to prevent slippage. Not only the finishing of the surface after hardening, but also the choice of forms can affect surface texture. Sometimes inside faces of forms are chosen specifically, so that the concrete will have a certain finish, without requiring any further finishing operation. Over-troweling of concrete surface during finishing must be avoided. Overtroweling will cause surface to be more susceptible to erosion, weathering and spalling. Evenly finishing concrete paving or exposed slab surfaces is important to prevent water from accumulating in dented locations.

Roller Compacted Concrete (RCC)

RCC contains the same ingredients as traditional concrete such as cement, water, aggregates, sand, but it is a drier mix, and contains less cement than structural concrete. It it used for pavements, large slabs on grade, parking areas, in dam bodies and needs no forms, vibration, finishing or reinforcements therefore it can enable very fast construction schedules for the volume to be poured. It is placed by first pouring it on compacted grade and then finishing and compacting with rollers similar to placing asphalt pavements. Lower cement content also means lower hydration heat during placement of high quantities, which is an advantage for mass concrete pours such as dams. RCC can sometimes also be called as lean or dry lean concrete. It is frequently used where large volumes are placed for cost efficiency and other advantages. It is low cost, durable, dense, strong, has very long life and needs little maintenance, and can be grouped under rigid type of pavement, as listed in Chapter 4. For example it does not easily deform under heavy and concentrated loads and very resistant to fuel spills and deterioration, has high freeze - thaw resistance, can withstand high temperatures.

During construction of RCC, proper compaction is critical to ensure all characteristics mentioned above are obtained. As in earthwork compaction, here the compaction also continues until the concrete meets density requirements, such as 98% of maximum density of modified proctor test. In the field, degree of compaction can be measured with nuclear density test (See Soil mechanics>Soil Testing). Each compaction layer can be anywhere from 4 to 10 inches thick. After construction, as in regular concrete, RCC concrete must also be cured for the cement to hydrate, but the curing period to support vehicle loads is far shorter, which is a big advantage for opening a road to traffic quickly. Curing can be done by water or applying sealer compound, similar to regular concrete, but more attention must be paid here to not allow water to escape, as this is already drier mix. Before pouring the actual RCC, a test section may be advisable or mandatory depending on project location, type and size, to better foresee the finish and strength characteristics, the need of joints and their gaps, for desired result.

STEEL CONSTRUCTION

Steel is the most commonly used construction material after concrete. Steel beams, columns, angles, channels, tubes, plates, bolts, decks are connected to each other to build limitless variety of structures and various structural components of any size. Steel is not only used to build structures but also as miscellaneous metals that serve as support framework for installations of other trades.

Comparison of Structural Steel vs. Concrete

The advantages of steel construction with respect to reinforced concrete construction include:

- Steel is much stronger than concrete, both in compression and tension. Compressive and tensile strength of steel are both very high and can be taken as equal to each other. Steel does not need additional reinforcing. In general, structural steel with strength values of anywhere from 200 to 400 MPa are used, which is roughly 10 times that of concrete. Codes and standards give range of values to use. For example, in Eurocode 3 - Design of Steel Structures, Part 1.1, in Table 3.1, for hot rolled structural steel, nominal yield strength values of anywhere between 215 to 460 MPa, and nominal ultimate strength values between 360 to 570 MPa can be found.

- In addition to its strength, steel is ductile and tough, which means high energy absorption during seismic events. Steel shows a lot of deformation (warning) before failure, whereas concrete is brittle, have little energy absorption capacity and fails suddenly after little deformation. Steel's higher strength, and ductility are its main advantages over concrete.
- Being stronger, steel can allow considerably thinner and lighter structures than concrete. Its strength to weight ratio is much higher. Lighter buildings perform better during earthquakes as we have seen under earthquake section. This means less burden on the structure and foundations. Using lighter, thinner members also means more rentable or sellable space too.
- Steel is manufactured in highly controlled environments in factories, which allows us to control its properties precisely, whereas concrete is cast in place on site, thus exposed to a lot of variables all the way from mixing and batching, until delivery and pouring. Precast and prestressed-precast concrete are exceptions to this comparison.
- Once fabricated and delivered on site, steel members can be erected fast. Unless highly developed and coordinated methods are strictly and efficiently followed, cast in place concrete construction is generally slower, mostly because of the curing time of concrete.
- Obtaining unusual shaped structures with steel is easier than with cast in place concrete.
- Modifications and repairs to an existing structure can be done with relative ease in steel structures, because simply removing and replacing steel columns or beams is simpler than cutting, preparing, forming, reinforcing and pouring concrete for a similar size and location.

Disadvantages or challenges of steel construction with respect to reinforced concrete include:

- Steel can have long lead times to manufacture, and sometimes long distance deliveries are required. This makes steel construction highly dependent on steel manufacturer's ability to prepare shop drawings, get approval from structural engineer, fabricate and deliver on time. To coordinate and plan these stages properly from the beginning is critical. Steel frame interacts with a lot of other trades, therefore extensive dimensional coordination is needed. Other trades will depend on and follow steel frame dimensions, *(both by looking at the approved shop drawings and also especially by field measuring the installed frame)*, which still makes completion of shop drawings and installation of steel critical. Other than structural frame drawings, also miscellaneous metals shop drawings are prepared for the steel that supports various equipment and nonstructural components, which needs coordination. Miscellaneous metal shop drawing delay have the potential to affect all subsequent trades. General contractor oversees this coordination. If architect, structural engineer, general contractor and steel subcontractor proactively work together, the approval, manufacturing, delivery and installation cycle does not necessarily mean disadvantage, but nevertheless poses risk.
- Steel loses its strength upon exposure to fire, and therefore must be fireproofed. Fireproofing can be applied on site, which adds to the cost and schedule, plus the added thickness to be considered for dimensional coordination. If we want to save the thickness for fireproofing, intumescent paint can also be applied but it is not a cheap material. Concrete is very resistant to

fire (Note: If concrete is exposed to fire for an extended period of time, there may also be negative effects, such as spalling of concrete and the cracking of frame due to expansion).

- Being a lighter material for the same strength is not always a positive. As concrete buildings are heavier, they are inherently more resistant to very strong winds than steel buildings that have the same surface area *(for earthquakes, the situation is the opposite as discussed)*.
- Steel is vulnerable to corrosion, and proper maintenance and precautions are needed. Concrete is much more water resistant than steel, although it also requires measures for improving water resistance in highly exposed conditions. The most important condition for reinforced concrete is to protect its rebar from contact with water or humidity through cracks.

As for environmental considerations, both steel and concrete have minuses and plusses. Producing steel and erecting on site has a high energy footprint and therefore not very friendly for the environment but the same thing is also true for concrete construction. For steel one plus is the ability to recycle previously used metals as steel is highly recyclable. In fact, most material used in steel production comes from recycled steel. For concrete, some waste materials can also be included such as blast furnace slag, fly ash, but the degree of recycling is much lower.

Structural Steel Shapes

Steel members are commonly classified according to their cross sectional shapes as shown below:

- W section, has a cross section that resembles the letter "H" and used as columns or beams. That is why they are also called H-beams. Their flanges are wide, so they are also called wide flange (W) beams. Often W beams have flanges and webs of equal thicknesses.
- M section has the cross section that resembles the letter "I" and used as columns or beams. The main difference between W and M sections are that for the same dimensional range, W beams are heavier and the flanges and web are thicker. W beam flanges have significant dimension in comparison to their web height. M beam flanges on the other hand, are narrower. Because of their narrow flanges, I sections have low torsion (twist) tolerance, and therefore used less frequently when loads are present along their length.
- S section, can be considered as a middle shape between W and M shapes, and the inner flange surface is sloped.
- Channel shape resembles the letter C and shown with letter C. Channels are for supporting much lighter loads than the W, M, S shapes, because their cross section is not symmetrical. If their compression flange are unbraced, they can easily undergo lateral torsional buckling. They are often used to connect stronger members, which don't demand heavy loads from them, and for smaller miscellaneous structures (i.e. a support structure for a glass canopy in front of a building) or parts of structures.
- Angles are also weaker than the W, M, S shapes, and same points as C shapes apply. They are also used around openings in concrete or as lintel in masonry for example, along with channel shapes. Legs of angles may not be equal to each other.

- Structural T sections also have similar purposes and limitations as C and angle shapes.
- Hollow Structural Section, (HSS), are used as columns or posts or for various structural or mechanical applications. They can be circular, rectangular or square in shape. Their main strength is their compression, tension and torsion resisting ability but they are less efficient under flexure.

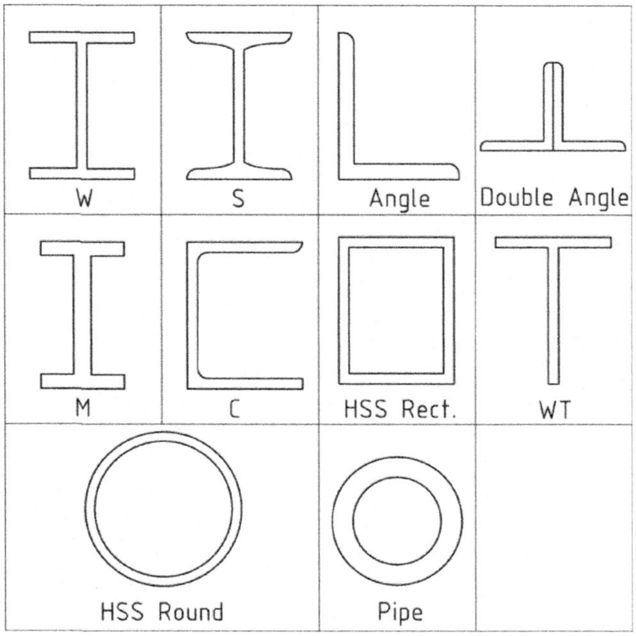

Figure - Steel 1

- Pipes look like HSS round but their purpose is mainly to carry fluids or used in mechanical context. As per in AISC steel manual, steel pipe and round HSS are made to conform to different ASTM standards. Pipes can still sometimes be used as structural members and listed in AISC manual including for compression and flexural loads.
- Plates / Flat sections, are just simple rectangles or squares, and they are used together with other shapes as strengthening or connecting pieces *(i.e. plates are welded to the side of the web of beams, which are called stiffener plates, they can be used as continuity plates to extend beam flanges into the column web etc...).*
- Custom sections, even curved shapes can also be made from steel. Architects can design appealing structures, utilizing steel's strength and workability into any shape. Of course, special sections require more complex structural design and dimensional coordination, more complex connections, longer fabrication times, slower installation, and these mean higher costs and longer schedules, therefore standard sections are used to the greatest extent possible.
- Other than the main structural shapes listed above, steel also has other uses such as,
    - metal decks which are used as slabs and often receive concrete topping and connected to beams by shear rods,
    - steel sheets, which are used as sheet pile walls,

- checker plates, which are used as flooring or stair material in many locations where mechanical equipment are present and in other places for aesthetical purposes.

All these steel shapes come in various geometries, thicknesses and grades (strengths). Steel Construction Manual by American Institute of Steel Construction (AISC) provides detailed properties for all items above with all their varieties and design criteria to be followed.

A truss girder roof made of structural steel elements such as I, C shapes, angles, and plates for connections. The elements of girder trusses at the bottom part of the photo wait for lifting and assembly in place but ground assembly and then lifting and placement on top of concrete columns can also be done depending on project circumstances.

Structural Steel Design Considerations

Steel design is a major branch of structural engineering. Some fundamental points are listed below:

- Steel members are designed according to strength and serviceability limit states. Strength limit states such as yielding, buckling and rupture ensure that the structure will not collapse. Serviceability limit states ensure that the structure functions properly, without causing discomfort, which can be listed as drift, deflection, deformation, vibration, wind motion, thermal expansion and contraction, connection slippage, such as described in ANSI/AISC 360-16. Strength and serviceability limit states were also mentioned in structural design approaches section. They are checked for both construction phase and for the finished structure.

- As far as their function, steel members can be listed as beams, girders, columns, bolts, rods, shear plates, bearing plates, base plates, bracings, truss members, bolts, anchors, embeds.
- As far as how they take on forces, steel members can be classified as:
  - Compression members, which work mainly to compression (such as columns)
  - Flexural members which work mainly under bending stresses (such as beams)
  - Tension members which work only in tension
  - Members that work under combination of the above
- Design of compression members are done both considering the yield and buckling failures. After a critical load is reached, a member under compression will buckle, which will lead to a stability failure as discussed under structural stability, even if the yield stress has not been reached. To prevent buckling, compression members' slenderness ratio (member's effective length divided by its radius of gyration) is limited (i.e. 200 as in ANSI/AISC 360-16 section E2).
- Design of tension members is done for yield and rupture. The safety factor for rupture is higher than the one for tension yield, due to more variety of failure possibilities, (for example, for Allowable Stress Design (ASD), the safety factor for tension is 1.67 but for rupture it is 2 as per ANSI/AISC 360-16, Section D2). For tension members, buckling is obviously not an issue, but to limit any unwanted vibrations and provide some stiffness a maximum slenderness ratio limit is still used, although it is much higher than that of compression members (i.e. 300 per ANSI/AISC 360-16, Section D1). Similarly, any slight eccentricity of loads, or deviation from straightness does not matter for tension members, because tension works to reduce eccentricity, which is the opposite of compression, where, eccentricity may cause even more eccentricity and failure.
- A beam is a bending (flexural) member which carries transverse loads along its length perpendicular to its main direction. By flexural members we mean members under bending. The loads are usually distributed over the length, but may also be concentrated at certain locations. In steel construction, when we say beam, we usually mean H or I beams, but other sections including box or tube sections, built-up sections, or even unsymmetrical shapes are also possible. Buckling of flanges and web, lateral torsional buckling, yielding of steel itself (compression or tension yielding at upper or bottom flange, shear yielding of web) may occur as a result of bending. The most critical of these must be found through structural analysis.
- Design of beam-columns, which are members subjected to both compression and transverse loading. These are designed differently from members under bending or compression only.
- There are also members which will be subject to torsion only, although it is not common.
- Design for shear, which is done for members that have shear stresses in their plane of web.
- Design of braced frames. Bracing significantly increases the lateral load resistance of a structural steel frame by resisting its swaying towards sides. A braced frame is similar to a truss which is vertical, so, its strength comes from the axial force capacity of its components. Braced frames are discussed in Chapter 2 under Types of Gravity/ Lateral Load Resisting Systems.
- Bracing of beams, which are provided by cross members connected to beam webs or a concrete floor slab that serves as a continuous bracing to the supporting beams underneath.

- Design of connections, such as the ones that take shear as pin support, or moment connections that take both moment and shear. Load and displacement requirements demanded of connections will be known after structural analysis after which the connections can be designed accordingly. Connections can have limited amount of inelastic deformation, as long as it doesn't cause bigger problems. ANSI/AISC 360-16 section B3.4 allows self limiting plastic deformations.
- The column web, where the column meets the beam, is called the panel zone. Panel zones are further discussed in Chapter 2 > Lateral Load Resisting Systems > Moment Frames.
- Beam to column connections affect the lateral resistance of a moment frame. To make steel moment frames more ductile so that they can consume more energy during earthquakes, the beam column connections must be strong. This means, both the panel zone and the connection detail that contains all the plates, bolts and welds as applicable.
- As the column beam connections must be strong, the splices of columns between floors need to be located far from connections, towards midway of the floor height. This is similar to splicing of reinforcing steel at mid floor height, in case of cast in place columns.
- Other types of connections include anchor rods and embedments, column base plates. The column base plates must have sufficient thickness and size, so that they can take on the large moments from columns. If they are not embedded level into the slab or foundation, the gap between the bottom of the plate and the foundation is completely filled by a nonshrink grout *(grouting must be done right after erecting the column, before it is loaded, so that the moment at column bases can be resisted. Negligence of doing this timely can easily lead to moment failure at the column base during or after construction).*
- Design of composite construction of steel and concrete, such as an H-column encased in concrete, concrete encased in a hollow steel shape, or concrete poured over steel deck (also called composite metal deck). Note that the concrete is still reinforced. For more on composite concrete metal decks, see Chapter 2 > Building Frame and Structural Members > Slabs.
- Built-up members are made from multiple pieces. This allows infinitely more flexibility in design, to obtain any geometry we want efficiently. Built-up members can be made in two ways. The pieces can be joined closely together, or they can be made from widely spaced flange components, which are connected by intermediate members to serve as web. This way we can create a beam or column exactly as we want, which can be much larger, (and for that overall size more efficient), than a single manufactured I or H beam or column.
- Design of diaphragms and collector elements. Slab collector elements are also mentioned in Chapter 2 for reinforced concrete. Whether RC or steel building, the purpose of collector is the same, which is to transfer forces between the diaphragm and the vertical elements.
- Design of welds and bolts in connections. Bolts have replaced rivets since at least several decades. Welded joints must be made to prevent brittle failure. Welds need to have equal or greater yield strength than the connecting members' yield strength.
- Fabrication and installation imperfections, residual stresses, yield strength variations are accounted for in applicable codes (i.e. as stated in Eurocode 3, Part 1.1, Section 2.4.2).

- Fatigue can be critical for steel under cyclic, repeated loads, such as vibrations from machinery, moving crowd, exercise activities, traffic... Fatigue does not need to be taken into account for earthquake or wind effects for typical building structures, per ANSI/AISC 360-16, Section B3.11.
- For buckling analysis of members, worst conditions in all planes must be checked, including lateral torsional buckling. Not only columns but beams, plates and pieces used to make built up members can also buckle.
- Floor vibrations in steel buildings due to the reasons above is checked for serviceability.
- Steel frame dimensions should satisfy standard tolerances set by such as AISC, but the requirements for other trades must also be considered, especially masonry.
- The stability of the entire structure and its effects on members must also be checked, such as the P-Delta effect - this of course applies to concrete structures as well. For steel structures however, stability issues not only for the finished structure but also during construction are also particularly important. Temporary bracing must be provided during construction as necessary.

AISC publications, such as the Steel Construction Manual, ANSI/AISC 360-16 Specifications for Structural Steel buildings, ANSI/AISC 341-16 Seismic Provisions for Structural Steel Buildings, provide detailed information and criteria for steel structure design concepts that are followed in the US.

Constructability must be considered during design. The Engineer of Record (EOR) will design the structure, and then leave details of connections and shop drawings to steel contractors' structural engineer. The specialty engineer then prepares shop drawings and submits (through general contractor) to EOR's approval. To facilitate easier preparation of shop drawings and construction in the field, the EOR can follow practices such as the ones listed below. Some of these may structurally be less efficient (too conservative) but still more cost effective:

- Connections have the biggest impact on the speed of installation and a major factor affecting the efficiency of overall design. They should be as simple as possible considering their fabrication and installation. Complex, angled geometries should be avoided to the greatest extent and should only exist when really necessary. The fewer the pieces, the simpler the connections, the better, as long as it is functional and safe. Sometimes difficult connections cannot be avoided such as for structures with complex architectural design.
- Repetitions are good, (of course on the conservative side) even if this may mean slightly below ideal structural efficiency in some locations, when the relative ease of fabrication and installation is worth it. A cost benefit analysis should accompany the engineer's judgment here.
- Making column sizes slightly more than needed instead of adding plates to achieve connection strength and stiffness and strong column weak beam criteria can be more cost effective overall, even if not the most structurally efficient and more conservative than needed. Of course this must not cause conflict with architectural aspects.
- Field welding, which is more expensive than shop welding, should be minimized, and pieces should be designed to allow shop welding to the greatest extent. Bolted connections may also be considered, as it may proceed faster than welding and do not require welding inspections.

- Construction stages, the incomplete state of the structure and the temporary loads should be accounted for, even though erecting the structure is the responsibility of the contractor.
- Member loads obtained from structural analysis should be given to the contractor. This enables easier, more efficient and economical connection design and shop drawing preparation by the contractor. For example the specialty engineer can then design the connections to actual forces, which may be less than the full force that the connected members are able to support.

Construction Phase Considerations for Structural Steel

Once the project is designed and the structural steel members are fabricated, the materials are shipped to the jobsite and installation starts. In addition to the items we already mentioned within the context, let's list a few more regarding the construction phase:

- In construction, questions that emerge during installation of any trade, including steel, are submitted in the form of RFI - Request for Information, which are sequentially numbered and effective means to track issues on a project. Due to field conditions, and also the steel structure interacting with many other trades, it is common to have frequently emerging questions in the field regarding the installation and coordination of steel. Good communication between the architect, structural engineer, general contractor and steel subcontractor is needed to quickly and effectively resolve these. On a structural steel building project involving many trades, a good portion of the RFIs will be originated as a result of steel installation and its coordination.
- There are some critical things that need inspection during steel structure installation, which are determined by local codes applicable to a project. Some common items to be inspected are welds, bolts, connections, and overall compliance of frame with construction documents.
- The erection sequence, means and methods and temporary stability of the structure during erection is the contractor's responsibility. The structure during construction phase can be the most vulnerable to failure and the contractor must have clear understanding about his means and methods of erection. The structural engineer of the project may still make remarks on the drawings for items such as shoring of decks, bracing, erection sequence, loading sequence etc.
- As steel members are delivered to the jobsite, the delivery and staging area and their coordination with installation activities and other trades must be well organized ahead of time to avoid delays and inefficiencies. After that, being able to lift and place members in an uninterrupted manner is important for efficient work flow. Good coordination of crane usage *(which comes at an expensive rate)* versus steel installation is needed. The larger the project, the earlier "ahead of time" means, which may even need to take place even before construction starts on large projects. During design, the design team should get together with the construction team to discuss potential issues. Construction is much more expensive phase than design phase *(which is true for all trades)*. Cost impacts due to poor construction coordination and not sufficiently considering constructability issues will easily surpass the cost savings from even the most efficient designs. This of course applies to all trades in construction, but for structural steel, the cost and schedule impacts can be very large.

MASONRY CONSTRUCTION

Masonry construction dates back to thousands of years, which enabled building of most historical structures and it is still a very popular system and material of construction. In our age, instead of being used as the only structural system as in the past, it is rather more often used in combination with steel, reinforced concrete or even wood frames in many different settings and as part of the structural system or not. The most typical use is as walls, which can be as reinforced, unreinforced or even prestressed, load bearing (structural) or non load bearing (nonstructural). They can serve as partition walls, facade walls, shear walls, basement walls, shaft walls, retaining walls and as various enclosures.

Masonry is an old type of construction but the properties of blocks and construction methods are now considerably different than in the past. Especially older unreinforced masonry structures in developed countries are now seismically retrofitted, if their use continues.

Masonry units can be made from clay (such as blocks or brick veneers), concrete (such as Concrete Masonry Unit (CMU) walls from normal or lightweight aggregate) or stone (manufactured or cut and dimensioned from natural). In this section from now on, we mainly mean masonry blocks such as CMU walls, unless noted otherwise. For brick veneer used at facades, see Chapter 2 > Facade section.

The advantages masonry construction include:

- The design and construction is relatively more straightforward and flexible, and can accommodate last minute changes easily. Submittal production or review is less of an issue than structural steel.
- Masonry blocks have reasonably high compressive strength, almost in the range of concrete.
- Facade can be of veneer, CMU or stone. The CMU blocks can also be made as colored, if used for facade. In general, choosing CMU, stone or brick masonry as facade or outside wall, almost always achieves good, elegant looks.
- Masonry blocks have excellent fire resistance and can be used as fire walls.
- Masonry blocks are extremely durable, especially if natural stone. They are resistant to weather and need low to no maintenance. Historical structures that still stand for hundreds or thousands of years are made of stone. This means, when we build a masonry structure, we will know that it will have long life span and the resale value will be much higher than light framed structures in the long term.
- Using masonry can be energy efficient for buildings, as masonry has high thermal mass, which means, it can absorb a lot of heat, and release it slow. With proper design, masonry buildings tend to stay cool in summer and warm in winter. In addition to high thermal mass, masonry has good insulating value. Designed correctly in the right environment, this can mean a lot of savings in the long run. Except in a hot and humid climate, using masonry construction can be energy efficient, if proper design practices are applied, which also considers other factors such as windows, location and orientation of building, insulation, tree shading from outside...
- Masonry provides good sound insulation.

- Masonry is resistant to harming effects from water. A masonry block just absorbs the water and slowly releases it back without losing strength or durability.
- Masonry structures are heavy, therefore they are not sensitive to wind and tornado forces as light wood or steel framed structures. During earthquakes, although the added mass means more burden on the structure, masonry walls can also dissipate larger amount of energy than light framed buildings.

In this building, insulating masonry blocks are used for both exterior and interior. The structure is a moment frame building with concrete shear walls. These thick masonry blocks everywhere in the building (even if not intended as shear walls) will still help lateral force resisting system during a seismic event but at the same time considerably increase the building mass.

Various Structural Considerations

Structural System and Configuration:

Using masonry walls also as structural walls, instead of just non load bearing walls, will reduce the burden on structural frame (whether steel or concrete). Masonry walls are strong in their plane and can resist earthquake forces well when the lateral force is in the plane of the wall, therefore, in addition to carrying vertical loads, they can be used as shear walls in different configurations, similar to concrete walls. Therefore the walls that will be used as structural walls must be adequately arranged to resist the loads in their plane, to meet the stiffness and strength criteria both globally and locally for each wall. The design should start with preliminary proposed masonry wall configurations (and therefore stiffness and strength distribution with respect to demand) and continue by refining the design until the final wall configuration is acceptable.

Masonry walls can be used as a primary lateral force resisting system of a building, or used as a hybrid system with steel or reinforced concrete frame, to contribute to the strength and stiffness of the structure, and provide extra load paths, by properly orienting the walls so that the stiffness balance of the structure will be maintained. This applies to unreinforced walls too for compressive stresses, which is something CMU blocks are good at. When used as a hybrid system, lower burden on structural frame will enable us to make it lighter, but the masonry and frame working together will also mean more complex design for the frame and the walls, different load transfer mechanisms between the masonry and the frame, therefore a cost benefit analysis must be performed. With masonry it is also possible to create stiff box configurations.

Infill masonry walls (walls that fill an opening of the structural frame) influence lateral stiffness of the structure by behaving as compression struts, which must be accounted for during structural modeling and analysis. For earthquakes where the displacements are small, this can be a beneficial addition, but for stronger earthquakes when the expected displacements are large, and if these walls suddenly fail, this can mean sudden added burden and unexpected behavior / irregularity on the structural frame. At the same time it also means that the wall dissipated some of the earthquake energy. The relative stiffnesses of the walls and frame is relevant here. If the wall is too strong with respect to frame then the wall can remain undamaged while the frame may be damaged (such as when the frame was not well/efficiently designed or constructed). Partial height walls can also cause short column effects. At this point you may want to check Chapter 2 > Building Frame and Structural Members > Building Irregularities section, and try to visualize which irregularities could also be caused by masonry infill walls. Criteria such as: maximum lateral floor deflection limits, requirements to account for forces that act on the columns by these walls, more strict rebar details, usage of isolation joints between the infill walls and the frame are imposed by codes to reduce such risks.

Masonry walls can be arranged in various configurations, such as a line or T, U, C, I (with flanges) or L shapes. Depending on the load paths in the building and the desired behavior, these can be made to act together or separated by joints.

When a masonry wall is used as shear wall, it can mainly work in flexure or shear. As in Reinforced Concrete (RC) construction, shear failure, which is brittle, is not preferred, and flexural failure, which is ductile, is preferred for masonry construction as well, as much as the architectural design allows us, although this is not mandatory. Flexural failure is preferred because in that case the failure is by the plastic yield of rebar, which gives warning and absorbs a lot of energy while deforming considerably and can withstand cyclic reversing earthquake loads much better, while shear failure is brittle and occurs as a sudden diagonal crack on the wall. Walls governed by shear also attract higher seismic lateral forces than flexure dominated walls, and this affects the other members in the entire structure by increasing demands on them (which they may really need to overcome if shear dominated walls are inevitable). Therefore, a flexural dominated wall will have higher R factor (see Earthquake section) to allow us to divide the design base shear by a higher number, which will reduce the demand forces on also other members in the structure. Here of course we must not forget that this comes at the expense of greater deformations, but for a masonry wall building this criteria is not often critical. A wall will have both flexural and shear capacities, but one of these modes will govern. Often having shear dominated masonry walls is unavoidable because of required geometries in architectural plans. The macro conditions to fail in flexure rather than shear are the same for masonry and a RC construction, such as:

- Aspect (height / length) ratio of a wall. A shear wall that is too short with respect to its length will fail mainly from shear. In other words, a wall must be "tall enough" with respect to its length to increase the likelihood of flexural failure.
- Another criteria is axial load on the wall. As the compressive load on the wall decreases, the likelihood of flexural failure will increase.
- The rebar details as far as the percentages, diameters, and how it is distributed throughout the wall is another major criteria. With these we control the minimum and maximum strains we achieve throughout the wall, to satisfy flexure criteria. In a wall with higher horizontal to vertical reinforcing ratio, flexural failure is more likely.

- The wall cross section itself such as its thickness and whether partially grouted or not. As the wall thickness increases, and when we fully grout voids instead of partially, the wall's shear strength increases which makes flexural failure more likely.

These criteria all affect each other, and depend on what level of ductility we want. The criteria to ensure flexural failure is clarified in codes, such as TMS 402/602 in the US, OR Eurocode 6 in Europe. Similar to concrete shear walls, we can make masonry shear walls as ordinary, intermediate and special as per codes such as TMS 402 and ASCE 7, where the ductility criteria is the strictest and R factor *(see earthquake section)* is highest for special walls, which we ensure by proper detailing of reinforcement.

We mentioned M-N interaction diagram under Structural Engineering > Structural terms section. This plays important role in masonry shear wall design by allowing us to remain within allowable combinations of axial load and moment, thus design bending reinforcement accordingly.

We can also make tall masonry shear walls as coupled walls (See Chapter 2, Figure Buildings - 4), with similar logic to RC shear walls. Again, the coupling behavior may be achieved by beams or slabs.

If an infill wall is intended as nonstructural, it should have gap with frame (filled by flexible material), to avoid unaccounted behavior OF structure. The attachment can still resist out of plane movement.

Out of Plane Loads:

Although masonry walls are strong in their plane, they are weak against out of plane loads. When the load is out of plane, they transfer this load to vertical framing elements, walls in perpendicular direction, and beams and diaphragms. The out of plane load can be an external load from wind or user caused loads or inertial earthquake load. The thicker the wall with respect to its height, the stronger resistance against out of plane loads. Even if attached, there is still a maximum height to thickness (slenderness) ratio specified by codes. For example per Eurocode 6, the slenderness ratio can not be more than 27, for walls loaded mainly vertically.

The wall and its reinforcing design for out of plane loads may often govern the design which may already satisfy the criteria for in plane loads - the loads which the walls mainly support when acting as a lateral force resisting system.

Reinforcing:

Load bearing masonry walls are often reinforced, which take on the tension force, adds to the strength, provides ductility and improves serviceability, similar to the purpose of rebar in concrete. The vertical reinforcing at masonry walls is for bending, also called flexural reinforcement. They ensure that the walls behave as a unit and do not break and separate into sub parts during earthquake, and also provide force transfer between the wall slab, foundation and the roof. After placing reinforcements, the interior voids of the blocks are filled with grout. Rebar is needed especially at the exterior walls, where wind and seismic loads can be significant, especially if the walls are tall.

Rebar congestion can more easily happen for masonry walls than reinforced concrete because of the limited space inside the block voids. This not only makes placement of rebar difficult to impossible but also makes it more difficult for the grout to flow in between and fill all voids.

Similar to reinforced concrete, providing confinement rebar in compression zone greatly increases the compressive capacity, but for masonry this is harder to achieve in the field due to geometrical restrictions inside the blocks. For this reason, we may make tension and compression ends of a wall out of reinforced concrete that acts together with the masonry.

Nonstructural partition walls may or may not need reinforcing as per codes, depending on the wall height, block thickness and seismicity of the area. For a load bearing wall, different criteria applies.

CMU can also be made as prestressed to enhance its strength. Prestressing makes the wall blocks press on to each other and ensures a more stiff overall. With similar logic, masonry walls which bear higher loads from above, may need less reinforcing than the ones bearing lighter loads from above because compression makes a masonry wall stronger, stiffer when behaving as a whole wall unit.

CMU blocks and bricks are very brittle. This is why, even with the reinforcement, masonry structures have small tolerances to differential settlement of the footings or story drifts during earthquakes, and they may start to crack quickly.

We can also make post tensioned walls, which can increase the ductility and out of plane load and bending moment capacity. It also reduces cracking.

Openings:

The openings on the wall such as window openings will significantly reduce masonry walls' in-plane resistance. Excessive size and number of openings in the walls, and the openings that are too close to slab connections or adjacent wall corners will prevent proper flow of force during earthquakes, and must be avoided. Too much / too large openings may deem a wall as nonstructural, which would be structural otherwise. It will also prevent the wall from supporting arch action behavior, which could have naturally developed within the wall otherwise. This is not to say that we may not have multiple openings on a wall, as long as we comply with criteria set in codes. As a result of openings, the wall's behavior will be more complex to analyse, such as, a lot of openings can cause either horizontal or vertical strips of the wall have more effect in the wall's behavior, unwanted brittle shear failures in certain sections, and the wall to behave as a coupled wall even if it was not intended during a seismic event. Walls with a lot of openings can be analyzed with FEM using nonlinear methods as long as the relative strengths of sections and areas above openings can be represented appropriately in the model.

Although steel lintels are often used over openings, masonry lintels are also possible, which, unlike a steel lintel, can also take advantage of arching action. In fact, with proper design, masonry lintels should be preferred for compatibility with the rest of the wall, and eliminating the schedule coordination issues. Prefabricated or precast lintels are also possible, although less common.

Other Design Considerations:

When movement joint is used for the wall to work separately from the frame, the wall must still have enough support for its stability during earthquakes, since the seismic acceleration can come from any direction (and not just from the plane of the wall but from perpendicular direction too, where the walls' lateral resistance will be close to nothing). The more rigid the structure, the lower requirement will be for the separation joint.

As they are made of concrete, CMU blocks shrink over time. For clay bricks this is reversed and they permanently expand due to moisture in the long term. There are also deformations due to temperature differences. These effects must be considered in the design by providing adequate vertical expansion and contraction joints in proper places and also by horizontal rebar. Through these joints, the design can be fundamentally affected, such as the aspect ratios (was discussed a couple pages back) of the walls and how the seismic forces are distributed within them.

Coordination with Other Trades:

As always, construction stages must be considered and coordinated with other trades, surroundings. These include steel or concrete frame, thermal and sound insulation, dampproofing.

Finished building is designed to be stable but construction stages must also be kept in mind, which may present unaccounted critical conditions. Stability of walls during construction will need to be ensured by either temporary braces or by coordinating progress of structural frame, slabs versus masonry wall progress, as the frame and diaphragm serve as bracing to the walls. For example we cannot keep building perimeter walls over a certain height if the slab of a steel frame building is not in place yet which would brace the walls, unless we use temporary braces, which can also be up to a point and will mean cost. The regular practice is that the frame is built first, and then the masonry walls are infilled inside and or built around the perimeter, which means safer and faster construction. Also, just like for the finished structure, for the construction stages as well, providing alternate load paths are preferable.

Mortar and Grout

Mortar and grout are important components of masonry walls. First let's list points that apply to both:

- Both are made from cement, water and sand but with different properties.
- Similar to concrete, their strength and durability is important.
- Workability during placement is critical as we place them in narrow voids of blocks.
- We want to avoid or minimize their shrinkage effects over time.
- For both, we may use admixtures.
- They may be mixed at batching plant or on site.

Mortar:

The quality of mortar mix and how it is applied (quality of workmanship), including the coating if any, also affect the overall strength and performance of a masonry wall. Masonry mortar adheres to the blocks and develops shear and some tensile resistance between them. Therefore good adhesion is essential. Similar to concrete, the mix can have various properties, such as general purpose, lightweight, thin layered, which can also be enhanced or modified by using admixtures for various purposes such as for joints, plastering, veneer base, bedding (the main mortar bed that bricks or blocks sit on) or pointing (repairing or finishing masonry mortar joints).

Grout:

Grout should not be confused with mortar. Mortar is applied by trowels between the blocks or bricks, to make them stick together. Grout however, is used to fill the core voids in hollow masonry blocks, to wrap around the reinforcement bars so that the rebar can work together with the blocks and it is protected from corrosion, and also to make the blocks work as a whole wall unit. Grout is stronger than mortar. Its strength needs to be close to the strength of CMU blocks and it is assumed to work as one with the blocks. Grout is applied by pouring it into the hollow core of the blocks. For higher workability, better flow and filling the voids completely, the slump should be at a certain minimum value (which is much higher than of concrete for instance) but this may cause too much shrinkage, so optimum must be found including using admixtures. During grout placement, blocks that allow horizontal flow of grout makes grouting easier. Grout can be fully or partially filled into the block voids. Fully grouted means all cells are filled with grout, and partial means only the cells with rebar are filled. Although obviously there are cost advantages to using partially grouted walls, fully grouted walls are stronger. By making full grout, we increase the wall's stiffness and strength and make flexural failure easier.

SITEWORK

Sitework is the general name given to a set of activities that take place outdoors, around the main structure itself, or, some projects can entirely be sitework such as construction of a new public park.

Sitework can include items such as:

- Earthwork such as excavation, backfilling, grading, compaction
- Underground utilities such as water supply lines, storm and sewer drain systems, electrical, fiber optic utility conduits
- Drainage components
- Landscaping
- Site electrical and lighting
- Asphalt or concrete paving
- Site concrete such as sidewalks, pedestrian ways
- Curbs, channels
- Striping and signage of parking and other areas as applicable
- Stabilized slopes, retaining structures
- Water fountains, pools
- Site equipment such as fences, gates, bollards, bike racks and other miscellaneous site features

Site coordination is a challenging task which must be well planned and executed, with continuous communication between parties, to avoid delays and extra costs. All site activities must be coordinated not only between each other, but if the sitework takes place around construction of structure(s), with those activities too, while also considering items listed below. The coordination is an ongoing, dynamic process, because as the job progresses different locations will be complete, and the work locations and materials needed by the trades will change.

A public park project which includes water features, landscaping, underground utilities, pedestrian walkways, slope stabilization, retaining structures, site lighting, sports tracks, fences, various site features, all of which were coordinated with each other during construction.

So, considering the activities above, the following items as applicable, which fall under construction execution plan, will continuously need to be coordinated, between the general contractor and all trades as the job progresses for both time and location:

- Site traffic plan, temporary access roads, ramps, walkways and parking
- Crane locations and pads, and their radii of operation with respect to existing or new structures, which is called the "rigging plan"
- Equipment and material deliveries and staging areas
- Underground utility locations and progress as they will cross access roads. Especially when the pipes reach near structures, heavy coordination is needed between the trenching and pipe laying operations and other trades that may work on the structure.
- Dewatering
- Location of temporary facilities such as jobsite trailers
- Excavation and backfill and compaction locations, stockpile management
- Scaffolding locations
- Waste disposal locations
- Heavy equipment travel routes which may require temporary structures over installed utilities
- Noise reduction, dust control, site access measures and coordination with neighboring facilities

As in any type of construction, also in sitework, when awarding subcontracts, the work scope must be coordinated. For example, who pours the footings of perimeter fencing; fence subcontractor or the site concrete subcontractor? Who makes the excavation for underground utilities; earthwork subcontractor or utility subcontractor? Who makes the connection of building utilities to site utilities? and so on...

Some of the other common coordination items to pay attention during sitework are also listed under earthwork, underground utilities and pile foundations sections of this chapter, which may be good to review at this point, for a better overall picture in mind.

SURVEYING

Surveying is the measuring of geometrical aspects and performing layout of a construction project. Due to importance of dimensional accuracy, to have a competent surveyor or surveying team on board is of utmost importance for any construction project. The surveying scope given in this section also depend on the project type, size and location. The ones listed here are general, regardless of the project type. The surveyor's tasks before the design of the project are performing measurements on site, locate the site and its main components, and based on these, prepare the site surveying drawings. These drawings typically include items such as:

- Limits of adjacent lots, property line, project site and the proposed project component footprints located with respect to official benchmarks, existing features.
- Established control points by the surveyor for horizontal and vertical control. GPS data, which is used often now, is also calibrated with the control points.
- Gridlines to be used for the project.
- Elevations of natural ground such as contour lines (lines connecting points of equal elevation)
- Existing features on site, such as existing buildings, streets, fire hydrants, monuments, fences, manholes, sidewalks, light poles, etc...
- The true North and the project North and the angle between them, if any. Usually the project North is taken differently than the true North, aligned with one of the main directions of the project components, for convenience during design and construction.
- Drawing scale

The surveyor's tasks after the design of the project, and during construction include:

- Mark benchmarks, set control lines, stakes, measure and set distances, heights, angles, grades, elevations, slopes, curves in order to locate and measure existing, and or layout new items.
- Examples of items to be laid out and located include: Excavations, structure footprints, footings, piles, utility pipe runs and manholes, culverts, retaining walls, bridge substructures and superstructures, piers, abutments, stabilized slopes, construction joints, steel base plates for columns, concrete formwork, road centerline and cross section, subgrades, shoulders,

superelevation (see Chapter 4), curbs, sidewalks, clearing and grubbing limits, site features and more... These are all tied to grids and horizontal and vertical control points for the site.

- Perform as-built measurements of the installed items such as the ones above as applicable, continuously update the records with the measured data, make calculations as necessary and produce as-built drawings, maps, cross sections, diagrams as applicable.
- Perform calculations for distance, area and volume to calculate quantities of completed portions of various work scope items to date. For example in projects involving mass earthwork, surveyors data is periodically needed for billing purposes.
- Stake markings should be removed and replaced as necessary if they start to interfere with progressing work or get destroyed by construction activities. The survey markings should be at locations that cannot easily be disturbed by construction activities later.

Considering the scope above, it is clear that the surveying crew needs to be on site from the beginning until the very last stages.

Surveying measurements are traditionally done with equipment called "theodolite", tacheometer (a type of theodolite) and measuring tapes. In the last few decades or so, theodolites were gradually replaced to a great extent by "total stations" which can roughly be defined as the electronic version of theodolites, which also include electronic distance measurement (EDM) system. Total stations allow easier, quicker measurements, can also make calculations, store and transmit data to external applications such as CAD programs.

Another branch of surveying is photogrammetry, which involves taking aerial photos, from which, not only measurements can be made but also 3D views of the area can be obtained. It is used as applicable, as a complementary method to classical field surveying. For example, for unsafe or difficult to access locations, photogrammetry can be very useful. Photographs provide a broad view of everything in the area, including terrain info, artificial features, which are also measurable. Photogrammetry is used in many fields other than construction as well, such as city planning, military, agriculture, forestry, geology, navigation, utilities and infrastructure, traffic studies, disaster planning and more.

In our information age, surveying goes beyond just measuring things. A frequently included feature in today's surveying systems is the GIS data. GIS stands for Geographic Information System. Through this, a lot more layers of data can be accompanied with the bare surveying data, to have a more complete picture about all aspects of the project site. GIS is also used in so many fields, such as the ones counted in the previous paragraph for photogrammetry.

Global Positioning System (GPS), helps locate points on earth's surface by receiving information from a minimum of three satellites. Since its launch in the United States in 1970s, the coverage and precision has always improved and the costs decreased. Apart from countless other applications, GPS data can also be used in surveying such as determining coordinates.

Even with today's complex systems and technology, a surveyor will often check his or her work though the use of simple geometry. For example even if a large rectangle was laid out in the field, it still makes sense to just measure the diagonal and compare its square to the sum of squares of the two sides.

UNDERGROUND UTILITIES

Storm and sewer drain lines, municipal water lines, electrical lines, fiber optic lines, gas lines can all be grouped as underground utilities. Transmission pipelines which are large scale construction projects are separated from this definition for our purposes here and will be discussed separately in Chapter 11. Storm and sanitary sewer and water distribution systems will be discussed in more detail in Chapter 7. In this section, underground utilities are meant in general, and only from construction point of view.

Construction Coordination

The route of underground utilities intersect with almost anything else on site, therefore coordinating this work with other tasks is the very critical in sitework coordination. They should either be completed earlier than other work, which is the ideal situation, but if this cannot be achieved, their gradual installation sequence must be carefully coordinated with other trades. Their timely completion of submittals, approvals, material fabrication and deliveries are of utmost importance. For coordination with other activities, the width of the utility lines should be taken as the whole trench plus the stockpiled earth material plus even the moving equipment around the trench as necessary.

Before digging for any utilities, the project team must make sure that there are no existing lines that are in use at that location. The location of existing underground utilities needs to be obtained as early as possible from local jurisdiction. The existing utilities must be exposed with hand tools without any sharp edges, to avoid strong impact of construction equipment. The impact on an existing utility line from a construction excavator can cause damage not only at the location of impact, but also at connections because of transmitted vibrations. After the location of the utilities are field verified, this must also be compared with the drawings for those existing utilities to make sure that both field verified lines and the drawings are not missing anything.

The elevation of the trench bottom must be verified before laying the pipes. If the manholes are the correct size and placed on correct elevations, the end points of the trench should be at correct elevations, but regardless of that, after the trench excavation is complete, the elevations must be checked along the trench at its ends and every certain interval. After placement of pipes and backfilling of the trench, the discovery of wrong elevations can be disastrous for the project budget and schedule. Especially if there are design updates during the course of the project, the project team must keep track of these changes and effectively communicate the final design elevations and layouts to the field.

On sites where heavy equipment will travel over utilities, temporary ramps to protect utilities may be required, which must be coordinated with other tasks.

A relatively new technology, Horizontal Directional Drilling (HDD) may be used on some projects where the utility pipe is advanced horizontally underground, similar to tunnel digging, without having to open a trench. This eliminates a lot of coordination issues or solves problems where a trench could not be dug for a certain reason. It can also be called directional boring method. This method can save cost and time because the following items are eliminated:

- disturbing the existing surface,
- digging the trench, backfilling and compaction,
- the relatively more number of labor,
- wait of other trades, business or public activities because of and open trench.

A disadvantage of this method is the increased difficulty to avoid existing underground utilities.

TEMPORARY CONSTRUCTION

Temporary construction items are installed in order to facilitate construction of permanent items, by providing a safe, temporary working environment to workers, or by temporarily supporting the incomplete portions of the structure before the permanent structure is adequately built to take on loads. Below we mainly discuss temporary structures and in the end only list a few nonstructural items.

The main standard used in the US for loads that apply to temporary structures is SEI/ASCE 37 - Design Loads on Structures During Construction. This standard not only applies to temporary structures, but also to partially completed permanent structures during their construction phase (as per SEI/ASCE 37-02, Section 1.2). For the latter *(construction phases of permanent structures)*, each distinct and critical partially completed state of the structure will also require a temporary structure in its different stage, with different loads and even supports *(the most typical example is bridge construction phases)*.

Temporary structures is the responsibility of the contractor but they also require engineering input. Especially for temporary structures that are supposed to brace or carry the incomplete portions of the structure during construction, strong coordination is needed between the structural engineer of the project and contractor's engineer. This temporary carrying or bracing can be in various forms as needed including formwork and shoring, reshoring, braces, cables, even temporary trusses and more. The timeliness of this coordination during planning phase, before the actual construction starts is critical, especially for larger scale projects. Not only the load carrying aspects but also a well planned construction stage sequence along with temporary structure erection and dismantling plan is also made in this stage. The stage of transferring loads from temporary structure to the permanent structure after its completion must also be studied.

In addition to the temporary structure's engineering analysis and design, paying attention during their installation and dismantling stages are also critical to avoid collapses, which happen much more frequently than collapse of permanent structures. Items to pay attention include avoiding overloading (which is easier to overlook than for permanent structures), providing proper bracing, guy wiring etc..., following the required erection sequence and procedures per the applicable codes, standards and the designer of the temporary structure, and maintain structural stability *(discussed under structural engineering section)* at all stages of erection and dismantling.

Similar to permanent structure design, for temporary structures all possible loading combinations must be checked and the most unfavorable ones must be satisfied per applicable codes and standards.

Formwork

Formwork, also called "forms", is where the wet concrete is poured into and takes its final shape. Formwork design and installation takes a lot of effort. The selection of formwork system to use affects the construction schedule and cost. Types of formwork systems include the traditional wood posts supporting the deck and beams, prefabricated panelized systems, flying form systems and more. The selection depends mainly on what is built, construction schedule, the unsupported span to be passed, cost. Designing and especially erecting formwork can cost around half of the whole concrete structure. Forms must be designed and erected in as simple way as possible, because erecting them takes a lot of manpower. Also, the design should allow reusable and standard size forms to greatest extent.

In the US, the publication called ACI-347 - Guide to Formwork for Concrete is typically followed, and each project can have specific requirements on top of that.

Forms are designed for worst case, which is when the concrete is still liquid and applies all its weight without any strength of its own, which is approximately the first 20-30 minutes of concrete placement. After that, the most critical stage is over. For structural slabs, although in expedited projects with special precautions forms may be stripped as soon as in several days, which is not common, in many projects they generally stay at a minimum of 2 weeks. Earlier stripping than the design strength period is subject to EOR's approval. For all other components, form removal times can be shorter, which all depend on particular project circumstances. Too early removal of formwork under structural slabs and beams will cause sagging, and even if it may not cause immediate failure, it will affect the load carrying capacity negatively afterwards.

Forms for an elevated slab must not only withstand the weight of concrete, rebar and the forms itself, but also must be able to resist other loads, such as lateral forces before and during concrete placement, wind, equipment impacts, wet concrete impacts, live load of workers, vibrations...

Wind force is multiplied by a greater factor as the estimated construction period increases. The most critical stage for wind load (lateral or uplift) is when the forms are erected and waiting for the concrete to be poured.

The stability against lateral loads can be achieved by bracings that are placed every certain interval as support in both perpendicular directions. In addition to stability, the forms must be very stiff as well, and not bulge when wet concrete hits the forms with all its weight. Also, the joints between the panels must be secured tightly, in order to prevent wet concrete from leaking.

In Figure 1 below, the main components of a typical deck formwork is shown. Based on the expected slab loads, the thickness of the sheathing, the size and spacing of joists and stringers and finally the shores and bracings will be designed. If it was a shear wall or column formwork for instance, the wet concrete's lateral pressure (since it will behave as liquid) must be found and the sheathing (vertical in that case) and its supporting wales, ties and studs (not shown here) would be designed. The sheathing material is more often plywood than wood.

Forms must be oiled inside, in order to remove them easier later. Different textures of concrete can be obtained with forms, if the final concrete surface is supposed to be exposed, which can be an architectural preference and can save money on finishes, while still maintaining aesthetics. Before the pour, inside of the forms must be inspected so that they are clean of any debris, dirt, or free standing water. Inside of forms must <u>not</u> be made of moisture absorbent material that would absorb the moisture of fresh concrete, as that moisture is necessary for cement hydration. During inspection, greater attention must be paid to areas with complex geometry and or not straightforward, in order to make sure that the forms are built correctly.

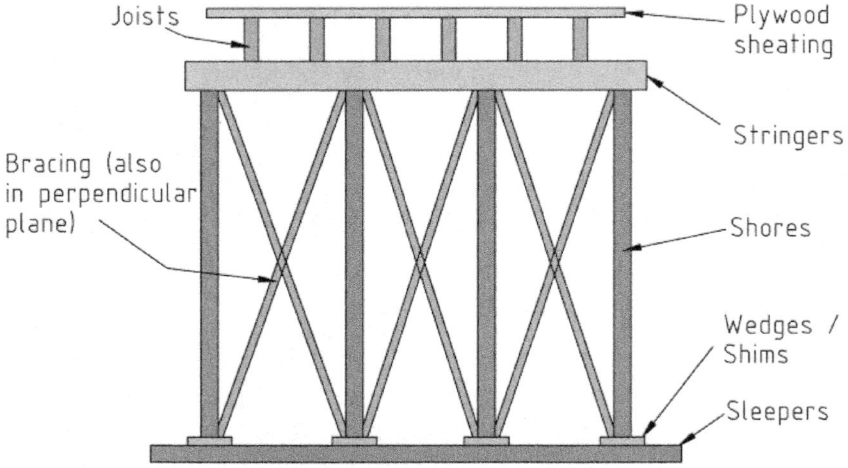

Figure - Formwork 1

How the concrete pour is performed is also important. Concrete should not be poured from too high and the vertical placement rate should not exceed a certain limit, to allow at least the previously placed concrete to start hardening and not to exert the full liquid pressure to the forms for the entire height. Liquids exert pressure always perpendicularly to the vessel they are contained in. At the bottom, where the wet concrete height is the most, there is the most pressure. No matter what we calculate for this value, there are also minimum pressure values that must be used per code. Therefore it is necessary to restrain the column or wall formwork tightly not only at their bottom, but also at every certain interval along its height as well, by attaching clamps all around it, to laterally wrap it.

Traditional type formwork for a RC apartment building of several stories. The uppermost floor is waiting to be poured. The floor below was poured but forms and shores were not yet removed. One floor below has several reshores at edges where sagging is most critical, but apparently sufficient time has elapsed

from pour that most interior reshores were removed. Pay attention to the distinction of terms "shore" and "reshore" here, as will be explained under its own subsection a few pages later.

A considerable portion of cost of a concrete structure is the cost of formwork. And for formwork, the largest part of cost is the cost of labor to install and remove it. Therefore, during structural design phase, in addition to the main goal which is to design an efficient structure, the formwork installation phase cannot be disregarded otherwise this will easily negate the savings from even the best design. Therefore, at the expense of slightly adjusting the structural design (i.e. adjusting geometry, rebar details or concrete strengths at various locations), as long as code and architectural requirements are met, greater savings can be achieved on some projects by making the formwork design the most efficient. For example, keeping the variation of column and beam cross sections, shear wall pilasters, beam drops below slab, openings in walls and slabs to a minimum and making everything as uniform and straightforward as possible will help reduce formwork labor and materials. Selecting of the optimal formwork system which will be easiest to work with, which can be reused at multiple locations to the greatest extent, and as always coordinating the efficient work of crew are of course also important.

In the photo on the left, column forms at the uppermost deck can be seen and the tower crane is lifting the form for a shear wall, which will be poured together with columns. After the columns and shear walls (the vertical elements) are all poured, formwork for the deck and beams are erected, while the newly poured vertical elements continue to gain strength. After the deck and beam are formed, its rebar will be installed. So the poured columns and shear walls, together with the shores under the deck and beam forms, must be able to withstand the dead load of deck form, rebar, slab's wet concrete plus the concrete pouring equipment and crew weight during pouring of the slab. Also see how the tower crane is supported at slab edges by discrete type of bracing. The slab edges must be able to withstand these forces as well which may require additional rebar details.

Bracing

Braces are used for both temporary or permanent structures to laterally restrain and stabilize them. From structural stability point of view, we may group braces as:

- Relative Bracing: The braces that affect the movement of adjacent members. For example on a scaffold, we can have just one vertical bay cross braced, and it can hold the other unbraced bays. The permanent vertical bracing of buildings are also similar to this type in concept.
- Discrete Bracing: The brace is provided at only certain discrete points to the element to be braced, to limit movement at those points. For example, we may attach a scaffold or a vertical

frame of a hoisting manlift to the permanent structure at various discrete locations which is an example of discrete bracing for the scaffold or manlift frame. See photo under Hoisting below.

- Continuous Bracing: The brace provides continuous uninterrupted support along the braced member's length. For example when we have concrete filled metal deck on top of steel beams, the whole deck serves a continuous brace for the beams under the deck (the deck is attached to the deck through many shear studs along the top beam flange).
- Lean-on Bracing: This is when one vertical member is braced with the support of another similar member, which are connected by horizontal members, which means the vertical members just lean on each other.

The analysis of the different bracing types as in the cases above fall under structural stability subject. Each of these cases require considerably different approach and equations for solution, which take into account the member lengths, stiffnesses, moment of inertia of the members, support types and locations, expected deflections based on geometry, as always in structural stability calculations. With braces we can support scaffolds, excavations, which makes the system far more efficient.

Scaffolding

It is a temporary structure which allows workers to climb and make installations at high elevations. Scaffolding design is part of construction execution phase so they may be designed by construction engineers as well. Large scaffolds can show quite complex behavior and not only their member or connection strengths but also the overall structural stability must be analyzed. Scaffold design can get very challenging as the size and complexity grow. Let's list some important points:

- The design should take all steps of erecting, using and dismantling of scaffold into consideration without the scaffold collapsing or moving.
- Bracings, attachments to the structure, member connections, buckling of legs are critical.
- The manufacturer provides the data for members and connections.
- The scaffold must have adequate connections with the actual structure itself, and not with the nonstructural architectural exterior facade. During this process the actual structure's design details also need to be checked to make sure that it can support the scaffold and also no part of the structure itself will be subjected to excessive loads. The scaffold should be braced to the structure preferably at every floor. Coordination of scaffold design with the structure as far as support attachments is critical whether new or existing structure renovation work.
- The self weight of the scaffold, weight of workers on it, tools and equipment and construction materials are considered to varying degrees of usage intensity and different phases of usage.
- Wind plays an important role in scaffold design. This is not only because scaffolds do not have strong foundations fixed to the ground like a permanent structure, but also they are very lightweight compared to a permanent structure, but they can encompass the whole facade area. Wind load depends on whether the scaffold has cover or not, geometry of surface behind the

scaffold, wind direction, aerodynamic coefficient of scaffold. Wind tunnel test may also be required. Even for no wind condition, a certain amount of lateral load must be considered.

- Snow load probability of occurrence increases as the scaffold stays longer.
- Hand calculations may suffice in many instances, while in more complex cases computer software including FEA solutions may be utilized.
- During work on scaffold, connections must be maintained as shown on the approved design drawings. Modifications as a result of job circumstances will require reevaluation of the connections and their attachments to the structure and the scaffold itself.
- Where the scaffold will stand, whether the ground or a lower part of the structure, must also be assessed for its adequacy to support the scaffold weight plus the live loads on it.
- The ground must provide firm support to scaffold. In some difficult cases, temporary slabs, excavation and or backfill or ground improvement may be necessary.
- Similar to formwork, scaffolds must also be inspected, including periodical inspections.
- Components must be designed to allow easy storage, erecting, maintenance and dismantling.
- For ease of erecting and dismantling, the design of connections are especially important. They must be as light and simple as possible and able to be erected with simple hand tools preferably by only one worker. Simple connections not only mean savings from cost and schedule, but also leave less room for error.
- Scaffolds can also be used to support temporary roofs.
- The scaffold components should be resistant to corrosion.
- As the building height increases, suspended scaffolding, instead of erected scaffold will start to make more sense.

Hoisting

Construction elevators, commonly called in the field as "manlifts" are used during high rise building construction, to aid in vertical transportation of workers, equipment, construction materials and tools. These are very helpful and cost effective means of drastically reducing work load of cranes (which are much more expensive to operate) and take care of a great deal of vertical transportation during construction. They can easily be installed as soon as a few building floors are completed. This is much earlier than the installation of permanent elevators. They are usually installed on the outside of the building by attachments as in a discrete type bracing as we listed above under bracing, at the edges of each floor slab, and/or also at the elevator shaft slab edges inside of the building. Until the very last stages of construction they continue to operate and dismantled when material deliveries are almost or entirely complete and permanent elevators including freight elevators are operational. After they are dismantled, permanent freight elevators take over.

Shoring and Reshoring

Freshly poured concrete needs as much as 2 hours to completely harden, but this doesn't mean it is strong yet. It needs 4 weeks to gain its full design strength. When we build multistory cast-in-place concrete buildings, this strength gain period, which affects the formwork removal date, becomes especially critical for keeping the construction schedule on track and it is almost always far too long of a period than the construction schedules can accommodate. If we want to save time and remove the forms earlier than 4 weeks, we take measures to support the slab load while it is still not at its full strength by distributing its loads to lower floors which had more time to gain strength. Through proper design and construction, we may achieve 2 weeks or even better floor pour cycles.

Shoring is straightforward to understand. It means supporting the formwork of the slab that will be poured, by posts. Let's say, we will pour the slab of 5th floor for example. We place the posts on level 4, to shore and carry the formwork and wet concrete of level 5. That is shoring.

Reshoring means, after that level 5 slab is poured, hardened and gained enough strength to carry just itself, removing the shores under it, letting the level 5 concrete slab to start carrying its own weight, in other words allowing it to deflect (sag) a little bit under its own weight, and right after that, without waiting and letting no time dependent sagging, reinserting supporting posts under level 5, but this time not too tight to cause supporting the slab again by posts, but tight barely enough so that the posts can barely stand there by friction. Now, these are called reshores. The reshores do nothing at this instant, other than to prevent time dependent sagging of level 5. They will start to be useful when additional shoring load will come from above later, when we will start shoring and forming and pouring for level 6. When that happens, that load will not be taken by level 5 slab itself, which will still not be at its full strength yet, but by these reshores right under level 5, which will transfer the load from level 6 to level 4, which had time to gain strength, and the reshores under level 4 will transfer some of the load to level 3 and so on... So floors 4 and below are now sharing this new load from level 6. This way, when a new slab (level 6) is poured over that slab we just let to carry only itself (level 5), we do not yet allow any extra load on the recently finished slab (level 5) from above, bypass it, and transfer the extra loads (from floor 6) to stronger floors below (level 4 and down) right away.

Generally the reshores may continue for at least a couple and as much as several floors below recently poured slab, to distribute its load to more slabs below. So to summarize, reshores do not carry the weight of the slab they are immediately under. They are inserted after that slab already carries itself. Reshores only carry the additional load that will come later when a new upper slab is being poured.

ACI 347.2R-05 - Guide for Shoring/Reshoring of Multistory Concrete Buildings, is the main standard that is followed in the US for shoring and reshoring. It describes the process in detail, and presents a lot of important points, a few of which can be summarized as below, which may vary between projects:

- The ground slab is assumed as rigid, if reshores continue down to this level.
- We must place shores and reshores aligned with each other from floor to floor.
- Shores are assumed as infinitely more rigid than slabs. This means, shoring and reshoring posts only transfer the deflection from one slab to the lower, without deforming themselves.
- Slabs that are connected by shores, share the load in proportion to their stiffnesses (since stiffness attracts load, as we mentioned before).

Bottom line is, we want to proceed as fast as possible with floor pours, without sacrificing strength and serviceability. Proceeding with form removal and reshoring too fast and loading fresh slabs too early may not only be unsafe but it can also cause cracking in concrete, permanent sagging and weaker slabs, even if no short term failure occurs. Waiting too much will be structurally okay but it will affect the job schedule negatively *(and therefore cost because of longer overhead duration, consequential delays etc...)*, therefore the owner, architect, engineer and the contractor must find the optimum together.

For shoring, the weight not only includes the weight of fresh concrete but also the live loads during construction, such as the weight of an entire concrete crew and all equipment they use, and of course the formwork materials. These can weigh as much as half of the slab as an additional load to fresh slab weight. The live loads during construction phase can exceed the design live load for the entire life of the structure, so, it is important to analyze and know the values and distribution of live loads during construction and formwork loads (in addition to dead load of the slab itself, which is straightforward to calculate from drawings), versus early strength gain of concrete and the strength of the lower slabs for shoring and reshoring operations. Understanding load paths and the sequence of shore removal is also important. Post tensioned slabs can present additional challenges.

Above we described the most common and straightforward meaning of shoring in general, which involved vertical upwards progress of construction of a multistory building. But shoring can be done in any way necessary to temporarily support anything else too. For example, when constructing complex shaped buildings or other structures and for some bridges, the construction progress is structurally far more complicated than just vertically advancing a multistory building. Before a complex shaped structure takes its final shape and starts to act together, the construction stages will also have their own unique set of loads and supporting requirements. In such cases, we will need to use temporary braces and shoring as necessary. These can be large items to design and build as the size of the permanent structure grows, which will also need modifications or additions as the construction progresses for each critical loading situation, which will not only account for self weight of the permanent structure but also construction loads and the weight of the temporary structure as applicable. The displacements in such structures, including differential displacements among components, will also need to be monitored as the construction progresses and necessary modifications to the shoring done as necessary.

Cofferdams

A cofferdam is a temporary structure, basically a box with open top, which is built to keep water and soil out of an excavation area, where a permanent structure such as a bridge pier in a river will be built. So the cofferdam is built in water, and then the water is pumped out from inside. This way, a dry working area is obtained to start construction work for the permanent structure.

Sheet pile walls are installed first by penetrating them into the river or lake bed, and then welding them to each other. Then the concrete is poured at the bottom of the enclosure, while there is still water, which is shown in the figure below. This concrete, which is called the sealing or tremie concrete, serves as bracing to the walls later when the water is emptied from inside *(as there will be great pressure difference between inside and outside)* and with its weight it also ensures that the empty box can resist water uplift pressure. There can also be piles installed before pouring this concrete slab. Water inside the cofferdam should not be pumped out before the foundation seal concrete below gains strength, and the upper bracings inside the walls are installed as seen below.

For everything in construction there are a lot of safety requirements, but for cofferdams these are especially more stringent and more difficult to comply, because inside cofferdams there will be workers surrounded by water. Not only all materials and structure are designed per these strict requirements, but also the performance of the structure is constantly monitored as the construction progresses.

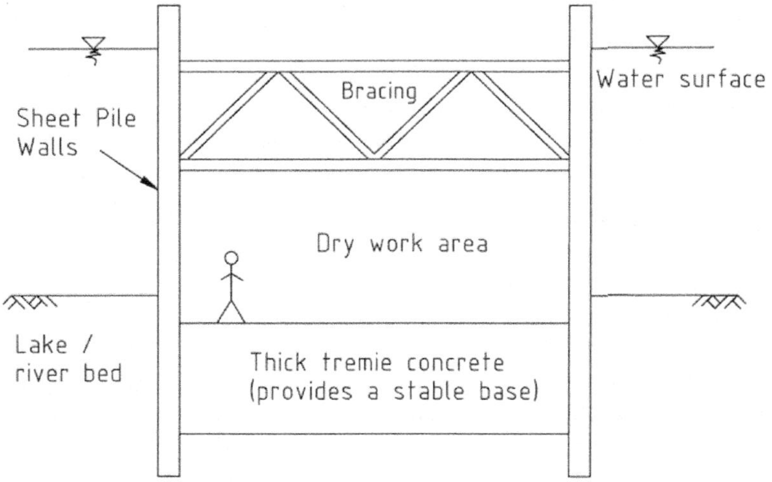

Figure - Cofferdams 1

It is much more challenging and costly to build a structure in water than on land; therefore cofferdam installers try to do as much of their work on land as possible before moving into water. The cost and difficulty of building a cofferdam increases almost exponentially, as the depth increases.

A cofferdam is a challenging structure to build. Although temporary structures such as shoring, formwork and scaffolding are designed usually by construction engineers, designing cofferdams requires soil mechanics, hydromechanics, and structural design knowledge together. Construction in water adds to the challenge, and it is difficult to keep the exact design dimensions during construction due to large dynamic water forces acting on the walls, and modifications may be needed in the field.

The forces that act on a cofferdam include the ones listed below, which are shown in the figure below:

- Hydrostatic water pressure on cofferdam walls, which is lateral, and the uplift force to the base.
- Seepage force of water from below.
- Dynamic wave and / or current forces from water.
- Lateral earth pressure from soil itself and also from the surcharge force of the weight of heavy construction equipment nearby, which is there to install the walls.
- The loads that will be exerted during permanent construction phase.
- In cold climates, ice forces.
- In case of dented sheet pile wall cross sections, in addition to the water forces described above, there can also be drag force by water. The more dented the wall surface, the larger this force.

- Impact forces that may occur during construction activities.
- Scour is not a force itself, but it negatively affects stability when it causes decrease of the support of the wall from the sea or river bed.

The figure below summarizes the forces we have just mentioned in the form of a free body diagram. Item "a" in the figure, which can be ice, wave and current forces, can not only act at the surface, but deeper too as a distributed force (still mainly horizontally, but not necessarily uniform or triangular like item c, hydrostatic force). It is shown only at the surface for clarity. Also see, how the rate of lateral load increases when we transfer from "c" to "f". This is because, the force "c" is only due to water, but "f" is due to saturated soil which has higher unit weight, and therefore exerts more intense stress.

Due to these forces some of which are dynamic, and many variables involved, a cofferdam can fail in many different ways including moving or tilting of the entire assembly or some of the walls, buckling or deflection of walls, welding failure of walls, bracing failure, water leakage, scour, failure of tremie concrete from uplift, sliding from walls, boiling of soil from below...

Figure - Cofferdams 2

In the figure, the letters represent the following:

a: water current, ice, wave forces
b: weight of tremie (base) concrete
c: hydrostatic force
d: seepage force, uplift force,reaction force from riverbed
e: bracing reaction
f: lateral earth pressure of saturated soil (soil–water)
g: lateral reaction of base concrete

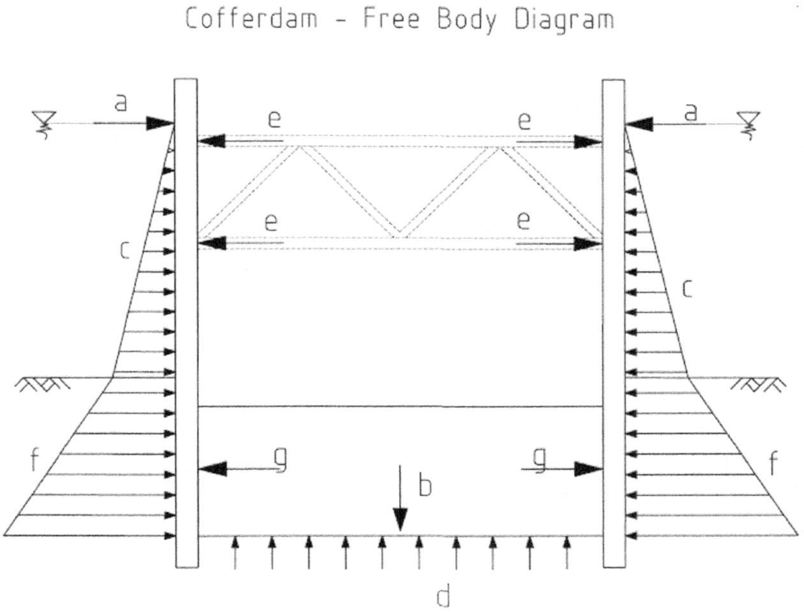

After the cofferdam is complete and the permanent construction is also finished, it is necessary to remove the cofferdam, and this also must be done with care, to prevent removal operations and water forces from damaging the permanent structure. Sometimes, the bottom of the sheet pile walls that extended below the riverbed or sea bed is simply cut off to remain in place permanently, as the removal of those may disturb and weaken the soil around the permanent foundation.

Caissons:

In pile foundations section, we described one meaning for caisson, which meant drilled shafts. The other meaning of caisson is a retaining structure or foundation, which can also allow underwater construction, and can become part of the permanent structure. Caissons can be grouped as:

- Open caissons, which do not have any bottom or top and driven into the ground by self weight, while the soil inside the caisson is removed.

- Box caissons, which can be part of permanent foundation (see example project, Chapter 3).
- Pneumatic caissons, which are open at the bottom and closed at top. Inside there is compressed air to create a dry workable condition.
- Excavated caissons

Non-Structural Temporary Construction Items

We described structural temporary construction items so far. Now, let's also list some of the nonstructural temporary construction items below:

- Setting up temporary site utilities such as temporary power, lighting, water for construction activities, jobsite trailers.
- Placing temporary fences around the construction area, build temporary access roads.
- Erosion and sediment control. Methods include placing sandbags or gravel bags around the site, placing gravel at access points to the site to prevent soil removal by vehicle tires, silt fencing, plastic fences, sediment traps, soil binders/stabilizers, covering soil surface, establishing temporary turfs and more.
- Pour slab on grade for crane, which are also called as "pad", if required

These items must be well coordinated with construction activities during the course of the project and must be clearly indicated in construction site plan for all phases of work.

With this, we finish Chapter 1, which is also the first part of this book. So far we introduced and discussed various terms and concepts in civil engineering, design, construction and structures in general. From now, while introducing different structures, we will often refer to fundamental knowledge presented in Chapter 1, which should also reinforce understanding of these concepts.

CHAPTER 2 - BUILDINGS

About buildings, one of the first places that comes to mind of many, may be Manhattan, New York City. A view of downtown Manhattan from New Jersey, across Hudson river.

Buildings are the most common type of structures therefore we start the chapters about structures by looking at buildings first. A building can broadly be defined as a structure that encloses and separates an area from the outside for an intended type of occupancy or purpose, such as office buildings, skyscrapers, warehouses, malls, residential buildings, factories, gymnasiums and so on...

Buildings are commonly identified by their number of floors and height. As of 2023 the tallest building in the world is Burj Khalifa, in Dubai, United Arab Emirates, with a height of 830 meters (2717 feet) to its tip, with 163 stories. This is also the world's tallest freestanding structure.

Buildings are primarily designed by architects per the owner's requirements as far as their overall size, shape, appearance and function *(you may refer to Chapter 1 > Civil Engineering > Role of Architects subtitle, to see a brief description of Architects' role)*. But their structure and foundation are designed by civil engineers, while coordinating with architect and other consultants. Here we will discuss the buildings from civil engineering (structural and geotechnical) viewpoint.

TYPES OF BUILDINGS BY STRUCTURAL MATERIAL

According to structural materials, buildings may be grouped as:

- Cast in Place Reinforced Concrete (RC) Buildings: RC buildings provide more mass, damping value than steel buildings and more resistant to movements. They can also have pre-tensioned or post-tensioned slabs. In this chapter we will mostly mean cast in place RC building when discussing concepts, but sometimes steel or precast buildings as well.
- Steel Frame Buildings: Due to steel's very high strength, very tall buildings are often built as steel frame. Many other low rise, commercial buildings are also of steel. Steel buildings are lighter than RC buildings for a given footprint and height, which means advantages such as less burden on the structural system and the foundations and slightly increased usable interior space area. Being lighter makes them more sensitive to wind than concrete buildings.
- Precast Concrete Buildings: These are usually up to several stories tall buildings, although taller ones are also seen. The most critical item here is designing and making the connections / joints properly between walls, slab and roof diaphragms, beams and columns.
- Masonry Buildings: Short buildings made of masonry blocks is also possible with their particular advantages. Masonry is often used together with steel or concrete frame.
- Wood Frame Buildings: For light and short buildings, wood is a popular material in many parts of the world. In the US, especially residence homes and 1-3 story shopping centers are often constructed from wood. In this book, wood material and wood construction is not discussed.

It is also possible to see buildings made of a combination of these by making them work together. Combining them lets us benefit from the advantages of each but requires more thorough coordination during design and construction. All of these materials and their design and construction aspects were discussed in Chapter 1, except wood. In this chapter we will mainly focus on reinforced concrete and discuss many more concepts about buildings and also add to what we introduced before.

TYPES OF GRAVITY / LATERAL FORCE RESISTING SYSTEMS (LFRS)

Structures carry us and anything else on them, because of their ability to support weight. When a structure collapses, it is because its ability to resist gravity loads disappeared. Then why do we discuss so much about lateral loads *(as you have seen mostly under Earthquake section in Chapter 1)*, even much more than resisting gravity loads? Because the inability to resist lateral loads will cause failures in the structural system which can also destroy its ability to support gravity loads. Therefore, gravity and lateral load resistance ability of a structure are inherently inseparable from each other. Designing for lateral loads is far more challenging task than the straightforward task of designing just to support gravity load. In addition we must remember that lateral loads are often dynamic, unlike static gravity loads. A structure must be designed in the form of a complete system as a whole, to allow uninterrupted transfer of loads by having adequate load paths for the worst possible loading conditions. All this, is achieved by using various types of structural systems. For buildings, we can group gravity - lateral force resisting systems (LFRS) mainly as:

- Moment Frame Systems
- Braced Systems
- Bearing Wall and/or Shear Wall Systems

This is the broadest type of logical grouping and may be too simple, but sufficient for an introduction. In all systems, the goal is to dissipate energy in a controlled manner during seismic events, and can show a balanced, symmetric behavior, while at the same time supporting gravity loads. The degree to achieve this goal varies with each system and its sub type.

These systems can also be divided into subcategories. In ASCE 7-16 Table 12.2-1 for instance, you can find a much more detailed categorization with subcategories of each, based on the materials used and how they are connected, along with their seismic coefficients and seismic design categories. It is a recommended exercise for the reader to take a look at that table at this point, and compare the coefficients (which were discussed in Chapter 1) and limitations of different systems with each other, which would help visualize each system better. For example, R values of moment frame systems, as seen under Group C (Moment Resisting Frame Systems), are high. This means that moment frames enable us to divide the base shear by that larger number and they are flexible systems *(see earthquake section in Chapter 1 for more on R value)*. Or for example as seen in Group A, there are both seismic category and height limitations on shear walls which work as bearing wall system, but, if we have shear walls working with moment frames as a dual system (Group D), not only all seismic categories are possible but also there is not even a height limit on most of the subsystems shown under that category, because dual systems have certain benefits as will be explained below.

Also in the same table, notice the terms "ordinary", "intermediate" and "special" listed under each system, whether it is a shear wall, moment frame, braced or dual system. See how much difference it makes as far as the coefficients and the allowable seismic categories and heights we can build to, when we detail for special vs. ordinary versions. For example with special versions of each system, we can use higher R values, which allows us to design to lower force demand values, but this comes at the expense of more ductile and deformation capacity detailing requirements. Going from ordinary towards special system, seismic energy dissipation capacity by resisting multiple load cycles without losing significant strength or stiffness increases, while the likelihood of brittle shear or other undesirable modes of failures decrease. For example for reinforced concrete structures, ACI 318 also makes the distinction of ordinary, intermediate and special, and in ACI 318-19, section R18.2, as we go from ordinary to intermediate to special, the requirements for detailing and proportioning increase to achieve increased deformation capacity. The systems below were explained by keeping in mind how a special version should perform, but many items may also apply to intermediate or even to ordinary versions.

Structural system is highly dependent on how the architect envisioned the building in the first place, after which, the engineers do its design. Thus the structural design also seeks the optimum between architectural and structural aspects and how will they blend together while architect and engineer coordinate. As always, structural aspects broadly mean the criteria for strength *(load carrying capacity)*, stability, and serviceability *(i.e. seismic or wind story drifts, slab sagging and vibrations, long term displacements... as applicable)* are addressed for the whole structure.

Other than earthquake and gravity forces, wind may also govern the design, which becomes especially more critical for taller and or lighter buildings. For tall buildings, wind may cause large story drifts and vibrations which may be structurally okay but create discomfort for occupants, therefore they must be adequately controlled by the lateral force resisting system (LFRS).

Also important to remind that when we say lateral force resisting system, we automatically mean diaphragms too, which connect all the vertical members in LFRS together and very effectively brace them at every floor level, transmit force to and from them and basically makes possible for everything to work together as a 3D frame.

Moment Frame Systems

In this system, the lateral forces are resisted by the moments created at beam column junctions and the bending of members as seen in the figure below:

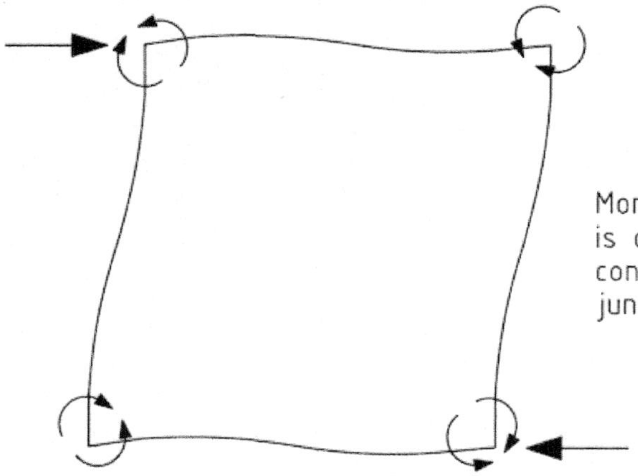

Figure - Buildings 1

Here we make the connections as strong as the members. A moment frame can take on a lot of deformation before failure. Structural steel or reinforced concrete (RC) can both be used to build moment resisting frames.

Various Aspects of Moment Frames:

Let's first list some general points for moment frames that can apply to all frames, whether RC or steel:

- The continuity of frame is of utmost importance, which must be continuous everywhere within the LFRS both horizontally and vertically. The more continuous frame, without interruptions or sudden changes or weaknesses, the better earthquake resistance. Moment frames should have as regular configurations as possible. We should have distinct frames that run parallel to each other in both perpendicular directions and the distance between frames should be equal or close. Also keep this in mind when reading building irregularities section later in this chapter.

- Strong column - weak beam concept as discussed in Chapter 1 is the main energy dissipation mechanism of moment frames where beam ends can yield but columns stand. If we have weak columns, this will produce weak stories and cause damage to concentrate to specific floors. Strong columns will also help distribution of drift more uniformly along the height of the building. Strong column - weak beam is a major criteria that can govern the design of moment frames, among other criteria such as drift, overall strength, P-Delta effects.
- As mentioned above, we want inelastic yielding mainly to happen at the ends of beams as hinging due to bending while the columns remain mainly elastic, although inelastic yielding in columns and column bases, and in case of steel frames, panel zone shear yielding are still possible. To achieve yielding at beam ends, we must first design the beam ends to their flexural capacity *(i.e. for RC beams this means rebar yield capacity is reached under bending before concrete crushes)* at the expected yield locations to yield first, thus avoiding brittle beam shear. In other words, we want the shear strength of the beam to be greater than the shear demand.
- Due to being flexible which means having large R values (high drifting, ductility, and energy dissipation), moment frames have smaller base design shear force during earthquakes, which means lower lateral demand from the foundations.
- Since moment frames are more flexible than other systems, they tolerate stresses from differential settlements generally better as far as the structure itself - but these larger deformations can damage items such as non structural infill walls, facade.
- When shaken, tall moment frame buildings generate large compression forces especially at the exterior perimeter and especially corner columns. This is because perimeter and corner columns are the furthest from center of mass, thus their eccentricity is more, so they take on more axial force than center columns which also vary more between both extremes of a shaking cycle, in comparison to center columns. Tension would usually be cancelled by the weight of the structure above but the compression from shaking will add on top of gravity compression load. Large compressive forces on these columns increase likelihood of brittle shear failure (this is also true for shear walls, see for example Chapter 1 > Masonry > Various Structural Considerations). Strength can be the governing criteria for such columns.
- The beam lengths (bay lengths) should not be too short or too long. Too short beams can force connections to take on too much shear which causes brittle failure instead of a ductile flexural failure taking place on the beam end. Too long beams will be too flexible which will necessitate larger beams (and in turn larger connections and columns) to control drift and or P-delta effects and satisfy strong column-weak beam criteria.
- When there are infill nonstructural masonry walls in a moment frame, they must be separated from the structural frame by flexible joints as needed and or regularly distributed to avoid soft story, short column or torsional irregularities as they may contribute to lateral resistance. In some cases they may work with the frame. Also see masonry section in Chapter 1 about this. Ramps in parking structures have similar effects. For example, a ramp between two or more levels may act like bracing element and take on considerable seismic force. They also must be regularly placed, to avoid irregularities or be separated by movement joints.

- Composite action of beams and slabs acting together may have considerable impact to overall lateral stiffness and strength of the structure. This statement is true whether the beam belongs to moment frame or gravity framing *(see the next list below for more on gravity framing)*.
- Structural steel column splices or RC column rebar lap splices must be located away from potential yield zones and close to inflection points where moment is zero (see Chapter 1 > Structural Engineering > Various Terms). This is true not only for columns that belong to moment frame but also gravity columns because they also undergo the same deflections as the columns in moment frame and while doing so, experience moment. Failure of column splice locations is undesirable, because it can lead to sudden, unpredictable response. They should be considered as force controlled and have sufficient capacity to resist load demand without yield.
- The behavior of lowest floor columns has great importance in determining earthquake response of the entire structure, as they also support the lateral loads from upper floors. The effect will increase from P-Delta effects. Plastic hinges can form at the bottom of these columns and walls, which can be shown by rotational springs. The rotational stiffness of these springs can be limited to a certain value, to put a limit on the resisted moment there and represent the flexibility of column to foundation connection. Significant portion of earthquake related collapses occurred as a result of lowest level columns failing.
- Wind becomes increasingly more critical as the height of a building increases. A building should respond to wind elastically, therefore we must still satisfy the strength criteria for wind, which might govern a tall structure's design. In addition, the floor accelerations can be considerably high especially at upper floors as a result of wind shaking, which must also be controlled. Note that the taller the building, the more need for dual system, instead of moment frame alone.
- Relatively more straightforward structural designs are possible with moment frames. Absence of bracing and shear walls increases usable area, which means more revenue from the building, and more flexibility in architectural design of the interior spaces and even exterior appearance.
- Although there is no rule, moment frames can be more expensive to build than shear wall or braced systems because we may need larger members and more expensive connections.

Gravity Frame and Its Interaction with Moment Frame:

A moment frame building doesn't necessarily mean all beams or columns work as part of the lateral force resisting system (LFRS). That was more common in the past. In current designs, some columns and beams may simply be there to support gravity loads only. Let's list some important items about interaction of gravity frame elements with moment frame system:

- Although resisting moments is not their duty and their connection is not designed for resisting moment but only shear, gravity frame members do have some and contribute to moment resistance. This can also help better distribution of forces to entire structure. But these contributions can be ignored as a conservative assumption *(it may be considered if a detailed analysis is wanted for precise performance level evaluation on top of collapse prevention)*.
- Gravity frame must work compatibly with moment frame. Gravity columns not part of LFRS must still be able to accommodate building sway deflections, while supporting gravity loads.

- Some members working for just gravity loads is a good thing, because in that case we avoid too much of the frame members requiring seismic detailing (not cost efficient), and also prevent moment frame system columns to carry too much gravity load, which would necessitate strength (capacity) design for them and lead to larger members and more stiff overall structure (instead of designing for displacements), which is also not economical. So, LFRS working together with gravity system where we find an efficient balance between the two is better.
- Gravity framing during earthquake can have less predictable and nonductile failure modes as well, but this is insignificant as long as the LFRS (in this case the moment frame) is doing its job.

Beam Column Joints in Moment Frames:

Moment frames function through their beam - column joints. Several important points about joints are listed below. Again, these apply to both steel and concrete moment frames.

- Moment frames, especially joints, must be structurally tough, stronger than the beams and columns connecting to it, and able to sustain repeatedly reversing seismic stresses in the inelastic range without significant deformation. The joint should also be stronger in shear.
- The beams are designed to yield in flexure first at their ends, right adjacent to beam-column junctions, to achieve strong column - weak beam behavior (in both perpendicular directions). As the stresses keep reversing during earthquake, the yield location on the beam should not significantly progress otherwise the deformations and rotations will accumulate.
- The only yielding for columns (again in flexure) may happen at their bases at foundation level because that is where the column meets even a stiffer element - the foundation *(as opposed to joining beams where columns do not yield but the beams which are weaker, yield).*
- The beams and columns should have proper cross sections to be able work together as frame, which means, both columns and beams take moment. Strong column - weak beam doesn't mean that the beams should be too weak. They should have some strength, to contribute to framing action, but still not too strong to cause column to fail first. If beams are too weak, then the columns below will mainly work as if freestanding columns tied with pin supported bar at their tops which will cause columns to take greater moments and disturb frame action.

Drift Considerations of Moment Frames:

Moment frame systems are more flexible than braced or shear wall systems, which means they tend to have larger story drifts and periods of vibration. So, their behavior is often controlled by drift, which means a moment frame must be stiff enough, while still having adequate strength *(this is why, adding even a small amount of shear walls will reduce the stiffness burden on moment frames, as in dual systems, but in this section we list items for moment frames alone).* Again, these items apply to both steel and concrete moment frames. Some of the important considerations about drift are:

- Drift is not only a result of elastic action of the structure but also inelastic deformations after yield. As mentioned under earthquake section in Chapter 1, drift considering inelasticity is obtained by multiplying the elastic drift amount with a deflection amplification factor C_d.

- The overall drift, and the drift between adjacent stories must be under a certain limit for design earthquake or wind, whichever governs.
- High story drifts are more difficult to control, which may also negatively affect nonstructural components such as facade, partition walls, finishes, operation of machinery, MEP connections.
- Large drifts can cause secondary moments and P-Delta effects, which must be considered for structural stability and may govern the design.

Items Specific to Reinforced Concrete Moment Frames:

Specific to concrete moment frames, important considerations include:

- Connections are heavily reinforced by placing more beam stirrups and column ties with smaller gaps in between, to provide better confinement. This drives the detailing and installation costs at joints up, but it is very critical for a concrete moment frame to function properly.
- Cracked sections of members should be taken into account, which can affect the forces and deflections within the structure considerably.
- Beams and columns must be poured monolithically to provide strong connection and adequate moment transfer between each other.
- Also see: Building Frame and Structural Members> Columns, Beams, and, Chapter 1>Concrete>Reinforcing Steel.

Items Specific to Structural Steel Moment Frames:

Specific to steel moment frames, important items include (also see Chapter 1 > Steel Construction > Design Considerations):

- We can use various details for column - beam connections. It can either be a connection detail prescribed by code (which were extensively tested and reviewed before and available for use by anyone who wants) and applicable to that situation, or, it can be a proprietary connection as long as it is / was demonstrated that it can qualify (which can be a long and expensive process if it was not done before or may be required anyway for that project, involving costly tests in labs and extensive reviews). So unless we have a large and or landmark project with special geometries where code prescribed connections will not work, using new types of connections are generally avoided.
- Lateral torsional buckling of moment resisting beams must be prevented. The slab above the beam provides continuous bracing to the beam upper flanges but we may also need to use separate lateral bracing elements to brace bottom flanges.
- The beam ends where plastic hinging is wanted are sensitive areas. No other trade should make unauthorized attachments to these areas, to not cause unexpected failures. The metal deck above the beam can be welded instead of using shear studs at these sensitive zones.

- 1994 Northridge earthquake in California, USA, showed engineers some unexpected fractures and behavior in steel moment frame connections. After that, the codes have been updated.
- A few notes about panel zones:

  Column web at the location where beams connect to the column flanges is called the panel zone. Depending on the detail used, the beam top and bottom flanges can be extended into the column web through the panel zone by using continuity plates, which makes the panel zone bound by column webs on the sides and continuity plates at the top and bottom.

  Panel zone performance is of primary importance for steel moment frames. Panel zone is at the center of moment transfer between columns and beams therefore it must resist large shear force which can result in failure. This failure in fact dissipates energy which reduces the burden on beams and columns framing into that joint, therefore upto a point that would not cause significant decrease in strength or stiffness, and the behavior is mainly ductile, it is acceptable.

  Because of supporting large amount of shear, panel zones must be designed to provide adequate shear capacity. In general the shear depends on the dimensions of beams and columns and applied bending moments. We can increase panel zone shear capacity by adding stiffener / doubler plates to column web. This also makes the frame more stiff overall and reduces story drifts, but of course the balance of loads with beams and welds must be watched as mentioned next. Note that instead of plates, a heavier column section may also be used which will result in a conservative frame but may end up lowering bottom line cost of the steel frame since adding plates is labor intensive. Cost benefit analysis is needed.

  The shear yield of panel zone should happen nearly at the same time with flexural yielding of beam or beams so that they balance each other and we ensure a ductile failure from beam yield. Weld capacities should also be considered. Welds should not fail unexpectedly in brittle manner, before the expected panel zone and beam yield.

Interested readers may also take a look at author's article about RC moment frames at: http://constructmagazine.com/structures/buildings/lateral-force-resisting-systems-lfrs/rc-moment-frame-structural-analysis-and-behavior-of-members/

Shear Wall / Bearing Wall Systems

In this type of system, earthquake forces are resisted by the lateral stiffness of shear walls. Shear walls generally provide the highest lateral resistance, and they are most stiff against lateral forces, among the three main systems listed here. As stated above, when working dually with moment frames, very tall buildings are possible with this system (and with outriggers, even taller ones). Shear walls can also work as bearing walls but in that case the system is mainly a gravity load resisting system. We discussed shear walls for masonry in Chapter 1 and will also do so for RC in this chapter.

Braced Systems

Cross bracings made of steel *(or steel concrete composite)* which work in compression and tension, can provide significant lateral resistance. Note that many different bracing shape configurations also exist than in the figure below. In comparison to shear walls, braces are considerably more flexible and lighter, but they allow stiffer systems than moment frames alone.

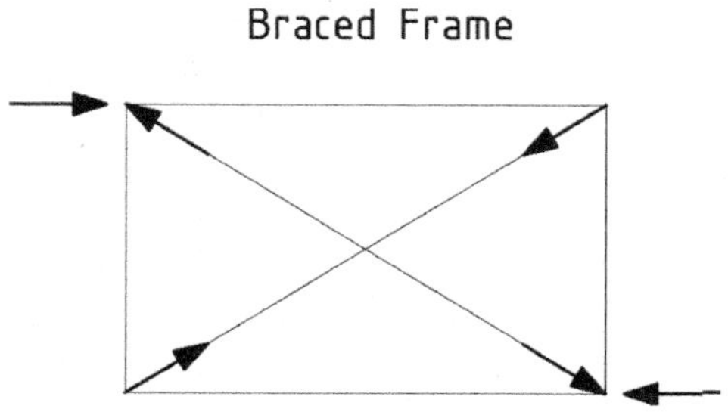

Figure - Buildings 2

The main structural frame must not yield and it must remain elastic as the braces yield, so the goal here is for the braces - no matter what type of bracing we use - to yield and fail first, in other words, yield in tension or buckle / yield in compression first but protect the structural frame. So the braces serve as "fuses" for the structure. This is not to say that the structural frame will remain without deformations during this process. As the braces deform and yield, inevitably there will also be corresponding displacements in the frame. So the deformations of the braces vs. the frame's elastic action to accommodate those deformations must be compatible. The participation of frame will considerably increase after failure of braces, but upto that point a considerable amount of energy will be spent by the inelastic yielding of braces. As always, overall capacity, story drift and structural stability criteria are to be satisfied, among others. When used with moment frames, the toughness of moment frame will combine with the laterally stiff braces, which would be desirable. Braces can be made from steel double angles, tube sections.

Braced systems have advantages such as:

- They provide more flexibility to architectural design with respect to shear walls because slender members allow more space.
- By having relatively larger R values (more drifting and ductility and inelastic energy dissipation) than shear walls, they allow smaller base design shear force in earthquakes with respect to shear walls, which means less lateral demand on the foundations.

Braced systems have disadvantages such as:

- Architecturally we must still accommodate space for the braces, whereas for moment frames there are no such requirement. This aspect is a little offset by our ability to use lighter beams and columns with braced frame systems, than in moment frames.
- Steel members and connection detailing and construction costs are usually higher.
- Story drifts are also difficult to control, as in moment frames.
- Braced frames can suddenly develop a high degree of torsional irregularity, when a brace on one side of the building fails, while the other(s) are still adequately resisting loads, even if the braces were identically and symmetrically built.

Several important design considerations for braced frames include:

- When brace or its connection at a particular story fails, this quickly turns that location into a weak spot. Failure concentration on a single area should be avoided, and be spread to as large area as possible, to avoid too much burden at a single location and formation of weak stories, P-Delta effects.
- Bracings and their connections must be designed to avoid brittle but allow ductile failure.
- Placing the braces at the building perimeter is the most effective in countering overall torsion of the building. Same goes for shear wall systems.

The points above are general, and should be kept in mind while reading the types of bracing systems below. You may also want to revisit Structural Steel section in Chapter 1 at this point.

There are several types of braced systems in common use, three of which are listed below. Other configurations also exist.

Special Concentric Braced Frames (SCBF):

This is a common bracing system for steel or concrete-steel composite buildings located. In this system the bracing members' center axes meet or nearly meet with each other, therefore the word concentric. In other words, the braces, gusset plate, column and beam connections are made such that the braces will only have axial load.

Above we said that for all braced systems, our goal is to make bracing system to fail first but protect the structure. In SCBF, this is achieved by axial yielding and buckling of braces for the most part, which only makes sense because as we said above, mainly axial loads exist in these braces. And we also said that special systems are made ductile with higher toleration for drifts. In SCBF this is achieved by detailing beam-column and brace-brace connections in a way to allow this. The failure should be limited to yielding in tension or buckling in compression but connection failure which can be unpredictable is not desired, therefore the connection capacity must exceed that of braces. Compression buckling is also a sudden and difficult to predict after critical load is reached as introduced in structural stability subsection in Chapter 1. This critical load generally corresponds to a lower threshold than the yield capacity of that material unless the column under compression is very thick and wide. Therefore in SCBF, which have slender bracing members, we can expect compression buckling way before tensile yielding can happen. In addition, buckling can be activated in small amount of story drifts. The negative effects of compression buckling is balanced by sizing members properly to account for this situation and by tension yielding in another brace. This balancing is the only way we can achieve more efficient, stable energy dissipation in CBF, which would be unstable (because of buckling) otherwise. There are also minimum and maximum compression limits to design for in each direction of earthquake. The balancing of tension yield vs. compression buckling is achieved by using diagonal bracings in symmetrical pairs opposing each other. Multiple configurations are possible, such as X bracing, V or inverted V bracing, multistory X bracing, diagonal bracing.

Buckling Restrained Braced Frames (BRBF):

In this system, similar to CBF described above, the members are axially aligned but here the difference is that the buckling of steel core is not allowed and the earthquake energy is absorbed by yielding of the brace member only. The restraining of buckling makes a big difference on the end result of how the braces behave and effect the building. The connections to accommodate this behavior are obviously also different than CBF. It is a newer system than the other bracing types but through its performance it gained a reasonable amount of use and now fairly well documented including code references.

The restraining of the core bracing member, which is a steel plate, is done by enclosing it in a steel tube filled with concrete, so that the center core plate cannot buckle, but because its cross section area in the middle is kept below a certain limit, it will yield instead, at that location. The core which we want to yield is unbonded with surrounding concrete for axial loads, so we ensure that we control strength based on the yield of steel core only. To make a brace of this complexity would not be possible in the field, therefore BRBF braces must be fabricated by specialized manufacturers in the shop. This may arguably be a downside in comparison to CBF, because CBFs can be made in the field by a larger number of familiar crews, however BRBFs have important advantages, which include:

- By not allowing buckling, we can obtain higher capacity from the brace, as the yield strength for a slender member such as a brace is higher than critical buckling load. We can obtain a fairly close compressive and tensile yielding strengths which also means symmetric behavior both in tension and compression. So there is no excess, "wasted" tensile strength as in the case of CBF.
- As there is no buckling, the unfavorable effects of buckling is eliminated. Instead we have a much better controlled axial compression or tension yielding, which means a balanced, more efficient hysteretic response under cyclic lateral loading. Because of the symmetric, controlled behavior under cyclic loading, we may also get the benefit of strain hardening (the increased brittleness from strain hardening would not be significant in comparison to the added strength because until reaching that yield range, a good amount of energy will already be spent in a ductile manner in multiple load cycles).
- Yielding is more ductile behavior than buckling therefore in comparison to CBF we have higher ductility, energy dissipation and inelastic strain capacity which means the same advantages associated with it which are mentioned in various parts of the book.
- The advantages listed above enable us to design the bracing connections, welds and the overall system more efficiently which may not only result in cost and schedule savings but also enable more flexibility in facade and architectural design and coordination.
- Damaged braces can easily be replaced after an earthquake.

Eccentrically Braced Frames (EBF):

In eccentrically braced frames, unlike CBF, the bracing axes do not meet but there is some eccentricity distance in between (at least about a foot, upto several feet) the braces or a brace and the adjacent column. This linking portion is called the link beam or link element which works by bending moment. Let's list some important points about EBF:

- EBF is a newer system than CBF (but still several decades of common use).

- The fact that we can have eccentricity for braces, allows more flexibility in architectural design in comparison to CBF.
- The link element is critical for EBF performance. During earthquake it tends to rotate which forces it to yield in moment or shear or mix of both, (depending on the link length), while plastic hinges emerge, acting as a fuse by ductile behavior, dissipating energy. This is a better controlled behavior than buckling of CBF and the behavior under cyclic loads is more regular. While the link beams yield, the other members of the system which are the braces, columns and the beams other than the link beams are designed to remain in their elastic zone.

Side Note: Think of the linking element as a line or stick that connects two triangles from their corners, or a triangle to a post. Under lateral forces, the triangles will be stable in themselves and remain in their elastic zone while the stick will be much more prone to rotating and bending.

- Because of the effect of linking beams, EBF are more ductile, flexible than CBF, giving a building a higher R value. This also means larger story drifts, longer building period and higher P-Δ effect potential but lower demand on the structure (lighter members) and foundations (reduced base shear). EBF systems can act as a mixture of moment frame and braced system taking advantage of the flexibility and ductility of moment frame, and the stiffness of the braced system.

Dual Systems

Some buildings have more than one of the structural systems, which creates redundancy and also help reduce demands for one another, therefore desirable from structural engineering point, although it may not be the most cost effective. Each system takes on the load in proportion to its stiffness and the codes regulate the portion each system will carry, to make sure both systems work effectively.

When shear wall is used together with moment frame system for example, the building can enjoy shear wall's high stiffness, and at the same time the large ductility and energy dissipating capacity of moment frame as long as each system takes proper amount of load. Add to it the redundant/alternate load paths that both systems create.

Redundancy can also be created even if only one system is used, by having more than one load paths through more than enough members in the system, but having two different systems working together is still preferable. In nonredundant systems, the analysis must carefully consider the effects of failure of each member or connection would have on the structural stability of the entire structure.

For very tall buildings, outrigger systems (see below) are added at the perimeter to be used together with a main shear resisting core, which significantly increases the lateral resistance of a building.

Outriggers

These systems are for significantly improving overall lateral resistance of especially tall buildings, by connecting the lateral resistance system such as shear walls at the core of the building to columns /

lateral resistance system at the outer sections of the building which enables all members contribute to the system effectively and work together, which provides a lot more lateral resistance than their total, if they worked separately. This also means, both inner and outer sections of LFRS can be made lighter. Outrigger systems also help lowering lateral drift and reduces moment at the building base. In buildings with reducing floor area at every certain height interval, which is common for very tall buildings, the force transfers between each section can also be through outriggers. Outriggers must be strong and therefore made of very deep members, which is more suitable to be placed in floors such as mechanical floors that contain mechanical equipment. Outriggers also help remedy differential shortening of columns by enabling load redistribution between inner core and outer columns.

BUILDING FRAME AND STRUCTURAL MEMBERS

Now let's list building structural elements by highlighting some important points about each. We assume cast in place reinforced concrete building here. Codes (i.e. ACI 318, IBC, Eurocode Chapter 2 and so on...) specify the design criteria, such as allowed dimensions, length to width to height or thickness ratios, opening sizes, displacements, connections, rebar details and so on. This section will be studied under the following subsections:

Foundations - Page 376
Columns - Page 376
Beams - Page 380
Slabs - Page 381
Shear Walls - Page 392
Stairs / Ramps - Page 396
Balconies - 396
Shafts - Page 396
Building Irregularities - Page 397

Foundations

Foundations were discussed in Chapter 1 as a main section. As mentioned, they are grouped into two as shallow and deep foundations. For buildings both systems can be used. The foundations of a building should be stronger than the columns and shear walls that they carry, and the connections between the columns/shear walls and the foundations must be strong enough to transmit not only gravity but also shear forces and moments from columns/shear walls to the foundation during earthquakes.

Columns

Columns are the vertical elements that carry the building. They mainly carry axial gravity loads, but during earthquakes and high winds, they not only continue to resist gravity loads but also lateral forces and a lot higher moments than normal too.

We mentioned columns many times before in the book within context. Some of them are mentioned again along with additional important points below:

- The column locations and preliminary cross sectional dimensions are predetermined in architectural drawings but as a result of structural analysis, there may be slight modifications.
- Column should be adequate to resist all loads, satisfy criteria such as strong column weak beam, deflection, buckling, P-Delta effects and accommodate the required rebar.
- One of the primary things to check about column and shear wall sizes and locations is if there will be torsion of the whole structure during earthquake. Columns and shear walls resist lateral forces together. The entire lateral resistance resultant force provided by columns and shear walls must be at or very close to the center of mass of the floor area, (because the inertial earthquake force is considered to act at the center of mass of the floor), in order to prevent torsional moment to twist the structure as a whole.
- All columns should bear on foundations directly. Columns which bear on beams create a high degree of irregularity. Also see building irregularities subsection later.
- Short column effect must be avoided. See various structural terms section in Chapter 1.
- To satisfy strong column weak beam condition, the total moment carrying capacity of columns at a column beam junction has to be x many times (i.e 1.2 times) stronger than total moment carrying capacity of beams at that junction. Increasing the column cross sectional area will make the column stronger, but at the same time this will cause the column to take more lateral load, which must be coordinated with lateral resisting system, to find optimum.
- Earthquake energy should be dissipated at maximum number of plastic regions as possible. This is why we want the yielding at the beam ends instead of column ends. See below.

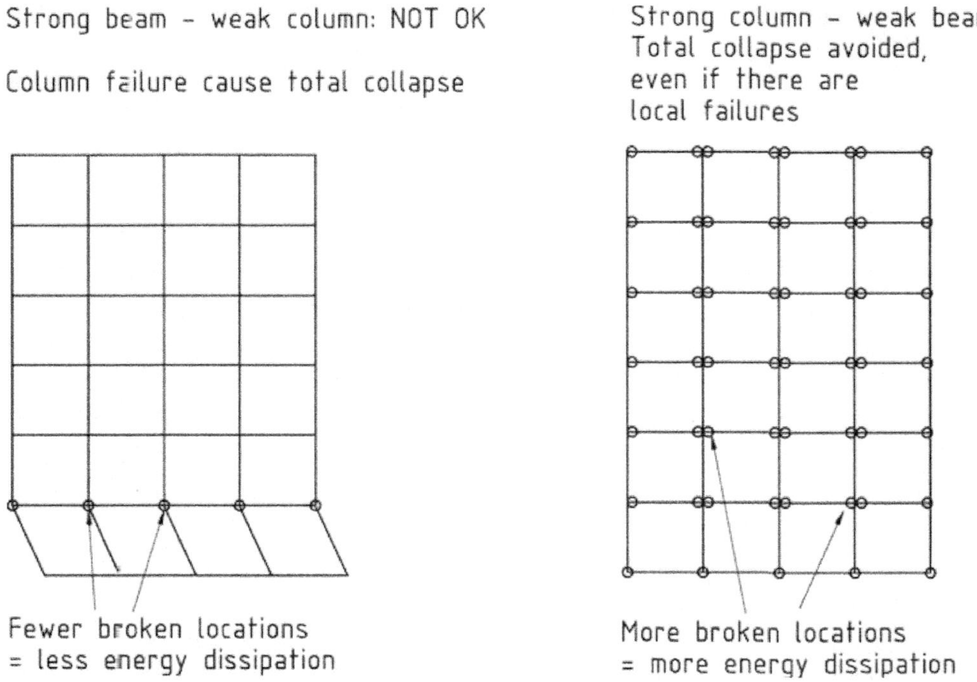

Figure - Buildings 3

In the figure above, the building on the left has strong beams and weak columns. Not only column failure is worse than beam failure, but it also causes fewer locations to yield, which means less energy consumption. It causes weak floors, concentrating damage to those. The building on the right has strong columns and weak beams, and as columns remained intact, drift is much more evenly distributed over the floors, and the building stands, even if the beams at each floor may be damaged. Also by allowing more number of locations to break, much more earthquake energy was consumed. During earthquake, the building on the right will also sway to the sides, and it will be increasingly more flexible if beams fail progressively, but it will not fail as the one on the left. In reality we can still have some damage in columns as long as they do not cause significant strength loss.

- Certain live loads can be reduced, with the logic that when the influence area of a member increases, the possibility of all live loads will exist on that member at the same time decreases. For example an interior column which can carry loads from all around it, has lower probability to experience all live loads from everywhere around it at the same time. But a slab will certainly experience the live load that it was designed for, so the live load on interior column will be reduced more than reducing it for the slab. This is simply reducing possible overdesign.

- Where columns meet the foundations take the greatest shear and moment as a result of lateral earthquake loads. In the figure above on the right, also notice that we showed hinges at columns only at foundation level. This is because foundations are stiffer than columns but beams are not since we design for strong column weak beam. We mentioned this when discussing moment frames, under lateral load resisting systems.

- When two or more columns are combined on a footing, their total moment resistance will be much more than the sum of the individual columns.

- The reinforcing bars should continue up past the floor level, spliced at mid height between floors, instead of terminating and splicing at column beam joints *(same is true for beam rebar which must also be continuous through beam column joint unless it is edge beam)*. This also makes construction easier when tying rebar and erecting upper floor forms.

- The column shear capacity must be adequate to resist lateral earthquake forces. We need to avoid shear failures which are brittle. This can be ensured by providing adequate column cross section area, using adequate grade of concrete, and with proper rebar detail. Under lateral loading, the highest load demand on a column will be at its top and bottom ends. And the shear capacity of the column will depend on the moment capacity of it at these top and bottom ends. As the column gets shorter for a certain cross section area, shear becomes more critical.

- Columns are members under compression and anything that is compressed, shortens. The taller a structure, the more load and the more shortening, for a given column stiffness. No matter how strong concrete we specify for columns - *and as a building gets taller we do specify stronger concrete for columns* - there will still be shortening. The effect accumulates as the building height increases (also add to this the effects of creep and shrinkage).

  When all columns and shear walls shorten the same amount, the process is easier to manage. But when they shorten different amounts, we now have a problem called "differential column shortening". This may lead to undue stresses on other members of the structure including beams, slabs and outriggers if any, and nonstructural items such as partition or facade walls, ceilings, floors, MEP items. The height of columns and shear walls on a floor are equal. Therefore, from $\sigma = E \cdot \varepsilon$, the difference in strain of columns of a story can only be a result of difference in stress (σ), or elasticity (E), but not original height. In other words, for vertical members, we must keep these two things, σ and E, the same, or close enough across members. To make stress in vertical members close to each other, we must make the layout such that some vertical members will not take on too much stress, in other words, we must uniformly distribute stresses between them. This can be done by avoiding placement of columns too close to shear walls since columns will take on higher stress as their cross section area is smaller, by not making column sizes too different from each other and having adequate load paths for uniform load distribution, including outriggers. Also keep in mind that interior columns will take on higher compressive stresses than perimeter columns *(as far as the permanent gravity stresses - during earthquake this changes, but that is irrelevant to long term differential shortening we are discussing)*. To keep the modulus of elasticity (E) balanced among members, we need to make sure that the same design concrete mix is poured for all members. Shortening of columns at lower stories will be more than higher floors and the cumulative shortening can be as much as several inches for tall buildings. For concrete core buildings, the perimeter settling may be more than the core. For composite buildings of steel and concrete, analysis of differential shortening is more complex as two materials shorten differently.

- In addition to differential and overall column shortening, which happens in short term as elastic shortening, there are also the long term effects of shrinkage and creep, which can also add up for columns and walls. The overall shortening effect of these affect the structure itself, load paths, facade, elevator guide rails, interior partitions, floor levelness, MEP supports, pipes and conduits, slab and other components... These must be taken into account for construction sequencing, and for finished structure, for tall buildings.

- Although we use rectangular or square columns more often than circular columns due to ease of construction and fitting easier architecturally at corners or under beams, circular columns are inherently more structurally advantageous such as:

    - They have better buckling and bending resistance, due to higher moment of inertia with respect to a rectangular or square column of same cross section area. The resistance is also the same with respect to any direction, unlike rectangular cross section.
    - They do not have corners that cause stress concentrations.
    - They have better torsional resistance.
    - As lateral rebar we can use spiraling ties, which are better than rectangular or square ties.

The structural advantages of circular cross section are not surprising, because the structural carrying systems in nature, such as plants, trees, also take benefit of circular columns. If architectural and constructability worries did not exist, the use of circular columns would provide higher strength and stiffness over rectangular or square columns for overall structural performance. The problem of not fitting circular columns under the beams can be solved by providing a flat slab (see the subsection for slabs below).

Beams

After we determine locations and preliminary sizes of columns, we can proceed with beams during modeling. The beams span from column to column as they must transfer the load they take from slabs to the columns. At this point you may also want to revisit Chapter 1 > Concrete > Reinforcing Steel section where important points about beam rebar were mentioned.

The beams are assigned preliminary sizes, usually to match the column widths that they connect to. These sizes are finalized after structural analysis. Beams will generally have positive moment in midspans and negative moment (will bend upwards) at column locations as discussed before.

Like all structural members, beams must also satisfy certain minimum dimensions per code, even if structural analysis may allow lesser dimensions. For example they must be over a certain minimum depth. The end conditions of the beam is the primary factor that affects this. For example in ACI 318-19, Table 9.3.1.1, a cantilever beam, which is most prone to bending, must have a depth of at least 1/8 of its length, while a simply supported beam can have 1/16 depth to length ratio, and a beam with both ends continuously supported *(so that also the portions outside of the beam can also contribute to its bending resistance)*, can have even lower depth/length ratio. As far as their width, it is most often limited to the column width for constructability reasons, although it may be wider.

When we make strong column weak beam connections, we must do this by making beams fail in tension (as a result of bending), because tension failure is ductile but shear failure is brittle (and beams are not likely to fail from compression). During tension failure, there will be a lot of rebar deformation, which gives us plenty of warning and consumes a lot of energy during earthquake.

The most common slab type is the one that rests on beams. Beams and slabs are poured monolithically, and they are heavily connected with rebar. Therefore, when we analyze a beam, we consider it working together with a certain strip width of the slab, as the slab will effectively increase moment of inertia thus bending resistance of the beam.

Sometimes it is unavoidable that mechanical or plumbing ducts intersect beams. These must be coordinated in the beginning between architectural, structural design and MEP. If, due to insufficient coordination or unforeseeable circumstances, penetration to beams are unavoidable, special detailing is necessary. Multiple smaller openings should be preferred over a single larger opening. Mechanical engineer must also consider options of creating more ducts with smaller sizes than a single large one.

Secondary beams connecting into primary beams without column support should be avoided, which causes complicated, indirect load distribution. It causes larger deflection, torque, bending moment, and shear in the beams. During earthquake the push from the secondary beam causes lateral load directly to the middle of the primary beam. If we cannot avoid this, we may try to place an extra column below the beams junction or eliminate the secondary beam altogether and increasing the slab thickness. If none of these work architecturally, then proper rebar details must be used.

Coupling beams (also may be called as link beams) are short beams that connect shear walls together, which increase the sum of lateral stiffness from the two walls combined by making them act together. They are very useful to consume energy during earthquakes when they yield, acting as "fuses".

There are various coupling beams. The most straightforward is the one reinforced similar to moment frame beams. There are also other types such as diagonally reinforced, steel beam encased or not encased in concrete. Strength, stiffness, ductility, energy dissipation capacity, span length and beam cross section dimensions limitations, construction difficulty *(as these walls must have sufficient attachment to work with shear walls)*, and as always cost and schedule are among the factors that affect the choice of the coupler beam type. Important to note that coupling behavior may also be achieved with slabs.

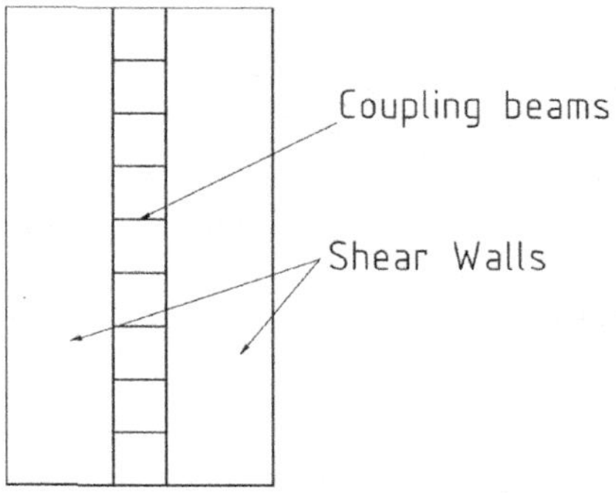

Figure - Buildings 4

Span to depth ratio of a beam is important to determine if flexure or shear will have more effect on it. Long beams are flexible and they can bend and fail from bending which is a tension failure. They can be reinforced regularly similar to what we showed in Chapter 1. Shorter beams however, do not have enough length to bend, and they fail in shear, which is brittle. For example the coupling beams are short, in other words, their span to depth ratio is small, so their behavior is governed by shear much more than a regular long beam, so we must have sufficient shear reinforcing in them, such as stirrups or diagonal reinforcing (*if you think of the cube as in Mohr's Circle, here the tension is mostly diagonal due to shear, similar to the two ends of the beam in Figure - Concrete 2 in Chapter 1, because in this case the middle portion of the beam under flexure is very short to nothing in length*). As beams' span to depth ratios increase, the efficiency of diagonal bars decrease (although stirrups are used as discussed).

During earthquakes, after rebar yields in tension under cyclic stresses, strain hardening will occur, which can give beams additional strength (though reduced ductility) and cause moments along the beams be redistributed.

For exterior beams, especially if long spans, beam bending deflection due to live loads become important for facade, which needs to be coordinated with facade system requirements *(there are also other deflections for facade such as due to seismic story drifts, temperature, shrinkage and creep of concrete, differential column shortening)*.

Slabs

The floor slabs, except slab on grade, are called as structural slabs or elevated slabs. Structural slabs not only carry their self weight and the live loads over them, but also significantly contribute to the overall integrity of a building by serving as diaphragms by connecting to vertical elements. In addition they increase capacity of beams by working with them. Below we discuss structural slabs in several subtopics. You may also want to revisit Chapter 1 > Concrete > Reinforcing Steel section where some important points about slab rebar were mentioned.

Slab Continuity and Designing for Gravity Loads:

The first thing we design slabs for is the gravity loads, which will usually be sufficient for it to work as diaphragm as well. Traditionally a slab is designed as a group of single or equivalent beams, which are commonly called as strips. The girder, beam, column, wall supports below the slab can be idealized as pins, axial or rotational springs depending on particular case. Although there are assumptions involved here, these methods provide acceptable results for relatively straightforward structures or low to medium seismic zones. As the complexity of structure and seismic force potential increase, usage of FEM *(see Computer Aided Analysis in Chapter 1)* makes more sense.

Slabs usually do not work as isolated areas enclosed by beams, but they have continuity over the beams, so they affect adjacent slabs on other sides of the beams. Continuity is better for the overall slab strength. See Figure - Concrete 4 in Chapter 1, where we showed a continuous span of slab over multiple beams, as they work together. The carrying capacity of one section of slab will be higher like this, as other pieces also contribute. So, unless there are unusual cases, such as difference in elevation, we design the slab of an entire floor as a continuous slab, by extending rebar between sections. In addition to improved dead load and live load carrying capacity, this will also ensure proper diaphragm action. So the thickness of a slab depends on factors such as the slab area, live load, the connection conditions of slab edge and the continuity of the slab over multiple spans, in order to satisfy capacity, displacement (and vibration, which is important for slabs) criteria.

The continuity of the slabs can automatically be assigned by structural analysis programs (which would be proper as this is the case almost always), or the engineer can identify noncontinuous locations specifically. If we were to design manually, we could take a typical slab strip similar to a beam under dead and live loads, and then apply that design to entire slab area as long as the conditions are similar.

Diaphragm Action, Chords and Collectors (Drag Struts), Lateral Load Design:

Slabs not only carry vertical loads over them but also contribute to the integrity of the whole structure by serving as diaphragms, which has additional and very critical duties such as:

- connecting vertical elements of lateral force resisting system (LFRS) around them to each other to make them work together by transferring lateral loads and moments between them,
- act as centerpiece of the frame to work in 3D with orthogonal planes supporting each other,
- bracing the vertical elements and prevent their buckling,
- during strong seismic events when some vertical members may be weakened, redistribute forces to stronger members
- ensure adjacent frame bays have compatible displacement.

Diaphragm action needs to be provided at each floor level of a building. We introduced diaphragms in Chapter 1, under Structural Engineering>Various Terms.

Under the action of lateral effects, axial forces, bending moments and shear forces will exist in a diaphragm. Lateral loads may mainly result from wind, earthquake, lateral soil pressure at basement levels, and the rare case of horizontal force component of any sloped columns. The total lateral load of a diaphragm includes the loads caused by these plus any transfer forces between vertical elements while performing diaphragm action. The analysis of distribution of lateral forces on a slab while working as diaphragm is more complex than analysis of it under only gravity loads.

To perform the critical duties listed above, the diaphragms must have sufficient strength and stiffness, in other words they must be rigid enough. For this they should be designed as linear elastic, unless there are reasons that nonlinearities in slab behavior may occur (i.e. too high transfer stresses, cracking of slab, relative slab stiffness to lateral stiffness of vertical elements being close, irregularities/too much openings in slab). In other words, ideally we do not want any plastic yielding in slabs, and leave that task to other elements of the LFRS.

Typical cast in place concrete slabs can usually be considered as rigid slabs and their flexibility can be neglected. This simplifies calculations. If slabs were flexible, the full transfer of lateral force would not happen between slab and vertical elements. Flexible diaphragms cannot for example distribute lateral load fully if building is under torsion, which may cause too much stress on a limited number of locations. The more rigid the diaphragm, the better force transfer occurs, however, modeling a slab as rigid where it may be stiff (but not rigid) can cause overestimation of load transfers, which can be more pronounced at slabs with large transfer forces. The thicker the slab and the higher strength concrete we use, the more rigidly and less flexible it acts. With respect to a rigid diaphragm, a flexible diaphragm would increase the building period, and decrease the earthquake force in the structure, as it contributes to a more flexible structure. Under lateral action, the edges of the diaphragm will tend to bend as seen below, since no diaphragm is perfectly rigid. A flexible diaphragm can be visualized as in the exaggerated figure below, similar to a very large horizontal beam under bending, where the loads act horizontally and one side of the beam is in tension and the other is in compression.

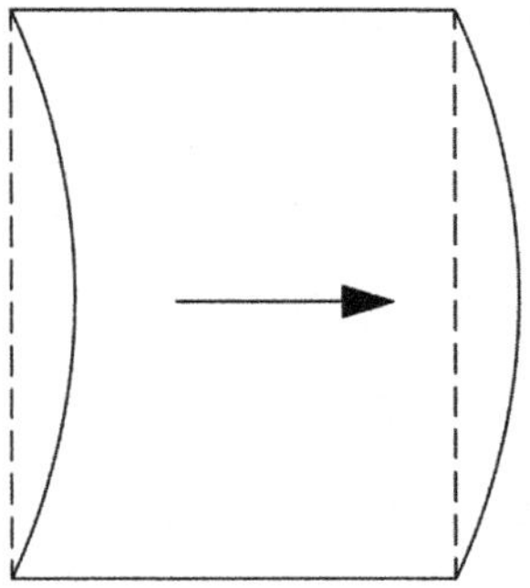

Figure - Buildings 5

So, at diaphragm edges, reinforcing steel must be provided to take this tension which will alternate sides during shaking of the building. These are called <u>chords</u>. So in the figure above on the right edge of slab we will have tension chord and on the left side a compression chord, which would alternate as the shaking continues. Chord reinforcing must also be placed around large openings in the slab and continue for an adequate distance into the slab, or when we have irregular slab shapes, chord reinforcing will not be just at the edges but again will continue into the slab for a certain distance.

> **Side Note:** For ease of visualizing, think of drag struts as strong, stiff "sticks" placed inside slab, that run between or along the vertical members *(of LFRS)* to horizontally connect vertical members to each other and of course to slab and "drag" the vertical members laterally. So by these strong, stiff "sticks", we achieve lateral load transfer between vertical members through slab.

The diaphragm will distribute lateral forces to vertical elements in proportion to their lateral load resisting abilities. During earthquake or wind, the distribution / transfer of inertial lateral forces between the diaphragm and the vertical elements (shear walls, columns) directly takes place between slab to vertical element connections and also by <u>collector</u> elements within the slab or by beams. Collectors are not separate elements, but formed by placing extra reinforcing steel <u>within the slabs</u>, working as linear axial compression and tension zones that transfer the diaphragm lateral forces to vertical elements and vice versa, meaning, they also distribute forces from vertical elements to diaphragms, and in those cases they work as a <u>distributor</u>, which is for redistribution of forces among vertical elements. Collectors may also be called <u>drag struts</u>. So they either "collect" or "drag" the shear forces in the diaphragm from or to vertical members. For this action to happen, in addition to adequate rebar detail, the slab must also be strong and stiff enough (i.e. by having adequate thickness, concrete class etc...), if there is no beam at that location. Collector width can continue well into the slab from the face of wall, to prevent rebar congestion (as they are heavily reinforced) and to provide adequate strength.

To adequately design drag struts, we must have a good understanding of both magnitude and load paths of the transfer loads through the slab between vertical elements. Therefore they must provide sufficient strength and stiffness at the correct path and essentially maintain their elasticity, for the loads to be transferred between vertical elements of LFRS. Stiff chords and collectors together also contribute to a higher overall stiffness of the building as a whole.

When a shear wall continues full length along one side of the slab, then no collector is required there because the slab to wall connection itself can perform the task of force transfer, as long as this common edge is heavily reinforced and connected into each other. Connections between diaphragms and the vertical seismic framing elements must have sufficient strength to transfer the maximum possible diaphragm shear forces to vertical elements. The reinforcing steel from shear walls must be bent and extend sufficient distance into the slabs, to prevent tearing of diaphragm from walls during earthquake.

Collector layout must not interfere with conduits and embedded items in the slab, which must be coordinated between a number of parties including structural engineer, architect, general contractor, MEP subcontractors at the very least.

In very tall buildings where higher mode effects are important, for certain floors modes other than the first vibration mode may be critical as far as the peak acceleration affecting there, and the diaphragm design should consider this. Dynamic analysis methods (Chapter 1>Earthquakes), can capture these.

Diaphragm, chord and collector concepts described above also apply to concrete filled metal decks. Unfilled steel decks, which are used commonly at roof levels, also serve as diaphragms.

Basement Slabs (and Walls):

The figure below is drawn to show the importance of basement level slabs (diaphragms). In these levels, similar to upper levels, lateral earthquake force from the core shear walls are transferred to slabs, but unlike the levels above ground, here the slabs can transfer lateral forces also to the basement walls, which provide far more additional support as they lean on earth, plus form a large box frame that encircles the perimeter together with basement level diaphragms. So, at the basement level due to this higher capacity, considerably higher force transfer occurs between the slabs and vertical elements.

For a situation as in Figure 6 below, during an earthquake or strong wind, the upper section of the structure behaves like a cantilever beam socketed into the ground, even in the absence of piles. This is called the backstay effect. Here because of the support from basement perimeter walls as described above, and the walls and slab forming boxes, larger shear forces can exist at the basement levels than upper levels. This is good because of the greater support at the basement levels, but locally it can cause large stresses in basement level diaphragms, and the locations where they meet with shear walls and basement walls which make backstay effect possible by transferring these large lateral forces to basement walls. Therefore, these slabs, especially the ground level slab, can be made thicker than others, to achieve this strength and rigidity. And where they meet shear walls or basement walls, slab openings which cause interruption and irregularity of this force transfer, should be avoided to the greatest extent, even more carefully than avoiding these at upper level shear walls and slabs. We said the ground level slab especially because that is where sudden change of footprint of building occurs (as in Figure 6 below), so it is as if this slab forms a force couple *(see Chapter 1 > Statics)* together with the mat foundation, to counter the overturning effect of the building during wind or seismic event. The similar force coupling effect between the mat and the slab where footprint suddenly changes could be experienced by the same building in Figure 6 during seismic event, even if it was entirely above ground without basement levels, and the backstay effect could still happen (although to a lesser extent), because the area of lower levels are much larger than the levels above which means the overall lateral stiffness is still much larger than the upper levels. So, much larger diaphragm force in the ground or podium level slab can exist. To relieve this force form the diaphragm, one option is to separate the slab by movement joint from the building but in this case the whole purpose of backstay effect, in other words, the support from basements may be eliminated and suddenly the shear wall effective height increases and the building becomes more flexible therefore it may not be the best option.

The presence of basement walls and the lateral force transfer from them to the surrounding earth releases considerable burden from the foundation to resist moment during earthquake, but the stiffnesses of the basement diaphragms, basement walls and the passive resistance from earth must be estimated fairly close, otherwise their effect may be ignored which is a conservative assumption because if you remember from soil mechanics section under lateral earth pressure topic, soil deformation amount to produce full passive resistance is a lot. In other projects, we can represent the support from surrounding earth by springs. Ignoring passive resistance can increase foundation costs.

For basement walls lateral earth pressure becomes more critical than seismic forces as the distance between columns and height of wall increase, which may necessitate thicker walls.

Basement levels affect the natural period of the building and the whole seismic response and load paths. By making some levels underground, due to backstay effect, period of the structure shortens.

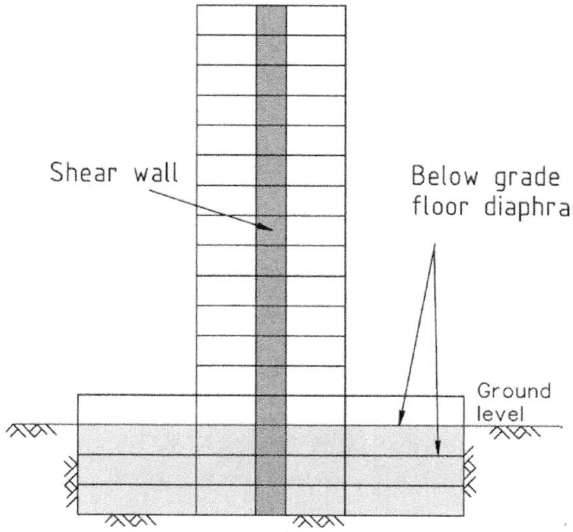

Figure - Buildings 6

If soil structure interaction (SSI) effects *(see earthquake section in Chapter 1)* are considerably present at the foundation, this will also effect the forces in the entire structure including diaphragms. The potential for SSI increases by basement walls acting together with mat foundations, in comparison to mats just sitting close to the ground surface.

On projects where we need to backfill outside of the basement walls, this may need to be delayed until the basement level slab or slabs are in place, to provide bracing from the inside.

One Way & Two Way Slabs:

A slab is considered as two way slab if its length to width ratio is smaller than 2, otherwise it is considered a one way slab. We described it under Chapter 1 > Concrete > Reinforcing Steel section.

Flat Slabs (Flat Plate Slabs):

A structural slab does not necessarily need to have beams around. Sometimes a slab may be designed to have a uniform thickness, and can rest only on top of columns directly, which can be called flat slab or flat plate slab. So here, on top of columns there is just one piece of uniform thickness deck, without separate thicknesses for beams, as the slab contains the beams within itself. Displacement for serviceability, punching shear capacity is important for flat slabs. If shear caps on top of columns are used, the load is spread to a greater area which will result in increased punching shear resistance and therefore the thickness of slab can be reduced. Advantages of flat slabs include:

- Schedule and cost savings can be achieved by using less formwork, which may be a significant part of a concrete structure cost and schedule, because installing formwork for a flat slab is much easier than installing for a floor with beams.
- Columns can be more flexibly laid out, which becomes more important when we have irregular column layout. The columns can still work with the slab itself as a frame (even if we have no beams) with proper detailing, to contribute to the lateral force resistance of the structure. This can include shear walls too.
- Savings can be achieved from overall building height, which in turn means savings from important trades including facade, MEP, interior finishes, and / or more available space for interior items. These are again, due to absence of beams (even if the slab itself may be thicker).

Composite Concrete Metal Deck Slabs:

This type of slab is mostly used in steel framed buildings, but it can also work in a concrete framed buildings, resting on concrete beams, concrete or even masonry walls.

Composite metal deck slab's load carrying capacity is through the composite action of concrete topping and the steel deck itself, which can create stiff and strong slabs, which are generally thinner than regular cast in place slabs, which means less burden on the structural system and foundations. As the compressive strength of concrete and the gauge of steel increases, the load carrying capacity and stiffness increases, for a given slab thickness. The concrete topping and deck are normally tied to each other by steel shear studs, which are welded over supporting beams before pouring of concrete, to prevent them from sliding over each other and obtain greater total capacity while working with each other and also with the supporting beams. The most efficient combination of deck gauge and supporting beam spacing (as it affects freespan length) for a given structure must be found. The beams working together with the slab is called the composite action, which greatly increases the moment of inertia of the beam and therefore bending capacity. Composite action is not only for carrying gravity loads but also can be beneficial during lateral load analysis of the building, especially for drift calculation of moment frames. Like all other types of structural slabs, metal deck slabs can also act as diaphragms.

During construction phase, before the concrete pour and during pouring before the deck gains full strength and takes advantage of its composite action, the stability must be ensured by any temporary means necessary. Deflections must also be controlled. The ponding of wet concrete must be avoided during concrete pour, which can make the deck sag. These can be the governing situations for deck design as far as the deck steel gauge and supporting beam design, or temporary shorings may be needed alternatively, although not common.

One of the big advantages of composite metal decks is to eliminate slab formwork entirely, as the metal deck itself also serves as the form during pouring of slab concrete. In addition, most of the time there is no shoring necessary. These two aspects can save a great deal of time and money. The cost savings from elimination of shoring can be offset a little by the increase in the supporting beam geometry and thicker deck gauge, but this is not much and it also gives us higher stiffness. In addition we always have the schedule savings from elimination of shores and not interfering with construction activities below.

For reinforcing, usually welded wire mesh is used. It is provided to resist the tension or shrinkage cracks, contribute to load carrying capacity, which also makes thinner slabs possible. Where the deck continues over beams, in other words in negative moment sections, rebar is needed at the top, otherwise, since concrete cannot carry any tension, the deck on both sides of the beam will turn into simply supported slabs, with hinges over the beams.

For longspan steel beams, steel girder bridges, the beams may slightly be cambered to account for slab deflections, in order to obtain a level final slab but the cambering depends on a few variables (such as span length, expected loads, beam cross section and strength, temperatures) which is also hard to quantify exactly and should not be overdone and therefore cannot be applied always.

Lightweight concrete is often used as topping slab on metal decks because it has better fire rating than regular concrete due to its porous aggregates, which also provides good thermal insulation. Lighter weight also means lower load on the building frame and foundations.

Metal deck can be together on the same floor with cast-in-place slabs. For instance in tall buildings, the core can be concrete shear walls and slabs, surrounded by steel frame, with metal deck slab.

Waffle Slabs:

A waffle slab is made of a thin concrete topping slab beneath which narrow concrete ribs run in both directions, to give it the appearance of a waffle, when looked at from below.

There are certain advantages of waffle slabs including:

- They provide a lighter but stiffer geometry, because the unnecessary concrete sections of slab that would provide little contribution to bending strength are eliminated (i.e. *the moment of inertia of the section for bending resistance is not very different although we use less material*).
- More stiffness also contributes to the rigidity of the slab. Considering diaphragm action, this means more rigidity and better distribution of lateral loads as discussed above.

- They provide good vibration resistance, which becomes important in hospitals, laboratories.
- Longer spans can be passed without columns, due to stiffer slabs.
- It can be left as exposed and gives aesthetical soffit appearance.

Waffle slab of a parking structure, which is sometimes preferred for these type of structures. See the wide band beam in the middle of the photo in the left-right direction, which can be commonly seen in these type of slabs (the other alternative can be making wide column heads without having these beams). These slabs work as two way.

Disadvantages include:

- They will cause greater floor to floor heights, given the same ceiling clearance we want inside. This means a taller building, for the number of floors we want. This means greater total height of columns, exterior cladding, conduits and pipes for the same building function. A taller building will also need to have a stronger lateral force resisting system and foundations.
- They have high rigidity, so cause loss of ductility of the building and frame action. Buildings with waffle slabs should have shear walls. Even so, waffle slabs are not the best for tall buildings.
- Forming the slab before pour is difficult. They are formed by placing hollow box shaped forms on top of a flat form deck.

Deflection and Vibration of Slabs:

For slabs, serviceability (deflection and vibration) criteria is often more critical than strength. This is mainly due to the fact that certain nonstructural, architectural elements such as the facade or interior partitions will not tolerate deflections beyond a certain limit and excessive deflections and vibrations will create discomfort for the building occupants. The criteria to limit deflections and vibrations are established by codes and standards such as the IBC or ACI 318.

Cast in place concrete buildings are best at absorbing vibrations. As the span length of a slab increases and its thickness decreases (which means lower flexural stiffness), the amount of vibrations increase. Span length and thickness are the main factors, although the behavior should also be checked by a finite element analysis that considers the structure as a whole. To quickly consume vibrations, the slab should have good damping, which is not only affected by the slab length and thickness *(wider and thinner slabs mean less damping)* but also by nonstructural components above and below the slab.

For certain buildings such as healthcare facilities, or the buildings near strong vibration sources such as rail lines, it gets more difficult to meet code and standard requirements. Exercise rooms, machinery rooms or corridors with a lot of pedestrian traffic are among other sources of floor vibrations.

Openings in Slabs:

For various reasons, such as to run conduits, ducts or shafts, stairs, skylights and so on, it is unavoidable to make openings in slabs. The number, size and arrangement of openings in the diaphragm must be designed to not interrupt the diaphragm action. These openings should ideally be placed as close as possible to the center area of the slab and their number and sizes be kept to a minimum. Depending on the size of opening, framing or extra reinforcing around the opening may be required. If several openings are too close to each other, they might need to be treated as a one large single opening. Analyzing the effect of openings in slabs may be more complex especially for lateral force load paths during seismic events, which must be taken care of with proper chord and collector design. Finite element analysis will also be suitable to analyze such slabs. Having too much openings in a slab will prevent it from working as a rigid diaphragm (and even from working as a diaphragm or slab).

For metal decks, the main thing to pay attention is that the metal should remain in place and not be cut out before the concrete fully cures and gains its strength, otherwise the metal's supporting capacity will be reduced with negative effects on deflections or structural strength.

Pour Strip / Delay Strip / Closure Strip:

When we pour slabs, in certain situations we exclude a strip of the slab from the pour, to be poured later. Pour strips (also called as closure or delay strip) are the long strips of slabs that temporarily remain without concrete, that run for the entire slab width, in order to allow for shrinkage of the poured sections on both sides of the strip. This is especially necessary for long span and wide slabs, and even more critical if the slab edge movements are restrained by shear walls.

The longer a pour strip stays open, the better for allowing more shrinkage to take place, which should be around at least several weeks, but this may create hardship for the progress of other construction activities, not only on the slab around the unpoured strip, but also below it because we usually cannot remove the forms and shoring below, or can only remove it partially only upto a certain distance to strip edge, so that we still support the unpoured strip and poured slab segments' edges on both sides, until the strip is also cast, gains strength and connects two previously poured segments adequately. This affects the entire construction schedule by delaying a lot of

> **Side Note:** In very large PT slabs, expansion joints may be preferred instead of strips, so that the two sections of the slab (and the structure) will not be connected at all. But of course expansion joints have their own challenges as discussed in Chapter 1.

other activities especially when multiple floors are affected. Strip location(s), width and length, for how long will they be open without concrete, the design of the slab itself, and shoring and reshoring design must be well coordinated between the structural engineer and the contractor. In fact, several weeks are also often not enough to allow adequate amount of shrinkage to take place, which can be at least 1-2 years, but in practice we cannot leave the strip unpoured for that long. Therefore, in addition to the partial relief provided by pour strips, often we may also need to provide permanent relief details such as slip joints.

Pour strips are especially used in large post tensioned slabs, when a slab is too large to be poured and tensioned in one piece. For PT slabs, in addition to shrinkage, the shortening of the slab - both short term (elastic) and long term (creep)- in its horizontal plane can also be remedied by using pour strips. The rebar from both poured sides are extended into the strip and lapped on each other, before casting concrete in the strip. Width of the strip, which should be at least around one meter or more, must be enough to allow sufficient rebar lapping length and also to place and operate the PT equipment.

Before casting pour strip concrete, the edges of the poured sections on both sides of the strip are roughened for better bonding and water saturated to not absorb the water necessary for hydration of wet concrete in the strip, which is also similar to what we discussed under concrete repair topic.

Slab on Ground (Slab on Grade):

This is typically a non structural slab, which means, it is not elevated and rests on earth, but not on beams or columns. This means the design is different; for example, ACI 318-19, Building Code Requirements for Structural Concrete, does not apply to slabs on ground, even if the slab may be for a building, except when the condition stated in section 1.4.8 applies, which basically says, when the slab transfers loads from the structure to the soil, in other words, performs a structural function / behaves as structural slab. Slab on grade is frequently abbreviated as "s.o.g." in project documents. Similar to foundation design, when designing s.o.g we need soil parameters from geotechnical engineer. ACI 360R - Guide to Design Slabs on Ground is a widely used or adopted source in the world, for s.o.g design.

Loose topsoil of the ground must be removed first. Then the soil is graded and compacted. If the soil has sufficient bearing capacity, the slab can be built directly over it, even if it is clay. Or, we can excavate a little more to remove and replace clay with gravel to support the slab better. In worse soil or more sensitive cases, other ground improvement methods may be necessary. In both cases, after the area is compacted and brought to final grade *(considering the thickness of slab itself plus the sand layer that will be placed)* and ready to receive the slab, any electrical conduits or plumbing pipes that are supposed to remain under the slab are also laid, along with their slab penetrations, a sand layer is placed, a moisture barrier (vapor retarder) is laid to prevent ground moisture reach to the slab, then the rebar is installed and finally the concrete is poured.

At locations where large point or continuous loads exist, such as from individual columns or walls, the slab may be thickened. For example, it is common to thicken edges to support exterior walls.

There needs to be control joints at every certain interval in each perpendicular direction on the slab, to prevent random cracks because too large slab area will crack no matter how well it was designed or constructed. Also, slab on grade on expansive soils must be separated from the structure with expansion joints.

Slab on grade bears on the soil below therefore it requires lower amount of reinforcement than a structural slab. The reinforcing bars can be as welded wire fabric (WWF) or simply regular deformed bars. To provide higher stiffness and durability, s.o.g. can be made as post tensioned (PT) especially if it will be on expansive soil or needs to have stricter displacement tolerances and better load carrying capacity (which becomes especially useful for industrial slabs but many residential developers also require PT slabs anyway). The duty of PT here is different than PT structural slab as was discussed in Chapter 1 however. In the case of structural slab, the tendons are responsible for the slab's load carrying capacity, therefore its vertical position within the slab varies, to take on tensile loads where necessary as bending direction of the slab varies between center slab and column locations. In the case of a slab on grade however, the slab is already supported by the ground, which carries the weight of the slab and live loads over it. Therefore, the duty of PT here is not to carry slab loads but to cause a compressive stress up front in the plane of slab and make it stiffer against the bending from both live loads from above or expansion of soil from below, which means it is better to run PT tendons at mid height of the slab thickness everywhere. This compression also makes the slab more resistant against cracks due to temperature or shrinkage. After concrete pour, post tensioning of tendons should not delay too much, as the cracks may start to form in the concrete slab, which already has lower amount of rebar than normal, so, as soon as we determine that the slab is strong enough, we should perform the post tensioning and compress the slab in its plane, which generally corresponds to several days in practice. During post tensioning, the compression in the plane of the slab itself will vary from slab edges towards middle slab sections, due to spread of loads within the slab and the friction from grade below. Post tensioned system costs money but this may be offset by the thinner slab and lower amount of regular rebar we can use, and of course by the better long term performance in most, if not all cases.

The slab can also be made as ribbed, meaning, the foundation walls are not only at the perimeter but also run at interior areas at every certain distance, in both perpendicular directions, *(so it becomes somewhat similar to a waffle slab as we described above, but this time it rests on grade)* which increases the stiffness of the slab significantly but adds to the cost. Since the stiffness is increased, the slab thickness can be lower than a uniform thickness slab without ribs. The gaps between the ribs must not be too wide otherwise the slab stiffness will vary greatly between ribbed and unribbed locations, which is not acceptable. The rib footings of the slab and the slab itself need to be poured monolithically for composite action to be achieved, otherwise, a lot of connection rebar will be needed in between. This system can also be post tensioned.

As stated in the first paragraph, slab on grade is not meant to serve as a structural slab. If it will serve a structural function, then the criteria in ACI 318 will be valid, for what function the slab is performing. As an example, per ACI 318-19, Section 18.13.3.2, for structures in certain seismic categories, if the slab on grade serves as a lateral load transfer mechanism between vertical elements during an earthquake, then they should be designed as diaphragm, per the criteria in there.

Pouring slab on grade can be harder than it seems, especially if we are trying to achieve tight tolerances, such as a factory or a gymnasium floor. In such cases, laser guided trowels can be used.

Although it is common practice to first pour slab on grade that would remain interior of a building and then proceed with higher construction, in rare cases we may need to delay the slab even if the building structure moves ahead. For instance, a delay in design for pipes running under the slab might be a reason. In such a case, we will not have vertical access to the area to conveniently pour the slab through a pump hose vertically, but will need to hold the hose inclined, which can only be up to a point because of the risk of segregation. In such case, although it will be slower, wheelbarrows can be used.

A slab on ground should not be confused with a mat (raft) foundation, which was discussed in Chapter 1 > Foundations. Mat foundations transmit the loads from building columns and shear walls to the ground or piles below, so they are massive structural members. They are also far more thicker than s.o.g. and have heavy reinforcing.

Shear Walls

Shear walls are structural walls made to resist axial, shear and moment loads and their combinations. They not only carry gravity loads permanently, but also receive lateral forces from slab diaphragms during earthquakes and transmit them to foundation. These are concrete walls that go all the way up from foundations to the top level just like columns. So shear walls are similar to columns but much wider and therefore can significantly carry more load in all directions. Their large width enables them to take on most of the lateral loads in their plane. Buildings with shear walls have great resistance to earthquakes. Shear walls are also the best way in limiting lateral displacements of a building, but this also means we highly depend on elastic strength and there is limited structural ductility and slow energy dissipation, unless we use them as a dual system with ductile moment frames *(See ASCE 7-16 Table 12.2-1, Group D. Notice that it says that the moment frame must resist at least 25% of the seismic forces, which is the way we make sure to utilize the ductility of the moment frame. Notice how a shear wall - moment frame dual system enables us to build for much stronger seismic categories and taller buildings and gives us higher R value)*. Use of shear walls can eliminate the use of heavy moment frames, which not only saves structure costs but also enables wider spaces inside the building.

A composite building of concrete core and steel framing. The concrete core serves also as shear walls, to provide lateral strength against wind and earthquakes.

Shear wall must start from foundation level and bear directly on foundations and continue all the way up. Shear walls attract a lot of lateral force so their connection to foundations must be very strong.

Even introducing a few of these walls in each perpendicular direction, will relieve the moment and shear burden on the entire columns of the building greatly, even at locations far from the shear walls. The overall lateral deflections of the building will also significantly decrease. In strong earthquake zones, shear walls are almost always used for buildings above a certain height. One of the most effective ways of seismically strengthening a structure is addition of shear walls.

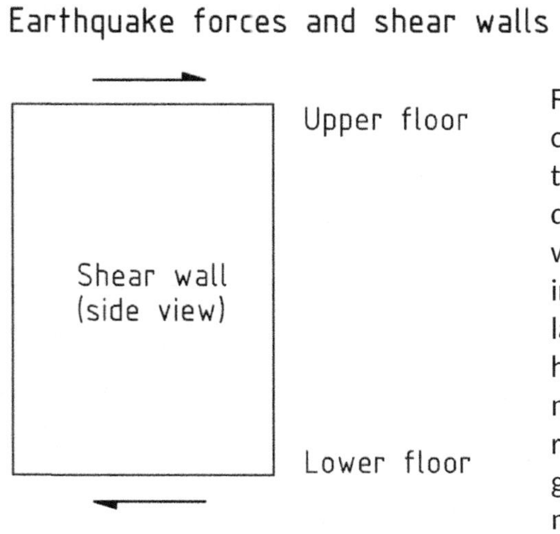

Figure - Buildings 7

Figure 7, shows a wall under shear. Short shear walls, also called as squat walls, have great resistance to flexure therefore they more likely fail in shear, which governs their design. Similar to deep beams, these walls can be analyzed with strut and tie method which was discussed in Chapter 1, in other words, they form diagonal strut when resisting lateral forces. Slender concrete shear walls on the other hand, are much taller with respect to their width (which means high aspect ratio), and have much lower flexural resistance so they work in bending and their design will be governed by flexural deformations rather than shear. The more slender the wall, the more significant its flexural deformation. Here the bottom of shear wall needs to be the strongest and detailed for more ductility, where the most moment occurs and some ductile flexural hinging can be accepted. See Figure 8 below. Flexure dominated walls are desirable because they are ductile and can endure high plastic deformations where a lot of energy can be dissipated during seismic events. So we must design the slender walls with adequate shear strength and prevent their shear failure. Walls failing in flexure behave similar to beam-columns. If we cannot have slender walls, short walls are also possible which will fail in shear. As the axial load on the wall increases, likelihood of shear failure increases. We must ensure flexural failure with adequate rebar confinement. As always in RC design, we must proportion the rebar to yield first, before the crushing brittle compressive failure occurs in concrete. In this case it is the rebar that takes on the flexural stresses.

Figure - Buildings 8

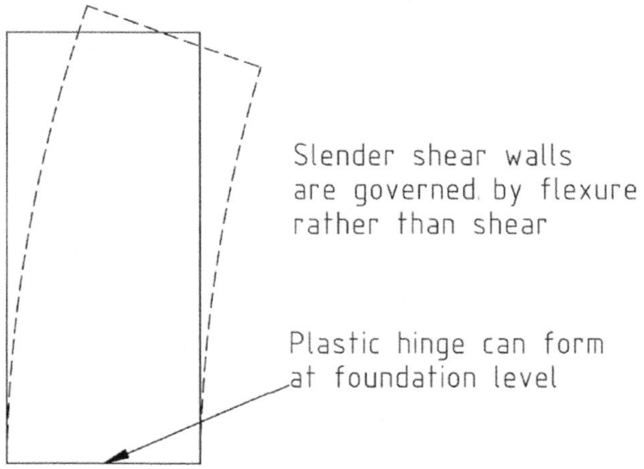

In designing shear walls, structural engineer complies with the overall architectural intent of the building as far as the wall layout. The shear walls' total stiffness in one direction must be sufficient to not only counter lateral loads in that direction, but also to limit lateral drift. Not only the total stiffness, but also how that stiffness is distributed on the floorplan is also important. The most important thing is that the geometrical center of resisted lateral forces on horizontal plane, in other words, the geometrical center of shear walls and columns on horizontal plane, must coincide or be very close to center of mass of the floor on horizontal plane, otherwise the building may undergo torsion as a whole, which is an important cause of structural failure. (*Imagine a table with four equal legs fixed to the floor. You push the table from the top at the center, and it will deflect but no torsion will happen. Now assume on one side of the table you add an additional leg, (or make one of the legs thicker or replace it with a leg made of stiffer material) and fix it to the floor too. When you push again from middle, as stiffness attracts load, this time the side where you added the leg would take on more force. This will cause unequal deflection of legs, which will lead to torsion of the table top as you still push at the center - actually the earthquake will shake from floor and act as an inertial force at center of mass, so nothing will "push" from side).* Symmetrical arrangement will usually provide this match very easily. Other configurations need a closer look. If the building undergoes torsion, such as when lateral stiffnesses are not well balanced, this can further increase the lateral load on shear walls. The best arrangement to prevent torsion is to place sufficient shear walls at the building perimeter, even better, at the corners, as long as architecturally acceptable (*which is also true for braced systems as mentioned before)*. Here also note that the perimeter walls are also subject to a lot of axial forces due to overturning effects of the building, therefore the columns and walls towards the center should take on adequate amount of gravity loads. While doing so, they will also take on lateral loads, therefore a balance must be found. When performing computer structural analysis, this is easily done by the structural analysis software, which can identify torsional irregularities and adequacy for gravity load carrying capacity among many other checks.

Let's list several other design considerations for shear walls below:

- We can add flanges at both sides of the shear wall *(so that its cross section resembles an I-beam with short flanges when looked from top)* which allows it to take on considerably more moment. This arrangement allows more ductility of the wall too, as we can also add more reinforcement *(similar to column ties - remember column ties also provided ductility in addition to higher compression and moment capacity)* at both flanges. Of course it must be architecturally compatible and forming and rebar will have minimal cost impact.

- Two or more walls can also be combined in orthogonal direction to work together such as L, T, I or C configurations. This increases efficiency. Overall building torsion must be considered as explained above by balancing lateral stiffness with other shear walls and columns.

- There is a limit to cross section length to width ratio for a vertical member to be considered as a shear wall. Below that limit, that vertical member is considered as a wall pier or even a column with smaller ratios and must be detailed accordingly.

- Shear walls are usually between 10-14 inches thick. Each inch of added thickness increases the wall's stiffness considerably and relieves rebar congestion and accommodation issues.

- Openings such as for windows, doors, ducts may be placed on shear walls up to a limit. For the shear wall to be considered a solid wall, there is a limit to sizes of openings and how close they can be to the edge as per codes. If the opening is too large or too close to the edge of the wall, its integrity may be compromised, and its ability to act as shear wall can decrease. Openings and discontinuities can lead to stress concentrations and too much can lead to unpredictable seismic response and flow of force therefore as much as architectural demands allow, openings should be placed as little and as regularly as possible. The further the design deviates from this, the more detailed analysis will be needed. FEM is useful to simulate such complex behavior.

- Shear walls mainly have reinforcement to resist tension and compression at both edges, and in the middle horizontal ties to counter shear. There must be sufficient wrapping of bars in both directions. The openings, (which can not be too large or too close to the edges of the wall), must be enclosed by extra rebar around them. The transverse ties mentioned for columns in Chapter 1, Concrete > Reinforcing Steel, have the same duties in case of a shear wall. Short shear walls have great flexural strength therefore tend to fail in shear in a diagonal manner, therefore this should be resisted by both adequate vertical and horizontal shear reinforcement.
- Shear walls must have sufficient connections, in other words, contact length, from all sides where they meet floor slabs and it may not be a good idea to build them right near open shafts such as elevator or mechanical shafts. The load from slabs to shear walls must adequately be transferred, and then to the foundation. This is the main usefulness of a shear wall. Otherwise it just serves as a wide column when no lateral loads exist.
- Two adjacent walls with space in between can be combined by coupling beams (which we also mentioned under beams subsection above), which makes the walls' total capacity greater than their sum. These coupling beams are short beams that can break during strong earthquakes and dissipate a great deal of energy safely. The purpose of these beams are to act as "fuses" and reduce burden on walls, and their failure will not cause structural collapse. Where we have coupler beams, walls can be made slightly thicker to accommodate coupler beam rebar. As these beams' length increase, they can spend more energy by working more in flexure and less in shear, which means lower amount of shear rebar. Analyzing forces in a shear wall - coupling beam system is considerably more complex than analyzing a single wall.
- The location with the most demand from the wall during seismic event is where it meets the foundation. A limited amount of plastic hinging may be accepted here. This can happen in addition to coupling beam failure. The taller the building gets, the more likely dynamic effects will start to influence upper level floors and for very tall buildings, shear wall yielding can also happen at upper stories due to higher mode shapes being effective.
- It is also common to have basement walls (perimeter walls below the ground level). See Figure 6. Basement walls make the building as if socketed into the ground. This particular behavior is different than shear wall, which takes advantage of the soil around, but contributes to earthquake resistance. In addition to also taking lateral forces, they also reduce the height of the building that is exposed to earthquake by their height at the basement level, as they can move together with the ground. For perimeter basement walls below grade level, also see Slabs > Basement level slabs (and walls) section, several pages before.
- Although for moment frames we can model column bases at foundation level as pinned, for shear walls we need to provide fixed connection to foundation (foundation rocking and soil structure interaction effects may still happen).
- Axial load vs moment interaction diagram is important to determine shear wall capacity, as we also discussed under masonry section in Chapter 1.
- And as far as the materials:

- Shear walls are built not only in cast in place concrete but also in composite concrete and steel buildings such as the one in the photo above. Here, the shear walls are concentrated at the core. These also have the other purpose of serving as elevator or stair shaft.
- Shear walls can be of precast concrete too, although not common. In such case, if the joints between the two adjacent panels vertically are stronger than the precast panels, this will be a similar case to a cast in place concrete shear wall, and the failure will not happen at the connections. Or we can allow failure at the joints. In high seismic zones, strong connection type is preferred.
- Shear walls can also be of large steel plates, although this is not common. During earthquake, although they will buckle, at the same time they develop tension which can provide lateral resistance and good ductility and considerable energy dissipation.
- Shear walls can also be of masonry. Also see Masonry section in Chapter 1, where we discussed masonry shear walls.

Stairs / Ramps

In addition to analyzing within the structure, the stairs must also be solved separately by isolating them from surroundings. It may be necessary to place additional beams especially at stair landings which are in between main floor levels. When doing this, attention must be paid that these additional beams will not cause a short column effect for the columns they are connected to. Ramps, which can be seen in parking structures, span longer and wider than stairs. Stairs and especially ramps should be designed to not cause unintended lateral stiffness or serve as unintended, sloped connection between adjacent levels which can be complex to accurately analyze. Under lateral loads, these can cause problems such as torsion, short column effects, unintended load paths and force redistributions, vertical element bypasses, unintended modification of diaphragm behavior.

Balconies

Balconies can be designed as slabs but one edge of the balcony will float in the air, unless there are columns supporting at the outer edges which makes it a regular terrace structural slab. We can place two cantilever beams at both sides of the balcony, which will be supported from the building structure.

Shafts

A shaft is on opening in the floor slab, which continues all the way from the bottom to top level. Shafts can be for stairs, elevators, air ducts, trash chutes and whenever we need an opening in the slab for something that runs vertically. The edges of the slab around the shaft openings must additionally be reinforced. If the shaft is large, we may also need chord reinforcement.

Building Irregularities

Above, we listed the main structural elements of a building. Let's also introduce building irregularities, as it affects the overall structure. Also see seismic retrofitting section later in this chapter, to see how some of these irregularities can be remedied during seismic rehabilitation.

The owner and architect decide the size and shape of the building. From structural engineering standpoint, the structure needs to be as simple, symmetrical and uniform as possible for structural efficiency and to be able to better foresee earthquake effects, however, architectural concerns may often prevail and not so simple or even complex structures may be designed. The most important factor for regularity of a building is how the architect designed its overall shape in the first place. If a building was very irregularly planned by the architect in the first place, even the best engineering design may only be up to a point. For some structures, such as landmark structures or the ones that are built for prestige, the cost effectiveness of structural design and construction can be of second importance to the owner. In those cases, as long as it is determined that the structural safety is not sacrificed, engineers are still able to come up with designs to accommodate almost any configuration that the architect wants, at the expense of increased design and construction costs and complexity. Therefore, close communication between the architect and the engineer must begin during initial planning and design stage. Even with good design and construction, buildings with irregular configurations may suffer more damage than buildings with regular configuration during earthquakes. Especially in high seismic zones, buildings with high degree of irregularities should not be built.

What we mean by regularity in general is that the mass, strength and stiffness of the structure should be distributed as evenly as possible everywhere, structural frame is continuous and uniform vertically and horizontally without abrupt changes and discontinuities, and the frame should allow loads on the structure to be transferred down to foundation as simply as possible through the shortest path. Generally, building irregularities can be grouped as horizontal and vertical irregularities.

Irregularities in a structure can cause negative results such as:

- interruption of load path, which can become longer and more complex
- torsion of entire structure
- concentration of forces in certain areas which can become unevenly distributed and can cause overloading and failure of certain members which can propagate to larger parts of structure
- different drift demands between different parts of the lateral force resisting system
- higher demands on the foundation system
- lower energy dissipation efficiency

Building irregularities are listed in country, local jurisdiction seismic codes along with their criteria to qualify as regular or irregular for various aspects. The codes include checks for various irregularities versus the actions that need to be taken upon discovery of such irregularities. The actions include penalizing the design by using higher forces to comply, using more stringent analysis methods or simply not allowing an irregularity for a certain case. To analyse a building with irregularities, we must make a more detailed model of the building, to capture the effects of irregularities on members.

Now let's list various common aspects to achieve structural regularity in a building. Some items that can also fit in this list were already mentioned in the subsections above where we described each type of structural element. Or conversely, some items listed below could also be written under the corresponding structural element that we described in the previous few pages. In other words, the subject of building regularity is inseparable from the discussion of the structure and its members.

- For overall shape, square, rectangular or circular footprints are best. Even if symmetrical, +, L, T, H, U shaped footprints can still behave as irregular during earthquake due to response of building wings. In addition, stress concentrations can happen at the corners where the wings meet, plus the torsion, and the combined the effect can be even more complex. The +, L, T, H, U shaped building wings can be separated from each other structurally by using expansion joints with adequate gap (by also considering anticipated floor drifts toward each other), or they can be joined by using stiff elements such as shear walls, or splays.
- Even if the building is rectangular, making one of the dimensions too large in comparison to other two should be avoided. For example, too tall building in comparison to its base, or too long building on one side w.r.t. width will need more complex details and stronger foundations.
- Buildings on a slope with front and back sides that are considerably different in height with trapezoidal sides can also be considered as irregular.
- A weak story has lower strength and a soft story has lower stiffness than the other stories, both of which must be avoided. This means, floors should not have significantly different mass and/or stiffness, relative to each other. For stiffness, the sum of vertical element cross sections are compared between adjacent floors, and the difference must be below a certain limit. Floor heights should be as equal as possible, and building elevation should look as regular as possible. Considerably different floor sizes, which are called as setbacks should be avoided, especially if an upper floor is larger than lower floors. Tall ground level floors, such as lobbies, are inherently less stiff - add to it that the most earthquake load is at the ground floor. Columns which do not satisfy strong column - weak beam also increase the risk of weak stories, where that story can collapse leading to total collapse. Strong columns help better, more evenly distribution of story drifts through the building height.
- The more continuous framing in a building, the better earthquake resistance. Structural frame and lateral load resistance system must be continuous and distributed as uniformly as possible, all the way up and all around the structure. In two perpendicular directions, separate frames should exist with strength and stiffnesses and also overall ductilities that are very close to each other. All interior columns should preferably be surrounded by beams on all their four sides. We can not make separately working structural frames that are located arbitrarily. Beam and column layout should be regular, as evenly spaced as possible, and need to continue all the way through the width and height of the building to greatest possible extent. Redundancies in the frame system are preferable, so that when one load path fails, another can take over. Secondary beams connecting into primary beams without columns below are not preferable.
- Overall torsional irregularity must be prevented. Shear walls should also be placed as symmetrically as possible with respect to the geometrical center on the horizontal plane, in other words, lateral stiffness center and mass center of the floor on horizontal plane should be as close as possible to each other, to avoid overall building torsion. A single shear wall should

not take too much of the lateral load. Of course there is also contribution from columns but in case of shear wall systems, the walls take much greater lateral load than columns. If there is torsion, this will not only be as a single torsional rotation but also in the form of torsional vibrations, which must also be minimized. The best way to resist torsion is to arrange shear walls as much as possible towards the perimeter (rather than core), as also mentioned before under shear walls. This will also help improve the mass participation amount during vibrations to sides, if some of the mass were not participating due to too much overall torsion. To be on the conservative side, a certain small torsional eccentricity is still assumed during analysis to account for uncertainties in mass and stiffness distribution.

- Shear walls must continue without interruption from topmost floor all the way down to the foundation. This also applies to columns, although they can sometimes bear on beams, which is highly undesired. Column and wall stiffnesses should be uniform.
- Between two different sections of the building, stories that meet with different elevations cause short columns. This also disrupts the continuity of continuous framing.
- The strength and stiffness of perimeter should be as uniform as possible all around. If one side of a wide building is considerably heavier than the other side due to live loads, this can also cause torsion, even if the structure is regular otherwise.
- Slabs should be with as little as openings as possible and be able to adequately transfer loads between vertical members. If the openings are too much, then we cannot consider it as a single rigid diaphragm anymore and the sections of that slab may need to be analyzed as separate diaphragms. For the +, L, T, H, U shaped buildings mentioned above, the regularity also applies to diaphragms, when a part of diaphragm protrudes too much and the regular shape is lost. Separating building sections by expansion joints will remove the diaphragm irregularity and allow sections to be looked at separately but expansion joints should be used with care.
- Ramps in parking structures will contribute to seismic resistance acting as braces but can also cause irregularities.
- Footings should be at the same elevation as possible.
- Cantilevered beams at the interior of the structure should be avoided as it disturbs load paths. Too long cantilevered slab sections will also create irregularities.
- Expansion joints should be used only when structurally unavoidable, such as preventing irregularities or obtaining more balanced arrangement of structural members, but not when we simply decide a section of the building is "too long or wide". In case of using expansion joints, precautions must be taken to prevent pounding effect of separated sections on each other during earthquake. Expansion joints should allow relative movement of building blocks for not only one but all directions.

In ASCE 7-16, Tables 12.3-1 and 12.3-2 various irregularities are listed along with their quantified criteria. FEMA 454 also discusses many irregularities.

STRUCTURAL MODELING AND ANALYSIS OF BUILDINGS

Some fundamental aspects of structural analysis with computers was covered in Chapter 1 > Computer Aided Analysis, which you may refer to as needed. Now we discuss more specifics related to modeling the structure in computer. Although this is presented under the chapter for buildings, the majority of concepts described below are applicable to structural modeling and analyzing of any type of structure.

In this section we will discuss:

1) General input for performing analysis
2) Modeling the structure
3) Running analysis
4) Interpreting analysis results, refining and finalizing the design

The steps above can briefly be described as:

1- Obtain overall geometry of building preliminary design from architect. Enter general information about the structure and members, enter reasonable properties for materials. and set general options of analysis, which can proceed simultaneously with step 2.

2- Assume preliminary structural member sizes based on experience and by also keeping in mind the architectural drawings. Construct the model of the structure, with all members and supports along with their properties, and imposed loads. Modeling can either be done manually from scratch based on architectural plans, or the architectural plans may be imposed into the structural analysis program, which also depends on software, and then the model can be continued using that. This will help avoiding big errors in matching of building axes and overall dimensions.

3- Run an appropriate analysis method such as the ones described in earthquake section in Chapter 1.

4- Interpret the results and refine the design as necessary. Repeat steps 1-4 until all results are acceptable, which means, the capacity is more than the demand, the displacements are within tolerances everywhere and all other criteria such as the regularity criteria set by codes are satisfied. Make sure that all results such as displacements, stresses, settlement values, story drifts, slab deflections, connection details... make sense both globally and for parts of the structure and members, even if per computer analysis everything checks out.

Now let's discuss the steps above one by one and the whole process in greater detail...

General Input for Performing Analysis

Before we can make any analysis, we need to input general information about the structure in the structural analysis program. In these programs, many of the items we mention below come with their relevant options, in different formats. So this step also includes setting general options of the analysis. At a minimum, the following info should be entered:

- The design code to use. Commonly used structural analysis programs have the relevant design codes already predefined in them. After we finish modeling and run analysis, the program checks things against the chosen code criteria.
- Structural materials, such as concrete, structural steel or precast concrete and their types or grades. For example in the US, most of the time concrete can be as 3000, 4000 or 5000 psi for different locations in the building structure and rebar grade as 60. Then the program will bring the previously defined properties for that material, such as unit weight, stress strain data and modulus of elasticity, coefficient of thermal expansion, Poisson's ratio, rebar grade. We can modify these if necessary, or enter information based on our specific project materials. After we define the materials, we can assign them to the relevant members in the structural model.
- Typical sections that we will be using, for precast or structural steel members as applicable. Many standard shapes exist as predefined in structural analysis programs, such as the different steel or precast concrete shapes we showed in Chapter 1. Again, after we define the sections, we can assign them to the members in the structural model that we will construct.
- Earthquake related data, such as:
    - seismic design category
    - earthquake design parameters such as the damping value, importance factor
    - depending on method, earthquake acceleration data to produce earthquake loads
- Soil material, its modulus of subgrade reaction, unit weight, angle of internal friction, cohesion. Foundation information. A structural engineer should not start the modeling of a structure before talking to the geotechnical engineer who prepared the geotechnical report for that site or project (geotechnical engineer should also get the structural engineer's input while preparing the soil report).
- Gridlines and coordinate system.
- Any special inputs other than the default values for wind and temperature effects.
- Load combinations to use for dead loads, live loads, wind load, snow load and so on as discussed in Chapter 1, which are already shown on codes. We define load analysis situations (cases) for these combinations depending on what we want to analyse where, in which case.
- When we finish our structural model, we can assign external loads with their numerical values to act on members or joints, either as point or linear or area loads. There are also the dead, live, wind, earthquake, snow etc... loads. These enter into the load combinations and load cases we created in the previous step to be used in our analysis.
- Type of earthquake analysis to use, such as linear, nonlinear analysis, and their options. As discussed in Chapter 1, structural analysis is not just using live and dead loads, but often also the earthquake analysis, which is much more complex.

Modeling The Structure

The engineer takes the preliminary architectural drawings and imports it to his/her structural program but these cannot directly be used in structural analysis. They must be converted to a usable model for structural analysis, which is called the structural model or the mathematical model of the structure. Importing the drawings is mainly for having perpendicular axes, column locations, elevations and the footprint, so that gross dimensional matching errors are prevented, and also to see where the architect intended the columns, slabs, walls, stairs, balconies to be, along with their preliminary sizes. If we did not have any preliminary architectural drawing available for some reason, or want to start from scratch anyway, we should enter members in the following order: axes, then foundations, columns and walls, then beams and slabs. So in modeling we "draw" our structure in the analysis program but this is not a typical drawing. This is called a mathematical structural model because it will have all members with their dimensions, connections, nodes, support types and locations, member properties such as strengths and stiffnesses... so it is a mathematically understandable and solvable entity for structural analysis, which represents the mass and stiffness of the whole structure in 3D space located in an x, y, z global coordinate system. As discussed under computer aided analysis section in Chapter 1, the model can mainly be a hinge model, fiber sections or finite element model or a combination. We can make entire 3D models only by connecting 1D line and 2D surface elements in three dimensions, as was explained under discrete models. There are always multiple ways of modeling a structure and its components, so there are no "musts" but preferred and better ways. No matter which approach is used, when solved, the model must represent the behavior of the structure with acceptable efficiency and accuracy in 3D under applied loads; this basically means, based on the model the analysis should be able to produce stresses and displacements everywhere that satisfies the code and any performance criteria that we are after.

The simply supported beam example in the very beginning of the book under statics topic, is also a structural model, even if it is the simplest possible. The examples under indeterminate structural analysis methods in Chapter 1, or the random structure examples you saw under structural stability section are also structural models. A building structural model will involve far more members, loads and supports. A simple structural model of a precast concrete bridge is also shown in Chapter 3.

General Points in Making a Structural Model:

Below is a list of some important items to consider, when modeling any type of structure:

- A model should start as simple as possible and then refined. It should include no unnecessary details. This will not only be useless but also will increase the risk of errors and calculation load unnecessarily. The model should represent the real behavior of members as close and simply as possible. This also makes it easier to check. As long as the model has sufficient details, the easier we can verify it by hand, the easier it is understood by everyone, the better the model.
- The load carrying system of a building should always be modeled in 3D even if symmetrical orthogonally.
- A model should generate accurate solutions and allow us to meaningfully interpret the results. After we interpret the results we can further refine our model if necessary.

- Models must be able to represent the particular type of analysis we are using and predict the behavior of the structure during earthquakes adequately. To represent the structure's behavior during earthquake is a critical aspect of modeling.
- We must make sure to assign the data that we entered during general inputs such as properties, cross sections, to proper elements of our model.
- The model must show all loads acting on the model at their proper places on members and/or joints. Assigning the loads visually will help avoiding errors, but they can also be assigned in tabular form to enter data faster and then checked visually afterwards.
- From the stiffnesses of individual members and structural configuration, the program will calculate the stiffness matrix of the whole structure. As the members yield and other nonlinearities happen, the overall stiffness of the structure will also decrease, which is taken into account if a nonlinear method is used.
- If any interaction of a structural element with a nonstructural element has a nonnegligible effect on the behavior of the structural element, this effect must be defined in the model.
- When modeling RC structure, we normally model concrete members as if homogeneous, and based on the forces, moments and displacements that we obtain from analysis, reinforcement details are produced. The program will assign equivalent stiffness value for stiffness of concrete and rebar combined. When using fiber type models, rebar can separately be accounted for.
- For some structures, especially for bridges, staged construction becomes important, each of which must be modeled separately.

Modeling Nonlinearities:

Nonlinear analysis basics was introduced in Chapter 1, earthquake section. Nonlinear analyses and modeling are much more complex than linear. Several important aspects are listed below:

- To represent nonlinear behavior, we can use different types of modeling such as: Nonlinear rotational spring hinges, nonlinear axial springs of finite length, nonlinear fiber sections and continuum finite elements.

  For example to nonlinear rotational spring hinges, we can assign nonlinear moment vs. rotation relationships. These zero length hinges connect beams and columns to each other which are represented by elastic frame elements (as in discrete modeling we introduced in Chapter 1, computer analysis section). So it means we take care of nonlinearity at these hinges while frame elements are elastic, which greatly simplifies things and at the same time adequate in most cases. In these hinges, inelasticity is concentrated, while in finite length hinges and fiber sections inelasticity can be modeled as distributed. Continuum finite element modeling represents everything at the most fundamental level as a continuum of small elements as also introduced in Chapter 1, computer analysis section.

  And as reminders:

  - Upto yield point hinges are linear. Nonlinearity begins after yield point.

  - Hinges should be added only where necessary and nothing more. Each extra hinge is a burden on calculations, and interpretation of results, and may cause inaccuracies in results.

- Common nonlinear aspects to consider include (also see earthquake section of Chapter 1):

  - Concrete: Crushing, cracking, shrinkage, creep
  - Masonry: Crushing, cracking
  - Structural steel beams, columns, braces: Yielding, fracture, rupture (tearing), local buckling of web or flanges, overall bending buckling, lateral torsional buckling
  - Structural steel bolts, welds, connectors, plates: Cracks, fracture, yield, slippage, rupture, plate bending or buckling
  - Reinforcing steel: Yielding, bond slip, splice failure, buckling
  - Prestressed reinforcing steel: Relaxation
  - Anchors: Bond slip, yield
  - Temperature effects
  - Seismic Isolators and Dampers: Their force-deformation characteristics
  - Part or overall structure: Deformed geometry of members and part or whole structure and corresponding secondary effects such as P-Delta, structural instability such as buckling of columns, hinging in beams and columns, nonlinear damping
  - Foundations: Foundation rocking, SSI (Soil Structure Interaction) effects where the structure is not considered fixed to the ground but rest on nonlinear springs, yielding of foundation soil, emerging gaps between foundation / basement walls and the soil.

  Correct modeling of these items as far as location, size, strength and stiffness and other relevant properties is important. The items listed above must be represented in their correct locations in hinge or fiber models, or if FEM is used, they must be appropriately defined into elements so that the stresses and deformations occur as close as possible to real behavior. As long as properly modeled with reasonably accurate input, these nonlinearities can represent the behavior adequately upto ultimate strengths (but after that, it is difficult to track the quick degradation and collapse of the structure, where the structure is already considered as failed).

  As a result of these type of effects (of which usually few are significant), stiffness of members, moment curvature relationships can change as the damage of the structure progresses during an earthquake. The overall stiffness of the structure gradually decreases, which leads to increased overall flexibility, increased deformations and amplified secondary effects under sustained cyclic seismic loads. Also note here that, as far as wind, even though the structure vibrates, it is designed to remain elastic, not undergo nonlinearities as listed above.

- As in any type of modeling and analysis, the model should start from a simple one, gradually evolving to a more complex one as we enter new nonlinearities and make refinements. Making a simple model at first is useful to quickly get an idea about the structure's overall behavior.

- Nonlinear analyses are more difficult than linear ones because:

  - More items are modeled
  - Entering a higher number and more detailed properties. It is more difficult to accurately enter the nonlinear properties
  - It is more difficult to represent the exact location of nonlinearities

- It is more difficult to control the behavior of the nonlinearities. The nonlinearities can be far less controllable and predictable

- Nonlinear stress strain curves for each material type should be defined. Force deformation or drift curves (commonly called as "backbone curves" for columns, beams, shear walls, coupling beams) need to be entered. Overall, the complete hysteretic behavior, loading and unloading characteristics and the effect on strength and stiffness reduction of each element must be entered. There can also be the effects such as loads from multiple directions, strain hardening, cyclic loads which must be accounted for as applicable. When assigning these to members, usually there are shortcuts to use, and no need to do it one by one for the entire structure.
- The expected performance criteria also determines the degree of nonlinearities allowed.

Modeling External Supports:

When modeling external supports, important points to consider include:

- The first thing we must do to model supports is to locate their coordinates and then define them, by telling the program along which of the x, y, z axes the joint is not free to rotate or translate (move). The restrained translation and/or rotation means, an external reaction force and/or moment at that joint, in other words, an external support.
- Some supports are made flexible on purpose. For example, the beams of a precast concrete bridge may rest on rubber supports (base isolators) which bear on abutments as shown in Chapter 3. The base isolators allow lateral movement during an earthquake, thus avoiding transfer of a high amount of lateral load to the pier(s), and in turn, to foundations. These supports are modeled with their proper stiffness values when considering movement in horizontal direction (i.e. for horizontal direction they have flexibility with a low stiffness value but for vertical direction they will have much higher stiffness value).
- Flexible supports are not necessarily only the items purposefully made flexible, such as seismic isolators. The parts of the structure itself, or any external supports can also behave flexible, even if they are made of solid materials. For example when modeling a slab as a beam strip, we may assume the actual beams, columns or shear walls carrying the slab as rigid or flexible supports below the beam strip, depending on expected behavior of components.
- Supports are usually assumed to rest firmly on rigid surfaces and not moving, but in reality this is not the case, and even if we neglect, they move. If these movements are not negligible we must include these effects in our model.
- It is very important to have correct type of supports in the model at their correct locations and directions. Even if a small error in these may cause significant differences in the calculated internal forces, moments and the final deformed shape. Fortunately however, it is easy to catch these errors, just by looking at the model again if something doesn't make sense in the results, because a missing support can cause extraordinarily large movement, or a support placed where it shouldn't be, may prevent the movement of a large part of or the whole structure *(Imagine for example deliberately placing an external support to the top floor of the building, as if bearing on air, which will prevent the entire lateral earthquake movement. This would also cause*

considerably different results for all deflections and stress distributions of the entire structure, which is easy to detect).

- To model foundations, different paths can be chosen, applicable to different situations, such as:
    - The structure can be assumed fixed to the ground, entirely resists the moments at column to footing junctions which produces the shortest periods, highest forces caused by seismic action in the structural frame, but this also means lowest deflections and vibrations of the building. Assuming fixed base condition can also serve to make up the upper limit of an analysis in our minds for forces, from where we may proceed to other methods. For shear walls, fixed connections to foundations is appropriate.
    - For moment frame and gravity columns, instead of fixed, another way is to assume pinned connections at the column base at the foundation, which is the most flexible option, as pins only resist shear and axial column force but not moment. This results in lower member forces in the structure, but higher drifts, which is a disadvantage, and building periods. In reality there is moment resistance which may need to be accounted for, otherwise we can end up with too conservatively stiff designs to restrain flexibility and therefore drift. Not designing for moment also enables lighter detailing. When grade beams are modeled around the pins, they also affect the results within the building and the building period and help achieving moment resistance at column to footing locations. The foundation to column details must provide necessary moment freedom if pin connection is assumed.
    - The structure can be assumed to bear on springs which represent soil per code criteria and then the reaction forces obtained at the foundation are used in a separate model. Effect of foundation rocking may need to be included in some cases.
    - In other cases when we have weak soil and heavy and or rigid structures, we might need to take SSI (soil structure interaction) effects into account as discussed in earthquake section in Chapter 1. We can either make separate analyses for foundation and the structure, or together, which would be more complex. Depending on the project and the type of analysis we may need to make few cycles of structural and geotechnical analysis after each other, since the foundation spring constants used in structural analysis affect the structural analysis results and the loads transmitted to the foundation, and these loads affect the geotechnical analysis of foundations which then affect the soil spring constants and so on.

Representing Members with Elements:

Models are made of members such as beams, columns, walls, foundations and slabs that are represented by certain elements, which are connected to each other and to external supports. During making the model, even before running the analysis, the engineer should visualize in mind how the structure would respond to forces in general, and start building the model and arrange members to capture expected responses. There are always multiple ways of modeling. Use of certain elements types and modeling approaches can also be mandated or recommended by codes or standards for given situations. In the list below the goal was to list general points about modeling members with elements. Some of the specifics of modeling of members can be found in other places in this chapter or Chapter 1, while discussing such members.

- The commonly used structural analysis programs come with a predefined set of elements. The commonly used elements are:
  - 1D frame elements, also can be called as beam elements, or line elements,
  - 2D elements such as plate, shell or membrane,
  - 3D elements also called brick or solid elements,
  - Axial or rotational spring elements,
  - Rigid links
  - Other elements include: Cables, base isolators, viscous dampers, and even gap elements

  The elements mentioned above also have subcategories each, depending on how they are supposed to react to loads and how they are to perform in the model.

- To these elements, stiffness properties are assigned as linear or nonlinear, as applicable and per our needs. These properties are either clearly known from common material data or must be found from previous experience, or experiments. It is important to assign these properties as correctly as possible because they not only affect the members they are assigned to but the whole structural behavior and how loads and stresses are distributed.

- We can define 2D surfaces such as slabs, shear walls, steel plates with 2D elements. These are often confused with each other, so let's try to clarify some of their important differences:
  - Membrane Element: Membrane elements are flexible elements that can only resist tension in their plane. So in other words they resist loadings though their axial stiffness only. They do not have bending stiffness and cannot take any bending. A camping tent cloth is a membrane for instance. A membrane is basically a 2D version of a 1D cable or string. Even if they do not have bending stiffness, membrane elements can still take on transverse loading. For example with the help of a friend let's say you pull a table cloth from its four corners so that it is in tension, and another friend places a box on the cloth. The table cloth can carry the box, not because the cloth can take bending, but because it is in tension which transmits the weight to supports, which are your and your friend's hands in this case.

  - Plate Element: Plates have bending stiffness and can work in bending. Plates are thin, flat members but have thickness. Plates can be classified as thin or thick. A thin plate's length and width is much more than its thickness and for plates shear deformation is neglected. A plate's edges can be free, hinge supported or fix supported. Now imagine a box is on a table. The table's top is the plate element here. It carries the weight of the box through its bending resistance. So similar to a beam in bending, the table top's bottom part will be in tension and upper part in compression, and it would sag, although this will be negligibly small for us to see. Similarly a reinforced concrete slab can be modeled as plate or shell (or just rigid), depending on which one fits better in our analysis.

  - Shell Element: Shells can take both membrane (axial) and plate (bending) forces, so they are like the combination of membrane and plate elements. Shell elements may also be curved. They can also have layers *(also called as sandwich model)* to simulate nonlinear behavior. With shells, we can represent any surface such as shear walls, basement walls, slabs, bridge decks, curved surfaces in 3D... Shell or plate elements can also be used for beams with wide flanges. Shells can also be grouped as thin or thick, and similar to what we said for thin plates, shear is neglected for thin shells.

- For the sake of example only, if we assigned slabs as membrane type in a building model *(which would be improper since slabs work in bending)*, diaphragm action would be absent, the total period of the building would be greater, the building would be much more flexible and deflect more, in comparison to assigning the same slabs as plates or shells.

- If a structural analysis program can automatically make members from an imported architectural drawing into the analysis program, this saves time, but still all these elements must be checked and modified as needed, and additional elements added as necessary by the engineer. At the least, the imported drawings can be used for tracing and starting the axes.
- Columns, beams, trusses, coupling beams, struts, braces can be represented by 1D frame (line / bar) elements, with their properties and cross section geometry assigned. So it is as if the whole beam or column cross section is concentrated into that line element.
- Shear walls are modeled as shell elements or by using line elements as well with concentrated hinges at each level with certain stiffness. Tall shear walls can be modeled as fiber elements along their length.
- Slabs can be modeled as plates or shells or as 2D finite element mesh. The slab connection to vertical elements for the diaphragm action of the slab should be represented in the model. If the diaphragm is not rigid we should model it with 2D finite elements, otherwise as rigid slab. If the slab geometry is simple, it can also be solved as representing as stiff or rigid beams.
- Floor masses can be represented as lumped masses at each floor for earthquake analysis.
- Spread footings can be represented as fixed or by axial springs. Mat foundations are normally modeled as finite element mesh. Piles can be represented by vertical and lateral springs. The subgrade can be represented by springs. Geotechnical engineer input is required.
- When the centerlines of frame elements such as beams and columns do not intersect, these distances (eccentricity) must be accounted for in the geometry of the model.
- Adjacent surface elements should work together. When the surface elements are connected only at their corner nodes, it means the edges will still work independent from each other, when in reality they shouldn't. This means we must make surface elements small enough for better precision or the program must have the ability to make adjacent elements' edges work together, not just at their corner nodes.
- Joints can be represented either as rigid, which means infinitely stiff, or, stiff with a certain finite value. The beam column connections for example, may be represented by rigid connections, since beams and columns are cast monolithically.
- For concrete and masonry, we should include the case for cracking of members, along with their expected locations. Cracks will reduce the overall stiffness of the structure. In some cases cracks can be represented even before any loading on a structure due to shrinkage of concrete. Cracking causes the load resisting (effective) cross section area of the members to decrease. Columns and beams may have reduced cross section, and the slab elements may have reduced

thickness due to cracking. These should be accounted for in the model by considering cracked section properties.

- For braced frames, compression buckling or tension yielding need to be represented at proper locations in proper sequence per design criteria to initiate at their respective strengths. It is hard to predict and model connection failures, therefore this must be avoided during design.

Running Analysis and Interpreting Results

After the previous steps are complete and the draft model is constructed, it will be analyzed, by using one or more of the analysis methods we covered under earthquake section in Chapter 1.

In this step modeling mistakes may prevent the analysis from progressing which must be corrected by the engineer. Mistakes such as forgetting supports, placing a hinge too close to supports, entering critical info incorrectly, program not being able to mesh a specific area properly... are modeling errors which may prevent the analysis to even run or finish. These kind of mistakes must all be manually corrected no matter how advanced programs are used.

After the modeling mistakes are resolved, the analysis is performed. This time the program gives us many results including external reaction forces and moments, forces, stresses, deformations, rotations and deflections for each member and the structure, the deformed shape of the entire structure (including higher modes), natural period, mass of the structure and floors, base shear..., and presents them in tabular and graphical forms as applicable. Although the analysis was able to progress, many errors may be generated in this step, this time not because of the inability of running the analysis but because of unacceptable results of the completed analysis that do not comply to the criteria established by the selected code, or some warning messages which the program is able to automatically identify at once. More specifically, the things to correct here can include: Relative and total floor drifts, secondary effects such as P-Δ, overall or partial stability issues, column and wall slenderness, loads vs. capacities in beams, columns, slabs, walls, and member joints, overturning of the whole structure, irregularities *(i.e. irregular floor plan, torsional irregularity, insufficient diaphragm continuity, weak and soft stories, vertical element discontinuities - also see building irregularities section earlier)*, mass participation ratio, strong column weak beam and more...

After the error and warning messages are also resolved, the engineers' job is still far from finished. Even though the computer analysis is now able to run and every error and warning messages are taken care of, this surely doesn't automatically mean that everything about the analysis was performed as intended. Taking care of warning error and warning messages is an important step but in the end this simply means that we have just constructed a model and performed analysis per the criteria defined in the program which doesn't generate errors anymore, but this doesn't necessarily mean that everything went according to structural engineers' design and analysis intent. The engineer or the engineering team must do further checks and verifications, and try to make sense of the results, even after all error and warning messages are taken care of. In other words, simply constructing a nice structural model and having the program perform an analysis without errors is not enough to say that the structural design is complete and the plans can be submitted. The results, all results, must make sense to the engineer, be acceptable to the engineer in the first place.

Of course, this checking and verification of results doesn't mean that the engineers should hand perform the complex mathematical work that was performed by the program - this is neither possible nor beneficial. What must be done here is that the engineers should perform certain tasks to verify and make sense of the results. As we said at the start of this section, this means making sure that all results such as loads, displacements, stresses, settlement values, story drifts, slab deflections, connections... make sense both globally and for parts of the structure and members, even if per computer analysis everything checks out and no more errors are generated by the program. For example:

- Viewing the deflected shape under load combinations is essential both on global level and for members as necessary. Also depending on program capabilities, performing visual animation of the displacement results is also helpful to see if they make sense. Both the static (not moving) and animated (moving) deflected shapes are shown in exaggerated form for clarity and seeing things easier. Vibration mode graphical display should be looked at. The shapes and magnitudes of all these, can help us identify various things at once such as gross modeling errors, critical members, if something is disconnected, how members work together partially or in the whole structure... An experienced engineer will know what to expect generally from the analysis and can often tell easily if the structure was modeled incorrectly, after looking at the overall deflected shape, deformations and magnitudes, when something doesn't feel right. Doing this step properly will reduce the time required for performing the rest of the steps below.
- Performing hand calculations, to get an idea about key aspects, and comparing these with computer results. Basically, the engineer should be able to hand verify any results that is produced by the program, in other words, the engineer should be able to simplify and hand check any aspect of the model as he/she sees necessary. Hand and computer results should not be "too far" from each other based on the engineer's judgment. If they are too far, something may really be wrong, which can be a sign to investigate the model further such as the dimensions, supports, loads, members, connections. More specifically we mean items such as below (and we should emphasize that the engineer's experience goes a long way for all of these):
    - Check global reactions in the first place. For example under gravity load, the total vertical reactions at the foundation, should be equal to the building weight. Check the symmetry of the reactions and how they are distributed to see if they seem to be distributed as expected or if a reaction seems too high or low and there is irregular distribution.
    - Look at how forces are distributed and flow in the structure globally and within members. Check for signs of irregularities.
    - Check the member results such as displacements, force and moment diagrams, for columns, beams, slabs, shear walls, bracings, connections, footings... manually and see if they seem proper. For example see if a member is getting too much or too little force or moment than expected. Nowhere in the structure should be left without looking at it or checked. First checking a typical beam, slab, column and making sure it seems ok, may greatly reduce the time required for further checks. Check critical locations, places with stress concentrations, for example hand verify a typical location with punching shear.
    - Reinforcing steel details for RC buildings or steel details for structural steel buildings can now automatically be generated by structural analysis and design programs in minutes,

which took days or weeks in the past manually. The engineers must review all these details generated by the program.

- Sometimes a good way of checking things and identify potential problems is, after saving the backup of the completed analysis file, make a copy for testing, and perform certain tests on that copy by introducing intentional disconnections or loads anywhere as necessary and view the results and see how the analysis would react or if it will react at all or what messages will it generate if any.
- A large, complex model should be checked with an independent second modeling and analysis.
- For nonlinear analyses, at least one engineer who is well experienced and trained in nonlinear analysis must review and interpret the results who, in some cases may also come as an outside consultant.

The software is ultimately a computer program and the engineer must be knowledgeable about the programs capabilities, assumptions, limitations, theoretical background that is utilized, graphical and tabular results, and detail drawings that are generated by the program. Consulting the user manual of the program is often necessary. And anytime a new software is utilized, or even a new version of an existing software is installed, it should first be checked independently by in-house multiple users if possible, with respect to a project with known results. Performing analysis on a real project without adequate in-house testing of the software first, no matter how widely used that software may be, is not recommended. The primary purpose here is not necessarily to find mistakes in the program but rather to see if it will produce results correctly and as intended for the particular needs of the project team based on their way of modeling and using the program.

Based on these, the preliminary member sizes and arrangements assumed in the beginning of the analysis stage can now be refined to meet capacity and performance criteria.

After corrections, another round of analysis is performed and further refinements can be made. Because of this loop, and the preliminary design model affecting analysis and the analysis affecting the modeling are inseparable steps.

If needed, analysis for critical construction stages must also be performed especially for structures such as bridges or structures with irregular geometries. Remember, during construction, the structure is not able to work as a whole yet, which might present unbalanced, inadequately braced / supported loading situations, all of which must be taken into account, for different stages of construction.

When finished, the designs go to architect for coordination, along with drawings from other consultants such as mechanical, electrical, plumbing, IT and others. It is only possible to see certain coordination issues only after all drawings are put together. At this stage, in addition to combining the different sets of drawings together, we should also never forget about the construction stage and the real world. Many things that seem to be working during the design phase and on the plans, may not work in the field, which means input from field engineers and superintendents must also be obtained at this stage. If you are a design engineer, you are not responsible for the exact means and methods that the contractor is going to use, but still, you cannot design with disregard of the construction phase. Often, suggestions and certain notes about construction methods and construction stage will also be added to project drawings and specifications, which would be quite similar or even identical to a point for the same type of projects. These notes are there for purposes such as clarifying scope for parties, communicating design intent better, to give warnings or simply to present options.

CONSTRUCTION PHASE

Finally, after all the designs are finished, coordinated and they are "complete", the finalized drawing set goes to the officials to be approved, and after approval, the general contractor can pull a permit to start construction. If you are involved in construction in any way, you might have already guessed why the word "complete" is in quotes. This is because many things can only be seen or clarified during construction, which would mean (hopefully minor) modifications to these designs. During construction, typically the contractor will have many questions to the consultants who prepared the drawings.

As we will see in Chapter 13 - Project Management, the design of drawings, making project budget and schedule falls under planning, and construction stage falls under execution part of project management. We mentioned project management at this point because for a construction project, all of its concepts start to work together in this stage. So, the skills of a construction project manager, although essential in every stage, is needed the most during planning and then construction phase where everything goes on at once. The better the planning stage is performed, the less burden will be on the project budget and schedule and the project team during the construction phase.

Construction Coordination

Coordination of different trades is of utmost importance during construction of any type of structure - not just buildings. We already mentioned or discussed many concepts applicable to construction of buildings in Chapter 1 and in this chapter, in their proper places. Now let's list some other common construction coordination issues for buildings:

- The first thing we need to do is to survey the building footprint and then excavate. During this stage, the sitework and utilities can also be surveyed and started but this is not always the case at least around the building, because we will need working space there. This space is for cranes, or deliveries and staging area and for other equipment, and therefore depending on site size and layout, we can start sitework and site utility activities after at least the structural work of the building is complete. On many jobs sitework can also be done in phases, to finish locations where can be done first and then leave the ones which conflict with other work for later.
- Coordinating the footing and slab inspections with the progress of concrete work may be difficult. The inspector may not come to look at it before the forms are complete, but that stage is actually too late to schedule a concrete pump. Therefore we will need to be in constant communication with the concrete crew and the inspector who will inspect the work. Most

inspectors will need about a day of notice at a minimum to check the work, so we need to have good estimates about the actual schedule of work at given points in time.

- MEP (Mechanical, Electrical and Plumbing also including Fire Alarm, Fire Protection, Security System, IT) coordination drawings must be prepared and approved as soon as possible, which must be coordinated with architectural and structural sets, (especially beams). This is one of the heaviest coordination tasks in design of a building and should ideally be complete before ceiling work starts. Architect, structural engineer, MEP engineers, general contractor and his MEP subcontractors should all participate. Various 3D software also gained popularity since over a decade, such as BIM (Building Information Modeling), some of which can also take into account time (which makes it four dimension) but on smaller projects just CAD drawings coordination is still common. Contractors for each trade will require the CAD drawings from architect to start their own shop drawings. Usually the mechanical contractor is the first to start, who works around the architectural set (which is ideally already coordinated with structural at that point), as his ducts occupy the largest volume. After that the other MEP and architectural trades coordinate. This includes not only the MEP items themselves and their conduits and pipes, but also their supports, such as hangers, support frames, seismic braces which make the process even more challenging. In some locations the slab or beams will need to be penetrated by these conduits. At other locations there will be heavy concentrated loads, which must be coordinated with the structure and architectural items.
- Work scope conflicts and questions between trades must be resolved during contract award stage, which otherwise will likely cost time and money by being a change order afterwards.
- For a steel building, coordinating steel shop drawings with almost everything else is necessary, and we make sure that the shop drawings, after approval, are distributed to every affected trade. Then the fabrication and delivery of the structural steel will be the critical activity for the whole schedule (and before that, the submittal preparation and approval stages).
- The goal of coordination before construction is to minimize the unforeseen issues but it is unavoidable that some issues are seen and solved during construction. For these, the contractor, who is in coordination with his subcontractors, will advise the architect about the issues and the architect will come up with a solution together with other consultants. For each item, there can be multiple variables that needs input from different parties. Let's give a dimensional coordination example. Let's say we are locating cooling towers on a steel framed roof. First of all there must be enough clearances around the towers. Second, the pipe locations vs. the drain locations vs. the steel beam locations at the roof must be coordinated. Third, the legs of the towers will need to bear on the proper locations on the beams. If this coordination was not done during design stage, it will be done as the work progresses. In many cases, installed structural elements will also be field measured by the trades who come later, before they start their own work, regardless of prior coordination.
- When making construction schedules, it is important to know when we will be able to enclose the building with the exterior shell, in order to separate inside from the outside weather and start interior work. For each floor, all interior activities ultimately depend on that date. For example if you see a tall building structure going up many floors without facade installation following it from several floors below, this usually means not the best project management, or it may even mean financial issues about the project if not facade work is seen at all. Because after

the structure is erected, there is no physical obstacle for the facade to not follow the highest erected / poured floor from 4-5 floors below, on which all interior progress depend on.

- For slab on grade, very soft soils should be excavated as necessary and replaced altogether with crushed stone aggregates and sand. During this, rainy periods would present extra difficulty and delays until the soil dries, as the wet soil (mud) cannot support the construction equipment. The deeper the weak soil layer, the more difficult and lengthy the process becomes. To make the process faster, aggregates may need to be poured over the area, just to support the construction equipment, but regardless, there may be no other way than to just wait for the drying of the area to continue with excavation.

Construction in Congested Urban Locations

We build our buildings more often in congested urban areas than in wide open spaces where nothing or little prior improvements exist. Let's list some typical points to consider and coordinate for construction of buildings in congested urban locations:

- The businesses, residences, public buildings, streets... in the area will still be operational. All construction activities, material deliveries must be coordinated with these, not only for them to remain operational and accessible but also to protect them from the effects of construction. If we need to temporarily use a part of these, this will create further challenges, costs and coordination need. For example in very congested sites without any room for delivery and staging of materials, or construction equipment excess, we may need to get permissions to use existing sidewalks, parts of streets, and even part of existing facilities. These needs will change as construction progresses and different activities take place, all of which must also be timely coordinated. Inadequate or late coordination may cause large delays and cost impacts.
- As we are talking about urban areas, in most cases we will also need to demolish an existing structure where the new structure will be built. So, most coordination and protection mentioned in the item above for construction, can also apply to demolition with different requirements. In tight locations where the noise, vibrations and impacts must be very low, demolition can only be done slowly with excavators that can reach high or even manually, instead of massive implosions or wrecking balls, which means cost and schedule impacts.
- Protection of existing often mostly means, protection from vibrations, noise, dust generated by construction and demolition activities. The working hours and even days may also be limited because in many cases disturbance from at least some noise or vibrations, although minimized, is unavoidable.
- Pedestrian traffic will also often need to be redirected and or protected. Temporary covered walkways or even tunnels may need to be built.
- The excavation will need to be vertical with tiebacks, due to space restrictions. Tieback locations should be coordinated to avoid interference with existing underground features such as utilities, tunnels or foundations.

- Any construction exceeding a certain height will need approval from aviation authorities (in urban area or not). For example in the US, FAA approval is needed when height of construction exceeds 200 feet or when the construction is too close to an airport, per FAA criteria.
- To serve as reference if future claims are filed by the existing facilities owners, surveying of existing facilities is needed, prior to start of any excavation work for the new construction.
- The utilities for new construction must be connected to existing utility system. This may create temporary disruptions of utility services in the area. In some cases temporary or permanent relocation or support of existing utilities may also be required. Utility trenches will need to be opened and closed fast.
- Sometimes there may be metro tunnels, large utility lines etc... running right below the proposed structure. Foundation locations must be coordinated with these during the design. In construction phase, in addition to sticking to those tight tolerances, the effect of foundation construction are constantly monitored at the existing underground structures. These underground structures can not only affect the design of foundation but may also affect the superstructure above, including eccentric design of certain parts of the structure.
- Surrounding streets will need to be cleaned frequently to remove dust and dirt generated by construction activities and traffic.
- Deliveries to the site, crane operations etc... must accurately be coordinated with construction activities to minimize the required access and staging area.
- Parking for employees who work on the project must be coordinated with the neighborhood. Various other measures is needed if street parking will not be available for the project team.
- Access of excessively large equipment or vehicles will need special permits not only for surrounding streets but also for the entire route and only certain hours may be available.

Crane Operations

For high rise building construction we will need cranes, which are very costly to rent, especially if crawler crane. Here are some typical points to consider when renting a crane:

- The type, boom length info and the required certifications, such as OSHA and others.
- We will need to provide a level surface for crane and before that make sure that the ground is strong enough for the weight of the crane, and if it is not, we might need to excavate and fill with stronger material and compact. In some cases, just a concrete pad will not be enough and even piles may be necessary which will require a foundation design on its own.
- The crane provider should include all mobilizing and demobilizing on site in his scope. They must have all the permits required during mobilization, demobilization or transportation. In tight locations, the crane company may need to close the traffic temporarily on an adjacent street when erecting or dismantling their crane. We must coordinate not only with the city officials but

with neighbors and other trades as well. This may affect a lot of things for construction activities during that time period for access of everyone else.

- We must make sure to clarify show up time, meal periods, overtime, holidays etc... because cranes are very costly and it is important to know who is going to bear what costs.
- Utilizing manlifts should definitely be considered to reduce the workload of cranes.
- For single, 1 day operations such as lifting large pieces, we need to coordinate the delivery date with the crane time allocated to it, since other trades will also be needing the crane.
- We need to make sure that there is enough working space for the crane. It requires much more area than it may seem at first. We must consider the boom length and the operating radius on a rigging plan and draw everything with proper scale on paper, also considering the dynamic nature of ongoing construction activities.

SEISMIC RETROFITTING **OF EXISTING BUILDINGS**

Seismic retrofitting is modifying an existing structure and or foundation to increase its ability to resist earthquakes. This improvement can be directed if per current codes the structure is deemed deficient or it can be voluntary by owner when a structure already satisfies the collapse prevention criteria per code, in order to bring its performance to higher levels. Both cases improve the value of the structure.

Any type of structure of any material can be seismically retrofitted, such as buildings, bridges, dams, tunnels and so on. Each needs considerably different solutions, but the main common goal is to increase energy absorption capacity of the structure. Almost all items listed below in the following paragraphs, are ultimately towards this goal, but, here we must not forget that adding new elements or increasing member sizes, thus increase in overall stiffness and mass of the structure also mean an <u>increase</u> in the earthquake demand on our structure, because a stiffer and heavier structure will have to withstand higher earthquake load in itself. Therefore, these additions are done such that, the increase in strength and energy absorption capacity of the structure will be sufficient to counter the increase in earthquake force demand, plus, make up for prior deficiencies.

Per ASCE/SEI 41-17 - Seismic Evaluation and Retrofit of Existing Buildings, we can aim for different performance levels during seismic rehabilitation (and of course for new construction too and the criteria in ASCE 41 is also applicable to new buildings). From lower performance (more loss) towards higher performance (lower loss) these levels can be listed in order as: Collapse Prevention, Reduced Safety, Life Safety, Damage Control, Immediate Occupancy levels, each of which require different objectives in rehabilitation, where the cost and schedule of retrofitting increase towards higher levels. For example, collapse prevention level means our goal is to prevent the building from total collapse, which will

> **Side Note:** We should analyse relative cost and schedule impacts of different performance levels. While better performance cost more, the difference may not be significant, even between collapse prevention to immediate occupancy levels, considering the fact that the design and construction teams already mobilize.

save lives, but that's about it and the building cannot be used anymore and has no remaining safety margin before collapse. Immediate occupancy level means the building still maintains its strength and stiffness that it had before the earthquake and can be continued to be occupied after the earthquake. The other levels lie in between. Each level has their damage level criteria defined in detail for different members and types of buildings.

Seismic rehabilitation earthquake analysis is more difficult than the analysis for a new structure, because instead of just making a new building work, here we must make a new system / new additions work together with an existing system that was built many years ago. Important things to consider include the ones below, most of which also apply during seismic design of new buildings.

- It is very desirable that the lateral force carrying system has redundancy, to create alternate load paths in case of local failures. If a building lateral resisting system lacks redundancy, adding of an additional lateral resisting system to create alternate load path (redundancy) is preferred over strengthening the existing system.
- For overall strengthening, the most straightforward and common alternatives is to add entirely new shear walls or braced frames which would run the entire building height. These can be much more stiffer than the structure they are added to, so any new irregularities, stress concentrations must be avoided and the connections to existing must be made strong. When shear walls are added, corresponding footing must also be added. In doing so, to reduce eccentric loading during earthquake, this new footing can be extended to connect with adjacent column footings, to take advantage of compression from columns there and to work together with the rest of the foundation system (for the latter, grade beams can also be used).
- Moment frames, if used, can only take on loads and be useful after the existing brittle elements of the building are damaged, since they are more flexible than shear walls.
- Sometimes parts of the building should work separately, which are easier to analyse, as long as they are adequate within themselves. In such case we must completely separate those sections, by movement joints. We must pay attention that the two parts will not have pounding (hammering) effect to each other (hit each other during earthquake), by providing adequate distance in between, which may also mean increasing the gap distance as we go up the floors. Pounding is more dangerous when floor elevations of adjacent buildings do not match and the slab of one building crashes the columns of the other at mid floor height.
- We can add entirely new frame adjacent to / outside existing frame to work together. In such case, it must be ensured that the newly added frame should be connected properly to existing slabs to transfer load adequately to and from slabs.
- We may add a whole new structure adjacent to an old one, so that they brace each other. This can be very effective and entirely eliminate the need for any further seismic modification. Another advantage is that the usage of existing structure might still continue during the addition of the new structure. Although expensive, if new extra floor space is already needed by the owner, and there is available area, this might be the best option.
- Simply decreasing the overall mass of the structure is a very effective solution to reduce the overall earthquake demands, if it is acceptable to the owner. Reducing of mass may include demolition of upper floor(s), replacing a heavy roof with lighter one, replacement of heavy

interior and exterior walls, heavy balconies with lighter ones, moving heavy equipment or storage loads from roof level to lower floors.

- We must check if there are any columns that would create short column effect, and take one of the measures as discussed under short column title, in structural terms section of Chapter 1.
- Seismic dampers or base isolators can be added. Especially adding base isolators is a highly challenging task that can be done by a small number of highly specialized firms.
- Nonstructural elements that may interfere with the building frame can be removed.
- For the soil and foundations, separate analysis is required. Soil materials may need to be improved, additional footings may also be required, or existing footings may need to be enlarged. Foundation strengthening and ground improvement were discussed in Chapter 1.
- We may make a member fail in flexure, which is ductile, instead of shear, which is brittle.
- The sequencing of construction is the responsibility of the contractor. But in seismic rehabilitation projects, if the removal and installation of certain members may affect the force distribution in the structure, then the structural engineer may dictate a certain sequence. The same logic also applies to new construction as well of course.
- After an analysis of the existing structure, it may not always be to our advantage to simply increase the strength of some members that we find to be the weak link. Strengthening them may sometimes cause too much load transfer to some other members during a seismic event, which were not designed for such loads. In such cases, instead of strengthening those weak members, we may just choose to increase their inelastic deformation capacities.
- Irregularities and discontinuities in a structure cause load to take complex paths which is undesirable. We must ensure that there is continuity of the structural frame everywhere and load paths are as straight and simple as possible. Irregularity in geometric sense can be as vertical or horizontal irregularity, or it can be as stiffness, mass and strength irregularity. Some examples to achieve continuity and regularity, are listed below. Also see Building Frame and Structural Members > Building Irregularities section under this chapter.
    - Columns and shear walls which do not continue all the way from roof to foundation must be continued to extend all the way. For shear walls this is a must. For columns there may be exceptions, which is still highly undesirable.
    - Parts of the building which are supposed to work together but tend to work separate from each other must be connected. In such cases we must make sure that the earthquake response of the connected assembly will behave as expected, including the foundations that will handle the combined effect of the connected parts.
    - Overall torsional irregularities must be corrected, to prevent torsional rotation of building, such as by adding bracings, shear walls (which must continue through the entire height) to balance lateral stiffness distribution on stories.
    - If a story is softer / weaker than the others, it can be strengthened with bracing or shear wall, but we must also watch for the continuity of load path with other floors.

- If we prefer just removing the irregularity, we must remember that this may affect the functionality and overall appearance of the building and for historical structures this may not even be allowed. In fact, all items listed above will be done with more difficulty when retrofitting historical buildings, to cause minimum disturbance.
- In cases when two adjacent buildings are connected by common elements, their effect must be analyzed and the connecting element may need to be separated, to prevent the buildings transmitting unwanted forces to each other.

- Items above applied to buildings of both steel or concrete *(or even wood but in this book we do not discuss wood)*. Specific to concrete buildings, important items include:
  - Jacketing of beams, columns, shear walls, joints with new reinforced concrete layer, which increases their cross section *(also see Concrete > Concrete Repair in Chapter 1)*, energy absorption capacity and strength against all loads. In such case, the continuity of added rebar between floors must be ensured. Jacketing by steel or carbon fiber reinforced polymer is also possible, which can significantly increase the lateral strength and drift capacity of column without failure. Especially fiberwrap jacketing can be efficient, considering their negligibly small added weight and cross section, employed effectively in many situations requiring strengthening of various members.
  - Post tensioning existing beams, columns or joints externally.
  - If there are members that are negatively affected from rebar corrosion, these must be clearly identified before seismic evaluation. If the rebar corrosion is at an advanced stage in some members, those may entirely need to be neglected from analysis, as the bond between rebar and concrete may not exist for those members anymore.
  - Improving defective existing reinforcement and then covering it with new poured concrete so that it will have full bonding with the existing section and the rebar.
  - Shear walls with inadequate flexural or shear strength must receive appropriate detailing. New shear walls can also be added as mentioned above.
- Specific to steel structures, important items include:
  - It can be difficult to know the strength and toughness of existing members, welds and connections. The testing methods depend on the project. The older the structure, the harder and more critical this becomes.
  - When we add new frame to work with an existing frame, the connections between the old and the new frame and new load paths must be analyzed.
  - Moment resisting connections may need to be strengthened to force plastic failure away from the joint location, since the joints are more critical than the beams.
  - Bracings, concrete or masonry infills can be added. These must not cause torsional irregularity for the whole floor. Among these options braces have the least weight, can still provide good stiffness and they are likely better compatible with a steel building.
  - Weak connections must be analyzed and introduced into the structural model.

Having accurate information about existing conditions, such as obtaining accurate as-built drawings and project records of the original structure is very important in seismic rehabilitation. In addition, as-built measurements are critical both prior to analysis and also during installation. Various tests may also be needed on the existing structure. Various unforeseen items may be encountered during installation. Therefore, the designer of the retrofit system and installer must work closely at each step, probably more in comparison to during a new construction *(which is usually relatively more straightforward)*. After we are satisfied that we have accurate and sufficient information about the existing structure, we can perform our analysis accordingly and decide what to do, which also depends on the cost of demolition plus new construction vs. cost of retrofitting. If we choose the option to retrofit, depending on the size and complexity of the structure, and the elasticity, stiffness and regularity that we will obtain, we choose a proper earthquake analysis method that was discussed in Chapter 1. The costs and benefits of each approach to retrofitting (target strength, stiffness, performance level, costs of construction) must be compared and finally the proper analysis method must be applied.

DAMAGE ASSESSMENT of RC BUILDINGS

So far we always talked about designing and building things, but it is always also good to see something from another angle too. Knowing about structural damages will provide further insight into understanding structures, which is the purpose of this section. Damages can occur in a building due to earthquakes and various other sources. Our focus here will be Reinforced Concrete (RC) buildings.

Post Earthquake Damage Assessment of RC Buildings

In previous section, we listed items for seismic strengthening, to better resist a future earthquake. In this section we discuss assessment of past earthquake damage that has already happened. These two things should never be confused with each other. The goal of damage assessment is only to list damages and to know if the structure can still serve.

Similar to anything else about structures, also for a damage assessment task, country / state / local requirements must be verified, and competent professionals must be consulted. The most important thing when evaluating damage after an earthquake is the engineer's experience. No methodology, measurement or equipment, no matter how great, can be a complete substitute to an experienced engineer's overall visual assessment.

Below are some items to be considered for post earthquake damage assessment. The source of some of these may also be structural overloading prior to earthquake.

- If overall damage seems too high, a decision may immediately be made from outside to call that building unfit for use, without even entering into the building. Too much floor displacement or overall rotation of the building are typical causes. If the building passes these criteria, and it is determined as safe for entry, assessment from inside can begin. Engineers can asses the floor with the greatest damage first which is usually the ground floor level or any soft story that might have occurred. If that floor has heavy damage, then it is not necessary to assess the rest and the building can be declared as unfit for use.

- The damage level of individual members of load carrying system and lateral force resisting system and also the overall state of the structure will be assessed. Then depending on the local criteria, these together will give the complete picture and form base for a decision of whether the building can continue service or not, and if necessary the extent of necessary repairs.
- Cracks from different forces occur differently such as:
    - Bending failures are ductile. They start from the edges of members and continue towards the middle, with a decreasing crack widths.
    - Shear failure is brittle, thus fast. Shear cracks appear as diagonal (see Figure Strength of Materials - 14 in Chapter 1), whether they are on columns, beams or walls. If the member was forced from single direction, it will appear as one diagonal, if forced from two directions, it will appear as two diagonals crossing each other, similar to an "x".
    - Torsional cracks are also brittle. They appear on opposite faces of a member as diagonal cracks and in opposite directions.
    - A tension crack can happen in a column, if it worked against uplift force. For example more commonly from moment created during earthquake such as from a force couple in two adjacent columns or less commonly from overall overturning effects. These appear as horizontal cracks. A tension crack on a beam will appear vertical in the middle sections.
    - Compression failure is brittle. The concrete will just be crushed in compression and from outside cracks may appear vertical. The concrete cover outside of rebar will pop outwards.
    - Multiple forces can also cause cracks at the same time, which would make the cracks show mixed characteristics from the effects as listed above and make them difficult to identify.
- For a concrete member the lowest level of damage is thin cracks. Then as damage level increases cracks get wider, larger and rebar starts to yield and the concrete cover may start to see some damage. After this, concrete cover is entirely broken, and falls off either by itself or by a small external force. And finally the rebar bends or buckles and the concrete core is also broken which means total damage. This paragraph can apply to all types of cracks from forces.
- Earthquake related damages can further be categorized as structural and nonstructural.
- To determine the condition of a concrete member, an experienced engineer can use a hammer. From the sound of blows, he/she can have a feeling of the integrity of the member. A clear ringing sound will indicate a healthy concrete, while a dull sound can mean damage.

Note: An experienced engineer assessing the building may also informally make useful comments based on experience about future seismic strengthening of the building, if that building still seems fit for use, but of course only the actual structural analysis after preparing a complete analytical model of the existing structure will formally determine what is really needed.

Damages from Various Non-Earthquake Sources

There may be other damages which are not earthquake related, due to other prior causes. To be able to correctly identify these is important to make proper judgment about the building's state of damage and suitability for future use. Non earthquake damages include:

- Plaster cracks. To make sure this is the case, plaster must be removed and concrete exposed.
- Some cracks can be due to rebar corrosion. Corrosion cracks follow rebar in concrete. See corrosion subtitle in Chapter 1 > Concrete > Reinforcing Steel. Corrosion also causes the bond between rebar and concrete to weaken. Thus, even when it is not in advanced state, an earthquake may trigger failure and completely remove the bond and cracks in concrete cover will occur due to this.
- Corrosion and foundation settling related cracks can sometimes be traced to water leaks. In steel buildings, water can also cause damages to structural members. Water leak in a structure almost certainly means structural damage to varying extents in the long term.
- There can also be cracks due to insufficient concrete cover around the rebar.
- Differential settlement of the building over time. We listed the type of damages differential settlement may cause. See Chapter 1 > Foundations > Foundation Repair and Strengthening. Foundation related damages can also result from heaving of soil.
- Temperature cracks.
- Shrinkage cracks.
- Differential column shortening related cracks.
- Vertical restraint related cracks in post tensioned slabs.

FACADE (EXTERIOR SKIN)

Although facade is not part of the structural load carrying system of the building *(except exterior load bearing concrete or masonry walls that are also part of the structural system)*, it still makes sense to discuss it now while we finish the chapter for buildings, because firstly, facade is a major component of a building, and secondly, the design of facade itself, how it will be attached to the building structure and interact with it, how it is affected by the movement or displacements of the structure, how it will withstand loads of wind, rainy storms and temperature changes still involves a great deal of structural engineering input at all stages. So, in design of facades, specialty structural engineers experienced in facade design *(who work for a facade subcontractor)* play primary role, and he/she must interact with the architect and the structural engineer (EOR) of the project, in order to coordinate with the structural loads, displacements and at the same time meet architectural intent, cost and schedule restrictions.

Times Square New York City, one of the most popular touristic places in the world. Here many buildings have large and heavy digital advertisement screens attached to their exteriors. These must be coordinated with the building structure and the facade both dimensionally and for strength, considering all possible loads, displacements, vibrations and environmental and thermal effects.

Completion of exterior skin of the building is almost always a critical item in the schedule, because only after the enclosure of the building the interior trades can start or proceed further.

Facades can broadly be grouped into curtain wall systems and masonry system. Masonry means CMU or clay blocks (but mainly CMU) which was discussed in Chapter 1 under masonry section, or brick veneer which will be discussed here. Block masonry were discussed in Chapter 1 because in most cases it is also a structural component, while brick veneer is strictly something related to facade only.

A facade can be of a single system or a combination of multiple systems. The variations in details are limitless to provide a unique and pleasing look to the building which must also function properly, to meet the facade goals as listed below. The facade system must:

- effectively separate the building interior from outside, and allow occupants to comfortably use the building,
- protect its integrity and not easily crack or fall off from inside or outside effects, work seamlessly with the main structure under load and displacement effects,
- protect the interior of the building from noise, wind, water and thermally insulate it,
- allow adequate sunlight into the building,
- be durable. It is easy to have a nice looking facade but making that durable without the need of frequent maintenance and functional problems is even more important.
- provide sufficient viewing of outside for its occupants.

Here the curtain wall also acts as storefront at the ground level lobby of a high rise building. Pay attention that the curtain wall is also working in coordination with the main entry canopy steel frame above, and the wall on the right which later receives large stone panels, and the canopy is also supported by the framing of stone wall. Facade elements at an entry are particularly important for a building's image and added value but long term durability of components is important.

We will discuss curtain wall and masonry facade systems under separate subsections below, but now let's mention a few general points that could apply to all facades in general:

- Facade must withstand not only its own weight but also dynamic wind forces, seismic forces and transmit these properly to the building structure.
- The actual displacements as a result of the following must be known:
    - Loads on the building, such as the elastic deformations from gravity loads, story drifts, building torsion from earthquake or wind in both directions, which get more critical for taller and more flexible (i.e. moment frame), more irregular buildings and higher floors, exterior beam or slab edge sagging which gets more critical with longer freespans.
    - Creep and shrinkage, column shortening related long term displacements.
    - Temperature changes (expansion and contraction cycles) which is more critical for steel buildings.

  All of these displacements listed above will result in applied forces on the facade system. As the building shape gets more irregular and complex, estimating these becomes more challenging.
- Corners of the building deserve special attention due to movements and displacements from both perpendicular directions from all reasons listed above. Movement joints in the facade or special corner sections or both will be required at corners.

- Interior walls, ceilings, roofs (including parapets), MEP trades, fire insulation, miscellaneous metals (i.e. steel supports for various items, handrails of a balcony and similar things) must work well with the facade in order to maintain the goals of the facade system we listed above, and their design and facade design can affect each other, which must be coordinated by the architect as early as possible which also needs input from structural engineer, general contractor and facade subcontractor. This item is more critical for curtain walls than masonry because masonry absorbs deviations in the field as necessary, easier than a curtain wall system.
- Connections that connect the facade to the main structure is designed by the facade specialty engineer. These connections must not only adequately hold the facade but also not cause local damage to the structure at attachment locations. The structural engineer of the project (EOR) on the other hand, is responsible for the structure to withstand the loads that are introduced by the facade to the structure.
- Depending on local codes and particular project location, the acceptable level of damage to facade during seismic events vary. Preventing injuries from falling of facade is a major common goal, while other damage may be acceptable.

Building a mock-up will also be needed in many cases, such as on a small portion of the exterior frame, which may be helpful in identifying potential issues and provide an actual physical sample to review by the owner and the architect, before proceeding with the full design and construction.

The items above apply to facades in general and should be kept in mind at all times while reading the two subsections below, which will not be repeated. Now lets cover more specifics.

Curtain Wall Systems

Curtain walls can be as framed, glazed curtain walls (including storefronts), EIFS (Exterior Insulated Finish System) or other less frequently used systems. They can also be used in combination with brick.

The name "curtain wall" is because it is as if these are hung to the building structure like curtains, covering its exterior, separating from outside. To do this "hanging", while also protecting the building interior from outside effects, making sure that the curtain wall itself is strong enough and can accommodate movements, temperature, water and other effects is challenging.

A curtain wall system is made of various components <u>working together</u>, which can include the framing, glazing, attachments such as anchors, embedded plates, metal panels, joints and sealants, spandrel panels, insulation, finishes and other components as applicable, which are attached to the edges of the structural frame and the backing wall, which can be of studs or masonry.

Due to a lot of components working together which must achieve multiple facade goals as we listed, a curtain wall system is one of the most challenging components of a building to design and build and must undergo a large number of tests and verifications, which include: load, air infiltration, water penetration, thermal (resistance, conductivity...), fire resistance, light, acoustical, impact, condensation tests and more... and standard details and procedures are available for all.

Curtain wall installation starts at lower floors and continues upwards as seen in this typical example in the photo on the right. Not only this naturally follows the construction of the structure but for concrete or concrete core buildings it also allows shortening of upper level columns (and shear walls) more due to gravity loads by allowing more time, before anything gets permanently attached to the exterior. This becomes more important as building height increases.

In the previous page we listed the trades that facade can interact with. Even after the submittal approval, before manufacturing the panels and or glazing in the factory, field measurement of the structural frame is strongly recommended or necessary, and usually this is also mentioned in submittals as well. Even so, minor changes may still be needed in the field.

Even if all panels were fitted without problems, they must also create a smooth surface together. This can be checked by looking up from the bottom of the building, and also by observing the reflection of light on the finished facade, from different angles of sunlight and by holding a strong light from below at night. In other words, the goal is to align the panels on or within the frames correctly along all x,y,z axes, and with correct distance from the slab edge. As a simple geometric check, diagonals of frame must be checked and be equal to or very close to each other in length.

Sealants must be able to accommodate movements, adhere well and be compatible with their surroundings. They must be the proper type for the joint in which it is applied. Sealants can be exposed to shear, compression and tensile stresses to varying degrees depending on where they are placed and the selected sealant must be able to work well under the expected stresses.

To hold the curtain wall, at the slab edges, drilled anchors or embedded plates with anchors can be used. Anchors must be strong and stiff enough, and they must also have enough embedding into the slab. Embedded plates are laid out and placed around the slab edge before the concrete pours. In both cases, the slab edge must be strong enough to accommodate facade loads, which should not be an issue because of the minimum slab thicknesses that have to satisfy structural criteria anyway.

In buildings where architectural aspects play a heavy role, there may be two framing systems such as concrete interior with steel exterior or steel for both, where the exterior steel framing can be arranged in complex geometries as if a separate structure - which is still supported by the interior core structure - to shape and support the exterior of the building. Coordination becomes even more difficult in such cases. This situation is somewhat similar to inserting an extra layer to be coordinated between the main core structure and the facade, which complicates things exponentially. Here there can also be differential footing settlement issues if two systems are supported by separate footings.

As mentioned under wind load section in Chapter 1, as the floor height increases, the effect of wind loading will be greater, and wind can also act as a suction force, which must be taken into account.

When considering thermal effects,

- Overall expansion and contraction of frame must be accommodated for supports and panels.
- The fact that aluminum expands and contracts much more than glass must be considered.
- Glass is very brittle, unlike metals. Anything that may raise the temperature of the glass excessively must be avoided or the system must be able to tolerate it.

Brick Veneer

Brick veneer refers to non load bearing masonry units used at the exterior. Brick veneers do not contribute to the strength or stiffness of the structure and can be attached to the structure either by adhering or mechanical anchors to the backing wall such as steel or wood studs, concrete wall or CMU wall. Where anchors are used, the veneer bricks can be thicker. When masonry is used as exterior envelope, it must obviously withstand external effects.

Attachment to Structure and Allowing Displacements:

The veneer bricks at the exterior can not continue from ground to the roof level without interruption all the way up. At every certain vertical distance, they are separated by horizontal gaps (joints) in order to allow relative movements with the structure due to general reasons listed under facade section, a couple pages back. To that list here we should also add expansion of clay bricks due to moisture. The larger the vertical distance between joints, the wider the joints should be. To achieve this gap, wherever we require the horizontal joint, L shaped shelf angles (also called relieving angles) are attached to the structure itself, such as the slab edges of the building, on which the veneer bricks of that level stand. And below the angle, masonry from floor below is terminated with a certain gap, which makes up the joint, which is filled by compressible material and then sealed from outside for watertightness. To clear any possible confusion here: The bricks at the lower levels could still withstand the whole compressive stress of bricks of many stories above, even if we did not interrupt the wall at every floor. Weight and the compressive strength of brick is not the issue here *(unless we are talking about a very, very tall skyscraper which would most probably not have brick specified at its exterior in the first place)*. The purpose of shelf angles therefore, as mentioned, is to allow <u>relative displacements between floors</u> due to movement, expansion or shrinkage. The maximum vertical distance allowed between joints, if any, depending on various factors, are found in applicable codes, such as TMS 402 in the US. The design of veneer and shelf angles must take into account the self weight, wind and seismic loads and it must closely be coordinated with the backing such as CMU or steel studs, thermal and water insulation, structural frame, sealants and flashing.

In addition to horizontal, there can also be vertical joints, again to prevent differential movements and cracking. In addition to having them every certain distance, they are also needed when two adjacent portions of the wall are supported on different members or sections of the structure, to eliminate cracking when differential movement occurs between the two supporting members. Vertical joints can also be used for masonry blocks, which we discussed in Chapter 1, whether they are used at the exterior or interior. However, a horizontal joint is specific to brick veneer at the facade for the purpose we discussed above. A horizontal joint it is not used for CMU walls, whether structural or not, which would eliminate the whole purpose of constructing CMU walls in the first place and weaken it.

As opposed to concrete masonry units, which lose moisture and shrink slightly over time, clay bricks absorb moisture and permanently expand in the long term. Even very little, this will happen and clay bricks will always be larger than the time of their manufacturing. Due to having so many bricks together, these very little volume expansions accumulate and cause wall to tend to expand in both horizontal and vertical directions. Because of this tendency to expand, if bricks are to be used over concrete backing, such as CMU walls which tend to shrink over time, it is advisable to use bricks that have previously manufactured at a sufficiently earlier date and absorbed some moisture already.

Water, Moisture, Air Resistance:

The water that may enter to a brick veneer system mainly enters between mortar and brick joints, and not through the bricks themselves. Even so, this water is still resisted by a waterproof layer. At the bottom, flashing is installed to direct the accumulated water away through weep holes. For the same reason, flashing is also needed over shelf angles, wall penetrations and lintels. There can also be a vapor barrier layer usually installed, to prevent moisture from reaching to interior, or reduce it to negligible amounts. Moisture inside the system can also cause corrosion of masonry ties.

CHAPTER 3 - BRIDGES

Bridges are among the oldest types of structures which are built to pass a certain span. They are built to cross any gap, such as a valley, river, strait or road that would not be possible or too difficult to cross otherwise. Ancient bridges consisted of stones, that mainly worked by arch action. After the invention of reinforced concrete, high strength steel cables, and wide availability of manufactured steel members, today it is possible to build much longer and taller bridges, pass longer free spans without piers (*for bridges, columns are called piers*).

Free (unsupported) span distance is one of the main identifiers of a bridge. The longer the free span, the harder to build. We can immediately judge a bridge by looking at its free span, before anything else.

As experienced bridge engineers would say, a bridge should look elegant, natural, and simple, in other words, beautiful to human eye, no matter the span. As long as the target functionality and quality is achieved, the best design is not when you have a lot on it, but when you cannot further simplify it, which is a principle that applies to any field. What we unknowingly call as "beautiful" is probably such simple designs. We all have a sense of load. Upon looking at a bridge, it should give us the impression that it can carry the loads naturally and as simply as possible, with minimum number of and as slender as possible components, where each component appears to carry load and their arrangement gives the impression of natural flow of forces. And indeed this usually means it is structurally efficient. For example if most people think that a bridge doesn't look appealing to the eye, then it is fairly high possibility that it is structurally inefficient. As can be inferred from these, especially designing large bridges is one thing where structural engineering gets as close as possible to art. The main items to consider for the aesthetics of a bridge include:

- The bridge should not be higher than necessary. This is not only structurally inefficient and costlier but also does not look the most appealing to the eye.
- The overall proportions of freespan, total span, central span and each of the spans, and their proportion to the depth and height of the bridge (including towers if any) must be considered and they must not be disproportionate in their proper place.
- The entire structure, such as its color and texture should blend in with its surroundings, it should look in harmony with natural landscape.
- We should have the minimum number of piers as possible. The piers need to be as slender as possible and they should preferably divide the span of the bridge into odd numbered spans where the span lengths should also be equal or as close as possible to each other.
- The approaching road should transition to the bridge smoothly, blending with overall design.
- Deck facade, soffit, slopes on both sides, abutments, tower shape and facade if any, cable configuration if any, must blend with the overall design. The deck must also appear as slender as possible.
- Shadow and light effects can also be used to better emphasize the desired features for the components and objectives listed above.

For buildings, architects are the lead designers and it is he/she who decides the geometry and look. Structural engineers design around that, to make it work. A bridge however, is almost entirely a structural item and in its design structural concerns play much larger role by far, with respect to architectural features than for buildings, because in the end bottom line is that we must pass a certain span over a certain terrain no matter how we want the bridge to look. For example building structure costs may probably only account for anywhere between 10-40% of the buildings' total cost, but for a bridge the cost of the structure is probably almost never less than 90% of the total and often approach to almost 100%. Due to aesthetical reasons we mentioned above however, for larger and more important bridges, although the structural engineer specialized in bridge design can be the lead designer, architects with knowledge in bridges are also usually involved. For common shorter span bridges however, it can just be the engineer.

The longest freespan bridge in the world as of 2024 is in Çanakkale, Türkiye, which was officially opened in March 2022. It has an unsupported length of 2023 meters (6637 feet) between towers. The second longest free span bridge is the Akashi Kaikyo Bridge in Japan, with a freespan of 1991 meters between its towers, and the third one is Yangsigang Yangtze Bridge in China, with the freespan length of 1700 meters. The longest bridge in the world, as far as the overall bridge length, is the Danyang Kunshan Grand viaduct in China, between Shanghai and Nanjing, opened in 2011, with a total length of 165 km (just over 100 miles), which carries high speed trains. Large bridges are perceived as among the most prestigious and iconic structures by the countries or the cities they are built in, such as the Golden Gate Bridge in San Francisco or the Brooklyn Bridge in New York, USA.

To choose the type and size of bridge to build, many factors are considered including *(as for any structure, also see the general factors listed in construction section in the beginning of Chapter 1)*:

- geometrical requirements and constraints,
- aesthetics,
- traffic type and load (current and future traffic demand), impact to regional traffic,
- traffic disruption during construction, if any,
- cost of construction, operation and maintenance,
- duration of construction,
- frequency and impact of maintenance,
- income or cost savings through the use of bridge,
- available equipment, materials, design and construction capabilities,
- public opinion, legal and political considerations, socioeconomic impacts,
- environmental impacts.

Bridges can be classified in a number of ways. Although it is possible to see different ways of classification in different sources, below are two of the logical ways of classifying. One method is by the materials used and the other method is by the structural system. We list both below.

AN EXAMPLE PROJECT: LONGEST FREESPAN BRIDGE

We will now introduce 1915 Çanakkale Bridge, the longest freespan bridge in the world, located in Çanakkale - Türkiye, connecting Europe and Asia to each other. Reading about a large project can also help learning many important aspects of bridge design and construction. *Note: Some parts of the information below might be followed a little easier if a minute is spent on internet looking at the general location of this bridge on the map.*

View from Asian side, looking North.

Overall Geometry

Middle freespan length between two towers is 2023 meters, making this world's largest freespan. The freespans on both ends are 770 meters each. So the total net length of the bridge is 2023+770+770= 3,563 meters. With the addition of approach viaducts of 365 meters on European side and 680 meters on Asian side, the gross bridge length becomes 4,608 meters, between the Asian and European land connections to highways. The height of the towers from sea level is 318 m each (and with inclusion of symbols on top it becomes 334 m), which also makes them the tallest towers in the world for a bridge.

Determination of Location and Length

The bridge is located at Çanakkale strait (Dardanelles in English) in Türkiye, where Europe and Asia joins and Mediterranean Sea connects to Black Sea, therefore it is at a location of extreme importance. Determining the exact route involved many criteria similar to the general criteria as we listed before. Let's look at some specific ones:

Choosing one of the narrower parts of the strait was obviously needed. Most parts of the Çanakkale strait would necessitate a longer freespan which would not be feasible.

View from European side, looking North. Try to visualize the span distances written above w.r.t photo.

In choosing the route, the slope and strength of the ground material was also important. On many narrow sections of the strait, the sea floor is either too steep and / or lacks adequate strength. Therefore, although the bridge is at one of the narrow locations, it is not necessarily located at the narrowest section of the strait, but the ground conditions were determined to be the most favorable at its current location after a detailed study by the transportation department.

Natural and historical significance of the nearby areas to the proposed route was also important, in order to create the least disturbance. The southern parts of the strait actually allowed shorter bridges, but at the European side there are highly regarded national park, monuments, memorials and historical sites. If the bridge was built there, the connecting highways which also carry heavy trade transit trucks and many tourist busses would greatly disturb the area. And on southern part on Asian side there is the Çanakkale city center, which would also be disturbed from the connecting highways and not only that but also the entrances to the bridge from that side would see heavier traffic and possible congestion.

Considering mainly the criteria above and some less significant ones, the current location was chosen. Apart from choosing the bridge location, this very long freespan length was decided on *(instead of building a third tower in the middle of the strait)* because:

- Çanakkale strait lies between Europe and Asia. It is an important sea trade route connecting Mediterranean sea to Istanbul and Black Sea. Such a busy sea traffic should not be interrupted.
- The long span allows avoiding the deeper sections of the sea bed for a tower foundation. These areas may also be of less suitable material for foundations. Consider also the exponential growth of the cost of foundations as the water depth increases. Of course here we should also remember that the longer freespan suspension bridge will require taller towers carrying more load each, which need higher capacity foundations, but this cost increase was easily compensated by placing the foundations at much shallower water depths.

- There is also the aspect of constructing a symbol, an icon just like in many other cities in the world - the longer a bridge and its freespan, the more iconic structure and prominent symbol. When it was found that the freespan length could be around 2 km, it was then decided that it should be 2023 meters *(see Symbols section a few pages forward)* in order to push the world record - which at the time belonged to the Japanese Akashi Kaikyo bridge with 1991 meters.

One unavoidable fact was that the bridge is located close to a highly seismic zone, only several kilometers away from the North Anatolian fault, a major and very long fault line that can produce very strong motions, which has similar characteristics to San Andreas fault in California. The bridge and its connecting viaducts have a seismic design return period of 2475 years for preventing collapse.

Impacts to Environment, Social Life and Sea Traffic

Environmental, social, economical and traffic impact studies were performed during initial stages, for both construction and operational phases of the bridge.

Socioeconomic Impacts:

The positive effects of this bridge is not only regional but also national. At Çanakkale, it reduced travel time between Europe and Asia sides from a minimum of an hour with sea ferries, which could sometimes be a few hours during high traffic or adverse weather, to just several minutes by car. This not only impacts the regional traffic but also national and international land trade and passenger transportation. Other Europe-Asian land connection is further up North in Istanbul, which was already connected by three bridges and an immersed tunnel, but those are far and Çanakkale bridge removes significant load from heavily burdened Istanbul traffic and brings much needed stimulation to this region. With this bridge, the whole Western and South Western part of the country and all the valuable touristic, agricultural and industrial areas in the region have much better connection with Istanbul, the most developed city in the country and also to western neighbors of Türkiye, Greece and Bulgaria and subsequently to the rest of Europe. In other words, the highway route from west / southwest of the country to the European side of Türkiye is literally shortened by at least a couple hundred kilometers on average, without needing to pass from Istanbul anymore. Also the historical silk road which spans from China to West Europe, has now a second connection at Asia-Europe border, in addition to Istanbul. The highways connected to this bridge will also be integrated with nearby airports, train stations and ports, which will greatly stimulate growth in the region. Forget all the socioeconomic stimulus, even the savings from fuel consumption and travel time alone would in the long run justify building this bridge.

Night view from Asian side, looking South.

Environmental Impacts:

Impact to marine life had to be taken into account. The activities that were found to have potential impacts to marine life were identified and appropriate measures were taken. This included pausing construction activities of foundations as necessary, which is a noisy operation that also causes vibrations. Some marine species which lived close to the tower base location were moved to safer distances. As another measure over 1.5 million trees were planted under the greater highway project, of which the bridge was part of.

Sea Traffic Impacts:

During the 5 year construction duration, over 250,000 sea vessels passed the strait. To minimize risk of sea accidents, all work areas over water and the construction vessels were equipped with signal emitting devices. During deck lifting operations, which consisted of lifting very large and heavy deck segments all the way up to the hanger cables over the strait, sea traffic was temporarily stopped for safety. In many other instances although the sea traffic was not stopped, it was slowed down.

Project Management

Construction started on March 18th 2017 and finished on March 18th, 2022, exactly one year before its planned completion date. Completing the world's longest freespan bridge 1 year earlier than planned at a strait with strong wind, current and waves, with most activities over water, during a worldwide Covid pandemic was indeed a project management success at the highest level.

The project was truly an international one. The capabilities and knowledge of many nations made it possible. A total of around 30,000 people worked on the project on one aspect or the other and at different times during entire design and construction periods, from many different countries. Although the workforce and site engineers, mostly civil, were to a great extent from Türkiye, there were also international talent on site, brought whenever necessary. The general contractor was a joint venture of two South Korean and two Turkish construction firms who operated as one unified team. Main structural design was performed in Denmark. Architectural design was by Turkish firms. Independent design verifications were performed in England and Norway. The main and hanging cables were manufactured in South Korea and tested and bundled in China. The steel shafts, bridge decks, and steel towers were made by Turkish firms. Expansion joints were made in Germany. Tower cranes were obtained from Australia. Wind tunnel testing including erection stages for the deck, tower and full bridge were performed in Canada, Denmark and China. Caisson placement system was provided by Dutch firms. Fluid dampers and flexible bridge bearings were obtained from Italy. Of course in addition to these there were hundreds more suppliers or subcontractors either local or from many different countries. Construction engineering, design of temporary construction aspects and temporary structures were performed through collaboration of the contractors, the structural designers and any subcontractors or suppliers as applicable.

Like in any construction project, so many different pieces working precisely together as intended and timely scheduling of installation of all components with highest quality standards and without cost overruns and were critical but here things were at the greatest scale. This also meant utilizing the best manufacturers, bring the best people and crews, and accurately coordinate their tasks. Very large and expensive construction equipment were used, such as the largest deck lifting crane in the world, and each moment of the operation of these large machines needed to be efficiently utilized.

A significant challenge during construction was emergence of Covid pandemics. For such a large project with thousands of workers, and under a deadline to finish, strict measures were taken to prevent an outbreak and run the project on schedule at the same time, which was successfully managed.

A close view from Asian side, looking North.

Foundations

Dredging and Piles:

To establish a foundation base for the towers, first the soft material on the sea bed was removed and then the sea bed was flattened by dredging. After that steel piles with open ends of 2.5 meters in diameter were driven. These piles not only transmit load to a stronger layer below, but also serve as ground improvement. Remember from Chapter 1 that the driven piles also serve as ground improvement by displacing and making soil denser around them. A total of over 300 piles were installed, ranging between 21-45 meters according to ground depth. These steel piles are called inclusion piles. An inclusion pile is something not directly connected to the foundation but installed in the soft soil layer as ground improvement beneath actual foundation, to improve load carrying capacity.

Crushed Stone Base:

After that, crushed stone layer was placed over the improved ground, which serves as base for the rectangular caissons. In other words, the bridge caissons that sit above this stone layer are not connected to the inclusion piles below this layer. This ensures that considerable energy will be dissipated during a strong earthquake or a large ship impact here at this crushed stone layer before even reaching to the structure. It was designed this way as an extra layer of safety for very strong events, not because it is designed to move at each earthquake or impact. Therefore, this layer also serves as a natural seismic isolator, in addition to helping more uniform load distribution to the improved ground below. The stone base is about 2-3 meters thick and needed to be laid very uniformly with high precision to provide a level surface for caissons. This levelness was critical in order to prevent tilting of the towers, to prevent nonuniform load transfer and settlements, and also to prevent nonuniform movements during a strong earthquake. Installing this stone layer uniformly with high precision in the order of several centimeters at the sea bottom required a special robot. After the crushed stone layer was complete, it was covered with an armor layer, for scour protection.

Caissons:

Beneath each tower, there is one giant rectangular caisson to serve as tower foundation, placed over crushed stone layer described above. In addition to serving as a strong foundation, the caissons also brace both legs of the towers together from bottom as can be seen in the photos.

Both rectangular caissons are 74 x 83 meters of width and length, 20 meters tall, and each is made of 50,000 m3 of concrete weighing more than 65,000 tons. They were first made right near the sea, on land at the European shore, by excavating a dry working area (also called as the "dry pool" or "dry dock"). Both caissons were partially completed in dry pool side by side which took 9 months until each reached to around 50K tons, and then they were floated by filling the dry pool with water and then pulled towards the sea, to the "wet" pool, to be finished there. To make sure that the flooding water lifted such giant caissons in the dry pool, before the caisson construction started, the flat surface of the pool was trenched with many channels, for flooded water to be able to fill in later and lift the caissons from below. If the flooding water did not lift the caissons, there is nothing else in the world to lift such weights. After flooding water lifted and floated the caissons, the sea side of the flooded dry pool was excavated and removed as necessary and the caissons were pulled by boats towards wet pool in sea.

The reason for first building in dry pool and then completing in the wet pool was that, making them completely in dry pool would mean a deeper excavation for the dry pool for the full height of the caisson, plus the more weight that would be needed to be lifted by floating water. This way, a lot of dry pool excavation was saved. The excavation and floating are not the only issues here, but we are talking about a dry pool area right near the sea and as the height of the separation between excavated dry area and the sea grows, the difficulty to make and support this separation grows exponentially.

The wet pool was an area in the sea with a pier, where it took 5 more months to complete the caissons. In the middle of each caisson, steel shafts which would form the bottom of the tower legs were installed. After the caissons and shafts were completed in the wet pool, the caissons were pulled by boats to their final location and sunk down to around 45 meters on Asia side and 35 meters on Europe side, to rest on the crushed stone base. The interior area of each caisson is not just empty like one big hollow pool. It is divided into many cells / compartments by concrete walls, over which a concrete cover was poured. These cells were gradually filled with water in a precisely controlled way in order to perform the sinking operation uniformly without allowing any shaking or tilting and also to place them in their precise location with a very small allowable tolerance of much less than a quarter meter. This was very slowly but successfully completed well within tolerances, the operation taking at least more than a day for each. The sinking operation had been carefully planned and simulated in computers way ahead of time, also taking into account the currents, waves and wind in the strait which had all been especially measured and recorded for a year ahead of time.

To protect the rebar in concrete of caissons from sea water action, a very special concrete mixture was used in order to guarantee a durable life of well above a hundred years. To obtain such mixture, the project team had to test a lot of trial mixtures in collaboration with universities. After the pour, even the smallest shrinkage cracks were sealed.

Towers

Towers heights are 318 meters from sea level, (and with addition of symbols on top, they become 343 m) which makes them the highest towers for a bridge in the world. Erecting the towers would obviously be easier on land, but this alternative could not be chosen, because in that case the towers would have to be too far apart, which would almost double the freespan distance needed to around 4000 meters, which would not be feasible with current technology, unless a third tower in the sea was added, which defeats the purpose *(plus that would be in deeper section of the strait)*. Therefore, even though building the towers over water presented significant design and especially construction challenges, the decision to place them in the sea had to be taken.

One of the towers (Asian side tower) zoomed from the photo a couple pages back.

The towers are made of steel, mainly because it allows faster erection speed and offers lighter weight for a given load carrying capacity. The segments are chamfered box cross section shape, with width and length varying between 75 to 110 meters and height of 11 meters.

The towers carry a total load of 36,000 tons from weight of the main cables (18000 each), and 55,000 tons from the deck, and also the loads that could occur due to thermal, wind and earthquake effects.

After the caissons including tower leg bottom shafts were successfully placed, thick concrete plinths were poured on the shafts, to evenly distribute the load from tower legs down to the footing. After this, the precast concrete tie beams between the tower legs at the bottom *(see photos above and below)* were also installed, which ensures that the legs brace each other and firmly attached to the caissons as a whole, for an overall frame behavior.

As an example of one of many challenges about construction over water, to keep the crane stable while lifting heavy pieces is not as easy as for a crane that operates on land, because in this case we do not have a firm crane pad foundation that conveniently bears on the soil as in land construction. All we have to support the crane is a barge that just floats on water. This problem is solved by constantly making up for the load imbalance by pumping in and out water at certain parts of the barge (which is called "barge ballasting"), in order to keep the barge level while lifting very heavy pieces. To keep the floating barge level becomes especially difficult during when not only lifting is required but also during precise placement of the large and heavy pieces to their place in the air, which was just the case while installing tower tie beams and lifting tower pieces.

To erect the towers, the largest capacity tower crane in the world had to be built especially for this project. The crane capacity was increased to be able to make each block piece larger and therefore substantially reducing the need for field welding and bolting (which is more difficult, expensive and takes longer to do in the field with respect to doing in a factory, as discussed in Chapter 1, under Structural Steel). The heavier pieces also had the advantage of being more stable under wind action, which was never ending. Then lifted blocks were first bolted and then welded each other. With this crane, a load of over 155 tons was lifted to a height of 318 meters during point of completion which set yet another world record.

After the towers were erected full height, other smaller cranes were also placed on top of the two towers for various lifting operations. Erecting and dismantling of these large cranes at 318 meters was also a challenge due to constant high winds.

Potential ship impacts to the tower was another major design criteria for the towers and the foundation. The design was made to account for a greater force than the impact force of the largest and fully loaded container ship impact.

There are mass dampers at the towers to dissipate energy from vibrations caused by strong winds or earthquake. To these mass dampers, sensors are connected to detect motion and act accordingly to dissipate energy and reduce vibrations. Note that mass dampers can also be without sensors which is also common, but adding sensors and actuation system can increase their effectiveness, as long as the counter action to movement is quickly and effectively provided by the dampers.

At the towers, between the deck and the tower there are fluid dampers. At the approach viaducts, the deck is supported by elastomeric bearings.

As seen in the photos, the tower tie beams are horizontal, but another choice was to make them as x-braces. That option would result in a stiffer, less flexible response to earthquakes, but at the same time it would increase the seismic force demand on the structure (remember, stiffness attracts force).

Anchorage Blocks

All tension force in the main cables are resisted by the massive concrete anchorage blocks at both ends of the bridge, which were made using around 90,000 cubic meters of concrete and weighing over 230,000 tons. The maximum expected force from main cables are around 50,000 tons but the blocks were made to resist much more than that.

In suspension bridges the huge reaction force needed to counter the tension from main cables is primarily provided by the body weight of the blocks and in turn the friction of the block and the soil. There may also be additional sources of reactions such as: the passive resistance of any soil that may be in front of the blocks, weight of any backfill soil that may be placed over the blocks, lateral resistance of any piles or diaphragm walls that may be constructed first as a foundation for the blocks. The goal is primarily to find an optimum for excavation and concrete quantities and minimizing construction schedule, while safely countering the immense tension forces from the main cables. In this project, in addition to the friction between the block and the soil, on Asian side, where the ground was weaker, diaphragm walls were also added below. On European side, where the ground conditions were more favorable, it was sufficient to add weight of backfill over the block, without needing a diaphragm wall.

The sides spans of the bridge on both Asian and European sides are 700 meters each. These long side spans allowed placing both Asian and European anchorage blocks further away from the shore, where the ground conditions were more favorable.

Main and Hanger Cables

After the towers and the concrete anchorage blocks at both ends were complete, the main cables could be pulled. Each main cable which is about 4368 meters of uninterrupted length, and made of fibers wrapped together. The ultra strong steel fibers that make up the main cable have a strength of 1960 MPa, which is almost 5 times that of ordinary reinforcing steel used in common construction. 127 steel wires make up one bundle, and 144 bundles make up the main cable. The main cables of the bridge is made of 162,000 km of high strength wires totally.

During manufacture of both main and hanger cables which are all different in length for different locations, extreme dimensional accuracy was needed because there was no chance of adjusting lengths later in the field. The dimensions were all based on not only on the final geometry of the bridge but also the calculated sags and lengthening, in other words all anticipated displacements, expansions and contractions in the cables and the bridge deck and towers, which varies not only from structural loads but also due to wind and temperature which can greatly vary across the year and also seismic forces when needed. The main cables were pulled by making use of catwalks on each side and temporary guide cables, also called hauling rope that were installed between the towers beforehand. Only catwalk itself was actually a big challenge, which was an important milestone.

At anchorage block connection, the bundles that make up the main cable were separated into multiple parts, to distribute the tension force to the concrete block at multiple points, instead of applying all force at one concentrated point. After the main cables were finished and tied to anchorage blocks, the hanger cables were installed, to which deck segments are attached.

Deck

Bridge deck is made of twin steel split box girders segments with steel deck on top, which are 24 meters long and just over 45 meters wide, with max thickness of 3.5 meters, each of which are supposed to carry 3 lanes of vehicular traffic each way, for a total of 6 lanes. Most important thing to note here is the 9 meter air gap in the middle of the deck for the entire length of bridge. On the sides of these segments, there are 3 meter wide walkways for maintenance.

The deck segments were made on land, and then lifted piece by piece, and attached to hanger cables and then to each other first temporarily by bolts and then permanently by welding. As each piece was very heavy and large, storing, transporting to final location and final attachments had to be done with great care to avoid any unwanted deformations, deflections and tilting anywhere.

Remember that in Chapter 1 under wind load section, we mentioned that wind tunnel tests are needed for especially large structures. For a long bridge like this, this becomes the most critical. The deck also includes a 9 meter air gap in the middle as mentioned above. Both sides of the air gap is a symmetrical box girder, to carry 3 lane traffic for one direction, connected to each other by steel cross girders at every 24 meters to act as a single deck. The 9 meter air gap is there for aerodynamic reasons, in other words, it was determined after wind tunnel tests that this gap was needed in the middle to prevent the deck taking a lot more force and causing much higher force and vibration demands on itself and the cables and the tower. The entire shape of the deck, including the exterior faces and the interior faces around this gap, and the wind screens *(which are for weakening wind force acting on vehicles and reducing risk of accidents and operational disruptions, at the expense of more wind force demand on the structure)*, was determined as a result of detailed wind tunnel tests which also included different erection stages. For a long bridge like this, wind becomes the governing criteria, even when the bridge is located close to a major earthquake fault line like this one. During wind tunnel tests it was shown that the bridge could withstand winds upto 280 km/hour, while the main design wind speed at the region is 110 km/h. Even though wind screens were used, the traffic on the bridge without limitations will be upto a certain wind speed which is a standard practice for suspension bridges. Then gradually as the wind speed increases, larger vehicles will need to wait.

The project location is one of the windiest in the country. Wind was an important factor always during construction, by affecting the lifting and placement of the large and heavy deck segments. Lifting operations could only be performed when wind was weak which affected the schedule.

The deck's bottom elevation is close to 73 meters from the strait sea level. Because of this, the maximum clearance height for the strait's sea traffic was decided as 70 meters, which is plenty for almost any sea vessel. And the available width for the sea traffic was decided at 1600 meters, centered within the 2023 meters freespan distance, which is again more than enough for even the most congested sea traffic conditions at the strait.

To lift deck pieces which weighed about 1300 tons each up to 73 meters high over the sea, a floating crane of 106 meters high with 5000 tons lifting capacity was used which made it the world's highest capacity floating crane. The moving lifting portals that are attached to main cables were also used for lifting. There were a total of 87 deck segments which took 2 years to lift and finish the deck.

To keep the balance of cables and towers, at midspan the deck pieces were started to be installed from the middle towards the sides. And at side spans, pieces were started to be installed from ends towards the middle span, to achieve overall symmetry and avoid unbalanced forces during construction.

At the two ends of the bridge, where the deck is connected to approach viaducts, there are over 16 meters wide expansion joints to allow movements and rotations in all directions while withstanding large forces. There are also smaller expansion joints between the ends of approach viaducts and the connecting highway ends from both Asian and European sides.

Photo taken from Asian side, showing the 700m side span and the approach viaduct on European side.

Approach Viaducts

At both Asia and Europe ends of the bridge, there are approach viaducts *(see photo above)*, which serve as links between the main highway and the bridge. These were made of post tensioned box section girders, something we mentioned under Chapter 1> Concrete>Prestressed Concrete. Viaducts are joined to the connecting highway and the bridge deck by expansion joints. They sit on bearings which are able to slide. The viaducts were built by incremental launching method, *(see Construction Methods section later in this chapter)* which not only saved time but also allowed traffic below to continue without disruption. As the soil beneath both viaducts was weak and the upper soil layer had liquefaction risk, both ground improvement and deep bored piles were used at the same time. *(See Chapter 1, Soil*

Mechanics for liquefaction. Remember we said the bridge is not very far from a high seismic zone, and the soil here is obviously saturated, being near the shore, which was also sandy.)

Operation and Maintenance

On the bridge there are over 1000 sensors which continuously give feedback to the operations center about deflections, vibrations and the bridge structural behavior. The maintenance operations can also be optimized this way. Maintenance accessibility was also a concern during design. Among the maintenance activities corrosion prevention is of utmost importance as the bridge is made of steel.

Project Funding

The project was built by a system called "Build, Operate, Transfer", where the design and construction financing was entirely provided by the contractors, with zero government spending, but in return the contractors get to collect toll revenues for a certain number of years (16 years and 2 months), with an agreed minimum of 45,000 vehicles per day toll revenue guaranteed by the government, where the difference is paid by the treasury when the actual traffic is lower. As of writing of this section in 2024, two years after opening, the actual traffic yet seems considerably lower than the agreed minimum, which is a subject of criticism because it means higher amounts paid by the treasury and in turn the taxpayers. The author also observed the low traffic even during day time while driving on the bridge. However, as explained above under social impacts section, the traffic will certainly increase, because of the great stimulation provided by the bridge to the area. There is no doubt that this bridge was needed here, with many direct and countless indirect benefits, but its cost seems likely to remain a topic of debate, at least in the near future.

Symbols

Like many mega-structures of the world, 1915 Çanakkale bridge also has national symbols built into it. But the importance of this aspect for this structure is high, because it is located near an utmost historically important location from near history. It is to such extent that, the bridge is regarded also as a monument as much as a bridge and the symbols reflect even in overall geometry, which should also be mentioned here, as below...

Located at the junction of Europe-Asia-Mediterranean Sea-Black Sea, Çanakkale strait is one of the most strategical points in the world and it was the battleground of one of the fiercest wars in history with global effects. During WW1 it was attacked by British and French combined navies, including soldiers brought from colonies, forming the strongest attack force of the time to pass the strait, capture Istanbul and win the world war quickly, and also to deliver help to Russian Tzar who was facing internal opposition. But against the odds, the Ottoman Empire, once a superpower but in decline since two centuries and in its final years, managed to defend the strait almost with its last breath. Çanakkale was impassable. The battles continued for 10 months on land and sea with casualties from both sides totaling around 500,000. The height of towers, 318 meters, symbolizes March 18th 1915, the date of sea victory for Ottoman Empire, Turkish Republic's predecessor. The freespan length of 2023 symbolizes the 100th anniversary of founding of Turkish Republic and also the planned completion year. The bridge's red-white colors are from Turkish flag. The figures that look like bullets on top of the towers are 75 tons and 20 meters each, and symbolize a cannonball that was lifted by a soldier to be loaded into cannon when the loading crane was broken during battle. The piers of the approach viaducts were made to resemble the monument of martyrs, located at strait's entrance. The battlefields also saw the entry of Mustafa Kemal Atatürk into history scene as a successful commander, who was a young lieutenant colonel at the time, who took the right initiatives at various critical battle stages that had effect on the course of war.

Therefore it is not a surprise that the bridge was completed one full year earlier than its planned date in 2022 instead of 2023 even during a Covid pandemic and under unending winds and currents, and with great success on all fronts, as for the nation and for everyone involved in the project it was not only building the longest freespan bridge, but also a monument that had to be completed with excellence, out of respect to history.

As stated several pages back, this section aimed to introduce many important aspects of bridge design and construction, by giving examples from a mega project. Now let's proceed with bridge categories and then we will look at various design and construction aspects.

BRIDGE CLASSIFICATION - BY MATERIAL

Concrete Bridges

Concrete bridges are mostly made as precast, but it is also possible to see cast in place. Precast concrete bridges are very commonly used especially for highway systems, where they can be designed as beams or cantilevers as far as support method during construction. In fact, precast concrete beam is the most common bridge type. The main advantages of precast as explained in Chapter 1 are, better quality of material, accommodating tighter work schedules, elimination of formwork, and the main disadvantages are the necessity to transport large and heavy materials to site, setting up logistics, lifting and properly connecting heavy pieces in place. The unsupported spans of regular precast systems can be as much as around 45 meters, but this is enough to make these bridges a logical choice for almost any highway or even railway systems at the majority of the locations, because multiple spans can also be created by simply adding piers to cross much longer distances such as large valleys, rivers or lakes, as long as we can place piers to support each span. With post tensioning, longer free spans are possible.

Steel Bridges

Steel bridges are also very common, although not as much as concrete bridges, and for the same type of system, they can cross relatively longer unsupported spans than reinforced concrete.

Composite bridges are made of both structural steel and concrete working together. A typical composite bridge is the one that carries the slab with steel girders and beams *(that is why we categorized it under steel bridges)*, and on top of that a metal deck slab is filled with reinforced concrete, where the efficiencies of both materials are utilized together. The concrete slab that fills the steel deck also serves as a continuous bracing to steel girders.

A steel bridge that supports two lane traffic, over a small valley park. Main girders are arches, over which columns are supported, carrying the composite (concrete poured over steel) deck. Near the bridge also see the tiered slope stabilization and supporting each tier by retaining walls which further stabilizes the slope as necessary because on top of the slope notice the surcharge load from buildings. The retaining walls are then decorated with plants which blend with the surrounding park's appearance.

Timber Bridges

They can be built where loads are not heavy, such as pedestrian bridges or bridges with light vehicular loads and traffic, where small free spans are also sufficient and aesthetical look of wood is desired. Short free spanned wooden bridges were built in the past for supporting vehicles or even trains.

BRIDGE CLASSIFICATION - BY STRUCTURAL SYSTEM

Beam (Girder) Bridges

These can be thought as the most basic and most common type of bridges, as they are often adequate. They can be of steel or concrete or composite (steel and concrete) or timber. Concrete beams are usually pre-tensioned or post-tensioned, although cast in place is also possible for smaller bridges. The beams can be of any cross section, such as I-section, T-section, or box shapes as needed. Post tensioned box girders can have largest and most efficient cross sectional area in which a standing person can easily fit inside and they enable longest spans for beam bridges.

Beam bridges can be single or multiple beam spans end to end, which bear on intermediate piers at each end - so one pier carries two adjacent beams end to end, separated from each other structurally by expansion joints of usually a few inches to allow possible lateral movements of beams. The number of spans can be as many as needed. Usually for highway bridges, each span length of a pretensioned beam can generally be anywhere between 10-45 meters (approx. 35 to 150 feet). In other words, at maximum of every 150 feet or so, which is typically much less, there must be a column or pier that will support the ends of the beams. As long as this is done, the overall bridge length can extend indefinitely, by just adding more piers and segments one after another, such as for instance, the Florida overseas highway. Figure 5.4.3.1 of LRFD for Highway Bridge Superstructures Reference Manual, July 2015 Revision by FHWA shows varying beam types with cross section depths ranging from 36" to 72", which allow unsupported spans that vary between 40 to over 150 feet, but notes that with some extra steps such as post tensioning and spliced joints, this can even be longer by 10% - 20%. For bridges that need longer free spans, other systems must be used such as truss, cable stayed or suspension bridges.

Precast beams seen from bottom. Beam spacing could increase, which means less number of beams if larger beam cross sections were used, which engineer optimizes during design.

Truss Bridges

Trusses were introduced in Chapter 1 under statics section. Relative to beam bridges, usually larger unsupported spans can be covered and heavier loads can be carried with truss bridges. This is because, when truss triangles are connected to each other and form a structure, together they can act like much deeper beams and provide structural efficiency, without the full weight of a solid beam of same depth. Also see photo under Chapter 1>Statics>Trusses for a view from different angle.

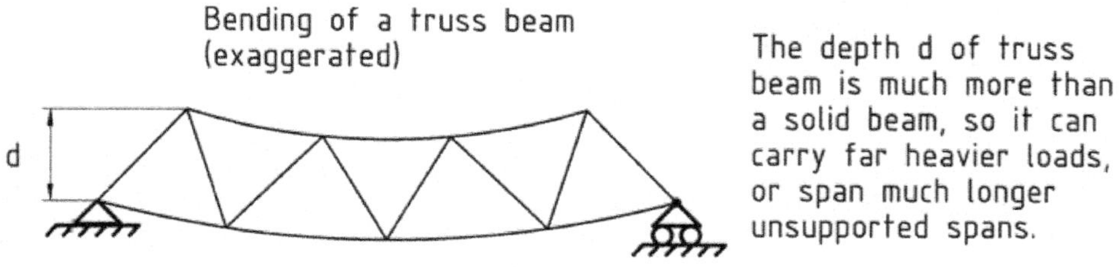

Figure - Bridges 1

Cable Stayed Bridges

Cables which directly carry the bridge deck, are hung from towers. Even longer free spans than truss bridges can be crossed with these bridges. The cables, which directly connect the bridge deck and the towers, carry the bridge deck through their vertical force components. Their horizontal force components cancel each other out, by the cables on each side of each tower

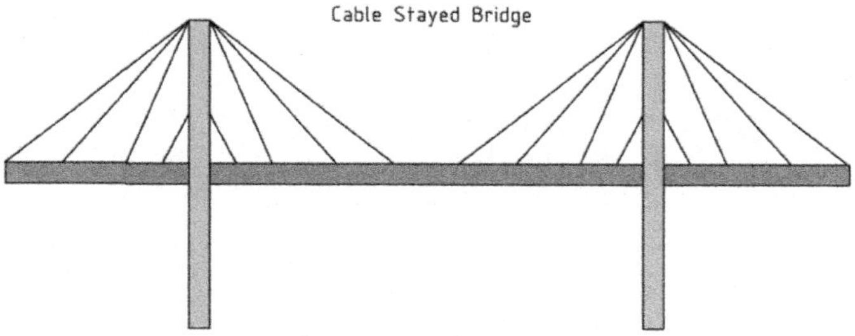

Figure - Bridges 2

Suspension Bridges

Suspension bridges are efficient structures considering the materials used versus the span passed. This system also uses cables, but instead of directly tensioning the cables that carry the midspan, a main carrier cable is suspended between the towers first, which support the other cables that carry the bridge deck. The longest freespan bridges in the world are of suspension type. These are more flexible than cable stayed bridges with higher periods, both in bending and torsion due to longer length and the way the cables support the deck *(and because of this, if higher stiffness is needed, they can be supplemented by cable stayed system)*. As the free span increases, the design and construction becomes increasingly difficult, the resistance to cyclic and dynamic loads such as earthquake and wind become more critical and the cost and project duration increases rapidly. The effect of wind in the design of suspension bridges is among the highest of any type of structure. The vibration modes (see Chapter 1, earthquake section) that the wind and earthquake can create in different axes can add up to many modes, (*theoretically, this is true for any type of structure of any material and size, but here we mean the modes that are not negligible*), more than even the number of modes for a very tall building.

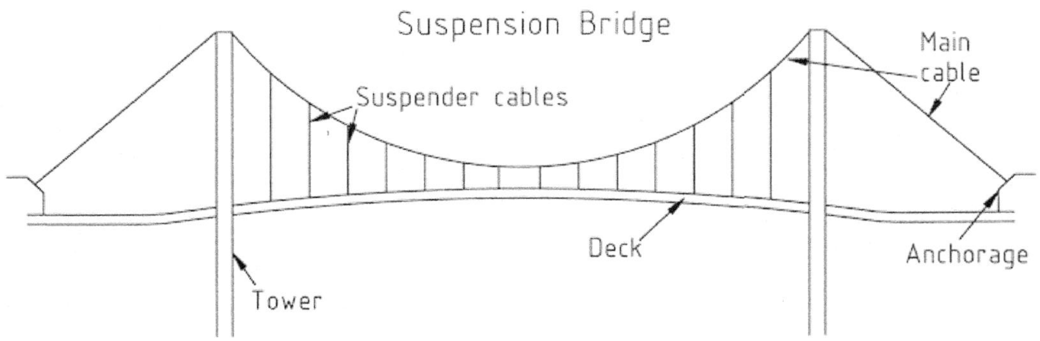

Figure - Bridges 3

The slight concave up shape of the deck that we see in most suspension bridges, is not (enough) for the arch action, but it is largely due to the fact that it is more comforting to the eye, and looks more appealing, considering that suspension bridges are large bridges which also serve as icons.

Arch Bridges

This is the oldest type of bridge, and it could be constructed with stone, when there was no reinforced concrete or structural steel technology. Arch action is also discussed under structural engineering section of Chapter 1, and Chapter 12 - Historical Structures.

DESIGN

Let's list some important points to consider in bridge design. These points apply to all bridge types as listed above, unless noted otherwise or it apparently belongs to a specific bridge type or types.

- Bridges are critical structures that must continue to serve to vehicular or rail transportation needs, even after earthquakes or floods. By saying vehicles, we mean trucks, because truck loads are much higher than car loads and represent the worst case. The most unfavorable possible truck configuration must be taken into account, which also means finding the worst case by making a moving load analysis over the bridge. Railroad bridges are built with even stricter criteria than the ones that carry vehicles. Especially the bridges for high speed trains have the strictest criteria, both for foundation and the structure, with extremely small tolerances for displacements and vibrations, which means very stiff and stable structures.
- The following loads are analyzed and are combined in various load combinations as applicable:
  - Loads from the self weight of the bridge with all its components.
  - Earthquake loads, for the worst possible case and load combination. We determine a return period such as 2500 years, set performance objectives that consider expected damage

levels and serviceability based on bridge importance category. As explained before under earthquake section in Chapter 1, the primary input into the seismic design of the structure is the peak ground acceleration, PGA, short (S_s) and long (S_1) period of the spectral acceleration coefficients, which depend on the seismic zone the bridge is at, which can be obtained from contour maps in AASHTO LRFD seismic design guide.

- Vehicle loads, such as HL-93 truck load as defined by AASHTO, and the uniform lane load.
- Impact loads when vehicles jump over bumps over the deck, vehicular or vessel collisions.
- Generally, the more flexible a bridge *(i.e. the higher its period)*, the more critical wind loads. For example, per AASHTO Bridge Design Specifications, section 3.8.3.1, bridges with span to depth ratio greater than 30, all cable supported bridges, and bridges with periods of 1 second or more are considered sensitive to wind. In other words, generally, the longer the bridge, the more critical the wind load on it.
- The edges of long bridges can be designed to minimize aerodynamic effects such as aeroelastic flutter, vortex shedding resulting from wind, as discussed in Chapter 1. There will also be uplift forces. Wind tunnel tests will also be required for long bridges. Even if all other conditions are satisfied, wind analysis may easily make a design unsatisfactory, therefore it may be preferable to analyse wind effects first.
- Other load types as applicable are: Ice loads, lateral centrifugal forces of vehicles while turning curves, blast loading, jacking forces
- Moving traffic load analysis must also be performed, to find the worst possible instance that would create the largest forces in the bridge members and largest reactions on supports. The concept of influence lines is useful here, which was introduced in Chapter 1. Of course, to move loads over the geometry of the bridge means a lot of calculations for each case, which were done by hand in the past, but now with computer software.

- Deformations due to various reasons can also cause additional forces, even secondary effects on a bridge structure, which can be used in loads combinations together with the ones from above. These deformation load effects include:
  - Uniform change in temperature of the whole structure: This causes the entire structure to lengthen or shorten. Bearing supports, deck joints, top of piers where bearings are fixed must be designed to tolerate this, otherwise there can be large additional stresses. The amount of elongation or shortening depends on the original length, the material properties and the change in temperature. Contrary to intuition, the coefficient of thermal expansion of steel and concrete are very close to each other, such as 0.0000065/F for steel and 0.0000060/F for normal weight concrete and 0.0000050/F for lightweight concrete, which prevents the problem of different elongation or shortening, which is a very useful fact for design and construction of composite steel-concrete and reinforced concrete structures.
  - Temperature gradient (temperature that varies within the structure): This happens when the temperature of a part of the structure changes more than the other parts, which causes that part to change in length more. For example during sunny weather, the deck can heat up

more than the girders below, which can cause the deck to expand more than the girders, thus an upward bending shape of the bridge.

- Creep & Shrinkage: As discussed in Chapter 1, these are time dependent slow deformations in concrete. Especially for prestressed, cable stayed bridges they can be significant and may govern the design. Shrinkage has different types and speeds, as mentioned in Chapter 1.
- Settlement of foundation: Settlement or differential settlement of foundations can cause large stresses in the structure. The more ductile the structure, the less severe will be the stresses caused by differential settlement.
- Post tensioning forces: In post tensioned bridges, the deformations caused by post tensioning can create additional forces.
- Fatigue effects must be considered for part or whole of the structure, when load cycles that occur on a regular basis are present. This applies not only to steel but concrete bridges too.

- In the US, AASHTO LRFD Bridge Design Specifications and AASHTO Standard Specifications for Highway Bridges are used for obtaining load combinations with the loads mentioned above. There are also other sources for more specific purposes such as the materials, pedestrian bridges, railway bridges, utilities.
- For cable stayed or suspension bridges:
  - In suspension bridges, the immense tensile force in the main cables are taken by the concrete anchor blocks at both ends *(see Figure 3)*, to which these main cables are attached. These huge concrete blocks resist the upwards force component from the cables by their weight of tens of thousands of tons, and the horizontal force component is resisted by the friction between the bottom of anchor blocks and the ground, plus the passive resistance *(see Chapter 1 > Soil Mechanics)* of earth in front of the block.
  - Cable changing and failure conditions for cables, vibration of cables due to all other loads including wind must be analyzed. Note: For suspension bridges, the main cables must be designed such that no failure or the need to change will occur for the structure's entire life.
  - All cables must maintain tensile force at all times including earthquake or wind conditions *(in other words we cannot allow a cable's tension drop to zero and carry no force).*
- Due to heavy loads, the foundations under bridge piers or towers must be very strong. So, usually pile foundations are used, but shallow foundations are also possible, if justified. The input loads for foundation design are provided by the structural engineer, as a result of the structural analysis of the bridge if foundations and structures are analyzed separately. Both the superstructure and the foundation can be analyzed together as well. After we know the forces and moments coming from the bridge pier based on structural analysis, we can find the number of piles, pile sizes under each pier by calculating the skin friction and tip resistance as you can see in free body diagram Figure 6. In many cases, soil structure interaction (SSI) effects must be looked at. Uplift and scour effects, flood water levels must be considered.
- Due to maintenance reasons, and to be able to easily detect any damages and perform repairs with least disruption to flowing traffic as possible, things such as bearings, expansion joints,

anchors, cables and other parts of the structure should be easily accessible. Accessibility for structural inspections and repairs and for corrosion protection is also important.

- For curved bridges, center of gravity vs. center of shear resistance must be looked at, to prevent torsion (rotation of bridge around a vertical axis) during earthquake.
- Bridge deck must be waterproofed to protect the structure below. Effect of de-icing salts must be accounted for. Especially at expansion joints waterproofing is more delicate and critical.
- For steel bridges, some of the important considerations are:
    - Critical items to consider for ensuring deck stiffness include the thickness and splices of deck plates, stiffeners, cross beams and main girders, their connections to each other, differential deflections.
    - Main girders may be cambered, to allow for sagging under loads, which should also consider the erection sequence of girders. Design and construction should seek to minimize and tolerate differences in the actual camber amount of girders, both from the expected values and also with respect to each other, which may occur due to various reasons that may affect geometric accuracy or relative stiffnesses and behavior under loads.
    - Even though the finished structure braces itself, during construction before the deck is poured, structural stability must be ensured. For example, long span girders must either be braced or have sufficient lateral torsional buckling stability which especially becomes critical during deck pours, with the weight of wet concrete, deck pour crew and their equipment and vibrations while the concrete is still wet, without any strength.
    - For arched steel bridges, in and out of plane buckling of the main arches must be considered. Buckling is discussed in Chapter 1 > Structural Engineering > Structural Stability. Buckling of arches is different than of columns. It happens in the form of snap through buckling, where a significantly different buckled shape occurs suddenly.

As mentioned before, bridges can be grouped in different ways, based on the materials they are made of or the structural system they belong to. Below, to serve as example, some important design points from the simplest and most common bridge type, precast concrete beam bridge is given.

Design Steps of a Precast Concrete Beam Bridge

The general steps to design a precast bridge may be summarized as:

- Determine the total bridge length depending on the gap that needs to be passed. This length will also dictate how many spans and therefore piers are needed.
- Determine the traffic load and type. This will dictate bridge width and the load it needs to support. Railroad bridges are designed to heavier loads than highways. This means, deeper or stronger or more number of beams (girders) need to be used, and piers and foundations must be designed to heavier loads, or the unsupported spans must be shorter with more piers.

- With experience and data from previous designs, assign a reasonable number of beams with reasonable depth to carry the bridge deck. This is the preliminary design stage.
- Now you can draw a cross section of the bridge, with the precast beams, and the cast in place deck on top. This section as in the figure below can be typical for a bridge carrying heavy loads, such as train loads, where the beams are densely positioned.
- Determine the deck elevation which will dictate pier and abutment heights. The bridges on rivers for example, must have decks above highest flood water levels. Flowing flood water + sediment forces can be huge and this is the most common cause of bridge collapses.
- After we have the overall cross section, height of piers and length of bridge, we calculate the self weight of the bridge by simple volume calculations, and come up with the dead load.

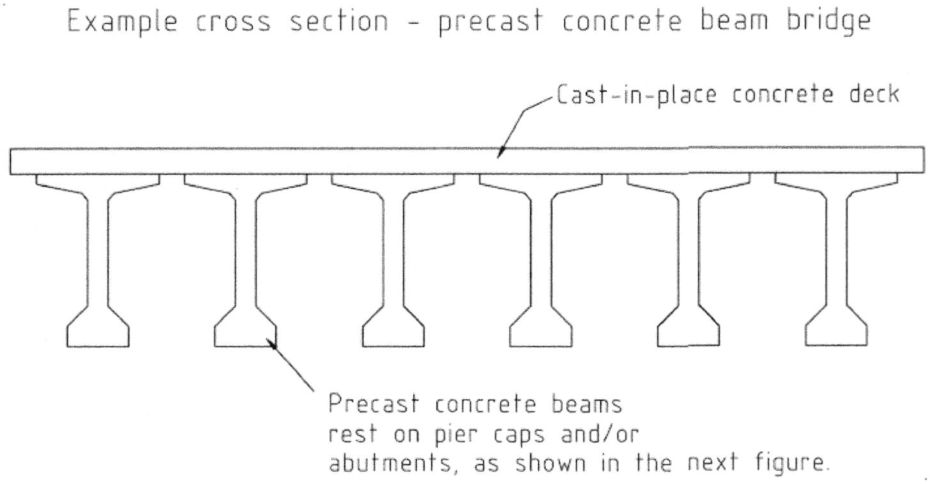

Figure - Bridges 4

- Now we can draw a free body diagram based on dead loads. A typical precast beam bridge and its free body diagram may look like in Figures 5 and 6 below. After that the bridge is analyzed with an applicable method for relevant load combinations from codes, such as dead and live load, earthquake, and wind loads. If the strength *(demand vs. capacity, with applicable safety factors)* and serviceability requirements are not met for the part or whole of the structure, the design is refined as needed.

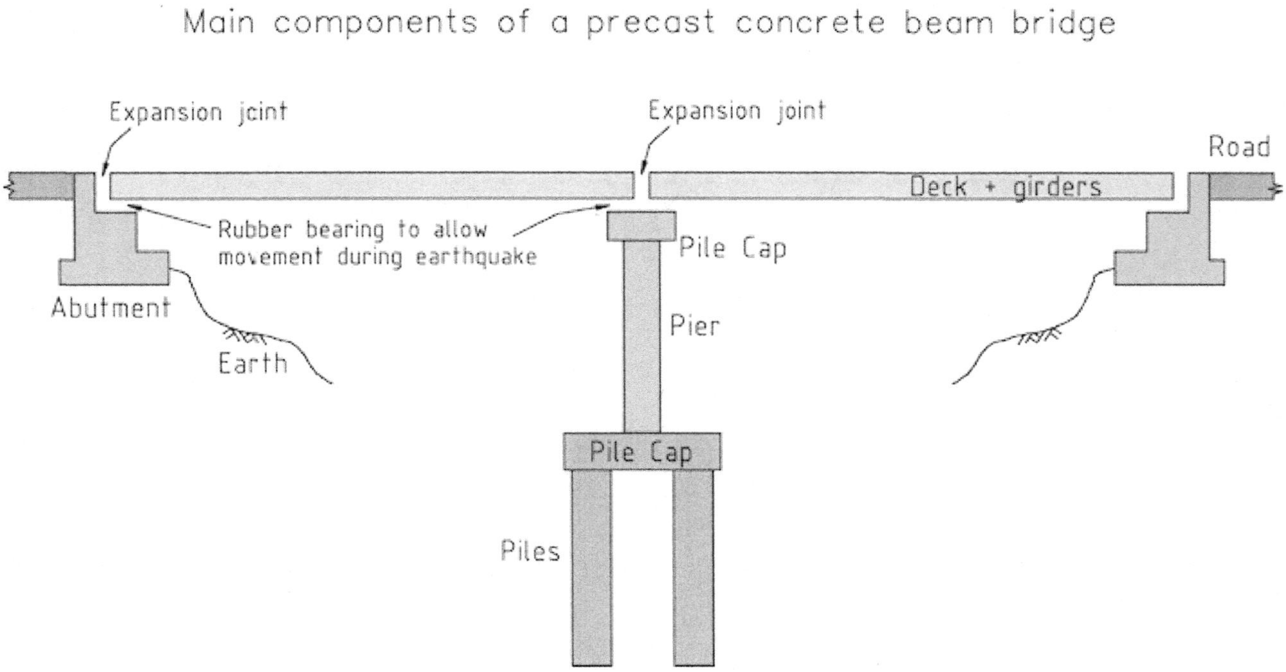

Figure - Bridges 5

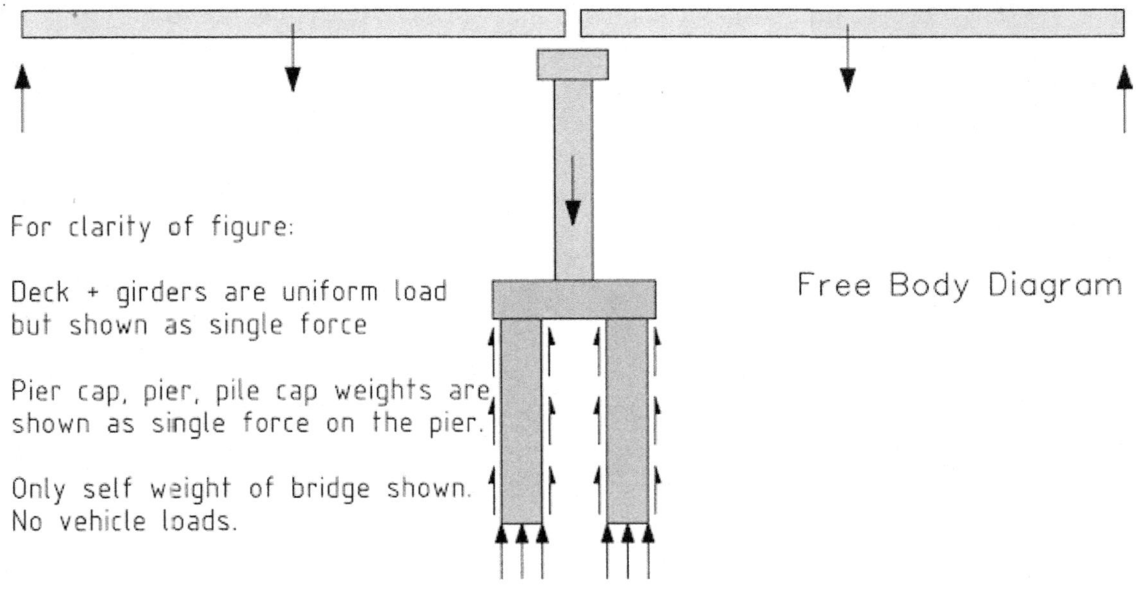

Figure - Bridges 6

We use stiff rubber bearings, for flexibility and to dissipate energy during earthquakes, which function as base isolators for beams. The gaps between beams are filled by expansion joint.

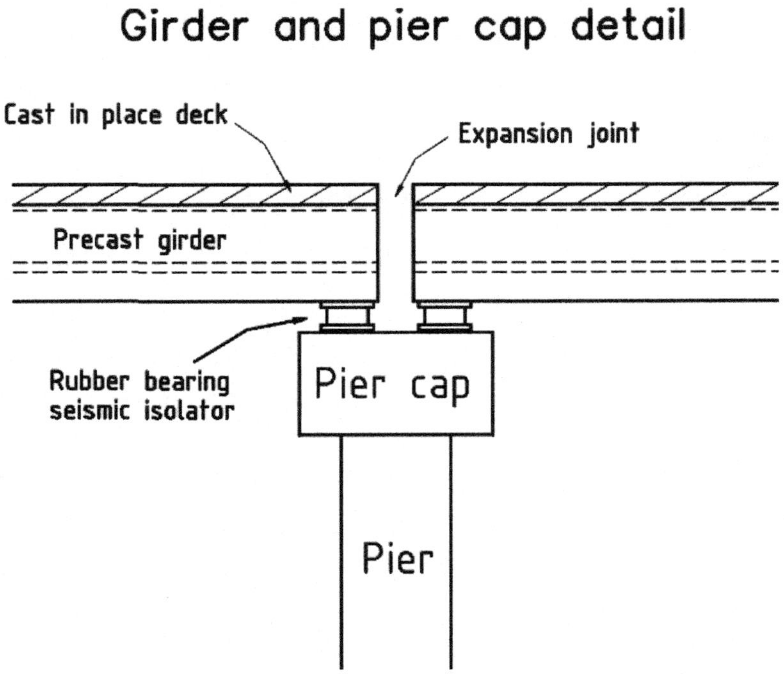

Figure - Bridges 7

In Figure - 8 below, during earthquake the beam may move towards both sides horizontally as it rests on rubber bearing. When the beam touches the abutment, we may or may not ignore the abutment stiffness. If we do not ignore it, then we consider the force generated by abutment plus the passive earth resistance because after breaking the abutment the beam will push towards the soil behind. If we did not allow the abutment to break, then the force would be reflected back to the beam and consequently towards the middle of the bridge and then to the pier which may be undesirable.

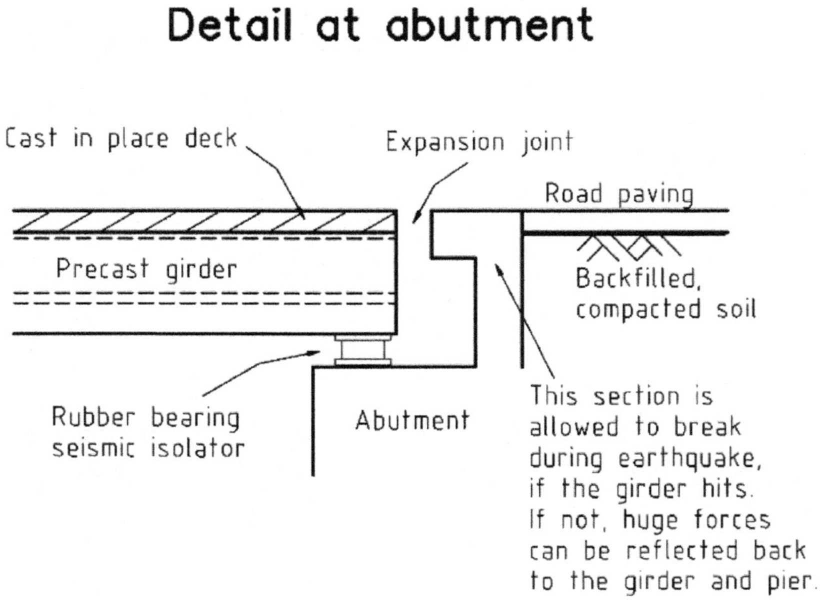

Figure - Bridges 8

The structure is analyzed using a proper earthquake analysis method based on worst combinations of dead load, live loads, earthquake loads... from any direction. Upon running analysis, the reaction forces at supports, such as abutments and piles, internal forces of members, shear and moment diagrams and displacements are found everywhere. Per these results we can make revisions until all are acceptable per the code criteria.

Bridge design is one of the most challenging parts of civil engineering, where dynamic loads such as vehicle loads, wind, temperature effects, and if the piers are located in water, hydrostatic and hydrodynamic forces, ship impacts, and depending on seismic zone, earthquake forces also come into play and each of these can have great effect on the design. Bridges are also critical structures that they must be continue to serve after an earthquake or do so after minor repairs. In addition, when building bridges, the crossing traffic (be it vehicular traffic or sea traffic) must sometimes be interrupted and the operations must be coordinated. Then there is also the aspect of the load combinations during construction phases, which become the most important for bridges with respect to almost any other structure. Constructability is another issue, because in most cases we need to use large equipment which the operation must be well coordinated with everything else. The larger the pieces we build, the more critical and expensive things get, including the cost of mistakes. Therefore a bridge design involves so much more than designing a structure only for the loads after finish of construction.

Example Modeling of a Precast Concrete Girder Bridge

In the figure below, you see a simplified version of how we can model a precast concrete bridge as far as the beams and piers, that is used in computer analysis. Dividing the structure by nodes like this gives us accurate enough results, and we do not need to represent the bridge as an assembly of millions of nodes as a 3D mesh. As discussed in Chapter 1, Computer Aided Analysis section, this is a discrete type modeling, which suffices here. We use adequate number of 1D elements to represent girders, pier cap and pier (and 2D elements for deck which is not shown here) to represent the displacement and stress responses of the structure with sufficient accuracy everywhere. The figure below seems as 2D but the model is actually a 3D model, by having more girders in parallel into the paper plane you are looking at. In the figure below, nodes in the program are located by their x,y,z coordinates, and linked by rigid links or beam elements *(and surface elements too in case of the slab but we did not show x-y plane here)*. We created a skeletal structure in 3D, made of beam elements and rigid lines, which are connected by nodes, such as at every 1 meter or an appropriate distance, which are all free to move in x,y,z directions, except at the external supports as seen at the bottom of the figure and the abutment.

The rigid links in the figure are provided in the locations and directions where the deformations are negligible. For example, the three rigid links at the pier cap first take the loads from beams (which sit on seismic isolators) and transfer it to the center of the pier cap, and then the lower rigid link is to make up for the distance between the center of the pier cap and above the pier. We can do this because the deformation of pier cap in xz plane (the plane of paper) is negligible. If we did not include the rigid links here, and represented these by deformable line elements, that would introduce unnecessary complexity into calculations without giving us any additional useful information.

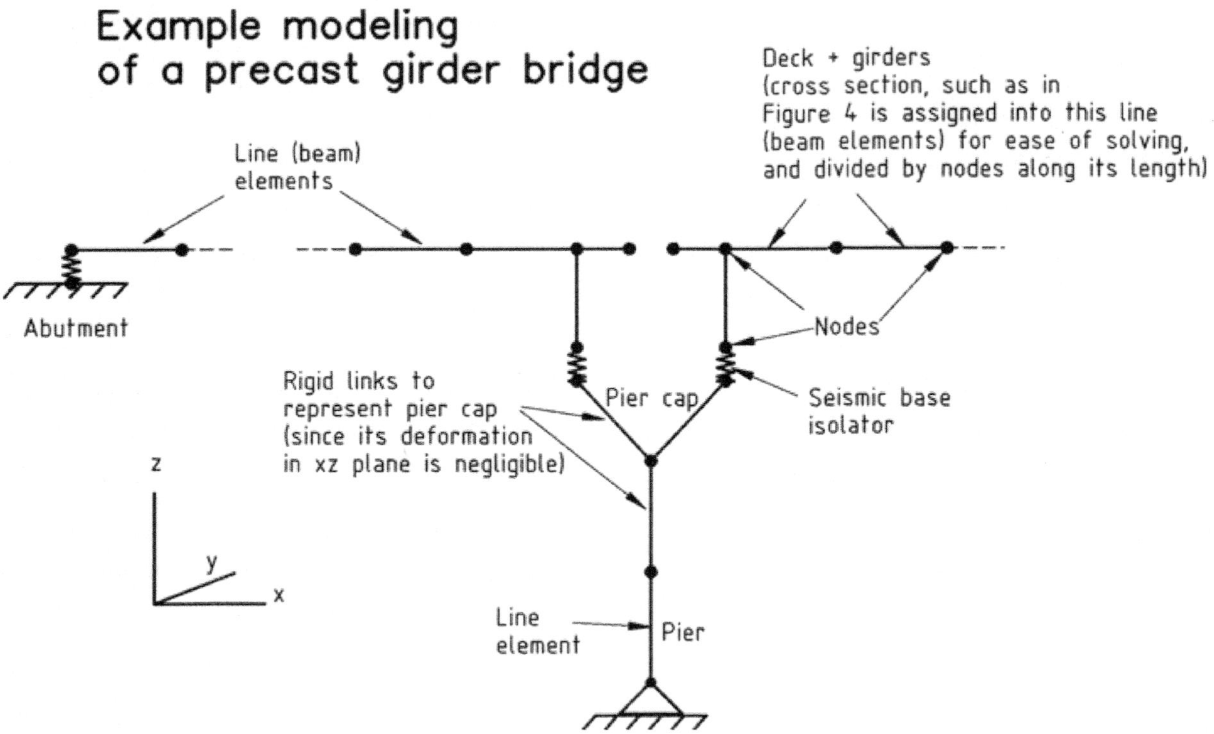

Figure - Bridges 9

The beam and deck assembly as seen in Figure 4 is assumed to be condensed into a line (line elements that make up the beam), and then that line is divided by nodes along its length, between the pier and the abutments. Condensing the deck + girders assembly means, we still enter the actual cross section dimensions and material properties into the analysis program, but assign this whole cross section to a line in x axis between the pier and the abutment, which is divided by nodes along its length, such as at every 3 feet or 1 meter. Unlike the rigid links for the pier cap, these lines are beam elements, because their deformation in xz plane is not negligible. We assigned not only the cross section of deck + girders assembly, but also the (nonrigid) stiffness value to these beam elements for instance. So whenever we need line elements where deformations are not negligible, we can use beam elements that can deform and bend. Similarly for representing the pier itself we used line (beam) element as above, which was oriented of course vertically this time.

If we showed the model from perpendicular plane (looking from yz plane), which would mean showing the cross section of deck+girders, then we would represent the geometry of cross section by rigid elements in yz plane, since the deformation in yz plane of deck+girder assembly is negligible for us, at least in comparison to its deformation in xz plane. When showing the cross section, under each girder we would show springs to represent seismic isolators at abutments and pier caps.

For representing the concrete deck, we can use 2D plate elements above girders. Each plate element will have nodes at their corners. After imposing the loads and then running the analysis, we can see displacement and loads for each node.

The girders sit on rubber seismic isolators at the abutments and pier cap (*the abutment on the right end is omitted in Figure 9*). The rubber seismic isolators are shown by springs. In the program, we assign stiffness values for these springs in x, y, z directions. For example, in z direction, the stiffness of the spring will be very high, to represent the fact that it has very little flexibility in vertical direction and would deform very little, whereas for the horizontal x or y directions, its stiffness will be far lower, which means, the girder+deck assembly will be able to move horizontally during an earthquake.

At the bottom of the pier, if you pay attention there is pin support. This prevents build up of moment there. It means we allow that concrete pier may crack at the base during the earthquake, provided that we still prevent collapse of the bridge. We could also use fixed connection if we want it to resist the moment and accept no damage, which would require strong moment resisting detail and cross section.

Abutments

Abutments are multi purpose structures that support bridges from its ends. They transmit and distribute the loads from the bridge to the ground, retain earth behind, provide the transitioning between the bridge and the terrain at the ends of the bridge, accommodate the long term displacements from the factors we listed, such as temperature, creep, shrinkage and post tensioning.

Abutments can be supported by piles or spread footings with cantilever walls, and or compacted fills. The fill material must be made of nonexpansive soil. If the abutment is backfilled before the bridge structure is placed on it, this may represent a critical loading case for the abutment, due to lateral earth pressure of the fill tending to overturn the abutment. Since we have retained backfill behind, abutment design is also similar to retaining wall design. Weep holes, perforated drainage pipes must also be provided to drain water from retained earth. We showed abutments in figures 5,8,9 of this chapter.

CONSTRUCTION METHODS

Construction sequence of bridges require analysis, because we are talking about something that will hang in the air not only when finished, but also during construction without the benefit of completed structure acting as a whole. Therefore, every critical stage must be analyzed, which must also consider construction stage sequences, and this can be as challenging as designing the finished bridge itself. Bridge design software can also include construction phase design modules. In the US, AASHTO LRFD Bridge Construction Specifications is used for temporary construction loads, in conjunction with the AASHTO LRFD Bridge Design Specifications that we mentioned earlier. There are also other guides just for temporary works, such as the formwork and falsework by AASHTO. Several common construction methods for bridges are listed below. These can be named differently by different firms or countries, some can be combined, and there are also many intermediate methods to fit a particular project. The first four methods below are for decks and apply to different type of bridges as appropriate. The last one is about installation of cables, which only applies to cable stayed and suspension bridges. For foundations, when crossing a river, lake or sea, to construct the piles, we can utilize cofferdams.

Span by Span Casting / Installation (Segmental Construction)

A straightforward method, where we build one segment of piers, beams and decks, then move on to the next segment, and so on. This method applies to both steel and concrete bridges. Unlimited number of segments can be built this way to cover very long routes. A viaduct is a bridge which is composed of several or many spans that crosses a terrain such as a valley or wetland, is a good example of this method. The girders are lifted to place by cranes and slowly and precisely aligned at each others end for all three axes. For curved girders, where the bridge or viaduct makes a curve, the process of correctly picking, lifting and placing the pieces is even more challenging. The length of segments are not only limited by structural but also by constructability concerns.

Incremental Launching Method

Similar to span by span method except that the sections are built in place incrementally by sliding a form, which can be efficient in narrow areas with difficult access for forms. This method causes the least disruption to surroundings and can be applied in narrowest locations with smaller assembly area, or at environmentally sensitive areas. Typically 15 to 30 meter sections of segments are installed at a time, after which the form is pushed forward in the air for the next segment. After finishing sections, the members can be post tensioned. It can also be applied as balanced cantilever as described below.

Heavy Lifting and Placement of Bridge Sections

In this method, large, heavy prefabricated bridge segments are raised in place by lifting. It can be applied on beam bridges and it is the only method for suspension bridges to erect the deck, after the cables are installed. Where the bridge crosses over water, the pieces are first carried on water, and then pulled upwards by ropes attached to the main cables.

Balanced Cantilever Method

Formwork or lifting bridge pieces can be a challenge. The higher the bridge deck, the greater the cost and construction difficulty for forms or lifting. Difficulty increases in rough or congested terrain, environmentally sensitive places, long spans, weak soils, or when crossing over existing highways or railways. As a solution, the pier or tower is built first, and then beams are spread outward equally towards both sides without forms or the need to lift pieces, as seen in the figure below.

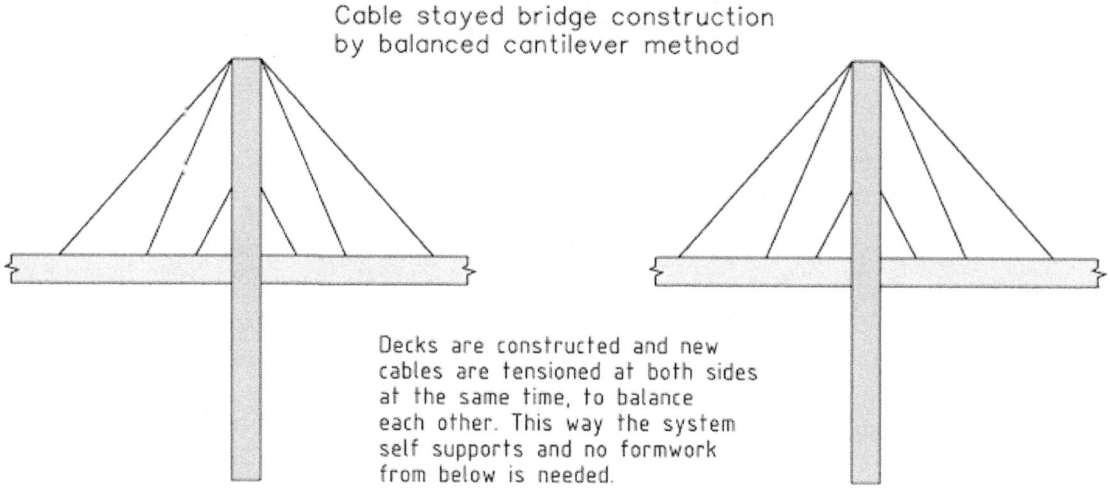

Figure - Bridges 10

This method can be used for free spans upto around 250 meters. Precast systems or casting concrete in place by using sliding form are both possible so the cantilevers can be made of box sections, precast beam sections, trusses or by incremental pouring. Both beam and cable stayed bridges can be built with this method (the figure above shows a cable stayed example). During construction, as the cantilevers proceed, new segments are added at equal rate on both sides to balance each other and finally the free ends of beams from two adjacent piers meet in the middle and get connected in the air.

When advancing cantilevers, the deflection of the free end is important. The construction must progress such that the downward deflection of the free end is accounted for, which means, at first the advancing cantilevers have to be made higher than their final intended elevation. If cast in place method is used, time of concrete pour vs. rate of downward defection of the free end is also taken into account and each step must be separately analyzed.

Installation of Cables

Cables are the primary load carrying components of suspension and cable stayed bridges. The main cables of a suspension bridge are first raised up and tensioned between the previously erected towers. To the main cables, the suspender cables are connected, and to them the deck segments are lifted up and attached to each other. For cable stayed bridges, the deck and cables proceed together, as in the balanced cantilever method described above. To both sides of the previously erected towers, the cables are installed and connected to deck symmetrically, and then they are tensioned to their design forces. After all of the bridge components are installed, the forces on the cables are checked and adjusted as necessary.

OPERATION and MAINTENANCE

Most bridges in the US are designed for a lifespan of about at least 50 years. Continuous monitoring, inspection and maintaining of the tens of thousands of bridges in the world is an ongoing process.

The concern for maintenance actually starts during the design phase of the bridge. The design should not disregard the ease and speed of maintenance during the life of the bridge. Bridge repair could mean disruption of transportation to surrounding communities, therefore repairs must proceed rapidly. Bridge inspection and maintenance is actually a cost savings measure to extend bridge life for many years, because building new bridges are much costlier. Maintaining bridges include structural repairs, scour protection, debris cleaning, snow and ice removal, cleaning deck drains, bank restoration, painting and so on. The maintenance tasks also differ greatly between each type of bridge according to its material and structural system. Regardless of their sizes, bridges must be inspected at every applicable interval. In addition, the performance is monitored as far as displacements of the structure, and settlements of foundations. The larger the bridge, the more complex monitoring measurements are made and recorded.

The most common reason for bridge failures is the running water damage, especially in developing countries where land development plans and flood risk assessment may be inadequate.

During operation, one thing to pay attention in winter months is that, the bridge paving freezes earlier than roads on soil because the cold air can also flow under the bridge deck. The frequent use of deicing salt in the long term can cause damages to paving and drains.

Bridge Inspection

Inspection of bridges is important to make sure that they operate safely. As an example, bridge inspection categories can be classified as below, according to NYDOT bridge inspection manual:

- General bridge inspection: These are regular inspections that are performed at a maximum interval of every 2 years.
- General diving inspection: These are done underwater, at a maximum of 60 months intervals.
- In-depth inspection: These inspections are performed on an as-needed basis.
- Special inspections: Also performed on as-needed basis. They can be for flood watch, after a flood, earthquake, or an impact, or to inspect a specific structural or other various types of repairs. Special inspection may also be performed in place of general inspections.
- Inspections on bridges that are closed to traffic.

All components of bridge are subject to inspection, including decks, superstructure, bearings, rails, substructure, culverts, joints, abutments, wearing surface and protective systems, approach slabs.

During an inspection, commonly detected issues may include:

- Scour and undermining
- Poorly aligned stream channel
- Various concrete damages such as efflorescence, delamination and spalling
- Bearing not fully supporting the deck because of misalignment
- Impact damage
- Tilting of abutments or wingwalls
- Deck deficiencies
- Abrasion of deck where the wheels touch the most and exposure of coarse aggregate
- Cracks in deck
- Spalled concrete which can expose anchorages, and rebar, which can cause corrosion
- Corrosion of metal decking
- Corrosion of steel columns beams, bracings, plates, which can also lead to section loss
- Buckling of steel columns, beams, bracings
- Distortion of steel girders due to excessive loading, fire, impacts, section loss
- Welding cracks
- Cracking, excessive sagging of girders
- Worn footings at river beds
- Differential footing settlement, that causes stresses, cracks, displacements at upper members, or excessive overall settlement

Scour

Scour effect for bridges mean the excessive washing away of soil and sediments in a riverbed from the base of a bridge pier, wall or even pile cap due to water current, which strips the bottom of piers from soil and leaves them exposed, and may cause the bridge to cease functioning or even total collapse. Scour effects must be analyzed for every structure built on rivers, such as cofferdams, bridge piers and also for structures that lie on flood planes, for the worst scour condition that can happen during the life of the structure. To prevent scour, a blanket of crushed rock or large size gravel, or a fabric filled with rock can be laid on the riverbed at the bottom of bridge piers or cofferdam walls. Pile foundations are preferred over shallow foundations, to decrease the risks from scour.

To analyze scour potential, flow depth and velocity distribution profile of the river directly upstream of piers or abutments must be understood. This modeling can be done by specialized computer software, and the analysis is based on river water location, depth, velocity and flow angle measurements with respect to bridge piers. Scour calculations need good engineering judgment, as for every different geometry of the piers and abutments in combination with different soil types, flow rates, natural or modified riverbed cross sections, river flow route, results can be very different, and it is not possible to have equations to cover all possible conditions.

CHAPTER 4 - HIGHWAYS & TRANSPORTATION ENGINEERING

A highway that runs a long distance between mountains. It is not separated for opposing traffic, which can be mainly because mountainous regions present greater challenge in every way, and add to the cost

Of the five major branches of civil engineering that we listed in the very beginning of the book, so far we discussed four, which were, structural, geotechnical, water resources and construction branches in various places and depths. Now we look at transportation engineering. Discussing highways will be similar to introducing transportation engineering itself, hence the title of this chapter.

Transportation engineering is a major part of civil engineering. Even just by looking at the chapter titles of this book, out of ten chapters that introduce different types of structures or systems, five of them are related to transportation engineering discipline: Chapter 3 - Bridges, Chapter 4 - Highways, Chapter 5 - Railroads, Chapter 6 - Tunnels, Chapter 10 - Airports and Airfields. Also consider supplementary components when building roads, such as stabilized slopes, retaining structures, drainage systems... From these, it is easy to see that transportation engineering is always heavily coordinated with all other four disciplines of civil engineering.

Roads are structures that are built directly on earth to carry moving people and vehicles. We said roads are structures, because they also carry things and have their own foundation. Design and construction of roads is one of the most frequently performed tasks in civil engineering. When we say roads in this chapter, we mainly mean highways and freeways but not the streets or narrow roads in urban areas with traffic lights and pedestrian crossings etc... although many items in this chapter would also apply to them. Railroads will be discussed in the next chapter.

Below we will cover some of the fundamental concepts in transportation engineering and highway design and construction. This chapter is organized as follows:

- Traffic Demand Calculations
- Geometrical Design
- Supplementary Components for Highways
- Construction

In the US, the main references that are followed for roadway design are the references published by AASHTO - American Association of State Highway and Transportation Officials, and FHWA - Federal Highway Administration. These provide the standards and criteria for any type of roadway anywhere for all aspects of design. These are such useful standards that they are also used in various other countries, who base their codes significantly on these. *"A Policy on Geometric Design of Highways and Streets"* by AASHTO, also known commonly as the *"Green Book"*, is one of the primary sources that the transportation engineers follow in the US for geometrical criteria and a lot of other design aspects. The state departments of transportation in the US have also published their own manuals mostly based on AASHTO and FHWA sources, which are also useful references.

TRAFFIC DEMAND CALCULATIONS

Highway design starts with determining needs for the proposed route in terms of number of vehicles per year. Traffic volume can be defined as the number of vehicles that travels through a point on a highway for a given duration. Some common terms for measuring traffic volume are, Average Annual Daily Traffic (AADT), Average Daily Traffic (ADT), Design Hourly Volume (DHV). The design will not be based on average but peak volumes. Peak hourly volumes can be derived from AADT. In the "Green Book" we mentioned above, ADT, which is called as the most basic measure of the traffic demand, is defined as the total traffic volume in a given period, divided by the number of days in that period, expressed in whole days, where, the number of days should be greater than one and less than a year.

The peak traffic flow, flow density and speed are interrelated. At a given optimum traffic flow, there is the optimum density and corresponding speed. For very fast speeds, the density is very low, as well as the total flow volume. Then as the density increases, total flow volume also increases but speed decreases, until we reach optimum flow volume. After that, if we try to increase density further, the traffic will just be slower and the optimum flow will decrease. So for example for the greatest densities, the speed will be very low, and the flow will also be very low.

After we have the traffic demand numbers as estimates or the actual numbers from traffic measurements over an existing route that connects the same start and finish points, we can determine the capacity and type of highway that is needed to be able to handle the existing volume plus the expected volume increase during the proposed life of the route. With this information, we can choose the highway type and start geometrical design. The traffic volume information is also necessary to determine the right program for traffic handling operations which we will not discuss.

For traffic and highway capacity calculations, *Highway Capacity Manual* by the Transportation Research Board of National Academies of Sciences is a valuable source.

GEOMETRICAL DESIGN OF HIGHWAYS

A separated highway connecting major cities. Design of exact highway geometry is in fact a challenging task. It involves consideration of quite a few variables simultaneously, as discussed below.

When we know the required highway capacity, we can decide the number of lanes, as well as the speed we must design to which must handle that peak traffic. The next step is to impose the proposed route on map, and decide the best layout by also keeping in mind the road width and the required number of lanes, and the other constraints we must satisfy for the speed we are designing to. For instance, a highway which is designed to a maximum speed of 70 mph will have stricter geometrical constraints than a road designed for 45 mph.

The most straightforward route to build a highway would obviously be to just draw a straight line from point A to point B. But this is almost never possible. The overall layout in the first place is affected by stakeholder input from many different fields and macro factors such as:

- project cost and schedule,
- access control to and from the highways vs. adjacent properties,
- macroeconomics,
- environmental and noise concerns,
- administrative, land development, legal and social issues, right of way, pedestrian concerns, obstacles to be avoided,
- topography, soil conditions, hydrological and drainage which fall on the side of civil engineers,

After the overall layout is agreed on, the final road geometry is determined by the factors below. All these items below affect each other simultaneously, to finalize the horizontal and vertical layout, so we try to find the optimum balance between them:

- Design speed
- Earthwork quantities
- Ride comfort
- Sight distances
- Clearances to horizontal and vertical obstacles
- Road cross section
- Drainage components
- Horizontal Layout, Vertical Layout

Therefore, we will discuss this section, geometrical design of highways, by studying each of the items in the list above as subsections, which will influence horizontal and vertical layout.

Design Speed

Design speed is selected based on factors such as traffic demand, overall slope of the area, topography, highway type and the extent to which surroundings are developed, such as rural or urban areas. The design speed tends to increase when the terrain is more level (flat), with fewer points of access and with less development around the road, in other words, in rural areas. The selected design speed is the primary ingredient that keeps the geometric features of a roadway in balance by determining the maximum and minimum values, such as the sight distance, vertical and horizontal alignment, lane and shoulder widths, superelevation, roadside clearances, maximum gradients, and that is why we list it under geometrical design section. Conversely, if, when designing the highway we find out that for a certain speed we are not obtaining enough sight or stopping distances, we may need to reduce the design speed in certain sections, but we must be sure to still achieve the targeted road capacity.

The road can be curved only to a point, for a given speed. The higher the speed, the larger the radiuses need to be. But sometimes, a certain curvature must still be given to the road, to follow natural terrain as much as possible, or to avoid a certain obstacle, which would cause a sharper curve than a certain speed requires. In these cases, the cross section of the road can be tilted to one side, so that the vehicles will not be thrown off the road. This tilting amount is called superelevation, which we will discuss. The options for design speeds, minimum curve radiuses, sight distances are given for each particular situation and type of highway in AASHTO "Green Book".

Earthwork Quantities

We seek to minimize earthwork. This can only be possible if we laid out our road in such a way that it follows the natural terrain as closely as possible (again, assuming we are not building bridge or tunnel). Minimizing earthwork means, minimizing the amount of cut and fill we need to make.

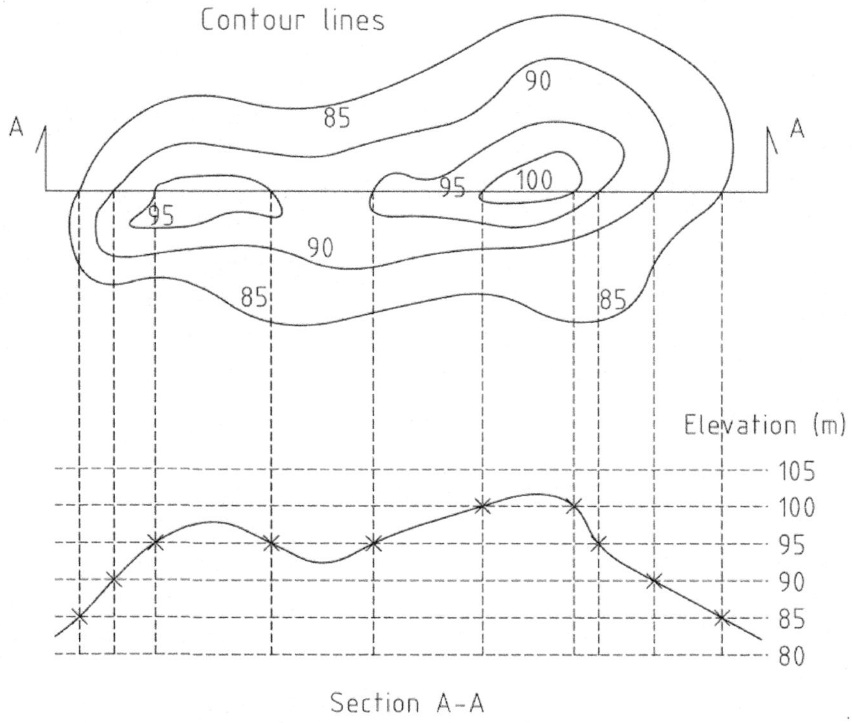

Figure - Transportation 1

Contour lines are made by joining points of equal elevation on the natural terrain map, as seen above. The top part of the figure is the contour lines map, looking from top, and the bottom part represents the side view, through the cross section we marked as A-A. As you can see, where the contour lines get denser, the terrain is steeper, and where contour lines are far from each other, the terrain is more flat. Now, to see where we cut and/or where we fill, let's impose a highway cross section on A-A:

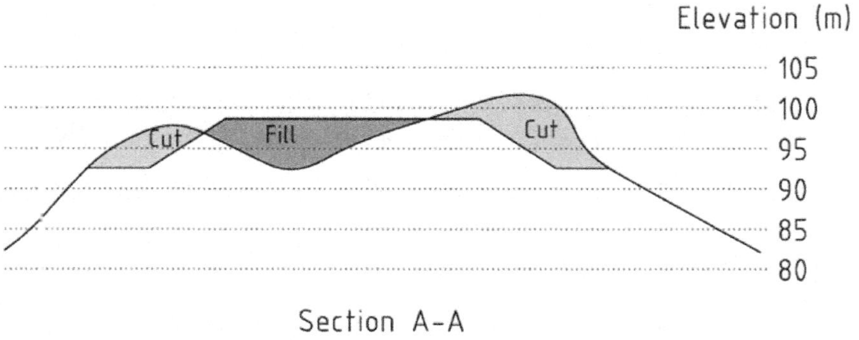

Figure - Transportation 2

Actual cut and fill would be deeper than shown here, to allow enough depth for placing subgrade. We also need to remove at least the top 6" of natural top soil in any case, even if the cross section is a fill.

Figure 2 above shows only one cross section. There may be thousands of cross sections if we consider a long highway, and draw a cross section every 50 or 100 m.

After this point, calculating earthwork volumes is simple geometry. We take the distance between two consecutive cross sections, let's say 100 m, and then multiply that distance with the average area of the fill and average area of cut between those two cross sections, to find the total amount of fill and cut in cubic meters, for that 100 m section. Different methods are also available to calculate this volume.

We prefer cut and fill amounts to cancel each other to the greatest extent, so, we try to follow a route that would minimize the difference between cut and fill. For example, if we shifted this road a lot towards left, and everywhere would be fill, then the fill amount would increase but cut amount would decrease or even be zero. And if we continued this way for all cross sections, we would end up with a lot of fill that we need to import from outside and compact under the road. Or similarly, if our cross section had much more cut than fill, then we would end up with a lot of extra earth that we need to haul away. So, minimizing earthwork means, making cut and fill amounts as close as possible to each other so that we can use the cut material for filling, although not all cut material will be suitable for backfilling and we will still need to bring material from outside. For each interval, we calculate the earthwork amounts as described, and come up with cumulative earthwork amount, which can be represented by simple graphs that are called "mass diagrams", which display the cumulative earthwork amount on vertical axis and the stations in meters (in other words, distance) on horizontal axis.

As mentioned in the beginning of geometrical design of highways section, factors such as the design speed, earthwork quantities, ride comfort, sight distance, clearances to obstacles, road cross section and drainage affect each other simultaneously, so the optimum must be found. For example, we may come up with cross sections that seem to be minimizing earthwork, but, they may not work when we also consider the horizontal and vertical layout required based on design speed and sight distances. Then if we shift the whole cross section to a new location, our earthwork amount may be increasing considerably and we may need to think about further revising it and so on...

Ride Comfort

The general performance goal is to ensure smooth and efficient flowing highways which would be comfortable for the drivers, by avoiding excessive or poor combinations of curves. This also ensures more pleasing appearance, as well as reduced travel time. For ride comfort, we should consider:

- Alignment should generally conform to the natural terrain, and sharp curves should be avoided as much as possible. Curves should also not be used right after long straight sections.
- In addition to single radius curves sometimes two curves with different radii can join each other to make a compound curve. This can make the design easier but it should not be overdone, and the radius difference should be kept to a maximum such as 50% of each other per AASHTO.
- If one curve must follow another curve in the opposite direction, they should not be used immediately after another, and there should be at least some straight distance (also called "tangent"). Similarly, two curves in the same direction should not be made if they follow each other too closely.

- Sometimes, in case of extremely long, flat and straight routes, the designers may give the road a very light curve and its opposite every now and then, to make sure that drivers are following the road attentively. However, the resting stations, with their signage provided at every certain distance also serves this purpose.
- Horizontal and vertical alignment are inseparable and should be designed simultaneously as per terrain conditions. Combining vertical and horizontal curves generally creates a pleasing ride and look, but this should be upto a point over a given distance and without sharp changes.
- Older driver's performance must be taken into account, by using a conservative values, such as the reaction time for breaking by providing adequate sight distance, adequate light for night time, road markings and signs that are bright and large. The road and its components should not confuse drivers and force them to process too much information in a short amount of time, in other words, things must be kept simple.
- Acceleration and deceleration capabilities of vehicles are among the primary factors to be considered. Poor performance vehicles should be accounted for, so, a passenger car rarely controls the design, because it can perform better.

Of course, for all items above and the similar ones that we did not mention here, the required highway capacity and the design speed must be maintained at all times. Again, when satisfying ride comfort requirements, we must also watch for all other criteria that we listed under geometrical design simultaneously, to finalize the horizontal and vertical layout.

Sight Distance

Sight distance is the length of roadway which the driver is able to see for various purposes as listed below. The geometry of the road should allow adequate sight distances to drivers, therefore this is a major factor that affects the horizontal and vertical layout of a road. Types of sight distance are:

- Stopping Sight Distance
- Passing Sight Distance
- Intersection Sight Distance
- Sight Distance on Horizontal Curves
- Vertical Sight Distance

Stopping Sight Distance:

We must allow adequate distance for the driver to brake and stop before reaching a stationary object on the road, when he/she travels with maximum design speed. This distance includes the distance that the vehicle travels before the driver can react to hit the brakes, plus the distance traveled while applying the brakes. For sloped road sections, we also consider the effects of gravity. For example when the vehicle travels downhill, stopping is harder so we add a certain distance based on kinetic energy calculations and friction. Similarly, when driving upwards the distance will be shorter.

Passing Sight Distance:

It is the sum of the distances that the car travels before and during passing the car ahead, continuing a little straight after passing, and then completely moving back to right lane. For multilane highways, where there are two or more lanes in each direction, and opposing traffic are separated, it is not necessary to consider this.

Intersection Sight Distance:

It provides adequate distance to the driver so that he/she sees an entire intersection ahead, and reacts to avoid collision. In this sense it is similar to stopping sight distance.

Sight Distance on Horizontal Curves:

When turning a curve around a mountain or hill, we may not have enough sight distance for a given speed as the mountain itself would block the view. So, just for this reason, to provide adequate sight distance to drivers, the section of mountain can be cut inside the curve. For situations that we cannot remove the obstruction, such as a building which is there to stay, we must lower the speed to reduce the sight distance needed, otherwise the route will need to be adjusted in greater scale.

Sight distances must also be considered on vertical curves, which also affect the vertical layout. Height of driver's eye, height of object, sight obstructions are factors to consider. Sight distances are shown in detail for all possible situations in AASHTO "Green Book".

Clearances to Horizontal and Vertical Obstacles

The road cross section must remain at a certain distance from horizontal and vertical obstacles, which obviously affect horizontal and vertical layout.

Road Cross Section

We can list the following aspects when describing a typical highway cross section:

- Superelevation
- Lane width and number of lanes
- Shoulder width
- Roadside clearances
- Foundation: Base, sub base, paving
- Drainage related components

Now let's look at these items we have just listed. A typical basic road section may look like below:

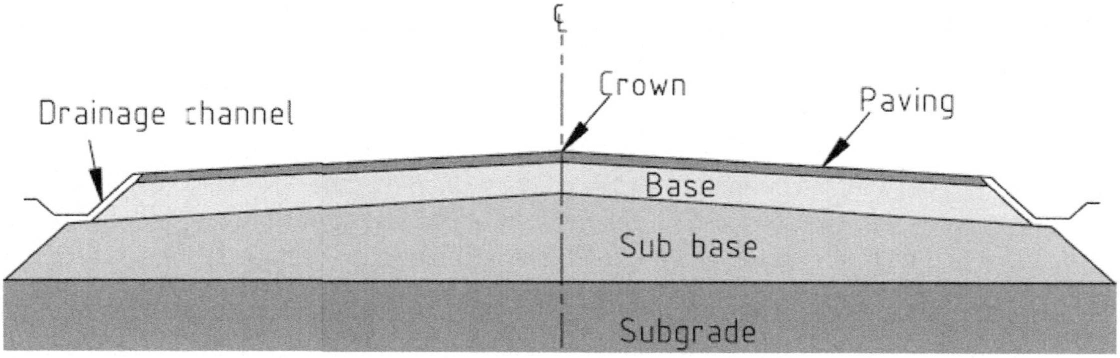

Figure - Transportation 3

The center crown is slightly higher, so that the rain water is diverted outwards to drainage channels.

Superelevation:

Superelevation is applied to road cross section on curves when necessary. It is the tilting of the road to one side, to counter the effect of the centrifugal force which tends to push vehicles outward when turning curves. It is calculated based on the speed and radius. The higher the speed, the sharper the curve, the more superelevation must be. Friction between the car tires and the road also counters the outward push of centrifugal force. Superelevation is needed when this friction is inadequate. That is one of the reasons why you do not see tilted roads often in urban areas, as the speeds are lower, but you see them on rural areas where speeds are high. There are also other reasons to this such as urban roads frequently intersecting cross streets, drainage issues and the need to meet adjacent grades. So in urban areas, the centrifugal force is only countered by tire and road friction, on a flat, not tilted surface.

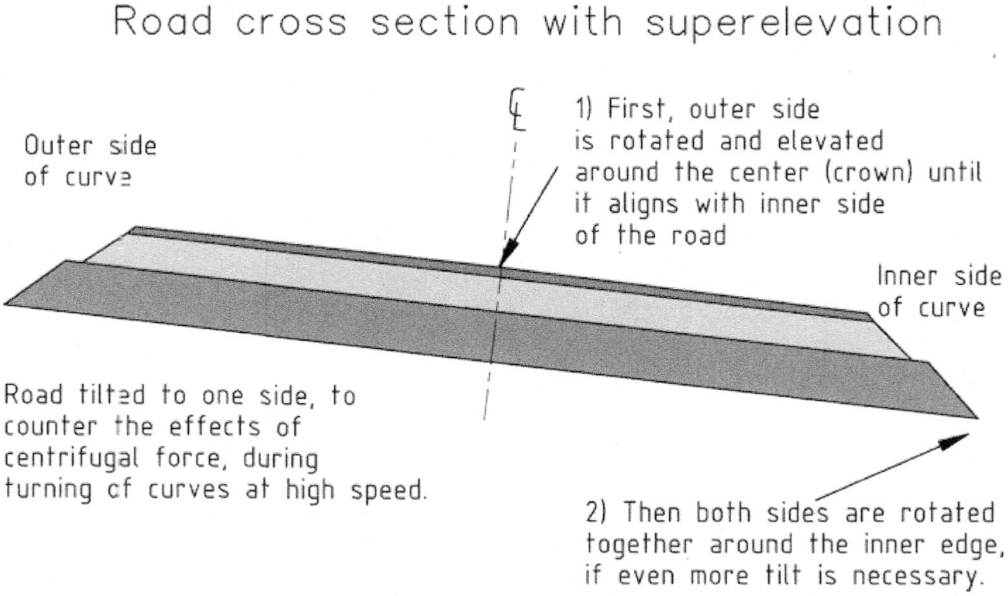

Figure - Transportation 4

On highways with superelevation, the transition from the normal cross section to superelevation must be done only gradually, per specified values. From a flat cross section (*actually it is not flat but has a crown as in Figure 3*), the transition begins even before the curve, and then it reaches the full superelevation cross section after the curve starts. There is a maximum tilting limit, (usually around 8%). Too much tilting would cause problems at low speeds, and in icy and snowy conditions.

Tilting of road starts with outer side by making center crown as hinge point, and when outer side is aligned with inner side, we can continue to tilt the whole section together (as in Figure 4) if needed. The curvature and superelevation should also be aesthetically appealing. After the curve is finished, superelevation is gradually eliminated and the road is returned back to its horizontal cross section.

Side friction factor, which is a measure of the friction between the tires and the road, is an important criteria to consider when designing superelevation. There are limits to sharpest curve that is possible without a superelevation, and the superelevation rates for different types of roads, radiuses and speeds, are shown in detail in the AASTHO "Green Book".

Lane Width and Number of Lanes:

These are determined based on the required highway traffic capacity, speed limits, truck volumes, available land and natural terrain conditions. In the US, the highway lanes are usually 12 feet wide. In Europe, it is generally slightly lower value. When the lane widths are reduced, design free flow speed must also be reduced. There are different types of lanes such as through lane where the main traffic flows, curb lane, shoulder lane, right / left turn lanes, parking lane.

Shoulder Width:

Shoulder is the extra clear space in addition to lanes of the highway on one or both sides of the road cross section. They are for emergency and otherwise should not be used. Similar to lane widths, a reduction in shoulder width results in reduction of design speed, but as the number of lanes in one direction increases, this reduction is less significant.

Roadside Clearances:

Per FHWA, a roadside clearance is a cleared area near the traveled road, for the use of errant vehicles, to allow the driver to stop safely and regain control, it can also include shoulders or slopes or run-out areas and no vertical elements or obstacles must be present in this area, such as curbs, walls, barriers, signs and signals, trees, power poles and so on...

Subbase, Base:

Pavement cannot just be placed over earth. A road is also a structure which supports loads. Therefore it must also have foundation. This foundation rests on the subgrade earth material, which can be a cut or fill section or both. Foundation consists of compacted sub base, and on top of that base material.

If the natural ground material is weak, it may necessitate extra cut and backfill with stronger material. The more we excavate, usually the stronger soil, and it will also mean less stress from the road above because of distribution of loads to a larger area as depth and width of foundation increases.

Drainage Components

Drainage and road layout affect each other. Drainage can especially control vertical curves, slopes, shoulders or even embankments, depending mainly on topography of the area. After the main layout and the geometry of the road is chosen, the specifics of the drainage system can be designed. Drainage in fact is a major aspect of all road designs.

Water must be directed away from the road surface quickly. There are a few reasons for this:

- Safety of vehicles, since accumulated water on the road will make driving difficult.
- Preventing water from seeping through the paving and reach the base course and below as the water will weaken those layers. As discussed in soil mechanics, water will affect shear strength of soil which will weaken the road foundation and may cause unwanted settlement. If the fill material is plastic soil such as clay, it will expand when water seeps in. There can also be freeze-thaw cycles in cold weather, which may cause even more deformation.
- Running water has energy, which can create considerable force. The force exerted by flowing water increases exponentially, as the velocity of water increases. Therefore, we cannot allow water to increase to a large volume and velocity that may cause erosion of the fill material, soil along ditches or below culverts, or wear exposed surfaces, the paving itself.

Problems caused by water due to improper drainage are among the most common causes of road failures such as paving, embankment or subgrade, slope, and supplementary components.

As discussed under Hydrology in Chapter 1, storm water drains from basins into outlets of each basin, by a natural water flow pattern. Therefore, the design should stay as close as possible to this natural pattern. Doing so is also important to avoid unfavorable conditions downstream, which may lead to problems, including environmental impacts, claims. In or around floodplains, flood water elevations must be considered during planning. Drainage system must also prevent downstream erosion.

Drainage of roads does not only mean surface water but subsurface water too. For subsurface drainage, not only the water that seeped into the road foundation system but also the water that can rise above groundwater table with capillary action must be considered. By surface, we mean paving surface here.

Therefore, we can classify road drainage into three parts as:

- Surface drainage
- Subsurface drainage
- Cross drainage (such as crossing stream)

Many different details and combinations exist for all of the drainage items listed below, for different road geometries, hydrological, topographical, groundwater conditions.

Surface drainage items:

Roadside Channels:

At the edges of the road, channel drains (roadside channels) may be placed, to take the surface water on the road *(as road cross section is already sloped towards the sides as in Figure 3)* and convey it to the storm drain system. In cut sections, these channels also take the surface rain water from sloped terrain. The location and geometry of channels and roadside clearance must be coordinated.

Curbs, Chutes, Toe Drains:

Curbs are used at the edges of roads to contain water flowing from the crown of the road at the road edge. In urban areas, this keeps water away from adjacent properties and provide a limit for the pavement. On rural highways, this prevents water to flow onto the road fill and enter into the road foundation, which would both weaken the foundation and also cause erosion of fill. The water that is contained at the edge is then diverted away from road surface by sloped drain chutes that run over shoulders at certain intervals, to toe drains. Chutes and toe drains are also used for draining surface water at sloped embankment or cut sections on one or both sides of highway.

A chute for draining water from a highway slope which is also strengthened by a gravity wall.

Inlets:

Inlet capacity determines the rate of water removal from the gutter, and the amount of water that can enter into the storm drainage system. Inadequate inlet capacity or poor inlet location will cause flooding on the road surface. Storm drain inlets collect runoff water and convey it to storm drainage system through pipes. We must avoid making drain pipes turning sharp angles, or complicated pipe routes that have a lot of turns. Common inlets for highway drainage can be classified as grates, curb opening, or combination of grate and curb opening. These can be poured in place or come as prefabricated units. At locations with bicycle traffic, grates should be bicycle safe. At low points of sag vertical curves, there should be extra inlets to take on possible overflows.

Temporary Drainage:

For many projects, temporary precautions must also be taken for the proper drainage during construction period, during inclement weather, before the permanent drainage system is installed and functioning, otherwise the accumulated water can cause delays or even damage, such as mud or heaving soil, or excessive water damaging part of built sections.

Subsurface drainage:

Subsurface drainage is for keeping water away from the road foundation *(either by seepage of rain water or in cut sections where the terrain on one or two sides of the road is higher than the road elevation, it can be groundwater seepage too)* as soon as possible, keeping the possible capillary water rise below the road subgrade, draining the water away that was collected by the roadside channels and so on. Filter material, such as geosynthetics, longitudinal perforated drain pipes surrounded by stone / sand (which is permeable) can be installed in the ground. If there is clay subgrade, the drainage will not be as effective, and an impermeable cut off layer to isolate road foundation can be installed.

Cross Drainage / Culverts:

Cross drainage is needed when the road passes a river or stream. In such cases, bridges can be built, but in many cases culverts will be enough, which are a part of the drainage system. Highway should be planned to comply with the natural drainage pattern as much as possible at these locations. Culverts can be considered as minor bridges, which carry the road above and allow water to pass underneath. A culvert has a clear opening of generally 20 feet or less. Culverts must be designed considering structural, geotechnical, hydrology, hydromechanics, construction and maintenance factors. They can be made of masonry arches, concrete slabs with abutments and /or piers, or concrete box sections.

Flowing water is a very common cause of all bridge failures. For culverts, which are minor bridges built on water crossings, this is even more critical. The opening of culvert must be able to handle the highest stormwater flow rate. Culverts must be aligned with the centerline of the stream and also be at same grade or slope for minimal resistance to water, otherwise, unless precautions are taken, erosion will occur all around, even in the absence of floods. In case of not aligning the culvert with river, we must make supplementary components to divert the river flow into the culvert at entry point, and then divert the water back to the river direction at culvert exit point. The road should be as perpendicular as possible to the stream flow, in order to keep the culvert length at a minimum.

In designing culverts, depending on the expected water levels at the inlet or outlet, the design can be inlet or outlet controlled, whichever is critical. Here "control" means, the discharge water flow capacity that the culvert will need to handle. For inlet controlled culverts, the geometry of the entry point of culvert, the depth of water at the entry, the type of inlet edge (such as square or mitered edge) are of primary importance. For these type of culverts, water does not fill the entire cross section of the culvert. For outlet controlled culverts however, the water flows at full cross section of the culvert, either through the whole or partial length of the culvert. This means, in addition to the factors we mentioned for inlet, the length of the culvert through which the water flows, elevation of water at the exit point, slope and roughness of the culvert's inside are also important. Also, the settlement of embankment around the culvert alters its grade, which must be accounted for during design.

Horizontal Layout

We layout the roads on horizontal plane with station points, which serve as the coordinate system for the entire length of road. Horizontal curves are made from segments of circular arcs. Now let's discuss the two basics of horizontal layout: the stationing and the circular arc.

Stationing:

The route is laid out on plans and in the field surveyed by marking stations, called stationing, which helps tracking distances. See Figure below which shows a stationing example with horizontal curve on the top part of the figure and its corresponding vertical position at the bottom part, for same stations.

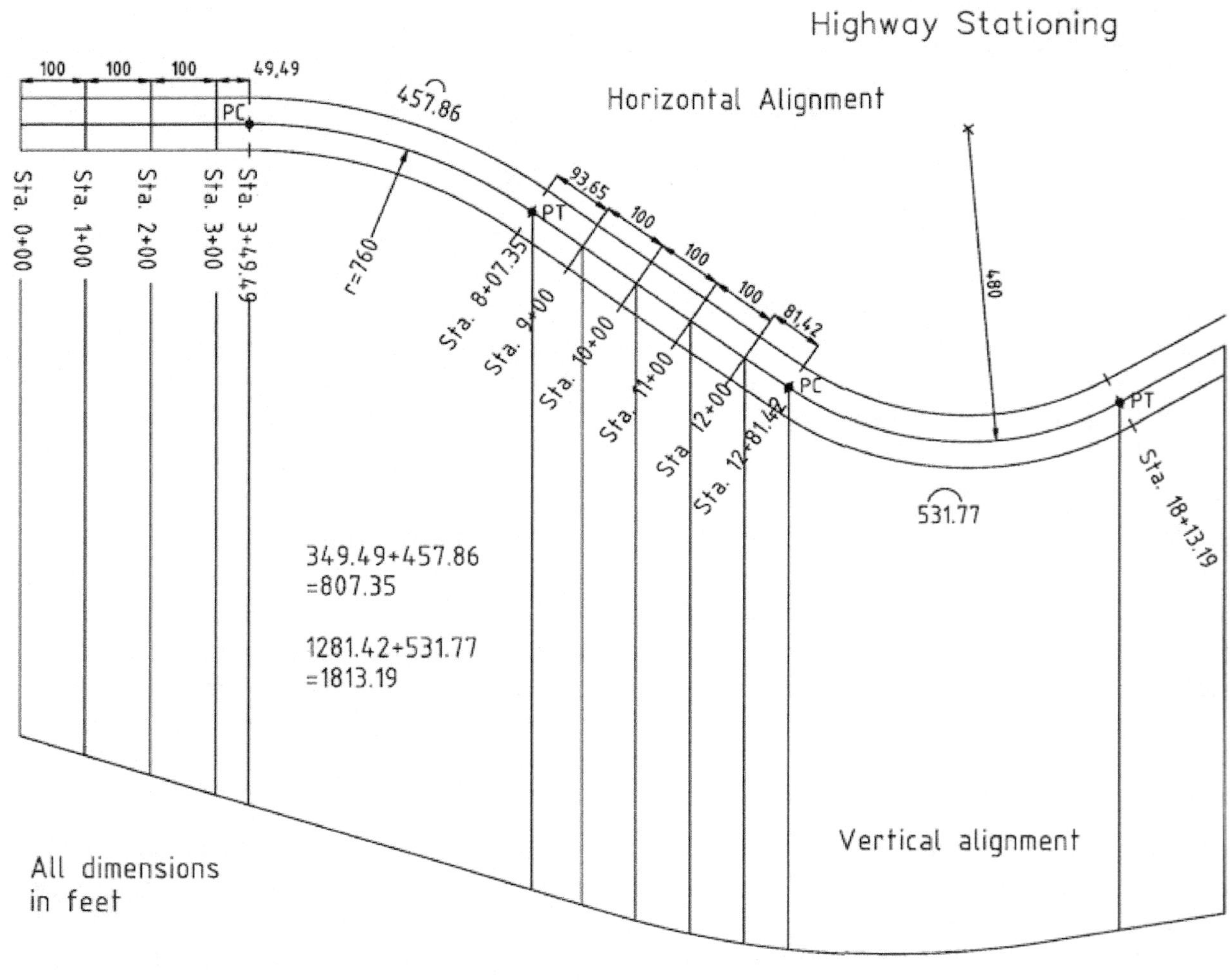

Figure - Transportation 5

The station 0+00 indicates the start location of the project. After that, every 100 meter or 100 feet increment, depending on what unit system is used, the stations are marked accordingly, both on the plans and in the field. For example, the station of 18+13.19 indicates 1800 + 13.19 feet = 1813.19 feet, from the beginning of the project, as in the figure above, where imperial unit system is used. For a metric format, expression is somewhat different, although the basic logic is the same. For example, the station 1+ 813.19 means, 1 km plus 813.19 meters = 1813.19 meters from start point.

The stations, as seen above, follow horizontal alignment - they do not follow vertical alignment - and therefore, roads are actually always longer (in other words, a car's wheels travel longer) than the road stations indicate, due to the fact that stations exclude the increased distance caused by vertical curves. If we tried the stations to follow vertical curves, this time in horizontal layout it would cause shift, therefore we choose horizontal plane for stationing.

As seen above, vertical curves are measured in horizontal distances, and they follow the centerline of the road in horizontal. The point of curvature (PC) and point of tangency (PT) which are also shown above *(look at the centerline of the road)*, are the start and end points of curves respectively, which can be seen on horizontal curve as below. Pay attention to the measured distances vs. station numbering in the figure above. The centerline distance covered is equal to the increase in station numbering. Also notice the distances between the first PC to PT and then from PT to next PC vs. station numbering.

Horizontal Curves:

The figure below shows how circular arc shaped horizontal curves are laid out for a road:

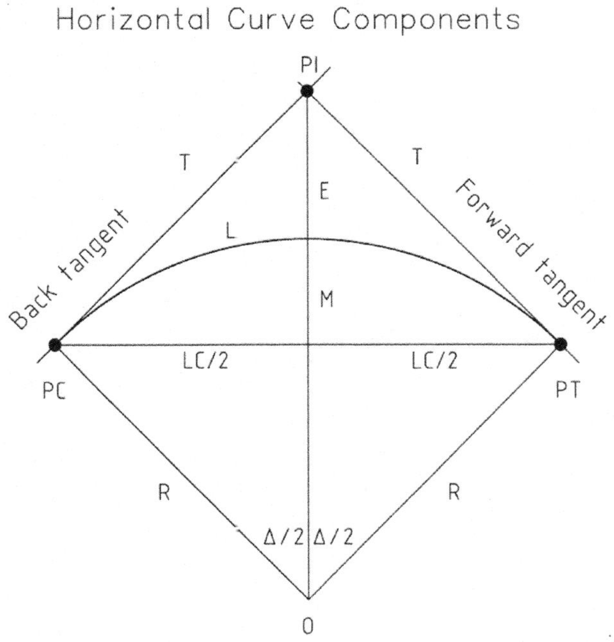

Figure - Transportation 6

In this figure:

Δ: Intersection / Interior angle (may also be shown as I)
PC: Point of Curvature
PT: Point of Tangency
PI: Intersection of tangents
L: Length of curve (arch length)
T: Tangent length
E: External distance
M: Middle ordinate distance
R: Radius of arc
LC: Long chord (may also be written as C)
O: Center of arc

The relationship between these quantities are given by the geometrical formulas below. Note that the formulas consider Δ to be in degrees. If it was radians, these formulas will differ. The shape above is an arc of a circle, so the formulas below are actually just common geometrical formulas for a circular arc.

$L=2\pi R\Delta/360°$

$M=R(1-\cos(\Delta/2))$

$LC=2R\sin(\Delta/2)$

PT Station = PC + L

$T=R\tan(\Delta/2)$

$E=R(\sec(\Delta/2)-1)$

PI Station = PC+ T

And let's clarify a something that is often confused. When you start with the station point of PC, and add the chord length L (which is curved) to it, you reach to the end of the curve, point PT. But now look at PI, where tangents intersect. To find the station of PI, as you would expect, you must add T, to the station value of PC. But then, you cannot just continue from PI to PT with stations. Because then you would have station of PT=PC+T+T, but we said PT=PC+L, which is obviously smaller. So again, when we want to find the station point of PT, we just follow the road centerline, starting from PC. Not tangents.

One parameter we did not show above to avoid confusion is the degree of curve, which is the angle created by a curve length of 100 feet, and shown by D. It is only used on projects with imperial units.

Vertical Layout

Vertical layout of a road is primarily influenced by the topography of terrain to minimize earthwork. In addition, other criteria must also be considered for vertical layout, such as: keeping vertical clearances, stopping and passing sight distances at crest curves, undercrossing sight distance at sag curves, paving friction, drainage, in addition to minimizing earthwork.

To put it simply, to change elevation of roads in vertical plane, we make slopes, and to connect these slopes, we make vertical curves. Let's call two consecutive slopes (grades) as g1 and g2. If we did not use vertical curves, g1 and g2 would join each other with a sharp edge, which would not only be unsafe and uncomfortable to drive on, but also unsightly. The vertical curves which are convex upward are called crest curves, as if the crest of a hill, and the opposite are called sag curves. A vertical curve can geometrically completely be defined if we knew the curve length, and the two grades, g1 and g2. Vertical curves are more complex than horizontal curves because a horizontal curve is simply an arc of a circle, while vertical curves are not. A parabolic equation is needed to define a vertical curve. We cannot simply make circular arcs for vertical curves because in this case we must join two adjacent grades with different slopes to each other. Therefore, the grade change over distance plays the key role for vertical curves, which requires more complex shape than a circular arc.

In the figure below, you can see a typical crest type of vertical curve and its components. For sag type curve, the equation is the same, except that only the grades g_1 and g_2 will change signs. The equation of the curve below is defined by the distance y from tangent, which is a function of x. Also, the distance L, is not the true distance that the car tires travel but the horizontal distance, as we also explained under stationing subtitle. In the equation, x starts as zero, and its values are in terms of stations for its distance, in other words horizontal distance from PVC, the start of the curve.

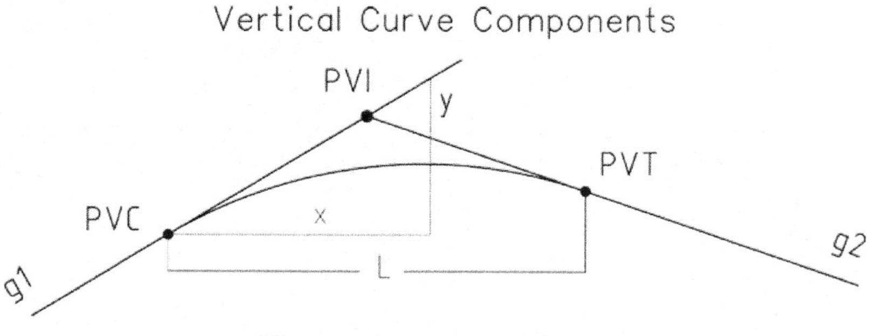

Figure - Transportation 7

In the figure above:

y: distance y from tangent (equation of vertical curve)
x: distance of a certain point from PVC, <u>in terms of stations</u> (therefore, on horizontal plane)
y is expressed in terms of x as:
$y(x)=y_{PVC} + g_1x + (g_2 - g_1)x^2/2L$
y_{PVC}: starting elevation at PVC
PVC: Point of vertical curvature
PVI: Point of vertical ntersection
PVT: Point of vertical tangency
L: Length of curve (horizontal)
g_1: Initial grade
g_2: Final grade

Many vertical curves are symmetrical about their central point, called the vertex. There is a limit to the grades we can have on a road, based on its usage. For example per AASHTO "Green Book", the maximum allowed grade on a 70 mph (110 km/h) highway is 5%, while for a road of max speed of 30 mph (50 km/h) a maximum grade of 7% to as much as 12% is possible. Vertical curves should be arranged such that the frequent ups and downs resembling a roller coaster ride, two vertical curves in the same direction (sag-sag or crest-crest) separated by short tangent line should be avoided for aesthetics, as well as the comfort of the ride. In cut sections, sag curves should be avoided unless the section is sufficiently drained, which would add to the cost of drainage work.

SUPPLEMENTARY COMPONENTS FOR HIGHWAYS

Bridges and Viaducts

When the road must cross deep and wide valleys or rivers, a specialized structure must be built to make this possible, which are bridges and viaducts. They were discussed in Chapter 3.

The photo on previous page shows a precast concrete bridge overpassing a highway, which is the the most common type used. The photo on the right shows the abutment closer, which is built on a stabilized slope, which is done more firmly right below the abutment as the slope also carries the bridge load there. And on the sides it is done by shotcrete, where it is sufficient to make the slope only to hold itself together.

Retaining Structures, Stabilized Slopes

Roads often continue near hills, mountains, slopes, from where rock and soil must be prevented to fall on to the road. This is achieved by retaining walls and or stabilized slopes, discussed in Chapter 1.

In the photo above, see the tiered excavation performed on the slope on the right side. The front part still needed some extra stabilization by a sloped gravity wall. On the left side, the slope is made and covered with shotcrete. The thin long line running down the slope on the right side is the drain chute, as it is important to protect slopes against accumulation of water, as discussed in Chapter 1, Soil Mechanics. For a closer look at the chute and retaining wall, see photo under chutes, few pages back.

Tunnels

Sometimes, it is less feasible to make the road to follow the side of a hill or mountain, or cross waters by bridges. In such cases, drilling through the mountain or below the river, lake or sea may make more sense, which is accomplished by tunnels. We introduce tunnels in Chapter 6.

Noise Walls

These walls do not support anything but just lower the noise from freeway to surroundings. Although they are freestanding, they must still withstand vibration, earthquake, wind and impact forces.

Ripraps for Bank Protection

These are placed on adjacent slopes, to protect against erosion. There are three kinds:

- Placed Riprap: rock placed on prepared surface as a well graded soil mass
- Keyed Riprap: rock placed on prepared surface by impact pressure
- Grouted Riprap: rock placed or keyed on a prepared surface by filling voids with grout

CONSTRUCTION

After the road layout, cross section and all other components are designed, the construction phase can begin. To build roads in the field, mainly three things are performed in order:

- Layout
- Earthwork
- Road cross section: Placing sub base, base, paving, drainage and supplementary components to build the road and its components

Layout

Layout, which is the first step in construction, includes the following:

- As shown in design documents, laying out the centerline of the road, pinning stakes on the ground to make up the centerline.
- Layout of slope and reference stakes on both sides of the centerline, which are the stakes that define the road clearing limit, which means the limits of the road plus any required earthwork. These are laid out in 30 meter or 100 feet intervals, per FHWA.

- Layout road cross section at every certain interval such as 100 feet, 50 meter or 100 meters, and intermediate cross sections as necessary.
- Laying out the road approaches.
- Placing grade finishing stakes.
- Marking culvert locations, including inlet, outlet, inlet basin stations, culvert dimensions.
- Layout of bridges, retaining walls, reinforced soil slopes.
- Layout of borrow and waste sites for earthwork.
- Layout of other miscellaneous items, such as guardrails, sidewalks, curb and gutters, turfs, utilities, trenches and drains, channels and ditches.

Earthwork

In Chapter 1, we discussed earthwork in general. This time we look at it in regards to construction of road foundation. First, clearing and grubbing of plants and trees, removal of various obstacles and structures (such as manholes, pipes, culverts to be abandoned) must be completed within work limit.

Surveying stakes are not only placed to indicate location, but they also show the final proposed grade of the road. Based on this, the excavation crew knows how much to cut or fill at a particular location. After staking is complete and direction and limits of the roadway are laid out, the natural ground must be corrected as necessary, which are shown on the cross sections that we discussed before. In Figure 3, we showed the layers of a typical road foundation. Now let's expand it a little:

Through earthwork, the ground is brought to the bottom level of the subbase, as seen in the figure. The very bottom layer can either be in the form of an embankment over natural soil, such as fill placement areas, or it can directly be natural soil, such as in cut sections.

The first pass is called rough grading, where the soil is handled in mass quantities. Second pass is done with equipment that can more precisely grade the soil. This is step is called fine grading.

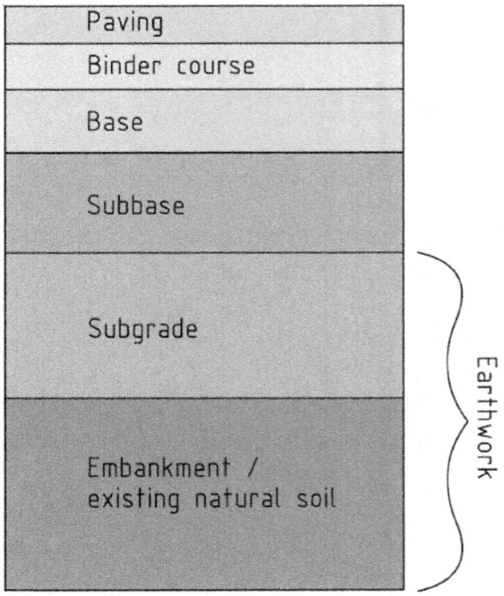

Figure - Transportation 8

At fill locations, the suitable soil is brought from approved sites (if the cut soil is usable, it is preferred) and placed. The soil is compacted, and then another layer is placed and compacted and so on. The soil is placed in layers rather than one thick layer for better compaction. The maximum layer thickness is 1 foot per FHWA standard specifications FP-14, but for the embankment soil that falls outside the roadway cross section limit, it can be raised to 2 feet. Usually the soil is compacted up to 95% of its maximum density, a value determined with proctor test as discussed in Chapter 1.

Where rock is encountered, blasting may be needed. At locations where natural slopes seem to be risky, slope stabilization measures must be taken. Slope stability analysis should also be performed for large amount of fills with considerable height where proper drainage must also be provided.

As earthwork operations proceed, at the required depths, geosynthetics such as geotextile filters, geomembranes, drainage pipes, moisture barriers are also installed as applicable, for reinforcing, drainage, filtration and separating layers from each other (not shown above).

In some cases, the subgrade materials may not be of enough strength and may be sensitive to excessive settlement, swell, frost-heave action. In this case soil stabilization must be performed such as with chemical admixtures, fly ash, hydraulic cement, lime, water, as will be discussed.

During earthwork operations, proper erosion control procedures must be followed.

Placement of Sub-Base, Base and Paving

Roads are also structures because they also have foundation (base and sub base) and carry loads. As seen in Figure 8 above, after the earthwork is complete, the following layers are placed in order:

Sub base > Base > Binder course > Surface course (paving)

On top of paving, sealers can also be applied based on climatic and usage conditions in order to provide better skid resistance and serve as waterproofing.

Similar to compaction requirements for earthwork, sub base, base and paving (if asphalt) courses are also compacted to specified density values. Placement of materials must proceed in a coordinated manner with the drainage installation. Now let's describe each of these layers:

Sub-Base:

It consists of natural gravel or crushed stone, so it also acts as a drainage layer. Sometimes there is no need to place sub base, if the subgrade below is strong enough. In some cases cement can also be applied as binder material. During installation of geosynthetics, as mentioned above after Figure 8, they must be protected from long time exposure to sunlight (as ultraviolet radiation causes harm) and also from direct construction traffic over it, and must be covered with base as soon as possible.

Base:

Base must be made of very good quality material, since it will be right under the pavement and be subject to high level of loads. Various materials can be used such as:

- Bitumen treated aggregate, for moisture resistance, which is called a treated base.
- Granular base, which is made of soil with particle sizes of coarse to fine
- Crushed stone or rock, aggregates, compacted together with finer materials

Binder:

Binder course is asphaltic material that is placed between the base and the paving in order to bind both layers and also to make sure that loads from paving to base are more uniformly distributed.

Paving:

Paving layer provides a smooth driving surface, friction to tires and also acts as a waterproof cover for the whole system. Paving must resist the effects of weathering by the effects of water, sun, hot and cold weather, cracking, wheel friction. Paving can be classified as rigid or flexible paving. Both rigid and flexible paving contain aggregates. For rigid pavement, the binder material is cement, and therefore we have the cementitious concrete. For flexible pavement, the binder material is bituminous, and it is also called asphaltic concrete.

A rigid paving, thanks to its stiffness, deforms very little, and transmits tire stresses almost uniformly to base and sub base. They are more expensive, and generally used on routes with heavy vehicle loads, harsh environmental conditions such as freeze thaw cycles, relatively worse ground conditions, which would have required more soil stabilization if flexible pavings were used. In some cases, asphalt surface material can be laid on top of concrete, for better wear resistance. Crack control can be achieved by control joints which are cut after the pour, at every certain interval, where plain rebar can be used to transmit stresses between adjacent plates, and the top of joints are covered with flexible joint filler. In case of full reinforcement, joints may be avoided. Concrete pavings have much longer life than asphalt pavings on average. Another type of rigid paving is the roller compacted concrete, which was discussed under Chapter 1 > Concrete > Roller Compacted Concrete section.

Flexible paving, as the name suggests, flexes under high loads. The tire loads are transferred in a much more concentrated manner to the base and sub base, which means the foundation needs to be thicker. Flexible paving is easier and cheaper to build and maintain.

Soil Stabilization

Soil stabilization is blending or mixing the soil with certain materials, to obtain a fill material with better properties for a highway or airport paving. Through soil stabilization, strength and durability of soil is improved, plasticity index (PI) and swelling potential is reduced, better gradation *(meaning, the soil with different size particles together)* is achieved. During construction, the stabilized soil also serves as a stable working platform (for heavy construction equipment), which can also withstand rains much better than unstabilized soil.

Soil stabilization is not the same as ground improvement, although they both serve the purpose of strengthening and stiffening the ground and make it less permeable. Soil stabilization is blending and mixing of the backfill soil that we compact at the surface. Ground improvement means improving the quality of the existing ground which may also be at much larger depths and below structures.

Stabilization mixing can be done in two ways, through additive (chemical) or mechanical methods. Mechanical method is mixing of soil with two or more grain sizes in order to achieve better graded soil with small and large particles so that the smaller particles can fill the voids between larger ones and the soil can be compacted better.

Additive method is improving the qualities of the soil by mixing it with certain chemical additives, which can include a combination of cement, lime, fly ash, bituminous materials and more. To give an example about mixtures, per FHWA Standard Specifications for Construction of Roads and Bridges - FP-14, the unconfined compressive strength requirement for a lime + soil mixture is 100 psi, whereas Cement + Fly Ash + Soil or, Fly Ash + Soil mixture is 200 psi. Per FP-14, fly ash is applied with dry method, which means just enough water to obtain the proper moisture content for the mixture. Lime and cement can be applied either by dry method or slurry method (wet method). Slurry method means obtaining a mixture of slurry consistency and then applying it. The materials must be spread as soon as the mixing is complete and then compacted to 95% of optimum dry density. After the application, the compacted mix should be cured with sufficient moisture until the next layer is placed on top of it.

Environmental effects, if any, of soil stabilization must be studied and understood before application.

Construction Coordination with Existing Traffic

In most bridge and highway construction work, where there is existing traffic, traffic diversion and control is critical. Separate temporary diversion routes or even bridges might be needed and removed. These routes must be maintained to remain open during operation, such as removal of soil or snow, watering for dust control, repairs. It is also common to work at nights where some portions of the permanent or temporary work must be finished or repaired quickly. The lighting system must be coordinated for such cases. Erosion control is also important, which needs regular maintenance. Some highway or bridge contracts can include very high liquidated damages to contractor, in case the work doesn't finish timely, because everyday of unfinished work creates a burden on public transportation.

Apart from coordinating field operations, effective coordination with the affected properties and jurisdictions is also necessary.

CHAPTER 5 - RAILROADS

Given the same tractive force, much more load can be pulled over rails than on an asphalt or concrete, due to far lower friction between steel wheels and steel rails, than between vehicle tires and a road. Although their initial and maintenance costs are higher than highways, railroads are more efficient means of transportation than highways in many ways. The direct and indirect benefits and efficiencies resulting from rail and metro transportation are countless, in terms of macro economy, environment, efficient use of land, energy, public health, time and cost savings and many more... In some developed European or Asian countries, people can access to almost anywhere with railroads, often including to even small towns, and the cities have extensive metro networks. Railroads can be classified mainly into two as the tracks that carry freight or passenger trains. Freight trains are much longer, heavier and generally slower than passenger trains.

In the figure below, a typical cross section for a railway is seen:

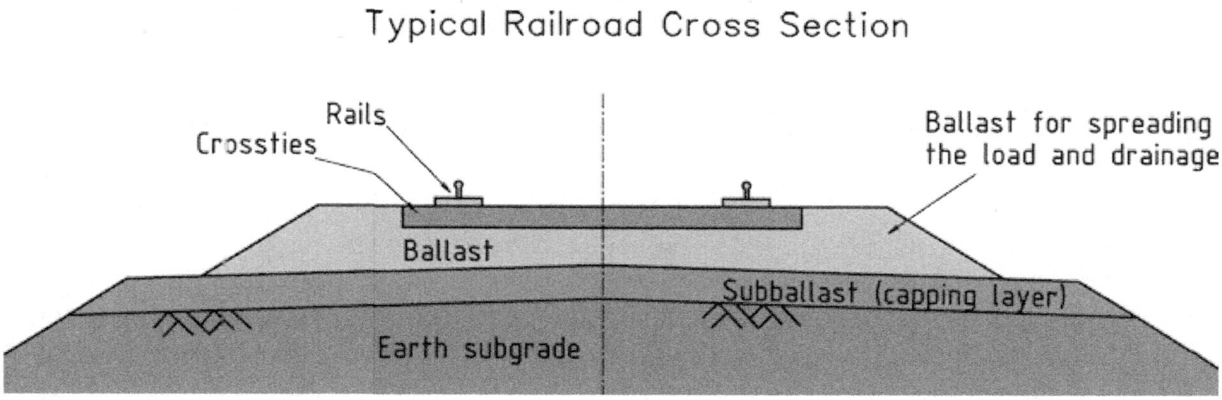

Figure - Railroads 1

The train load is first spread by rails and crossties, and then the ballast and finally the subballast.

Crossties, which has a spacing of usually between 19" - 24", can be of wood or concrete, and used to:

- Transmit train load from rails to ballast, and spread it over a much larger area, reducing stress
- Serve as a base for the rail anchors to attach
- Ensure the stability of rails by serving as foundation and bracing

The ballast layer further distributes the load from crossties, stabilizes them, absorbs vibrations, and serves as a strong, elastic foundation. Ballast is made of angular stones, which means high shear strength, and in turn, bearing strength of the overall layer. It also allows water to drain easily as it is very porous. On the sides longitudinal drainage can also be placed depending on the terrain conditions.

A geotextile filter fabric can be placed between ballast and subballast and subballast and supporting earth, thus protecting the system by preventing migration of finer particles and also for improving lateral drainage to remove water much faster from the system. By using geotextiles, it is also possible to save some subballast thickness. At weak subgrade, given long enough time, subballast and earth material tend to mix, under the action of heavy, dynamic vertical loads, which will cause compaction, deform subballast and cause pockets. In the long run, pockets can also start to collect water, which will accelerate deformation and decrease load bearing capacity. Installation of geotextiles will prevent this process by separating ballast from subgrade, besides improving filtration and drainage.

Ballast, crossties and rails.

In addition to most of the criteria we discussed for highways, various other items must also be considered when designing railroads, such as:

- Geometric layout principles of railroads is generally similar to that of highways, but with more variables and stricter tolerances. For example, horizontal curve radii must be much larger for railroads. Maximum allowed slopes and superelevation are lower. Clearance requirements, and curve transition lengths are far greater than highways. All these stricter requirements greatly contribute to much higher costs and longer construction schedules in general, especially over rough terrain, because highways can turn sharper curves, climb steeper slopes and therefore go along better with natural terrain, while railroads cannot adapt so easily and often require longer routes, bridges, a higher number of viaducts or tunnels in order to remain in their relatively more straight and flat route, with higher clearance requirements. Therefore, when determining the main route and the station points, more criteria must be satisfied. The bridges, viaducts and tunnels for railways will also generally be costlier than the ones for highways.

- The standard distance between the two rails on a railroad track in the US and many other countries in Europe, Middle East and some in Africa and Asia are made as 1435mm (4' - 8 1/2") but various other standards also exist. As this distance gets larger, more passengers and freight can be carried, but at the same time the construction costs are higher.

- Crossties can be of steel, wood or precast prestressed concrete. Wood crossties were gradually replaced by the other two. Steel ties are strong and flexible. The ones that are most commonly used today are the prestressed precast crossties.

- Grade crossings: Constructed where railway meets highway. Before the grade crossing is reached, any existing superelevation as a result of a previous curve must already be completed and two tracks must be at the same elevation. Storm water must be directed away from the crossing to prevent dirt accumulation.

- Signal systems: Used for safe and efficient movement of trains. Can be manual or automatic.
- Special track work: Turnouts, slip switches, intersecting rails are where trains can be directed to different directions by adjusting the tracks at intersection. A bolted rail crossing is fixed, does not change train direction, it is simply an intersection of rails in two different directions.
- Noise and vibration control: Strong and excessive vibrations from fast traveling trains will eventually weaken the railroad structure so vibration control is an important criteria in railroad design. In addition, trains can cause significant amount of noise and vibration for nearby structures if any. Therefore depending on the location, noise and vibration control measures are needed, which may also needed to be applied to the foundation of adjacent structures such as base isolators and or ground improvement. Not only the design and construction but also the maintenance of tracks also affect the generation of vibration and noise.
- Locomotives and wagons impose heavy loads on the foundations. Not only heavier loads in comparison to highways must be carried, but also strong vibrations must be prevented, which is critical especially with faster trains, due to very high momentum of heavy wagons when they move at high speeds and even small impacts due to vibration will mean a large forces on the railroad structure and back on the train at those speeds. Similarly the impact kinetic energy is very high which must be absorbed by the foundations. Therefore, not only the geometric layout but also the foundations of railroads are designed to stricter requirements. Settlement of rails can be very dangerous for fast moving trains, and the tolerances are very low.
- Because of the reasons above, railroad bridges must be built for larger forces and stricter tolerances for displacements and vibrations than bridges that will support trucks and cars. The acceleration or braking of a train, lateral impacts of wheels to rails will exert great forces on a bridge structure, which will affect not only the deck but also the piers and the foundations. Track to bridge interaction affects must be considered. Dynamic effects must be considered for high speed trains or for bridges with long free spans. All these must not only satisfy structural safety and deflection criteria but passenger comfort criteria as well. It is important to standardize railroad bridges (and tunnels) for overall efficiencies to economy.
- Corrosion: Preventing corrosion of tracks.
- Stations: Design and construction of stations can be considered as separate and specialized projects. Main stations also have a lot of storage tracks for unused wagons.
- For high speed rails, rubber shock absorbers can be installed below the tracks.
- For locomotives that use electricity for traction, substations and power distribution is major electrical work, which must also be coordinated with the construction of the railroad itself.

The photos show track repair of a surface metro system. The concrete paving around the rails were to be replaced, and it was removed. To make up for the lost support, the rails were braced on both sides and to each other, so that they maintain their position. After this, new concrete will be poured. Welded rebar mesh are placed between the tracks extending a certain distance outwards, as reinforcement. Although this detail is different than the typical detail figure we showed for railroads, and ballast and crossties are absent, the goal is the same, which is to firmly stabilize the rails in place by means of a strong foundation. The stabilized and compacted soil underneath serve in place of the subballast that would be found in a typical detail. The photo below shows the same rail line several feet back, where concrete pour continues. See that it is poured upto the bottom of rails for supporting them. Concrete, which is strong and rigid, here effectively replaces and does the same duty as ballast plus the crossties that would be found in a typical cross section. The concrete mix design used here should also ensure a very durable final product, in order to withstand environmental conditions and wear for many years.

CHAPTER 6 - TUNNELS

Twin tunnels on a major highway route, where each is used for one way traffic. Also see the stabilized mountain section slope right above the tunnels.

Transportation tunnels are underground structures that are built for passage of vehicles, trains or passengers, which we will cover in this chapter.

Tunnels are also dug commonly for water conveyance, and for various other purposes such as even the particle colliders. In fact, the longest several tunnels built in the world so far are for water conveyance and a considerable portion of aqueducts usually run underground, contrary to common knowledge, but, when we use the word "tunnel" in this chapter from now, it will only mean transportation (vehicular and railway/metro) tunnel, although many of the aspects of these tunnels inherently apply to tunnels for other purposes too. Transportation tunnels have much larger cross sections than tunnels for other purposes. For example, a tunnel diameter of 10 feet would be a very large one for an aqueduct, but this diameter would not suffice even for a small transportation tunnel.

Tunnels are in many situations better alternatives to bridges or longer length roads when crossing physical barriers such as mountains, hills, rivers, roads or man made structures. When completed, they have no adverse traffic noise or environmental effects to surroundings, they are entirely invisible and hidden underground which does not disturb the natural scenery, they do not occupy and disturb valuable land area in urban locations except their entry and exit points.

We will study this chapter under the following sections:

- Primary planning
- Geotechnical aspects
- Tunnel geometry
- Tunnel classification
- Bored tunnels
- Cut and cover tunnels
- Immersed tunnels
- Jacked box tunnels
- Water drainage and waterproofing
- Interior finishes
- Other tunnel components

As of 2023, the world's longest tunnel (ignoring aqueduct tunnels) is the 57 km (35 mile) Gotthard Base Tunnel through the base of Swiss Alps. The tunnel is made of two main tubes, each of which are 57 km long, small connector tubes between the two main tunnels at every 325 meters, and three vertical access shafts, drilled from the mountain surface to the tunnel for service and emergency purposes. With respect to highest point of the mountain, it is also the deepest tunnel, with 2300 meters maximum depth. 80% of the excavation took place with Tunnel Boring Machines (TBM) and the rest mainly by drilling & blasting, and the total length of all tubes add up to 152 km. It was built between 1999 and 2016, with average daily work force of around 2500. The tunnel was constructed for high speed rail transport and reduces rail travel time between North and South Europe by 30 minutes. 65 passenger and 260 freight trains can travel through the tunnel everyday.

PRIMARY PLANNING

Many factors are considered when choosing between a tunnel or bridge, or between a tunnel and going around a mountain via highway. These factors include the topography, construction cost and duration, alternative routes, long term life cycle maintenance costs, life expectancy, environmental and ecological impacts, aesthetics and visual intrusion, future benefits from saving land or water space, water conditions, navigation, cost savings by using the tunnel and administrative issues. Tunnels usually have much longer economical life expectancy than bridges or roads, such as 100-150 years.

For primary planning, the "Technical Manual for Design and Construction of Road Tunnels by Federal Highway Administration, FHWA-NHI-10-034" is an important source that is followed for highway tunnels in the US, where many terms and concepts discussed here can be found in much greater detail.

GEOTECHNICAL ASPECTS

Tunnels differ from other structures because the geotechnical environment is part of the design of the structure itself, as far as its stability, strength and construction. Since tunnels are built underground, the most important thing to know when starting the design is the geological properties of the proposed tunnel route. We discussed preparation of geotechnical reports in Chapter 1, which you may want to review again at this point. A tunnel continues for long distances, and therefore in addition to applicable items that are typically included in a geotechnical report, a geotechnical study for a tunnel should also include items such as:

- The geological map that shows the geological features of the tunnel route such as: Valleys, fault lines, sinkholes, karstic terrain, thermal water and gas, and extensive information on rocks such as the rock types, discontinuity (joint) types, spacing and orientation, weathering, block size, strength, color, texture, hardness, mineral composition, layering directions, roughness.
- A more detailed study on groundwater, for design water levels for all expected situations including the existing groundwater flows and the changes of water levels both during and after the construction of the tunnel, identifying potentials for groundwater inflows.
- Anticipated behavior of soil/rock in response to construction operations and their sequence.
- In some cases test shafts may need to be drilled to obtain better information about ground conditions. For better groundwater data, aquifer pumping test can be performed.

These site characteristics determine the design and the method (which also means the schedule and cost) of construction. In some projects there may be more than one method that will be appropriate for a tunnel since the ground conditions can change along the length of the tunnel.

Geological mapping of the tunnel route continues as construction progresses. This work must be well coordinated with the progress of construction. To eliminate or minimize unforeseen conditions during construction phase, which have the potential to affect the design, a thorough geotechnical study in the preliminary design phase is necessary.

Once a tunnel is excavated, a certain portion of soil or rock above the excavated hole (i.e. "ceiling" of the hole, see figure below) will now be in tension, tending to fall. Rocks can hold themselves better, while soft soils will collapse almost immediately. This will affect the tunneling method and the linings (structural support layers), and if necessary any ground improvement or soil/rock reinforcement to be applied and its type. The tunnel does not and can not carry the thousands of tons of rocks of a hill or mountain in a vertical column directly on the top of the tunnel all the way up to the surface. The more we go up, the more portion of that rock in that vertical column is squeezed laterally by the surrounding soil/rock and the area that tends to fall gets smaller as seen below. The size of the area in tension depends on shear strength of the soil / rock. For hard rock for example, the area that tends to fall would practically reduce to zero. As the soil gets weaker the size of the area will increase.

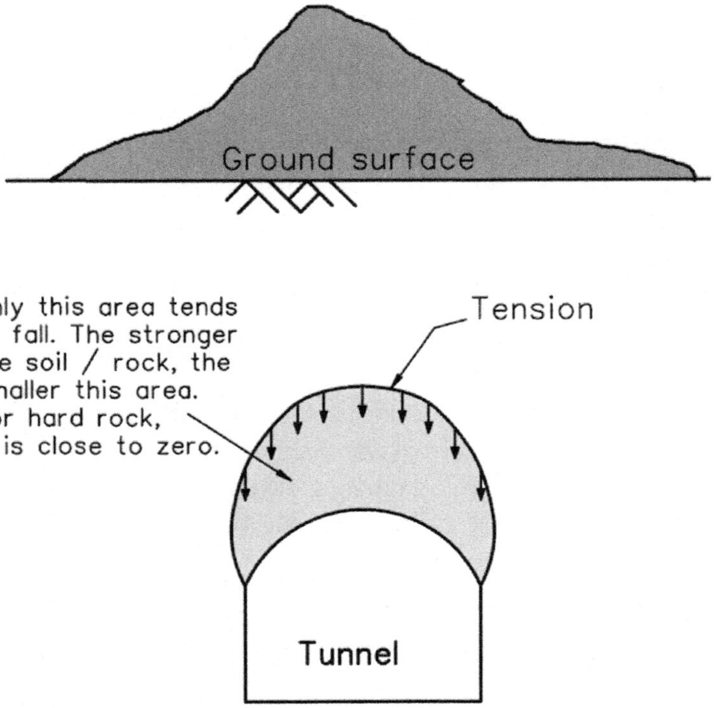

Figure - Tunnels 1

Most tunnels, by their shape, heavily utilize arch effect. The arch shape of the tunnel, enables all that huge weight of the soil or rock on top of the tunnel to get redirected to go around the tunnel.

TUNNEL GEOMETRY

The three main shapes of tunnels and their common methods of construction are:

- Rectangular - suitable for cut and cover and immersed methods
- Circular - tunnel boring, drill & blast methods
- Horseshoe - drill & blast method

Tunnel length should be kept as short as possible primarily due to the very high costs. For vehicular tunnels this is also for having sufficient sight distance. At the end of long tunnels, a slight curve may be required to make it easier for the drivers to gradually adjust eyes to daylight brightness. Tunnel exit and entry portals have some other special requirements in addition to the tunnel's.

AASHTO Technical Committee for Tunnels define road tunnels as enclosed roadways with vehicle access that is restricted to portals regardless of type of structure or method of construction. This also means, highway tunnels are viewed as a continuation of highways, with at least equal traffic capacity of the road it is connecting to. Geometrical requirements for highways would also apply to tunnels, however due to much higher costs in case of tunnels, typically the minimum requirements are used for tunnels to minimize costs, such as for design speed, superelevation, shoulder width, horizontal and vertical clearances, lane width, sight distance, grade and cross slope, without sacrificing safety or the required traffic capacity. For vehicular tunnels, too low ceilings or too narrow lanes to save from cross section costs should be avoided as this tends to slow down drivers hence the traffic. The feeling of confinement in a tunnel is already uncomfortable for some drivers.

In addition to having adequate size to allow the required traffic flow, the cross section of a tunnel must also take into account elements such sidewalks/curbs, ventilation and its ducts, lighting, drainage system, traffic controls, fire detection, fire protection pipes and components, communication, emergency system and exists, surveillance cameras. These must fit efficiently into the cross section and be kept to a minimum to minimize costs, without sacrificing safety or functionality. Of course similar or more stringent principles apply for railway tunnels.

To keep the cross section and therefore the costs and construction schedule to a minimum, there are many factors, from which, several are listed as examples below:

- The design of the roadway plus tunnel system should be such that there will not need to be signage inside the tunnel.
- Traffic merging or diverging situations within a tunnel should be avoided.
- Long, steep slopes should be avoided as they necessitate climbing lanes for heavy vehicles.
- Typically the shoulder width in tunnels is reduced.

With all geometrical, geotechnical, structural, traffic, environmental, cost, schedule and other constraints, the actual geometry of a tunnel can be determined after a number of iterative loops.

It is also common to see two tunnels dug in parallel, for each traffic direction, instead of one bigger tunnel. It is much easier and economical to dig and structurally support each of the smaller cross sections than one large cross section, even if in the case of two smaller holes we will dig twice the length with respect to one larger diameter tunnel. To support the larger diameter hole would require much bigger excavation machine and a stronger and more expensive structural support system. In addition to structural issues, from simple geometry, the total excavated volume is less in the case of two smaller tunnels, if you think about two small circles drawn side by side with radius r, and placing them inside a larger circle with radius 2r. The larger circle will have twice the area, thus volume, than the sum of two smaller circles. The ventilation, lighting, drainage requirements in two options need to be compared. The total circumference would be equal for the two options which affect the surface area of liner, (which can be of concrete, shotcrete, precast concrete or steel) and waterproofing. For long tunnels, two tube system is also preferred for safety reasons, such as in the case of fire, for evacuating passengers from one tube to the other, by making small connection tubes between main tubes at every quarter mile or so. Thus, all variables that we discussed so far in this chapter are taken into account and the decision is made for one larger hole vs. two smaller holes.

TUNNEL CLASSIFICATION, AND SUPPORT SYSTEMS

AASHTO "Green Book" classifies tunnels into two main categories: Tunnels dug by mining methods and tunnels constructed by cut and cover methods. According to FHWA-NHI-10-034, main types of tunnels according to methods of construction are:

- Tunnels bored in rock
- Tunnels bored in difficult ground
- Tunnels bored in soft ground
- Cut and cover tunnels
- Immersed tunnels
- Jacked box tunnels

In the list above, we can combine the first three (bored tunnels in rock, difficult ground and in soft ground) under one section as "Bored Tunnels". Bored tunnels are the traditional and most widely used form of tunneling which we will discuss in greater detail. It is also called as the "mining" method.

BORED TUNNELS

Bored tunnels are the traditional and most widely built type of tunnels. We will look at them in several subsections as below:

- Loads and modeling
- Excavation
- Ground improvement and groundwater control
- Drilling & Blasting
- Surveying and measuring
- Initial and final lining (Tunnel structure)
- Bored tunnel classification per rock/soil type
- New Austrian Tunneling Method (NATM)

Loads and Modeling

The loads that act on a tunnel system include:

- Lateral and vertical loads exerted by the ground.
- The dead loads of liners, finishes and other components.

- Static or flowing water pressure, considering the worst case future levels and flows. This can only be disregarded for the finished tunnel, if an effective permanent drainage system can be guaranteed. In principle, the load analysis should also look at the case without water, regardless of the actual water effects and drainage in the field.
- Seismic loads.
- Load effects of shrinkage and creep.
- Load effects of temperature variances.
- Load effects from nearby cavities.
- Prestressing loads, such as from prestressed anchors, pretensioned connections.
- Loads acting on the ground surface, such as the dead loads from the structures, traffic live loads, vibration loads, including future expectations for all.
- Time dependent effects, such as the hardening of shotcrete, rearrangement of loads of the surrounding earth mass in the long term, future loads from rock due to weathering.
- Grouting pressures.

For analysis, a mathematical model of the tunnel should be prepared, to be used in a 3D finite element analysis. The model will include all these loads as applicable, the ground vs. tunnel interaction, material strengths, soil parameters, groundwater conditions, effect of existing underground features (such as large utility lines, other tunnels...) and foundations nearby. This will not only analyse the tunnel and determine the loads and deformations of the tunnel system, but also calculate the stresses in the soil and estimate the settlements at ground level, and the effect on existing structures' foundations, which becomes critical in urban areas with nearby streets and structures. Critical intermediate construction phases are also analyzed. The model and design may be modified as the tunnel progresses, upon the results of surveying checks and encountered ground conditions. Note that design of other types of structures also get updates as the job progresses but that may be due to coordination issues only. For the tunnels however, making updates to the design during the course of the work is an inherent part of the design, as we progress through the natural and non homogeneous ground material, although usually the same principal design will be used for the most part or entirely. The ground investigation in the beginning of the project is performed to minimize these changes.

Before the common usage of 3D finite element methods with powerful geotechnical software, the analysis above for loads, deformations and settlements was traditionally done based on soil mechanics calculations based on the soil type, tunnel geometry and depth, groundwater conditions and surcharge loads from existing foundations. For tunnels that are sufficiently deep, small diameter, and or far from significant existing nearby loads, this method is still adequate. It is still useful in any case, in any situation, to at least provide common sense check and preliminary point to start with.

Excavation

Tunnels are dug in mainly two ways:

- Excavation
- Drilling & Blasting

We discuss excavation now *(also called "mechanical excavation")* and blasting in the next subsection.

Tunnel Boring Machines (TBMs):

Tunnel Boring Machines, in short, TBMs, are highly automated, very large and expensive construction equipment, with a rotating cutting wheel at the front, and the screen and all other equipment that follows. It is basically an underground plant, which excavates and advances in the soil horizontally.

TBMs are used mostly in two cases:

- Boring long tunnels, where economies of scale can be achieved due to high advancement rate of these machines once they are set up. As the length of tunnel increases, using TBMs makes more sense. The only exception may be for locations with the hardest rocks, where drill and blast method may be more effective, but if there are concerns about disturbance to surroundings from blasting, then TBMs especially made for hard rock can also be used.
- Boring tunnels in collapsible soils *(i.e soft ground or below groundwater table and soil/rock is permeable/fractured)*, where it is impractical or more expensive to prevent the collapse/flood with other methods, as TBMs can also have their own protection screen (similar to a tube) that supports and protects the newly excavated hole from inside, which advances behind the cutting wheel in the front, as seen in Figure 2 below.

The type and size of TBMs vary. Large TBMs are the most powerful, largest and most expensive construction machines. The large ones can be over 200 meters (over 650 feet) in length, and weigh more than 2000 tons in total with all components. As an example, the TBM that worked on Gotthard Base Tunnel that we mentioned, which is the world's longest tunnel, was over 400 meters in length and weighed over 3000 tons, with cutter head diameter of 9.5 meters (31 feet). Just the mechanical assembly of large TBMs on site and the electrical and mechanical setup to operate the machine is a big task, with miles of power and low voltage cables.

TBM digs the circular cross section tunnels by rotation of its cutting wheel (also called cutter head wheel, drill head), which has many teeth on its face as seen in Figure 2. This rotating wheel is pushed and pressed against the soil to excavate with a contact pressure as high as several thousand psi. As the machine digs at the front, the excavated rock/soil which is now ground to small pieces, are transferred from the front towards back, and then with a conveying system, it is transferred further back, to be removed from the tunnel. As the ground is dug, the cutting wheel and protection screen assembly is pushed forward, and typically precast concrete rings are installed to make up the tunnel structure. This forward push is made by hydraulic thrust cylinders, which bear on the faces of the last installed panels, as seen in Figure 2.

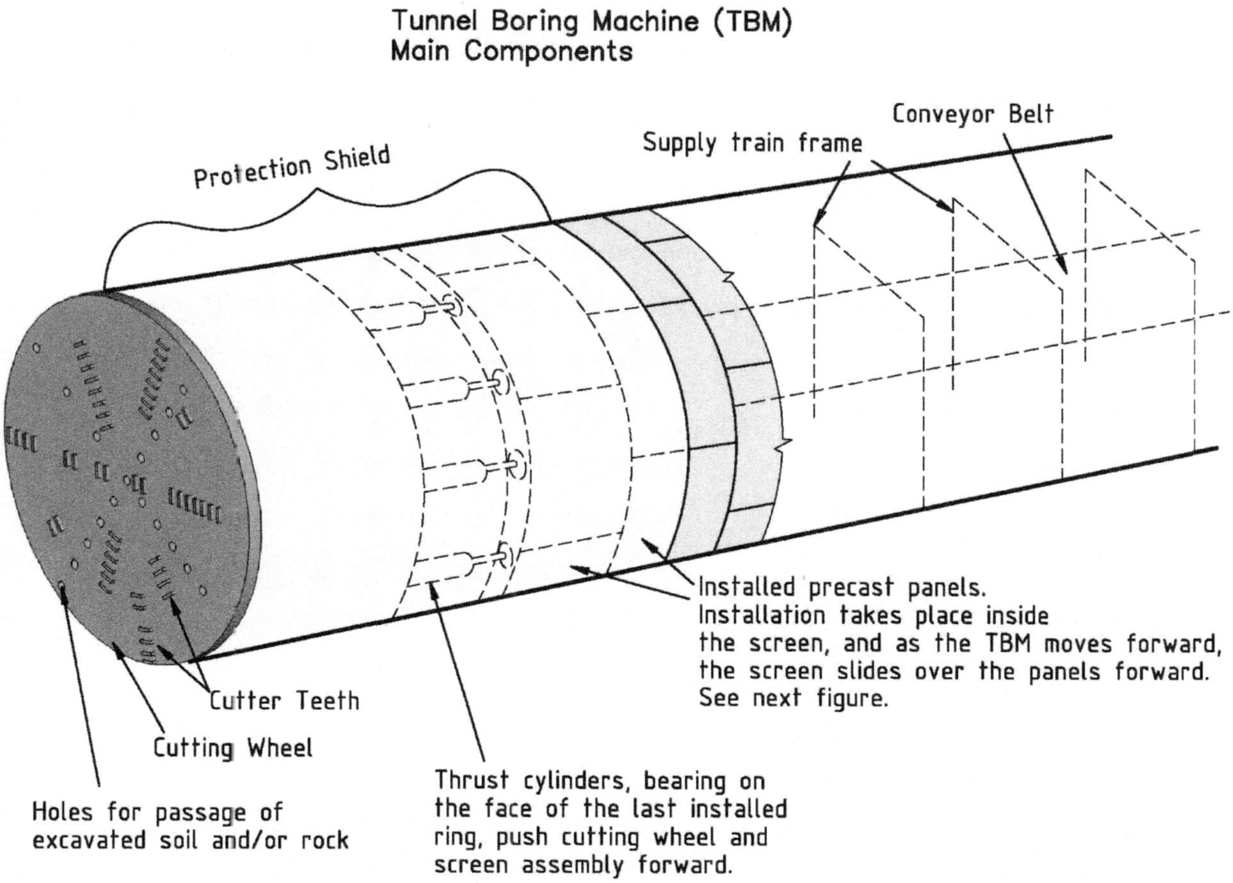

Figure - Tunnels 2

These rings are made of several arc shaped pieces to make up the whole 360 degree ring as seen in the figure. Each arc is placed one by one with automated ring erector system, which works in coordination with the hydraulic thrust cylinders. The cylinders are pulled back at the specific location of the ring that is being installed, and then pushed to start to bear on the newly installed piece. The final piece of the ring is shaped as a slight trapezoid, to be placed the last, as the keystone *(see Chapter 1, various structural terms, arch action)* of that ring. As the rings are placed, and the protection screen slides forward over the installed precast panels, grout is injected to the space between the earth and the outer face of the panels as in Figure 3, to make them stable in the excavated hole and also make the rings take on the load from the earth above uniformly.

Above you see a general sketch of a TBM, that shows its main components as:

- Cutter head wheel (cutting wheel), which digs the soil/rock at the front by rotation and applying pressure, which contains many cutting teeth made from hard elements. The wheel also has holes, to transfer broken soil/rock to its back side. The layout of teeth and holes vary for each machine. The wheel pushes the soil with tens of thousands of KN force, which varies according to soil conditions, as it rotates and slowly moves forward.
- The wheel also has sensors, to monitor the contact pressure, it has injection system to inject foam to bring the consistency of the soil more suitable for excavation as necessary.

- The screw conveyor (not shown) behind the shield, carries soil to the conveyor belt, which transfers the soil all the way back towards the entrance of tunnel, from where it is removed.
- Protective shield. Inside this shield, there are hydraulic pushing system (thrust cylinders), rotation system motors, precast ring segment lifting and placement (segment erector) system, precast segment feeder that supplies precast blocks to the erector. This shield also protects the excavated tunnel from caving in, in case of soft soils and or soils under water table. In that case it is also called earth pressure balancing shield.
- The set of trailers (also called supply train or trailing gear etc..), include the following:
  - the steel frame that serves as the structure to hold everything together
  - conveyor belt
  - central control cabin for the machine operator
  - mechanical and electrical equipment: pumps, ventilation systems, hydraulic units etc...
  - laser measuring system
  - rail system for transporting the crew, panels and everything else,
  - storage for soil conditioning agents which can be added to the soil during excavation
  - all necessary facilities for the TBM crew's needs.

To start using TBMs, especially in urban areas, a vertical launch shaft is first dug into the ground, which can be as deep as few hundred feet depending on the tunnel entrance depth needed. TBM parts are lowered into the shaft, where they get assembled, to start excavation from there, by moving forward on previously installed rails.

TBMs are expensive machines with long setup time, (i.e. at least few weeks, depending on size) that make sense mainly for long tunnels, where economies of scale is possible, and by using them, the largest lengths of tunnel digging per day can be achieved. The great deal of automation means higher equipment but lower labor costs, and as the distance gets larger this efficiency increases, so the tunnel must be "long enough". Due to a great variety of project conditions, giving an average value of "long enough" is not possible to state because it varies between countries, projects and local costs. If we really want to guess an average figure here, may be saying a tunnel length of around 2 miles will not be unreasonable. For majority of projects in the world, the tunnel lengths are not that long, and setting up a TBM is not the best alternative. In such cases, smaller equipment called roadheaders, and/or drilling & blasting are used. Drill and blast method will also be used where needed on any tunnel length or sometimes almost entirely. One disadvantage of TBM is that it can not accommodate varying tunnel cross sections. One of the most significant challenges in tunneling is that it is not possible to know all exact conditions in the ground beforehand. Therefore, a TBM is designed for the worst geological condition that is expected throughout the length of the tunnel. Even though this may mean overdesign for most part of the tunnel, there might be no way around this.

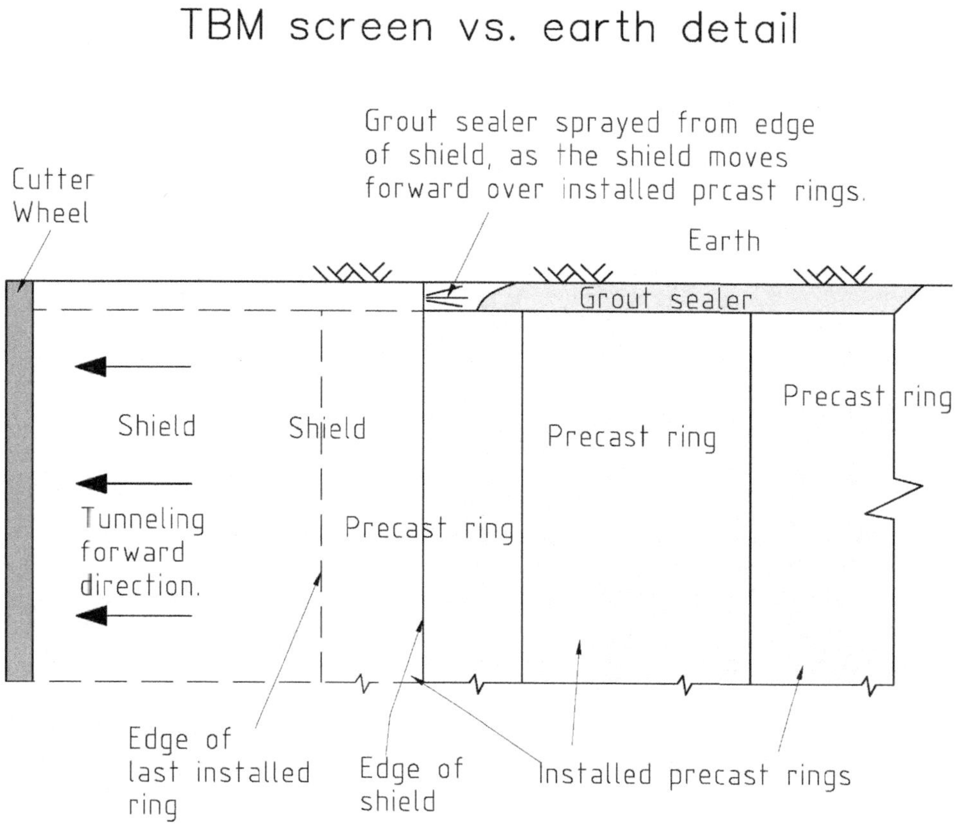

Figure - Tunnels 3

Roadheaders:

Roadheaders are far more smaller, flexible and cheaper than TBMs and can immediately get to work once they are at a jobsite. In comparison to TBM or blasting, they produce far less vibration or shock waves. Therefore, in smaller, shorter tunnels, critical urban areas or sensitive locations, which together constitute a great majority of the tunnels in the world, they are usually the first alternative to choose.

Roadheaders have a special arm in the front that can rotate in 3D and at the tip of the arm is a cutting head with a lot of teeth/spikes (called cutter picks) on it, to scrape and excavate rock/soil.

An advantage of roadheaders is that they can dig a tunnel in any cross section shape (TBMs can only dig circular), and also can make a tunnel turn sharp curves, while for TBMs, the curve making ability is limited. Due to this, even for the projects that TBMs are used, for fine digging at critical shorter sections, roadheaders are also used.

In the figure below, see the typical main components of a roadheader. The cutting head rotates and excavates while the rotating blade picks up the fallen earth material and pushes it backwards towards the conveyor belt, from where it can be loaded to a truck and hauled away. This system increases efficiency, by not interrupting the roadheader to wait for a loader to pick up fallen material from the ground and load to a truck.

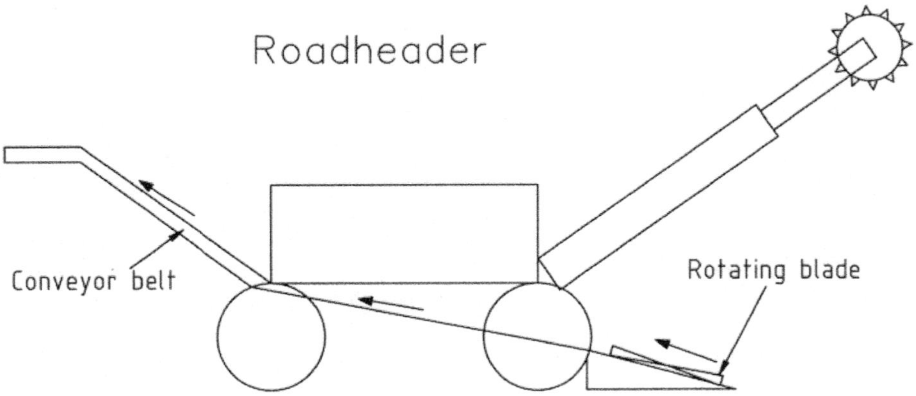

Figure - Tunnels 4

Electrically operated road headers are preferable due to elimination of exhaust gasses and reducing burden on tunnel ventilation system. Newer models of roadheaders also have automatic sensors at the tip where the pressure between the machine and rock is measured automatically and after the operator brings the arm to the location to be excavated, the machine does fine adjustment to the arm's position automatically for optimal pressure, which not only lengthens the economic life of the excavator and its parts, but also ensures digging at the most effective rate.

In addition to roadheaders, earthwork equipment can also be used when appropriate, such as the hydraulic excavators. Hydraulic hammers are also used for breaking and fine trimming rocks.

Ground Improvement & Groundwater Control

In certain weak ground conditions, ground improvement (see Chapter 1) may be needed before the excavation work can proceed, to prevent collapses, settlements at the surface, as a result of tunnel excavation. Since the tunnel is usually deep in the ground, doing ground improvement from surface in this case may not be practical, so it is done from within the tunnel, as the tunnel progresses. In the case of TBMs, this can also be done by the TBM machine injecting grout from front as the machine proceeds, where it fills around its front face with grout, which not only makes the ground stronger but also less permeable for groundwater control. The ground being stronger prevents surface settlements during and after excavation, and it also means less burden on the tunnel structure, both during and after construction, as the soil can hold itself better, which means a lighter structural system will also be sufficient. Performing ground improvement will not cause significant difference in excavation effort as the machines can also easily dig through the improved soil as well *(only hard rocks are difficult to dig with machines, which are much harder than improved ground).*

Groundwater can be controlled by drawing water from filter wells along tunnel length, or slurry walls or ground freezing. TBMs' protecting screen can also keep water out of the excavated tunnel.

Drilling & Blasting

For bored tunnels, blasting methods can sometimes be more economical than excavating, especially for hard rocks, if we are at a location and/or depth that blasting will not cause disturbance to surroundings.

To perform blasting, first the surface is prepared and holes are drilled in it, to insert explosives and because of this, this method is also called as "drilling and blasting". After that the explosives are detonated. After the blast, there is a wait time for fume dissipation and debris removal (debris is also called "muck"). Then the excavated part of rock is supported and the entire cycle is repeated. Because of these preparation and wait times inherent to this method, drilling and blasting is usually a slower method than excavating. Blasting can be used together with roadheader excavation if the ground contains both hard and soft rock.

Blasting causes great noise and vibration, so especially in urban areas, great care and skill is needed, and the tunnel must be at a certain depth below the surface, which varies for each project. Depending on the job location, a minimal amount of blast vibration can be allowed to be felt on the surface, but in that case the blasting may be limited to certain daytime hours. Sometimes blasting will not be allowed at all, and the only feasible option will be excavation.

This method has the advantage of being flexible and it can easily be used to open varying cross sections as needed. It has relatively lower equipment costs, and fast mobilization is possible.

One disadvantage of blasting is that it is less precise than excavating. For example, it can cause excessive loosening of surrounding rock, some of which can cause larger than required voids. The average precision of excavation with TBMs and roadheaders are a few inches at most, while for blasting this could be as much as a couple feet. This means more shotcrete must be used to fill the voids.

Surveying and Measuring

The planned layout of bored tunnels is prepared during the design phase, but the actual layout can only be finalized as the work progresses. In our age we have such precise equipment and established practices that, even for long tunnels of many miles, when the excavation is started from both ends, they are able to meet in the middle by a very small error margin. The reason of starting from both ends instead of one end is mainly for a faster construction schedule but for very long tunnels it also reduces error in the main alignment, because starting from two known points and meeting in the middle is prone to less amount of error than starting from one known point and going all the way full length.

Surveying during construction is not only for laying out the geometry but also to constantly check the stability and accuracy of the tunnel, which is continuously performed as the construction progresses. For example the distance between two fixed points across the cross section and length of the tunnel is often measured, in all three dimensions, such as by electro optical distance measuring, which detects any possible deformations inside the tunnel face, with the help of pins placed on the face of shotcrete. This allows determination of support locations on an as needed basis. Inclinometer measurements to determine slopes, extensometer measurements to know material elongations or shortening (strains) are also performed. If larger than expected deformations emerge, additional supports will be placed. In unfavorable ground or when passing under critical structures above the ground, measurements are performed at closer intervals. As the tunnel progresses, the sections that are left behind are measured and recorded in increasingly longer time intervals such as once a week or month. When it is finally determined that the deformations at those areas are finished and the tunnel is stable, measuring operations can end.

Other types of measurements include, seismic measurements, ultrasonic measurements to determine loosening zones, in-situ stresses, measurements of water pressure in the ground and the pressure between the ground and the extension of tunnel. Measurements can also be performed above ground, to check effects on ground surface, nearby structures and environment, as far as vibration and settlement. After tunnel completion, if future changes in geological, hydrological conditions or tunnel and ground / mountain interaction is expected, measuring can be ongoing. In that case, measurement devices must be placed such that they will be functional and accessible in the long term.

Initial and Final Lining (Structure of Tunnel)

Support systems vary greatly for bored tunnels, depending on ground conditions and tunnel diameter. Even along the length of the same tunnel the support may vary, due to possibly differing ground conditions, although usually one method will suffice for the most part of, or the entire tunnel.

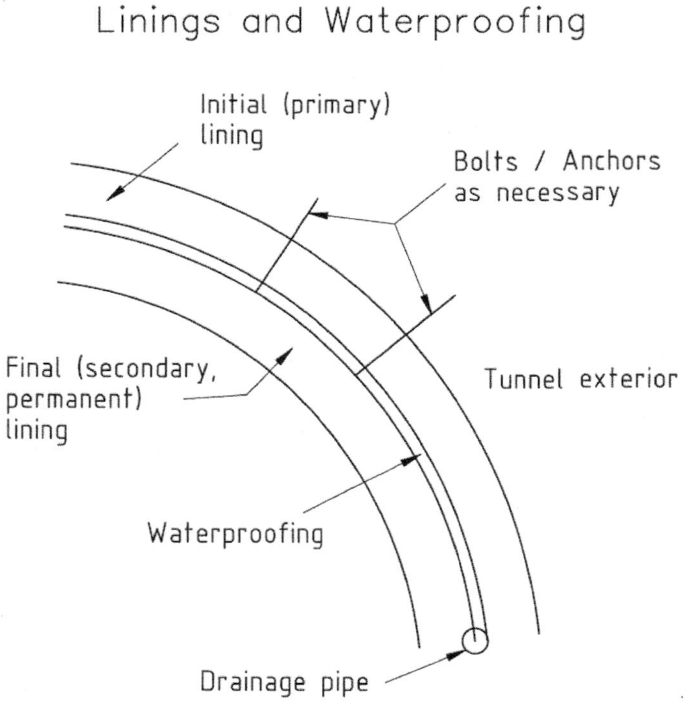

Figure - Tunnels 5

Tunnel lining makes up the structure of a tunnel, which supports the earth at the exterior of the tunnel, and separates inside of the tunnel from exterior. Although a single layer of liner structure is possible, often there are two main layers, also called as liners or linings, such as the initial and final liners.

Initial lining means, the outermost structural support layer of the tunnel, that directly touches and supports the earth / rock outside of the tunnel. Final lining is the inner layer. Waterproofing can typically come between these two liner layers.

How these layers are made, vary for each project. A typical profile in general sense may look like the one shown above, with reinforced shotcrete as the initial lining to stabilize the excavation, geotextile and membrane waterproofing in the middle, and finally, the final lining as shotcrete or concrete again. The reinforcement for shotcrete may be by steel fibers, welded wire mesh, steel arch frames, lattice girders. As extra support, ground anchors and bolts etc... can also be used. Waterproofing can also be applied at the exterior of the tunnel too, at the outer face of initial lining. For another case, we may have precast concrete panels as the only lining, with its inside face covered with protective coating and the waterproofing is again outside, or inside the initial precast lining we may have cast in place concrete as final liner which is a common case for shielded excavations.

Lining material can be made of:

- Concrete (plain or reinforced)
- Shotcrete (plain or reinforced)
- Precast concrete segments (reinforced)
- Steel plates

Reinforcing of concrete or shotcrete include:

- Reinforcing steel
- Steel mesh
- Lattice girders
- Steel fibers
- Steel arches

Anchoring, doweling, bolting items include:

- Rock bolts, dowels
- Ground anchors
- Micropiles
- Pipe umbrella

A combination of the items listed above can be used for each tunnel. In addition to these, ground improvement may also be applied to make the ground less permeable and stronger, and exert less burden on the structural support system on the tunnel.

We determine the liner cross section and its geometry, anchoring/bolting and its type and scope, and if any ground improvement is needed, in order to support the loads from the soil/rock and other loads as discussed under "Loads and Modeling" subsection. When entering data into the model, since soil is a natural material, some simplifications and assumptions must be made in certain situations, as long as we are satisfied that we obtain sufficiently close representation of the soil and tunnel interaction. As a result, force and moment diagrams on the tunnel liner, soil stress and deformation values and graphs are obtained. This process is repeated until satisfactory results are found for the given cross section of the tunnel and liner. The computer programs use FEM applications to solve these. This design forms the basis of how the tunnel will be built, but, as the tunnel progresses, depending on the actual soil/rock conditions and deformation measurements it can be modified as necessary, also to take advantage of the self supporting ability of the earth as much as possible, which we will talk about under "New Austrian Tunneling Method" shortly.

Tunnel linings are also structural systems but they are different from overground structures in the sense that they directly interact with the soil and support it, instead of bearing on it. This interaction behavior and contact with the soil determines their deformation and load carrying capacity.

We listed the main liner materials above. Now let's take a look at each:

Cast In Place Concrete:

Cast in place concrete linings are used for any hard or soft soil type, and can be plain or reinforced. The main condition of applying concrete liner is that, excavated hole must first be supported before, such as with shotcrete, which is why very frequently we see shotcrete as initial and concrete as the final lining.

For concrete, to achieve high early strength is important to be able to strip forms quicker, but the excessive temperature rise during cement hydration must be prevented. Since this also depends on the surrounding temperature, the mix can change according to time of the year. The temperature in deep and long tunnels can climb up to uncomfortable levels both for the crew and concreting operations which will put burden on the ventilation system.

Concrete is usually placed against a waterproofing membrane. This membrane must be protected during placement of forms and rebar. During the pour, the intended position of the reinforcement steel must continuously be checked, to make sure that it doesn't move.

As the tunnel progresses forward, the concrete is poured in sections and the gaps between sections must be grouted. This operation must be done before rise of water level, if the water table fluctuates in the area. Then these locations are sealed by joint tapes. Pouring in sections is also beneficial to prevent cracking, and it also allows formwork to proceed in sections and becomes more practical for the sequencing of construction and concrete production, delivery and pouring.

Concrete can be used with any excavation and initial ground support method. It can take any shape, and with this, it can correct the irregularities of an excavation. Its surface can be smooth finished which provides a good base for finish materials. It has high durability which means low maintenance, less traffic interruption and cost savings over the years.

Among the disadvantages are that the placement of concrete in tunnels is difficult due to limited access and most importantly, it requires formwork. Vibration, an important part of concrete placement, can be done only up to a point here. Insufficient vibration may cause voids in the finished product, so we will need higher slump (see Chapter 1, concrete section) values than usual, which will have a negative effect on strength. Another alternative may be SCC - Self Consolidating Concrete as discussed in Chapter 1. If reinforcement is used, there is a relatively higher risk of corrosion in comparison to over ground structures, due to underground water and if higher slump concrete is used. The minor cracks that may form because of long term redistribution of loads from the earth above, can also affect water, thus corrosion resistance negatively.

Except for short tunnels, concrete is supplied from on site mixing plants which enable continuous and more cost effective concrete delivery. For small jobs however, delivery by concrete mixer trucks from off site concrete plants is more efficient.

For formwork, a sliding form over rails is used for faster operation. When sliding over rails, the form has the ability to slightly move down and retreat from sides, and when it is brought to final position it is raised and pushed outwards again, through hydraulic adjuster pistons, to make the final shape of the tunnel. See a general figure below that shows components and concept in general for a possible configuration, which of course will vary for each project. The flow of concrete must be maintained with a minimum proper slump value, without sacrificing strength (hence admixtures or SCC will be needed) because of the need to pump through hose and fill in the forms as seen below.

Figure - Tunnels 6

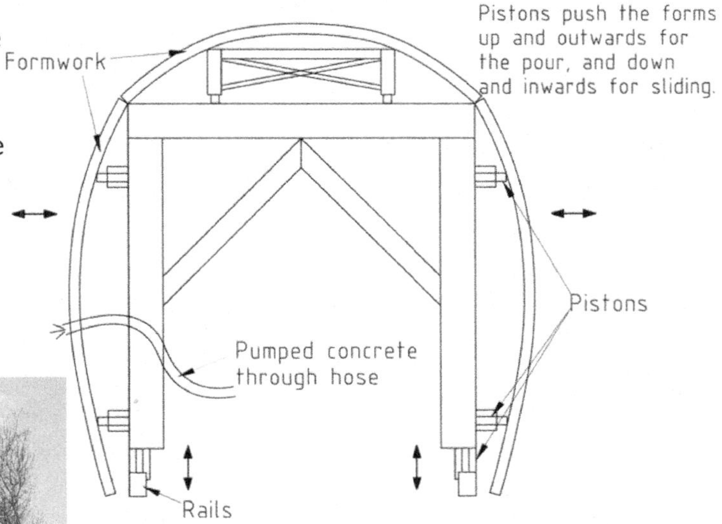

A tunnel formwork for an urban metro line. It of course has differences from the general figure above, but the main idea is the same.

Shotcrete:

Shotcrete can be used both for initial and final lining, but as discussed above, it is usually preferred as initial liner to initially support the excavated earth and to stabilize it, to facilitate natural redistribution of loads by acting together and deforming with the surrounding earth while still being supported with anchors/bolts/reinforcing, to prevent loosening of soil/rock, before final concrete layer is applied. Thus its high early strength is also important. This means, the excavated soil must be able to support itself to some degree at least for a short while, before the shotcrete is applied and starts to harden. This is not an issue for rocks, and many soils are also able to provide such short term support, except very weak soils or when there are significant water leakage problems, in which case, ground improvement and/or shielded excavation will be necessary, if that tunnel route cannot be avoided.

Before the start of applying shotcrete, test areas may also be required, to represent the actual typical conditions. The test must be done early enough for the results to be observed. Workmanship quality when applying the shotcrete through nozzle is of utmost importance. Shotcrete can also be used to replace concrete altogether, such as, when access is difficult, tunnel geometry is complex and formwork will not be practical, or the soil/rock above is fairly strong and can self support to a great extent.

Precast Concrete:

Precast concrete liners have the main advantage of ease of installation and handling inside the tunnel which also does not require any formwork or curing as in the case of cast in place concrete. Similar to cast in place concrete, precast linings also provide good foundation for tunnel finishes and enables a low maintenance, durable, stable structure. Among the disadvantages are, that the precast segments and the reinforcing steel must be made and installed to tight tolerances, segments must be stored and protected from damage while handling and be placed with specialized equipment which all take place in a tight space. As in the case of cast in place concrete, underground water, and corrosive chemicals and the potential minor crackings due to load redistributions pose the risk of corrosion.

The precast liner is made of 360 degree rings, placed one after another, to make up the whole tunnel, and each ring is made of several segments, as we also showed in Figure 2. After the ring's completion, pressured mortar is injected between the outside of the ring and the soil/rock, as in Figure 3 to make it stable and behave together with the soil/rock, to facilitate more uniform load distribution from the soil outside. The gaps between the rings are also filled with cellular concrete. There are gaskets on all sides of the panels, both circumferentially and against the previous ring provide sealing against water leaks. Any curves that the tunnel must follow can be achieved by adjusting ring segment dimensions. Precast panels are frequently used with TBMs.

Steel:

Steel can also be used as lining material although it is not that common. It can be used as both the initial and final layer. Similar to precast concrete, between the plates, gaskets limit the flow of water, and they must be manufactured within tight tolerances. They require specialized equipment to erect.

Steel plates has the similar advantages to precast panels that they require no forming and curing and can be easily handled inside the tunnel and they provide a stable ground support. The main disadvantage is the low corrosion and fire resistance and they will require concrete cover.

Bored Tunnel Classification per Rock/Soil Type

We listed 6 main tunneling methods, under the section "Tunnel Classification and Support Systems". Three of those methods were bored tunnels which were classified according to soil/rock type. The prior subsections under bored tunnels, which were, Loads and Modeling, Excavation, Ground Improvement and Groundwater Control, Drilling & Blasting, Surveying, Structural Support & Lining all apply to the three subsections below with some variations, which we will not repeat. We will only mention some remaining points for bored tunnels specific to each main soil type in subsections below.

Bored Tunnels in Rock:

Rocks can be of different quality and therefore support methods vary. For a high RQD (good quality rock with no or minimal fractures) rock, even no support may be fine and the drilled rock surface can be left like that, except weak spots, but still wire mesh can be used to catch small falling fragments. For weaker rock surfaces with more fractures, shotcreting and bolting may be applied and for even more sensitive cases, heavily reinforced concrete can be placed at the rock face. Deeper rock strata are usually stronger as they have gone through less to no degree of weathering in the past.

A common failure mechanism for rock tunnels is the wedge failure. As seen below, when one wedge falls, it can cause other pieces to fall too, until enough pieces fall so that the surface stresses balance, or, up to the point of collapse of the entire tunnel. Therefore, these wedge movements are restricted by shotcreting and bolting. As RQD goes up, the likelihood of the wedge failures, and the convergence of the drilled surface towards inside decrease.

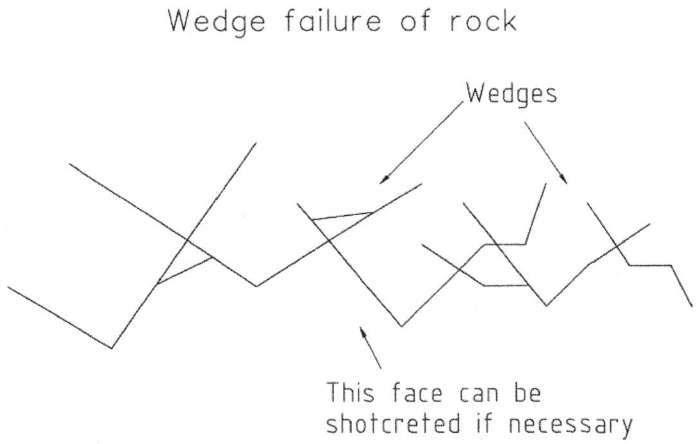

Figure - Tunnels 7

Bored Tunnels in Soft Ground:

By soft ground we mean any type of weak soil regardless of its type, but not rock. When digging tunnels in soft ground, it must first be understood that it is not possible to avoid deformations altogether and this is expected. Rate of entry of water (if any) inside the tunnel also determines what method will be used. Dealing with water in soft ground is one of the biggest tunneling challenges.

In tunneling, generally some deformation is desirable and necessary, for the loads to be distributed favorably to the support system as we mentioned before. It is up to the engineering design team to determine up to which level this deformation will be allowed, which will determine where and how strong the supports will be. In soft ground however, these deformations may be excessive or the tunnel may just tend to co lapse easily. Thus, in most cases in soft, weak soils, for easier and faster installation, a uniform support is provided to cover everywhere based on the worst case, unless those worst case locations are very rare with respect to the rest of the tunnel, in which case they deserve to be dealt differently than the rest of the tunnel with additional supports.

Methods for soft ground include shield tunneling, which is a very old method invented in early 1800s, and is performed by advancing a shield inside the tunnel as the excavation progresses. This method is so effective that it was used in all soft ground cases until a few decades ago, which was then started to be replaced by the TBMs, where the shield and the digging machine became parts of one machine working together. Other methods to support the tunneling are newer. One of them involves providing compressed slurry and air inside the excavation, to support the tunnel hole from inside. Newer, more sophisticated methods also emerged, which require higher level of control and specialization.

Digging tunnels through soft ground is more expensive than digging through hard rock, because of the need for stronger support. The difficulty and cost of digging and the liner increase even more, if the tunnel is below the water table, where expensive techniques must be used.

Bored Tunnels in Difficult Ground:

The main sources of difficult conditions are heavy loading from above, sandy soils below water, weak clay type soils, adverse fracture orientation of rocks, fault zones, presence of water and water flow (which require significant amount of dewatering), obstacles and constraints such as in urban areas. Combining these with the variety of soil types and tunneling needs, each case is unique and must be dealt with on a case by case basis.

New Austrian Tunneling Method

We already discussed many principles in tunneling that would also apply to this method, similarly, it can be called as other names too, such as the shotcrete liner or sequential excavation method. This is a tunneling method that was originally developed in the mountainous country of Austria in 1960s and gained widespread acceptance everywhere in the world since.

This method basically involves taking advantage of the natural support strength of rock, by allowing it to deform and mobilize to the greatest extent possible, and then by applying the initial liner and anchors to stabilize and help rock support itself to behave like a ring, by making the liner and rock work together. This is ensured by allowing deformations and stress redistributions in a controlled way, by constantly monitoring the deformations as excavation progresses, and creating more support (i.e. with rock bolts or dowels, ground anchors) only where necessary, instead of reinforcing or supporting the entire excavation or by using heavy structures such as ribs and lagging etc..., thus utilizing natural shear strength of the ground and the arch action (of rock plus shotcrete) to the greatest extent possible. Therefore, although more sophisticated measurement is needed to monitor deformations, overall it still means saving money from cost of permanent structural support. Of course, this doesn't only depend on what we see during construction, without prior analysis. Before construction, during geotechnical analysis, these intermediate steps which impose varying loads at different points at different stages (construction stages) is taken into account as much as possible. Thus, what we expect from geotechnical analysis before construction, is complemented by field monitoring during construction, and the temporary and permanent supports are installed accordingly.

The application of the initial liner may be immediate, or it can be delayed a little to allow for the excavated ground to deform by itself, which is determined on a case by case basis, depending on the stiffness of the rock. When calculating this delay, the time for the shotcrete to harden and gain strength must also be considered.

The deformations we mentioned in the paragraph above are primary deformations, which take place in a relatively short time. There are also secondary deformations, which may continue for many months, which are then supported by applying the final liner structure. So, after the optimum condition of rock/shotcrete assembly is achieved by some (below critical) deformation and stress redistributions, then the final liner of concrete can be applied for permanent support, which will now have to support less load, had we not utilized the natural self support of the rock above.

The "ring" we said above means, the bottom part of the tunnel (called invert) also needs to be closed to complete the liner (similar to the keystone of an arc) to make it behave like a complete ring.

After shotcrete, waterproofing membrane is applied, and after that the final reinforced concrete liner, as shown in the typical section in Figure 5.

NATM is more suitable for shorter tunnels in comparison to TBM tunneling, but due to much more number of shorter tunnels, it is commonly used in many tunnels. The excavation can be done by drill blast and/or roadheaders. It is a flexible, lower cost method, but the rate of progress may be slower than other methods, due to constant need to monitor and applying supports individually as necessary, and also excavating only in short length increments, due to the need of supporting the excavation. This method can not be used in soft soils that may immediately cave in, or in permeable soils below water table. In those cases the methods with screen would be more appropriate as discussed.

CUT AND COVER TUNNELS

This is another common method which can be grouped into two as bottom-up, and top-down methods.

In bottom-up method shown below, first retaining structures are installed into the ground, then the ground is excavated between these walls where we obtain an open excavation, supported by these retaining walls on its sides. Another option is to just start excavating and as it gets deeper, installing anchor walls. This is especially common in urban locations due to limited area instead of making a sloped excavation. If the size of the construction area is not a problem, then open cut excavation can also be made. After that, a tunnel box section is placed, and backfill is placed over it. The bottom-up method is the more straightforward, conventional method, where the excavation is made first, then the tunnel box is placed and then everything is covered by backfilling. This method has several advantages. It allows applying waterproofing at the outside face of the tunnel. Inside of excavation is easily accessible during construction work. Drainage can be conveniently placed outside of the structure and water can be diverted easily with this system. The main disadvantage of this method is that the ground surface can not be restored back to original until after the construction is complete. For the retaining wall, secant piles or large soldier piles can be used, and braced from inside as the excavation progresses. Cut and cover method is the most frequently used method for shallow tunnels (i.e. usually up to around 60 feet of depth), where excavating from above makes economic sense.

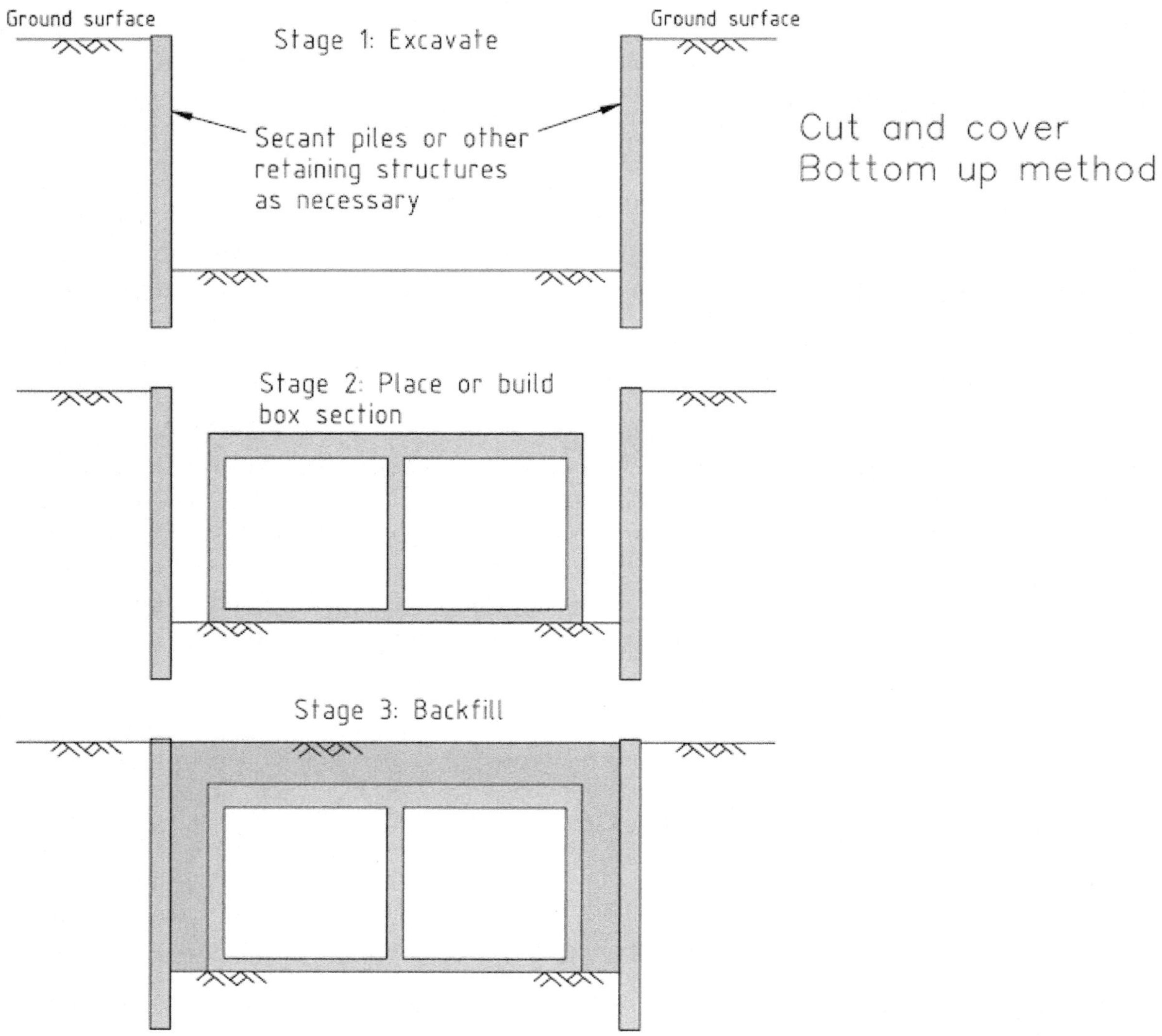

Figure - Tunnels 8

Another method of constructing cut & cover tunnels is called the top-down method. Here, first, secant pile walls are inserted into the ground. After that, without making any excavation yet, the top of the piles are connected by pouring the deck, which will serve as the roof. After that the excavation continues below that roof deck, and the tunnel is constructed. Therefore in this method, unlike in the figure above, the retaining walls on the sides also serve as the side walls of the tunnel.

The main advantages of top-down method are:

- The ground surface can be restored back to its original position far more quicker
- Using retaining walls as the permanent walls of the tunnel can also mean cost efficiency
- Roof is constructed directly on the ground before excavation, therefore saving formwork
- Roof deck also serves as a very effective bracing to the retaining walls at the two sides, reducing the amount of tiebacks required to brace the walls

The disadvantages of top-down method are:

- Waterproofing can not be installed outside of the tunnel, since the retaining walls which was inserted into the ground also serve as the side walls of the tunnel itself,
- More complicated structural connections between the deck, slab and walls is required, which may also mean more potential for water leakage,
- There is the risk of exterior walls (which are drilled into ground in the beginning), extending into the required clear space of the tunnel if they deviated from vertical during installation,
- There is more limited access to excavation and construction operations since the top is closed with the roof deck and the excavation proceeds under it.

In both methods, the walls on the side need to extend considerably below the bottom of the excavation, to be stable, as also seen in the figure above.

IMMERSED TUNNELS

For under sea/lake/river tunnels, the two alternatives are bored tunnels or immersed tunnels. If the tunnel must be below the sea/lake/river bed, we bore the tunnel, which we discussed. If the tunnel can just sit on the sea/river bed, we use immersed type of tunnel.

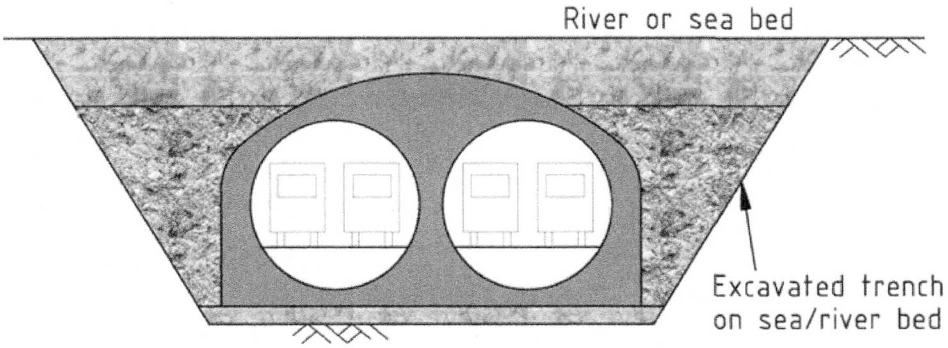

Figure - Tunnels 9

The figure above shows a possible configuration. The steps in making immersed tunnel can be listed as:

1. Large hollow box segments of precast concrete, as seen above, are made on land. These are very large sections which are heavily reinforced. Figure above shows a possible cross section.
2. Once complete, they are floated, slid to water similar to ships, (their ends are temporarily sealed so they can float like ships), and pulled by ships to bring them over their final location.
3. They are sunk to their final installation location on to the trench which was already dug and prepared by dredging beforehand. There may be water currents, waves during this operation,

which makes this task highly challenging where the pieces must be perfectly aligned on the trench and right next to each other.

4. The sections are connected to each other underwater and sealing is removed between them.
5. Backfill is placed.

The segments must be made to withstand not only the permanent load effects after the finish of construction, but also the loads during fabrication, transportation and placement. If the tunnel is located in a seismic zone, flexible seismic joints must be placed at certain locations, *(which are much more complex systems than seismic joints on bridges)* such as at both ends of the tunnel, which must be determined per seismic analysis.

The top of tunnel must usually be at a certain distance below the original water bed, as in the figure above, to allow for proper backfilling and restoration of original water bed, but if the hydraulic conditions are favorable, it can be left higher too, with an underwater protective embankment over it.

Since they sit on water bed and not go below it, immersed tunnels can be constructed at shallower depths than bored tunnels when passing water bodies, which saves costs not only for tunnel but also for highway (or railway) approach. They may also be very cost effective alternatives to bridges in certain soil conditions, for a given span distance. The main disadvantage of an immersed tunnel can be that they may disturb the natural sea/lake/river bed, which can negatively affect ecology in the area, and also may impact navigation in certain conditions.

The trench for an immersed tunnel is usually excavated by a clamshell dredger. Dredging is done in mainly two stages, which are, removing the bulk material to roughly form the trench and then fine trimming it. The environmental impact of trenching must be assessed and the appropriate excavation method and precautions must be applied. Construction schedule may also affect the design.

Typically gravel/sand is placed at the bottom of dredged trench as foundation layer, and after the tunnel segments are placed and connected, the backfill on top is done with gravel and stones. Details will vary for projects, based on hydraulic and geotechnical conditions. Backfilling method and materials and the armor material on top must be selected to withstand effects of erosion and scour. Backfilling operation must not damage the structure and or waterproofing. Backfill must be placed evenly in layers to prevent uneven pressures on the structure. In seismic zones, foundation, backfill method and material must ensure that no excess pore pressure to cause liquefaction will develop during an earthquake.

For immersed tunnels, there is the risk of ship anchor impact / dragging or ship sinking events, if the tunnel top is close to original river bed or even at a higher elevation. These should be considered during preliminary design. Bored tunnels are not exposed to such risks.

JACKED BOX TUNNELS

This is a method specifically for construction of shallow rectangular tunnels below critical structures or railways, highways etc... that allows for construction of tunnel without disruption of services above. It works with the same general logic of pipe jacking, where pipes are pushed into the ground without opening trenches, by drilling at the front face, which is also called microtunneling in utilities industry. Here, instead of pipe segments, the tunnel segments, such as concrete boxes are pushed, by drilling the ground in the front. This method is suitable in soft ground for shallow depths and short tunnel lengths.

With this, we finished discussing all six main tunnel types that we listed before under tunnel classification section. Now we will list other topics in tunneling, such as waterproofing, interior finishes and other tunnel components, that would apply to all tunnel types.

WATER DRAINAGE & WATERPROOFING

Water Drainage

The design of water drainage should be such that the entry of water into the tunnel such as by floods, surges, tides, waves or their combination from storms through the entry and exit portals should be prevented in the first place. To achieve this, proper design of entry and exit grades, the shape of walls around portals are required to prevent water entry. FHWA publication NHI-10-034 recommends 500 year flood level be considered as design water level, which can be obtained from hydrological studies.

For draining underground water entering into the tunnel, the perimeter of the tunnel is waterproofed, and the water is directed to the drainage system made of channels and pipes placed at the bottom of the tunnel, along the center and/or side lanes, to prevent water accumulation. Drained water can be diverted to be stored in water storage tank or tanks and then discharged to the drainage system. Sump pumps at low points, water/oil separators and control systems are also installed as applicable, with easy access for maintenance. The drainage system should be sized to handle the surface water, leakage and fire fighting water, but FHWA NHI-10-034 allows that the fire and storm will not occur at the same time when sizing the system. The drainage system should be made of noncombustible materials for fire safety. Water drainage details vary based on soil, tunnel design and the underground water conditions.

When designing water drainage, it must also be understood that water drainage during construction and after the completion of the tunnel must be approached differently.

Waterproofing

Tunnels must be watertight, even if they may be above groundwater table, because there will always be water coming from rain or snow. Design of waterproofing system depends on soil and hydraulic conditions, tunnel lining design, construction methods, geometry and layout. Waterproofing system should also take intermediate construction stages into account, as well as the finished condition and operation and maintenance. During installation, if waterproofing is laid first and rebar is placed over it, care must be taken to not damage the waterproofing, and a final inspection of waterproofing is needed before concrete pour. A typical waterproofing system can be made of geotextile layer, membrane waterproofing sheets, water barriers or it can be sprayed on. Ground improvement methods can also considerably reduce soil permeability around the tunnel. There are two types of waterproofing, which are open waterproofing and closed waterproofing system. Each has its own uses and advantages.

Open Waterproofing Systems (Drained):

In this system water is allowed to enter the tunnel drainage system and then it is drained. Therefore, the tunnel does not remain submerged in water here. The tunnel vault area is protected with waterproofing barrier as an umbrella at the perimeter, which directs water towards the bottom of the tunnel. This system is used where infiltration rates are low, such as rock tunnels. There is none to little hydrostatic load to be resisted in this system, as the water cannot build up and is allowed to drain. This is usually installed between initial and final lining layers, as also seen in Figure 5. As the hydrostatic load is considerably reduced, final (permanent) lining and invert can be built more economically. Since this system drains the water, it can also be called as drained system.

Closed Waterproofing Systems (Undrained):

In this system water is not allowed in the tunnel system at all. Higher loads must be resisted here, because water is not drained, it just builds up at the exterior and the tunnel remains in submerged condition, thus the system can also be called as undrained system. These systems must be built where there is significant water, or in permeable soils, where lowering of water table will not only be too expensive and impractical but also may cause surface settlements.

INTERIOR FINISHES

After placing the final liner, interior finishing materials such as metal panels or porcelain can be applied. These have different properties for sound attenuation, ease of cleaning, maintenance, lighting reflectance and visibility.

OTHER TUNNEL COMPONENTS

The systems listed below allow a tunnel to function properly. They are not there only for after completion, but also essential components which are always needed during construction too.

Ventilation System

Ventilation constantly provides fresh air for safe and comfortable tunnel usage. The required air quantity is calculated based on the traffic value, size and length of the tunnel, maximum allowed polluted air threshold, fire safety, smoke removal. After the volume is calculated, the required number and size of fans is placed. High carbon build-up inside tunnels is not only harmful for health and unsafe but also causes carbonation in concrete, as explained under concrete section in Chapter 1.

Fire Protection System

Fire in a tunnel can be catastrophic. The smoke accumulates and temperatures rise fast. Fire protection is not only required for safety of construction workers or passengers, but also for protecting the tunnel structure. At the entries of long tunnels, firefighting vehicles can also be stationed, as the fire damage may cause big loss of life or injuries, and economic losses, which increase as the tunnel damage increases. Limiting or prohibiting vehicles with certain type of fuels are also enforced in some tunnels.

Power Supply System

The temporary power required during construction needs long lengths of power cables to be installed, to feed the TBM and/or other construction equipment. For permanent construction, power supply system is needed for fans and equipment that needs to operate inside the tunnel.

Lighting System

Tunnel lighting is necessary for safe and comfortable usage of the tunnel. As these lights must always be on, high efficiency, low running cost, low maintenance, long life, and durable light fixtures are required. Different lighting patterns are required for interior, backup, entrance and emergency lights.

Communication and Control System

Emergency items (telephones, push buttons, alarms and egresses at every certain interval etc...), wireless systems, CCTV systems, sensors for monitoring enable a tunnel to function properly. Lighting, communication & control, and power supply conduits are installed per electrical design requirements.

CHAPTER 7 - WATER SUPPLY & COLLECTION SYSTEMS

In this chapter, we will look at:

- Water distribution systems such as the municipal water supply systems
- Water collection systems, such as storm and sanitary sewers

In Chapter 1 we discussed water resources engineering, that would serve as foundation to what we will discuss here, which you may want to review at this point if you need to.

Water collection and distribution systems are expensive investments and therefore should be designed not only considering today's demand but also future demand as well. Replacing these systems is a costly undertaking.

For both types of systems, first the demand must be determined. This is a very broad subject which involves not only civil engineering but also other subjects such as land use policies, demography, level of development and more, which will not be discussed here. After the demand is determined, design can start. Some of the important points to consider that applies to design of both water distribution and collection systems include the items below.

- Material, length and slope of pipes
- Pumps, with their locations and their power
- Storage tanks or structures
- Valves with their locations and types
- Ground contour maps
- In areas where existing construction is present (which is often), existing underground utility and other prior construction items on maps.
- Minimum and maximum allowed earth cover amounts for pipes. Minimum cover is needed for pipe to be below frost depth, to protect it from higher weight loads closer to the ground. The minimum cover amount is often already obtained when we give slope to the pipes for them to satisfy a certain slope over a certain pipe run length.
- When sanitary sewer pipes run under the same street as storm sewer or municipal water pipes, sewer pipes must be placed at a lower elevation. This is to prevent any leakage from sanitary sewer getting into storm or municipal lines, and also lower elevation of sanitary lines ensures that the dirty water coming from the street or buildings can enter into the sewer lines easier due to larger elevation difference as the system works with gravity.
- Infiltration of fine grained soils into pipes through pipe joints is a problem. Watertight flexible joint materials for rigid pipes, and, watertight coupling bands in flexible pipes can be used.

- Buried pipes cannot be stable just by backfilling on top, no matter how well we compact the soil, since the running water in the pipes cause thrust and vibration. In time, inevitably the pipes can push the soil around them, move, vibrate and cause joints to break. For this reason, concrete thrust blocks are used to keep the pipes stable, to prevent vibrations and relative movement and separation of joints. Thrust blocks are placed at pipe joints and at intermediate points as needed. They must be large and heavy enough to counter forces imposed by moving water in pipes, without causing any deformation or settlement in the surrounding soil. Although they are also placed where lines run straight, they are especially important at elbow locations, where there is a change in direction or slope, to counter the change in momentum from fast moving water. When designing thrust blocks, an optimum must be found between size and gap of placement. Bearing surface area of blocks must be large enough to prevent high stress concentration on the surrounding soil and therefore soil deformation.
- During design and construction, ease of future maintenance of the system must always be kept in mind for every possible aspect.
- The pipes should not be located under highways but under pedestrian sidewalks, in order not to cause major disruption to vehicular traffic during future repairs or failures.

WATER SUPPLY SYSTEMS

It is a complex task to determine the amount of water that will be needed for a populated area. One of the best places to start is to look at previous systems that work for a similar case. The demand should be estimated and met for the next 20-30 years.

Water supply (also called distribution) systems distribute water from sources such as dams, rivers, wells, springs or lakes to the places where it will be used for domestic, commercial and industrial purposes. They can include open channel flow or pressurized flow or both, which were introduced in Chapter 1. From now we will mean municipal water supply systems, for water distribution systems.

After withdrawing water from the reservoir, it is treated to prevent waterborne diseases and to ensure better hygiene. If water is of adequate quality, the treatment can be only physical, such as separating sediments. Less clean waters are also treated chemically. And the least hygienic water are treated also biologically, in addition to physical and chemical treatment.

As shown in the figure below, after treatment plant, water is distributed to water storage tanks for each sub area, and then to lateral branches that are in different distribution zones in the sub area. For each zone, booster pumping stations are used as necessary to maintain the required pressure.

In the figure below, each water main that supplies laterals is an independent distribution area with its own pressure zone, each having its own pumping station and storage facility. The storage at each zone is necessary to maintain the required pressure in that zone because water demand varies by seasons or even days, or hours. At night, when the demand is low, storage tanks are filled with water that come from the reservoir, and during the day when the demand is high, the demand can be met by the supply from the mains plus the previously stored water. Therefore the water that we use, comes from the storage of our own zone. The capacities of these storages vary according to expected demand and required system pressure. Instead of a dam reservoir, water can be withdrawn directly from a river or well or spring in less common cases. In case of spring, it is often not necessary to treat water in plant. Well water may contain minerals and may need to be treated.

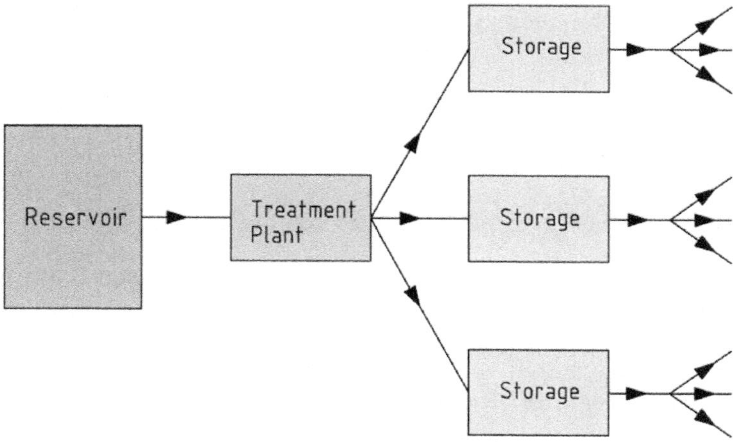

Figure - Water Supply & Collection 1

Water supply systems are complex, divided into many zones with their own pumping stations, storages, water mains, thousands of pipes and valves of different sizes, fire hydrants... A municipal water distribution system should meet the conditions including:

- Adequate pressure at the final destination where it is used. Usually, a pressure range of anywhere between 50 to 100 psi is maintained within the pipes of the municipal system.
- For high rise buildings, the pressure range of municipal system will not be enough therefore additional pumps will be needed at each building to elevate water to high floors with enough pressure. Without these pumps, the overall municipal water system pressure would need to be maintained at high levels, which would be too much for the whole system to handle.
- Fire fighting water line should be distinguished from domestic water line, as the fire line needs much more pressure.
- Excessive pressure in the system is not allowed, to prevent leakages and damage.
- The system must be watertight, to prevent leaks and entry of foreign materials into water.
- Failure of one pipe should cause interruption of service to none or minimum number of possible locations downstream. For this, grid network type of systems, which run parallel should be preferred, instead of branch systems in series, as we will discuss.
- The system must have backflow prevention,
    - to prevent contaminants entering into the system at connection points,
    - at points where there is risk of rise of water levels due to floods,
    - at points where the pipe ownership changes for cost and liability purposes,
    - and just to increase safety of the system.

Energy Principles in Water Transmission

In Chapter 1, under water resources engineering, we introduced Energy Grade Line (EGL) and Hydraulic Grade Line (HGL) concepts and also open channel vs. pressurized flow. We also covered water head concept under seepage topic of soil mechanics section, again in Chapter 1. Those will now serve as foundation to this section, which you may want to revisit if you need.

In water supply systems, water is moved in two ways: gravity (open channel), or pressurized flow, or both, in different parts of the system.

As discussed before, EGL is made of:

EGL: Elevation head (EH) + Velocity head (VH) + Pressure head (PH)

The equation above forms the basis of design calculations in water distribution systems.

As water flows in pipes, EGL decreases due to frictional losses and also due to the elevation losses when water flows downward by gravity. At locations where we need to elevate water, and / or increase its pressure, we can use pumps, which cause a sudden increase in EGL.

In the past, when there were no pumps, the only way we could convey water was by gravity. It meant, the source where water was drawn had to be considerably higher than the entrance elevation of city water supply system. Such sources were the springs in mountains. Ancient Roman aqueducts used this principle. Alternatively, the cities were simply founded near rivers, without requiring water supply lines or aqueducts which were difficult to build. Today, we use both gravity action and pumps to maintain the required pressure, velocity and elevation in water supply lines.

During transmission of water, sudden pump stops or starts cause sudden water impacts in the system that cause shocks of stress in pipes and valves, which propagate at the speed of sound in water and can cause damage to pipes and valves. To prevent this, balancing tanks are used, as seen below, which contain compressed air on top of water, which compensates these sudden pressure increases or decreases and protects the system.

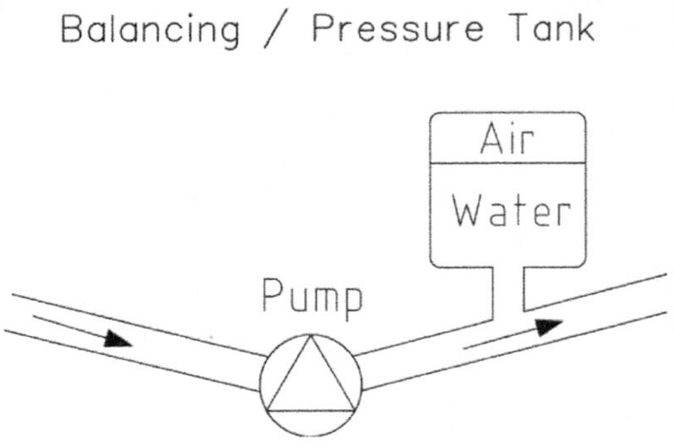

Figure - Water Supply & Collection 2

The compressed air at the top part of the tank pushes water down to the pipe when the system pressure drops, and when the pump works again, the pump pushes the water upwards in the tank and compresses the air on top again which happens in a smooth manner instead of the sudden shocks if the tank was not there. The storages that were shown in Figure 1 also serve this purpose, in addition to storing water for a region. Pressure tanks in much smaller scale are also used inside the buildings plumbing system, with the same logic.

The cost of distributing water goes down if we increase the pipe diameter that provides water from the source to the system, such as the maintenance and energy costs. But larger diameter pipes mean higher construction costs and longer construction schedules. Pipe material is another important variable, which affects the frictional losses in the system and the tolerable pressures. So engineers seek to find an optimum point for pipe diameters, considering the costs and also the water demand. Of course, when doing all this, the distances and elevation differences from point A to point B (and any intermediate points in between as necessary) are known in the first place.

In Figure 3 below, the reservoir at point A is at higher elevation than the city at point C, so, water can run by gravity. The reservoir at point E is at a lower elevation than the city, so we need pumping, to pressurize water and give it enough energy, to reach to the city. See how the energy grade line (and hydraulic grade line) suddenly jumps up at the pump station. Points B and D are arbitrary; they can happen due to topography. But assuming we ignore any frictional losses, pay attention to how they do not affect EGL & HGL. This is because, even if the pipe slope is downwards (and elevation decreases), now the velocity of water increases as gravity moves water down. So from equation EGL: EH + VH + PH, elevation head decreases while velocity increases, which is why EGL is not affected at points B and D. The only gradual decrease that you see in EGL is because of general frictional losses in the system.

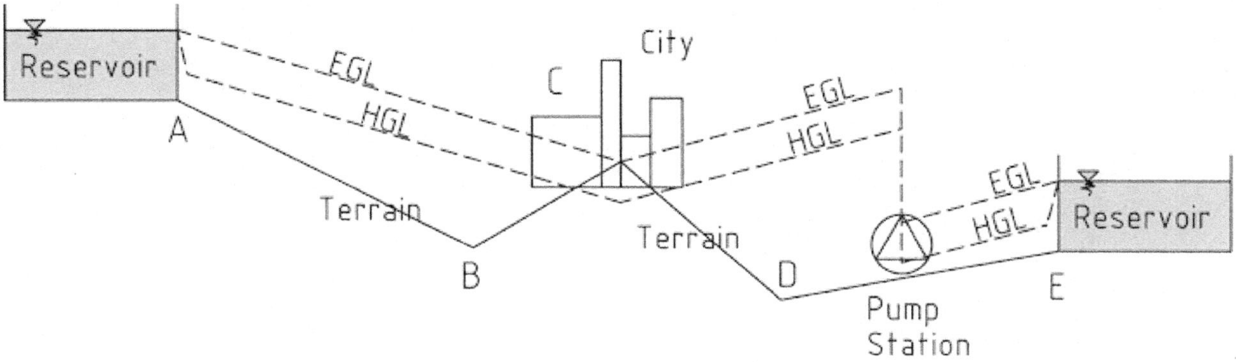

Figure - Water Supply & Collection 3

Also, in the figure above, notice how HGL runs parallel but lower than EGL. This is because, as discussed, HGL = EGL - velocity head. So, at any point in the reservoir, HGL = EGL as there is no running water there and the velocity is zero. But as soon as the water leaves the reservoir and enters into the pipe, it will have velocity. Look closer to the shape of HGL above, right where water leaves reservoirs. At those points, the EGL still does not get affected but HGL immediately drops to a lower value as shown in the figure, because now we must subtract that nonzero velocity head from EGL, to obtain HGL.

Types of Water Distribution Systems

The way water mains and laterals are connected, such as in branches or as a network, depends on many factors but in general smaller places can be supplied with branches while the larger areas should be served by grid networks. There are mainly three types of water distribution systems:

- Branch pattern with dead end
- Grid pattern
- Grid pattern with loop

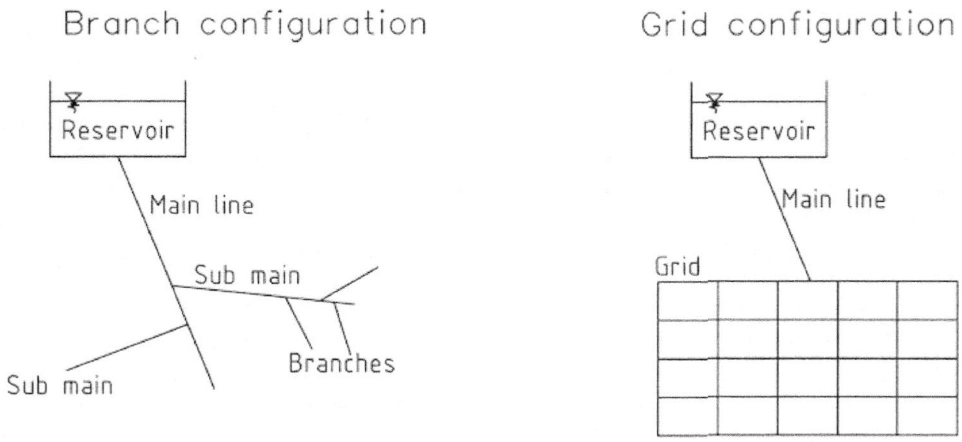

Figure - Water Supply & Collection 4

Branch system has the following characteristics:

- It is connected in series to each other and therefore if there is a problem in the main, all downstream locations remain without water.
- They need cleaning more often.
- It is the simplest, most economical and most straightforward method of water distribution. Calculations are simple.
- Requires smaller number of cut-off valves.
- It is relatively older type of practice.
- There are a lot of dead ends where sediments cannot flow and thus accumulate. At these locations bacteria can also grow. To prevent this, they must often be cleaned. It is also difficult to maintain desired chlorine levels at these dead ends.
- Fire fighting water is limited because there is only one water main and the required pressures for domestic water and fire fighting water are different.

Grid system on the other hand, has the following characteristics:

- Water comes from many different directions and therefore kept fresher and accumulation of sediments and bacteria as in branch pattern will not happen or be very limited.
- Alternate routes are available in case of failure of one line.
- Fire water can effectively be supplied. During fires, water flow can be adjusted by valves and more water can be concentrated for fire fighting.
- Water can reach all locations with head losses kept to minimum.
- Designing and building this system costs more than branch pattern, because of laying more pipe length and more number of valves in total. Design is more complex.
- Some grids have loops to increase water pressure in certain parts of a city. Loops should be located such that, as the city grows, sufficient water pressure can be provided to new locations.

Hydraulic Analysis of Distribution Systems

Distribution systems can be hydraulically analyzed in different ways. Three most common methods are:

- Dead end method
- Hardy Cross method
- Equivalent pipe method

With these methods which are based on the energy principles as discussed, between start and end points of a system (of which we know the locations and elevations), we calculate pipe lengths and elevations/slopes, manhole locations and invert depths, flow rates, operational pressures, head loss in members, pipe diameters, storage tank elevations, pump requirements and all necessary components.

Plumbing

After water reaches its final destination, such as a building, the municipal water supply completes its part. From that point, water is distributed by the plumbing system inside the structure. In older buildings it was more common to elevate water to a tank at high elevation by pumps, which provided pressurized water flow by gravity. This system has gradually been replaced for the most part by pressurized systems with pressure tanks and pumps. Water pressure values in the plumbing system depend on the country or even different jurisdictions but mostly range anywhere from 40 to 80 psi. Excessive pressure will damage the system. If the main pipes are allowed to exceed this value, then the lateral lines must be protected with pressure reducing valves. During pump starts and stops, a sudden surge occurs in the system. The pumps should be designed to minimize these and the pipes should be designed to withstand these. High rise buildings should have their own booster pumps, as the municipal system pressure at the building entry may not be enough to provide water to high floors. Plumbing, which is designed by mechanical engineers, is one of the major trades in commercial and residential construction that also affects the installation and schedule of at least several other trades, which must be coordinated by construction project managers. After water is used, the waste water is removed by the sewer system. Some water coming out of sewer treatment plants can be used again as nonpotable water after treated, such as for industrial or irrigation purposes. We will discuss sewer systems below.

Aqueducts

Aqueducts are water channels that convey water. They can be on, above or below ground, have open or closed channel flow, although the first thing that comes to our minds when we hear the word aqueduct is the elevated aqueduct above ground with open channel flow.

Aqueducts were constructed since ancient Roman times, or even hundreds of years before them, such as the ones constructed by Egyptians, Persians, Indians, Middle Eastern and may be other civilizations. They can be considered as the first structures related to water distribution and plumbing. Romans used them for carrying irrigation water to farms, municipal water to cities, which could cover many miles of distance, the longest being a 57 mile aqueduct that supplied water to Rome.

We have aqueducts now too. For example, the Catskill Aqueduct, which is one of the three main supplies of freshwater to New York City, brings water from a distance of 120 km (75 miles), from Catskill mountains to Yonkers, from where it is distributed to the city by a pressurized freshwater distribution system. California Aqueduct, which is by far the longest in the world, conveys water for a distance of 700 km, from Northern to Southern California and supplies water to as much as 25 million people.

WATER COLLECTION SYSTEMS

Water collection systems collect <u>storm</u> or <u>sanitary</u> sewer water and transmit them safely to a treatment plant or a large water body as applicable. Storm and sanitary sewer systems can be combined with each other or built separately.

A combined sewer system collects domestic, public and industrial waste and storm water runoff together. A combined system may not always be adequate in times of heavy rainfall. It is more economical to convey sanitary and storm sewer systems in the same pipes, which was more common in the past, but today the regulatory requirements are stricter, and separate systems are usually built.

A separated system carries the sewer and storm water separately, with dedicated manholes and pipes for each. Sewer and storm pipes are usually placed on opposite sides of a street, but if they must run on the same side, they can use the same trench, but still a minimum distance between them should be maintained and the sewer lines should run lower than storm lines. Both systems' starting points should be from as high elevation as possible (*but at the same time maintaining minimum earth cover on top*), which will minimize excavation amount for the whole system. Due to larger volumes, storm water lines are on average larger than sewer lines. The storm water collected by dedicated storm lines can be transmitted to lakes, seas or rivers, without any treatment.

For both storm and sewer systems, manhole locations, sizes and types, invert elevations and pipe runs are the primary components to design. For both systems, a manhole is needed at street junctions, or when a change in diameter, slope or direction of the pipes are necessary. Where pipes run straight, the distance between manholes can be increased. Sometimes, if curved pipes are used, to overcome horizontal alignment or slope differences, some of the manholes (and their cost) may be eliminated. But for the most part curved piping is not used in the world, because it is harder to survey and install them in comparison to digging straight trenches and laying straight pipes, and it is also more difficult to locate curved pipes later in the ground, when the need arises in the future.

During construction of storm and sewer lines on public streets, vehicle traffic through that area will be interrupted due to open trenches. During construction on a jobsite that involve other components such as buildings, many other trades will be affected. Therefore, trenching, placing manholes, laying pipes and backfilling must be completed in rapid pace. This means, all manholes, pipes and materials must be timely on site, they must be placed and connected at a reasonable speed and the inspector of the jurisdiction must be there to inspect the open trenches, pipe and manhole connections, before the backfill. Just like the soil compaction for roads, this compaction work is also tested in place to verify that it satisfies the requirements (i.e. 90% or 95 % of maximum dry density of soil). The statements in this paragraph also apply to municipal water supply systems too, except that they do not have manholes. Now let's discuss each system separately.

Stormwater Systems (Storm Sewer Systems)

Storm sewer systems (also called stormwater systems or storm drain systems) collect surface runoff water resulting from snow or rainfall, and direct it to water bodies on earth. The surface runoff is received through inlets, conveyed through conduits (pipes) to an outfall point. Therefore, to design and build stormwater systems, the most important thing to know is the amount of surface runoff water that the system must handle. As mentioned in Chapter 1, under hydrology, as urbanization increases, the surface area that water can penetrate into the soil decreases which causes more rainwater to run as surface water, which puts more burden on stormwater systems.

Pipe sizes of stormwater systems increase gradually in the direction of flow. For example, where water is first collected, a small pipe is adequate. But later downstream, when additional water from different branches join together, larger pipes are needed. This is the opposite with respect to a fresh water supply system, where the pipes have larger diameter at the source first, and then they gradually become smaller as they approach towards final user location.

The largest quantity of water that a storm sewer system is supposed to carry is found by considering the worst storm that may occur in the area for the expected life of the system. This would be a rare, big storm that happens every certain number of years. This number of years is called the return period or equivalently, the frequency of storm, or storm frequency, which is taken as the worst case. What storm occurs in which period comes from hydrological data. But of course, if a very high return period storm is taken into account, the capacity of the system must be higher, which means costlier construction which also takes longer time, so engineers must find the optimal balance between the available budget and the acceptable risks and demands, at the same time complying with code. Higher importance areas, or the areas that would have worse damage in the event of a flood, can have longer storm frequencies, meaning, they are designed for less frequent, larger storms.

Only knowing the storm intensity for the worst case storm is not enough. Equally important is the surface area on which this rain will fall. The larger the area, the more water to be drained. This area of rainfall is called basin, or tributary area, as discussed in Chapter 1, which are naturally distinct from each other depending on the natural ground, hills, mountains or slope of the area. Basins and sub basins must be taken into account during design.

When the surface runoff is more than the storm sewer system can handle, flooding will occur. If the storm sewer system does not have sufficient capacity for discharge, even a 2 year storm can cause floods. This is why in poorly planned cities especially in developing countries lacking properly regulated growth, streets may get flooded every time there is an intense rain for even 15 minutes.

To reduce and delay runoff, in urban areas, certain measures can be taken. For example:

- The most important thing to do is, to increase the percentage of green area in the cities. European cities present a good example about this. According to European Commission website as of 2021, European cities became %38 more green in the last 25 years, and 44% of Europe's urban population live within 300 meters of a public park.

- Green roofs capture significant amount of rain water, before they let the excess water out into the storm system. Green roofs also have other benefits. They are good thermal isolators, provide a green rooftop breathing area for building residents by also increasing total walkable area in the cities, can grow vegetables and fruits even if small quantities, considerably decrease heat reflection from buildings which helps cooling down of cities during hot summer months, provide a pleasant view, and capture carbon. Green roofs should be encouraged everywhere.
- Blue roofs also work to help to reduce storm water runoff, by capturing some water to temporarily pond on the roof, then by slowly letting it drain into the system through controlled flow roof drains. This delay is provided by weirs placed at roof drain inlets.
- Subsurface systems such as storage vaults or tanks can store considerable amount of water before it is released into the city storm sewer system.
- Rainwater recycling systems, which can be as simple as a barrel connected to a downspout, or much larger underground tanks and cisterns. The water obtained from here can be used in the building such as for water closets, urinals, laundry systems, cooling towers and so on, thus not only helping reducing the runoff but also water demand from the city supply.

After the quantity of rainfall and the tributary area is identified, the next step in the design is to locate the receiving water bodies for identifying the outfall or overflow points.

Then, a preliminary horizontal layout of pipes are drawn, which should be able to catch water from the entire tributary area, but at the same time keeping the total pipe length minimum. The most efficient designs are made when the pipes approximately follow the natural drainage flow patterns. So basically what we are trying to do is, to imitate how nature drains water, but in a controlled way. If we divert too much from the natural slopes, things would get costly, as more excavations and pipes will be required.

After locating pipes horizontally, their sizes and slopes are determined. This is the hydraulic design of pipes, which uses energy principles we described. Locations of storm inlets are also determined in this step. Inlet sizes are determined by the amount of surface runoff water they are supposed to receive.

When designing pipe layout and determining slopes, too small slopes, such as 1 in 400, should be avoided, as it is difficult to construct accurately and at the same time difficult to maintain efficient gravity flow (such as a speed of 2.5 feet per second, which is usually needed to prevent accumulation of suspended particles in the pipes). Pumping should be avoided, but slope and length limitations may make it necessary in some cases. Size and slope of pipes can be revised many times, until design criteria per codes are met. It must be remembered that many branches of pipes interact with each other here, so in the design the whole system must also be considered, in addition to individual branches, and this is why the design may continue many cycles, iterations, until all individual and overall branches satisfy the design criteria. It is not necessary to lay storm sewer lines under every street. They are necessary only when the street is long enough that creates sufficient amount of rain water to run due to its slopes. The whole system is finalized with pipe runs and slopes, pipe sizes and materials, connections, storm inlet and manhole locations, sizes and elevations of manholes, pipe inlet and outlet elevations to manholes. It is not necessarily the goal to minimize the pipe lengths 100% of the time. If shorter pipe lengths would mean more excavation, then sometimes slightly longer pipe routes may be followed.

A water channel that is used to convey storm water with open channel flow. See how it is stepped in many locations, in order to dissipate energy of water and keep its velocity down. It also serves as an aesthetical water feature this way, and also cools down its urban surroundings during summer.

A storm drain system can be designed as an open channel flow or a pressured flow system. For pressurized systems, special care must be taken when designing and installing pipe joints, to withstand pressure. For a storm drain running in full, the flow capacity increases with pipe diameter, slope and decreases with roughness coefficient of pipe material. Even a small increase in the diameter of pipe will increase flow capacity greatly, as it is proportional to its square. If the slope of pipe doubled, the capacity would increase by 1.4 times only. If the roughness of pipe is doubled, flow rate will decrease by 50%. Oval pipes have better flow rate per area ratio than circular pipes, but, circular shapes are more economical to make and therefore still preferred. A square box section will have the same efficiency as a circular conduit. Pipe to manhole connections should be made from flexible joints, as the settlement of manhole over time may cause joint failure in the future otherwise.

Sanitary Sewer Systems

Sanitary sewer systems collect, transmit, treat and dispose wastewater. Treatment step is necessary if the collected water does not satisfy certain environmental criteria, because ultimately sewer water must be disposed off to natural water bodies on earth.

Sewer system design is similar to storm system design in the sense that first the required capacity is determined, then preliminary layout of pipes are drawn, and then pipes are given proper slopes and sizes, manhole locations and elevations to optimize lengths and excavations are determined.

Here the required capacity (demand) means the total sewer generated in the area, not only by present but by estimated future developments too. The proximity of available disposal areas and/or sewage treatment plants or the cost to build those, environmental considerations have the greatest effect on the design, other than the demand. Industrial sewage is generally not allowed to mix with public sewer system due to its negative effects on biological sewage treatment process, but if mixing cannot be avoided, industrial waste should be treated first. Enforcement and practices vary between countries.

Sewer can be carried by gravity or pumping or vacuum action. Gravity is preferred as much as possible due to lower costs and higher efficiency. But in some cases pumping is needed such as when basement elevation of building is very low with respect to street level, or when conveying wastewater through highly variable, sloped terrains. It would not be economical to try to run an entire sewer drainage system by pumps. This is to the contrary of freshwater supply system, where water is distributed to users by pressurizing with pumps almost all the time. Flow by gravity means, pipes normally flowing partially filled, per design. Pipes running full reflects the worst case condition, which does not always happen.

The site sewer plan in Figure 5 below shows the most basic components of a sewer system. The manholes are numbered as per the needs of the project and the arrows indicate the direction of flow. Pipe diameters are smaller at the branches, and they become larger at the main lines.

After collection and treatment, ultimately the treated sewer water is discharged into the sea. In the past, the discharge was done at the shore, and from single point. Today, the discharge is done in deep sea, by extending the sewer line a certain distance to open sea, such as half a mile. At the discharge point, the water is let out from several points. Ultimately salty sea water consumes the sewage.

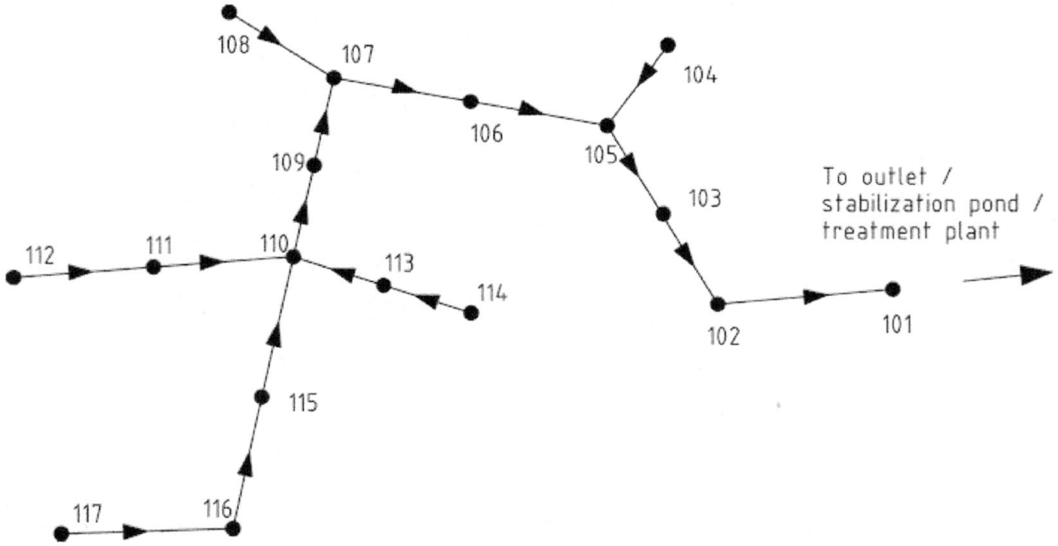

Figure - Water Supply & Collection 5

Sewer Pipes:

Sewer pipes can be made of different materials as listed below:

- Plain or Reinforced Concrete: Wide range of structural and pressure strengths, nominal diameters and laying lengths are possible, but they are heavy. They should be made from sulphate resistant cement. They can be cast in place too, but mainly precast is used.
- PVC: Long laying lengths, lightweight, easy to fabricate, can tolerate excessive deflection, economical, low maintenance, but sensitive to organic chemicals in the ground.

- Cast Iron: High pressure and load bearing capacity, long laying lengths, but sizes are limited and they are heavy. Proper bedding is necessary with uniform weight and load distribution from above, as they can break. Sensitive to corrosion from sewage and acidic soils.
- Ductile Iron: Resistant to breaking, impact, wear and tear. They are machinable and have high tensile strength. Long laying lengths possible, but heavy, although lighter than cast iron.
- Corrugated Steel: They can resist internal pressure, impact load, and vibrations much better than cast iron. They are sensitive to failure when there are negative pressures in the pipe.
- Vitrified Clay: High resistance to corrosion from acids and abrasion, handles high velocities. As cast iron, they are brittle and need special bedding or cradling concrete in the trench.
- Brick: Were more common in the past. With brick, only large diameter conduits are built. Many old brick sewers are still in use. The failures that can happen after very long time are mainly due to disintegration of bricks and mortar. Construction in place makes any size or shape is possible. The disadvantages are high costs, large space requirements, slow construction.
- Glass Fiber Reinforced Plastic Pipes (GRP): These are corrosion resistant and lightweight.
- Other pipe types include materials such as, asbestos cement (not used anymore due to health risks), High Density Polyethylene Pipes (HDPE), Structured Wall Piping, UPVC.

If the flow rate is too low, suspended materials can settle and gradually decrease the flow area, so a minimum velocity must be maintained. This can be achieved by increasing the slope of the pipes but then deeper manholes may be required which means more excavation and more difficulty in manhole construction and higher costs and longer schedules. Optimum must be found. Also, the flow does not necessarily have to prevent accumulation of particles always. We can allow accumulation of a small amount during low flow periods, as long as the system is designed so that the peak flow will flush it out. One other benefit of maintaining a minimum velocity is to prevent slime accumulation at the submerged surfaces, and to prevent generation of Hydrogen Sulphide gas, which can attack the cement of concrete sewers. When sewer pipes get clogged, mechanical instrument brushing and water flushing can be used.

On the other hand, if too much slope causes more than the maximum allowed velocities within pipes, the slope will need to be limited, in which case series of drops at certain intervals by manholes may be used. Limiting the velocity to a maximum is necessary because too much velocity can damage pipe's interior surface, as the sewer water also contain suspended solid particles.

All these maximum and minimum limits were extensively studied and tabulated in standards. For example, minimum velocity depends on hydraulic radius of the pipe (as discussed in hydromechanics), Manning's roughness coefficient (a coefficient of friction of the interior surface of the pipe, which depends on the pipe material), specific gravity and size of average particle in the sewer water. The minimum and maximum velocities may be anywhere between approximately 0.6 - 3 meters/second.

The elevation of sewer pipe entry point to a manhole is called the invert elevation and it is taken as the <u>lowest</u> point of the <u>inside</u> face of the pipe. At every manhole location where the invert elevation reaches the deepest allowable limit, (since we cannot just continue to make pipes lower and lower all the time), we use a submersible sewage pump in the manhole to elevate the water again, and start the next pipe back from a higher elevation, which will carry water again with the help of gravity, by sloping downwards. When we raise the water, we raise it back to the minimum allowed distance from ground level, in other words as high as possible, in order to start the next pipe from as shallow depth as possible, to save costs on excavation. See figure below:

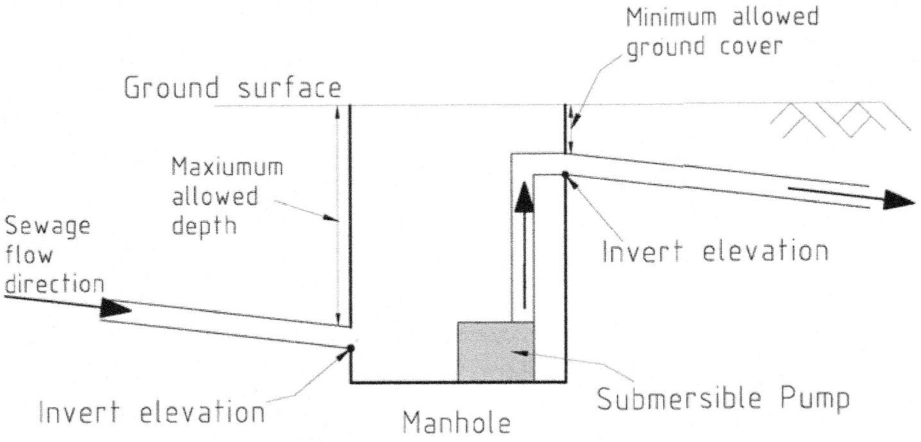

Figure - Water Supply & Collection 6

Another alternative to this may be to place the pump directly on the pipe which will let us running the pipe by upwards slope, eliminating the need to raise it by sump pump in the manhole.

Manholes:

A manhole serves as a connection point for pipes where there is a change in diameter, slope, elevation or direction of pipes. For sewer systems they can also be used at the end of branch or lateral sewers, to provide a means for cleaning or flushing. Manholes can be made of brick, cast in place or most commonly, precast concrete. There can be several pipes connecting to the same manhole, although it is usually two. The manhole geometry depends on the pipes connecting to the manhole such as the pipe sizes, the number of pipes, their connection elevations and angles, connection details. The structure of manhole design includes the following aspects:

- Backfill loads, surcharge loads from above, hydrostatic loads must be resisted
- Concrete must not remain too thin at any section (such as pipe entry holes), which may lead to breaking otherwise
- Round manholes have the advantage of utilizing ring action which allow the entire section to remain in compression, with no bending moment.
- The holes for pipes may be cast or cored into the structure

There can be biological activity inside the sewer system and therefore before starting any maintenance work, manholes should be checked for harmful gasses. Aerobic condition (presence of oxygen) needs to be maintained within the system, to prevent building up of toxic or explosive gasses.

Sewer Treatment Plants:

In addition to pipes and manholes, another important component of a sanitary sewer system is the design and construction of a sewer treatment plant or plants. For remote development areas that are far from the city or town centers on-site treatment plants can be constructed. But generally a central treatment facility where all sewage water is handled together is preferred due to efficiencies involved. Design of treatment plants is primarily the subject of environmental and mechanical engineering.

CHAPTER 8 - DAMS

Dams are among the oldest types of structures. Water resources on earth are distributed unevenly and they are subject to climatic conditions. To use this available water in a controlled and consistent manner to our benefit is critical for our survival. Dams are used to store water for various purposes and in many cases to create energy from that stored water at the same time. Storing water allows using it for domestic, commercial, industrial and agricultural purposes, so this was one of the main locomotives of development through our civilization's history.

Dam size definitions vary. According to US Bureau of Reclamation, a dam with a height of up to 100 feet is considered a low dam, between 100-300 feet as medium size dam, and more than 300 feet as large dam. According to International Commission of Large Dams, (ICOLD), a large dam is defined as a dam with a height of minimum 15 meters, from its lowest foundation to the crest of dam, or a dam from 5 to 15 meters, containing more than 3 million cubic meters. Most of water stored by dams in the world is stored by large dams. Also, according to ICOLD, most dams are single purpose dams, of which, 47 % are for irrigation, 21% for hydropower, 12% for water supply, 9% for flood control, 5% for recreation and less than 1% for navigation and fish farming according to 2020 data.

Within the last 100 years or so, in addition to using dams to store water alone, we also started creating energy from stored water, as it contains large amount of potential energy. The potential energy of water is turned into kinetic energy through turbines, which generates large amount of electrical energy at hydro plants. Per ICOLD data, hydroelectric plants can generate electricity anywhere from several hundred megawatts to as much as around 10000 megawatts, for a total capacity of 675,000 megawatts, and in total hydro plants generate roughly a quarter of world's electrical energy. This is also by far the largest source of renewable energy still, with around 90%, even though wind and solar electric generation are increasing their share.

Currently the largest hydroelectric plant in the world is the Three Gorges Dam in China, with a capacity of 22500 megawatts. It has a height of 594 feet and a length of 7770 feet, and the reservoir it creates has a surface area of 400 square miles, according to USGS website.

Other than irrigation, drinking water, domestic water, and creating renewable energy, additional benefits of dams include flood control by regulating river levels, water quality control, sediment control, reduced erosion, groundwater control and navigation, providing recreational and fishing area. Sometimes a series of dams on the same river (river basin) may work as a group, to further enhance or satisfy these benefits or requirements. Having so many benefits, dams can have large positive impacts on a country or region's social life and economy, and building them requires large investments.

Dams may also have negative impacts for the environment, which are listed under ecological impact of dams section later in this chapter. Therefore a good plan to consider all these benefits and negatives must be in place, which is a very complex task that involves a lot of variables from many disciplines, stakeholder organizations and people. This process can take long time, to determine to build the dam where, with what capacity, type, purpose, cost, economic life, and so on... Only after careful planning and funding, the design stage can begin.

DESIGN

Dams are structures that hold large amounts of water, which means enormously high pressures must be supported. A dam failure can have catastrophic consequences. The design of a dam primarily considers factors such as: Purpose/function, capacity, topographic and soil conditions, seismic conditions, environmental effects, available materials, flood control. Not only the dam itself, but the valley behind the dam that will hold water must also be able to support that future elevated water level. Dams may be grouped in a few different ways. Here we group and discuss them as embankment dams, gravity dams, arch dams and buttress dams.

Before designing the dam itself, the site and exact location to build the dam must be selected. The following items are considered or must be done for site selection:

- Using contour / topography map of the area, for different water levels water storage volume amounts will be calculated. On the contour map, the dam itself and the valley behind the dam that will hold water which will be the reservoir area for all possible arrangements of dam layout alternatives, spillway, diversion works, outlet works, construction access and other facilities as applicable are laid out, which are linked to a horizontal and vertical surveying control system established for the site.
- The site ground material should not contain a lot of faults and shears, otherwise, foundation improvement may be required. For arch and gravity dams the quality of rock is important and the soil is excavated down to rock level. For embankment (earth fill) dams, soil foundation is also possible as long as it has sufficient strength. Here a particularly important parameter is the seepage rate of water in the ground material, k value, as discussed in Chapter 1.
- Usually slope stabilization will also be performed on the valley behind the dam, most commonly as stepped slopes (due to availability of area), or other slope stabilization measures.
- Flood routing study, which means finding the relations between inflow hydrograph *(incoming water during storm)*, reservoir storage and dam spillway discharge is done for flood control.
- Since the reservoir of a dam may encompass a large area, existing improvements such as roads, canals, railroads, existing villages or towns may be affected, which must be relocated or abandoned which increases costs. Even the existence of one historical site which would remain below water after dam completion, may make a dam location unsuitable. Determining the exact dam location and the amount of water to be stored is a multi sided, multi discipline study, that involves many other variables than just engineering concerns, and may take long time.
- Ease of access to the site is also important for construction costs, considering that this is a massive operation that will continue for a long time which will involve millions of man-hours, equipment-hours, materials, administrative and management costs. New roads or even a bridge may be needed just for this.

After site selection is complete, design of the dam can proceed. The factors below, which can still affect the site selection (and could also be listed above), more specifically affect the design of the dam itself:

- water demand for industrial, municipal use and power generation, without shortages
- water demand for irrigation, for which, rare shortages might be acceptable
- river flow, which usually varies a lot between seasons
- the amount of flood flow
- groundwater conditions
- sedimentation studies
- water distribution losses, evaporation, reuse of return flows
- stream releases and storage requirements for fish and wildlife
- storage requirements for power
- recreational use
- legal water rights
- anticipated future improvements

After these factors, and the cost and schedule of construction and the cost/benefit are considered, the type and size of dam is decided and design can start. The ultimate goal is to store adequate water to meet the demand and the other purposes that the dam is designed for, even in dry seasons, for a planned number of years. The single purpose dams, such as the ones that are built only for irrigation, are usually small. However if sufficient funds and other resources are available, it usually makes more sense to build larger dams with multiple purposes, in order to benefit from economies of scale. Below is a general figure that shows a gravity dam cross section. By utilizing a gate at the top, even a little extra height of water may mean millions more of cubic meters of stored volume.

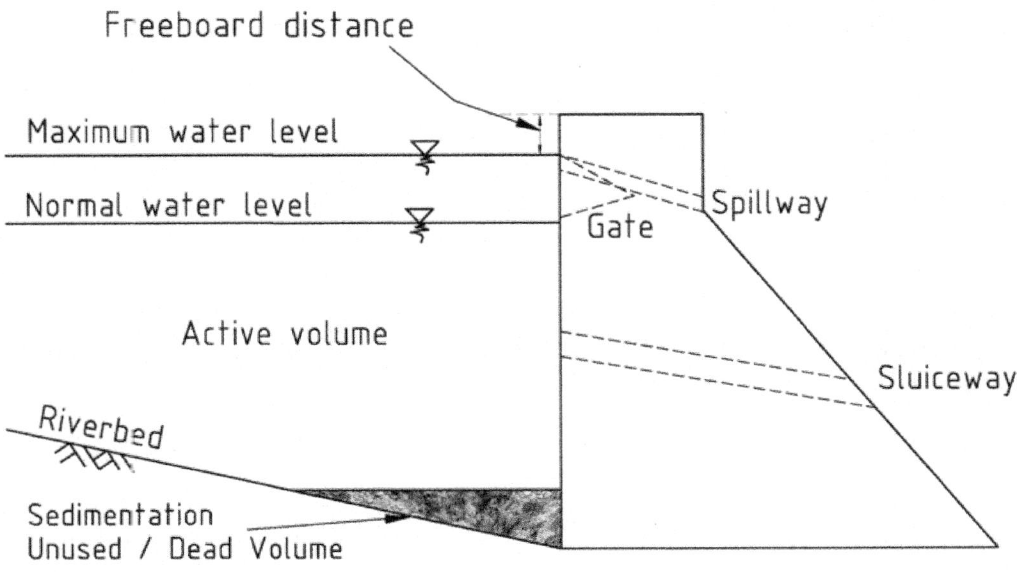

Figure - Dams 1

The forces that try to destabilize a dam are listed below. Also see the figure that showed a free body diagram of a retaining wall, Figure - Soil 15, and the Figure - Soil 39a in Chapter 1 that showed seepage of water in the ground below a dam.

The destabilizing forces on a dam are:

- Lateral water pressure: The hydrostatic force of water is the primary external destabilizing force that acts on a dam, that is to be resisted.
- Uplift force of water from below: This is another main destabilizing force, can be a significant problem if precautions are inadequate both at the foundation and in the dam body itself.
- Seepage forces in the soil: Seeping water exerts force on surfaces, as discussed in Chapter 1.
- Seismic forces: These are considered for both conditions when reservoir is empty or full, and in combination with other forces listed here, to find the worst case. Seismic force is caused by both the mass of stored water and the dam body itself, as an inertial force.
- Wave forces in the reservoir.
- Lateral earth pressure from sediments: As mentioned in Chapter 7, sediments in water can exert force on surfaces. In the early stages of a dam's life, this pressure is low when there is little sediment accumulation. As the sediment accumulation grows, its lateral pressure must also be considered, as the saturated soil unit weight is about 2-3 times more than that of water.
- Temperature stresses can be stabilizing or destabilizing, which must be considered. These are stresses caused by volume changes due to temperature variations. Varying weather conditions and water levels will affect the temperature of the dam body.
- Ice forces in cold climates.

The stabilizing forces are the weight of the dam and appurtenances, the friction at the base, the resistance of foundations such as the deep foundations, the resistance from the sides of the valley in case of concrete dams. These forces stabilize the dam against sliding, overturning and uplift. All these forces cannot act at the same time so we apply load combinations. The worst possible combinations of destabilizing loads must be found and applied in design, and the results must be satisfactory for structural strength, stability and for displacements and functionality. When we say the worst possible combination, it doesn't mean combining all loads that have remote possibility of occurrence at the same time. For example the maximum water level and earthquake should not be taken together in the same combination, which have an extremely small possibility of occurrence at the same instant. Worst combination means, taking the frequent or always existing loads, plus a rare event such as the earthquake. Functionality includes the criteria that the dam remains functional for its intended purposes. Leakages must also be limited and allowable leakage amounts are very low.

RIVER DIVERSION

Except for some small dams which can also be constructed on rivers in dry seasons without diverting the river or can be built inside cofferdam, or the dams that are constructed at a certain distance off stream for some reason, a dam's construction starts first with diverting the river completely so that the area is dry, foundation materials can be inspected and the unsuitable soil can be removed. This work can be very easy or very costly depending on project conditions, such as the river flow, valley profile, the dam itself, appurtenances, soil materials, construction schedule, cost, hydrological factors... River diversion can be done by first opening a tunnel or channel for the alternate river route, eventually joining it to original river path downstream. The river is then diverted through this new tunnel or channel, which is called the diversion channel.

DAM FOUNDATIONS

A thorough subsoil exploration must be performed before the design of dam foundations. We listed items to look for during subsurface exploration and preparing geotechnical reports in Chapter 1. As dams are large and heavy structures we also look at the foundation materials from a greater scale, for example, for rocks, the types, thicknesses, continuity and orientation of main joints, shears and faults, weathering depth and degree, and for the soil, permeability, types and other properties for both.

Foundation work starts with modifying the river bed. This means, after the river is diverted, the unsuitable materials for the foundation are excavated and removed, until an adequately strong soil or rock layer is reached.

The foundation of a dam must withstand huge loads in order to prevent unwanted settlement. This means, both the foundation must be strong and also the loads from above must be transferred to the foundation properly. While the load on riverbed was only the weight of flowing water before a dam was built, after the dam gets built now we must consider the load of several hundred feet high of earth or rock or concrete. This is a big difference, which necessitates ground improvement to varying degrees, such as injection of grout mix (i.e. cement, water, sand, chemicals) under pressure in to the cracks and deliberately prepared voids at the foundation and the sides of the valley. This not only strengthens the foundation and the sides of the valley but also makes them less permeable and resist seepage better. Watertightness can also be achieved by grout curtains or diaphragm walls, sheet piles below the dam foundation. Note: If the material is rock or soil that contains large proportion of boulders, driving sheet piles would be very difficult to impossible. After the dam is complete, water must be drained in a controlled manner under the dam, which must be slow enough to avoid erosion, but fast enough to prevent uplift forces of water. This can be achieved by drainage tunnels or boreholes. To all these, if we add the fact that the foundation soil or rock is a natural material which may have varying characteristics, the design of foundation system for a dam can be complicated.

For concrete gravity or arch dams, where the dam body must structurally be attached to the foundation, a stronger foundation is required in comparison to embankment dams. This means, deep foundation systems or making keys to the foundation rock are needed. To achieve this, loose material on top of firm rocky layer is excavated, until rock is reached. Unlike concrete dams, embankment dams (which are made of compacted earth and or rock) can also be built on softer foundation as they are more tolerant to settlement.

Stress analysis of the ground is done using effective stress parameters, because all water pore pressure will have dissipated after removing unsuitable clay soil (as *clay has low permeability, if it was present, pore pressures could have developed*). So, the total stress will be equal to effective stress for analysis.

Settlement analysis of embankment dams can be done by dividing embankment to horizontal layers and calculating the settlement as each layer is laid on. As a result, the center will tend to settle more, while outside the toes of the dam the ground may tend to heave. These values can be produced easily for all ground level locations with a finite element analysis. Large embankment dams that are built by laying and compacting many layers of soil to great heights can exert enormous stresses to the soil particles at the bottom parts of the dam, so that the sand particles themselves will be crushed, and their angularity, thus internal friction angle, Ø, will decrease, which might also need to be considered.

TYPES OF DAMS

According to the materials used, it is possible to group dams under two main categories:

- Earth / Rock embankment Dams
- Concrete Dams

Earth embankment dams can be of fill material made of soil or rock. Concrete dams can be as gravity, arch and buttress dams. Below we take a look at each: Embankment, Gravity, Arch and Buttress dams.

Embankment Dams

These can be classified into two: Earthfill dams, Rockfill Dams, or the combination of two, by adding both earth and rock. Soil can be treated with cement, to take on some concrete qualities, which is also cost effective due to large quantities involved. Rock and earth are readily available everywhere, and for this reason, nearly 75% of all dams in the world are embankment dams according to ICOLD data. It is always much cheaper to use local materials for millions of tons of placement, rather than importing from far. The soil to be used must be free of organic materials and have low plasticity. As the density of embankment materials are less than concrete, volume of embankment dams are larger than concrete dams, such as having lower slope and larger base, to achieve the adequate body weight. Slope stability, water seepage and overtopping are the most critical risks to be considered for embankment dams.

Earthfill Embankment Dams:

An embankment dam's slope must be sufficiently low in order to avoid slope stability problems. A slope stability analysis is needed for proper material selection, optimizing water content and determining the slope of the embankment.

Embankment dams must be impervious to passage of water. Using natural soil alone will not prevent water seepage. To ensure imperviousness in shallower dams, impervious materials can be selected and a sand drainage layer is used. For taller dams, as seen in Figure 2 below, a central core of impervious material such as clay is used, which must be well compacted. Around it, the regular dam fill material is placed. To avoid erosion of the core, a filter layer is provided around it. In some cases, geosynthetics such as geomembranes or geotextiles can also be used.

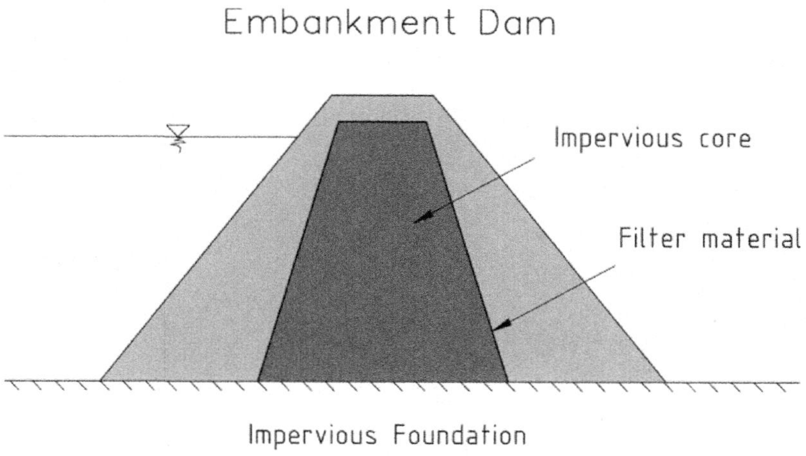

Figure - Dams 2

The function of the core in the figure above is to prevent not only seepage of water but also to prevent erosion of soil as well. If the foundation was not entirely impervious, a cut off trench can also be provided below the middle of the dam, to control seepage. If the internal impervious core was not present in the figure above, water would seep through the dam as in Figures 3a, 3b below, which means we could not have stored as much water as we wanted, and in 3a, seeping water would cause erosion. Figure 3a shows a dam without impervious core and without a drainage layer. The situation in Figure 3a *(the flow lines reaching to downstream face)* is not acceptable:

Side Note: Look at seepage flow and equipotential lines, and compare this with Figure - Soil 39a in Chapter 1.

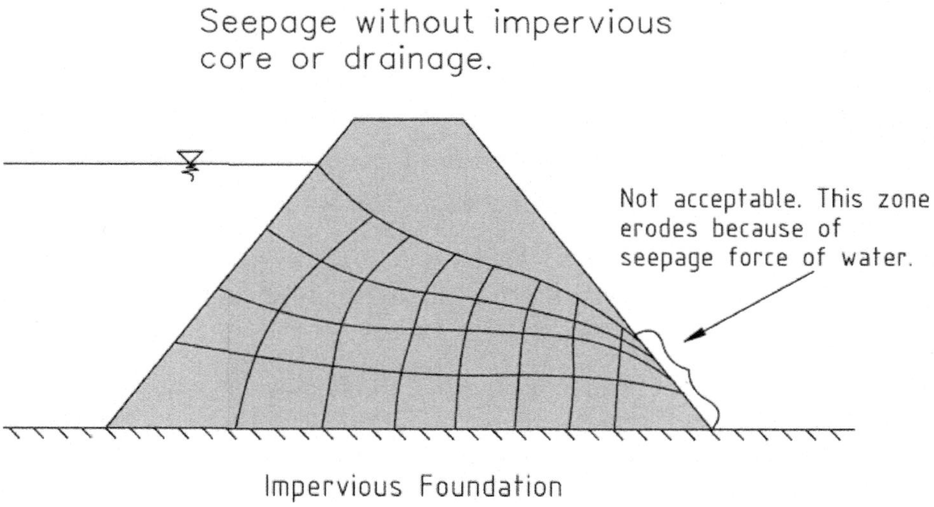

Figure - Dams 3a

Figure 3b below, although still no impervious core, shows a drainage layer this time, made of pervious material such as crushed stones, so that water can seep through. Like everything in nature, water also tends to take the easiest path, and therefore it would drain to this layer easier before reaching to the downstream face of the dam, which is what we want in order to prevent erosion.

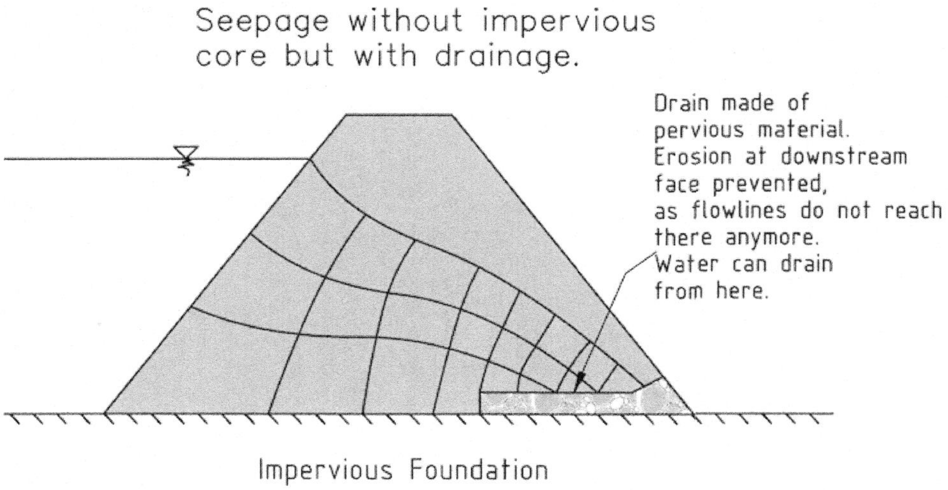

Figure - Dams 3b

The body of the dam itself is most subjected to erosion but the erosion can also stem from other reasons than the dam core or body. One reason is the poor design of foundation that does not consider erosion effects of groundwater. Another reason of erosion can be poorly designed pipe details, where erosion can take place around the pipes.

Overtopping is a great threat for failure of earth embankment dams, as the material on the top can easily be eroded during overflooding. To prevent this situation, spillways are constructed to allow extra water out. Overtopping and internal erosion are the two most common sources of failure of embankment dams by far, with respect to static instability or seismic failures.

Construction of especially large embankment dams require well organized earthwork operations. A large amount of earth material must be transported, spread in layers and compacted in an organized, efficient manner, otherwise schedule and budget impacts will add up. Rain or freezing may affect fill quality and work schedule. Placing the core and filter material is also a challenge, as it is done only in certain section of the dam fill, and layer by layer the location will be different because of the shape of the central core, which will complicate filling operations. For this reason, a cost per m^3 for filling and compacting of an earthfill dam will be higher than the cost per m^3 of excavation of soil.

Earthfill dams can be given more aesthetic appearances if their slope can be blended with natural environment, such as blending the fill on both ends with the nearby hills, placing vegetation on the downstream slope, and giving the fill a slight degree of curve.

Although embankment dams do not need to be attached structurally to rocky foundations as in the case of concrete gravity or arch dams, and can be placed on sufficiently strong soil, still the ground must be cleared off from any unsuitable soft materials and vegetation before the embankment work.

The critical stages for an earth embankment dam as far as slope stability are during:

- the construction stage,
- regular operational stage when there is the steady seepage of water,
- rapid drawdown of reservoir water, where the slope will remain wet, as mentioned in Chapter 1 soil mechanics section, under slope stability topic.

Rockfill Dams:

Other type of embankment dam is the rockfill type. This time the main filling material is rock, instead of compacted soil. Using rock allows steeper slopes than earth dams, because the angle of internal friction of a rock fill is higher than soil fill, which means it has better mechanical properties. This can allow overall cost savings, due to reduced overall dam body volume.

As the rock drilling and blasting equipment became more effective, in many cases rockfill dams also started to appear as a good solution. Another advantage of rockfill dams over earthfill dams is that the construction operation can also go on during rainy weather or icy conditions, since rock is far less affected by these factors in comparison to soil.

Similar to an earthfill dam, a rockfill dam can also have clay core for imperviousness. However for rockfill dams, imperviousness can also be achieved by an impervious layer at the upstream face. This slab is made of concrete or bituminous concrete (asphaltic concrete) and must be carefully designed to work together with the dam body as far as movements and to prevent cracking of the slab. Using an impervious external layer instead of placing a core allows easier construction and therefore cost and schedule savings because the whole dam body fill can be placed at once at higher speed, without having to worry about making an internal core.

The risk of failure of an earthfill dam vs. rockfill dam is about the same from overtopping, therefore, the spillway capacity for both earthfill and rockfill dams must be in the same range for dams that have similar storage capacities and flood risk.

Concrete Gravity Dams

Gravity dams resist lateral pressure through their own weight and by transferring lateral load from water to the sides of the valley and to the foundation. They are rigid structures, that can be made of concrete, or roller compacted concrete. Due to cost reasons, gravity dams are less common than embankment dams. The dam should be large enough that it can resist everything only by its weight, without allowing tension anywhere, so, all sections of dam should be under compressive stresses only, which is something concrete is good at. Also those stresses should not exceed the allowable stress of concrete, so there must be a minimum cross section for this too. If we allowed tension, rebar would be needed for such a huge volume, which would be very costly. To have compression everywhere is the case not only for gravity dams but also for arch and buttress dams as well.

In gravity dams, the concrete body of the dam itself serves as the impervious layer, and no central impervious core is necessary as in the case of embankment dams.

Gravity dams can be straight or curved. A curved gravity dam can be analyzed similar to an arch dam.

Gravity dams are not vulnerable to overtopping erosion like embankment dams, as there is no soil to erode here. They can tolerate overtopping water, unless the overtopping amount is so huge, and the pushing force gets so large that it knocks down the entire dam.

One disadvantage of gravity dams in comparison to embankment dams are that they need stronger foundations, preferably rock foundations, as these are dense, rigid structures, with less base footprint area, and need stronger support and attachment to the ground from below.

Because of huge volumes involved, for economical reasons concrete must be made from aggregates and sand that can be obtained either from the excavation itself or from places very near. By volume and weight, the largest ingredient in concrete is sand and aggregate, (See Chapter 1>Concrete>Aggregate) so obtaining aggregate and sand from nearby places is most important for cost efficiency. Then the aggregates and sand can be sorted and mixed in a large concrete mixing plant on site for quick and efficient delivery to the pour location. Cement may not be readily available always, because limestone and other minerals are needed to produce it, but it can separately be made at a further distance and transported to the site.

When pouring large amounts of concrete (mass concrete) such as a gravity dam body, hydration heat of cement is a big concern, especially in hot weather. Cement content should be kept to a minimum, provided that sufficient strength, durability and imperviousness of concrete is not sacrificed. Pozzolans or special cement (such as Type-IV cement) may be used, or the water to be used in concrete is sometimes mixed with ice or cold water. To obtain ice or cold water, a nearby refrigeration plant is also necessary. However, in no case ice should completely replace water and its amount should not be more than a certain portion of water volume. Another measure is cooling down aggregate by cool air blast.

Concrete can be moved from the mixing plant to the valley by various means, such as trucks, buckets, ropes, cranes, rails, conveying belts. Locating the concrete plant so that its location is optimized for receiving ingredients, but at the same time as close as possible to the body of the dam is important, for cost and schedule efficiency. Permanent access roads or even bridges may need to be built for this, and also for the purposes of general access and delivery of other materials to the site.

A concrete gravity dam is a large concrete structure but it cannot be built in one piece. It is built in 3D blocks of rectangular prisms, to obtain a single large mass. For a large dam, a large block can be around 10m in each dimension. Each block can easily weigh around a few thousand tons. The blocks can be poured so that the joints are keyed into each other by dents and bumps to make a more monolithic structure. This ensures stress transfer between blocks by shear. If the block joints are also grouted in addition to keys, the stress transfer will be done both by shear and moment between blocks. If the joints are neither keyed nor grouted to each other, then the structure cannot work monolithically.

As we said, we cannot just pour the whole dam body at once. Reasons of using blocks include:

- We can pour them one by one in a controlled manner. Each pour can last an entire day, due to large volumes involved.
- We can arrange the pour according to working times of crew, as overtime is more expensive.
- Eliminating large thermal stresses, by allowing thermal expansion or shrinkage between blocks due to temperature variations.

- Avoiding certain times of the day such as night time when setting up and maintaining artificial lighting for a large area would be costly, or, noon time when the weather is hottest which contributes to already high cement hydration heat.
- Downtime due to concrete plant maintenance.
- Waiting time for concrete ingredients delivery.

The joints that separate these blocks are called contraction joints, which prevent concrete cracking when concrete cools down and shrinks (contracts). In the middle part of these blocks, concrete can contain larger aggregates, and towards the edges, smaller aggregates.

During concrete pours, massive pipes will also be placed in the dam body to let water through, if the dam will also serve as a hydroelectric plant.

If, during structural analysis, any criteria such as overturning, sliding or uplift resistance is not satisfied, pile foundations can also be used but this is not common especially for large dams, and instead, keyways (making dents in rock) are provided between the dam body and rock below. So instead of expensive and difficult to build pile foundations, simply a heavier dam cross section is usually used, but still a good bond to the foundation rock must be ensured by making keys.

Arch Dams

Arch dams are concrete dams that are curved towards the upstream side of flowing water. This way they are able to transfer the lateral water pressure to the valley banks by arch action, which is the main source of their resistance to destabilizing force of water. Because of this reason, arch dams can be much thinner and require far less material to build than gravity dams. Arch dams are usually built in narrow rocky canyons, where the two sides of the canyon can serve as supports to the ends of the arch. So the dam puts the rock of the canyon into compression, which is something rock is good at resisting. Arch dam must also be firmly fixed to the underlying rock below, through its foundation.

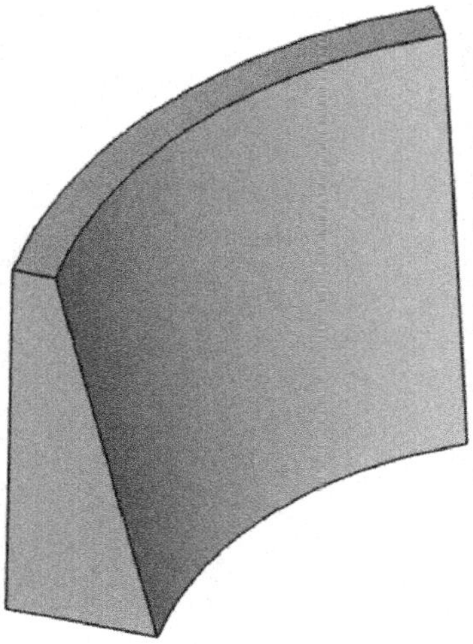

Figure - Dams 4

Similar to gravity dams, arch dams are also constructed in large blocks. Here the only difference is that the volume is much smaller, efficient, by taking advantage of the force redirection to the mountain by the arch action. The blocks are placed on top of each other, with different elevations each time, to make a stronger bond. Again, there is no reinforcement in these blocks, which would not only make the cost go up significantly and it is not needed either, since the whole section is in compression by arch action.

Lateral earth pressure and water pressure increases with depth. This means, the highest lateral pressure is at the bottom. Therefore, at the bottom the arch must have a sharper curve (less radius), to transmit that greater force to the valley banks. Gradually as the depth decreases towards the surface, the lateral pressure decreases, and the arch can have gradually larger radius. This is also to have a more stable base and it also gives the dams a pleasant looking shape but more importantly, it saves a lot of concrete too due to geometry.

The foundation of arch shaped dams is the most critical in comparison to other types of dams because here a relatively thinner structure must be fixed and held in place from bottom, as well as the sides. For better attachment at the foundation, keyways are made at the valley. The sides of the dam where it meets the valley require strong attachments due to high stresses involved, when the arched body transfers the compressive stress to the mountain such as by many anchors all the height in order to achieve sufficient bond. This is why, arch dams are built at locations with hard and good quality rocks, such as granite.

As explained under Structural terms > Arch action in Chapter 1, there needs to be compression in the body of the arch to avoid tension forces. The compressive stresses must also remain below elastic strength of concrete. As long as this is satisfied and there is sufficient bond between the body of the dam and the valley rock, arches can last practically forever, unless there are other effects to destroy the arch, since it is a natural way of transmitting load.

One major construction challenge in narrow valleys is to find adequate area for the contractor's plant, materials and equipment, and produce concrete near the dam to keep costs low. After the construction site is set on a suitable flat area near the dam, materials, workers and even the heavy equipment can be moved (flown over) to the exact dam section to be worked on by cable cars, high capacity cranes, or their combinations, which can be supported from one or two high sides of the valley.

Buttress Dams

Buttress dams can be considered as a subcategory of gravity dams, but more intricate structural design and construction is needed here, due to having a series of buttresses that run along the downstream face of the dam, to serve as supports. Because of this, buttress dams are not very common, unless other conditions exist to make them the best alternative.

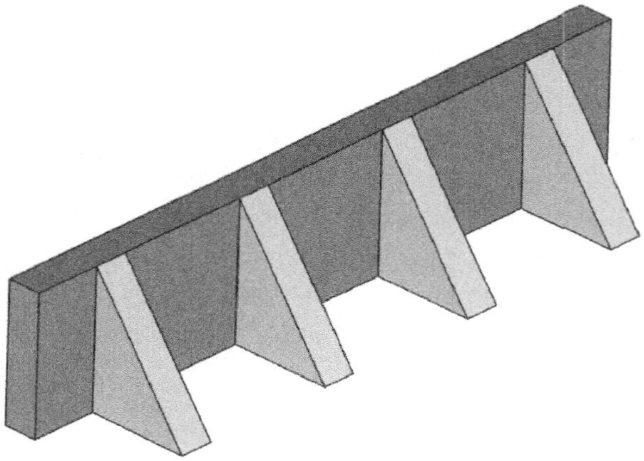

Figure - Dams 5

OPERATION OF DAMS AFTER CONSTRUCTION

Dams must continuously be monitored and maintained throughout their lives. However, dams also have relatively long lives in comparison to other structures. For example, it is possible to see some dams built in the 19th century which are still operational.

Things that are continuously monitored during operational life of dams include:

- Water levels at both sides of the dam
- Small displacements or tilt for both the overall dam structure and also the relative movement of large blocks (such as the concrete blocks of a gravity dam that we mentioned)
- Foundation settlements
- Strain due to high stresses developed in the dam structure
- Leakage amount
- Sedimentation (will be discussed)
- Pore pressure at drains. An increase in pore pressure indicates clogging of foundation drains, which can cause great destabilizing uplift forces at the foundation level

ECOLOGICAL IMPACT OF DAMS

Dams, which can be very large structures that store huge amount of water, can have significant impacts on the environment and even climate in their geographic location, and not all of them are positive. In fact, environmental study before deciding on a certain dam location and size is a very important part of the planning which affects the final decision. Large dams inevitably affect the environment. What we can do is to make this as positive as possible. Now let's list some of the ecological impacts of dams:

- Flood prevention may seem positive at first, but it may have harmful effects too, as it changes the existing water flow balance. For example, water that goes downstream will decrease.
- River valleys are habitats for many animals. Altering the valleys, in addition to a newly formed lake, may have negative impacts on wildlife and balance of ecosystem.
- Large dams can have effect on water quality, and the composition of water can change.
- Electric generation by hydropower, is renewable energy, which is extremely positive outcome.
- Dams close to mining locations can have significant siltation problems.
- Too shallow reservoir water can warm excessively which will be harmful to fish.

- Oxygen content of reservoir water may decrease with time and reaeration measures must be taken, such as during releasing water to the spillway.
- Stored still water can cause harmful microorganisms to easily reproduce.
- Dams can cause certain places such as fertile agricultural lands, recreational sites, historical sites and even villages or towns to remain under water.
- On some rivers, the entire river profile can be tapped with dams. This will turn the river into a series of lakes with successively decreasing elevations for which the environmental effects must be studied.

To keep reservoir water clean, the immediate area around the reservoir should not be polluted. This means, any recreational activity in these areas must be limited.

Other than environmental impacts, dams also have positive social impacts as it encourages more people and their families to live in the area for agriculture, or to operate hydroelectric plants, which prevents migration from rural areas to cities, creates industry and economy.

Although most of us may not see a river every day, our lives greatly depend on them, as they have major impact on ecosystems. Rivers provide fresh water, and that is why most cities were founded near or not far from rivers. Rivers are the connecting link in hydrological cycle of the world's waters, by carrying water from basins to oceans and seas and supporting all types of life in the meantime.

Dams which are carefully designed and properly located on a river will not alter the natural geometry of the river, and after storing the desired amount of water, they will continue to discharge and provide the necessary quantities of water to downstream side, which will continue to support life for both local use and natural habitat as before, and in some cases even enhance water quality and environmental conditions. For most dams, the stored water will be used for multiple purposes which adds great value.

VARIOUS TERMS RELATED TO DAMS

Conduit

A channel designed to carry flowing water under or through the dam body.

Freeboard

This is the height difference between the highest point (crest) of the dam vs. the water surface level at the maximum design height of the dam. So this distance is there for safety, in order to prevent any overflowing, also as shown in Figure 1 earlier in this chapter. Any settlement of dam foundation causes reduction in the freeboard distance, which must be accounted for.

Gallery

In concrete dams, it is an opening within the dam that provides access into or through the dam. Multiple galleries can connect power plant, elevators, pump chambers. Other than providing access into the dam for various purposes, galleries also provide drainage way for water, provide space for grouting and drilling the foundation.

Outlet Works

It is a combination of structures and equipment required for safe operation and control of water released from a reservoir to serve various purposes. They can be classified as river outlets, which regulates the flow of water into downstream side, irrigation outlets to serve agriculture, water supply outlets to serve domestic or industrial water needs, or power outlets which directs water to turbines to generate electricity. An outlet can also be multipurpose. Especially power outlets must be designed very efficiently both mechanically, electrically and also to minimize hydraulic losses, in order to generate as much electricity as possible. When the water is let out through the dam body, it must be designed in such a way that it will not harm the dam foundation. It can also go through tunnels, such as the ones that go around an arch dam, though the mountain rock.

Piping

It means soil erosion caused by seepage. Long duration piping may result in enough erosion to cause dam failures. Seepage forces must be analyzed for piping effects, especially at places with soil structure interaction, where the new structure disturbs the natural soil, even if the natural soil's measured seepage k value may be low. These locations must be treated for preventing seepage, such as by grouting. Reducing seepage and erosion also reduces the uplift effect.

Sedimentation

Without any human interference, rivers are in natural equilibrium as far as the sediments they carry. After a dam is built, the areas of low flow will cause the sediments to accumulate and negatively affect the natural balance dramatically. Due to sedimentation, dams in the world lose a small percentage of their overall capacity each year, therefore, sedimentation, which reduces the useful life of the dams, is one of the major concerns when planning dams.

It is not possible to eliminate sedimentation completely but we can take measures against it. Sediments can be cleaned in certain circumstances, with methods including dredging, flushing or water jet methods which can lengthen the service life of a dam significantly. Other than cleaning accumulated sediments, other sedimentation mitigation methods include, just making the total reservoir capacity of the dam larger, decreasing the amount of sediments entering into reservoir, redirecting the sediments to other locations, planting trees in the valley. In large dams, due to difficulties in extraction and storing of sediments in such large volumes, sedimentation will cause permanent irrecoverable effects.

Sedimentation will cause:

- adverse effects on environment.
- shortening the functional life of a dam, which will have negative effects on its energy production, drinking water supply, flood control, and will cause economic losses. Sometimes in worse cases, a dam's useful life can be as short as 20 years because of sedimentation.
- other sections of a river to flow with steeper slope, which will erode the river base both vertically and sideways.
- water levels to rise and in turn cause more frequent floods.
- underground water levels to rise and in turn lower soil quality in agricultural lands.
- negative effects to engineering structures such as bridges, roads, railroads on the way and along the river.
- coastal erosion.
- negative effects to river transportation and recreation.
- sedimentation elevation increase, which can also affect the outlet works design, that may necessitate a design to accommodate higher water intake levels, as sediment levels rise.
- delta formation at the entrance of the reservoir, if the carried sediments are coarse. If they are fine grained, they will just fill the reservoir.

Spillway

Spillway is a major part of a dam that controls floods by allowing water either from above or through the dam. Overtopping is one of the main causes of dam failures. Inadequate spillway design, blockage of spillway by debris or settlement of dam crest may be the causes of inadequate water flow through the spillway and can lead to dam overtopping. Spillways must be designed to consider the worst possible flood expected during the useful life of the dam, which can be over a century. This is called the spillway design flood (SDF). In some cases, to account for extraordinary floods, fuse gates or fuse dykes systems can be used to take care of the extra water only in the case of extraordinary floods, instead of designing the whole system for it, which can reduce overall cost of the system.

Many large spillways include gates, which are entirely or partially opened during flood times, but they are closed during normal operation, to maintain water level at the dam to its highest capacity possible.

Not gated spillways are also possible, to allow the water that will soon to overtop to flow through it. The advantage here is the elimination of maintenance requirements of gates, but the downside is that the maximum water storage capacity is lower. Spillways without gates can also be made as a labyrinth, to reduce the flow and energy of water. The gate and the spillway itself must be designed such that the total system is able to discharge sufficient water in a given duration. To give an example, without assuming any inflow, the system must be designed with sufficient capacity that it will be able to discharge 75% of the water between maximum design water level and the top point (crest) of the spillway within 48 hours, according to NY State Department of Environmental Conservation criteria.

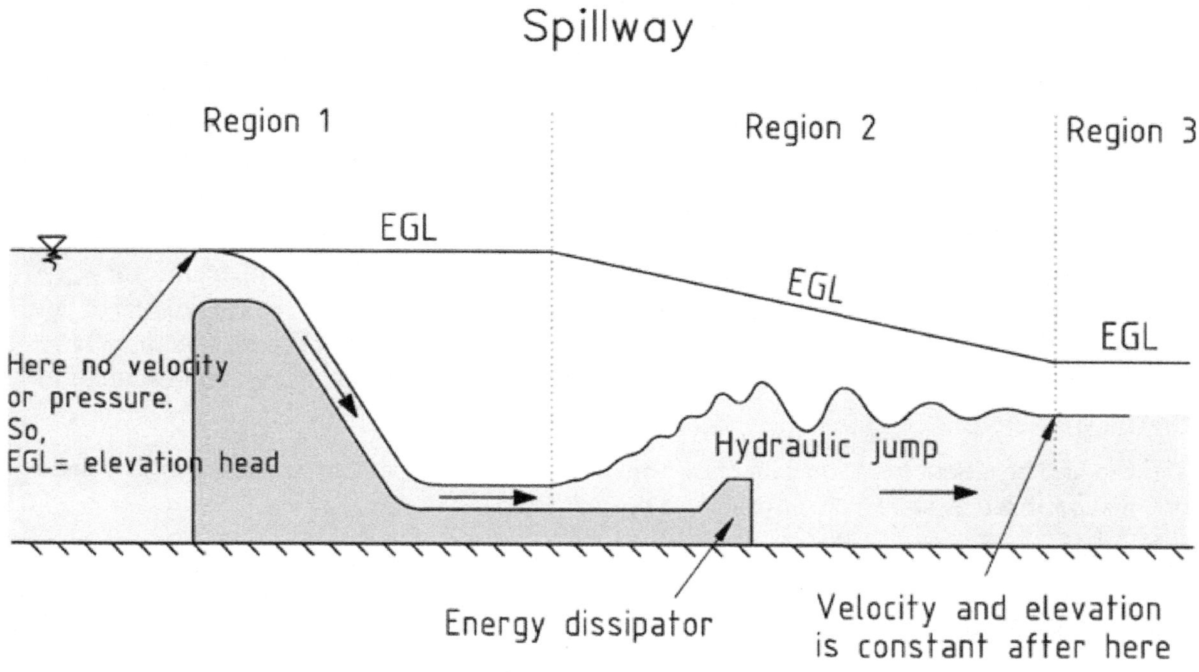

Figure - Dams 6

A spillway must also dissipate energy of water, otherwise the flowing water with high energy will cause soil erosion as soon as it leaves the spillway. At the toe of the dam adequate measures must be taken so that the falling water with high kinetic energy will not cause any erosion. To achieve this, energy breaking thick concrete sections can be provided as seen above.

As we saw before in Chapter 1 > Water Resources Engineering and Chapter 7,

Energy Grade Line (EGL) = Elevation Head + Pressure Head + Velocity Head

See the figure above. Just before water starts to flow over the spillway, velocity is still zero yet. So,

Total Energy (EGL) = Elevation Head + Pressure Head + 0

This value is constant anywhere in the reservoir behind the dam because as the elevation head decreases pressure head will increase (Also see Figure - Soil Mechanics 39a in Chapter 1). And on the water surface, pressure head will also be zero and EGL will just be equal to elevation head: EGL = Elevation head. That is why, EGL is exactly at the surface of the water over the entire reservoir area.

As seen in Region 1 of the figure above, as the water starts to fall down the spillway, its elevation decreases, pressure head is still zero, but the velocity increases this time, to keep the total energy constant. We ignore the minor energy loss due to friction here and show the EGL still continuing as a flat line, in other words, constant here.

In Region 2, at the toe of the dam, when the fast traveling water hits the energy dissipator, suddenly there is a big loss of velocity because of this impact, and significant energy loss, EGL decreases as seen.

Finally in Region 3, the water continues again almost straight, as its elevation and velocity is now constant, again ignoring minor friction losses there.

And why did the height of water suddenly increased in region 2? This is called a hydraulic jump. It occurs here because the decrease of velocity is so much and sudden that, water cannot adjust to lose its energy that quickly to make up for the very rapid loss of such an amount of velocity head. So, from the equation above, if we want to protect energy, when velocity head decreases, elevation head increases - although energy is not fully protected and there is a loss, as seen by the decreasing EGL in region 2 of the figure. You may also ask that water had much lower height before the hydraulic jump, so how can the height increase if the width of flow is the same? This is because we are not talking about volume but water flow amount here. The water flew with faster velocity in smaller cross section, which is equal to flow with lower velocity in larger cross section, since amount of water flow cannot change. Remember, water flow is equal to cross section area times velocity, Q=V.A as we saw in Chapter 1, Water Resources Engineering section.

Water can also be let out from the dam body from under reservoir water surface level. In that case, these are controlled by steel plates that can go up and down.

Auxiliary Spillway:

This is a secondary spillway that is built to operate only during extraordinary floods. If this is not provided, the single spillway must be able to handle the peak floods as well. Unlike the primary spillways, which is also called the service spillway, auxiliary spillways do not need to endure the sustained flow conditions. This also means, primary spillways must have adequate capacity so that the auxiliary spillways will not be subjected to frequent use.

CHAPTER 9 - COASTAL and MARINE STRUCTURES

Underside of a long pier, with a recreational area at its end.

COASTAL PROTECTION STRUCTURES

Considerable part of world's population live along shores. Design and construction of coastal protection and marine structures is important part of civil engineering, both of which are discussed in this chapter.

Coastal protection structures are built to make the coasts function as desired, in a controlled manner. They serve various purposes such as to protect the coast itself or the inner land from waves, to stabilize beaches, to serve as a platform for people or vehicles to safely and conveniently be present at the coast, to make more land available for use at the shore, increase the value of land, to serve as a platform for boats or even ships to approach, moor, load and unload cargo and passengers. Although they have such benefits, they can have adverse environmental impacts as well, which must be well studied. Therefore we must build these structures only when we really need them, otherwise the human ambitions to make use of and profit from valuable coastal land can turn into a big negative for everyone in the long run. Below, main types of coastal protection structures are introduced.

Sea Walls and Bulkheads

These are vertical or nearly vertical walls constructed at the landward end of beaches or coasts for reasons such as to protect people and property inside against waves and flooding from sea during storms, to overcome the elevation difference between land and beach, to artificially widen the beach to create additional recreational space, to separate beach from buildings and streets inland, to protect eroding coastlines and similar purposes. It is also important to know that some states do not even allow sea walls, bulkheads or revetments at ocean coasts, due to potentially negative environmental impacts.

The term "sea wall" may be confused with the term "bulkhead". A bulkhead is similar to a seawall but generally smaller and its primary purpose is to protect, retain and prevent sliding of the land behind to the sea, and protection against waves is low and of secondary importance. Along bulkheads ships can also moor, and in this sense they are similar to quays. Bulkheads are usually of corrosion protected steel sheet piles or protective treated timber which are embedded into the sea floor.

Sea walls on the other hand, are usually larger in size, and they are primarily for protection against waves, in addition to retaining soil behind. Therefore, for areas with large waves, sea walls are used.

Seawalls or bulkheads can be made as a cantilevered, gravity, sheet pile, soldier pile, gabion wall, and together with different geometries, details, shallow or pile foundations, with or without anchors, limitless number of details of systems can be applied depending on particular project circumstances. The shorter the height to be exposed and shallower water to be maintained, the easier to build them. For example a short bulkhead will not necessitate anchors and a cantilevered section can be built.

For sea walls, we consider items such as: Wave forces, current forces, water depth, tides, storm surges, seepage, seabed sedimentation, erosion and scour, retained soil pressure with or without water, and seismic forces as applicable. All dynamic, hydrostatic and uplift forces due to water effects are considered. Failures can happen from overturning, and sliding from the pressure of the retained earth behind, foundation settlement (sinking of the wall), global rotating slope failure, erosion at the toe of the wall, tilting or moving of wall from wave forces. For economical reasons, a reasonable amount and frequency of overtopping can be accepted, as long as this does not create any safety issue. Considering these constraints and the external effects listed above, sea walls need to have strong foundations. A gravity type wall must have strong soil below, to carry its weight without excessive settlement.

Seawalls and bulkheads are subject to erosion and scour at their toe, which affects their stability. To prevent this, ripraps revetments (discussed below) can be placed in front of the wall, which greatly dissipate wave energy. Geotextiles are used to prevent erosion of retained backfill material behind through seepage. The backfill material need to be coarse grained to be able to quickly let the water drain, before developing unbalanced hydrostatic pressure. Weep holes are provided to let the water out. Sea water can get to the back of the wall during wave overtopping. To entirely prevent wave overtopping would require larger and much costlier construction with higher environmental impacts.

Revetment

A revetment is placed on a slope such as the shore itself, dike or embankment to protect them against erosion, currents, waves and storm floods. They can also be placed in front of seawalls. They absorb a great deal of wave energy, instead of reflecting the waves which would cause erosion and slope's undermining. The top layer of revetment is the armor layer which is made of permeable and irregularly placed materials such as large concrete blocks, quarry stones or boulders, gabions, interlocking concrete slabs, timber, geotextile fabric bags. As they are placed more irregularly, the energy dissipation capacity also increases, therefore they should not be placed smoothly just because it may look better. Due to the varying erosion protection requirements and a large variety of coastal profiles and water effects, many different options are possible, one of which is shown below. Revetments made of precast concrete blocks or stone are the most common and most effective to absorb wave energy, and they are also very durable and easy to place, with almost no maintenance needed later. The armor layer stones or concrete blocks must be large and dense enough not only to resist the dynamic forces of waves but also to prevent the smaller size stones in the filter layer below from being lifted away. They can also tolerate large differential settlements than interlocked, uniformly placed concrete blocks. One disadvantage is, they are harder to walk on and don't look as neat.

In addition to the main material as stone, which is called the armor material, revetments should also have a filter layer and toe protection below the armor layer as seen in Figure 1. The filter layer is a layer of gravel or broken small size stones, which are considerably smaller in size than stones of armor layer, applied usually together with geotextile filters, placed between the underlying soil and armor layer. It not only drains and relieves the hydrostatic pressure of water below armor layer but also prevents soil below from being washed away, through the large gaps between armor layer stones or blocks. It also distributes the weight of armor layer to soil more uniformly thus prevents differential settlement. Toe protection prevents scour and erosion at the toe.

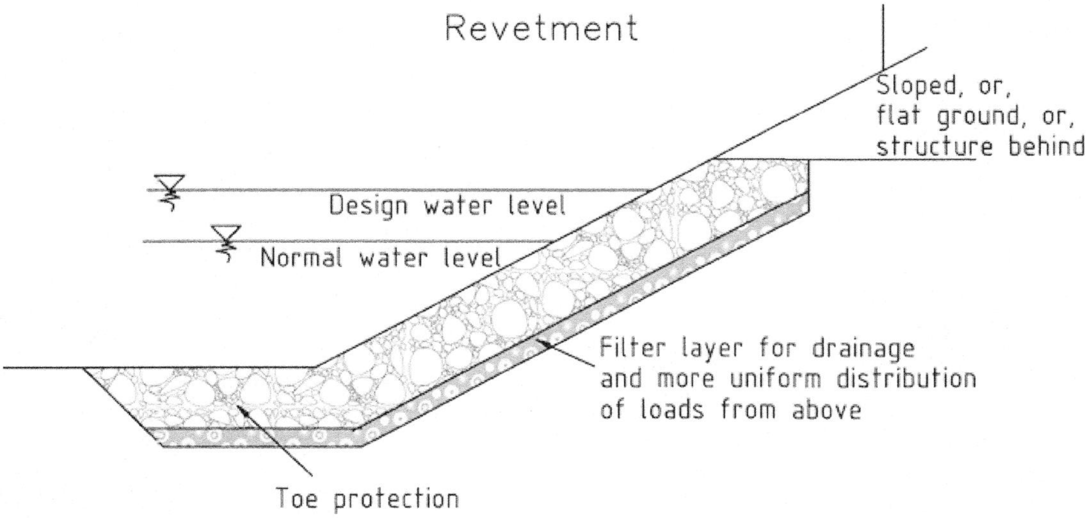

Figure - Coastal 1

The term riprap is used instead of revetment sometimes but although they do the same thing, there is a slight difference. Revetment is the system as described above, with the armor and filer layers. Riprap is simply the large size boulders placed on the shore for protection from waves, placed in similar fashion with the armor layer of revetment. So the revetment's armor layer can be made of riprap stone but not necessarily always - it can also be of the materials we mentioned couple paragraphs above.

Breakwater

These are built in the sea, to protect ships and boats by reducing or totally eliminating wave action. Breakwaters may or may not be attached to the shore. Typical breakwaters are made of boulders or randomly placed concrete blocks of various shapes, to dissipate wave energy. Due to the complexity of the situation created by beach topography, waves, tides, currents, winds and storms, it is not an easy task to determine the optimum layout of breakwaters and physical models may be worth the task, depending on the size of the project.

There are also floating type of breakwaters. Their advantages are that they are generally more economical to build, and have little to no adverse effect on aquatic life *(segmental breakwaters built in usual way as mentioned in the previous paragraph also have this advantage)*. On the other hand, floating breakwaters can only be effective for a certain range of wave period. Too lower or too higher frequencies of waves will transfer right through these structures *(remember, in Chapter 1, under earthquake section, when we discussed natural periods of vibration, we gave an example of a small boat and a large ship versus waves of different sizes, and only when the wave period gets close to the boat or ship's period it will rock the vessel which means the vessel is consuming wave energy)*. Therefore floating breakwaters' usage is far more limited. Mooring's connection to constantly moving breakwater blocks is another critical factor to consider during floating breakwater design.

A breakwater made of boulders, protects a marina.

Sea Dike

Sea dikes are usually sloped earth structures at beach shorelines that protect lower inner land from waves and flood. The sloped, raised core of the dike is protected by revetments, which is also called the armor layer, which can be made of stones, concrete blocks or slab, asphalt. The dike's slope stability, erosion resistance at the toe and even the core, core and armor material stability must be ensured. There can also be vertical dikes made of concrete. They should be checked against overturning, slippage, sinking, global failure, toe material erosion, undermining of material below toe.

A more natural and environmentally friendly version of dikes are artificially created stabilized sand dunes, which still serve the same purposes as dikes plus they are also considered as beach nourishment, where a new sand dune can be created by imported or locally dredged sand, or an existing dune can be made larger and reinforced, to stabilize it. In fact, whenever possible this should be a preferred method for protecting coasts, because not only they are more environmentally friendly (which is the most critical reason) and economical, but also they just blend better with the coast and allow recreational activities better than other systems. Their surface is reinforced and stabilized by grass and other suitable plants native to the shore, which provide reinforcing by their roots. One downside is that sand dunes need maintenance much more often, especially after storms.

Side Note: In general, for shore protection, we should always prefer natural arrangements that do not involve construction of structures, if we can.

Quay

It is part of a riverbank or coastline for ships to dock and moor alongside. So the quay is the shore itself, not a separate structure that protrudes from the coast as in the case of piers or wharfs, organized and improved for ships and boats to approach, and disembark. In this sense, it could also be listed as a marine structure in the next section, and therefore many of the design and construction points listed for marine structures in the next section also apply to quays. In a harbor or a wharf, there may be many piers and quays. Quays can be made of anchored screen walls made of driven sheet piles, tangent / secant piles of reinforced concrete, gravity blocks, concrete retaining wall sections, or as open decks (where there is no retained soil underneath the deck) supported on piles, all with their own applicable details and deck, deck support, foundation and backfill details, depending on different initial beach conditions and the proposed shore profile. Impact force potential from boats must be evaluated during structural design and protective measures should be taken to absorb energy. At the edge of quays (and seawalls), low parapet walls can also be built.

MARINE STRUCTURES (HARBOR WORKS)

These are structures built on the coast to accommodate ships and boats, so that they can moor and shelter, unload and receive cargo and people. For offshore structures, see Chapter 11.

Port

It is a large facility for ships of all sizes to load and unload. It can contain many piers, wharfs, breakwaters, sea walls, revetments, cranes, storage and repair facilities for goods and ships, rail stations for transferring goods inland and many substructures. A port can also contain one or more harbors, wharfs, piers. Ports are very large facilities that can influence the social life, industry, economy and the environment around them. A large port means a lot for a country's economy.

Harbor

It is a parking, shelter and storage space on a water body for ships and boats. Unlike ports which are only man-made, harbors can also be natural or semi natural, in addition to being man-made. Natural harbors are possible where there is suitable land or reef area at the coast surrounded by sufficient land so that waves will not be able to approach to ships or boats, or, their magnitude will be reduced down to an acceptable limit. A man made harbor will have breakwaters and seawalls. The term harbor is often confused with the term port. A harbor is smaller than a port. Harbors can also have sub facilities but not as extensive as ports. So, all ports can also serve the purpose of a harbor, but a harbor may not necessarily serve as a port. Harbors can be for commercial, military, refuge, fueling or other purposes.

Marina

A marina is a shelter and storage place for yachts and small boats only, so it is usually a smaller place than a harbor. A marina can be located within a harbor as well, but usually this is not the case. Marinas have high density docks, for many boats and yachts to tie up.

Wharf

A wharf is built on the shore of a harbor, which provides access to ships and boats, for them to moor and dock alongside for loading and unloading passengers and cargo. The term wharf is sometimes confused with docks or piers. A wharf may mean a pier parallel to the shore, or it can mean a larger facility which can include docks and piers. Within wharfs, warehouses, service buildings and other facilities may also be present for handling ships or boats. Wharfs can be of wood or concrete. They may contain navigable channels in between the piers. They can lie alongside the shore, and / or protrude towards the sea (when water depth is not enough at the shore for ships) and in that case connected to the shore through trestles supported by piles. The deck sections above water are supported on piles or pillars, which not only support weight, but also resist wave action and provide adequate lateral support for earthquakes or impact loadings from ships or boats. Wharves are built on river or canal shores too.

Pier

It is a raised walkway structure above water, usually made of wood that protrudes from the shore towards water, perpendicularly or with an angle to the shore, to allow ships to disembark and load and unload passengers while remaining in deeper water. Small berths may also be present for smaller boats, branched from piers, or when the pier reaches adequate depth, it can branch out perpendicular, forming a T or L shape. Piers also make possible for people to walk towards deeper sections of sea and do fishing, diving, and the edge may include small recreational structures. Piers are supported on piles, but in comparison to wharfs, these piles usually have wider spacing.

DESIGN AND CONSTRUCTION

Now let's list some important points in design and construction of coastal protection and marine structures that may apply to structures we listed so far in this chapter, but of course keep in mind that the importance and applicability of each item below varies for each structure, which should be fairly obvious when reading (i.e. some apply more or only to coastal protection structures, and others to marine structures). Design of coastal and marine structures can involve engineers from all civil engineering disciplines such as structural, geotechnical, water resources, transportation. In addition, oceanographers, meteorologists, geologists, urban planners also participate often. Note: Below when we say "harbor", it will mean all of the port, harbor, marina and wharf terms for convenience.

General Planning and Layout

Design constraints for both coastal protection and marine structures include:

- Beach profile (i.e. land and underwater contours, natural materials), water level and its fluctuations, effects of sedimentation and erosion, natural processes of coastal modification.
- Seabed geotechnical conditions
- Amount of dredging required
- Prevailing winds and wind forces, storms and even hurricanes. Storms can also effect the flooding of the harbor area. River discharge to the area must also be considered, if any.
- Wave heights, durations, periods and directions, which determine wave loads
- Water current strength, direction and durations. Also the water within the harbor area should be able to circulate and replace itself by natural means.
- Tidal effects
- Icy conditions
- Foggy days and visual aids such as lighthouse and radar
- Water quality, salinity (which also affects density)
- Environmental impacts of the structure on the surroundings
- Effect of environment on the structure, such as chemical and ecological processes, short term seasonal and meteorological effects, long term trends such as possible sea level change
- The existence or feasibility of constructing new land transportation access to the proposed harbor site
- ... and as for any type of project, the project budget and schedule, constructability issues, long term maintenance and operational costs, vs. the expected revenue, aesthetical concerns, presence of existing structures and features, technical capabilities, availability of materials, social, political and administrative issues.

And specifically for marine structures / harbor works:

- The required berthing area and layout depend on items such as the number, type and size of ships and boats, their mooring, navigation and maneuvering needs with certain safe distances, prevailing wind and current directions *(ships and boats should be moored heading towards prevailing wind and/or current directions as close as practically possible)*, the required surface area to accommodate the loading / unloading traffic and equipment installations on berths, the required storage and complimentary facilities (i.e. administration, maintenance and repairs, inspection, electrical, resting area...) between berths and within the harbor general limits, the required connections with the outside on land. Similarly, the elevation of decks must be compatible with approaching ships / boats to it.

- A harbor is divided to zones for large and smaller boats or ships for efficient usage of available area. The smaller ships should be located at the shallower, inner sections.

- The harbor entrance must be designed to minimize the entrance of waves while still be large enough to allow the ships to maneuver fully. And the inner boundaries of harbor must be designed such that it must dissipate the entering waves' energy as quickly as possible, without allowing significant wave reflections. The wave oscillations inside the harbor should be prevented to resonate with the natural period for wave resonance of the harbor itself, in order to prevent maximum oscillations, which could also resonate with the moored ships or boats and causing problems for them. These oscillations and all aspects of whole harbor system are also studied with physical and computer models *(better to do both because of the difficulty in adequately representing complex systems as these, which can still not be ideal)*, which also include all other effects such as tides, currents, storms, effect of vessel movements.

- Different layouts in the planning phase must be compared to each other while considering all factors indicated in this section through computer models. Construction of physical test models is necessary except for small projects.

Water Depth, Dredging vs. Ship / Boat Draft

Marine structures listed above are built by either digging out of a low area inland with access to a main water body, or, built into a larger body of navigable water, or a combination. Steep sloped sea bottom are preferred, to avoid or minimize dredging, but if this is not available then there must be dredging work and / or piers may be needed to extend towards water with sufficient depth. Soil at sea bottom might need to be removed by dredging, to open space for larger ships or boats, at least for a part of the port or harbor area. Squat effects (See Chapter 11, water canals) to ship draft must be considered as applicable. As an alternative, limiting the draft (how deep the ship hull sinks in water) of ships that can use the port can be considered.

The depth of water is affected mainly by waves, tides, currents and any other water flow, which are also dependent on different seasons. Erosion and sedimentation of sea floor is also a factor. The draft of a ship is affected by factors such as its weight, the load it carries and change in water density. All of these are considered together to determine a safe distance between the bottom of the ship and the sea floor in the harbor area. This affects the choice of the main location and dredging needs.

Just like when building highways earth is scraped and removed from the natural ground by earthwork operations to make the path for the road, dredging is the equivalent of it underwater. We mention dredging also in the case of:

- Opening path for immersed tunnels (see Chapter 6 -Tunnels > Immersed Tunnels)
- Clearing dam reservoir sedimentation (see Chapter 8 - Dams)
- Clearing and deepening the path for building and maintaining waterways (see Chapter 11 - Various Structures > Water Canals)

Dredged materials are called sediments. Dredging can be performed by a ship traveling over water and pulling away material at the same time, or by stationary cranes on shore, which have digger shovels attached. The fill material obtained from dredging can also be used for beach improvements such as sand dunes or coastal landfills, beach maintenance, creation of wetland and coastal habitat, or, if it will not be used, the site to place excess materials and its effects on surroundings must be clarified. Obviously, environmental concerns play a major role for the timing, method and amount of dredging operations, and also the final placement of dredged material, therefore they must be well studied.

The pier or wharf decks are designed to remain above the water even during high tides, so no buoyancy (uplift) forces of water are considered, except when/if there are lower sections, such as utility lines or vaults, which may stay fully or partially submerged if it cannot be avoided. Utility lines that run from the land to wharf area should have flexible connections at the junction locations, expansion joints on the deck, in order to tolerate possible movements. They should be located above water and preferably above the pier deck. The waves generated by ships must be taken into account when determining height (and also during structural design).

Entrance channel and harbor area should not be at a location where it will be filled with soil over time.

The depth of water influences the period of oscillation of water in the harbor, which affects what size ships oscillate more. Deeper harbors cause longer oscillation periods, which affect larger ships more, which may mean resonance for them *(similar to the example we gave for ships vs. waves in Chapter 1, Earthquake section, and taller, heavier structures having longer periods and resonate if the earthquake ground period is also long).*

Deterioration of water quality in harbors with deeper water (called "deep-draft") is less critical than in the harbors with shallow-draft, because deeper harbors have wider entrance to sea.

Loads

Coastal and marine structures are subject to a complex set of forces, including permanently and constantly varying dynamic loads. As for any structure, the structure itself, its foundation and underlying soil forms an entire system to be analyzed under these complex set of loads, which are listed below. Because of this, unless we are dealing with small structures with simple geometry, it is always beneficial to analyse them using numerical (FEM) methods, which we introduced in Chapter 1. In addition, coastal and marine structures also need hydraulic model studies, for more efficient design.

The loads that can act on coastal and marine structures and their foundations include:

- Constantly varying amount of hydrostatic force such as both lateral force and vertical uplift due to varying water levels and waves. There may also be differential pressure at the front and back. The maximum and minimum water levels and the dynamic forces that will be created are considered for the design of the structure and its foundations. The most unfavorable condition as far as the difference of water levels on both sides must be considered, such as the low tide in front of the wall, vs. the retained earth behind, plus the maximum height of water table that may occur behind the wall.

- Hydrodynamic wave and current forces. Resonance of wave forces and the structure must be avoided. Wave forces can be from non breaking, breaking and broken waves. These cause different force direction and intensities at different sections of the structure and must be modeled correctly. The height, period and direction of waves are important parameters.

- Live loads of vehicles, various components and people on decks.

- Seismic loads.

- At locations where loading and unloading of cargo to and from ships, any applicable truck, container and crane (plus container) loads are also considered for deck edges, which are multiplied by appropriate factors and affect the design of the deck and the foundation. As an example, the truck load can be the HL-93 truck which are also commonly used for bridges. The cranes are usually on rails which must be supported by considering the worst load combinations, including any breaking of piles in case of a ship impact. If closed form deck systems were used, the load from cranes may impose considerable surcharge on lateral retaining systems and bearing on the backfill below to result in settlement, therefore, to support cranes, tracks on open form decks directly supported on piles may be more preferable.
- Creep, shrinkage, thermal loads.
- Ice load must also be considered in cold climates. Moving surface ice can push the structures or create impact forces. During low tides, it can add significant weight by drag down, and during high tides, it can create uplift. Ice can have benefits too, such as covering the structure and protecting against severe winter waves but this aspect cannot be relied on.
- Wind loads on cranes which are supported on the deck.
- Earth pressure on the structure and foundations.
- Liquefaction potential must also be looked at, as these structures are built on saturated soils.
- Dynamic loads from sea vessels, including impacts, berthing and mooring. The design of mooring to berth supports must consider the ship sizes and weights, wind, tides, currents, waves from ships or sea. Moored ships can exert large loads on these supports and in turn the deck. Adequate mooring points, such as bollards and hooks along deck are needed to safely resist and distribute ship / boat loads and limit their movement. The moored ships or boats should be headed towards prevailing winds and towards the wave as much as possible, so that the forces on mooring lines are minimum. Berths should be directed as parallel as possible to the prevailing current, to minimize the force exposure.

  At the edges where ships approach and stop, there can be very large impact loads. The impact load of a ship on a berth structure can be resolved into a perpendicular and parallel force component to the deck. The parallel component is a friction force. This force depends on the ship weight and velocity, current strength and direction, wave and wind conditions. The worst case must be considered, as given in standards. From these, berthing energy is found. Velocity of the ship is the most important factor, which affects the berthing energy by its square (Kinetic Energy = $1/2\ mv^2$). This energy is then reduced by factors such as the energy absorbed by the fenders on the deck side, and the energy lost by ship's rotation ship when it impacts the deck unless it is a head on collusion. *(A head on collusion where the impact force passes through center of gravity of the object, is the most destructive as no energy is lost to rotation.)*

All of these loads can be significant. Add the difficulty of construction in or near water, designing and building coastal and marine structures are challenging, and performed by highly specialized firms.

Deck Design and Modeling

Various important items to consider for modeling and seismic design of wharf and pier decks on piles are listed below. Structural modeling was introduced in Chapter 2. Decks can be of open or closed type. Open type decks allow water to flow underneath and they are built on piles. Closed decks are built near the shore, they can be built on fill that are then surrounded by piles and or walls, to form a closed area, over which the deck is built. Below discussion mostly apply to open type decks.

Deck can be modeled as typical section strips, which can then be applied to whole deck, if suitable. Typically, deck rests on beams, which are supported by piles. They can also be made as flat plate slabs.

For vertical loads, to calculate the beam deck moments, the piles can be modeled as vertical springs under the deck. This can also enable calculation of pile axial loads. For lateral loads, the deck, pile caps and piles form moment frame. The piles contribute to the frame action by their bending stiffness.

High punching shear resistance must be provided by the deck against concentrated loads, which is best possible by using thick concrete decks.

For lateral loads, such as earthquake, mooring loads, wave forces, just to calculate a preliminary, conservative demand, piles can be assumed fixed at their bases, moving laterally, bending elastically *(similar to ELF method discussed in Chapter 1, where a building swings elastically, which causes the most demand on structure and the foundations because of no yielding)*. To be conservative lateral soil support may be ignored. After this, the effect of soil lateral support can be included by lateral springs, to obtain a more realistic, lower demand value. Lateral loads are distributed to each pile according to their relative stiffness, since as we mentioned a number of times before in the book, stiffness attracts load. What we described here applies only to decks supported by piles, with a distance from land, in other words open decks. For wharf decks touching the shore, in other words closed decks, there will also be great lateral support from the land side, such as the earth retaining structure there.

In Chapters 1 and 2, we discussed representing nonlinear items in structural models. Similarly any source of nonlinearities must be modeled for wharf and pier decks, such as the effect of cracking, any possible change in soil stiffness due to shaking of the foundation piles, the sequence and location of pile plastic hinge formations. For piles we find moment (from moment-curvature analysis), shear, displacement capacities, pile to deck interactions, P-Delta effects and represent in the model.

Where we use batter piles, the lateral load resistance is obtained much more easier but this means increased lateral stiffness as a whole which means higher seismic forces on the structure. The batter piles also impose high punching forces to the deck/pile caps during strong earthquakes.

A harbor deck can be designed as a ductile system and with strong beam (deck), weak column (supporting piles) concept, which is the opposite of strong column - weak beam concept that we discussed for buildings. Because here it is more important to keep the deck itself intact during earthquake. This means, plastic hinges will be allowed to develop at the top of the piles, where they connect to deck, (which must be simulated in the model with proper member stiffnesses and plastic hinges) but the deck and the people and structures on it will be safe, although the structure will have lost its serviceability which also depends on the design earthquake level. Battered piles resist horizontal loads better, but they may prevent development of plastic hinges and can punch through the deck or cause breakage where they join the deck *(as a remedy for this, flexible connections between the deck and the head of piles can be used)* and therefore each case must be analyzed for the particular project.

For concrete decks supported on piles, high punching shear (see Chapter 1, structural engineering, various terms) capacity is needed. This paragraph also applies to quays made as open deck systems supported on piles (and not as retaining systems).

Piers and wharf decks should be made with as few expansion joints as possible. Temperature variances, along with the loads imposed, the geometry and the materials of the structural system are inputs to the structural analysis, to determine the size and number of expansion joints. Both sides of an expansion joint should be supported on pile caps close to the joint, to minimize cantilevering distance. Also see Chapter 1, structural engineering, various terms section, for more on expansion joints.

As far as construction materials, concrete, steel or timber are used. High durability concrete can be the best option for high load capacity, long life span and low maintenance. Steel can also be used as long as it is protected from corrosion. Timber may be used as treated but under high loads, it is inadequate.

Various Items on Deck:

When there must be certain buildings on the deck, the interaction between the deck and the structure above as well as the combined seismic mass of the building and the deck must be accounted for, during the design of such building, and the deck below. A building structure on deck should be considered only when really necessary and preferably they should be located on the shore, to avoid many additional challenges of designing and constructing a building on deck.

There will often be cranes at the edge of the deck. These must be located as close as possible to the edge to extend their reach to the ships. In addition to the coordination of loads imposed on the deck, their location must also be coordinated with utility lines, which can also run at the deck edge.

Utility lines on decks, some may be fixed, create challenge when using expansion joints.

Bullrails must be adequately supported by the deck at the edge.

Depending on project, there may also be trucks or other vehicles need to be present on the deck. The deck layout should coordinate their maneuver space vs. other items on deck.

Any fire lanes as needed, must be coordinated with other items on the deck.

Proper drainage must be provided on deck, to divert rainwater to the sides.

Various Seismic Concerns

We discussed earthquake analysis methods in Chapter 1. They can also be used for marine and coastal protection structures.

In the case of closed form wharf decks or quays, retained systems must support the inertial force from all the mass of the retained fill material (plus any water) behind, during an earthquake. Because of this open decks supported by piles underneath can be preferred. If retaining systems are used, and the backfill material remains in water, it must be permeable enough to rapidly release increasing pore pressures during seismic event. To facilitate rapid evacuation of water, permeable fabric materials can also be used between the fill and the retaining system. The most earthquake resistant system would be a floating type deck (which would practically be completely resistant), isolated from the ground motion, but that of course brings its own design and construction challenges and limitations in use.

Liquefaction potential must also be evaluated for the overall section and behavior of fill materials, also considering existing soil *(Soil Structure Interaction and Liquefaction were discussed in Chapter 1)*.

Another effect of earthquakes are tsunamis. They do not have to be very large to cause damage. Even if a small tsunami wave reaches a coast, it can still be larger than wind or tidal waves, which may cause small to moderate damage at coasts and coastal structures such as marinas, harbors, can cause boats to crush each other and so on... Any small tsunami wave entering a harbor, may cause long lasting oscillations, as it may not easily leave the enclosed area. It must be kept in mind that the wave height of a tsunami will increase as it reaches to the shore, as we described in Chapter 1, water resources engineering section. Side *Note: Construction of walls all around coasts just for tsunami protection is not feasible and it greatly reduces our interaction with the sea. This is not practiced in the world except in rare locations. Instead, tsunami warning and evacuation plans must be in place.*

Geotechnical Concerns and Foundations

Designing and construction of coastal and marine structure foundations are more challenging than building foundations on land. Let's list some common items for consideration:

- Due to the need of resisting significant constant lateral dynamic and uplift forces, in addition to the gravity, seismic, earth pressure and other loads we listed, pile foundations or large and massive foundations that can resist these loads by their weight are needed.
- In deeper water sections where we need to construct decks, the piles need to be considerably long, not only to overcome the section above water plus the water depth but also to have sufficient embedment into the sea floor. This makes piles very slender and buckling of piles and or stability of piles and deck system may be the governing failure criteria. Increasing pile diameters will help piles to be less slender and resist lateral loads by bending.
- Piles experience large lateral loads and moments because of waves, earthquake forces, mooring, retained earth, tidal variations.
- Steel piles, including steel sheet piles, need corrosion protection at marine environment, which makes inspection during the life of the structure more difficult. Timber piles also need coating treatment to protect them from decay. For RC piles, the concrete cover over rebar must be adequate, and for durability the concrete must include cement / admixtures that will have greater resistance to salt water and the water / cement ratio of concrete must be low.
- In many locations that we want to build coastal and harbor structures, the soil may be considerably weak, often due to loose sediment transport from land.
- Backfill is placed below solid type decks. The fill must be permeable and drain water adequately. This backfill can also be subject to internal erosion due to waves and tides, if the water can seep through the wall, which must also be prevented by proper sealing, and arrangement of backfill material grain sizes. Before placing the fill, unsuitable, weak natural soil must be removed. The

fill material might also come from dredging if any, if it is made of granular soil of acceptable properties, which could mean substantial cost savings for backfilling.

- As coastal and marine structure foundations are in water, sometimes short term loads such as during or right after finish of construction may become more critical for soil carrying capacity, before sufficient soil consolidation occurs. This requires using undrained soil parameters, rather than drained condition (See Chapter 1>Soil Mechanics>Strength of Soil>Shear Strength of Soil).
- Possible erosion of sea bottom greatly affects the stability of coastal and marine structure foundations. To determine erosion potential, water velocity that drives soil particles, vs. specific gravity and the particle size distribution of soil needs to be known. The velocity of water is affected by currents, waves, and the movement of water from ship propellers.
- For toe protection against scour, the worst case, lowest sea level must be considered. Factors that affect scour include, wave breaking near toe, wave currents, wave run up and reflection, grain size of soil, seepage beneath the wall or revetment, water movement by ship propellers, effect of flowing debris. For cost reasons, design details may assume failure and replacement of materials at the toe after rare major events, instead of designing for it for the entire lifetime.
- Erosion and sedimentation of the coasts is an ongoing process which can never be fully controlled, mainly due to constant effect of winds, tides, currents, storms / floods and of course waves. The goal of any coastal improvement project therefore can only be to minimize its effects, accommodate natural processes, and stabilize the coasts to the greatest extent.

Construction Challenges

Building in or near water undoubtedly introduce extra challenges, versus building simply on land. Some of the typical issues are listed below in addition to the ones mentioned earlier:

- Any formwork for cast in place concrete, such as decks, must be installed over water, which means slower, more difficult formwork installation. For this reason, precast concrete should be used whenever possible for coastal and harbor works.
- As mentioned, piles must not only cover the height above water and depth of water, but must also be penetrated adequately into the sea floor which means very slender piles. Such piles must be installed in an environment with tidal and wave forces, from an installing equipment that may not necessarily be supported from ground but only floats on water.
- Similar to installing piles, for lifting other components, the crane can be on a floating platform but this makes stable work more difficult, slows down work rate and lowers lifting capacities.
- We may need to wait for low tides, which slows down construction.
- The safety aspects of work due to the items listed and not listed here are more stringent than construction on land which may create further costs and schedule impacts.

CHAPTER 10 - AIRPORTS AND AIRFIELDS

Airports are important projects that impact the economy, social life and environment around them. One large airport can mean a lot for its state or country. Airports include many structures in which a lot of specialized systems operate. Airports include large areas of high performance pavings, drainage and utilities, and special site requirements.

We covered buildings in Chapter 2. Although airport structures do not exactly look like the multistory buildings we discussed, still the concepts apply, because in the end they are also built as concrete and or steel framing and/or truss structures *(although airports often also have wide span roofs which are in many cases designed to make them look as iconic structures, which may involve more irregular and complex designs)*. We also discussed earthwork and highway paving in Chapters 1 and 4 respectively. Therefore, in this chapter we will mainly focus our attention to remaining applicable subjects for airports that are specific to them, and still fall into the realm of civil engineering.

Airports have a lot of highly specialized communication, information, security, baggage handling systems that fall under other engineering fields. Therefore, civil engineers working at airport construction sites, who will mainly focus on construction of structures and/or airfield paving area, must still know how and where these systems are installed on structural components, and their interaction with structure, architectural and other items for coordinating them as the general contractor.

PLANNING

Planning of an airport starts with determining the demand for passengers and cargo and estimating the required capacity, for a given region. Capacity estimation is both an art and science, because so many real life variables play role. Planning of airports are affected by many factors including:

- Demand of usage in terms of passengers and cargo
- Macroeconomics, political, administrative, legal, land use and social factors
- Project cost and schedule, and the expected revenue
- Environmental and noise concerns, effects on wildlife. Especially noise aspect can create issues with surrounding communities.

DESIGN and CONSTRUCTION

After planning is complete, which means the location, capacity, approximate area and the preliminary budget and schedule is determined, design can begin. Below, main components of an airport are listed. First few, such as the runways, taxiways, pavements, drainage, utility system are the civil work portion.

Runways, Taxiways

Number of runways and their sizes are primarily affected by the required airport capacity and the peak traveling volume, including all arrivals and departures that consider the number of aircrafts, their types, sizes. Number of runways and their lengths have the biggest impact on airport size and cost due to significant amount of earthwork, drainage and pavement work that will be required. The criteria for geometrically laying out and locating runways, taxiways, aprons, parking areas etc... are shown in sources such as US Federal Aviation Administration (FAA) Airport design document AC 150/5300-13B. Geometrically coordinating all these per a lot of strict requirements is a challenging task.

Let's list some of the important items about runway design:

- Runways should take advantage of prevailing winds, to run as parallel as possible to them. According to FAA, and European Aviation Safety Agency, a runway must be usable 95% of the time to have 95% prevailing wind coverage, and if this cannot be satisfied then another runway in different direction is required, so that the sum would make 95%. So in other words, the crosswinds should not exceed 5% per the allowable values given in FAA standard.
- Topography, such as mountains, hills, the terrain slope, soil conditions, manmade structures nearby also affect the design. These also affect the paving, earthwork and drainage work, which are two of the costliest items for an airport project.
- Runway lengths depend on the distance needed by the planes to land and take off. Design aircraft is the plane that will need the longest landing or takeoff distance (worst case). Takeoff distances are usually longer than of landing. Since the start of commercial flights about a hundred years ago, in the first few decades runway lengths increased due to increasing size and speed of aircrafts, but then the average tended to stabilize or even started to decease, due to efficiencies achieved in aircraft design. Runway lengths vary. For a very small airport, it can be only around a thousand feet, while for large international airports it might be over 15000 feet.
- Minimum distance of anywhere between 2500 to 5000 feet will be required between parallel runways per FAA, depending on the equipment and flight monitoring systems used.
- Runways, similar to roads as discussed in Chapter 4, must be sloped from centerline of the runway towards the sides, to prevent accumulation of and drain rainwater. As for longitudinal (main) direction of runway, ideally a grade of 0% should be made but FAA allows up to 2%. The limit on slope is also considering towing of planes if needed. The reason for not doing a 0% may be the cost savings from excavation or backfill by following the natural terrain as much as

possible, because runways are wide and long surfaces. But at the same time the extra gradient will increase the required runway length too, so the optimum must be found.

- Around the runways and taxiways, there are object free zones, where no object can be present on the ground. This is different than an obstacle free zone, which means an area where no obstacle can be present in vicinity of airspace where aircrafts approach or climb.
- Taxiways are the connector roads between terminal parking gate areas, runways and service parking locations. Taxiway geometry should be as simple as possible, without complex intersections to avoid causing pilot errors. Taxiways should be wide enough to allow the planes to be steered comfortably, by assuming that the cockpit will be on the centerlines, with a certain amount of allowance for wandering. Per FAA, at curves the generally accepted nose gear steering angle is maximum 50 degrees. Taxiway layout also depends on the wing tip clearances between multiple aircrafts that could be traveling close to each other.
- Apron, also called terminal apron or terminal gate, is the place where aircrafts park adjacent to terminal building while loading, unloading of cargo, boarding and unboarding of passengers, fueling, aircraft servicing. On aprons planes may also park short or long term. Wingtip clearance of the design aircraft, the position of how the planes will park or maneuver, affect the layout. Passengers often board or unboard through passenger boarding bridges at terminal aprons.
- A remote apron is where the planes can park and be serviced for long term. It can also have a hangar, where the planes can enter and exit, which is called a hangar apron.
- Remote aircraft parking positions are used for parking aircraft at a distance from terminal buildings. To and from this location passengers may be transported by bus or simply by walking.
- Vehicle service roads are for use of ground vehicles to move around the airside.
- The unused area (the area other than runways, taxiways, aprons) are called islands. They can be paved or not. If paved, then they must be clearly marked as unusable area to avoid pilot errors.
- Runway, taxiway, apron design should consider the effects of possible future expansions.

Pavements

Pavements for highways were discussed in Chapter 4 and the same general principles apply to airports except their own requirements, and with much higher load demands, which also means far more thicker pavings. Similar to highways, airport pavement can be bituminous/asphalt (flexible) or concrete (rigid), where concrete pavings are used for heavier loads. Airfield pavement design life, type, thickness, strength and stiffness are affected primarily by the types, sizes and weights of aircraft that will use the airport and the number of estimated annual departures per year. Other factors that affect pavement design include construction cost and schedule, environmental factors, readily available materials, ease of maintenance and future anticipated traffic change.

Pavement must be firm, able to support heavy and repeated loads from aircrafts including the heaviest aircraft, abrasion from wheels, deterioration from adverse weather. This means the subbase and subgrade must be able to support heavy and repeated loads. Airport pavings must be free of bumps and dents with strict tolerances, as this is essential for aircraft safety, but they must also be skid resistant to provide good traction for wheels, even under adverse weather with rain, snow or ice. For this purpose, grooving or another surface friction treatment method can be used. For aircraft weight, the departure weight is used in design, since planes are heavier with full fuel tank.

Soil stabilization can be applied to base and subgrade, as discussed in Chapter 4. Per FAA Airport Pavement Design criteria, in document AC 150/5320-6F, it is indicated in section 3.6 that a stabilized base course is required, if the aircraft in design traffic mix weighs 100,000 pounds (45,359 kg) in its maximum takeoff weight or more, which is referred to in both flexible or rigid pavement design sections. To give an idea, this weight is far below the max. takeoff weight of a commercial airliner carrying 100 people or more, which means, for commercial airports soil stabilization will be required.

For a given tire load, rigid (concrete) pavement distributes load more uniformly to the subbase and over a larger area. The slab itself provides the main load resistance, which means less stress on the subbase. A flexible (asphalt) paving will flex and transmit the same load in narrower, more concentrated area, which means more stress on the base and subbase and subgrade. If rigid paving is used, it is poured in sections, separated by control joints. Concrete pavings include reinforcing steel to make adjacent slabs to act together under loads, to transfer loads to the base without cracking and also preventing temperature cracks. Isolation, control, expansion and construction joints are used in various places throughout the entire area, between paving segments or between paving and other features such as drainage components. Types of joints were described under structural terms in Chapter 1.

Utility Systems

A significant amount of utility and drainage lines are required for airports, which include:

- Storm, sewer and industrial waste drainage lines
- De-icing lines
- Water supply lines
- Natural gas lines
- Conduits for electrical, in-pavement heating, in-pavement lighting, data and communications

All these systems require coordination with each other and the earthwork and paving work. In addition, utility buildings such as sewage treatment, pumping, electrical stations will also be required.

Drainage

Drainage was also discussed under Chapters 1, 4 and 7 and many general principles and practices also apply here, which will not be repeated. As always, the design of the drainage system starts with estimating the surface runoff amount. The system must be designed to require the least amount of maintenance. The impact on the environment and possible future expansions must also be considered.

Storm and de-icing drainage system is important for safety of planes on runways by removing water to prevent them from traveling on ponds of water, and also for long life of pavement by preventing water to seep through cracks and weaken the pavement, base and subbase. Removing water from airport pavement is more challenging than removing it from highway surfaces because of far larger paving area and the water must still be removed fast. Because of this, in addition to an effective drainage system, properly designed and applied pavement grates over these large surfaces are also critical.

Per FAA, AC 150/5300-13B, open channels and watercourses should be located away from runway approach and departure paths, to prevent a plane from falling into them if it travels off the runway during an emergency. Similar to highways, below the paving, sub drains are also used to drain the base and subbase materials. On runways, apron and taxiway areas, water is removed by drain grates, which convey water to storm drain pipes. These drains are installed level with the paving, to provide smooth ride for planes. Grates and their connections to paving must handle heavy, dynamic tire loads.

Icing is a problem in all airports except the ones located in warm climates. Additional variables must be considered for ice removal. For example the drain sizes must be larger, to accommodate for ice pieces that may block the flow, in addition, they must be more in number. When ice accumulates, deicing chemicals or heating are the most common methods to melt the ice. To install the heating system adds to the difficulty and cost of paving design and construction.

The components we discussed so far fall under the civil work components of airports. Now let's continue with other special components of airports.

Passenger and Cargo Terminals

For large airports, architectural aspects play significant role when designing passenger terminals, to serve as an icon. Only this aspect may prompt the owner, (i.e. government), to build a structure that may be larger and/or costlier than needed, as long as it can function as an icon *(for example a large international airport is the first place a foreigner sees when entering to another country).* Commonly wide span roofs, exposed steel and cable elements are preferred for such structures.

The expected number of passengers, the amount of cargo that needs to be handled, determine the number and size of terminals, which are coordinated with the layout of runways, taxiways, aprons etc... to fit them together within the airport limits, at the same time satisfying the code requirements. Terminals include check-in desks, security and passport control points, transferring passenger areas, baggage handling, shops, lounges, customs area, service areas, administrative offices, passenger movement corridors, departure gates and waiting areas, boarding bridges...

Construction coordination of baggage and cargo handling systems is challenging, as they interact with many trades, especially electrical, IT, security and architectural. For small airports, where only one or few flights are present, this is simple and the system runs between several points at most. But for large airports, the efficiency of cargo handling system is important for the overall operational costs of the airport and the airlines. In large airports, baggage handling is largely automated due to the number of check-in desks, airlines and destinations. Due to its complexity, cargo terminal and cargo handling system design may also be done by airlines directly. The system must also be flexible to future changes.

Air Traffic Control Tower

This is where air traffic is managed from, with complex computer and electronics systems. From construction point, these systems' installation must be coordinated with the structure, conduits, ducts, architectural systems. As always, coordination is not only geometrical but also considering schedule. An air traffic control tower's foundation is pile or very heavy mat foundation, to prevent overturning from wind and earthquake. Control towers are given high importance factor in earthquake design, as these are among the most critical structures that need to stay operational in case of natural disasters.

Communication, Security, Navigation, Luggage System

These are complex electronic and data systems that are also integrated with flight control management. Their installation is coordinated with utilities, architectural, structural components.

Support and Maintenance Facilities

These are buildings for purposes such as: aircraft and airport maintenance, firefighting, flight kitchens, engineering building, snow and ice equipment, fuel storage, hangars, and others.

Car Parking, Public Transport Access, Hotels

Parking structures, connections to public transportation such as metro or bus stations or metro lines, hotels must be built near airports, which can also be treated as separate projects.

CHAPTER 11 - VARIOUS STRUCTURES

In this chapter, we will look at several other structure types that we did not yet discuss, that might somewhat be considered as more specific ones.

AERIAL ROPEWAYS

Aerial ropeways (can also be called as cable cars, passenger cableways) can be very effective in transporting passenger, cargo or materials at mountainous areas, places with difficult access, congested sections of cities, ski resorts (i.e. including chairlifts) mines and various other industries and so on...

A ropeway for quickly transporting passengers from an urban location to a mountainous recreational area. In such situations ropeways offer indeed great efficiencies.

Ropeways have many advantages including:

- Effectively planned aerial passenger ropeways can be a viable mass transportation alternative in cities, although they are still mostly used in rural areas for recreational purposes.
- Ropeways have small negative impact on environment, such as the need for cutting some trees for towers, maintenance and emergency access roads, and during construction the negative effects may be more, but overall they are much more environmentally friendly than roads and using vehicles to travel between the same start and end points.

- Their initial capital costs per travel length are higher than roads but lower than underground metro. In the long run the efficiencies of aerial ropeways pay off. The life cycle operation and maintenance costs of ropeways are lower than roads or railroads.
- They are efficient way of transportation. In comparison to roads or rails, the same amount of passenger or material can be transported using less energy overall between two points.
- Aerial ropeways can also be alternatives to (or supplement) tunnels or bridges when there is the need for passing over a valley, river, lake, or mountainous areas, where it would be just easier to pass distances from air, and at locations where sensitive natural or historical etc.. features exist, if roads would disturb landscape.
- Their operation does not slow down in weather conditions such as rain, fog or snow, although during very high winds it can be interrupted due to passenger comfort concerns.

Disadvantages / downsides include:

- The reasons for them not being common in densely populated urban areas are mainly that, although they can just fly over the densely congested areas, they may visually be unpleasant, noise pollution, discomfort for the people and residents below and even privacy concerns.
- Under high winds, unlike road or rail systems, ropeways may need to be shut down.

Industrial ropeways to transport materials are quite different from passenger ropeways. They are specifically designed to handle the materials they are supposed to transport, with different safety requirements, and considerably different end stations. What we mean in this section when we say ropeways are the passenger ropeways. There are different types of systems of each, for different settings and purposes, with different capacities, but here we discuss them in general without covering details of each group. The capacity of passenger ropeways are measured in transported passengers per hour and the capacity of industrial ropeways are measured in transported tons of materials per hour.

In Figure 1 below, you can see the free body diagram of a section the rope. As you saw in the very beginning of the book, equilibrium, in other words, Total Fx = 0, Total Fy = 0, and Total Mz = 0 must be satisfied, from where we solve the support reaction forces from towers and the tension in the cable *(Note: We did not show the specific method to solve cables in this book)*.

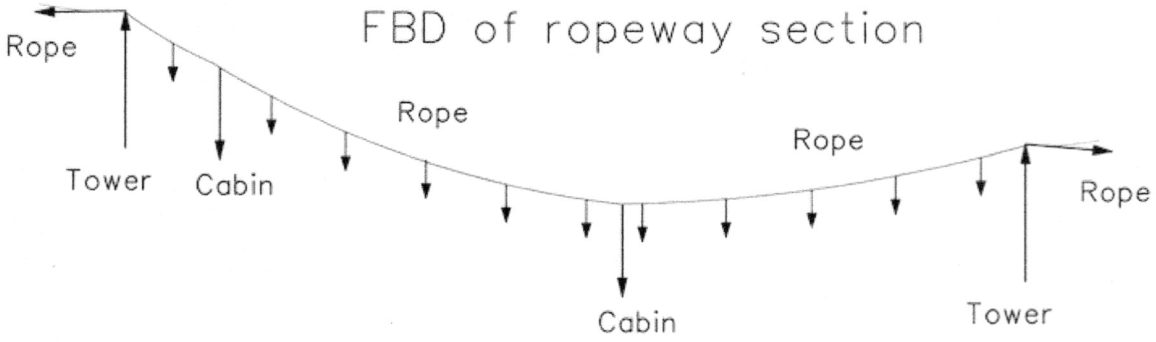

Figure - Ropeways 1

The main components of a ropeway are:

- Cabins (also called carrier, car, gondola)
- Ropes (cables)
- Towers (also called trestles)
- Tower foundations
- Stations at both ends
- Rope handling mechanism, grips, hangers at the top of each tower
- Machines and control systems at each station

The items below affect each other simultaneously during the design of ropeways:

- The general route of the ropeway, and based on this, the distance and elevation difference between two end stations.
- Number and height of towers and distances between them. As the distances between towers increase, the height of towers and the tension forces in the ropes must also increase, in order to keep this minimum clearance between the cabin and the terrain or other structures, because the cables will sag more. Therefore, any cost savings achieved with lower number of towers will be countered by the cost increases resulting from constructing each tower higher, stiffer and stronger, with larger foundations and stronger, heavier cables to resist higher tension forces. Sometimes, the vertical height clearance is satisfied simply by cutting a portion of the terrain, instead of raising tower heights or decreasing distances between towers.
- Length, number, maximum sagging, unit weight, yield strength, stiffness, thermal elongation of ropes, which are made from bundled strands of steel wires. The weight of rope acts as uniformly distributed load as seen in Figure 1 above. Due to high stresses and long spans, the elongation of the ropes need to be checked both for elastic and permanent deformations. The bending stresses must also be considered, as there is bending at the cabin locations and when the rope wraps around the bullwheel. Some systems have single (monocable), while others have multiple ropes in parallel. In multiple rope systems, there are suspension (carrying) and pulling cables. The rope(s) are kept in tension, whether the system is loaded or not, by counterweights or hydraulic systems. The tension force is constantly monitored and adjusted due to changes in loading and temperature conditions, so that the force remains within the design limits, to prevent too much cable sagging or too much tensile force. A factor of safety of anywhere between around 3.5 to 5.5 is commonly seen, for the tensile force on the rope vs. the yield strength and for ropes with different duties. Cable sags are measured vertically, at the midpoint of each section between towers. The tension in the rope due to self weight of the rope and due to weight of a cabin at a particular section is proportional to the square of the horizontal distance between the towers, and inversely proportional to the sag amount. A sag amount of anywhere between 1/30 to 1/50 of the span length is common.
- Potential dimensional conflicts of towers and / or end stations with the man made structures or features, railroads, transmission lines etc... This can especially be an issue in cities. Even if the tower locations or stations do not conflict with anything else, coordination must be made with appropriate authorities when the line runs adjacent to structures, roads, railroads and electrical

transmission lines or cross them. For example if crossing an electrical transmission line is unavoidable, the electrical lines might need to be buried underground.

- Distance between cabins, total number of cabins and number of cabins in each span. Cabin weights when full and empty to produce worst case loads for the towers, foundations and cables. Cabin dimensions.
- Structural analysis for overturning, buckling, vibration of the tower and the displacements.
- Settlement, differential settlement, overturning, sliding, uplift, rocking analysis of foundations. The foundations must resist to dynamic loads, which can also effect the supporting soil/rock in the long term. In addition to determining the soil/rock strength and properties, slope stability analyses must also be performed for the entire route to analyse the risk of slope failures, landslides or rockfalls when the towers are in place and operational with full loads.
- Structural analysis of rollers, bullwheel, grippers, hangers.
- Dynamic forces due to wind (which increases with height), due to service or emergency breaking, deceleration and acceleration of the loop. Dynamic forces cause swinging and oscillating motions of the ropes, towers, cabins, grips, hangers, vibrations in ropes. The cabins swing similar to a pendulum. The towers and ropes must not only withstand the dynamic loads, but the displacements must also be below certain limits, for the comfort of passengers. Therefore, extensive dynamic analysis is needed, in addition to static analysis.
- Seismic analysis of the towers and their foundations in earthquake zones. Seismic forces and other dynamic forces as mentioned above may occur simultaneously. The seismic design variables, earthquake analysis methods that we discussed in Chapter 1 are also used here.
- Fatigue loads, for a given number of loading cycles for every affected component.
- Any possible impact loads.
- As ropeways work with cables in tension, the route between two end stations must be fairly straight. From tower to tower, only a maximum of 1-2 degrees or less of alignment change may be allowed, and cumulatively a little more. Each tower and the components at their top must be designed to handle side loads in such cases. This limitation must be considered initially when planning the route. If larger angle changes are required, angle stations are necessary.
- Access roads for emergency, inspection and maintenance as applicable, depending on the size and nature of the project. An emergency plan must be in place to safely and quickly evacuating the passengers and return them to a station or safe location in case of an emergency, and the route must be designed to accommodate this requirement, so that all locations are accessible for both maintenance and emergency.
- In cold climates snow plus frost load on cabins, cables can add up to non negligible amounts.
- If there will be night operation, adequate lighting system must also be built, both for operation and also for emergency situations separately.
- Maximum and minimum air temperatures that are expected during the life of the system.

- Friction coefficient between the ropes and the bullwheels and rollers.
- Corrosion protection must be provided for applicable components. Lightning protection must be provided to the entire system. Fire protection measures must be installed.
- Mechanical, electrical design of rope driving system in stations, and the design of stations, with its own power, size and safety requirements (i.e. emergency shutdown, preventing rollback, overspeed and many more). The power system must be able to handle the acceleration and deceleration of the dynamically loaded system for the worst loading conditions and provide sufficient traction for the design operating speed. One contributing factor to energy efficiency of ropeways in comparison to roads or railroads is that the cabins traveling downhill will also help the hauling effort to pull upward moving cabins, as the system is a loop. This type of efficiency is not possible with roads or rail systems, except when a funicular system is arranged as two sided pulley system, where the opposite sides pull each other while traveling up and down, but in that case many benefits of ropeways are absent.

...and as for any type of project, the required capacity (demand) of the system, cost to benefit analysis, construction schedule, administrative issues and environmental impacts and other applicable aspects as listed under Chapter 1 > Construction section. As can be seen from the number of variables above, which affect each other simultaneously, design of a ropeway for a functional and cost effective system is a highly challenging and specialized task. The analysis must be performed for each component, each span separately and also for the entire loop working together.

A rope driving station

Chairlifts are frequently used inside ski resorts instead of cabins. Their loads are lighter which can enable smaller stations and towers. Surface lifts are even lighter and they follow the terrain slope, where the passengers sit and ski at the same time while being lifted higher to climb the slope.

FLOOD MITIGATION SYSTEMS

Humans historically settled near or close to rivers, streams, for readily available water, which is essential for sustaining life. Living close to rivers also bring the risk of floods however, since these are the natural drainage routes that the nature drew itself. The best flood mitigation measure therefore, is to avoid building in flood zones in the first place. Through hydrological studies and models, previous floodplain elevation data, these zones are extensively studied in many parts of the world, which receive or likely to receive human development.

For example, in the US, applicable local or regional entities such as municipal utilities or public works, regional flood control districts, regional planning commissions, FEMA regional offices, Federal agencies such as National Oceanographic and Atmospheric Administration, Federal Highway Administration, US Army Corps of Engineers, Natural Resources Conservation Service and US Geological Survey will have the appropriate data for a given area.

Per FEMA 551 - Selecting Appropriate Mitigation Measures for Flood Prone Structures, floods can cause damage through hydrodynamic pressure, erosion, inundation, debris impact or scour. Below we will discuss the flood mitigation measures, for structures that are likely to be exposed to floods. Per FEMA 551, flood mitigation measures include the following:

- Drainage improvements: such as widening natural channels to increase capacity
- Flood barriers: such as flood walls, levees
- Wet floodproofing: allowing water to flow through basement level
- Dry floodproofing: isolating of structure entirely form water
- Elevating the structure: making modifications to the structure to raise it entirely

Due to variety of geographical locations, river geometry and flow quantities, developments and alternatives, the whole picture must be considered and the optimum solution must be found. If none of the solutions above seems to work, then if the structure is a small and light structure such as a small house, it can either be moved away by wheeled vehicles or it must simply be demolished and built at a new location. Before all these however, the best solution of course, is not to allow development in the flood zone in the first place.

The flood protection depth is determined such as, from the expected base flood level, subtracting the lowest adjacent grade in the zone to be protected, and to that result, adding a safety distance which is known as a freeboard distance. The base flood level is determined from hydrological studies.

The flood velocity is affected by the terrain slope and features. Rough terrain with a lot of plants will absorb water well, have slow flood velocities and reduces the amount of floodwater in the first place. Impervious surfaces not only causes faster flood velocities but it also means more surface runoff as mentioned before, under hydrology in Chapter 1. The velocity of flood water increases in mountainous, sloped areas, and as the location is closer to the original stream or river flow route. Per FEMA 551, flood water velocities can be as high as 10 feet per second or even more in extreme cases, and even when it appears to move at a very slow speed such as 1-2 feet per second, it can exert very large forces on buildings. The calculation of exact flood water velocity for a certain location is difficult but estimated ranges of values are used based on typical previous data and hydraulic models. Especially for coastal areas accurate estimate is more difficult, due to more possible directions of water flow in those areas.

Flowing water not only exerts more force than static water but it also causes scour and erosion. The faster the water velocity and the heavier the structure, the more erosion and scour that the building foundation will experience. Now let's discuss each of the flood mitigation measures we mentioned above in more detail. Note that sometimes more than one method will be suitable.

Drainage Improvement

By drainage, here it is meant the water channels, rivers, streams that move surface water to lakes or seas. The improvement may mean enlarging culverts or cross sections of streams and rivers, straighten channels, building underground water storage areas. The major disadvantage of this method is that, while it improves the drainage capacity of the zone in question, it may create problems and disturb equilibrium in downstream or upstream of the river or stream. This is in fact similar to ecological impacts of dams that we saw before in Chapter 8, which might be considered as another means of managing floods. Diverting water and just changing its course can also be considered a form of drainage improvement, with similar ecological impacts to be considered.

Planting a lot of trees near the rivers can be viewed as a solution of flood mitigation as trees first hold some rain water before even it falls on the ground, increases water absorption and holding capacity of the soil, and also slows down water speed both by improving the surface roughness and through their own bodies. This solution has no negative impacts, but it may take a while for trees to grow and function as described here. Most likely however, this will not be a primary solution by itself, but a complimentary one, by reducing the burden on the other mitigation systems to be built.

Flood Barriers

Building barriers around rivers and structures is another flood mitigation measure. These include levees, flood walls, or temporary barriers. The main advantage of barriers is that they prevent water from reaching the structures altogether.

Flood Walls and Levees:

A flood wall is built along a river or a waterway to contain rising water levels during extreme rain events. Flood walls are designed to resist overturning, sliding, tilting, bearing, differential settlement, uplift failures. The forces to be resisted are the self weight of the wall, lateral forces from earth and water, hydrodynamic loads of flood water, wave action, earthquake loading, equipment loads during construction which are to be used to find worst cases. They can be of sheet piling (as secant piles), masonry, concrete or a combination of these and various cross sections, which can be as inverted T or I shaped, cantilever or look like a gravity wall. For aesthetical reasons, decorative bricks or blocks can also be used at top of flood walls. Flood walls can also be supported by elevating earth from behind. Pile foundations may also be needed, including battered piles.

A levee is a compacted earth structure, built at the edge of a river to prevent floods. We had introduced dikes in Chapter 9. Levees are similar to dikes, with the main difference that they are along rivers, instead of shores. Levees require more land area than floodwalls, as they are soil embankments sloped to reach to a certain height. They must be built by compacting the soil to high degree and they must be made of impervious soil such as clay. It is not acceptable to just pile the soil from the river towards the sides to make a levee. For stability and erosion resistance, the slope of levees must be low, such as 2 or 3 horizontal to 1 vertical, which means, a levee cross section base would cover about 6 times width of its height. For this reason, at locations with right of way restrictions, flood walls will be preferred. Since levees are made of earth materials, they are more vulnerable to erosion than flood walls. Therefore if there is not enough space, or if the water velocity is too high to create strong erosion effects, flood walls will be preferred, although more costly. As floodwalls require construction of a structure, their cost are usually higher than levees, except in cases where the soil required for the levee must be brought from far locations. One possible negative impact of levees and flood walls is that they change local natural drainage patterns downstream, which must be taken into account during design. Both levees and food walls require erosion and scour protection, as their bottom is constantly subject to flowing water which can undermine the structure and eventually cause failure.

The reasonable height of flood walls and levees are several feet. Beyond that, the cost of construction will significantly increase with each additional foot (when you also think about the foundation). Also for levees, there may not be enough space when we increase the height since it means increasing the base area too. If the height of flood wall or levee would not be sufficient for a certain location, then other flood mitigation measures discussed in this section can be considered, because overtopping, even by a small amount, is especially a bigger concern for levees, due to erosion potential at their top.

If flooding continues more than few days, even if the flood wall or levee contained the flood, water seepage through the soil from below the levee or flood wall can start to be a problem for nearby structures. For example, there can be considerable pressure on the basement walls and uplift force to the whole structure. Sump pumps placed around the structures can be used to remove this water. This seepage from soil can be prevented or reduced by installing cut off trenches or sheet piling, or extending foundation edge vertically down to serve as barrier to seepage below the flood wall or levee, or make grout curtains. The seepage through or around the ends of the wall must also be considered. In short, seepage control is a primary aspect of flood wall design.

The entire wall cannot be poured at once. Between sections there must be waterstops. Any utility lines that cross a floodwall must be separated from the structure to allow relative movement, with a proper expansion joint, and to prevent seepage by inserting waterstops. Groundwater which is close to ground surface and poorly controlled seepage below the foundation will cause pore pressure build up and reduction in bearing strength of the soil. Excessive pore pressure will even cause quick condition (see liquefaction in soil mechanics in Chapter 1). Toe drains made of perforated pipes in sand and gravel, which run parallel to the wall along the toe, or the larger trench drains can also be used to relieve seepage pressures. Land side seepage berms can be used which simply increase seepage path and reduce hydraulic gradient (*see soil mechanics, groundwater*), although they are not suitable in congested areas due to space restriction. Ground improvement to reduce soil permeability can also be applied. If rock is at very shallow depth right below the foundation, its joints may need to be grouted.

Temporary barriers:

There are also temporary, alternative barriers such as water filled tubes, temporary metal walls, disposable gates, sandbags. These can be stored and reused in future floods and they are far less cheaper than the permanent systems, although the protection they provide may not be as effective. They can also be used as supplementary measures to permanent barriers.

Wet Floodproofing

Deliberately allowing water to flow through the unoccupied basement level of the structure, balances water pressure inside and outside, which reduces flood damage, although for high velocity floods or wave action it is not effective. This is a very low cost method of limited use and effectiveness, since it simply involves allowing water to pass through. Before the flood comes, the basement area must be cleared. The items that cannot be relocated such as utilities, water heaters, electrical panels etc... must be protected. Pumping of water too soon after the flood is over, may lead to structural damage because there will still be saturated soil outside, which creates pressure difference. This is not an appropriate method if the flood will last longer than a day, due to potential of damage to the structure. Wet or dry floodproofing method are not suitable for manufactured homes, or for any structure in poor condition.

Dry Floodproofing

This is a low cost flood mitigation method which means sealing the structure against water from outside. For this, waterproof coatings, impermeable membranes, watertight shields can be used. Similar to wet floodproofing, this method is not effective against high velocity floods or waves. Dry floodproofing can withstand flood a little longer than wet floodproofing. This may not be a suitable solution for wood or metal frame structures, as they are harder to make watertight. It is also not suitable for any structure where the flood level is expected to be high, such as 3 feet depth per FEMA 551, because as the water level increases, larger hydrostatic forces need to be resisted by the basement walls, which may cause collapse of the walls. Sewer backup prevention must also be in place to prevent water from flowing backwards in sewer system during floods.

Elevating the Structure

Elevating is literally raising the structure above expected flood water level, by rising the height of the structure either in its entirety or by simply adding an elevated floor on top of the structure, so that its habitable areas remain above the expected flood water level. This can only be applied to small footprint, 1-2 story structures which are in good condition. Structures with basement, or especially slab on grade structures are more difficult to elevate than structures with crawlspace or columns/shear walls. Utility system components must also be relocated or elevated, and access to the entry of the structure will need to be modified as well.

OFFSHORE STRUCTURES

Offshore structures considerably differ from the other marine structures listed in Chapter 9, as they are built far from the coast, at locations where sea depth is much greater. They can be classified broadly into two as floating and fixed platforms. Floating vessels without foundations work with the principles of ship design which is not the subject of civil engineering and we will not discuss them here. We will discuss fixed platforms, which can be grouped as:

- Gravity platforms: These are made by fixing large concrete blocks to the sea bottom, which are then attached to the structures to support them. This method can provide the strongest and most stable support for platforms under most severe wave conditions. Another advantage of these platforms is that the support system is very durable and requires little to no maintenance.
- Jacket platforms: Built as traditional trusslike tower structures with pile foundations at the sea bottom. The truss structure is made of high strength, large and heavy tubular steel members.
- Tension leg platforms: The platform floats on water and is connected to fixed foundations at the sea bottom by steel pipe tendons that are in tension.
- Jack-up platforms: The platform structure is moved up and down by jacking units through installed vertical legs which are fixed to sea bottom. The advantage of these platforms is that once the platform finishes its work in one area, the legs can be jacked up, while the platform floats like a ship, and the whole platform is pulled to another location by another ship. The legs are made of high strength steel tubular members connected as trusses. Three or four legs are used. To find the load balance between legs while the platform is also affected by dynamic wave loads is critical. Offshore wind turbines can be installed this way.

Construction on or under water is much more difficult and costly than on land. Therefore, offshore structure components are fabricated and assembled on shore to the greatest extent possible, but this can only be up to a point since transporting large sections would also be difficult and costly.

For offshore structures, in addition to gravity, seismic and wind forces, we must also consider large dynamic wave forces, thermal forces and in cold climates ice forces. The effect of wind and waves are far greater for offshore structures, than the other marine structures listed before. Under these great dynamic forces, matching of period of waves and the structure (resonance) must also be prevented.

Since the structure itself is much elevated with respect to its foundation, this can create high moments at the foundation at the sea floor under the action of lateral wave, wind and seismic forces. The geotechnical survey and construction of footings are far more challenging and costly than the one done on land, because here we must first overcome the depth of water and only then perform the work, at the sea floor which is underwater, under great dynamic wave loads, that necessitates special equipment. Particularly for geotechnical survey, obtaining good quality samples is very difficult because of the reason above and also because of a high amount of water pressure difference between the depth from which the sample is obtained and the water surface. To precisely locate the desired foundation location is also more challenging in the sea, both for geotechnical survey and during construction. The properties of the soil and how it behaves under dynamic, cyclic loads and liquefaction are especially critical in the case of offshore structures. Both shallow or pile foundations are possible. In case of shallow foundations, they keep the structure stable with the help of their own self weight. They can be made on land and brought to final location by floating over water. Similar to land structures, for offshore structures shallow foundations are used where the sea floor is relatively stronger.

Different stages of construction will necessitate separate analyses, as the structure will be incomplete and its load resisting ability and self weight distribution will be different than the completed assembly.

Rust and corrosion protection play an important role in offshore structure design. For this purpose, stain free steel and / or cathodic protection methods can be used.

As design and construction methods improve, the number, size, types of uses and the depths over which we can build offshore structures will increase, and this is a growing branch of civil engineering and construction. Today mainly oil, gas and wind energy companies lead research and development in this area. The number of projects, and investment from other fields are also increasing.

ROLLER COASTERS

Roller coasters are inseparable and most significant components of amusement parks. Thousands of coasters in the world are operational today, in which millions of passengers ride every year. Roller coasters, which exist since well over a hundred years, have been improving all the time, and this trend will likely continue. The main reason that they are the most popular components of theme parks is probably because of the appealing look they have and our bodies being very sensitive to change in gravity and the fear against falling is a natural instinct we all have.

Design

Roller coaster design involves:

- The geometry of the ride and tracks
- Components (we will list below)
- Structural supports
- Foundations

Roller coasters have complex geometry in 3D space. As the coaster moves along the tracks, it continuously changes direction which causes dynamic loads on the tracks and the supporting structure in many directions in 3D.

1g-force, is the force due to earth's gravitational acceleration, g. So then 1g-force = mg, as in Newton's Law, F=ma. Human body is able to withstand up to a few times of g as a maximum (not necessarily downwards only), which is also kept in mind as a design constraint.

A roller coaster works by gravity. Potential energy (PE) and kinetic energy (KE) are continuously converted into each other as the coaster drops and rises, minus the constant loss of heat energy (HE) due to frictional losses. The train is first lifted an initial highest position, which may or may not have zero velocity at that instant. If zero velocity, this means zero kinetic energy and its starting energy is entirely potential energy. As you may remember from physics classes, the potential energy is equal to:

$$PE = mgh$$

where, m is the mass of the train, g is gravitational acceleration and h is the height of train's center of mass measured from a base elevation such as the lowest track elevation. As the train starts to move down on tracks, it will start to have velocity and this means KE, because kinetic energy is equal to:

$$KE = 1/2\ mv^2$$

where, m is the mass of the train, and v is the velocity. So, the PE will gradually be converted to KE.

And let's designate heat energy due to frictional losses as Q. So, we can write the conversion of PE and KE that we mentioned above between any points 1 and 2 as:

$$PE1 + KE1 - Q = PE2 + KE2$$

Which means, potential energy plus kinetic energy at point 1, minus heat energy due to frictional losses, is equal to the new potential plus kinetic energy at point 2. So, as the train moves along the tracks, its PE+KE gradually decreases as seen in the equation above, due to losses in heat, until it reaches the end of the ride. Note that it is the <u>PE+KE</u> that is decreasing <u>overall</u>, due to frictional (heat) losses. But during the ride, PE and KE individually will each rise and fall many times by transforming into each other, as the train climbs (PE increase, KE decrease) and falls (KE increase, PE decrease)...

Frictional losses as heat energy includes the frictional losses between the wheels and the tracks, the friction within the train itself due to turning of its wheels, and the aerodynamic resistance of air. Each of these are found by the formula Energy = Force x distance. Distance is the same for all, which is the distance traveled by the train. The friction force for each component is calculated separately and then (they may also be added first) they are multiplied by the distance to find the total energy lost due to friction. For example, the friction force between the train and tracks is dependent on the friction coefficient between the wheels and the track. The friction force within the train itself is another coefficient which should be provided by the train manufacturer. The air friction is proportional to the front area of the train including passengers, air density, the coefficient of air friction, and the <u>square</u> of the velocity, which means that the air friction force increases dramatically as the velocity increases. Then, after finding all friction forces, we find the total heat energy Q lost between points 1 and 2.

Then into the equation above, we insert the Q, the heights at point 1 and 2, gravity g, mass of train m, and the velocities v1 and v2 at point 1 and 2. This allows us to find the acceleration of the train, which then enables us to find force from F=ma, which is the force on the tracks due to acceleration of train. There is also the centrifugal force on curves. In addition to these two forces which always vary in direction and magnitude, there is also the weight of the train, which always acts downwards. So at any given time, the force that the train exerts on the tracks is the <u>3D vector sum</u> of the weight of the train, the force due to acceleration of the train at a given instant (if present), and the centrifugal force at that position (if present). Finding forces on the tracks and finally adding the self weight of tracks to it, enables us to draw free body diagrams at all critical locations and find total forces on the structure that carries the tracks, which then enables us to do the design for tracks and the structure, with proper structural analysis programs. As mentioned, acceleration force and centrifugal force can be in any direction in 3D, so there needs to be sufficient amount structural restraints in all possible directions in 3D to counter these forces, in addition to vertical to resist gravity loads. Finally, by adding the weight of the supporting structure itself, we find the total force and moment demands on foundations. The more spacing between structural columns, the longer free spans must be passed which means stronger supports and tracks must be used, so the optimum must be found. Dynamic, live and vibration loads, fatigue and temperature effects must also be considered. In the US, ASTM F 2291 outlines criteria to be followed when designing amusement rides.

A major constraint when designing the 3D geometry of the ride is the maximum possible distance a passenger can reach by his/her arms or legs in all directions, plus a certain distance for safety, while riding the train or carts. Nothing must be present in this imaginary border around the train as it proceeds, in order to prevent injuries.

The structure also includes a lift mechanism, to bring the train to a high point (give it enough potential energy) before the ride begins, and a catwalk platform, which are walkways installed near the tracks on certain locations, to safely remove the passengers in case of emergency evacuations, and also to enable maintenance. Tracks include breaking system to slow down the train.

The whole system is managed automatically, based on the inputs from sensors. Sensors are installed on many locations on the tracks and the trains to make sure that the ride continues without problems, such as monitoring position and keeping relative positions of the trains if there are multiple of them.

Apart from the roller coaster structure itself, the other important thing is the theme of the ride and the features around it, such as various shapes, figures, landscaping etc... to make it look more attractive. Some of these features also have their own structural supports. The geometrical design of the coaster structure and tracks must also consider these components.

Roller Coaster Operation

Roller coasters are generally the most expensive components of a theme park, which also bring the highest revenue. Therefore, without sacrificing safety, it is important to keep them operational, and they are inspected everyday. There are strict standards for inspection of every component of roller coasters in varying frequencies. During first trials after the construction is complete, everything in the beginning may be tighter, and the trains may go slower. The train must run smoothly on the tracks and stick to them well, to minimize vibration or impact forces to tracks as it moves with high velocity.

Wooden roller coasters were preferred more in the past and now they are less common than steel coasters, but they still exist. They are sensitive to weather and shaking more than steel coasters. The flexibility, height and speed that can be obtained by wood coasters are lower than steel ones. Humidity and dry weather can affect wood, and in turn the fitting of bolts, which may lead to shaking of trains, and this in turn will have negative effect on the structure over time, because of too much vibration and impact. Therefore wooden coasters require more maintenance than the ones made of steel, but this doesn't necessarily mean they have shorter lives, if they are properly maintained.

TRANSMISSION PIPELINES

Transmission pipelines are very effective means of conveying large amount of fluids, such as water, petroleum, natural gas for long distances, either on or off shore. By transmission lines, the need for using tankers or trucks are dramatically reduced or eliminated. Although long pipelines are very expensive endeavors, and incur maintenance and operation costs, in the long run they pay off their costs many times over, if we consider the savings obtained with respect to transporting by trucks or ships. They may run for thousands of miles, may cover entire continents or seas and often span over more than one countries and considered strategic investments for all involved nations and even for world politics. According to Interstate Natural Gas Association of America, (INGAA) website as of 2020, around 300,000 miles of gas pipelines in the US are in service, most of which run underground.

Onshore Transmission Lines

Transmission lines can run on and/or offshore. An onshore transmission line also includes facilities such as control and compressor stations, measuring and pressure regulation equipment which are facilities for smooth, safe and secure operation, but we will focus our attention on the pipeline itself here.

Design:

The first thing to determine is the pipeline route and its capacity. This depends on the demand, topography, cost, available technologies and materials, land use, political, environmental and many more factors, which require input or approval from a great number of officials, public agencies, private firms. After all inputs and approvals are obtained and the route and required pipeline capacity are determined, the pipeline itself is designed, with all its components such as the pipes, valves, storage, pressuring or control facilities, and of course the exact route in greater precision.

For transmission pipelines, steel pipes are used. The pipes may have different thicknesses or coating for different conditions such as operating pressure, diameter and environmental conditions. Most long distance transmission pipe diameters are between 20" - 40". Due to the long distances that the fluid travels in transmission pipelines, the operational pressure here is much higher than of a municipal water lines. Therefore, these pipes, their joints and valves must be made stronger.

The fluid transmitted also makes a difference. For example pipelines for water can be made of lower grade and thinner than steel pipelines, operating with lower pressure. Their connections are different.

As a pipeline crosses long distances, it may be inevitable that at least part of it crosses high seismic zones and even active faults. Things to consider for seismic design of pipelines include:

- Unlike other structures built at a single location, ground displacement criteria can become relatively more critical for pipelines because they span long distances over a large area, which may inevitably include crossing fault lines, where the ground slip may occur. So the major and minor fault lines locations and their possible displacement amounts with respect to the pipes are analyzed. The displacements of course, can happen not only horizontally but vertically as well, just like the movement of different fault types we mentioned in the beginning of earthquake section in Chapter 1. Not only the deformation of ground plates with respect to each other along fault lines, but also the strain in the ground over long distances because of earthquake waves from any direction must also be considered. These permanent ground deformations as a result of fault slips and ground strains can exert large forces on a pipeline.

- Earthquakes can also trigger slope failures, liquefaction, flooding / scour, lateral spreading of soil along the pipe route, each of which must be analyzed with respect to pipeline route as far as the stresses they may produce on the pipeline.
- Seismic ground acceleration is more critical for unburied segments than for buried segments.

Items listed above can either analytically be calculated by equations or simulated as the whole ground and pipeline system by computer using finite element methods by modeling the pipe and surrounding soil with appropriate elements, stiffness values in all directions and defining how the boundary works between the pipe and the surrounding soil. Buckling of pipe should be watched where compressive stresses exist. Joint and connection strengths versus expected forces should be checked. As far as tensile stresses it is usually less of a concern considering the high strength of steel.

To mitigate damages and economic losses and environmental issues that may occur in case of a pipeline breakage from seismic action, various measures can be taken. These can be applied separately or in combination, which are decided during the design stage to minimize cost and schedule impacts, while keeping the pipeline intact to prevent any leaks and - if possible - functional. These measures include:

- Aligning pipe orientation with respect to expected fault displacements, to minimize stresses on the pipeline. It is preferable to align to minimize compressive stresses (which can result from axial compression or bending), which may cause buckling.
- Place pipe in supports that allow sliding in specially designed vaults while crossing the fault line to minimize pipe displacement.
- Make high stress locations flexible.
- Increase thickness and strength of pipe material itself. Note that this causes decreased flexibility, so it must be done where flexibility is less of a concern than strength.
- Use movement joints such as expansion joints, slip joints.
- Install valve points to be able to quickly terminate fluid flow at expected breakage locations, in case the pipe failure occurs.

Fabrication of Pipes:

Pipes are fabricated in factory, to the required specifications per the design. Pipe manufacturing is the subject of metallurgical engineering and materials science.

Construction:

Construction of pipelines is a highly specialized task, performed as if a moving assembly line. An experienced crew can install around a mile of pipeline a day depending on the terrain and weather conditions. Below we list and describe general steps of a land transmission pipeline construction:

Surveying, Clearing Right of Way, Grading:

First, surveying is done to layout the pipeline and establish right of way limits. Any artificial or natural obstacles within these limits are cleared, such as trees, plants, boulders and so on. Then this area is roughly graded to bring it to proper elevations along the route and for straightening sharp slopes, so that the pipes will not have to be bent too much. Typically around 100 feet right of way is needed during construction, centered at the pipe location, with half distance to either side. *(Note: After construction finishes, the permanent right of way will be significantly narrower.)* At the end of the grading stage, erosion control measures such as silt fencing are placed around the work area.

Stringing:

Stringing is the distribution of pipes by trucks from manufacturing plants to their final location and arranging the pipes in position in their connection sequence, according to design plan.

Beveling, Bending and Welding of Pipes:

Beveling is preparing the pipe ends for welding, which is done on site.

In some cases, to fit to topography of the land, and to be able to follow the route curves, pipes can be bent slightly on site, by a special pipe bending machine *(this pipe bending is unlike storm or sewer lines where we use straight pipes and don't need to do bending or use bent pipes because in that case we can use manholes, from where pipe directions and elevations can be adjusted)*. The amount of bending cannot be much for one pipe, in order to prevent damage to pipe or coating, but we can bend multiple pipes in sequence, to make the pipeline follow any route and grade we want. The precise locations to bend which pipe to what degree is determined in the field, after the surveying crew vertically and horizontally lays out the pipe centerline vs. the ground, which is already roughly graded by then.

After that pipes are welded to each other, from both inside by a special welding machine, and outside. Tents can be used to protect the welding area in adverse weather conditions. Welding can continue from pipe to pipe until long sections of up to several thousand feet of pipes are connected each time.

Coating:

The purpose of coating is to protect pipes from corrosion. The pipes come as coated from the factory because applying coating there is more efficient, except a few inches at the ends of the pipe, in order keep the surface clear for welding. After the pipe ends are welded, coating can be applied to those sections as well. At this stage, the entire outside surface of the pipe is also inspected, to detect any possible damages or defects on coating and they are patched as necessary.

Trenching:

Trenches are excavated to bury pipes underground. In some cases cold weather is preferred for construction, as it makes ground harder and stronger to carry heavy construction equipment, such as the excavators and cranes. Harder soil makes trench excavation more difficult too but the excavation equipment such as backhoes and ditchers can handle this. In some rocky areas special rock excavating equipment or blasting can also be used.

To cross streams or rivers, different methods are used depending on the size and the environmental importance and sensitivity of the water body. Horizontal directional drilling (HDD) technique can be used to dig from underwater and to leave the stream in its natural state. This method can also be used when passing under highways in operation in order not to interrupt traffic. HDD method is not suitable in areas with sandy soils or where considerable existing underground utilities exist. Trenching depth should be such that the minimum required cover on top of the pipes is maintained. When crossing under streams or river, the minimum cover distance required above pipes is more.

Placing Pipes into Trench:

After the pipes are welded and the trench is ready, the pipes are then picked up by sidebooms and placed into the trench. Sidebooms are cranes that are specifically made to lift up pipes, with slings attached to their booms, while proceeding on their tracks. Many sidebooms follow each other to pick up the pipes, which are now connected by welds, which can sum up to several thousands of feet long in connected length. It is done this way to achieve better production rate. Sidebooms gradually lower the pipes into the trench.

Installing Valves:

Valves and faucets are installed at this stage, when the pipe is rested on its final position, but before backfilling. Later during operation, when necessary, the flow can be shut down to isolate certain sections of the pipeline by using these valves.

Backfill and Compaction:

Then the trench is backfilled and compacted, leaving only the end of the pipe in the open, for connection to next segment. Backfilling can be done with the material excavated from the trench, but care must be taken to not damage the pipe coating here.

Final Testing:

At this stage, section by section, the pipeline is tested by using pressured water, at pressure levels higher than the future operation pressure of the pipe to ensure that the pipeline will not leak.

Restoration of the Ground and Markings:

After the pipeline is operational, the construction right of way is turned into permanent right of way. The ground is brought as close as possible to its original contours, to maintain the original rain water drainage pattern in the basin. Soil stabilizing and grading are also done at this stage. Over the permanent right of way, trees or tall vegetation cannot be planted but still a ground cover natural to the area is maintained, to prevent erosion, and not much can be constructed except some parking areas, fences, driveways etc... and only if they meet special conditions. Near streams or wetlands, riprap stone or timber may be placed to stabilize soil. Permanent erosion control measures can also be taken.

The route of the pipe is marked with signage to make public aware of the pipeline in the area so that any unwanted damages by excavation can be prevented, and only valves and faucets will remain visible on the ground. The signage and ground markers along the route must be placed frequently enough, generally within line of sight distance of one another, and must clearly be visible to everyone in the area. Additional markers are placed along fences, road crossings, waterways etc...

Offshore Transmission Lines

Pipelines can also encompass seas or even oceans. A thorough surveying of route must be performed, first for the general route approximately and then in detail to know sea floor features both vertically and horizontally, all the way between both end points of the pipeline. In addition, borings at every certain distance must be performed at the sea floor. Not only the 3D topography but also seismic faults must also be identified, bearing capacity must be determined. SSI effects, liquefaction effects must also be studied as required. In other words, the pipes cannot just be laid on the sea floor before performing these steps. As a result of these studies, the pipe route may need to be modified, and / or certain extra cost and schedule impacts may be identified. Depending on the sea floor topography, we may need to backfill or excavate at certain sections, to provide a stable foundation for the pipes. At some locations, some unsupported length may be acceptable up to a point, such as when the pipe runs between two high points on sea bed, as long as the overall stability and integrity of the pipe system is not affected.

As discussed in Chapter 1, ocean waves have no effect beyond a certain depth. Therefore, in open sea, the pipes on deep sea floor will not be affected by the waves, however as they approach to the shore, wave forces will start to be effective, therefore, the pipes at shallow locations close to the shore must be buried under the sea bed, to protect them from wave and current action. An alternative is to protect the pipeline with concrete mattresses. In addition, sea traffic may also necessitate this.

Wave heights and lengths, water temperature and density, currents that act in 3D, density of the fluid transmitted, pipe material and density are all taken into account to determine drag and uplift forces that will affect the pipeline. Another critical phase is when the pipes are laid from the ship to the sea bed, where the tension forces in the pipe are also considered. During placement, different wave and current forces will act on pipes than the forces that would act when the pipeline is at its final location.

Connection and Laying of Pipes:

Similar to onshore lines, long offshore pipelines are also made of steel. In general, the steps of connecting the pipes to each other are the same or similar to land projects, but in case of offshore pipelines, the connection of pipes are done in the pipe laying ships, and instead of placing the pipes into a trench they are laid on the sea bottom. The pipes are delivered to the ship in large quantities. This means, the pipe laying ships can store a huge number of pipes in order to achieve economies of scale in production. After the pipes are delivered, the production stage begins, which is similar to onshore pipelines as discussed, such as, beveling pipe ends, welding, coating welded sections.

All these steps take place in ships, in much more factory like environments, on assembly lines. This surely saves time, not only because of assembly lines, but also because it can continue 24x7. Land projects are interrupted at nights because lighting a large area which would be for a night or two at most and coordinating crew and equipment during night time would not be feasible. The quality however is not necessarily better for offshore lines because land projects are also performed per the strictest standards with highly skilled and experienced crews.

After finishing all these steps above and obtaining connected pipes, the pipe assembly is laid on the sea bottom either from behind the ship by lowering them at an angle into the water, or from middle of the ship, by dipping them vertically. There are several methods for pipe laying, depending on the ship, seabed depth and the pipeline itself.

WATERWAYS (WATER CANALS)

Waterways / Water Canals are made to connect two larger water bodies for the passage of ships, such as the canals that connect two lakes, seas or even oceans. The two largest water canals in the world, Panama and Suez Canals, have global impact on world trade.

By building a canal, we are basically separating two land masses from each other, that were together before, and joining two distinct water bodies that were separate before. Therefore these are large projects with high environment impact, such as the effect on aquatic life, agricultural lands or forests. Dumping of millions of cubic meters of excavated and dredged soil *(and the constant working of trucks affecting existing traffic in the area for a long time)* and so on, need thorough feasibility studies and planning, and input from many stakeholders and the public. In addition to environmental impact, if the canal is close to a populated location, there may also be a lot of existing utilities, pipelines, roads in the area which may intersect the proposed canal location, which will most probably need to continue to operate. These will all need to be rerouted through future tunnels or the body of the bridges that would be built to cross the canal, which will all be other costly and lengthy projects each. The duration of construction of all these will also be long. After all these aspects the cost vs. benefit of the canal is carefully studied, and a great majority of the public's agreement is clearly obtained, only then, the decision should be made to go ahead with it or not.

After the general location and route of the canal is selected, its exact <u>length</u> is determined. Typically, this corresponds to where we can draw the shortest line over land between the two water bodies we want to join, to keep the length of the canal as short as possible, but this may not always be possible and a longer route may be necessary. The length is main factor in determining the main location, since every extra foot of excavation and construction of a canal costs a lot of money if we consider the whole cross section and all associated work. The canal must be kept as straight as possible, not only for minimum total distance, but also for easier navigation of ships. The required width and depth of the canal are also determining factors here.

For determining the <u>width</u>, the main criteria is to decide whether it will have one or two lanes, and the width of lanes which must be wide enough to accommodate the largest ship that is supposed to use that lane. The lane width must also include spaces at two sides of the largest ship possible, plus possible yaw of the ship due to winds and to allow maneuvers and avoid collisions. This is the width of the maneuverable lane. On top of it, we must also add the bank distances as seen in the figure below. Bank clearance is needed because without it the water flow on two sides of the ship would have different velocities, which would cause suction effect of the ship towards the bank. The width of the canal is the flat bottom portion. The entrances are wider due to many variables and navigation difficulties. The figure below shows a typical water canal cross section with two lanes, where:

BC: Bank clearance
ML: Maneuverable lane
SC: Ship clearance / Passing distance
A: Authorized depth (design depth, between water surface and the authorized depth)
B: Distance between underside of design ship to authorized depth (the gross under keel clearance)
C: Dredging + Maintenance Allowances

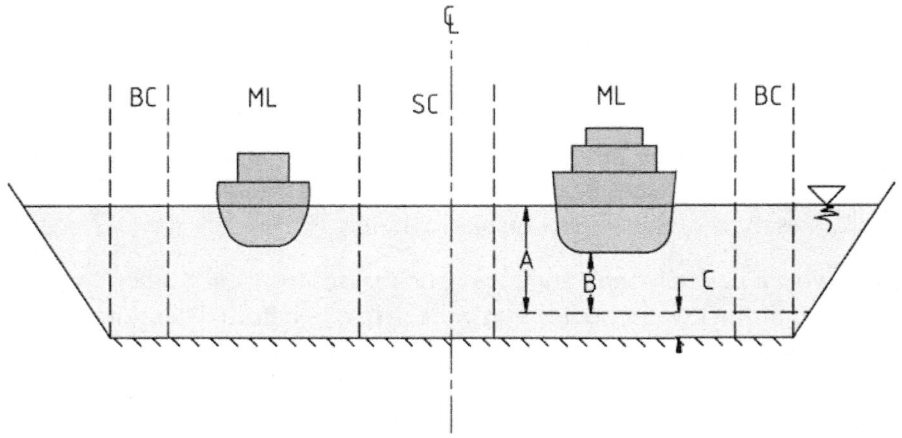

Figure - Water Canal 1

The water surface level in the figure above corresponds to average (mean) low water level.

The bank suction effect is not to be underestimated. For example in March 2021, Suez Canal, the world's largest, was blocked for about a week by a giant container ship, which lost control of its steering. A factor which might have contributed to the loss of control could be the bank effect, if the ship got too close. The blockage affected the world trade and cost billions of dollars of losses overall.

The depth of the waterway, is another major factor that influences its cost. Although increase in depth increases the cost of construction significantly, these can be offset by better use of present ships, allowing larger ships, reduction in delay times, which can all make a big difference in long term income and cost savings.

The waterway depth depends on factors including:

- Locations where water is naturally deeper is better, to reduce initial construction and also the maintenance dredging costs. When first opening the canal by excavation or dredging, slope stability and the effect on existing nearby structures, environmental effects must be checked. The site to place excess materials must be clarified.
- Locations which are not subject to excessive sedimentation and siltation are preferred. Maintenance dredging must also be done regularly.
- Existing water currents, water quality, prevailing winds all affect canal depth.
- Lower water density increases depth of ship in water (draft), which in turn affects canal depth. As the salinity of water increases, its density also increases.
- Closer to both ends, sea waves can be effective and cause up and down movement of ships.
- Squat effect must also be considered when calculating waterway depth. When a ship moves in deep water, the water displaced by the moving ship's hull can easily move around and under the ship. But when the ship moves in shallow water, that displaced water, when moving under the ship must move faster, because there is not enough space between the canal bed and the bottom of the ship. Per Bernoulli's principle, if a fluid moves faster, its pressure decreases. When the pressure of water beneath the ship decreases, its ability to lift the ship decreases, which causes the ship to sink more. Squat effect increases as the ship's speed increases or the distance

between canal bottom and ship decreases, or the ships distance to the side of the channel decreases. The shape of ship's hull also affects squat, because different hull shapes will displace water at different rates for the same speed. If the canal is too narrow, or there is another ship nearby, this will also increase squat because the water on the sides of the ship's hull will also move faster. Therefore the depth and width of the canal and the allowed maximum speed of ships are interrelated.

In some canals, water depths are known and can be guaranteed, while in others it cannot. Some canals are restricted, which means the entire canal is dredged, while others are not, and the water width continues beyond effective canal width that is being used.

At the end of a canal, there can be a harbor, to provide an accommodation area for ships. We discussed ports and harbors in Chapter 9.

WIDE SPAN ROOFS

Wide (long) span roofs are commonly used on structures such as warehouses, factories, malls, gymnasiums, airports and even stadiums... They typically pass widths of at least 20 m without intermediate columns. Having wide enclosed spaces free of columns have benefits in construction schedule and budget, and during the life of structure as more usable space, better movement, visibility and aesthetics inside. Wide span roofs are made as a combination of different structural systems as the ones listed below, so the items below usually are not used alone, although one can still make the majority of the roof structure.

- Truss systems such as:
    - Space truss
    - Beams
    - Decks
    - Arches
    - Bicycle wheel form
    - Truss and concrete deck composite systems

  Trusses should be assembled in the shop to the greatest extent possible, as allowed by the shop capabilities, transportation and lifting limitations.

- Precast prestressed concrete beams. Long span beams must have efficient cross sections while saving weight. For example, an ordinary RC beam of a building frame can be rectangular, but a long span beam must have more efficient cross section, such as I-beam for better bending efficiency, as most of the bending stresses are carried by bottom and top flanges of beam and as long as the remaining web section is sufficient for resisting shear. I-beams with holes in web are also used to allow for pipes and ducts to cross beams. In addition to geometry, using very high strength concrete is also a viable solution to obtain better efficiency.

- Structural steel systems.
- Cable structures of various forms such as cable nets and or suspension cable systems which are tensioned in 3D to take on forces from all directions, which can also give the roof a complex warped aesthetical appearance, while saving considerable weight. Cables can be carried and tension is obtained through towers (as in suspension bridges), and perimeter trusses. Membrane material such as fiberglass, tensioned membrane can be used as the main body of the roof, carried by cables. These can also allow light which is a big advantage.
- Long span arches and domes in shell form utilizing arch action. The compression ring in the middle and perimeter ring at the perimeter is critical for domes and arches. We discussed arch action in Chapter 1 > Structural Engineering > Various Terms.
- Fabric membrane structures that work by air pressure from inside.

During design of the above, although the specifics for each system greatly vary which are not discussed here, let's list some common things. Keep in mind, most are also valid for any structure, but for wide span roofs they can be much harder to achieve. This is because they are large footprint structures often with unusual or even complex geometries in 3D, made from a combination of different materials and systems as the ones listed above which rest on yet another different material as reinforced concrete footings / columns / towers / reinforced concrete or masonry walls.

- Long term effects such as shrinkage and creep for concrete components.
- Temperature effects for structural steel, truss, cables both for the elongation and shortening of members and also for the temperature differences across the structure.
- Long term settlement effects, differential settlement.
- Using high strength steel for any steel component (i.e. truss, structural steel, cable) is more critical than for most other structures, to save weight while passing long freespans.
- Unless we deal with a relatively lower span roof made with concrete beams, in some cases it is advisable or even necessary to perform wind tunnel test cycles while refining the design.
- Not only the self weight of the roof, but also items such as mechanical ducts and ventilation system, lights, sound system, large screens (i.e. screens hung from the roof in the middle), interior cranes, catwalks etc... as applicable to the project must be supported.
- Construction phases impose different conditions and loads on partially completed state of the structure. During construction, unlike the finished structure, different states of unbraced structure will exist. The load vs. capacity in members and structural stability due to these unfinished states must be analyzed. Another thing to pay attention is the lifting stresses that can occur in very large members during lifting by crane. This, and the stresses in installed members while the structure is incomplete can exceed the permanent stresses in the members in the completed structure. For installation, we may need to erect temporary structures as well.
- Rain and snow loads have greater importance for wide span roofs than for any other structure.
- A rectangular roof or a close shape can be designed as one way or two way slab.

- Wide span roofs also serve as diaphragms. We discussed diaphragms in: Chapter 1 > Structural Engineering > Various Terms, and, Chapter 2 > Building Frame and Structural Members.
- When steel is involved, as in any type of steel structure, the option of standardizing member sizes and connections, and reducing number of members and connections should be considered. The schedule and cost savings due to easier fabrication and installation vs. making the structure slightly less efficient should be weighed by an overall cost benefit analysis by the structural engineer, owner, architect (and contractor(s) even if not yet selected).
- There are also retractable roof systems, which can be installed on as large areas as stadiums and adds significantly to the complexity of the project. In addition to the concerns above, the weight and coordination of retraction system and retracted roof is critical in such cases.

A mall's widespan roof made of structural steel beams supported by several center columns and perimeter concrete. Primary design concerns include column buckling, beam deflections, thermal loads, and coordination with supporting concrete.

For wide span roofs, coordination between the main structural engineer and the specialty structural engineer(s) of the roof system(s) is critical at each stage of work. Designing and building wide span roofs is a much more challenging structural engineering undertaking than many other types of projects. Passing of large spans using highly efficient structural systems that are subjected to a great amount of load combinations not only for the finished but also unfinished structure is highly challenging. It requires timely coordination between all parties at every stage. Making everything work in 3D software beforehand should accompany this coordination. And of course as in any type of project, coordinating construction activities along with their submittal review and material ordering and clearly defining work scope of each party is essential. The erection method and schedule of the main roof structure should be obtained from the contractor as early as possible for review by all parties involved. As the span grows and the roof structure becomes more irregular, all challenges exponentially increase.

WIND TURBINES

For land turbines, hills are usually suitable locations, because of stronger winds.

Wind turbines generate electricity by converting wind kinetic energy into electrical energy through their propellers and generators. Wind is renewable, and benefiting from it is becoming more common every year, with new projects and research and development. Both the number of wind farms and the efficiency of turbines are increasing. Wind energy has great potential, which we are using only a small fraction of currently. According to the US Energy Department website, U.S. offshore wind has a potential of more than 2,000 GW of capacity, or 7,200 TWh of generation per year. For context, this is nearly double the nation's current electricity use. This means that even if only 1% of the this was recovered, nearly 6.5 million homes could be powered by offshore wind energy. Developing wind energy would also support many jobs, reduce greenhouse gas emissions and reduce power sector water consumption.

Site Selection

Wind farm location is chosen based on factors such as the prevailing and constant wind speeds of the area, ease of transportation of construction components, accessibility of site, soil conditions and for offshore wind farms, also the depth of sea bed. Wind turbines must be constructed at a certain distance from any obstructions that is on the way of prevailing winds in order not to lose any wind. For example, a location in the middle of a flat desert where nothing is around and transportation of materials are easy can be considered as a suitable location, but the side of a hill, where even though construction access is not ideal, may be better because of the generally higher wind speeds there.

Design

Foundations:

Land Turbine Foundations:

A wind turbine foundation, in addition to the weight of the generator, tower and the propellers, must also resist huge moments and lateral forces, thus, heavy raft or raft plus deep foundation are used. Due to changing directions of lateral forces, circular foundation shape is optimal, but it can also be hexagonal or octagonal. If no piles are used, the mat can be several meters deep. In any case, piles / pile caps / footings will need to be heavily reinforced, to resist large overturning moments, fatigue from cyclic wind loads. Soil with high plasticity must be replaced. In some cases we need to make other types of ground improvements as well, as was discussed in Chapter 1 > Foundations. Battered piles may also be considered. Estimating foundation and soil stiffnesses (for horizontal, vertical, rocking, torsional motion), any possible nonlinear behavior, and soil structure interaction (SSI) is important. We can analyse the foundations where we calculate the foundation stiffnesses by equations, or represent the foundation by springs and model and analyze by FEA, or model the whole structure plus the foundation plus the ground behavior with a complete 3D meshing for FEA, which can also simulate SSI effects.

The overturning moment must be checked in two places. At the bottom of foundation and also at the base of the tower structure where it meets the mat foundation. For the former case, we must make sure that everywhere in the soil remains under compression, as was discussed in Chapter 1, eccentric loading of foundations subsection. And at the tower base, we must make sure that the strength of materials can safely take on the overturning moments, plus the fatigue loads from millions of cycles over the lifetime of the structure.

Offshore Turbine Foundations:

For offshore turbines, in addition to the wind, there is also wave and current forces which increase the lateral force and moment significantly. The main factor determining the type of foundation for offshore turbines is the depth of the sea bed. For the shallowest depths, even a mat foundation can suffice, which can stabilize the tower with its self weight. For depths of greater than 30 meters, a combination of piles and another structure that will serve as a connector between the foundation and the tower can be used. The greater the sea depth, the more complex the foundation and the tower, and after a certain depth this becomes impractical, but most offshore locations suitable for wind energy generation are located where the sea bed is too deep. At those locations floating systems can be used.

Wave forces can be very significant, as we are talking about a windy area suitable for a wind farm in the first place. There is also scour and erosion risk at the sea bed level that must be analyzed during design.

Lateral forces we mentioned above for both land or offshore systems are not static but dynamic. The tolerances of movement and tilting for foundation are specified by codes and by manufacturers, for rotational, vertical and horizontal directions. In both land or offshore cases ground improvement may also be necessary.

Structure:

As far as the height, offshore turbines are taller and more powerful than land turbines on average. Offshore towers can be well above 100 meters above sea bed *(to the tip of the tower where propellers are attached)*. It must be remembered that the higher the tower, the stronger winds it can catch and therefore produce more power. Of course higher towers require deeper and stronger foundations and stronger tower body itself. For very tall towers, there might be approval issues with aviation authorities, such as FAA in the US. Some of the things to consider during design of wind turbine structures include:

- The towers can be built in the form of cages or pipes. As the base diameter and the wall thickness of tower segments increase, taller towers can be achieved as the stiffness of the tower increases. This also increases the mass. Both of these contribute to a higher natural frequency of the tower *(which is not necessarily always the best as we saw in earthquake section but we can adjust the natural period in this fashion to avoid or decrease probability of resonance, in other words, matching period of the structure with the ground period)*.
- A wind turbine tower is exposed to a very large number of fatigue cycles, higher than any other structure, because of wind and movements of propellers and generator.
- Towers are installed in sections, which consist of several segments of rings that are placed on top of each other which are secured with many bolts. The tower pieces are shipped to the project site in modules. Although we prefer to fabricate as large modules as possible, there are limitations such as fabrication capabilities, transportation limitations and obtaining permits for it, installation limitations, therefore installation in modules is inevitable.
- Blades are made of highly durable, strong and lightweight materials. The longer the blade, the more bending it will experience. Strength of blades are tested in factories by repeated cycles of bending tests with large loads. Excessive bending will also cause the blade to hit the tower itself, even if the blade doesn't fail, therefore the blades must be stiff enough, but not too stiff to experience brittle failure.
- Dynamic analysis of a wind turbine for earthquake must also consider things such as the rotation of the blades and any nonlinearities.
- Wind turbine towers can be considered as thin shell structures. Therefore, buckling becomes an important criteria in addition to strength, fatigue.
- There can be high amount of local, concentrated stresses in certain locations of the tower such as at penetrations, openings, bolts etc... which must also be checked.

Construction

Foundations:

For land turbines, the significant amount of rebar that is needed for mat foundations may cause rebar congestion or vibration difficulties during pour. Electrical cables that run under foundation are placed before the pour. After the foundation is complete, the top is backfilled and compacted, except the middle part that the tower will be placed. In remote areas where concrete supply may be an issue, the pour may need to be done in segments, which are separated by construction joints, and extending rebar dowels between sections. For offshore wind turbines methods vary greatly depending on sea bed depth and the foundation system to be used.

Structure (Land Turbines):

After pouring foundation, a crane is set. This must be a very tall crawler crane, which may need to be assembled on site. Very tall towers must be made and transported to the site in pieces. A piece may weigh around 100 tons and have length of 50 meters or more. By lifting the pieces, we mean lifting by crane vertically and set it to position so that each bolt and hole location matches, which is a slow process. First the bottom piece is placed and secured to the foundation. Usually this piece is not tall, but it forms the base for the upper pieces. A scaffolding can be placed inside the tower structure as a working platform for workers, which is raised as the tower goes up. After completely erecting the tower piece by piece, the generator is placed on top. This is a heavy piece and can weigh as much as around 100 tons. Rotors are assembled on site. A rotor would typically have 3 blades, and each blade is attached one by one to the center piece which serves as a hinge. After all blades are attached, the rotor is lifted and attached to the generator. Finally, the electrical and mechanical connections are made.

There is also a maximum allowable wind speed for the construction operations to take place (i.e. 9 m/s). This is not only because it is more difficult to perform installation during higher winds but also even a small amount of swinging of a very heavy load hanging from a very tall crane is dangerous.

Structure (Offshore Turbines):

As mentioned above, offshore turbines are usually larger. The pieces are brought to their final location by a special installer ship which both transports the pieces and then installs them, by using its onboard cranes. During installation, the wind speeds, wave heights and current forces must be below certain limits. After completion, the structure and the foundation must also be able to withstand all these dynamic forces throughout the entire design life, which can be far above the limits during construction.

An Offshore Wind Farm Example:

Let's illustrate the process and get a feel of some numbers, with an example from Europe, where offshore wind power generation is common. The offshore wind farm in our example, called DanTysk, is located in the North Sea, 70 km west of the shoreline of Sylt at Germany at Denmark border. Although not close to the shore, the depth of seabed on site is shallow, varying only between 21 to 32 meters, while at the same time having constant and strong winds, which were major factors in site selection. The site is 70 km^2 and includes 80 wind turbines of 3.6 MW each, with a total capacity of 288 MW.

The pieces were transported to their final location by ship, which also served as the installation platform. Precise locations towers were determined by an automated direct positioning system that positioned the ship, since surveying as on land was not possible. Seabed soil at each tower location was checked and confirmed before installing each tower, even though there was a prior geotechnical study.

To start installation, the ship transforms itself into a platform, by lowering legs down to the sea bed which penetrate to the soil for stability. Together, these legs totally lift the ship above the sea surface, while the loads on legs are monitored. This is done to provide stable, flat surface for installation, unaffected from wave action.

The piles were drilled first in to the sea bed as much as 30 meters. After placing the foundation, the transition bottom piece was attached to the foundation. On top of that, the tower itself, and finally the generator was placed. Finally the blades were attached one by one, to form the complete 3 blade rotor. The rotor blade diameter is 120 meters and the height of generator is 88 meters above sea level. All of these were performed by the large crane mounted on the ship, with a capacity of 500 tons. The highly skilled and experienced crew was able to install one wind turbine per day.

Operation of Wind Turbines

Wind turbines generate electricity when the wind pushes the blades, which are oriented diagonally to wind direction that causes rotation, which is then transferred to the generator via shafts and gearbox.

There is a lower and upper limit of wind speed for energy generation. The lower limit is called cut-in and the upper limit is called cut-out speed. The lower wind speed limit is usually around 2 to 4 m/s, and the higher limit can go as high as around 30 m/s. Usually the nominal speeds are between 10-15 m/s. A turbine can not indefinitely generate more power as the wind speed increases. There is an upper limit to its electricity generation capacity even if the wind speed kept increasing. This upper limit is called the nominal speed, where the turbine generates its maximum power it is designed for. The faster speeds can damage the blades and the turbine, by creating excessive voltage, vibration, so there are brake mechanisms in place that engage when the wind speed reaches the cut out speed. Between the cut-in and cut-out speeds, the blades turn 10-20 revolutions per minute typically. To generate electricity however, much more revolutions per minute is needed. This is done by the gearbox, which raises this number from 10-20, to more than a thousand per minute to transfer it to the generator.

Many wind turbines automatically adjust orientation according to the wind, to produce maximum power possible. This is done according to the wind data collected by the equipment installed on top of the tower, which then automatically adjusts the direction of the rotor accordingly, by a yaw motor.

The power that a wind turbine generates depends on:

- Number, orientation and shape of blades
- Length and width (surface area) of blades
- The height of the rotor from the ground or sea level
- Efficiency of the turbine, such as the capacity factor, which can be around 50% for the efficient turbines, although the newer ones can reach up to 60% or even more

The wind turbines can generate power anywhere from several hundred kilowatts to as much as well over 10 megawatts. The bigger the turbine, the higher cost efficiency. Many turbines are built together to make wind farms. Offshore wind turbines, which are usually larger and more powerful than the land turbines, can usually generate at least several megawatts each. Each blade can be around 100 meters long for very large offshore turbines. A turbine of this size can power several thousand homes. There are also small size wind turbines that generate electricity below 100 kilowatts, which can be used for single homes or similar locations. The higher the tower, longer the blade, the more wind it can harvest by sweeping more area.

Wind power generation are also done in many different configurations in smaller scales, including placing them on rooftops, or anywhere suitable on existing structures. For example, camouflaged wind turbines were installed on Eiffel Tower, which together can generate a total of 10 Megawatts, which is enough to power its restaurants, shops. The advantage of such installations are that although they generate small amounts of power, their accumulated benefits will be huge, if they are installed at a lot of urban locations. They do not need separate structures and can be installed on the buildings, and they do not need distribution power lines to transmit the energy they produce, as it is used right there.

Although wind is a renewable energy with many benefits, wind turbines have some disadvantages too. For land turbines the disadvantages seem to be more, although they are a lot cheaper. The main downside for both is that, turbines must be checked frequently, as they have moving mechanical and electrical parts operating outdoors. Wind turbines are usually designed to operate efficiently around 20 years, which is still shorter with respect to hydroelectric dams, nuclear or coal plants, but this aspect is also constantly being improved. These add to the cost, but the benefits increasingly outweigh the costs.

Comparison of Land vs. Offshore Turbines:

Some disadvantages which only apply to land turbines are:

- Turbines must be at a certain distance from obstacles, to catch the full wind. For land turbines sometimes this objective can not be fully achieved. Coordinating the location with existing developments and infrastructure can be a problem.
- Moving machine parts of turbine and the moving blade in the air create high level of noise. And to some people, appearance is not pleasant. Offshore turbines are far from people, which eliminates this problem in practice (although it may be harmful for certain animals in the sea).

The advantages of offshore wind turbines include:

- In general sea winds are stronger and less variable, because of no obstacles or rough surfaces to slow down the wind as on land. Also, winds over sea are stronger during daytime, and therefore offshore turbines can generate more power when the demand is higher. Land turbines usually generate more power at night, when the land winds are stronger.
- They can be larger, which means more power generation and more efficiency.
- Many people live along the coastlines. Offshore platforms provide electricity for these areas, thus eliminating the long power transmission lines on land from power plants.

Therefore, it is not surprising to see more and more wind farms constructed offshore, either with foundations at shallow depths, or as floating systems at greater depths, which are called Floating Offshore Wind Turbines (FOWT). FOWT systems can be tied with steel cables to the foundations below. Considering strong winds, possibly very large waves, current forces, all of which are also dynamic in nature, these connections must be very strong and be able to withstand endless cycles of large dynamic loads. Using FOWT systems greatly increases the area of sea that can be utilized for wind power.

Comparison of Horizontal vs. Vertical Axis Wind Turbines:

So far we discussed the more common type of wind turbines, which are horizontal axis wind turbines (HAWT), which means they rotate around horizontal axis (similar to a windmill). There are also the vertical axis wind turbines (VAWT), which rotate around a vertical axis. Now let's compare these two.

The advantages of horizontal axis wind turbines (HAWT) are:

- They can generate more electricity. This is due to having larger blades and higher rotation speeds, and the blades are fully utilized since their angle is adjusted to receive maximum wind. For VAWT, the blades at the opposite of rotation slow down the movement. Generating more power is the most important reason that HAWT are more common.
- They operate more stable in gusty wind conditions than VAWT. They vibrate less, because they are subjected to more steady winds.
- They do not need an initial push to start as in the case of VAWT.

The advantages of the vertical axis turbines (VAWT) are:

- They can generate power regardless of the wind direction, therefore no steering (i.e. yaw and pitch) is required. Thus, they can be used in low or turbulent winds which can vary in direction.
- They can better resist to extreme winds.
- They are lighter, which means lighter structure and foundation is needed. Their maintenance is also easier. They produce less noise than HAWT, as their blade rotation speeds are lower.
- They can be used in locations where local rules do not allow high turbines. Because of this, they can find more usage in urban environments. They can be placed closer to each other.
- New, innovative blade designs and systems emerge all the time, extending uses and increasing advantages.

CHAPTER 12 - HISTORICAL STRUCTURES

In a book that introduces structures, civil engineering and construction, it also makes sense to discuss historical structures, to give a sense of what people were able to build before our current knowledge and construction technology developed. Understanding the basic thinking of human mind enhances our understanding of today when equipped with more advanced tools and knowledge.

Among a huge selection of historical structures from many different nations lived on earth, some of which disappeared and others still survive, here we will focus our attention only on two of the very old but popular ones, which are ancient Roman Civil Engineering (about 2000 years old) and the Pyramids of the ancient Egypt (about 4500 years old), based on common historical knowledge and from civil engineering and construction point of view.

There are three types of forces - as we have seen in the very beginning of this book:

- Compression
- Shear
- Tension

and, the turning effect of force, moment.

Old structures built from stone and mortar were able to resist shear and mainly compression. But the ancient engineers did not have the technology such as reinforcing steel to make their structures to resist the other major force, tension sufficiently, which mostly means the tension from bending moment, when you think of a bending beam or slab. All they could do was to make their beams thicker so that they would not crack from the tension due to bending (since a deep beam bends less), which of course considerably added to the dead weight of their structures. And even though deep beams were used, the supports from below (such as columns or walls) had to be close to each other and large free spans could not be passed.

A very useful technique, utilizing arch action, emerged likely about 2000 years ago. In comparison to straight, thick beams we mentioned above, arches enabled passing of larger spans by using compression only, when the stones were arranged in an arch shape, just by pressing against each other. Arches not only enabled passing of spans but also saved a lot of weight. Many heavy stone structures would have collapsed under their own weight, without arches. A derivative of arch is the vault, which basically means, many arches continuing along a line, to form a corridor with arched ceiling. Another derivative is the dome, which is created by revolving an arch around vertical axis. Without using arches, vaults and the domes, most historical structures would not make it up to modern ages. We still use arches sometimes but it is now optional, used in combination with other arrangements or for aesthetical reasons to give a structure historical look.

Even without (sufficiently) resisting tension, ancient engineers were able to achieve great engineering feats, and built bridges, colosseums, aqueducts, temples with large domes, palaces, pyramids, water storage and supply systems, mines, tunnels and anything they needed, only by arranging materials to resist mainly compression and shear.

But there was only so much engineers could do, when they could not make structures to resist tension sufficiently, not to mention the absence of today's knowledge, technology, tools and equipment. In late 19th century, with the start of usage of iron and then steel bars in concrete, and also being able to start making structural steel members, the problem of not resisting tension and bending moment was solved. This meant a revolution in construction. It enabled wider spans to be passed, taller structures to be built which were also considerably lighter. It made possible not only frame but also truss structures.

In addition to structural, civil engineers/architects of the past also achieved success in surveying, water distribution, transportation, construction management, foundation systems, even with the limited equipment and knowledge they had.

The distinction between an architect and civil engineer started to appear in the late 19th century. Before that, architect and civil engineer meant the same person.

PYRAMIDS OF EGYPT

Pyramids are among the top most significant historical structures on earth. They were most likely built to serve as tombs for pharaohs. There are many pyramids that were built in ancient Egypt, over the course of a very long time. Khufu, the largest of Giza pyramids, is what we will discuss here. It was built in 26th century BC, for Pharaoh Khufu.

Giza Pyramids were the tallest man made structures for thousands of years. Khufu, the oldest and the largest of them, had an original height of 481 feet *(146.5 meters, but when the casing stone which covered the structure was removed over time, its height dropped to 138 meters)* and with a slope of around 51 degrees. Khafres pyramid, near Khufu, is a few meters shorter, but built on higher land, which makes it appear taller. Its slope is also slightly steeper than Khufu (around 53 degrees).

Giza Pyramids are one of the top architectural, civil engineering and construction successes ever, if we think it in proportion to available technology, knowledge and equipment at the time of construction.

Planning and Construction Management

The huge size of the project and the crew, required effective construction planning and management.

Estimates about construction duration about Giza pyramids seem to somewhat vary. Let's use 25 years for the Khufu pyramid, and for the entire Giza pyramids project site, let's use 80 years. Pyramid Khufu has around 2.3 millions stone blocks of limestone and granite. The limestone, which is the great majority of stones, weigh 2-3 tons on average. Granite blocks are far larger and heavier.

If we assume 2.3 million blocks, and 25 years, it makes 252 stone blocks everyday, for 25 years nonstop. Assuming there were delays, inefficiencies and not everyday was full work day, let's say 300 blocks for one full work day. Placing stone means, it is first quarried, trimmed, then transported to site, carried upwards, placed to its final location where its surface was also finished if it was outside. Using this number let's make a rough estimate here about the crew size. Let's assume it took a total of 50 men to work on a single stone, to do all the tasks we have just mentioned *(not that the same 50 men worked on it from beginning to the end as a team, of course the tasks were divided and every worker had specific locations and tasks only, but we are talking about overall effort, as in 50 men-days were spent for one stone)*. This makes 300x50=15000 workers per day, just for the tasks related to stone. On top of this, add more workers for all other tasks and the management of the job, and we might be looking at around 20000 people per day, provided that the assumptions here are reasonable. This number may seem high in comparison to even our large jobsites of today, but remember that the ancient Egyptians did not have the powerful construction equipment and efficient tools and the knowledge we have now. If citizens also participated in construction in addition to slaves, it meant even more careful consideration and planning for the worker's needs and health. There was also great success in transporting materials. The very large and heavy granite blocks came from very far, by boats along Nile.

Construction

"Alien intervention" stories are often heard about pyramids, due to their complexity, precision and size. However, ancient Egyptians were already skilled in construction before building Giza Pyramids.

Experience from Earlier Pyramids:

In addition to other types of structures, ancient Egyptians also built many smaller pyramids before Giza pyramids, which gave them experience. Three well known earlier pyramid examples are shown below. On the left, is the"Bent Pyramid" which started with a steep angle, but continued with a lower slope. This may be just to save work and time, or during construction they might have found out that the pyramid would not be stable with the initial steep slope. In the middle, is the pyramid called "Meidum Pyramid". And on the right, is the stepped pyramid. Early pyramids had only few levels of steps. Then, in time, more layers were added, to make them increasingly bigger, and look more like pyramids.

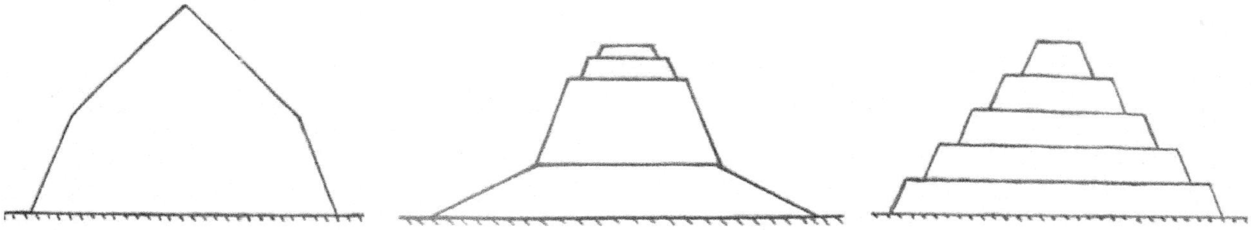

Figure - Historical 1

Surveying:

Egyptians had great surveying skills because they continuously needed to build things, establish borders between lands. Pyramids are dimensionally impressively accurate, considering that the ancient Egyptians only had the primitive equipment and knew only some simple geometry.

They might have used tools such as various frames and plumb bob, water level as seen below, ropes, groma (see Figure 5 in Roman civil engineering section) or other tools for measuring, setting lines, angles, and level surfaces.

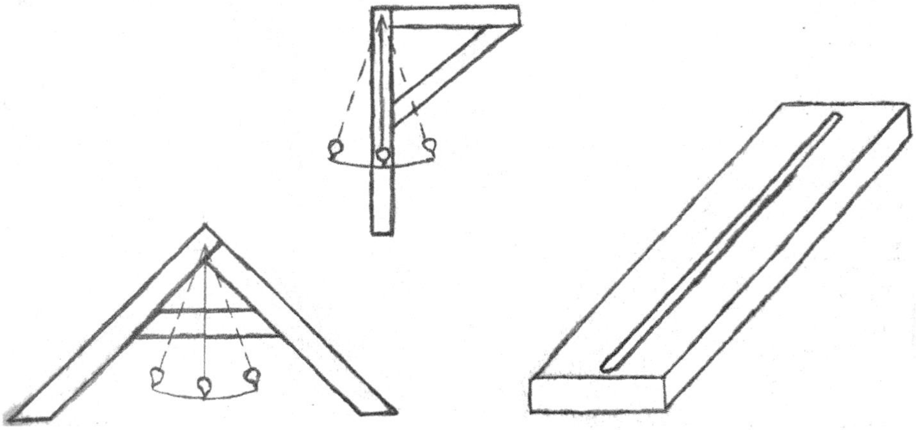

Figure - Historical 2

Each side of Khufu pyramid is about 755-756 foot, with very small variances. The three Giza pyramids corners pass through a straight line and each of their sides are parallel. In addition, the sides are practically in perfect alignment with north, south, east and west directions, even though they did not know compass. The base of pyramid is also level by a very small error. To obtain a level foundation base was critical for geometrical accuracy and stability because all layers above followed.

Quarrying Limestone Blocks:

As mentioned above, the great majority of stone blocks that make up the pyramid are limestone, which were quarried near the site, so they needed to be transported for a short distance only, which could be a reason for selecting that site. The blocks weighed a few tons each on average. As an additional note, the famous Sphinx near the pyramids were carved out of that limestone bedrock of the site.

Quarrying and Transporting Granite Blocks:

Inside the pyramid, the corridors, shafts, rooms and the King's chamber are of granite, which is a very hard rock. For it, they must have used more advanced techniques to quarry and trim than they did for limestone. Granite was not available at Giza, but was obtained from far and transported to the site by boats along Nile. They carried these blocks of up to 50-60 tons from quarry to the boat, transported very long distances along the river, and then unloaded and carried up to the final place.

Moving Blocks To Their Final Place:

After quarrying the stones and bringing them to the jobsite, they had to move them up. They might have pulled the stones over the sand on sleds, while pouring water to wet the sand. The wetting of sand was understood in 2014 when researchers from University of Amsterdam examined an artwork which showed a human pouring water in front of a stone sled as it was pulled. The team was able to test and confirm this. When water is added in front of the sled it not only decreases friction but also prevents piling up of sand. The added water must have been up to a point though, because as we saw under soil mechanics, too much water would decrease the bearing capacity of the soil by reducing effective stress and cause the sled to settle and sink.

To move stones up, one of the most believed theories is that they built temporary ramps which also rose as the pyramid rose in height, only for the purposes of construction, which were later removed.

A ramp that was discovered at a quarry called Hatnab, dated to Pharaoh Khufu, had at least a %20 slope and had stairs on both sides of the ramp with holes at every several steps, which were probably there to hold posts, for a rope and pulley system, as teams from the University of Liverpool and the French Institute for Oriental Archaeology in Cairo saw. See figure below that describes general concept.

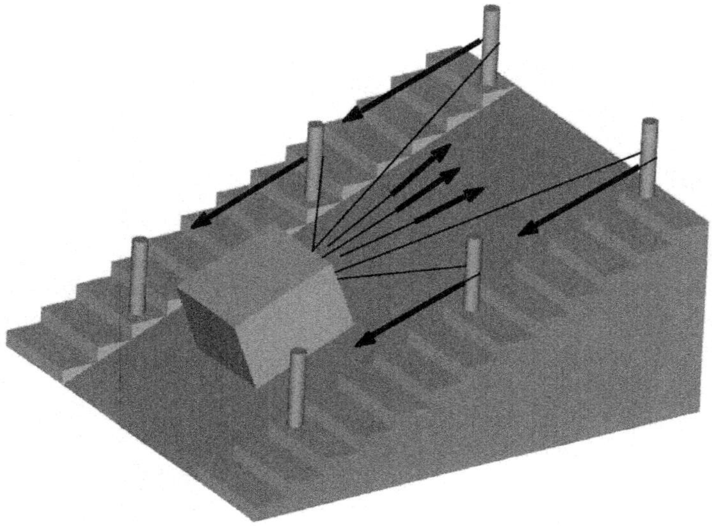

Figure - Historical 3

The advantage of pulling downwards, in addition to upwards was that this way the worker's weight force component were also being added to the pulling force, minus the friction between the rope and the posts. If this was indeed a ramp pulley system used for pyramid construction, it would have allowed more men pulling the stones, and ramps with steeper slopes, which would have required less effort and material than building ramps with lower slopes as previously thought.

Final Placement and Finishing Work:

After bringing the stone to its final location, levers might have been used to guide it to its final place. Outside casing stones were also fine trimmed to fit in their final place, with practically no gap remaining in between, and probably polished. So probably the pyramids originally looked like triangular white prisms with sharper edges, which looked even more appealing than today. Only a very small portion of stones at the exterior face of pyramid remained to date however, due to environmental effects such as the weakening effects of rain and temperature cycles and looting over thousands of years. The granite blocks in the burial chamber inside were also precisely placed and finished. Other interior stones however, were only roughly finished. Inside the pyramid, there are still sealed voids which have not been entered yet. Their purpose could be to save volume, or to hide items inside.

ROMAN CIVIL ENGINEERING

Roman Empire was one of the greatest empires in history which lived for a very long time. From its beginning in 8th century BC, Rome was a place for many people who could come from far and settle (*may be similar in a way to how the US started*), which allowed the newest knowledge to accumulate there, including construction. Many structures that the Romans built contributed greatly to the empire's power and were continued to be used even long time after the fall of the empire, including some that still stand today. In the sections below, we first give an overview and then discuss different types of structures that were built by Romans.

Overview

Romans built everything they needed for improving their civilization such as:

- Amphitheaters
- Water distribution systems including aqueducts
- Artificial lakes
- Basilicas
- Bridges
- Buildings
- Cisterns
- Dams
- Harbors
- Libraries
- Palaces
- Public baths
- Roads
- Sewer systems
- Shopping malls
- Stadiums (colosseums)
- Temples
- Tunnels
- Walls and forts

Many of these structures were better than those of other civilizations in their era. In addition to structures, they also built equipment and war machines that were larger and more advanced than of their neighbors, roads that enabled their legions to rapidly advance and trade taking place faster. All this construction work required a lot of workers, most of which were the captured slaves in addition to citizens or soldiers. Roman soldiers were also skilled in many trades, especially in construction and also making battle and siege machines and equipment.

The most important factors in Roman's success in civil engineering and construction included:

- utilizing arch action,
- their ability to make stronger, durable and hydraulic concrete,
- their advanced equipment relative to their time,
- skill in structural and foundation construction,
- skill in surveying,
- the abundance of resources and manpower.

Use of Arches:

One key factor in Roman's success in construction was utilizing arch action. A well built arch does not even need mortar or anchors to hold the stones, as the stones under compression can hold themselves together, plus carrying a considerable load from above, even for thousands of years. So in addition to passing larger spans than straight stone beams, arches were naturally able to support much higher loads, even for that larger span.

Romans used arches to make bridges, aqueducts, buildings, amphitheaters... They sometimes made the arches in tiers as arched rows on top of each other to form arcades, or continued the arch in perpendicular directions to create vaults in the form of corridors or they intersected vaults perpendicularly, to obtain large interior spaces.

Still, there was a limit to how high and wide the Romans could build their structures. This was due to structural (geometric) stability, foundation soil capacity, limited resistance to earthquakes and not being able to support any tension or moment. The limited earthquake resistance was still obtained by arranging the structures in a way to brace itself.

Because of the natural stability of arches, many ancient structures built with arches still stand, and the ones that do not, collapsed from other reasons than the arc mechanism itself.

Roman Concrete:

Mortar and concrete were used before Romans too, but Romans were the first to use it widely. Roman concrete, although weaker than our modern standards, was stronger and more durable than the concrete made by other civilizations of its time (and even long after that) and also could harden and didn't lose its strength underwater, which means it was hydraulic concrete.

Roman concrete included lime, sand, aggregate in the form of crushed stone or brick, water, and volcanic ash. They burned limestone and obtained a powder called quicklime. Then by mixing it with water and sand, they obtained mortar. Volcanic ash which is pozzolan (*Romans called it "pozzolana"*), was soft and excavated from the ground, but when it was added into mortar and reacted with lime, it enhanced durability and hydraulic property. Due to being hydraulic, Roman concrete was also very useful in water related structures such as aqueducts, baths, sewer and water distribution systems, foundation walls, harbors etc... and meant faster, stronger construction and more possibilities.

Romans did not use concrete only as applying mortar between the stone blocks but also by making structures made of concrete, such as the 142 foot diameter Pantheon dome in Rome, which was built 2000 years ago and still stands as the largest unreinforced concrete span in the world.

Use of Equipment:

Romans had great skill in making construction and battle equipment and machines. A few construction equipment and machine examples are compound pulleys which worked as cranes, water screws which were used for moving water up, pile driving rigs, groma (a precise surveying equipment to set lines).

Now let's discuss different types of structures that the ancient Romans built.

Water Distribution and Aqueducts

Roman Empire's water distribution system was way ahead of its time and by far the most advanced in the world. The huge amounts of water that was supplied to cities, especially to Rome, made larger populations possible. At some point it is estimated that Rome had a population of close to 1 million people which was far above the standards of its time, that enabled a strong empire.

When all were completed, there were totally 11 aqueducts to Rome, with a total length of at least a few hundred kilometers, which might have brought up to a million cubic meters of water a day into the city (estimates vary). That makes almost 1000 liters per person per day, if we assume a million cubic meters and a million residents, although this includes not only drinking and domestic water used by people but also outdoor springs and water features, irrigation of farms and gardens, public baths, commercial use, mills, mining operations, palace and the houses of wealthy people and so on, which makes up the most part of that amount, but it still competes with our modern standards *(for example in the US, per usgs.gov website, a person needs on average 80-100 gallons of water daily, for indoor home uses, which means about 250-300 liters/day)*.

When aqueducts reached the city, some of the water were divided into a few storage and distribution reservoirs and distributed to the city. Some aqueducts were built to serve only particular sections of the city. The pressure of water mains were maintained at sufficiently high levels. To many parts of Rome, a network of minor aqueduct branches reached.

The first aqueduct to Rome, Aqua Appia, was completed in 312 BC, after which more aqueducts were built as needed, until the 11th and last aqueduct Aqua Alexandrina, which was completed in 226 AD, so they were built gradually to meet increasing demand. The route of a new aqueduct was planned from the mountain spring to the city, which not only included geometrical concerns but also inputs from affected land owners. The longest aqueduct to Rome, Aqua Marcia ran for 91 kilometers, which ran 80 km underground and its final 11 km above ground on arches. Of these 11 ancient aqueducts to Rome, one of them, Aqua Virgo, still functions. Romans built aqueducts not only in Rome, but also in many other parts of the empire. For example one of the longest Roman aqueducts, Zaghouan aqueduct in today's Tunisia, ran for 132 km.

In an aqueduct, water flows by gravity as open channel flow (see Chapter 1), which means, a downwards slope needs to be maintained. But this slope must be very small, because of only a limited amount of elevation difference over long distances such as many miles. Romans achieved this very slight slope with highly precise surveying, and over highly variable terrain, which required the aqueducts to run both over and underground.

The overground arched aqueducts that comes to our minds when we hear the word Roman aqueduct were actually only a very small portion of the whole length, and these were only built when they had to cross terrain with too low elevation, such as valleys, or had to run the duct around an obstacle and digging a tunnel would be more difficult. These overground sections were often made of arches on columns, which carried the ducts and water. The water channels were made of bricks, concrete or stone and were waterproofed with hydraulic lime mortar that contained volcanic ash, as mentioned above under Roman concrete. The cross section of channels were often large enough to fit men inside, to perform maintenance and cleaning later. Top of the channels were covered, which prevented evaporation of water and also kept it cool and clean.

Overground aqueducts sometimes had more than one level of arches in tiers, which is called an arcade, which is partially destroyed but the columns and the arcade are still visible. Another example, the Proserpina aqueduct in Merida, Spain is as much as 100 feet high, and has a three tiered arcade, which is one of the most famous Roman aqueducts, is also called as "los milagros - miracles". The Segovia aqueduct also in Spain, is one of the best preserved, has two tiers and a height of 28 m (91 feet). Another example is the Pont Du Gard in France, again a three tiered aqueduct and bridge at the same time. Aqueduct itself plus water is heavy, and the Romans built the foundations that did not settle too much, or did not even undergo excessive differential settlement from one location to another which was a great foundation engineering success, in addition to structural. Some overground aqueducts also served as bridges or entry gates.

Underground portions consisted mostly of underground trenches and pipes if the slope allowed, or, when there was a hill or mountain, tunnels through them. The pipes could be of stone or lead. The tunnels included access shafts, which were used for material delivery, removing excavated materials, ventilation, and for inspection and maintenance after construction. When crossing low valleys, in addition to overground arched aqueducts mentioned above, sometimes an inverted siphon system (*a system where water can flow first downwards and then upwards, given that there is sufficient pressure difference at both ends*) that ran underground was used, if the slope allowed. The tunnels or conduits had arch shaped cross section on top to carry the loads from above. For underground conduits, at every certain distance, an inspection cover was placed. Running of water underground kept the water cool, and it also prevented entry of dust and dirt from outside, and might have decreased the risk of destruction or poisoning by enemies.

Aqueducts cost a lot, and after completion they needed inspection, maintenance, cleaning and operation effort by many people continuously working. This is why in many Roman cities, wells, cisterns were also used or the cities were simply built near rivers in addition to or instead of using aqueducts.

Without the equipment, knowledge and technology of today, building such aqueducts that brought large quantities of water from many miles by maintaining a precise and very low slope without settling too much under heavy loads which crossed a variety of land features with different soil conditions, both over and underground, and then to distribute it with sufficient pressure to many parts of the cities, is certainly one of the top civil engineering and construction feats ever, with respect to its own era. After the Roman Empire collapsed, water distribution success in these quantities was not achieved again in the world until modern times.

Baths

Romans built many baths across the empire, which greatly increased public hygiene. Emperor Caracalla baths, was the biggest public bath in Rome, which opened in 216 AD. The bath's exterior is still visible. It included many large rooms, swimming pools, bathing pools at varying temperatures, private bathing rooms and remained in use for over 300 years. Supplying many different places with varying temperatures of water and then to drain the wastewater meant a well designed plumbing and sewer system. The water was heated in boilers over furnaces, which circulated the pools by convection. Construction required coordination of masonry, stone and brick work, plumbing and water distribution, sewer system, structural and foundation engineering, architecture, and concrete. Baths were heated by warm air (heated by furnaces) circulating beneath the floors. The space needed for air to circulate was achieved by raised slabs, which beared on short pillars, instead of directly on grade. Using Roman experience, an upgrade to Roman style baths were later built by Selcuklu and then Ottoman Empires, called hamams, which had running hot and cold water, instead of static still water in pools as had been in Roman era baths, which remained as the most advanced widely used and public bathing system until modern era.

Bridges

Romans built hundreds of bridges mostly as part of their highways. Some Roman bridges still stand, some of which are even operational. The success was due to using the arch action so effectively which enabled great load carrying capacity, while reducing self weight and passing long spans, their hydraulic concrete which did not dissolve in water, and their skill in building foundations, including even piles. Applications in each project varied but often they built the piers in water by first standing over rafts on the river, from which they drove the pile foundations (see Figure 4 below). Over the piles they installed footings from thick wooden beams, which served as base for stone footings, and then stones. After that the stones went up as piers, also forming arches.

Romans constructed not only stone bridges which took longer time, but also wooden ones. One famous example is the bridge Julius Caesar's legions built to cross Rhine River to fight German tribes. They quickly constructed a bridge in the Rhine river which was wide and had strong flow. Boats could also have been used, but Caesar might have expected (or wanted his opponent to expect) future use, and / or scare them by building such a structure which seemed out of reality at the time. To build it, Caesar's soldiers first drove piles into the river bed. In the deepest middle part of the river, if we assume the depth was 25 feet, it means the longest piles had to be at least around 40 feet, to have enough length to be driven into the river bed soil and to extend above the water to hold the bridge. Figure 4 shows a Roman pile driving rig's general features, where a weight hammered the piles from top and drove them. Battered piles also served as lateral bracing to resist water flow. After the piles were driven, they could be connected by beams on top, over which, the deck could be laid.

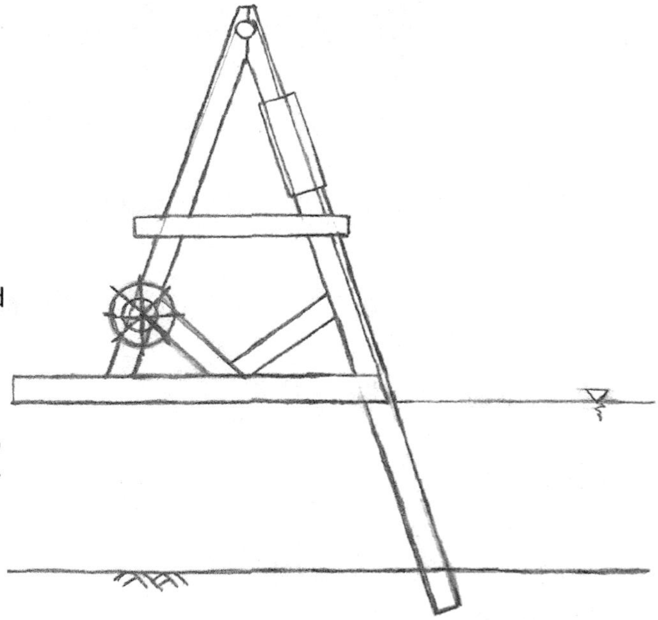

Figure - Historical 4

Dams

Romans built many dams, fed by aqueducts or rivers. They were built for irrigation, mining, river diversion, recreational purposes. Their dams were of earth or rock embankment or gravity type. They were waterproofed with hydraulic mortar.

Roads

Romans had a road network that expanded to all corners of the empire, from West Europe to Middle East and North Africa. Building this extensive network over a very large area had strategic importance. As once someone said: "*If you cannot go there, it is not yours*". Estimates vary but their total paved road length could have been as much as 100000 km, including posts along the way, to accommodate traveling soldiers. The total length of unpaved roads was far more than that and the entire road network reached to almost everywhere in the empire. These roads made possible for the Roman army to reach rapidly to war zones and trade and transportation to take place faster than anywhere else in the world. Of course one may think that other civilizations could also ride horses for long distances but still a straight and hard surface is better than an unpaved, longer route with many curves, and the difference would be pronounced even more because of bridges and tunnels that the Roman highways included, or during rainy weather, when the unpaved surface would be mud. The cumulative efficiencies would add up in the long run, both militarily and in regards to commerce, to contribute to overall strength of the nation. After the collapse of Roman Empire the roads they built continued to serve others for a very long time including middle ages, and many of them also greatly determined the routes for contemporary roads. Some Roman roads are still visible today.

Roman highway construction had many similar features and steps to what we discussed before in Chapter 4, Highways. The typical main steps to build a Roman road were:

- Determining the proposed route. They also utilized even older routes, if feasible. Apart from the cost and effort to build, the shortest possible length was the most critical for the marching armies or the speed of travel and commerce, as the road was traveled on foot or by horses. Thus, they tried to keep the routes as straight as possible, which sometimes meant building tunnels, bridges, retaining walls, excavation and backfilling, and stabilizing slopes.
- Surveying and route layout. In Figure 5, sketch of a common ancient surveying equipment called groma is shown, which allowed to set very precise straight and perpendicular lines.
- Clearing the area, removal of unsuitable topsoil.
- Grading and compaction of subgrade.
- Placing sub base, which was made of of boulders, gravel, rubble or sand, as applicable, which also allowed drainage.
- Placing the base, which included sand, gravel and crushed stone, mixed and compacted with mortar. The majority of the roads were unpaved, therefore this step was usually the last, unless stone pavers were used, which were made in areas with high traffic. The roads were sloped towards the sides with a crest in the middle, for water to drain and had drainage channels on

the sides. The side channels for water drainage were also placed in this stage. Also see Figure 3 in Chapter 4.

- Placing large stone pavers on top of the base and compact and firmly fix them onto the base. The stone pavers were made as smooth and level shapes and then placed tightly side by side, many of which lasted hundreds of years under heavy traffic.

Other features of Roman roads included:

- The roads widened near the city gates, to accommodate more traffic, just like the highways of our era which have more lanes when approaching major entry and exit points.
- The roads were marked with milestones at every Roman mile. The stones had information on them such as the mile numbering.
- As the road continued long distances, the construction materials could show some variance, which depended on the climate and the geology of the area.

Colosseums and Amphitheaters

The construction of the large colosseum in Rome was started by Emperor Vespasian, and completed in 80 AD under Emperor Titus. It is the equivalent size of a modern day stadium and hosted at least around 50000 spectators, and remained active for about 500 years. The structure was the highest Romans had built with a height of 57 meters, with lighter construction materials with increasing height. Construction of such a building also required sophisticated cranes which still worked with only muscle power. When completed, it became the tallest structure in Europe by far, and second to pyramids of Egypt in the world (which were already 2500 years old by then). It had many corridors, staircases, a plumbing and water delivery system and connection to sewer system. Beneath the arena there was a 2 story substructure called hypogeum, which contained a system of elevators and doors, worked by pulley systems that enabled wild animals and gladiators to suddenly be released on the floor. The colosseum in Rome is the largest of many amphitheaters that were built in the empire. The theater was partially destroyed by earthquake and some materials were looted over time.

Storm and Sanitary Sewer Systems

In the very early stages of Roman Empire in 6th century BC, the Romans drained the marshes and surface runoff in Rome into Tiber river by a large sewer system called "Cloaca Maxima". Roman sewer systems were made of large stone walled drains running under the cities. With precise surveying, very slight slope for open channel flow was achieved. The system, which drained both sewage and runoff water, served like a combined storm and sanitary sewer system of today.

This large drainage work enabled Rome and in turn the empire to grow. This is an excellent example of a well planned and built infrastructure system that does the right job at the right time for a nation.

Tunnels

Romans dug tunnels for various purposes such as to continue aqueducts or roads underground, when an obstacle such as a hill or mountain was met and it was less feasible to go around it. A tunnel digging technique was the Persian qanat method, where a series of vertical shafts were first dug from the top of the hill and then connected with the tunnel itself, which made layout of a straight route very easy, as it was first done at the surface. Other than ensuring a straight tunnel, the shafts also provided access and ventilation to the tunnel both during and after construction.

Temples

The best example to Roman temples is no doubt the Pantheon, completed around 125 AD. The structure and its large dome still stands today as the largest unreinforced concrete dome in the world.

The dome has a height of 150 feet at its center, and the unsupported span is 142 feet (with a 30 foot wide hole at the center). It pushes outwards at it base, which is an expected behavior of arches (see Chapter 1). The base walls that act as foundation walls also take that outward lateral load. The concrete mix was made gradually lighter, such as the lightweight volcanic pumice aggregate and the section gradually thinner, as the dome rose upwards. There are also recesses at the dome's interior, to decrease the load even further which also gave a much better appearance when looked from inside.

We said that the dome can be considered as an arch revolving around a central vertical axis. And for arches, we also said that the center piece at the top, called the key stone is the most important for completing the load transfer. But for the pantheon, there is a large hole in the middle, with an approximately 30 foot diameter. This is still possible, because, in 3D the dome can hold itself together with the help of a compression ring at its center, which is made of a series of stones, where all the revolving arches lean on. That ring holds itself together by compression of all stones by arch action, although this time the arch (stone ring) is a full circle.

And finally, below you see a sketch of the Roman surveying equipment, groma, which was used to layout straight and perpendicular lines with precision, for many structures that the Romans built.

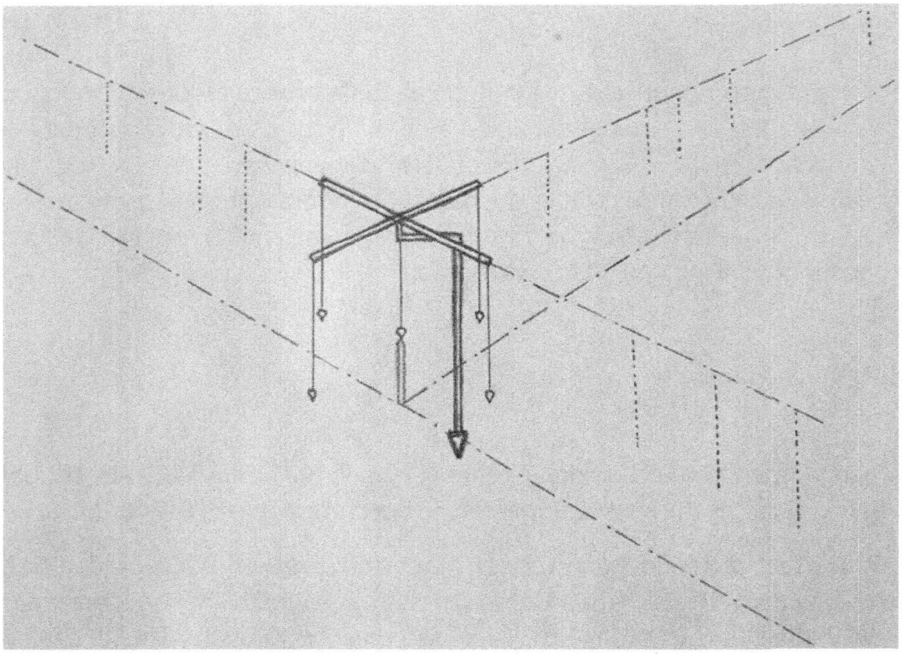

Figure - Historical 5

With this, we conclude this Chapter. As we said in the beginning of this chapter, it makes sense for today's people and engineers to understand what human mind can do when building something without the modern technology and knowledge, and hopefully this chapter gave some idea.

At this point, we finished all chapters that introduced civil engineering, structures and also their various essential design and construction aspects.

The last chapter we will study below, is rather a standalone one (but certainly not less important than the previous ones), introducing project management and scheduling.

CHAPTER 13 - BASICS OF PROJECT MANAGEMENT & SCHEDULING

Anyone in construction and civil engineering must at some point either manage or work on projects. A person's knowledge in civil engineering and construction may not be enough for success, if he/she had to manage a project but had poor project management skills or knowledge.

Now in this last chapter, we will look at project management, planning and scheduling, because everything we build is a project, which must be managed with proper project management practices.

To clarify first of all what this chapter is all about, let's first define the term "project" very briefly at this point, which we will discuss in more detail in the following pages: A project is a temporary work, undertaking, with definite start and end points in time, to deliver a unique outcome (result).

So in this chapter, which can be considered as a standalone chapter in relation to the rest of the book, we give an overview of project management, with particular emphasis on scheduling concepts. We say standalone because project management theory is applicable to any field and not just construction and civil engineering. Therefore, the contents in this chapter is independent from the rest of the book except the examples from construction that we use, which could also have been examples from another industry. In other words, even if this book was about electrical engineering or agriculture or space programs, the contents of this chapter would be the same, except the examples, because in all industries there are projects and the project management theory applies to all fields in the same way.

Let's repeat what we said in the very beginning of this book: Civil engineering has 5 main branches, where four of them deal with design, such as structural, geotechnical, water resources and transportation, and the fifth one is construction of these designs, which is construction management.

Construction management is known to be related the most with project management, and this has good reason because construction of a project is by far more expensive and involves more resources and more coordination than the design of that same project, in other words, most of the overall project resources and money are spent during construction, rather than design. But engineers in the other four branches, the design engineers, must also understand project management, not only because they can envision and follow the construction projects better when the construction is underway, but also because producing a design can also be viewed as a project in itself from the design office's point of view. And those design projects are managed by the design engineers. Therefore, all civil engineers, no matter which branch they work on, and anyone else involved in design and construction in other capacities, must understand what project means and how it works.

A project manager with a good understanding of project management theory, will effectively run the mechanism that makes up a working project. For example, he/she will include and coordinate qualified people with high technical knowledge and effectively integrate them into the process, no matter what level his/her own technical knowledge is *(still, in practice, people with high technical knowledge are selected as project managers but this was for the sake of giving a theoretical example)*. Being an effective project manager requires coordination, communication, intuition, practicality, follow up, organizing, initiation skills, in addition to having adequate technical background to understand and run a project. For construction project managers this is especially true, due to the amount of money and the number and size of variables involved and the speed of everything, which can place significant amount of stress on the project manager and his/her team.

This chapter presents some of the most essentials of project management knowledge and how they fit together in project management, focusing especially on estimating and scheduling, but mainly scheduling. We will cover scheduling in much greater detail, because it lies at the very center of project management by having relationships with the most number of processes, and ties everything together. A proper understanding of how the schedule works is key to understand how project management works overall. This is why this chapter is called "Basics of Project Management and Scheduling".

PROJECT MANAGEMENT

A project is a temporary (not ongoing or repetitive) undertaking, with a definite start and end date, a definite planned cost and planned time of completion and a definite unique result to be achieved with desired qualities. The previous sentence is fundamentally important and was underlined entirely for you to keep in mind the whole time while you read this chapter, which you will see why as you read.

In order to be considered as successfully completed, a project must satisfy the planned criteria in the previous sentence for:

- Schedule (Time)
- Budget (Money)
- Quality

These three items keep balancing (pulling) each other, in the sense that if we try to satisfy one or two, the other one or two will tend to slide towards the failure side. Proper project management practices seek to make all these three aspects succeed, which means, the job gets completed within the planned time and budget and the produced unique result - also called as deliverable or total work scope - meets the scope and quality criteria that were specified in the beginning.

The deliverable can be anything; a software product, a building, a tunnel, establishing a type of customer service system, producing the designs of a car or a bridge, making a space shuttle... Example: Manufacturing of telephones in a factory is not a project. It is not a temporary operation with a unique outcome. It is an ongoing, regular, repetitive operation with no definite start and end points. But producing the original working prototype of that telephone, is a project in itself. The prototype is the unique deliverable of the project which can be called such as *"making a prototype for the cellphone model xxxx-xxx which will be manufactured in 2023"*.

The start point of the project will also be the start point of the schedule. As example, for a construction project, this is usually taken as a milestone called Notice to Proceed (NTP).

In order to produce the final deliverable of the project, there are a lot of processes, all of which fall under the following 5 process groups (also called stages) per PMBOK® (Project Management Body of Knowledge) standards which was published by Project Management Institute (PMI), which are:

- Initiation
- Planning
- Execution
- Monitoring & Control
- Closeout

Let's briefly describe these five groups, both generally and also in construction context.

Initiation

Project emerges from the owner's need, who hires consultants. The main requirements to satisfy this need are outlined while stakeholders state their needs, to come up with a more precisely defined end result. The project team starts to be formed. Stakeholders include owner, architect, engineers, contractor, owners and residents of affected neighboring facilities, end users of the project, affected public agencies, and more as applicable.

Planning

Starting with the initial requirements from initiation, now the work scope and its qualities are clearly defined in detail and the project is designed, which can also proceed simultaneously up to a point. Based on these, the planned work schedule and budgeting are completed. The baseline, which means, a snapshot of the project schedule and budget for a given complete scope is created. The work schedule saved at this point in time is called the baseline schedule. Everything in terms of time and budget performance of the project is compared to this baseline later during monitoring and controlling stage. When making the schedule, costs and resources are assigned to each work scope item, which means a cost and resource loaded schedule. This is how we can compare the planned vs. actually earned values later during monitoring and controlling stage. The schedule is heavily used and updated during execution and monitoring and controlling stages.

Foreseeing all possible real world issues during designing is in practice not possible even for the most experienced and skilled design teams, even if they received good input from field people and stakeholders. There are simply no designs that "build themselves" in the real world of construction. Therefore, one of the main goals of the design team is to minimize the future changes, even if they cannot be eliminated entirely. As soon as the project documents are sufficiently completed for overall scheduling and budgeting purposes, through bidding and contract award, the party who will perform the project (contractor) should be selected, so that he/she can plan his/her own resources, prepare submittals, get approval and order long lead time items, and provide any possible useful inputs to design team who can further refine designs. These changes now will mean change order which may alter the contract award price if they modify the project scope awarded to the contractor, but even so, the contractor input is still valuable here because the price of making changes in the planning stage are still lower than the price of making changes during execution stage, moreover, this "change" is often something that was already needed for properly achieving the project goals, so it would have been in contract award price in the first place. Such inputs can also come during bidding stage from different contractors when the contractor bids are reviewed and before the contract is awarded, which is another reason that, as soon as the project documents are sufficiently complete, they should be sent out to contractors for bidding.

Execution

This is the part where the job actually gets done, in other words, deliverables are produced per the defined project scope. Therefore, construction falls under execution phase. It is the stage where most resources are spent. Project management team manages construction of the project, by trying to stay within time, budget and quality constraints, at the same time coordinating all activities. This "staying within constraints" is constantly checked with the monitoring and control process, listed below.

Monitoring & Control

Through this process it is continuously checked if the project execution is running and progressing per the plan, which means, the approved scope, schedule and cost baseline and the quality criteria that was determined in the planning stage. If it is not, then some of the tasks might be sent back to planning stage and be executed and checked and monitored again. So, in other words, after the initiation is complete, the three stages of planning, execution and monitor & control will keep running and repeating as a continuous loop, to evaluate the performance of the project while executing it, and corrective actions are taken as necessary, until the required project deliverable is obtained. Not only monitoring the existing scope but through this process, the work scope may also be changed, as long as the owner approves it, and then it also enters into this loop. The key concept during monitor & control process is the Earned Value Management which we will discuss.

Closeout

Closeout is making sure and verifying that the project satisfies the final objective, and finalizing all documentation and outstanding issues. When the construction of the project finishes, and upon completion of the closeout stage, the project ends. Repetitive operational items after that, such as operating and maintaining a bridge, dam, sewer system, building, will not be part of the project.

ESTIMATING

Estimating is determining the monetary value of the work, before it gets built. For this, the cost of individual activities must be found, which then makes up the cost baseline and the budget of the project. The earlier the estimate is made, the less accurate it will be, but regardless of this, it is needed as soon as the design documents are adequately complete, in order to proceed with budgeting and scheduling. Estimating requires good expert judgment often. There are a few main approaches to estimating:

Bottom up estimating: In this approach the cost of all activities are calculated in the most precise way as possible, by finding out the amount or quantity of the work first and then the resources that is needed to complete it and multiplying by unit costs. Therefore this is the most detailed but time consuming approach.

Analogous estimating: In this approach, the values from similar past projects are used to calculate the durations and costs of activities.

Parametric estimating: This is a shortcut method that uses unit values from similar past projects, by multiplying them with the overall quantities. For example the construction cost of a new office building can quickly, roughly be estimated based on the per square feet price of a similar office building project that was completed before, under similar circumstances. All we need to do is to multiply that unit price with the expected gross area of the new building.

In estimates, in addition to calculated activity costs, there are contingency and management reserves. Contingencies are for known unknown costs, such as paying for removal of a known existing obstacle. Contingencies are already included in the baseline, so they do not cause change in baseline when they are used, although owner approval will be needed to use them and move the funds from contingency to the appropriate places in the estimate and schedule. Management reserves on the other hand, are for unknown unknowns. They are not in the baseline, but part of the overall project budget. We will also explain this under the term Earned Value.

SCHEDULING

Schedule is the display of activities (tasks) that make up the project scope in time sequenced format, including their relations, dates, durations, milestones, constraints, start and end dates at a minimum. To function properly for performance measurement, *(which is done by monitoring and control process)* a properly made schedule should also include costs and resources assigned (loaded) to those activities. The activities that make up the project scope are grouped under Work Breakdown Structure (WBS), which should be <u>fully</u> represented in the schedule.

Of the 5 stages in project management that we mentioned, schedule is created during the planning stage, and afterwards it is one of the most heavily used documents of project in all stages. The schedule in project management is similar to the keystone of an arc, as everything in project management is directly or indirectly related to it and almost everything can be tracked with it, so it lies in the very center of project management. For example, from a simplified summary schedule with only 15-20 lines, and several columns showing only dates and activity relations, it is possible to derive hundreds of questions or key points about a project, in relation to many different aspects (such a schedule and some example questions are provided at the end of this chapter). This is possible because schedule connects almost everything in project management to each other. This is why, in this chapter which is about project management, our focus is scheduling. From here forward the subtitles below are terms and concepts related to scheduling, and necessary terms to know about project management, which still affect schedule.

The list below is in alphabetical order, so it can be read in any order. Almost everything in this list are interrelated. So you can also move from term to term as necessary. Sometimes the same information can be seen under different items, which may also help better understanding by seeing the same term in different contexts and from different angles. If you now proceed reading in alphabetical order and see a term that was not described before, chances are that you may find it in the rest of the terms.

Activity

The total project scope is divided into separate, <u>distinct</u> tasks called activities. Activities receive their codes according to Work Breakdown Structure (WBS). These codes should not be confused with activity ID numbers in the schedule. Activity ID numbers are simply created automatically by the scheduling program in sequence, as we enter the activities in sequence (which we can change manually if we wanted). WBS codes however, are the codes that show where an activity falls under the grouping and breakdown of the work scope, organized in logical hierarchical order. Therefore, the WBS code of an activity cannot be arbitrarily assigned or changed. Also see the term WBS.

Activities have durations, start and finish dates, and, in a proper schedule, they are also assigned resources, and sometimes even risks. Activities ideally should be linked by finish to start links only. Assigning lags during link (relationship) assignment is not a preferred practice, unless it seems that there is no other choice. And assigning negative lags are not wanted at all.

Activities should be <u>unique</u> for each task and be given proper, <u>descriptive</u> names. Several tasks should <u>not</u> be combined under one activity as a shortcut, but each of them should be represented by separate activities, although this must not be overdone in order not to congest the schedule and make it too complicated to track. For example, pouring the 5th floor concrete must mean pouring the 5th floor concrete, but it should <u>not</u> mean placement of formwork and rebar also. Other than being too broad, this would cause difficulty in assigning resources, and calculating start and finish dates. But with the same logic, pouring the 5th floor concrete should also not be separated into too much detail, such as, pumping the concrete from truck, vibrating wet concrete, troweling etc... Doing so, will not only gain us nothing, but it would also congest the schedule and would create too many lines and relations and resource assignments and makes tracking and making updates difficult.

Milestone:

A milestone is an activity with zero duration and without any assigned resources or costs, so it represents zero amount of work. Typically it is used for important events, such as notice to proceed, substantial completion, completion of a phase or location, end of the project and so on.

Summary Activity:

Related activities in a schedule can be grouped under a summary activity, and that summary activity can be placed under a higher level summary activity as well, where other summary activities may also exist, and this hierarchy can grow to as much levels as needed, based on the WBS. The larger the project size, the more level of summary activities. Although theoretically there is no limit to the number of levels, large projects may have around 6-7 levels of summary activities.

A summary activity has nothing assigned to itself, but serves as a container for the activities under it, in every way:

- For example, it starts with the earliest start of the earliest starting activity and finishes with the latest finish of the latest finishing activity of the activities under it, which are automatically calculated. Assigning start and finish dates, or a duration to a summary activity is wrong practice and will mean assigning constraints to that whole group of activities under that summary activity. If the constraint is wanted somewhere, which is still not an ideal thing to use, they should be assigned to individual activities as required, such as to the start date of first activity and or to the finish date of the last activity under the summary activity. The summary activity must also have no links assigned to itself, otherwise conflicts or circular references in schedule calculation can happen.
- The summary activity also represents the total of work, resources and costs of all activities under it, but no costs or resources should be assigned to itself. This means, the sum of all summary activities at the highest level will represent the entire project as far as the, total duration, cost and resources used, and the scope of work, with nothing more or less. This is called the 100% rule in scheduling.

Sometimes management may want to view a high level schedule, showing high level summary activities only. These should be obtained by working on the details of the individual lowest level activities, then collapsing them and keeping all calculations automatic, so that the summary activities are automatically updated. In other words, nothing should manually be entered into summary level activities.

Hammock Activity:

A Hammock activity is a supplementary activity which is not part of the project scope but has its own costs as overhead. A hammock activity has its start and end dates determined by other tasks but unlike a summary activity, a hammock activity has no activity under it. The only relation of a hammock activity to other activities are specifically the points A and B as seen in the figure below, and nothing else.

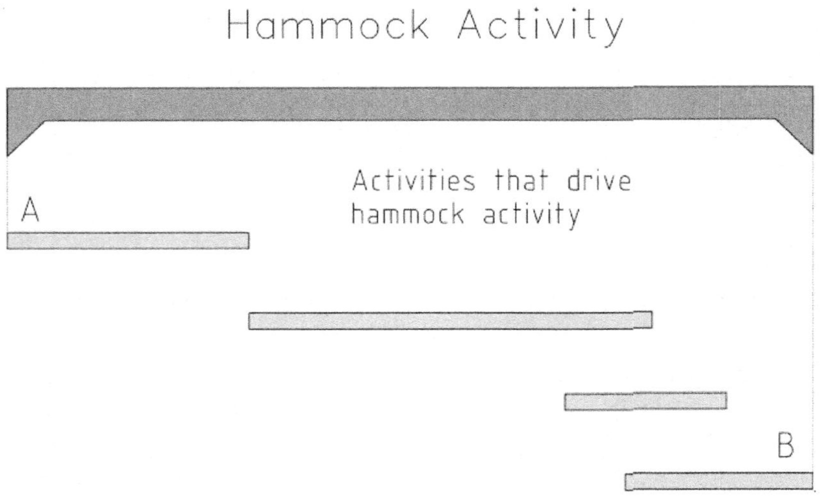

Figure - Project Management 1

Therefore a hammock activity does not have its own duration, but it has its own costs or resources. The resources and costs assigned to hammock are not included in the project earned value calculations, because they are not part of project scope, and similarly they do not produce any work. Therefore a hammock activity is a good way to represent supplementary, overhead type activities, which still affect the overall project budget, even though not the earned value. For example, for a facade renovation work, you need scaffolding outside. The rental of scaffolding starts when the first activity of renovation starts, and it can only finish when the last activity of facade work finishes. Whatever happens between points A and B for those activities, as far as their costs, resources, work amount, is irrelevant to scaffold, except the start date A, and the finish date B. A hammock activity can also span between two milestones. The only important thing is that it is logically tied to two points in time, which are A and B.

Baseline

As far as earned value calculations, a baseline represents an approved state of the project schedule, cost and scope, to which, all future progress is compared for performance monitoring, therefore it serves as a benchmark. Schedule baseline is the original approved schedule before any progress. Scope baseline is the originally approved project scope. This scope is represented in the schedule by dividing it into activities per WBS, and logically linking them. Cost baseline is the sum of all activity costs (estimates), plus any contingencies, as of baseline date. After addition of management reserves, it becomes the project budget. Costs and resources of activities are assigned in the schedule to the activities for a cost loaded schedule. Change orders, upon approval by owner, modify the baseline.

Budget

Budget is the total amount that the project will cost or is supposed to cost as estimated during planning stage. The approved budget, with all its broken down items, becomes the cost baseline.

Calendar

There may exist more than one working calendars in a schedule, with different working times and holidays. These can be assigned to an entire project or specific resources or activities, where progress can only happen on the working days of that particular calendar.

As a few examples:

- We can assign an outdoors calendar to outdoors activities which are affected by weather.
- A resource, if assigned a certain calendar, will only be available to work on the working days of that particular calendar which in turn affects the activity that that resource is assigned to.
- If we changed the calendar assigned to the overall project, the start and finish dates for many or all activities may change.

A common mistake is to skip assigning calendars to a schedule and let the program use default calendar. This may work for simple projects, but it usually won't, for the specific circumstances and time frame of that project. So the best is to assign a calendar to at least to the overall schedule and enter as much information as possible into that calendar as far as non working days, national holidays, etc... Instead of assigning the nonworking days in a calendar, we must <u>not</u> just increase the duration of activities, which would be against the logic of earned value calculation and tracking performance.

Change Management

It is the management, control and tracking of changes, which are represented by change order requests. A change order typically modifies the project schedule, scope or budget upon approval. And less commonly it can also modify a document, policy or plan, quality requirements, or it can even be to prevent a future negative impact. Typical causes of a change order include: A change to scope of work, unforeseen issues, owner or administrative delays. The cost of making changes are lowest in the beginning and increases as the job progresses.

Constraints

Activity constraints enforce certain conditions or dates on activities. From weakest to strongest they can be listed as:

- Zero Free Float
- Zero Total Float
- Finish No Later Than, Start No Later Than
- Finish No Earlier Than, Start No Earlier Than
- Start On
- Must Start, Must Finish

Constraints should not be used unless really needed, and in such cases the reasons must be clearly described and justified in the schedule narrative. Constraints prevent free calculation of the critical path, by imposing conditions and can even cause negative total float. Introducing a constraint to an activity can make it a critical activity. This may mean unrealistic dates for the rest of the activities. It also makes tracking resources harder. Too much constraints will make a schedule very hard to manage.

During making or updating schedules however, the scheduler can frequently use constraints temporarily, because this allows instant testing of the float for certain activities or milestones, or seeing the project finish date if a certain date for an activity happened.

Cost

Cost is the monetary value that is assigned to activities and also to resources. The total of these make up the sum of activity costs only. To make up the entire project budget, on top of this, activity contingencies, project contingencies and also management reserves are added. These are shown in more detail under the term "Earned Value Analysis" (EVA).

The total cost incurred by the project is slow at first but then it gradually increases. Cost spending reaches the peak rate during execution stage, and then starts to decline. The costs and resources (which also bear costs) of activities must be assigned to them in the schedule (cost loading) in order to properly track the performance of the job through monitoring and control process. The progress of spending time and cost is constantly monitored throughout the job and compared with their respective planned values. These are also commonly called as cost control and schedule control, or simply project control, as they are simultaneously performed, though EVA.

Cost control includes:

- Making sure that activities are assigned correct costs and they use their own funds as the job progresses
- Evaluating change requests and processing them for approval or rejection
- Evaluating work progress to understand the progress of resource (therefore cost) usage, which also means evaluating job performance by comparing the planned vs. actual usage

Critical Path

Critical path, as its name suggests, is a special path in the schedule, which would cause delay of the whole project if it delayed even a little bit. This path is formed automatically upon assigning links and durations to activities.

Total slack, also called total float, is the number of days that an activity in a schedule can delay before delaying the entire project. Free slack (free float) on the other hand, is the number of days an activity can delay without making its immediate successor critical.

From the definition above, it can be understood that the critical path has zero total float because any delay in critical path will immediately cause delay of the entire project. This also means, all other branches in the schedule will have positive total floats, since they are not as critical and can have some delay before causing delay to the entire project. If a noncritical branch delays more than its total float, then that branch will be critical now, and the previous critical path will start to have positive float, and so on... So, critical path has the longest total duration than other branches. In other words, we must spend at least that amount of time, equal to total duration of critical path, to complete the project.

In the schedule, all branches are formed though the link relations of the activities that we assign to them. Which of these branches will become critical, and how much total float the other branches will have, is an automatic calculation by the scheduling program, upon our update of activity progresses and recalculation of the schedule. Therefore per these calculations, the end date of the project can move freely but no matter which branch becomes critical, that branch will have zero total float. The only exception is when a constraint is assigned, which may cause negative float as it prevents free movement of the critical path, as we explained under the term constraints, which is not a preferred action unless needed or dictated by contract.

In a schedule, all activities, even if there are thousands, must have a successor, except the very last activity or milestone, to which all branches of the schedule will ultimately be tied to. If this is not done, individual branches will start becoming critical on their own, in other words, there will be more than one critical paths, which means more than one floating ends in the schedule, which is not a proper way of scheduling. See figure below. Activities 1, 2 and 3 have no successors, so the program automatically makes the branches leading to them critical, which means all those branches will have zero total float. Also note that we commonly show activities with rectangles and milestones with diamonds.

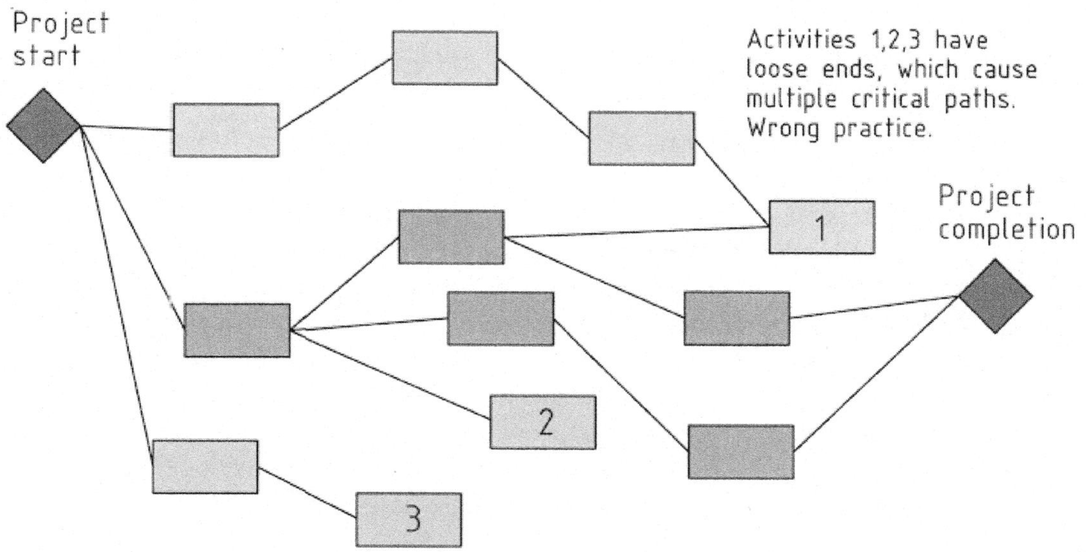

Figure - Project Management 2a

See correct practice below, where activities 1, 2 and 3 have successor. Here only the last activity, project completion milestone has no successor. This way there is only one critical path. We tied activities 1, 2 and 3 directly to project completion, but that doesn't need to be the case. They can have other successors too. All that matters is that ultimately all those must also be linked to project completion.

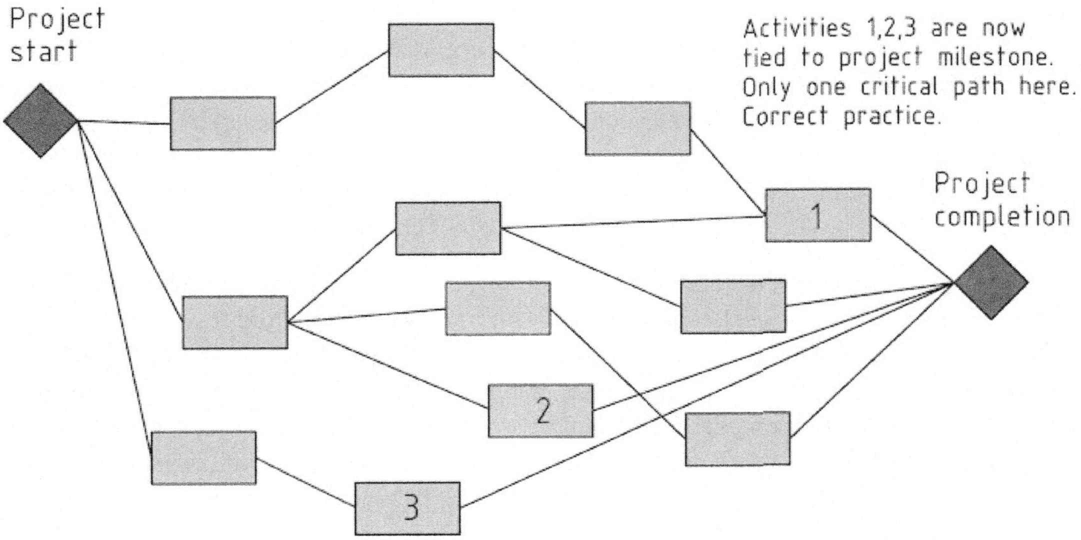

Figure - Project Management 2b

By successor, we mean at least one FS link as successor. If an activity has a successor with a SS link, and no other successor, it means that the successor only depends on the start of the predecessor, which only ties the start of the predecessor into critical path, but its finish will just float around without link to the project finish and will create another branch with its own end date, which means another critical path and finish date. A scheduling program may not show it as a separate critical path, if that loose end of the predecessor finished earlier than the project finish date. If you allow the program to calculate multiple critical paths, then it will show it as a critical path too.

Similar logic is true for project start and predecessors. All activities must have a predecessor, except one initial activity or milestone. We cannot have multiple activities serving as start point of project.

Activities have Early Start (ES), Early Finish (EF), Late Start (LS) and Late Finish (LF) dates. These represent the earliest and latest dates that the activities can start and finish. The early dates are calculated by forward pass, and late dates are calculated by backwards pass. We will not show details of their calculation but it is very straightforward and can easily be found in many resources. Forward pass starts with the project start date and continues forward. Backward pass starts with project end date and continues backwards. After the early and late dates are calculated, the total and free floats can be calculated for each activity. This is how we can manually calculate our schedule but it is almost exclusively done by computers now, and manual calculations are now only for educational purposes, which the computer programs in scheduling also follow. The important thing is that the early and late start and finish dates should be calculated based on schedule logic, but not manually entered. Manually entering these dates will mean enforcing constraints for these dates.

The total float is calculated as: TF = LF - EF or LS - ES. Free Float is calculated as: Earliest Early Start of immediate successors minus Early Finish of the activity. The activities on critical path have their total float and free float values equal to zero.

Critical Chain

Critical path method described above treats resources as if they were infinite. Critical chain method solves this problem by also considering limited resources, and delays tasks with overallocated resources. Overallocated resources are the resources that are expected to perform more work than they physically can, for a given time bracket. Allowing certain tasks to continue longer, so that they use resources slower, solves resource overallocation problem. This is called resource leveling, which may cause changes to critical path. Therefore, by using critical chain method, a more realistic version of the critical path schedule can be obtained. The resources on the critical chain are called critical resources.

Another significance of critical chain method is the placement of buffers at the end of certain branches (feeding buffer), or the entire schedule (project buffer). These buffers account for uncertainties, risks and limited resources. So instead of assigning longer durations to tasks to account for these (which is an incorrect practice anyway), here we give more realistic, realistic durations to tasks which reflect their real work value better, and we still do not sacrifice safety by using buffers. Instead of making activity durations longer, assigning buffers eliminates setting ourselves up for delays such as:

- Due to Parkinson's Law, when people are given time, they will take their time, and use it till the end, even if they might have finished it earlier.
- Even if the task gets finished early, its successor may still need to wait to start because it was not expected to start and its resources will not be available until a later date because of the unnecessarily longer date we had assigned to the first task. We said the resources may not be available because for example many companies use resources on multiple projects with a predetermined resource calendar, or the work space or permits may not be available for the next task yet, or simply the crew cannot be mobilized on time since the next task was not expected to start until a later date. So, time savings may not be reflected, but the delays will be.

So with critical chain method, we will have an advantage to finish earlier, and at the same time will have a more realistic schedule, due to resource leveling. Therefore, critical chain method evaluates the usage of buffers, for performance evaluation.

There are also resource buffers, in addition to feeding and project buffers. The resource buffers are assigned to resources in the form of not time, but in the form of resource such as people or equipment.

Other than resource leveling, there is also the term "resource smoothing". This does not delay the project unlike resource leveling, but it only levels the resources of activities only to the extent that their float allows, in other words by using their float time. As you might have guessed, this can be risky. This should only be applied if we know what we are doing and we are confident that this will not increase the risk of causing delays to the successors, because we are talking about using all available float here, which gets it close to a zero float, critical task. The benefit of resource smoothing may be to ease burden on resource overallocation (although it cannot fundamentally solve it as in resource leveling), and also in situations very sensitive to prompt cash flow, this will ease the burden on the firm. Another benefit may be easier field coordination with fewer people and possible benefits to the quality of work produced, depending on the nature of the task.

Critical chain is a relatively newer method which was introduced in 1997 by Eliyahu M. Goldratt, an Israeli business manager.

Data / Current / Status Date

When the schedule is updated each time, the current date is updated, the progress is brought to this current date for all activities, and the schedule is recalculated. The amount of work done, percent complete values, remaining durations, usage of resources and money are all updated. This is also commonly called as "making the schedule current". Current date is also called as data date or status date. After making progress current, and the schedule is recalculated, the progress as of this status date is compared to baseline in order to make earned value calculations, to measure project performance.

Current date is always brought to present day for regular progress update. For analysis purposes however, other dates can also be used, such as for forensic claim analysis.

Delay

Analyzing and including delays is an inherent part of monitoring and controlling process. When delay occurs, it is compensated by using appropriate contingencies or buffers in schedule.

The more time passes between making a mistake and discovering it, the costlier it will be to fix it and more delays it will cause.

Concurrent delay means multiple, simultaneous delays, which originate from different sources, such as the owner, consultants, contractor or "Force Measure". Proper documentation is necessary for assigning responsibilities later. Force Measure means, situations that could not have been reasonably predicted or controlled by anyone.

A delay may or may not be excusable to the contractor depending on its nature, local laws and the contract terms.

Anticipated or already occurred delays can be compensated for as described under the term Schedule Compression / Acceleration.

Delay Analysis

Delays can be analyzed in various ways. They can be categorized as backward looking, contemporaneous, or forward looking. The first one can be applied only after the delay occurred, the second can be applied simultaneously, and the last one can be applied even before it happens or as it happens, to make a best guess. The most frequently used delay analysis techniques are:

- Time Impact Analysis
- Impacted As-Planned
- As-Planned vs. As-Built
- Collapsed As-Built
- Window Analysis

Delay and claim analysis is the most complex part of scheduling, which usually means a lot of money to the parties involved, when considering the already high amount of costs in construction industry plus the litigation costs. Usually, parties tend to try to settle disputes through arbitration. Delay analysis must be performed by experienced teams in scheduling and project control.

Disruption

Disruption is similar to delay, and may be confused with delay often, to the degree that many people simply call it delay, although it is not the same thing. A delay happens because someone was just too slow or late, whether it was the owner or contractor. But disruption is slowing down of someone (usually the contractor), due to things that hinder progress of his/her work. This is more difficult to discern, especially in the construction environment where a lot of things happen at the same time, therefore detailed documentation is necessary, in addition to regular daily reports, where issues are only briefly written. As they are harder to document and prove, disruptions can be a source of disagreement between the contractor and the owner, such as when the contractor estimates a disruption amount and submits it in the form of a change order request.

Duration

Duration is the amount of time (almost always in days), that the activity is going to take to finish. It can either be entered directly from previous experience, or preferably calculated by using the number of resources, their production rate and the total amount of of work to be installed, which will give the duration. For example if a worker can lay 100 feet of utility pipes a day and the total amount of pipe to be laid out is 1000 feet, then it will take 10 days to finish the task, if he/she was working alone. When we add another worker to the job, then it will take 5 days. So,

Duration = Amount of work / production of assigned units (resources) per unit time.

Deliverable

As mentioned in the beginning of this chapter, a project is about producing a unique result. This is the entire work scope, also called the deliverable. It is the final result that the owner wanted, which the undertaker (contractor) must deliver to the owner, as specified in contract documents, (drawings, specifications and contract). The final deliverable is divided into meaningfully distinct tasks called activities, as grouped by the WBS.

In addition to tasks that are part of the scope (deliverable) itself, there are also supplementary activities that we perform on a project, which are not part of the scope, yet still needed in order to be able to perform the actual scope of work. For example, setting up the jobsite trailer, or surrounding the construction site with fences, laying cables for temporary power and so on. These items are part of general conditions. Although they are not part of the actual deliverable, they are in the necessary overhead that we need to spend in order to get the work done, so their usage and spending is also considered part of earned value as they are in the project budget.

There are also contingencies. They can be blended into the cost of activities and we can also have an overall contingency for the project. How contingencies can fit in overall budget is shown under next title (EVA) as the last item in the list there. Their way of entrance into EVA depends on the contract.

Earned Value Analysis (EVA)

EVA is mainly comparing the value earned, which means, how much work has been produced, versus planned value of work, as of a point in time.

In the figure below, you can see how the terms in EVA fit together. Earned Value (EV) is also called as Budgeted Cost of Work Performed (BCWP). It means the dollar worth of work which is completed to date. EVA is one of the most important concepts in project management, and forms the basis of evaluating performance.

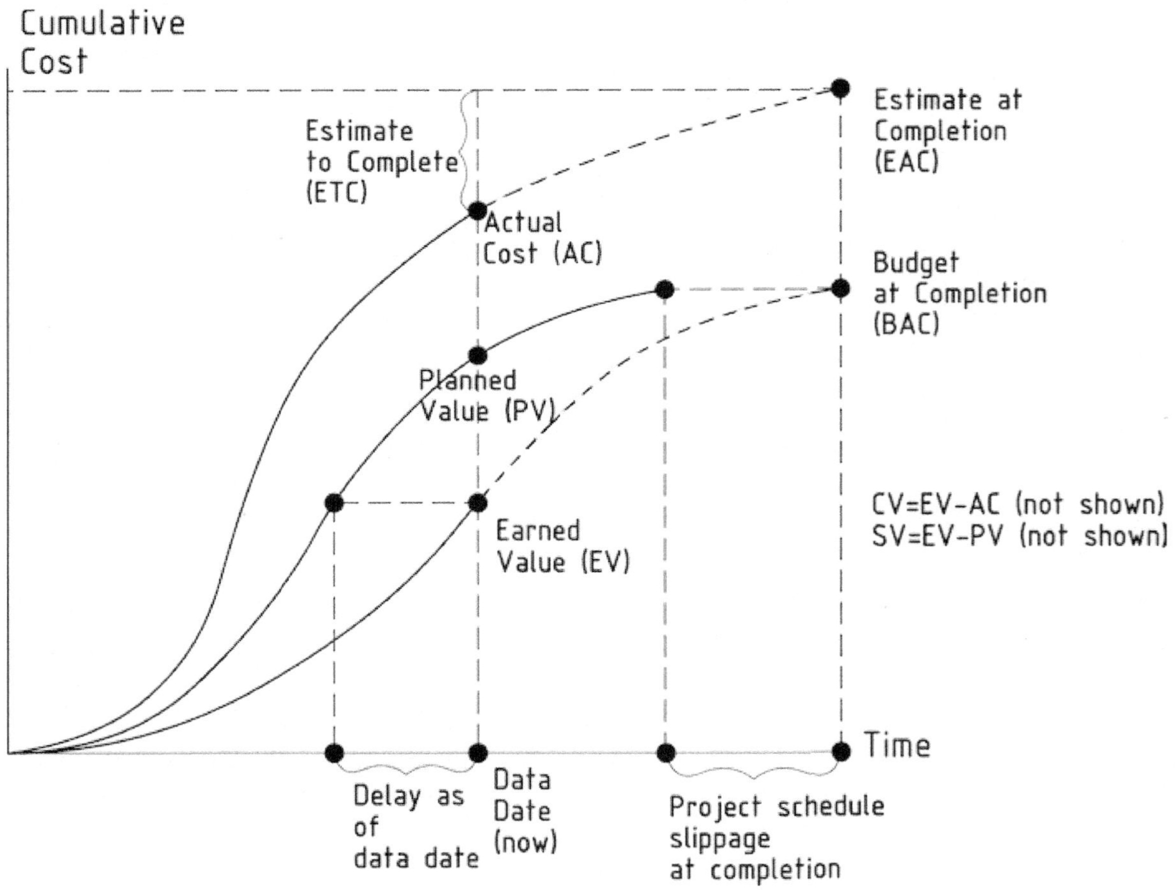

Figure - Project Management 3

Important items to note here:

- According to this graph, the work is behind schedule and above budget as of now. It is also estimated that it will be over budget when completed. The work is behind schedule because for the data date considered, originally we thought more work would be done, as you see in the PV (Planned Value) curve, but the EV for that same planned date, is less than that. It is also over budget because AC (Actual Cost) value is higher than PV for the current date.
- Data date is chosen as present date, but for claim analysis, other dates can also be chosen.
- Notice that EV eventually becomes equal to PV at the end of the job. Because delayed or not, the total work is the same, so, no matter how late or early, EV will be equal to PV eventually.
- See that the projected, future portions of EV and AC curves are shown as dashed lines, as they are only estimates at this point and may vary towards better or worse. But also notice that PV continues a solid line to the future, because it was already planned from the start, which will not change. That is the baseline which was already approved in the beginning.
- Management reserves, which are for unknown unknowns, are not included in baseline. They are in addition to baseline which means they are not a part of EVA. They can also be in terms of duration, not just cost. When they must be used, even if they are not included in the baseline, still a change order is needed, because this used money is now being added to the baseline.
- AC is also called as Actual Cost of Work Performed (ACWP)
- CV (cost variance) = EV - AC
- SV (schedule variance) = EV - PV
- VAC (variance at completion) = BAC - EAC
- SPI, Schedule Performance Index and CPI, Cost Performance Indexes are two of the most important performance indicators that managers look at. They are found as:

  SPI = EV / PV

  CPI = EV / AC
- EAC = BAC / CPI

  or

  EAC = AC + BAC - EV
- ETC (estimate to complete) = EAC - AC (assuming work will progress per plan). ETC can also be estimated if it is preferred.
- TCPI means, To Complete Performance Index. It is the ratio of the remaining work, to the remaining funds. So,

  TCPI = (BAC - EV) / (BAC - AC)

  or

TCPI = (BAC - EV) / (EAC - AC) *(if BAC can not be achieved and owner approves new EAC)*

- And although not displayed on the graph for clarity, the relationship between management reserve, project budget, contingency, and activity costs can be written as follows:

  Activity Costs + Activity Contingencies = Total of Work Packages

  Total of Work Packages + Project Contingency = Cost Baseline = Total of Control Accounts

  Cost Baseline + Management Reserve = Entire Project Budget

  Note that how contingencies are entered into EVA depends on the contract type and specific terms of the contract between the owner and contractor and the type of contingency.

Also to see an example of how we calculate PV, EV, AC, CV and SV, see the term "work", and also the exercises at the end of this chapter.

Gantt Chart

On a schedule, Gantt Chart is the format that we show activities as bars in rows. The bars are scaled according to activity durations, with their start and finish dates according to the proper chosen timescale. Early or actual start and finish dates, activity descriptions, IDs, durations, float amounts, link relations, % complete values, are also commonly shown and a lot more columns are possible to see depending on the need.

The column layout, bar format, level of detail will all vary greatly according to the program and the need of the project team. We can also choose to show two or three bars right on top of each other for each activity, which can show the planned and actual dates, and even revisions, if desired.

Gannt chart takes its name from Henry Gannt, an American engineer, who first used it in the beginning of the 20th century. An example Gannt chart schedule can be seen at the end of this chapter, from which a number of questions were developed.

Linking Activities / Activity Relationships

After dividing the entire project scope into meaningfully separate pieces called activities, we must establish the relationships between these activities, to obtain the sequences of activities from the start to finish. In doing this, except for very simple and small projects, almost always we will have more than one branch of activities that will connect the start and end points, which will be extremely complex for large projects. Proper linking with correct logic is essential for making a useful schedule.

Types of Links (FS, SS, FF, SF):

There are four kinds of activity relationships:

Finish to Start (FS) Relationship:

This is the most straightforward and natural relationship between two activities, which means, the successor can only start upon finishing of predecessor. A properly made schedule should preferably, ideally contain only FS relationships which show the natural flow of scope, which are broken down into meaningfully separate activities. In some contracts owner may not even allow any relations other than FS, but this may be hard to achieve and in some cases SS or even FF relations may reflect reality better.

Start to Start (SS) Relationship:

This relationship is assigned when we want the successor to start upon starting of the predecessor. So here, the successor does not need to wait for the predecessor to finish. There are usually lags added when SS relationship is used such as SS+5 days, which means, the successor starts 5 days after the start of predecessor, but this is not preferred practice, because we do not know how the predecessor will progress within that 5 days.

Pay attention that we said that the second activity does not wait for the first activity to finish. This means, the first activity's end date is not tied to anywhere (even if the successor has a successor). Activity ends being loose is not allowed and every activity in a schedule must have a successor, except the very last one. So at the very least, the end of the first activity must be tied to the very last activity (i.e. project completion), if not to anything else in the schedule.

Finish to Finish (FF) Relationship:

In this type of relationship, which is not often used, the successor can only finish when the predecessor finishes, but, now the start of the successor is not tied to anywhere, so at the very least it must be tied to project start, if not anything else.

Start to Finish (SF) Relationship:

This type of link should be avoided in a properly made schedule.

Link Dependencies:

The activities can depend on each other in different ways.

Mandatory dependencies result from the activities' real relations to each other physically. For example before erecting formwork, it is physically impossible to pour concrete. This is a mandatory dependency.

Discretionary dependencies are established at management's discretion.

External dependencies occur when external factors from outside of the project influence the activities or a certain activity of a project.

Resource dependency occurs when two or more activities are using the same resource.

Mandatory dependencies can naturally be shown by FS links, discretionary dependencies by FS links or constraints, and external dependencies can be shown by constraints or external links.

The link relationships are established in the beginning of the job, based on the project team's knowledge of the work. For example, the project team knows that we must first dig the trench before laying the utility pipes, and they establish the link as such. If, during the course of the work, some links will need to be changed, this may indicate poor planning and technical knowledge. Other type of changes such as delays, unforeseen things will happen, which do not necessarily indicate poor planning. They are already incorporated into the schedule in the form of contingencies and buffers.

Leads and Lags:

Leads bring forward and lags delay the start of a successor activity. Although it is common to see them in many schedules, usage of both, especially leads is not preferred, unless really needed. Depending on the owner or project sometimes they are not even allowed or only lags may be allowed up to a certain number of days, such as a few days at most. Both leads and lags are only a period of time and not activity. Therefore, there must be no resources assigned to them that would affect EVA calculations.

Using too much lags and during schedule updates constantly increasing or decreasing them to achieve certain dates is not proper scheduling practice, mostly because it will prevent us to see real float dates of activities (and not only activities with lags but everywhere). Plus they are harder to track and need manual update as they are fixed numbers which do not change by themselves.

Lags can be used in the following situations, when they are really needed:

- Sometimes even if the predecessor finishes, we may need to wait for a certain number of days for the successor activity to start, which means there is a lag. We must make sure that this wait period is really a waiting period and not a separate activity.
- We can use lags to avoid using too many intermediate activities which would not be meaningful to show, and on the contrary would congest the schedule and make it unnecessarily complex.
- They can be used in preliminary or summary level schedules on usually large projects as placeholders to represent activities far into the future which cannot be clearly known at present, to show that a certain estimated amount of time will also pass for those future activities to take place.

Instead of using lags, effort must be spent first to naturally represent everything simply by FS relationships without lags. A lag is shown in the schedule as FS+x, for x number of days of lag. When a lag is used, it applies to both early and late dates of the activities it is linking, which would enter into schedule calculation both when doing forward pass and backward pass. Lags should not be used to achieve dates. If starting on a certain date is really necessary for good reason for a given activity, then a constraint should be used. Lags should also not be used as safety buffers, as they are more difficult to track than buffers.

Lead is accelerating the schedule by bringing the start date of an activity forward. They can be shown as for instance FS - x days, where x is the number of days for lead. As this is a negative amount of time, it is not natural in the first place. They are used when it is wanted that the start of the successor should not wait for the predecessor to finish all the way. For example, formwork does not have to finish everywhere, in order to start installing the rebar. We can start rebar at the portions that the formwork completed. So the lead can be written as for example FS - 2, where we are saying that 2 days before finish of the formwork, we will start the rebar. But as we said above this is a negative amount of time which is based on a future event that has not yet even been completed, and we have no 100% guarantee that it will be. Therefore it is better to eliminate the lead and instead use a SS link plus lag, such as SS+3, where we will say that the rebar will start 3 days after the start of formwork. Better yet, we can eliminate using lag also, if we divide the formwork installation into 2 pieces for example as north and south area, and give only FS relation such as when we finish the south portion of formwork, we can start the rebar there while the north portion formwork starts. This must not be overdone however, in order not to congest the schedule.

Important Points To Consider When Linking :

Below are some important points to consider when linking activities:

- First and most important of all, the logic of links must be correct. During linking of activities, people with good technical knowledge about the project scope must provide input. Not having correct linking logic indicates poor technical knowledge about the job.
- Linking two activities as both predecessor and successor to each other will create an error called circular reference. We can not make activity A as predecessor and successor of activity B at the same time. Some scheduling programs do not even allow this in the first place.
- As we mentioned before, summary activities must not have any links themselves. Giving them links will cause schedule calculation errors and create circular references. Linking summary activities will affect all tasks under it too. The idea of having summary activity is to have it automatically be calculated by the activities under it in the first place. Assigning links to a summary activity will entirely defeat this purpose.
- Redundant links should be avoided. For example if activities A, B and C follow each other by the default FS relationships, it is redundant to place an FS link to make C as successor of A.
- As mentioned before, all activities except the very last activity must have successor. All activities except the very first activity must have predecessor. Not doing so, will cause multiple critical paths and will eliminate the purpose of the schedule.
- Path convergence, which means, merging too many activities into a single successor should be avoided. If any of these predecessors delay, the successor delays. Making one activity as successor of too many activities is not logical in the first place and implies some corrections may be needed. Some of the links may be directed to other activities, or another new successor activity can be created, if making that new activity makes sense, or may be too many activities were created in the first place and some of them can be eliminated. Merging a lot of activities into one activity also implies aggressive schedule because all these tasks must happen at the same time to not let any of them cause delay. **Figure - Project Management 4**

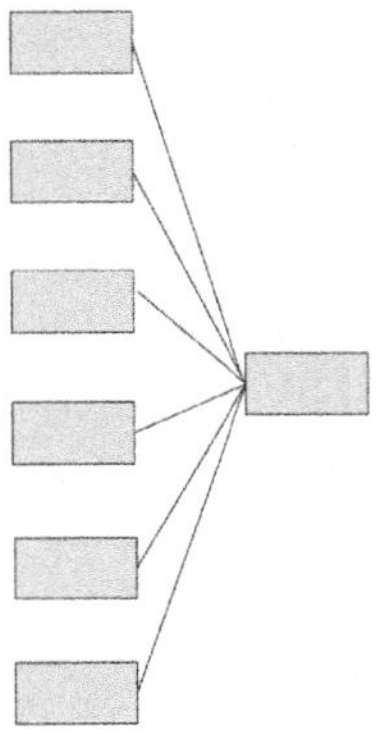

Means and Methods

These are the methods that the undertaker (contractor) applies in order to perform the work. In many cases, the undertaker can choose what method to apply and how, as long as he/she performs the work and delivers the result. Means and methods are not part of the project scope.

For example, on the drawings, it does not show site planning for where you will place the crane or how the vehicle traffic on site will circulate, how you are supposed to lift the steel beams, or how you are supposed to pour the concrete such as with buckets, or pumps and so on... These are called the "Means and Methods". Although they are not in the project scope, the owner still pays for these to the contractor, so they are still part of the overall project budget, integrated into activities as costs. Does it mean, means and methods is only the responsibility of the contractor who is building (executing) the job? Not entirely. The designers also have the responsibility to think about the construction phase and not disregard what might happen in the real world. On some projects, some of the means and methods may be suggested or required, which does not change the fact that they are not part of scope but still paid by the owner.

Means and methods are of course still subject to coordination or approval by whoever the undertaker reports to, in order not to cause any conflicts with others or negative effects to the project. For example, if we mean a subcontractor as the undertaker here, the general contractor coordinates or approves his/her methods. And from the owner's point of view, the general contractor is performing this work, and if that subcontractor's work is significant enough, the owner or his/her representative may also want to comment on it or approve.

Narrative

See the term Schedule Narrative.

Quality

It is a measurement of how well a deliverable meets the requirements of a satisfactory result, as specified in contract documents and any applicable codes and standards. Quality, time and cost always affect and balance each other. The goal of every project is to obtain the desired quality, without going over budget or incurring delays.

Resources

Resource means the needed labor, material and equipment to perform tasks. In order to measure the performance through EVA, resources must be assigned to activities, which is called resource loading. This will also ensure a more realistic schedule because it will enable us to see if there are periods when resources are overallocated. As a result, we may need to perform resource leveling, as described under critical chain *(scheduling programs can do this automatically, if all the links, resources, calendars and durations are properly assigned)*, which may increase the durations of certain activities and delay the project, or, we may need to assign more resources to the job. Both of these will mean higher costs, because delaying of the project overall will mean longer duration of overhead (and may be liquidated damages too) and assigning more resources will simply mean paying for more resources. Through resource loading these can at least be seen from the beginning and appropriate actions may be taken. Leveling overallocated resources may cause delays, but will ease the burden on the company's cash flow, if the firm's financial situation is sensitive to it. Assigning resources to activities, also enables us to see their cumulative costs are distributed over time, instead of seeing these costs just as a list in an estimate. In other words, it gives the costs an extra dimension, which is time. This cash flow vs. time graph is typically in the form of an S-Curve, as shown in Figure 3, under Earned Value Analysis (EVA). A resource calendar can also be followed during assigning resources.

As we mentioned under the term duration, the total amount of work (what we need to get done) for an activity (i.e. pouring concrete, laying utilities, building a brick wall...) are related to each other as: Duration of work = Amount of work / production of assigned units (resources) per unit time.

The three types of resources, which are, labor, material and equipment are discussed below. The formula above obviously applies to labor and equipment type of resource. The material resource simply makes up the physical item itself, which we can calculate by looking at the drawings and other applicable contract documents (i.e. we can have a 400 sf of brick wall, or 500 cubic yards of concrete and so on, which determines the amount of material resource we must use and its cost).

<u>The value we earn by performing activities, which we enter into Earned Value Analysis (EVA), is the total cost of usage of the resources (labor, equipment and material) to produce that activity.</u>

Labor:

Labor is a human resource, the people who actually perform the work such as the concrete crew, the welder, the brick layers and so on... (*but <u>not</u> the accountant for example, even if she is working on the same project because her work's cost falls into general conditions, which is part of general overhead for the project, so she is not a resource for performing activities here.*)

We wrote how duration of work is related to total amount of work above, as: Duration of work = Amount of work / production of assigned units (resources) per unit time. So per this relationship, it seems that if we increased the number of assigned labor to the work, the duration will keep shortening. But increasing labor to shorten the duration will only work up to a point. After that, assigning more labor will be redundant and produce no further benefits.

Sometimes the duration can be fixed. For example, performing a test on an installed component may take a certain number of days per code requirement, which cannot be changed, even if we decreased or increased the labor assigned to it.

When assigning labor units to activities, it is better to standardize as much as possible, for the labor with same pay level and skills, instead of assigning by people's names, which will make tracking more precise, but harder later, if we consider each person's vacations, sick leaves, rate of pay etc...

When executing the project, a smooth workflow must be ensured by the project manager, so that the resources can work efficiently. Abrupt starts and stops or too much fluctuations in work amount will decrease resource productivity. Although this is not very common, if it happens, it can be missed when making the schedule because it will not necessarily generate overallocation error. For example if, after loading the schedule with resources, we will need 20 labors in first week, then 5 in the next week, then 30 the following week, this may cause inefficiencies as far as mobilization and coordination of the crew in the field, even if it will not generate any scheduling errors.

Material:

These represent the materials which are consumed and become part of the project. They are measured in different units such as kilograms or pounds, tons, meters or feet, cubic yards, pallets, each and so on.

Even if material resource may not seem to be overallocated, the delivery date of these materials to the project site are tracked with the schedule. This also means ordering adequate amount of materials (which is usually by the subcontractor) and taking into account the submittal and approval dates of these materials, which may be tracked with the schedule as well.

Equipment:

Equipment are the machines that are used to perform work and build the project. Similar to labor resource, the same relationship between duration, work amount and equipment resource applies: Duration of work = Amount of work / production of assigned units (resources) per unit time. Again, increasing the amount of equipment assigned to the task will have benefits only to a point, after which it will be redundant and unnecessary cost (i.e. only so many excavators can work on excavating a pit).

Risk

Risks are uncertainties in the future that may affect the project performance. Although almost always regarded as negative, risks can technically have positive impacts too. The risks must be identified, analyzed and tracked. Tracking can be done with the schedule. This will also enable more realistic schedules and constant awareness about risks by the project team and stakeholders.

The most effective way of reducing risks is to know the job and properly represent the scope in the schedule in the first place. This means, having correct links, reasonable activity durations, proper breakdown of scope into logical activities, effective resource planning, using proper scheduling and performance measuring standards. Consulting stakeholders and getting their input during breaking down of the scope into activities, establishing a competent project team is also very important. After all proper practices are followed, there will still be outstanding risks, which must be shown realistically.

Risks can be in terms of time, cost, scope and quality. Of these, the first three are represented in the schedule and quality affects them. As a result of risk analysis, contingency funds of the project may increase or decrease, and a new critical path, which is called risk critical path, may emerge.

Negative risks or threats can be handled in various ways. Avoidance means eliminating the risk altogether, by preventing it from happening and all exposure to it in the first place. Mitigation (also can be called as limitation) strategy seeks to reduce the probability and negative impact of the risk. Transfer strategy means transferring the risk to another party, such as getting insurance, outsourcing payroll, customer service etc..., or including certain terms in a contract to transfer the risk to another party. This also enables the firm to focus more on its core competencies. Acceptance strategy means taking no action and dealing with the risk when it occurs, such as establishing a contingency reserve for it, which can be the most logical way when the cost of dealing with a risk is more than the benefit of avoiding it, which can usually mean this is for low probability and or low impact risks.

S-Curve

S-Curve is the shape of the graph that resembles the letter S, which is obtained when taking cumulative of variable A (such as cost) with respect to variable B (such as time), where variable A increases slow at first, then faster, then slower again, which makes S-Curve, as seen in the example below. Also see the similar curve under the term EVA. The keyword here, which causes the shape of graph is "cumulative".

	Cost / Month	Cumulative Cost
January	$ 10,000	$ 10,000
February	$ 25,000	$ 35,000
March	$ 40,000	$ 75,000
April	$ 50,000	$ 125,000
May	$ 80,000	$ 205,000
June	$ 70,000	$ 275,000
July	$ 80,000	$ 355,000
August	$ 60,000	$ 415,000
September	$ 40,000	$ 455,000
October	$ 20,000	$ 475,000
November	$ 30,000	$ 505,000
December	$ 5,000	$ 510,000

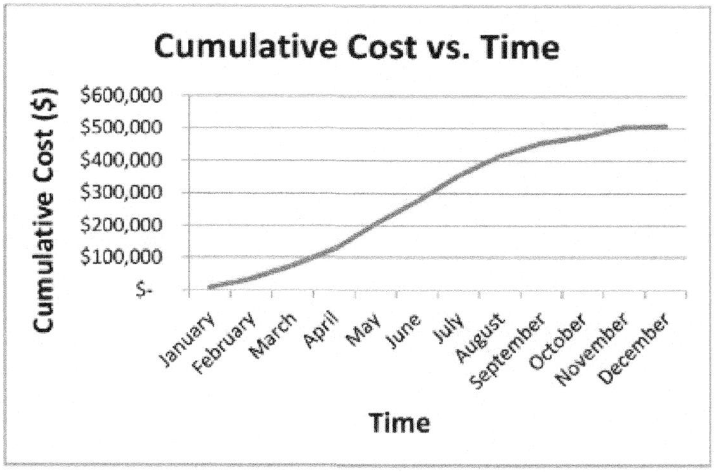

Figure - Project Management 5

Scope

Scope is the same as deliverable. Please see the term deliverable for the description.

Schedule Compression / Acceleration

As a result of delays to make up for lost time, or even before anticipated delays, or to simply run the job faster, we must take measures to accelerate the schedule. This is called schedule compression. Compressing schedule almost always has a cost, such as overtime work, increasing workforce, expediting delivery of materials. As a result of compression, coordination of work becomes harder. Sometimes, material deliveries, or external factors cannot accommodate the compressed schedule. Two main schedule compression methods are fast tracking and schedule crashing.

Fast Tracking:

It is making activities to happen more simultaneously than before, which makes coordination harder.

Schedule Crashing:

It means accelerating the speed of work by adding resources, reducing vacations, increasing overtime, expediting deliveries... It can be effective only up to a point, because it can decrease morale, reduce crew efficiency by placing too much burden, or too many people will simply not be beneficial after a point, which also depends on the nature of the task.

Other Measures For Schedule Acceleration:

We can list other ways to compress schedules as:

- Spending contingencies or buffers which were inserted into the schedule during planning stage. This spending should ideally have been accounted for during risk analysis before. Consuming buffers or contingencies actually does not compress the schedule, but simply prevents the delays from pushing the finish date, by absorbing those delays into contingencies or buffers.
- Splitting long duration activities into shorter activities in parallel. This is similar to fast tracking and may cause resource overallocation and make coordination harder.
- Reducing scope of work. This is not common or not even possible in many cases unless owner is willing to accept. It would be a negative reputation if proposed by the undertaker, unless there is also a justifiable reason such as value engineering, and even so, the contractor may still be responsible for delays up to that point. Otherwise, if the functionality of the project deliverable would decrease, it should only be proposed by the owner himself/herself.

One thing to remember here is that before taking any of the measures above, first of all we must make sure that there are no errors in the schedule. There may be constraints, lags, longer than needed durations for some activities, wrong links etc... If these are found, they will remove unnecessary delays.

Schedule Narrative

A schedule narrative is a supplemental document, attachment to a schedule, which includes explanations about the schedule, which can not be displayed on the schedule itself properly and in detail. Narratives must be written clearly to be understood by everyone. The items that need to be in a schedule narrative may also be specified by owners. Schedule narrative can include items such as:

- On the Gannt chart, the descriptions of activities cannot be long. In the schedule narrative, longer descriptions can be written. The WBS code for the activity must also be shown and the description given here must remain within the limits of what was described for that WBS level.
- Even if the resources are loaded to activities in the schedule, they can also be described in more detail here. The production rates and cost rates of resources, and per these, the basis of duration and work calculations should also be given.
- Critical path, general work sequence, critical dates can be described and discussed here in detail, such as the background information, underlying reasons and how it is likely to progress so that everyone can understand them better. If there are constraints in the schedule, they must be explained and justified here. Important project issues affecting the schedule, any coordination issues between trades, or between the project and the surroundings, any ongoing site mobilization activities and the affected trades and locations, impact of approved and upcoming change orders, outstanding RFIs, risks need to be described or at least listed here as far as their schedule effects. More detailed discussions can take place during owner-consultant-contractor-subcontractor meetings as applicable.
- Description of calendars that are used, such as their working days, holidays, shifts etc... and the reasons for assigning them to particular activities or resources.
- List/log of submittals *(submittals are items such as shop drawings, brochures, calculations, material samples, material data sheets and so on for the items to be installed)* should also be included or attached to the schedule narrative. The dates that they are needed from subcontractors, approval dates by consultants, fabrication durations and final delivery dates are important and many of them affect the critical path. All these dates must be in coordination with the dates of installations in the schedule. These can also be tracked in the schedule itself as non-work items or by a separate submittal and material fabrication schedule, which must then be connected into the main schedule. The issues that slow down the preparation of submittals, approvals, fabrication durations and deliveries should be described.

Any other item that may affect the work schedule in any way should also be included as necessary. As we said before, schedule is the keystone of a project and we can track everything with it. Schedule narrative is an excellent supplemental place for the project team to better communicate how the schedule works and will work, and to document and track things.

Updating and Reviewing Existing Schedules

Updating Existing Schedules:

Let's list some key points to consider when making periodic updates to schedules:

- Before all, the data date (current date), must be updated. This is almost always the current date, unless we are analyzing some scenario or a claim for a different date. After this, the schedule is recalculated where all incomplete work should remain to the future of data date, in other words, <u>no incomplete work can remain in the past</u> *(we cannot jump onto a time machine, travel to the past and work on an incomplete work in the past; we can only work on it in the present and future)*. This means for example, if an activity was planned to be complete by this date but it wasn't, then the incomplete portion of that activity will need to be pushed after the data date, so it means the activity just got longer in duration, in other words delayed.

- Progress of activities can be updated by entering remaining duration or % complete. If remaining duration is entered, the % complete value will automatically be generated.
- If updating of the schedule takes longer than a day, the data date should not be different for different activities. Either the first or the second... day should be taken as data date and all progress must be entered per that date.
- If the start and finish dates of an activity don't make sense, its links should be checked first. Any constraints should also be checked. All activities should be set to start as soon as they can per the schedule logic, unless a constraint is applied to them.
- Schedule narrative update should accompany each schedule update.
- At a minimum, the schedule update must be completed at each reporting period. But more often updates are also possible to better track the progress in the field and discuss during owner-contractor or contractor-subcontractor meetings.
- All schedule updates must be saved separately. In other words, there should be as much files as the number of updates. If we only keep updating one file, a lot of useful information during the course of the job will be lost. These would be especially valuable if there are claims later.
- Impact of approved change orders must be shown in schedule updates and be mentioned in the schedule narrative. A change order may also include a logic or other change of schedule.

Reviewing Existing Schedules:

Sometimes you will need to review schedules made by others. The items below represent some of the common mistakes or items that need to be revised the most often, in existing schedules:

- Logic of links must be checked. Activities without links must be assigned links. Overall common sense checks must be done for major activities and their predecessors and successors, to see if something may seem out of sequence. Improper usage of constraints must be found and deleted. Durations should be looked at, to see if they make sense. The schedule should also be collapsed to higher levels and checked that way. This shows the broader view at once and may help identifying major errors quickly.
- The completeness of the schedule should be checked as far as reflecting the full scope. Similarly, the items that do not belong in scope must be reviewed to verify that they do not have resources assigned and they remain there only if necessary or moved to schedule narrative.
- No incomplete activity dates should be in the past. If that is the case, it means the data date was not updated and the schedule was not recalculated.
- The float amounts must be reviewed to see if they make sense. Too long float amounts usually indicate errors in logic of links. If there are minus float amounts this tells us that there are constraints that prevent schedule from properly calculating.

- Percent complete values must be checked to see if they reflect actual conditions.
- Resource usage must be reviewed. Any overallocations must be leveled.
- There must be only one critical path. The absence of one, or presence of more than one indicate schedule logic problems. All activities must have predecessors except the very first one. All activities must have successors except the very last one. A milestone is also an activity so it is included in this definition *(see the term Activity, to see the description of milestone)*. The very first and the very last activity are usually milestones.
- Actual dates, when entered, supersede links. If an actual value is entered incorrectly, this can cause schedule slippage.
- Testing, punchlist, commissioning, signoffs and so on must be accounted for at the end.

Weather

Adverse weather can prevent or slow down the progress of most outdoor construction activities. Therefore, its effect must be accounted for in the schedule. The most proper method to show adverse weather days would be to include them in a separate calendar, which we may call as "outdoor work calendar", by assigning them as nonworking days as allowances. Then we can assign this calendar to outdoor activities which are sensitive to adverse weather, while indoor activities can still use the default calendar. Then as the dates progress and we see the actual weather, we either leave them as they are if adverse weather happened on those days, or change their dates for when they happened, or simply turn them into working days if adverse weather happened less than expected by either consuming those days or carrying to the future. The method we use and how to use it also can be specified in the contracts. Other methods include showing weather days as a contingency buffer at the end of the project or at the end of affected branches in the schedule, or to increase activity durations that might be affected from weather (but as we said before, accounting for future delays by assigning longer durations to activities is not correct practice).

The historical adverse weather data for every location in the US can be obtained from resources such as National Oceanic and Administration (NOAA). Adverse weather can be rain, wind, snow, freezing or even days with excessive humidity depending on the nature of activities. For example windy weather can affect tower cranes or formwork installment, while it has less effect on a concrete pour or earthwork, for which rain or snow would have more influence. Some days can still be counted as weather days, even if the adverse weather ends, such as the soil remaining wet and preventing construction equipment to travel or excavate, after a heavy rain.

Work

Work is the actual task to be performed, to complete an activity. So for example, we can have 1000 feet of pipe to lay or 500 cubic yards of concrete to pour. This is the work we need to do.

In scheduling though, work is represented as something that corresponds to the amount of effort spent by resources, just so that we can represent it as a value, in earned value analysis.

Let's study an example to explain the concepts. Let's say, a crew of 5 laborers actually worked for 3 days (8 hour workdays) to complete a wall. Pay attention that we underlined "actually worked" above, and keep it in mind, because our original plan might have been different, as you will see below.

The total amount of work for scheduling purposes to build the wall can be found with the fundamental formula we expressed before, which was:

Work = Duration x Resource units assigned to task

Work = (3x8) x 5 = 120 hours. This is the actual work done to build the wall. *(For example if we had only one of the workers above, it would have taken him 120 hours, which means 120/8=15 days to finish the work with the same productivity.)*

And if 1 labor-hour (commonly called as man-hour) costs $50, it means the actual cost, AC, of labor would be 120x50=$6000 here. This is the actual labor cost that we spent to build the wall.

Again, the 120 hours here represent the actual hours we spent in reality. But we might have planned it differently. For example, let's say that our wall has an area of 300 sf. And let's say 1 worker had a planned production rate of 25 sf of wall in 1 full work day (8 hour day). So, per this production rate, it would take 300/25 = 12 days = 12x8=96 hours for one worker to build this wall, per the original plan. See how the planned 96 hours can be different than the actual 120 hours that we spent in reality. So this 96 hours is the original planned amount for this work but in reality the workers worked less efficiently (one worker produced less than 25 sf of wall a day) and they spent 120 hours.

If we multiply this planned 96 hours by our labor rate, 96x50=$ 4800 is the total planned value of this work. This is the total value of work to earn per the original plan, upon fully finishing this work.

Cost variance, CV, as we described under the term EVA is:

CV = EV - AC

So we can calculate CV as: CV = $4800 - $6000 = -$1200. It means we are over budget by this amount because the value of the work per our original baseline plan (planned value, PV) was $4800, but we spent $6000 to achieve (earn) that much worth of work. *Note that we used EV and PV interchangeably here, because the work is completed, in other words, our EV ultimately became equal to our PV. If the work was not yet complete, EV would not be equal to PV.*

And what about schedule variance, SV?

For that we need to know the planned value, PV, at the end of 3 days. We said 5 labor actually worked 3 days above. But 3 days may not necessarily be equal to what we planned in the beginning as far as the original duration. For example, we might have decided employing 3 laborers in the original plan when we were making our schedule. To make it clear: The amount of planned work was not our decision. We calculated it based on quantity of wall and the worker production rate, which came to be as 96 hours. But how many workers to employ to build that wall is upto us and we wanted to do it with 3 workers during making our original plan. So, since we have 96 hours worth of work, it would take:

Duration = work / assigned units, so, Duration = 96 worker-hours / (3 workers x 8 hours) = 4 days (8 is used to convert hours to days) for 3 laborers to finish the wall per our original plan. So we had planned for 4 days originally, to finish this work. But we actually finished the wall in 3 days in reality. So it means, we completed the wall ahead of schedule.

Now coming back to calculating schedule variance... As we saw under EVA,

SV = EV - PV

We know our planned value for the entire work already. It is $4800. But we planned to earn that value at the end of 4 days. This means, at the end of 3 days, in our original plan : 3 men x 3 days = 3x3x8=72 hours work was planned. This means, 72x50=$3600. This is the PV at the end of 3 days per original plan *(you can also find it if you calculated it as 4800x3/4=3600).*

So,

SV= EV-PV =4800-3600=$1200. In other words, at the point in time (end of 3 days) when we originally planned to earn a value of 3600, we did better, and earned 4800 worth of value.

So this $1200 is our schedule variance, which is positive. A positive SV means work is ahead of schedule. So although we spent more money than we planned (as we have negative CV), we were able to finish the work ahead of schedule.

Do not confuse the concepts discussed here about work. Let's summarize: As we said first, work is actually the physical scope we need to perform, such as laying pipes or building wall. But in scheduling work is expressed in hours as the amount of effort spent by the resources, as in Work = duration x assigned units. Then we can calculate the dollar amounts (values) for these. So the dollar amounts you see here are the value of work. Not the work itself (which is the physical pipe or wall itself that we build). It is important to not confuse the concepts. These values, in other words, the values of work, are what we enter into EVA.

Again, the total value of work (planned value) was found directly from relating the physical work quantity to production rate of labor, which was found as 300/25 = 12 days = 12x8 = 96 hours and then 96 x 50 = $ 4800. This value of $4800 doesn't change no matter how many labor we plan to use and how productive they might be.

Also worth noting here that we did not include material costs in the values here for simplicity. But in reality, the cost of material would also be included here.

Work Breakdown Structure (WBS)

The total scope (deliverable), must be broken down into meaningfully distinct activities, in a properly structured format called Work Breakdown Structure (WBS). The total of WBS is therefore equal to the total of the scope (100% rule), which represents the highest level of WBS. The lowest level is the individual tasks themselves, in other words activities, which are also called work packages. WBS allows us to group and track tasks. It is basically as if assigning ID numbers to all activities, but in a structured way which is linked to project scope in a hierarchical way, and in this sense it is different than simple activity IDs in a schedule. WBS is similar to the structural frame, the skeleton of the project scope, and in turn, the schedule and also the estimate, since these must also represent the total scope. The WBS codes assigned to activities also gives us a standardized, efficient way of tracking and communicating these across the whole company such as between project management, estimating, accounting departments, among all project team members, and among all stakeholders as well.

As an example, a portion of a simple WBS can look like the one given below, within a building construction project estimate or schedule. It is quite straightforward. At the highest level which is level 1, we have the entire project scope, which is the building. Then we go one level down and we have 1.1 Structure, 1.2 Exterior Shell, and 1.3 Interior Finishes. See how the structure has minus near it, as it is expanded and we can see the details under it, but 1.2 and 1.3 have plusses near them, which means they are collapsed and we do not see their lower level summary and or individual activities under them. Lines such as 1.1, 1.2, 1.3, 1.1.2, 1.1.3.2 etc.. are all summary activities, although at different levels, and will be shown as summary bars with (preferably) different colors and thicknesses on Gantt chart. The ones such as 1.1.3.2.1 - Install Formwork, have neither plus nor minus near them, which means they cannot be expanded further because they are the lowest level, which means they are the actual activities (work packages) themselves, where the actual work takes place, gets done. So the entire project is represented by activities like them, all of which reside under their own summary (parent) activities. Only these actual activities are the ones to which we assign resources, links and constraints, as mentioned earlier in this chapter. They would also be shown with different color and/or lighter or smaller font than summary bars, which can be adjusted per user preference.

-1-BUILDING PROJECT NAME

    -1.1-Structural frame

        +1.1.1-Ground floor

        +1.1.2-First floor

        -1.1.3-Second floor

            +1.1.3.1-Columns

            -1.1.3.2-Beams and Deck

                1.1.3.2.1-Install Formwork

                1.1.3.2.2-Install Rebar

                1.1.3.2.3-Pour Concrete

        +1.1.4-Fourth floor

    +1.2-Exterior Shell

    +1.3-Interior Finishes

As can be seen above, WBS provides a structured way of organizing activities, by assigning them codes in proper hierarchy, which is an excellent way of serving as a standard for everyone. It makes not only scheduling, but also estimating and budgeting easier, because the cost of a summary activity is simply made up of costs of all activities under it. In other words, the scope and cost of entire activities make up the total scope and cost of the summary activity above them and in turn the project, nothing more or less, which is called the 100% rule. The summaries being the sum of everything under them also serve as useful check points during making schedules and budgets and updating them.

There are more than one way to break down the scope and form the WBS. For example the WBS shown above is formed by organizing the main components of the building as: structure, exterior shell and interior finishes. But on another project, the project scope can very well be divided by location. On another one, it can be divided by the main installed components. This is determined by organizational and project needs.

There should also be a WBS dictionary, which will describe in detail what each code is about, the required resources and responsibilities, how it relates in the hierarchy above and below it, and also to other codes in the same level or wherever the activity is linked to in the schedule. From project to project these descriptions can be copied to be used as a starting point, which is another positive aspect of standardization, but care must be taken here, and every code must be reviewed one by one, per that particular project and all appropriate adjustments must be made, because every project is unique by definition, even if there may be many similarities with past projects.

Sometimes during initial planning, not all details about all activities can be known, especially for large projects and for activities far into the future. Instead of entering activities with probably wrong estimates and having to update many times later, placeholder items with high level WBS codes called planning packages can be placed and these can be broken down further into activities with lower level WBS codes as the job progresses and more details are known. This is called rolling wave planning. By using planning packages, we can estimate total cost and duration of the planning packages, with appropriate links from both ends of the whole package. This way there is not a hole in the estimate or schedule and the packages serve as placeholders until things become clearer. Another benefit of using planning packages is that the packages which seem more critical (since they are also assigned links from their both ends) can be given more attention by the team and their details can be clarified sooner.

Example Schedule and Exercise Questions

Below is a highly simplified example schedule of an imaginary project that shows construction portion. For the purpose of answering the questions below, you may also assume that it is the entire project schedule as applicable. Try to answer the questions below first without looking at the answers, which should serve as a good wrap up of many concepts learned in this chapter. In predecessor and successor columns, if you only see an activity number, it means the FS (the most straightforward) relation.

ID	WBS	Task Name	Duration	Predecessor	Successor	Start	Finish	Total Slack
0	**0**	**Example Project**	**11 days**			**May 12**	**May 22**	**0 days**
1	**1**	**Building**	**11 days**			**May 12**	**May 22**	**0 days**
2	1.1	Foundation	2 days		3	May 12	May 13	0 days
3	1.2	Structural Frame	4 days	2	4SS+2d,5,7	May 14	May 17	0 days
4	1.3	Facade	4 days	3SS+2d	8	May 16	May 19	2 days
5	1.4	MEP	2 days	3	6SS,8	May 18	May 19	1 day
6	1.5	Interior Finishes	3 days	5SS	8	May 18	May 20	1 day
7	1.6	Roofing	4 days	3	8	May 18	May 21	0 days
8	1.7	Signoffs	1 day	4,5,6,7	16	May 22	May 22	0 days
9	**2**	**Sitework**	**10 days**			**May 12**	**May 21**	**1 day**
10	2.1	Layout	1 day		11,15	May 12	May 12	1 day
11	2.2	Utilities	5 days	10	12SS+2d,16	May 13	May 17	1 day
12	2.3	Backfill / Grading	4 days	11SS+2d	13,14	May 15	May 18	1 day
13	2.4	Signoff on Grades	1 day	12	14	May 19	May 19	1 day
14	2.5	Pavement	2 days	12,13	16	May 20	May 21	1 day
15	2.6	Landscaping	5 days	10	16	May 13	May 17	5 days
16	3	Project Completion	0 days	8,11,14,15		May 22	May 22	0 days

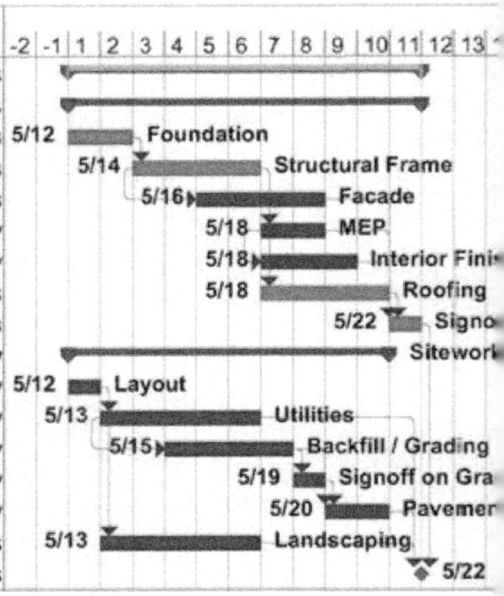

Figure - Project Management 6

1) How many days is the critical path in the schedule above?

The critical path here is: Foundation, Structural Frame, Roofing, Signoffs, as they have 0 Total Float (slack). Totally there are 11 days on this path. Note that this is also equal to total project duration.

2) By how many days can we delay the Landscaping activity, before it becomes critical?

It can delay 5 days before being critical. Similarly, Facade can delay 2 days, MEP can delay 1 day and so on, before being critical. Note that these numbers are equal to their total floats.

3) If you changed the predecessor of Activity 11 - Utilities, to 10FS+3, which choice(s) below are true?

A) The new project finish date would be 5/24
B) All sitework activities will now become critical, except the Landscaping activity
C) The Early Start (ES) of activity #10, Layout, would not change
D) It would bring the Total Float (Slack) Façade, to 4 days.
E) All of the above

Before changing the predecessor as in the question, Utilities had 1 day of total float. Now if we give a 3 day lag for its start, this would first consume that 1 day of float, and then, it would start pushing the schedule, which would be the 2 extra days. So now the project would finish on 5/24. So, A is correct. B is correct because now it will push all successors of Utility activity. C is also correct because Layout activity comes before Utilities and it will not be affected by whatever happens to dates of utility activity. D is correct because the total float of Facade work will also increase by 2 days, which will now be 4. So all choices are correct and the answer is E.

4) Assume each day of each task is performed by 1 labor only. How many worker is needed on Day 6?

We must look at day 6 and simply see how many activities take place on that day. There are 5 activities taking place on day #6. Since each day of each task is performed by 1 labor per the question, we will need 5 labor on day 6. If we had less than that, it would cause delay to one or more of the activities that day. (i.e. if we had 3 labor, only 3 tasks could be worked on, so 2 tasks wold delay)

5) Assuming each day of each activity is performed by 2 labors, how many manhours will it take to finish the Building portion? (1 workday is 8 hours).

We must first simply find out the total duration of activities for the building portion of the project. The total duration of all building related activities are: 2+4+4+2+3+4+1= 20 days. *(Pay attention that this is not how long the building portion takes, as some activities take place at the same time (stacked) on some days. We did not need that here.)* Since each day requires 2 labor, this makes 20x2=40 man-days. Then we simply convert man-days to man-hours as: 40x8=320 manhours, which is the answer.

6) Assume each day of each activity is performed by 2 labors, and 1 man-hour costs $100. What is the total Planned Value (PV), as of the end of day 4?

We must find how many work days are required by the end of day 4. If you draw a vertical line at the end of day 4, simply look at the total durations of activities before that. You will see that there are total of 12 days of activities up to the end of day 4. This makes 24 man-days since each day requires 2 labor. 24 man-days = 24x8= 192 man-hours as we have 8 hour workdays normally. And this makes 192x100=$19,200 as PV by the end of day 4. This is what we planned originally. We may spend more or less than this (AC), and / or earn more or less value than this (EV) by then.

7) Assume again as in the previous question that each day of each activity is performed by 2 labors, and 1 man-hour = $100. Assume that we performed $ 15,000 worth of work by the end of day 4, and spent $18,000 in doing so. Calculate the Cost Variance, CV.

Let's remember the formula CV=EV-AC. Doing "worth of work" means, earning that much value. So our Earned Value (EV) at the end of day 4 is $15,000. And the question already tells us that we have spent $18,000 to do earn this value, so this is our Actual Cost (AC). Then, CV = $15,000 - $18,000 = -$3000. Note that we did not need to know labor cost per hour or how many labor is needed for a day of each task here to find CV. The question already gave us EV and AC.

8) If, during the course of the job, we find out that we can start Roofing, 3 days after its predecessor starts, which one of the choices below is TRUE?

A) The predecessor link of Roofing must be revised to 3SS+3
B) The work scope is still the same
C) The project will now finish in 10 days instead of 11, and all sitework activities except landscaping will become critical.
D) This will not cause other relationships in the schedule to change
E) All of the above

All statements in the above are true. The answer is E. As mentioned before, note that assigning SS plus lag link is not preferred practice unless it is really needed. Here it means, we will start roofing 3 days after structural frame starts. This is not preferred because we simply do not know how structural frame will proceed, whether its progress in 3 days will be enough for roofing to start at the end of 3 days.

9) Which one below is FALSE about Activity #16, Project Completion?

A) This activity is a milestone
B) It does not need any resources
C) If we did not assume this schedule as the entire project schedule, then it would be more appropriate to call this activity as "Construction Completion"
D) Only the activities #8, #11, #14 and #15 drive this activity, but others do not

Choices A and B are true as this is a zero day activity which means it is a milestone and a milestone does not require any resources it is not a physical work. C is also true because a project also involves a close-out phase which is not shown here for simplicity. If we did not assume that this was the entire project schedule then what we have here looks like a construction schedule only. D is false here and it is the answer, because all activities, and not just 8, 14 and 15 can ultimately drive the last activity. Remember that we said all activities must have a successor ultimately linked to the project completion otherwise there will be free floating paths and multiple critical paths which is not acceptable practice.

10) Which one is FALSE about activity #11, Utilities?

A) Its cost is included in total sitework cost and its control account is sitework.
B) Its start date determines the sitework start date
C) If it delayed 2 days, the project would delay

The answer is B. Choice C is not false because of its start date, as Total Float is also TF = LS - ES, which is linked to activity 12 with SS link.

11) Which phase of the overall project below comes AFTER the activities in this schedule are complete?

A) Determining project requirements
B) Design
C) Feasibility study for the project
D) Project turnover

This schedule, which only shows the construction phase, is only a part of the whole project. There are also many other things to do before, during or after construction, which does not belong to construction. The answer is D. Phases in choices A, B and C take place before the construction phase.

12) In which process group(s) we will most often need this schedule?

A) Planning
B) Monitor and control
C) Executing
D) Both Executing and Monitoring and Controlling

A schedule is made during the planning stage but it is mostly needed and used during execution and monitoring and control processes. The answer is D.

13) During which process the scheduler filled out the Predecessors and Successors columns?

A) Define activities
B) Sequencing of activities
C) Create WBS

The answer is B

14) In the schedule above, to fill the contents of Task Name Column, which choice below shows the correct sequence of the processes that the project team followed?

A) Plan Schedule Management, Define Scope, Develop Schedule
B) Define Scope, Create WBS, Define Activities
C) Estimate Activity Resources, Define Activities, Develop Schedule
D) Define Activities, Sequence Activities, Estimate Activity Resources

Sum of tasks represent the work scope but the whole schedule is something more complex, with all the links, durations, resources etc... Therefore we do not talk about developing the schedule or assigning any resources here. If we can see this, we can eliminate choices A, C and D because they mention developing the entire schedule, estimate activity resources and sequencing activities. The answer is B. To write the task names, we must define the scope, break it down per WBS, and define the individual activities, as given in choice B.

15) If the links between the activities and their durations need frequent updates during the course of the job, which one below is TRUE?

A) A new baseline must be made after an approved change order and the project performance must be compared according to this new baseline.

B) Due to these updates, the overall duration of the project may change. Variances will be calculated according to new baseline.

C) Although the situation in the question may be understandable upto a point since in almost any project unforeseen things emerge, changing links and durations too much during execution (especially changing links) indicates inadequate understanding of the scope during planning.

D) All of the above

The answer is D.

16) The duration of Activity #11 - Utilities, was found by dividing the total quantity of installation, which is 500 feet, by the length that a typical crew can install per day, which is 100 feet, which produced the result of 5 days. What type of estimating technique is this?

A) Parametric Estimating
B) Analogous Estimating
C) Three Point Estimating
D) Most Likely Estimating

Choice A is correct.

17) A buffer period, which will be a part of the overall project requirements, but will not be included in this schedule, is called:

A) Management Reserve
B) Known Unknown
C) Contingency

Management reserves are not part of the baseline or earned value calculations. They are on top of them but from the owner's point of view they may still spend this money so for them it is part of the overall budget and schedule. Choice A is correct.

18) You are the PM of the general contractor and talking to the PM of façade subcontractor who you will likely select for the job. He lets you know that, they must have a signed document before starting facade installation, which must confirm that the installed structural steel frame complies with the designs and codes. He also says he will need 1 week's notice to schedule his crew, and if they cannot start on scheduled date, you must pay his crew for those dates anyway. He says that these points were written in their quote when they bid the job. You look at the schedule as shown above, and see that there is no way of knowing status of structural frame one week in advance of facade work, as the steel work would not even have started yet. Fortunately you are still at subcontract award stage and the job has not started yet. Which one below is a mitigation strategy?

A) Write a clause in the steel subcontractor's subcontract that, in the event of non acceptance of his work, the cost of delays of other trades, such as the wait of Facade installer crew, will be deducted from his subcontract amount.

B) Have a meeting with the structural steel and façade subcontractors, structural engineer and clearly identify risks. Take steps to either eliminate them or reduce their negative impacts.

C) Make a note of the risk but do not take any action and see how it goes.

D) Try to get another façade subcontractor, who can schedule crew with shorter notice or does not ask for such document.

Choice A represents a Transfer strategy. We can do this because we still did not award the subcontracts, per the question. Choice C is an acceptance strategy. Choice D is an avoidance strategy. The answer is B. We listed these under the term "risks".

19) In this schedule, how many activities require use of resources?

We have total of 16 activities, but activity #1 and #9 are summary activities, and they do not have resources assigned to them. Activity #16, a milestone, also does not require resources. All other activities are actual work tasks, which require use of resources to get done, and there are 13 of them.

20) What is the Early Start (ES) date of Activity #15, Landscaping?

ES is simply the calculated start date per the links and durations of activities. And in the schedule it can be seen as May 13th.

21) What is the Late Start (LS) of Landscaping activity?

In the previous question we saw that the early start was May 13th. The total float of this activity is 5 days. It means its start can delay 5 days before it gets critical. So its late start date is May 18th.

22) If you increased the duration of Task #11, Utilities from 5 days to 8 days, which one below is TRUE?

A) Since it has a Total float of 1 day, at first that day would be consumed, and then, there would be a 2 day impact to overall schedule.

B) Total float of this activity will now be 0 days

C) This particular change would not impact the end date of the schedule

D) Cost of performing this activity will increase

Choice A is wrong because the total float can be as LS-ES too and it is linked to activity 12 with a SS link, so, its TF will remain unchanged. It would only start to have zero float if we increased it to 10 days, instead of 8, because its end date is tied to nothing but project completion only, which was 5 days away from May 17, the finish date of this activity. So, choice B is wrong too. Choice C is the answer. As far as choice D, the cost may or may not increase. We cannot determine with given data. Did the duration increase because we now have to install more pipe? If so yes the cost will increase. Or the duration increased because we are now employing fewer workers to do the work? Or the production rate of workers decreased? Or there was an unexpected delay? If so the cost will not go up, but it will not decrease either because now the work remains the same. In the scheduling program, if you changed the duration you would also need to show to the program the reason.

23) Assume that the total Planned Value (PV) for Activity #4, Façade Work, was $4,000. At the end of the day May 18th, if the Earned Value (EV) for Façade work was $3500, which one of the choices below is definitely true?

A) This activity is ahead of its schedule
B) The activity will finish before its planned date
C) The successor of this activity can start earlier than planned
D) Costs are being managed, and money is spent effectively for this activity

The activity has 4 day duration. Assuming each day the same value is planned to be earned, the PV for each day is $1000, for a total of $4000. So at the end of the day May 18, which is the 3rd day for this activity, the PV is $3000. The question tells us that the EV for the same point is $3500. So we earned more than planned which means we are ahead of schedule *(although we do not know how much we had to spend to achieve, earn this much value)*. So, choice A is definitely correct. As far as B, although it is likely that we will finish early, we cannot be 100% sure. Choice C is also not certain because of the same reason and also because the start of the successor (Activity 8) is also affected by its other predecessors (5,6,7). Choice D is also something we are not sure about. May be we are ahead of schedule because we had to pay people to work overtime or used more people and spent more than planned. Or may be we just worked more efficiently. In the question how much we spent was not given.

24) Although we have different trades that require people with different skills and pay rates working on the job, for the sake of question, assume that each day of each activity requires 1 labor with same skills. Assume that there are 4 workers available for our project. Which one is TRUE?

A) We must perform resource leveling (see the term critical chain)
B) Starting day 5, more people will be needed
C) The overall construction duration will be longer because of resource leveling
D) A and C are true
E) All of the above are true

The answer is E. Starting day 5, more than 4 people will be needed, so we must perform resource leveling and the overall duration will be longer because of this.

25) If the WBS of activity #2, Foundation is 1.1, then what would be the WBS of Activity #15, Landscaping?

A) 1.14
B) 2.6
C) 15
D) Cannot be determined

WBS codes, thus the answer can directly be seen on the schedule, so the answer is B, but let's also explain. Foundation is under summary activity "Building." If the WBS code for Foundation is 1.1, then the WBS code for Building would be just 1, as it is one level higher. At this higher level, the next one is Sitework, so then its WBS code would be 2. Below Sitework, the activities will be assigned codes as 2.1,

2.2, 2.3 and so on. If you continue to landscaping, it will be the 6th activity listed under sitework and so its code would be 2 6. The same level activities in the hierarchy have the same number of digits.

26) What is the presentation format of this schedule?

A) Pareto Diagram
B) Activity List
C) Bar Chart
D) Network Diagram

The answer is C

27) If a buffer activity is placed after activity #7, what is this called?

A) Total Slack
B) It cannot be done
C) Project Buffer
D) Feeding Buffer

The answer is D

28) Originally, activity 6 was going to be linked to activity 5 with 5FS (which would be seen as only 5). However, the project manager, who worked on similar projects before, wanted it as 5 SS, arguing that there are no physical limitations to do it (in other words starting Activity 6 as soon as MEP starts), as long as we manage things properly and coordinate everyone involved effectively. What did he do?

A) Decomposing
B) Schedule Crashing
C) Fast Tracking

The answer is C

BIBLIOGRAPHY

Airport Design - AC 150/5300-13B, US Department of Transportation, Federal Aviation Administration. March 2022

Bridge Inspection Manual, Office of Structures, New York State Dept. of Transportation. March 2017

Building Code Requirements for Structural Concrete - ACI 318-19, American Concrete Institute, Farmington Hills, MI. June 2019

Coastal Engineering Manual- EM-1110-2-1100, US Army Corps of Engineers

Design of Gravity Dams - Bureau of Reclamation, US Department of the Interior, Denver, CO. 1976

Designing for Earthquakes - A Manual For Architects, Risk Management Series - FEMA 454, National Earthquake Hazards Reduction Program (NEHRP), Federal Emergency Management Agency, US Department of Homeland Security, Earthquake Engineering Research Institute, Oakland, CA. Dec. 2006

Airport Drainage Design - AC 150/5320-5D, US Department of Transportation, Federal Aviation Administration. August 2013

Code of Practice on the Design, Manufacture and Installation of Aerial Ropeways, 2018 Edition - Electrical and Mechanical Services Department, The Government of the Hong Kong Special Administrative Region. March 2018

Earthmoving Operations - FM 5-434, US Department of the Army, Washington DC. June 2000

Foundations and Earth Structures - NAVFAC DM 7.02, Naval Facilities Engineering Command, Alexandria, VA. September 1986

Ground Anchors and Anchored Systems - Geotechnical Engineering Circular No.4 - FHWA-IF-99-015, Office of Bridge Technology, Federal Highway Administration, US Department of Transportation, Washington DC. Authors: P.J. Sabatini, D.G. Pass, R.C. Bachus. Performing Organization: GeoSyntec Consultants. June 1999

Geotechnical Engineering Circular No.6, Shallow Foundations - FHWA-SA-02-054, Office of Bridge Technology, Federal Highway Administration, US Department of Transportation. Washington DC. Author: Robert E. Kimmerling. Performing Organization: PanGEO, Inc. September 2002

A Guide to the Project Management Body of Knowledge (PMBOK Guide) Fifth Edition, Project Management Institute, Newtown Square, PA. 2013

Guide for Shoring / Reshoring of Concrete Multistory Buildings - ACI 347.2R-05, American Concrete Institute, Farmington Hills, MI. July 2005

TMS 402/602-16: Building Code Requirements and Specification for Masonry Structures - The Masonry Society, Longmont CO. 2016

Guide to Concrete Repair, Second Edition - US Department of the Interior, Bureau of Reclamation, Technical Services Center. Prepared by: Kurt F. von Fay. August 2015

Guide to Tilt-Up Design and Construction - Cement, Concrete & Aggregates Australia, Concrete Institute of Australia, St Leonards NSW Australia. 2005

Guidelines for the Design and Construction of Stormwater Management Systems, New York City Department of Environmental Protection. July 2012

Guidelines for Design of Dams, New York State Department of Environmental Conservation, Division of Water, Albany, NY, January 1989

Haulage using Aerial Ropeways, Technical Memorandum No. 22. Transportation and Urban Projects Department, World Bank Study of the Substitution of Labor and Equipment in Civil Construction, 1976

International Building Code (IBC) 2018 - International Code Council Inc. Washington DC

International Commission on Large Dams - www.icold-cigb.org

Interstate Natural gas Association of America - www.ingaa.org

Kavramsal Su Muhendisligi. Authors: A. Melih Yanmaz, Nurünnisa Usul. Ortadogu Teknik Üniversitesi, Toplum ve Bilim Merkezi, ODTÜ Geliştirme Vakfı, Yayıncılık ve İletisim A.Ş. Yayınları, ODTÜ Yayıncılık. Ankara, Türkiye. 2006. *(Conceptual Water Engineering)*

Load Resistance Factor Design (LRFD) for Highway Bridge Superstructures - Reference Manual - FHWA-NHI-15-047, National Highway Institute, Federal Highway Administration, Arlington, VA. July 2015. Authors: Michael A. Grubb, P.E., Kenneth E. Wilson, P.E., S.E., Christopher D. White, P.E., William N. Nickas, P.E. Performing Organization: Michael baker International.

AASTHO LRFD Bridge Design Specifications (8th Edition) - American Association of State Highway and Transportation Officials, Washington DC. September 2017

Materials Testing - FM 5-472, C2, NAVFAC MO 330, Department of the Army, Washington DC. July 2001

Minimum Design Loads and Associated Criteria for Buildings and Other Structures - ASCE/SEI 7-16, American Society of Civil Engineers, Reston, VA. 2017

Windparkmontage vor Sylt - Millimeterarbeit bei Wind und Wetter | HD Doku - WELT documentary, April 2021. https://www.youtube.com/watch?v=BqRc_vAg9DY

Airport Pavement Design and Evaluation - 150/5320-6F, US Department of Transportation, Federal Aviation Administration. November 2016

Quantification of Building Seismic Performance Factors - FEMA 695, National Earthquake Hazards Reduction Program (NEHRP), Federal Emergency Management Agency, US Department of Homeland Security. Prepared By: Applied Technology Council, Redwood City, CA. June 2009

Approved Code of Practice for Passenger Ropeways in New Zealand - 1998 Edition Incorporating Amendment 1 - Occupational Safety and Health Service - New Zealand Department of Labour, Wellington, New Zealand

A Policy on Geometric Design of Highways and Streets - Fifth Edition, American Association of State Highway and Transportation Officials (AASHTO), Washington D.C. , 2004

Schedule Assessment Guide - Best practices for Project Schedules - GAO-16-89-G, US Government Accountability Office, Washington D.C. December 2015

Engineering and Design, Retaining and Flood walls - Engineer Manual 1110-2-2502, US Army Corps of Engineers, Department of the Army, Washington D.C. September 1989

Prestandard and Commentary for the Seismic Rehabilitation of Buildings - FEMA 356, Federal Emergency Management Agency (FEMA), Washington D.C. Prepared By: American Society of Civil Engineers (ASCE), Reston, VA. November 2000

Seismic Evaluation and Retrofit of Existing Buildings - ASCE/SEI 41-17, American Society of Civil Engineers, Reston, Virginia. 2017.

Engineering and Design, Settlement Analysis - Engineer Manual 1110-1-1904 - US Army Corps of Engineers, Department of the Army, Washington D.C. September 1990

Selecting Appropriate Mitigation Measures for Floodprone Structures - FEMA 551, Federal Emergency Management Agency (FEMA), US Department of Homeland Security, Washington D.C. March 2007

NIST (2016). *Seismic design of reinforced concrete special moment frames: A guide for practicing engineers, Second Edition*, GCR 16-917-40, NEHRP Seismic Design Technical Brief No. 1, produced by the Applied Technology Council and the Consortium of Universities for Research in Earthquake Engineering for the National Institute of Standards and Technology, Gaithersburg, MD.

NIST (2016). *Seismic design of steel special moment frames: A guide for practicing engineers, Second Edition*, GCR 16-917-41, NEHRP Seismic Design Technical Brief No. 2, produced by the Applied Technology Council and the Consortium of Universities for Research in Earthquake Engineering for the National Institute of Standards and Technology, Gaithersburg, MD.

NIST (2016). *Seismic design of cast-in-place concrete diaphragms chords, and collectors: A Guide for Practicing Engineers, Second Edition*, GCR 16-917-42, NEHRP Seismic Design Technical Brief No. 3, produced by the Applied Technology Council for the National Institute of Standards and Technology, Gaithersburg, MD.

NIST (2017). *Guidelines for Nonlinear Structural Analysis for Design of Buildings, Part I – General*, GCR 17-917-46v1, produced by the Applied Technology Council for the National Institute of Standards and Technology, Gaithersburg, MD.

NIST (2017). *Guidelines for Nonlinear Structural Analysis for Design of Buildings, Part IIa – Steel Moment Frames*, GCR 17-917-46v2, produced by the Applied Technology Council for the National Institute of Standards and Technology, Gaithersburg, MD.

NIST (2017). *Guidelines for Nonlinear Structural Analysis for Design of Buildings, Part IIb – Reinforced Concrete Moment Frames*, GCR 17-917-46v3, produced by the Applied Technology Council for the National Institute of Standards and Technology, Gaithersburg, MD.

NIST (2012). *Soil Structure Interaction for Building Structures, GCR 12-917-21.* Produced by the NEHRP Consultants Joint Venture, a partnership of the Applied Technology Council and the Consortium of Universities for Research in Earthquake Engineering, for the National Institute of Standards and Technology, Gaithersburg, MD. September 2012.

NIST (2014). Seismic design of special reinforced masonry shear walls: A guide for practicing engineers, NIST GCR 14-917-31, prepared by the Applied Technology Council for the National Institute of Standards and Technology, Gaithersburg, MD.

Specification for Structural Steel Buildings - ANSI/AISC 360-16 - American Institute of Steel Construction, Chicago IL. July 2016, Revised June 2019

Steel Construction Manual - American Institute of Steel Construction, Chicago, IL

Standard Specifications for Construction of Roads and Bridges on Federal Highway Projects - FP-14, Federal Highway Administration, US Department of Transportation. 2014

Technical Manual for Design and Construction of Road Tunnels - Civil Elements - FHWA-NHI-10-034 - Federal Highway Administration, National Highway Institute, US Department of Transportation, Washington D.C. Authors: C. Jeremy Hung, PE, James Monsees, PhD, PE, Nasri Munfah, PE, John Wisniewski, PE. Performing Organization: Parsons Brinckerhoff Inc. December 2009.

Toward a consistent design of structural concrete. Authors: Jörg Schlaich, Kurt Shäfer, Mattias Jennewein. Precast / Prestressed Concrete Institute (PCI) Journal / May-June 1987

Design of Coastal Revetments, Seawalls and Bulkheads - Engineer Manual 1110-2-1614 - US Army Corps of Engineers, Department of the Army, Washington D.C. June 1995

Eurocode 2, EN 1992-1-1 (2004), Design of Concrete Structures - Part 1-1: General Rules and Rules for Buildings, European Committee for Standardization, The European Union

Eurocode 2, EN 1992-2 (2005), Design of Concrete Structures - Part 2: Concrete Bridges - Design and Detailing Rules, European Committee for Standardization, The European Union

Eurocode 3, EN 1993-1-1 (2005), Design of Steel Structures - Part 1-1: General Rules and Rules for Buildings - European Committee for Standardization, The European Union

Eurocode 3, EN 1993-2 (2006), Design of Steel Structures - Part 2: Steel Bridges - European Committee for Standardization, The European Union

Eurocode 6, EN 1996-1-1 (2005), Design of Masonry Structures - Part 1-1: General Rules for Reinforced and Unreinforced Masonry Structures - European Committee for Standardization, The European Union

SEI/ASCE 37-02: Design Loads on Structures During Construction - American Society of Civil Engineers, (ASCE), Reston VA, 2002

Design: Piers and Wharves - United Facilities Criteria - UFC 4-152-01, US Department of Defense, 24 January 2017

Recomendaciones para Obras Marítimas - Recomendaciones para el proyecto y ejecución en Obras de Atraque y Amarre - Serie 2 - Obras Portuarias Interiores - ROM 2.0-11, Tomo 1, Tomo 2, Puertos del Estado, Ministerio de Fomento, Gobierno de España. 1st Edition: June 2012, Spain. *(Recommendations for Maritime Works - Recommendations for the design and execution of Docking and Mooring Works - Interior Port Works)*

Recomendaciones para Obras Marítimas - Recomendaciones Geotécnicas para Obras Marítimas y Portuarias - Serie 0 - Descripción y caracterización de los factores de proyecto - ROM 0.5-05, Puertos del Estado, Ministerio de Fomento, Gobierno de España. 1st Edition: November 2005, Spain. *(Recommendations for Maritime Works - Geotechnical Recommendations for Maritime and Port Works - Description and characterization of project factors)*

Türkiye'nin Dev Yapıları - 1915 Çanakkale Köprüsü - TRT Belgesel - https://www.youtube.com/watch?v=uE_8EXZOX7c

Mega Proje 1915 Çanakkale Köprüsü - TC Çanakkale Valiligi - https://www.youtube.com/watch?v=yjHV4n47vAA

http://www.1915canakkale.com - Official website of the 1915 Canakkale Bridge

Zusätzliche Technische Vertragsbedingungen und Richtlinien für Ingenieurbauten , Teil 5 - Tunnelbau, Abschnitt 1 - Geschlossene Bauweise, - Bundesanstalt für Straßenwesen, Deutschland. Germany, 2018. *(Additional Technical Contract Conditions and Guidelines for Civil Engineering. Tunneling, Closed Construction)*

TS EN 12811-1 - Geçici İş Donanımları - Bölüm 1: İş İskeleleri - Performans Gerekleri ve Genel Tasarım. Temporary works equipment - Part 1: Scaffolds - Performance Requirements and General Design, Türk Standartları Enstitüsü, Ankara, Türkiye. 2005

Türkiye Bina Deprem Yönetmeliği 2018, Deprem Etkisi Altında Binaların Tasarımı İçin Esaslar - Afet ve Acil Durum Yönetimi Başkanlığı, Türkiye Cumhuriyeti İçişleri Bakanlığı, Ankara, Türkiye, 2018. *(Building Earthquake Code of Türkiye, Turkish Republic Ministry of Interior, 2018)*

Wharf Design Criteria - Version 3.0 - Port of Long Beach, Long Beach, CA. February, 2012

Printed in Dunstable, United Kingdom

91343447R00380